D1659099

G. Radons, R. Neugebauer (Eds.)

Nonlinear Dynamics of Production Systems

G. Radons, R. Neugebauer (Eds.)

Nonlinear Dynamics of Production Systems

With a Foreword by Hans-Peter Wiendahl

WILEY-VCH Verlag GmbH & Co. KGaA, Weinheim

Editors

Prof. Dr. Günter Radons
Technische Universität Chemnitz
Institut für Physik, Theoretische Physik I
Komplexe Systeme und Nichtlineare Dynamik

Prof. Dr.-Ing. Reimund Neugebauer
Fraunhofer Institut für Werkzeugmaschinen
und Umformtechnik IWU Chemnitz

Cover picture
The photograph shows a working finger milling tool with spiral chips formed. The state space trajectories of the insert visualize the nonlinear dynamics of regenerative chatter which may perturb such machining operations. Courtesy: Gabor Stépán, Budapest.

Library of Congress Card No.: applied for
British Library Cataloging-in-Publication Data:
A catalogue record for this book is available from the British Library

Bibliographic information published by Die Deutsche Bibliothek.
Die Deutsche Bibliothek lists this publication in the Deutsche Nationalbibliografie; detailed bibliographic data is available in the Internet at <http://dnb.ddb.de>.

Printed in the Federal Republic of Germany
Printed on acid-free paper

Composition Steingraeber Satztechnik GmbH, Dossenheim
Printing Druckhaus Darmstadt GmbH, Darmstadt
Bookbinding Großbuchbinderei J. Schäffer GmbH & Co. KG., Grünstadt

ISBN 3-527-40430-9

Foreword

Since the 1980s both a distinct acceleration and an increasing interlinkage of technical and logistic manufacturing processes can be observed worldwide. This has led to phenomena which are often graphically described as "turbulent". Conventional linear models and approaches obviously no longer suffice to control the corresponding sudden and apparently unforeseeable process changes.

Prompted by works on chaos research in the mathematics and physics, a group of colleagues at the German Academic Society for Production Engineering (Wissenschaftliche Gesellschaft für Produktionstechnik WGP) posed the question whether the theories propounded by "chaos researchers" to describe non-linear dynamic systems might not also contribute to a deeper understanding of the behaviour of high-precision manufacturing processes, complex production facilities and cross-linked logistic processes.

To assess the need for action, a workshop on "Potentials of chaos research in manufacturing sciences" sponsored by the Volkswagen Foundation was staged in Hanover in 1994. About 30 participants from the manufacturing sciences, natural sciences and industry took part. The evaluation of this workshop and contacts with other scientists led to a joint proposal by Professors Wiendahl (Hanover), Weck (Aachen) and Lierath (Magdeburg) to initiate research with focus on the "Investigation of non-linear dynamic effects in production technology systems".

In the summer of 1995, the board of trustees of the Volkswagen Foundation agreed to set up this research program, which came to a conclusion at the end of 2001. The Foundation funded a total of 33 joint projects with up to six research groups involved. An important condition for the funding of a project was the collaboration of physicists and/or mathematicians with engineers. The Foundation granted a total sum of 13 mill. Euro for projects with this research focus, with 7.6 mill. Euro to go to the engineering sciences, 3.9 mill. Euro to physics and 1.5 mill. Euro to mathematics. In April 2003, the fourth and final symposium took place in Chemnitz.

This book contains selected and edited contributions to this symposium. The spectrum of the topics dealt with ranges from the modelling and optimisation of classical manufacturing processes, such as cutting, milling or grinding, to the development of innovative control processes for complex manufacturing machines and logistic analysis and modelling of production technology systems. These can now be better understood as regards their structure and dynamics, in particular their irregular behaviour and their characteristic complexity. Building on this, new concepts for their planning, design and control were designed.

In retrospect, it is clear that the initiative succeeded not only in gaining important scientific insights within the projects and on workshops, but also in promoting and strengthening the ongoing co-operation between engineers, physicists and mathematicians in this new field of manufacturing science.

The initiators and researchers involved would like to thank the Volkswagen Foundation for its generous funding of the projects, symposia and the publication of this book, and are also indebted to Dr Claudia Nitsch and Dr Franz Dettenwanger for their competent and helpful support.

Hans-Peter Wiendahl

Hannover, August 2003

Preface

After more than two decades of intense fundamental research in nonlinear dynamics, time has come to reap the fruits of this field. Applications in the area of production systems are possibly the most important and challenging ones. To enable progress in applying nonlinear dynamics to production systems, one needs the input from theoreticians, who are often affiliated with physics or mathematics departments, and from experts in the engineering sciences. Only a close cooperation between these groups can solve the many problems that arise from the ubiquitous presence of nonlinearities inherent in production processes and manufacturing techniques. This has been recognized clearly almost ten years ago by the initiators of the priority area "Investigations of Non-Linear Dynamic Effects in Production Systems" and the responsible persons at VolkswagenStiftung, the funding organization of this project.

Due to these efforts we are now in the lucky position to report on the progress and the many facets of this new research field. On occasion of the fourth and final symposium of this priority area held on 8–9 April 2003 in Chemnitz, Germany, we asked the members of the priority area and internationally renowned experts in the field to contribute to a book on "Nonlinear Dynamics of Production Systems". The response was overwhelming and enthusiastic and resulted in the current volume. This is the first book covering nonlinear dynamic effects in the broad field of production systems in such a comprehensive way. Of course, not every problem arising in one of the many different manufacturing techniques or production processes can be solved with the aid of nonlinear dynamics. And of the many cases where the inherent complexity and nonlinearity calls for such methods, we can only present a prototypical selection.

The content of this book is divided into five parts corresponding to different aspects or subfields of production systems. Part I is devoted to the dynamics and the optimal organization of whole production lines and general production systems. Classically, such problems have been topics in operations research. Recently, however, with the need for more flexibility and stability of production processes, the central importance of nonlinear dynamic effects has been recognized. Thereby a new field is emerging and the chapters in this part give an overview over these approaches. The aspects that arise are of interest for both the scientist who seeks interesting fields of research and the manager who wants to optimize his workshop. The largest section, Part II, is concerned with various mechanical manufacturing techniques. It reports on recent advances for the long-standing problem of machine chatter appearing in turning, milling, grinding and other mechanical machining operations, but also on new forming techniques. In addition, it treats various other important methods used nowadays to improve quality and performance of these techniques, such as the coating of tools. In a way this and the next part may be regarded as an update and an extension of the previously published Wiley book on "Dynamics and Chaos in Manufacturing Processes", edited by Francis C. Moon in 1998. Part III deals with certain aspects of the dynamics of machines and robots, which are relevant for or closely related to manufacturing processes. These range from nonlinear vibrations in forming machines and drives to the control of mechanical coordination tasks and the experimental identification of the friction dynamics in mechanical systems. This part also contains an obituary for one of our authors, František Peterka, who died unexpectedly while we were editing this book. In Part IV, non-conventional manufacturing methods, such as water-jet or laser-jet cutting and laser welding are treated. In many respects these advanced techniques complement the more traditional mechanical processes. It turns out, however, that the nonlinear dynamic phenomenon of pattern formation causes some problems in otherwise advantageous operating regimes. Possible solutions or at least new insights into these problems are provided. Pattern formation also plays a central role for many industrial chemical and electro-chemical processes which are treated in Part V. Often these processes are too complex to be understood in detail, but nevertheless the appearance and identification of certain patterns often helps to determine and control the state of the system, be it in a coal burner, or in an etching process or in a lead battery. This part also reveals that bifurcation and catastrophe theory are valuable tools in designing and controlling chemical processes.

In our opinion the contributions in this book demonstrate very convincingly the ubiquity of nonlinear dynamic effects in almost all aspects of production systems. As a consequence the understanding of such effects becomes an increasingly important pre-requisite for the further development of production techniques and systems. For the nonlinear-dynamics scientist this implies a continuing challenge from interesting real-world problems, and to the engineer it shows that nonlinear dynamics can provide promising approaches for the solution of his problems.

We want to thank all persons who made the publication of this book possible. First we thank the authors who cooperated so constructively and reliably also in our cross-refereeing procedure. Next, also in the name of all authors, we wish to express our gratitude towards the VolkswagenStiftung, Hannover, which funded not only much of the research presented here, but also the symposium in Chemnitz and previous ones from which this book emanated, and which in addition made it possible for this book to appear in color printing. Finally we thank Vera Palmer and Ulrike Werner from Wiley-VCH for their smooth and engaged cooperation, which made the editing of this volume a pleasant endeavor.

Günter Radons
Reimund Neugebauer

Chemnitz, September 2003

Participants of the “4th International Symposium on Investigations of Nonlinear Dynamic Effects in Production Systems” in Chemnitz, 8–9 April 2003

List of Contributors

Farid Al-Bender
Catholic University of Leuven
Department of Mechanical Engineering
farid.al-bender@mech.kuleuven.ac.be

Dieter Armbruster
Arizona State University
Department of Mathematics
dieter@source.la.asu.edu

Ralph T. Bailey
Babcock & Wilcox Canada, Ltd.
RTBailey@babcock.com

Michael Baune
Universität Bremen
Institut für Angewandte und Physikalische Chemie
m.baune@uni-bremen.de

Andreas Baus
Rheinisch-Westfälische Technische Hochschule Aachen
Werkzeugmaschinenlabor
a.baus@wzl.rwth-aachen.de

Arno Behrens
Universität der Bundeswehr Hamburg
Laboratorium Fertigungstechnik
Arno.behrens@unibw-hamburg.de

Andrzej Bodnar
Technical University of Szczecin

Ekkard Brinksmeier
Universität Bremen
Labor für Mikrozerspannung
brinksme@iwt.uni-bremen.de

Magnus Buhlert
Universität Bremen
Institut für Angewandte und Physikalische Chemie
buhlert@uni-bremen.de

Leonid A. Bunimovich
Georgia Institute of Technology
School of Mathematics
bunimovh@math.gatech.edu

Adrienn Cser
Universität Erlangen-Nürnberg
Lehrstuhl für Fertigungstechnologie
A.Cser@lft.uni-erlangen.de

C. Stuart Daw
Oak Ridge National Laboratory
Fuels, Engines and Emissions Research Center
dawcs@ornl.gov

Berend Denkena
Universität Hannover
Institut für Fertigungstechnik und Werkzeugmaschinen
denkena@ifw.uni-hannover.de

S. Diaz Alfonso
Instituto Superiore Politecnico, Havanna
Facultad de Ingenieria Chimiza

Thomas Ditzinger
Springer-Verlag, Heidelberg
Ditzinger@Springer.de

Reik Donner
Universität Potsdam
Institut für Physik
reik@agnld.uni-potsdam.de

David Engster
Universität Göttingen
III. Physikalisches Institut
D.Engster@DPI.Physik.Uni-Goettingen.de

Ronald Faassen
Technical University of Eindhoven
Department of Mechanical Engineering
r.p.h.faassen@tue.nl

Spilios D. Fassois
University of Patras
Department of Mechanical and Aeronautical Engineering
fassois@mech.upatras.gr

Ulrike Feudel
Carl von Ossietzky Universität Oldenburg
Institut für Chemie und Biologie des Meeres
U.Feudel@icbm.de

Charles E.A. Finney
Oak Ridge National Laboratory
Fuels, Engines and Emissions Research Center
finneyc@ornl.gov

Gerhard Finstermann
Johannes Kepler Universität Linz
gerhard.finstermann@vai.at

Thomas J. Flynn
Babcock & Wilcox Canada, Ltd.
TJFlynn@babcock.com

Michael Freitag
Universität Bremen
Planung und Steuerung produktionstechnischer Systeme
fmt@biba.uni-bremen.de

Rudolf Friedrich
Westfälische Wilhelms-Universität Münster
Institut für Theoretische Physik
fiddir@uni-muenster.de

Michael I. Friswell
University of Bristol
Department of Aerospace Engineering
m.i.friswell@bristol.ac.uk

Timothy A. Fuller
Babcock & Wilcox Canada, Ltd.
tafuller@babcock.com

Manfred Geiger
Universität Erlangen-Nürnberg
Lehrstuhl für Fertigungstechnologie
m.geiger@lft.uni-erlangen.de

Mark Geisel
Universität Erlangen-Nürnberg
Lehrstuhl für Fertigungstechnologie
m.geisel@lft.uni-erlangen.de

Carmen Gerlach
Universität Bremen
Institut für Angewandte und Physikalische Chemie
cgerlach@uni-bremen.de

Roland Göbel
Universität Dortmund
Lehrstuhl für Umformtechnik
Goebel@lfu.mb.uni-dortmund.de

Edvard Govekar
University of Ljubljana
Faculty of Mechanical Engineering
edvard.govekar@fs.uni-lj.si

Igor Grabec
University of Ljubljana
Faculty of Mechanical Engineering
igor.grabec@fs.uni-lj.si

Janez Gradisek
University of Ljubljana
Faculty of Mechanical Engineering
janez.gradisek@fs.uni-lj.si

Karol Grudzinski
Technical University of Szczecin
konrad@safona.tuniv.szczecin.pl

Maria Haase
Universität Stuttgart
Institut für Computeranwendungen
ica2mh@csv.ica.uni-stuttgart.de

Juergen Hahn
Texas A&M University, College Station
Department of Chemical Engineering
hahn@tamu.edu

Ernst-Christoph Haß
MIR-Chem GmbH, Bremen
Hass@mir-chem.de

Bodo Heimann
Universität Hannover
Institut für Mechanik
heimann@ifm.uni-hannover.de

Dirk Helbing
Technische Universität Dresden
Institut für Wirtschaft und Verkehr
helbing@trafficforum.de

Burkhard Heller
Universität Dortmund
Lehrstuhl für Umformtechnik
heller@lfu.mb.uni-dortmund.de

Axel Henning
Fraunhofer Institut für Produktionstechnik und Automatisierung, Stuttgart
henning@ipa.fhg.de

Helmut J. Holl
Johannes Kepler Universität Linz
Abteilung für Technische Mechanik
helmut.holl@jku.at

Alexander Hornstein
Universität Göttingen
III. Physikalisches Institut
A.Hornstein@DPI.Physik.Uni-Goettingen.de

Tamas Insperger
Budapest University of Technology and Economics
Department of Applied Mechanics
inspi@mm.bme.hu

Hans Irschik
Johannes Kepler Universität Linz
Abteilung für Technische Mechanik
Hans.irschik@jku.at

Jörn Jacobsen
Universität Hannover
Institut für Fertigungstechnik und Werkzeugmaschinen
jacobsen@ifw.uni-hannover.de

Karsten Kalisch
Universität der Bundeswehr Hamburg
Laboratorium Fertigungstechnik
Karsten.kalisch@unibw-hamburg.de

Tamás Kalmár-Nagy
United Technolgies Research Center
kalmart@utrc.utc.com

Holger Kantz
Max-Planck-Institut für Physik komplexer Systeme, Dresden
kantz@mpipks-dresden.mpg.de

Ines Katzorke
Universität Potsdam
Institut für Physik
ines@agnld.uni-potsdam.de

Matthias Kleiner
Universität Dortmund
Lehrstuhl für Umformtechnik
mkleiner@lfu.mb.uni-dortmund.de

Christian Klimmek
Universität Dortmund
Lehrstuhl für Umformtechnik
klimmek@lfu.mb.uni-dortmund.de

Fritz Klocke
Rheinisch-Westfälische Technische Hochschule Aachen
Werkzeugmaschinenlabor
f.klocke@wzl.rwth-aachen.de

Jan Konvicka
Mikron Comp-Tec AG
jan.konvicka@mikron-ac.com

Vadim Kostrykin
Fraunhofer Institut für Lasertechnik, Aachen
Vadim.kostrykin@ilt.fraunhofer.de

Alexei Kouzmitchev
Westfälische Wilhelms-Universität Münster
Institut für Theoretische Physik
kuz@uni-muenster.de

Jürgen Kurths
Universität Potsdam
Institut für Physik
Jkurths@Agnld.uni-potsdam.de

Vincent Lampaert
Catholic University of Leuven
Department of Mechanical Engineering
Vincent.Lampaert@mech.kuleuven.ac.be

Erjen Lefeber
Technical University of Eindhoven
Department of Mechanical Engineering
a.a.j.lefeber@tue.nl

Regina Leopold
Fraunhofer Institut für Werkzeugmaschinen und Umformtechnik, Chemnitz
leopold@iwu.fhg.de

Jianhui Li
Universität Bremen
Labor für Mikrozerspannung
jli@lfm.uni-bremen.de

Grzegorz Litak
Technical University of Lublin
Department of Applied Physics
litak@archimedes.pol.lublin.pl

Wolfgang Marquardt
Rheinisch-Westfälische Technische Hochschule Aachen
Lehrstuhl für Prozesstechnik
marquardt@lfpt.rwth-aachen.de

Hendrik Mathes
Universität Bremen
Institut für Angewandte und Physikalische Chemie
Hendrik@uni-bremen.de

Karl Mayrhofer
Johannes Kepler Universität Linz
Institut für Anwendungsorientierte Wissensverarbeitung
karl.mayrhofer@vai.at

Jan Michel
Fraunhofer Institut für Lasertechnik, Aachen
Jan.michel@ilt.fraunhofer.de

Martin Mönnigmann
Rheinisch-Westfälische Technische Hochschule Aachen
Lehrstuhl für Prozesstechnik
moennigmann@lpt.rwth-aachen.de

Francis C. Moon
Cornell University, Ithaca
Sibley School of Mechanical and Aerospace Engineering
Fcm3@cornell.edu

Alejandro Mora
Universität Stuttgart
Institut für Computeranwendungen
ica2am@csv.uni-stuttgart.de

Steffen Nestmann
Fraunhofer Institut für Werkzeugmaschinen und Umformtechnik, Chemnitz
Steffen.nestmann@iwu.fraunhofer.de

Reimund Neugebauer
Fraunhofer Institut für Werkzeugmaschinen und Umformtechnik, Chemnitz
Neugebauer@iwu.fhg.de

Markus Nießen
Fraunhofer Institut für Lasertechnik, Aachen
Markus.niessen@ilt.fraunhofer.de

Henk Nijmeijer
Technical University of Eindhoven
Department of Mechanical Engineering
h.nijmeijer@tue.nl

J.A.J. Oosterling
TNO Institute of Industrial Technology Enschede
Manufacturing Development

Andreas Otto
Universität Erlangen-Nürnberg
Lehrstuhl für Fertigungstechnologie
A.Otto@lft.uni-erlangen.de

Ulrich Parlitz
Universität Göttingen
III. Physikalisches Institut
U.Parlitz@DPI.Physik.Uni-Goettingen.DE

Frantisek Peterka †
Academy of Sciences of the Czech Republic, Prague
Institute of Thermomechanics

Karsten Peters
Universität Göttingen
III. Physikalisches Institut
karsten@physik3.gwdg.de

Stefan Pfeiffer
Rheinisch-Westfälische Technische Hochschule Aachen
Lehrstuhl für Lasertechnik

Arkady Pikovsky
Universität Potsdam
Institut für Physik
pikovsky@stat.physik.uni-potsdam.de

Peter Jörg Plath
Universität Bremen
Institut für Angewandte und Physikalische Chemie
plath@uni-bremen.de

Thomas Rabbow
Universität Bremen
Institut für Angewandte und Physikalische Chemie
rabbow@uni-bremen.de

Günter Radons
Technische Universität Chemnitz
Institut für Physik
Radons@physik.tu-chemnitz.de

Volker Reitmann
Max-Planck-Institut für Physik komplexer Systeme, Dresden
reitmann@rcs.urz.tu-dresden.de

Rüdiger Rentsch
Universität Bremen
Labor für Mikrozerspannung
rentsch@lfm.uni-bremen.de

Dimitris C. Rizos
University of Patras
Department of Mechanical and Aeronautical Engineering
Driz@mech.upatras.gr

Alejandro Rodriguez-Angeles
Technical University of Eindhoven
Department of Mechanical Engineering
a.rodriguez@tue.nl

Rafal Rusinek
Technical University of Lublin
Department of Applied Mechanics
raf@archimedes.pol.lublin.pl

Gerhard Schmidt
Fraunhofer Institut für Werkzeugmaschinen und Umformtechnik, Chemnitz
Schmidtg@iwu.fhg.de

Alf Schmieder
Universität Bremen
Planung und Steuerung produktionstechnischer Systeme
smi@biba.uni-bremen.de

Bernd Scholz-Reiter
Universität Bremen
Planung und Steuerung produktionstechnischer Systeme
bsr@biba.uni-bremen.de

Wolfgang Schulz
Fraunhofer Institut für Lasertechnik, Aachen
Wolfgang.schulz@ilt.fraunhofer.de

Oliver Schütte
Universität Hannover
Institut für Mechanik
schuette@ifm.uni-hannover.de

Udo Schwarz
Universität Potsdam
Institut für Physik
Uschwarz@Agnld.uni-potsdam.de

Gabor Stepan
Budapest University of Technology and Economics
Department of Applied Mechanics
stepan@mm.bme.hu

Uwe Sydow
MIR-Chem GmbH, Bremen
sydow@mir-chem.de

Kazimierz Szabelski
Technical University of Lublin
Department of Applied Mechanics
mechstos@archimedes.pol.lublin.pl

Robert Szalai
Budapest University of Technology and Economics
Department of Applied Mechanics
szalai@mm.bme.hu

Palaniappagounder Thangavel
Universität Bremen
Institut für Angewandte und Physikalische Chemie
Thangavelp@yahoo.com

Hans Kurt Tönshoff
Universität Hannover
Institut für Fertigungstechnik und Werkzeugmaschinen
toenshoff@ifw.uni-hannover.de

Ubbo Visser
Universität Bremen
Technologie-Zentrum Informatik
Visser@tzi.de

Nathan van de Wouw
Technical University of Eindhoven
Department of Mechanical Engineering
n.v.d.wouw@tue.nl

E. van Raaij
MIR-Chem GmbH, Bremen
info@mir-chem.de

Jerzy Warminski
Technical University of Lublin
Department of Applied Mechanics
jwar@archimedes.pol.lublin.pl

Frank Weidermann
Fachhochschule Mittweida
Fachbereich Maschinenbau/Feinwerktechnik
Frank.weidermann@htwm.de

A. Wessel
Universität Potsdam
Institut für Physik

Niels Wessel
Universität Potsdam
Institut für Physik
niels@agnld.uni-potsdam.de

Bert Westhoff
Universität der Bundeswehr Hamburg
Laboratorium Fertigungstechnik
Bert.westhoff@unibw-hamburg.de

Engelbert Westkämper
Fraunhofer Institut für Arbeitswirtschaft und Organisation, Stuttgart
wke@ipa.fhg.de

Hans-Peter Wiendahl
Universität Hannover
Institut für Fabrikanlagen und Logistik
wiendahl@ifa.uni-hannover.de

Jochen Worbs
Universität Hannover
Institut für Fabrikanlagen und Logistik
worbs@ifa.uni-hannover.de

Keith Worden
University of Sheffield
Department of Mechanical Engineering
K.Worden@sheffield.ac.uk

Jens Wulfsberg
Universität der Bundeswehr Hamburg
Laboratorium Fertigungstechnik
Jens.wulfsberg@unibw-hamburg.de

Contents

IV Non-conventional Manufacturing Processes 387

Part I

Dynamics and Control of Production Processes

The organization and management of production systems is a very difficult task because the decisions or the different policies of the manager result in different nonlinear dynamical systems for the production process. The nonlinearities induced by the applied rules and the complexity of the associated dynamics entail great problems in analyzing and predicting interesting features such as the throughput of the system. Often such problems are tackled by numerical computations using discrete-event simulations. The latter, however, become quite time-consuming already for moderate system sizes and suffer from the combinatorial explosion of the space of possible rules. In times of increasing demands with respect to flexibility in "turbulent" markets and rapidly-changing environments, such time-intensive procedures become a serious obstacle for the management of production processes. As a result there is an increasing need for appropriate models of the production process. They cannot only provide solutions for large-scale problems, but in addition one gains a deeper understanding of the possible phenomena occurring in the considered processes and thus one obtains a firm basis for decision making.

The first contribution by D. Armbruster reviews recent developments in modeling the organization and management of production systems and subsequently provides further perspectives of these approaches of nonlinear dynamical systems. The three model classes are i) production lines using principles of bucket brigade organization, ii) fluid models for production networks, which may result in chaotic dynamics, and iii) nonlinear traffic models for re-entrant production processes. Roughly, these classes are described mathematically by discrete dynamical systems, i.e. iterated maps, ordinary differential equations, and partial differential equations, respectively. The contribution of L. Bunimovich elaborates on the first class and its generalizations. Especially self-organization principles and stability aspects of such production systems are treated and a hierarchy of related models and their properties are discussed. The following two papers belong to class ii). The one by K. Peters et al. considers the theoretical and practical consequences of restricted buffer sizes, a feature which in real production systems is of great importance and may lead to chaotically varying output of such manufacturing systems. The contribution of B. Scholz-Reiter et al. develops a model for re-entrant manufacturing processes and corresponding dynamic control methods, which are implemented in a discrete-event simulation tool. E. Lefeber shows how the standard fluid model can be extended to incorporate throughput and cycle times resulting in a traffic model of class iii), and how the latter can be controlled. Also the last contribution in this part by D. Helbing discusses extensively the connection of production systems to traffic dynamics and the lessons that can be learnt from such analogies. This first part of the book demonstrates very clearly how on the one side the industrial manager can gain from nonlinear dynamics, and how on the other side such applications result in new and interesting problems in the theory of dynamical systems.

1 Dynamical Systems and Production Systems

D. Armbruster

Recent applications of nonlinear dynamics to three models of production systems are reviewed: (i) a re-organization procedure for bucket brigade systems is introduced. With such a procedure the bucket brigade self-organizes when a new worker is added whose skill level increases with time due to learning. (ii) Self-organization in chaotic networks of switched arrival systems is presented. (iii) A new continuum model of production flow through re-entrant factories leading to non-local nonlinear partial differential equations is discussed. Such a model may lead to fast simulations of supply chains as well as improved understanding of the dynamics of supply chains.

1.1 Introduction

This paper aims to give a review of some recent models for the organization and management of production systems, which were using ideas from dynamical systems theory. The topics that we will discuss are (i) the self-organization of workers along a linear production line using the organizing principles of a bucket brigade, (ii) chaotic dynamics in fluid models of production networks and (iii) nonlinear dynamics in a "traffic model" of a re-entrant factory. All these models have in common that the nonlinearity that makes them complex and interesting does not come from any nonlinear physical law. The nonlinearity typically is a result of policies or decisions that change the rules of the game. As a consequence, very often the underlying dynamical system is piecewise linear or even piecewise constant leading to hybrid dynamical systems. Unfortunately beyond some initial attempts [21, 37] there is no comprehensive theory of hybrid dynamical systems available at the moment, which leads in general to ad hoc methods of analysis. On the other hand, the existing models indicate the importance of further research into a general theory of hybrid systems and even further into a general theory of the dynamics of decision systems. Rather than reporting in detail on previous work, I will in general simply state the major results and focus on extensions and open questions.

1.2 The Bucket Brigade Production System

Operational control in a linear production line is extremely important to optimize production. Typical studies focus on a fixed and optimal worker allocation that balances throughput along the line. As an alternative there is a growing interest to study operational control in systems with multi-skilled, flexible workers that are allowed to search for their own optimal place along

Nonlinear Dynamics of Production Systems. Edited by G. Radons and R. Neugebauer

ISBN 3-527-40430-9

a production line. In that case, instead of fixed work allocation, control is generated through a set of policies that tells each worker what to do "next".

The Toyota Sewn Products Management System (TSS) is one of the most widely studied architectures of this kind. TSS is employed regularly by manufacturers of sewn products in modules that are used in the finishing and assembly of cut parts into a subassembly or finished garment. The production modules are typically U-shaped, and workers process garments as a team. In a TSS line, each worker picks up a task and processes (also carries) it at each station until he gets bumped by a downstream worker. There are no additional work in progress (WIP) buffers kept. The ordering of the workers has to be preserved but, other than that, workers are not restricted to any particular zones. When a worker comes to a busy station he must wait until the station becomes available; he also may not seek other work. The number of machines in a TSS line typically ranges from 2 to 16 with an average of 2.5 machines per worker [7]. In addition to the apparel and garment industry, TSS lines have also been shown to perform robustly in certain types of warehousing environments [5].

The term "bucket brigade" was coined by Bartholdi and Eisenstein [5] for TSS lines in which the workers are sequenced from slowest to fastest. The authors provided the first comprehensive analysis of the dynamics of such systems and showed that the bucket brigade is self-balancing; that is, eventually a stable partition of work among workers will emerge such that each worker repeatedly executes the same interval of work content. The basic assumptions in [5] are:

- work is continuous and workers can walk back and take over the work of their predecessors at any point;
- the work is deterministic and each worker has a velocity function that gives his/her instantaneous velocity at each point along the production line;
- processing times are deterministic; however, there is still variation due to the fact that workers have different skill profiles and, hence, different velocities at different tasks;
- workers walk back with infinite velocity (i.e. instantaneous reset of the line).

Such a setup will lead to a nonlinear dynamical system in the following way: consider N workers along a production line. We model the line by the interval $0 \leq \xi \leq 1$ with $\xi = 0$ denoting the start of production and $\xi = 1$ indicating the end of the line. Let $x^m(t)$ be the position of the worker m at time t along the production line and c_m her velocity. For the time being we assume that all velocities are constant along the whole production line. This can easily be changed.

From the time a new part is started on the production line by the first worker until the time that the last worker finishes her part, the time evolution of the positions of all N workers is given by

$$
\begin{aligned}
x^1(t) &= c_1 t, \\
x^2(t) &= c_2 t + x_0^2, \\
&\dots \\
x^N(t) &= c_N t + x_0^N.
\end{aligned}
$$

The time $\bar{t}$ to finish the next product is given by the equation

$$x^N(\bar{t}) = 1.$$

Hence

$$\bar{t} = \frac{1 - x_0^N}{c_N}.$$

The positions of all workers at the time $\bar{t}$ will be the starting point for the reset production line when all workers instantaneously walk back to pick up the part from their predecessor. Registering these positions leads to an obvious Poincaré map that relates the reset positions for finishing product n to the reset position for finishing the next product $n + 1$:

$$\mathbf{P} : \Sigma = \{(x^1(\bar{t}), x^2(\bar{t}), \ldots, x^{N-1}(\bar{t}), 1)\} \to \Sigma$$

$$\begin{aligned} x_{n+1}^1 &= \frac{c_1}{c_N}(1 - x_n^{N-1}), \\ x_{n+1}^2 &= \frac{c_2}{c_N}(1 - x_n^{N-1}) + x_n^1, \\ \ldots \\ x_{n+1}^{N-1} &= \frac{c_{N-1}}{c_N}(1 - x_n^{N-1}) + x_n^1. \end{aligned} \tag{1.1}$$

Bartholdi and Eisenstein [5] rigorously proved that the Poincaré map $\mathbf{P}$ has a globally stable fixed point if $c_1 < c_2 \cdots < c_N$. They also proved that the resulting throughput is optimal. Hence with a bucket brigade production rule, the production line organizes itself in an optimal way, obviating the need for intervention by management or for fixed work allocation rules. Recently we have extended this result in several directions [3, 4] which we discuss next. See also the related contribution by Bunimovich in this book [8].

1.2.1 Re-ordering

In the original bucket brigade setup, there is still management intervention in the form of determining the ordering of the workers along the production line. This is clearly a major problem since management will not always know the true ranking of worker speeds. In addition, worker speed will change not only stochastically, but also systematically. Workers will become faster (due for instance to learning) or slower (due for instance to health-related issues). In [4] we discuss the following re-order policy for bucket brigades: Whenever a worker gets blocked, he switches position with the blocking worker and continues to move ahead. At reset, every worker walks back until he encounters another worker along the production line. He will then take over the job from his predecessor, leaving the first worker to start a new product. The following lemmas will prove that, for a uniform bucket brigade (i.e. where the workers have constant speeds along the whole production line), this re-ordering algorithm will lead to a stable organization, corresponding to an ordering from slowest to fastest.

Lemma 1: A uniform bucket brigade that is not ordered slowest to fastest has an unstable fixed point.

Proof: Assume a balanced bucket brigade, i.e. every worker repeatedly executes the same interval along the production line. Consider two neighboring workers. They comprise a two-worker bucket brigade. Assume worker A has a speed of v_{A} and worker B has a speed of v_{B}. Call p the handover point from worker A to worker B. Then

$$p^{n+1} = \frac{v_{\mathrm{A}}}{v_{\mathrm{B}}}(c - p^n) \tag{1.2}$$

if worker A comes before worker B along the production line. The constant c describes the handover point for worker B and is unimportant. If $v_{\mathrm{A}} < v_{\mathrm{B}}$ the fixed point of the linear map (1.2) is stable whereas for $v_{\mathrm{A}} > v_{\mathrm{B}}$ the fixed point is unstable.

Lemma 2: A balanced uniform bucket brigade that has an unstable fixed point shows blocking.

Proof: The nature of blocking is that the worker that gets blocked slows down to the speed of the blocking worker. The system (1.1) with blocking can be written as

$$\mathbf{x}^{n+1} = \mathbf{A}^i \mathbf{x}^n + \mathbf{c}^i, \tag{1.3}$$

where $\mathbf{x}$ is the vector of all reset positions, $\mathbf{A}^i$ a constant non-singular matrix and $\mathbf{c}^i$ a constant vector. The index i counts the different instances of possible blocking and represents different initial domains that have more or less instances of blocking. As a result the system (1.3) is a piecewise-linear system. The definition of a fixed point implies that there is no blocking. We choose $i = 1$ to be that case. By assumption that fixed point is unstable. Since the dynamics restricted to the domain of the $i = 1$ system is linear, any starting point near that fixed point will eventually reach the boundary of that system. Blocking occurs at that point.

Lemma 3: A generic bucket brigade with the above reset rule will converge to the stable fixed point.

Proof: Since the system (1.1) is piecewise linear, there are generically no other attractors than fixed points completely inside just a single linear domain. Blocking can only happen if a trailing worker n has a speed c_n that is larger than the speed of the following worker c_{n+1}. In that case the worker order will be reversed. Hence the re-order rule can never switch a worker with lower speed into a higher position. This is a terminating process that leads to the stable ordering from slowest to fastest.

As a result of these lemmas, there is no management intervention necessary to balance the bucket brigade. As long as the order of the workers' speeds is uniform along the production line, the re-ordering rule will seek out the balanced line. Non-uniform bucket brigades (i.e. workers' speed order not constant along the production line) have the possibilities of multiple fixed points. Hence the above arguments will not work in general.

1.2.2 Non-constant Speeds

The assumption of a fixed worker ordering along a production line is too restrictive in many settings. This is true especially for production lines that have different types of tasks or for warehouse picking, where a worker will be familiar with one part of the inventory but very unfamiliar with another part of the warehouse. In our paper [3], we study two workers who have varying levels of specialization at different tasks. As as result no worker speed uniformly dominates the other over the whole production line. To illustrate the type of dynamics that happens in this case, we discuss a case study for the passing case.

Let's suppose we have two workers, A and B. For convenience, we refer to worker A as a male and worker B as a female. We scale work content and time such that worker B has a uniform speed along the production line which we set to 1. We assume that the speed of worker A is a function of his location on the line. To ensure that worker B's speed does not uniformly dominate that of worker A, we assume that worker A has speed c_1 on the interval $[0, X)$ and c_2 on the interval $[X, 1]$ with $c_1 < 1 < c_2$. A worker may overtake the next worker temporarily. However, a fixed worker order is assumed and we call the system unbalanced if that worker order is not true at the time of reset of the production line. Consider the case when worker A is placed at the end of the line. We define $x_{\mathrm{B}}(t)$ ($x_{\mathrm{A}}(t)$) as the position of worker B (worker A) along the production line at time t. Since the speed of worker B is equal to 1 for all portions of the work we find

$$x_{\mathrm{B}}(t) = t \text{ for all } t \geq 0 .$$

For worker A, $x_{\mathrm{A}}(t)$ will change as a function of where he started. Let x^0 denote the location that worker A took over the first job, that is, $x^0 := x_{\mathrm{A}}(0)$. If worker A starts at a point after the break point X (i.e. $x^0 > X$), then the location of worker A at time t will be governed by:

$$x_{\mathrm{A}}(t) = x^0 + c_2 t \text{ for all } t \geq 0 .$$

If worker A takes over the job at some point before X (i.e. $x^0 \leq X$) then we get:

$$x_{\mathrm{A}}(t) = \begin{cases} x^0 + c_1 t & \text{for } t < t_X , \\ X + c_2(t - t_X) & \text{for } t \geq t_X , \end{cases}$$

where t_X is the time it takes worker A to get to $x_{\mathrm{A}} = X$:

$$t_X = \frac{X - x^0}{c_1} .$$

It takes worker A $\bar{t}_1$ time units to complete the first job, which is found by setting $x_{\mathrm{A}}(t) = 1$:

$$\bar{t}_1 = \begin{cases} \dfrac{1 - x^0}{c_2} & \text{for } x^0 > X , \\ t_X + \dfrac{1 - X}{c_2} = X\left(\dfrac{1}{c_1} - \dfrac{1}{c_2}\right) - \dfrac{x^0}{c_1} + \dfrac{1}{c_2} & \text{for } x^0 \leq X . \end{cases}$$

Again we get a piecewise-linear map

$$x^{n+1} := x_{\mathrm{B}}(\bar{t}_n) = \begin{cases} \dfrac{1 - x^n}{c_2} & \text{for } x^n > X , \\ X\left(\dfrac{1}{c_1} - \dfrac{1}{c_2}\right) - \dfrac{x^n}{c_1} + \dfrac{1}{c_2} & \text{for } x^n \leq X . \end{cases} \tag{1.4}$$

The map is piecewise linear with slopes $-1/c_1 < -1$ and $-1/c_2 > -1$. The elbow of the map is at $x^n = X$ with $f(X) = (1 - X)/c_2$. Clearly there always exists exactly one fixed

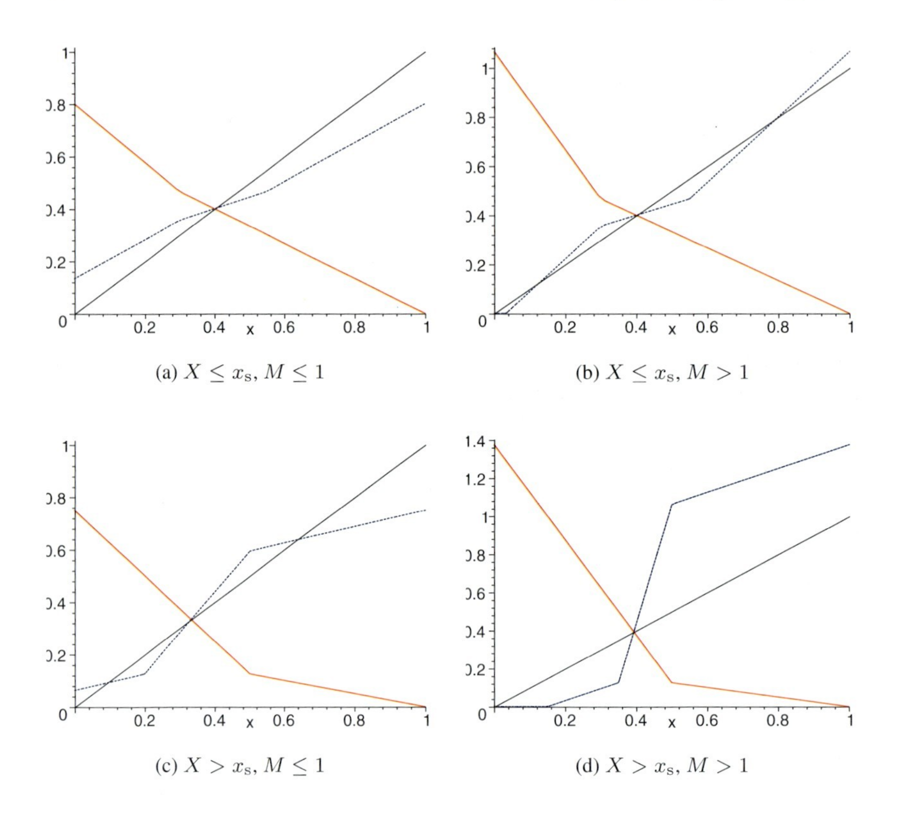

(a) $X \leq x_\mathrm{s}$, $M \leq 1$

(b) $X \leq x_\mathrm{s}$, $M > 1$

(c) $X > x_\mathrm{s}$, $M \leq 1$

(d) $X > x_\mathrm{s}$, $M > 1$

Figure 1.1: All possible dynamics for the passing case when worker A is placed at the end.

point of that map that we call x_s. If $X \leq x_\mathrm{s}$ (i.e. the slope at x_s is equal to $-1/c_2 > -1$), then the fixed x_s point is stable. If $X > x_\mathrm{s}$, the fixed point x_s is unstable.

Figure 1.1 shows all four possible cases for the dynamics of the bucket brigade. Each of the subfigures shows: (i) the Poincaré map $x^{n+1} = f(x^n)$ (depicted by the dashed line), (ii) $x^{n+2} = f^2(x^n) = f(f(x^n))$ and (iii) the $y = x$ line as the diagonal. Figure 1.1a and 1.1b show the case of $X \leq x_\mathrm{s}$, which results in a stable fixed point. Figure 1.1c and 1.1d show the case of $X > x_\mathrm{s}$, which results in an unstable fixed point. The critical quantity for the dynamics in both cases is M, which is defined as the time that worker A would need to get through the whole production line. Then, $M = f(0)$ can be found as

$$M = \frac{X}{c_1} + \frac{1-X}{c_2} \,.$$

Note that if $M > 1$, then worker B is *faster on average* than worker A.

- $X \leq x_\mathrm{s}$, stable fixed point:
  - Figure 1.1a shows the case with worker A faster on average than worker B ($M \leq 1$). We see that the stable fixed point attracts all initial conditions.

  - Figure 1.1c shows the case with worker B faster on average than worker A ($M > 1$). In this case, the fixed point is still stable, but there also exists a period-two orbit that is unstable. That is, all initial conditions inside the period-two orbit converge to the fixed point and all initial conditions outside the period-two orbit lead to a situation where worker B finishes her job before worker A does, which means that A is not fast enough at the end of the production line to maintain the ordering.

- $X > x_s$, unstable fixed point:

  - Figure 1.1b shows the case with worker A faster than worker B on average ($M \leq 1$). We see that the fixed point is unstable; however, there exists a stable period-two orbit that attracts all initial conditions and preserves the order of the bucket brigade.
  - Figure 1.1d shows the case $X > x_s$ with worker B faster than worker A on average ($M > 1$). In this case, the fixed point is unstable and all initial conditions eventually lead to a situation where worker B finishes her job before worker A does and, hence, the order of the bucket brigade is not preserved.

This case study can be generalized to the following theorem [3].

Theorem
The following results hold true for the dynamics
of two-worker bucket brigades with passing, regardless of worker ordering:

(i) If the worker at the end is on average the faster worker then either the fixed point is globally stable and attracts all initial conditions or the fixed point is unstable, in which case there exists a stable period-two orbit that attracts all initial conditions.

(ii) If the worker at the end is on average the slower worker then either the fixed point is unstable and all initial conditions lead to an unbalanced bucket brigade or the fixed point is stable and there exists an unstable period-two orbit. All initial conditions inside the period-two orbit lead to a balanced line at the fixed point; initial conditions outside the period-two orbit lead to a reversal of the finishing order of the bucket brigade.

1.2.3 Bucket Brigades and Learning

Bucket brigades are typically set up for relatively low-skilled work. Low pay and non-existing benefits often lead to a large worker turnover [16]. Hence very often an existing bucket brigade has to incorporate a new member who is unskilled and at the time a slow worker. As he gains experience, his skills improve and so does his production speed. Recently we have studied the implication of learning on the stability of a bucket brigade [4]. The most important feature of the models of bucket brigades with learning is the fact that the dynamical systems become non-autonomous. To be specific, we consider a well-ordered, uniform bucket brigade of N workers arranged from slowest to fastest: $c_1 < c_2 < \cdots < c_N$. One new worker will be added at an arbitrary place in the worker ordering. We adopt the re-order policy discussed previously. Hence workers can pass each other and handovers will always be done to the nearest worker upstream. We assume a starting velocity for the new worker: $v_l < c_i \;\; \forall i$, constant and a potential maximal velocity v_h. The worker learns only on those parts of the production line where he/she has worked.

Hutchinson, Villalobos and Bernvides suggested the following model for the increase in worker speeds due to learning [16]:

$$v_n = v_\mathrm{l} + (v_\mathrm{h} - v_\mathrm{l})(1 - e^{-t_n/\tau}),$$
$$t_n = \sum_{i=0}^{n} t_i.$$

Theorem
The bucket brigade will self-organize to the same throughput, independent of initial conditions (*initial worker ordering*).

Figure 1.2 shows the asymptotic velocities as a function of the positions along the production line of a four-worker bucket brigade with one new worker added. The constant speeds of the existing bucket brigade are

$$(c_1, c_2, c_3, c_4) = (1.3, 1.6, 2.0, 2.3),$$

the starting speed for the new worker is $v_\mathrm{l} = 1$ and the limiting speed for the new worker is $v_\mathrm{h} = 1.8$. We see that the new worker typically only learns significantly on a finite interval. That interval represents the part of the production line that he will cover as part of the work allocation for the stable fixed point of the bucket brigade. Note also that workers typically stay in the original order although occasionally they change order after a few iterations. Figure 1.2e shows such a case where the initial position of the new worker was last but he switched to second to last. This is reflected in some initial increase in velocity close to the end of the production line. It is instructive to place the new worker into a position where its limiting speed will lead to an instability of the bucket brigade. Assume the limiting speed of the new worker is significantly higher than the speeds of the workers around him. In that case, the new worker will cover larger and larger parts of the production line, shifting his region of high-velocity motion further and further out. Eventually the resulting bucket brigade will become unstable, leading to first periodic behavior and then to a re-ordering due to the fact that the old order cannot be sustained at reset. Figure 1.3 shows the transition points between the various workers as a function of time. We see that an initial order slowly becomes unstable to a beautiful period-doubling cascade until the system resets on a new order that dynamically creates a fixed point. What looks like a period-doubling cascade is really a transient that evolves slowly through the quasi-steady periodic orbits.

1.3 Fluid Models of Production Networks

There are currently two major approaches to simulate production flows: discrete event simulations (DES) and fluid networks. DES have successfully been used in large simulations for semiconductor factories (for instance [10, 12]) but typically are very time consuming. An alternative is the use of fluid models as discussed in [11, 19], etc. Fluid models come from traffic theory and were introduced by Newell [28, 30] to approximately solve queueing problems.

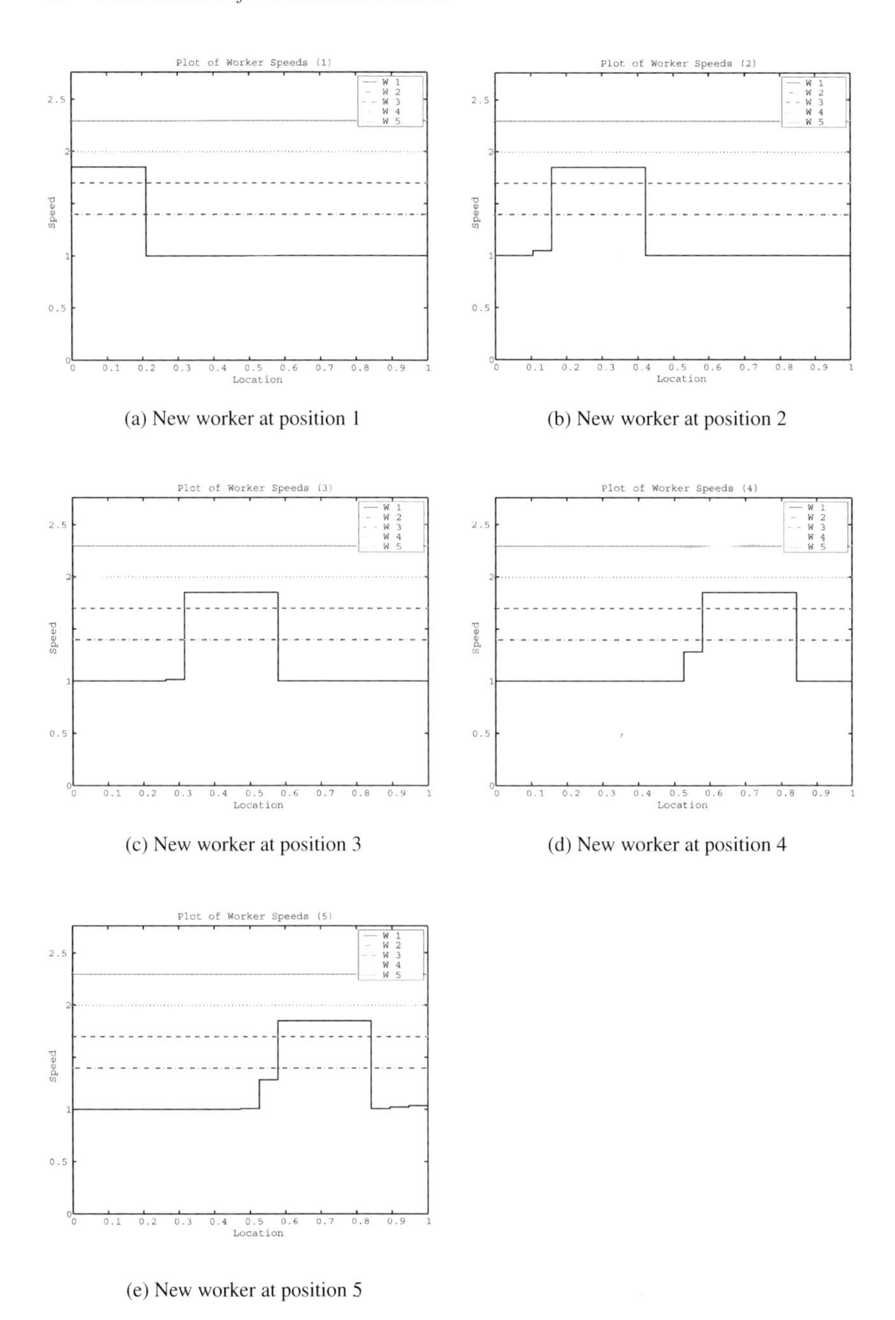

(a) New worker at position 1

(b) New worker at position 2

(c) New worker at position 3

(d) New worker at position 4

(e) New worker at position 5

Figure 1.2: Asymptotic velocity distributions for adding a new worker at different places in an existing bucket brigade.

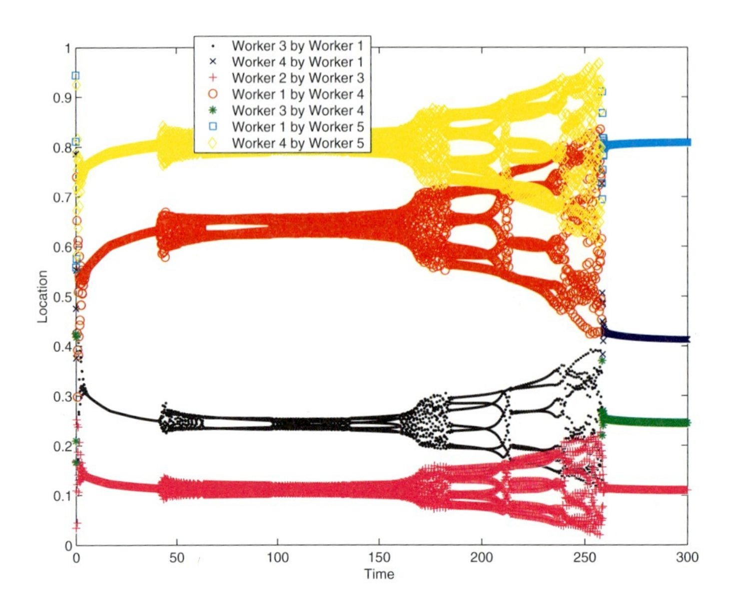

Figure 1.3: Handover points for a four-worker bucket brigade with one additional learner (worker 1).

They consider the length of a queue $x(t)$ as a continuous variable whose rate of change is given by:

$$\frac{dx}{dt} = \begin{cases} \lambda(t) - \mu(t) & \text{for } x(t) \neq 0, \\ 0 & \text{for } x(t) = 0, \end{cases} \tag{1.5}$$

where $\lambda(t)$ is the arrival rate and $\mu(t)$ the processing rate of the queue. Notice that while in general λ and μ are random variables and hence Eq. (1.5) represents a stochastic differential equation, the usual models assume nice distributions for the arrival and departure processes and hence use average start and processing rates. This basic building block for a queue can be connected to a *work-conserving fluid model* by feeding the outflux of each queue into other queues. The basic nonlinearity of this model stems from the fact that there is not a simple right-hand side to Eq. (1.5) but a discontinuous flow. In addition scheduling policies will influence the influx rate λ and the production rate μ. Dai and Weiss [11] have analyzed the relationship between the stability of the fluid model and the stability of scheduling policies for the associated queueing networks. Here stability of the fluid model is represented by the boundedness of the fluid variables for a given influx $\lambda(t)$ which is assumed to be less than the smallest processing rate $\mu_i(t)$ of any queue in the network. Stability in the queueing theory sense is given by a unique stationary distribution ψ for the underlying stochastic process describing the queueing network. Dai and Weiss [11] in particular showed that a queueing discipline is stable if the corresponding fluid model is stable.

The most well-known fluid model that showed chaotic dynamics is called the switched arrival system [9]. A flow model for N parallel machines and one switching server that distributes work over the machines is considered. Work drains at a constant work rate while

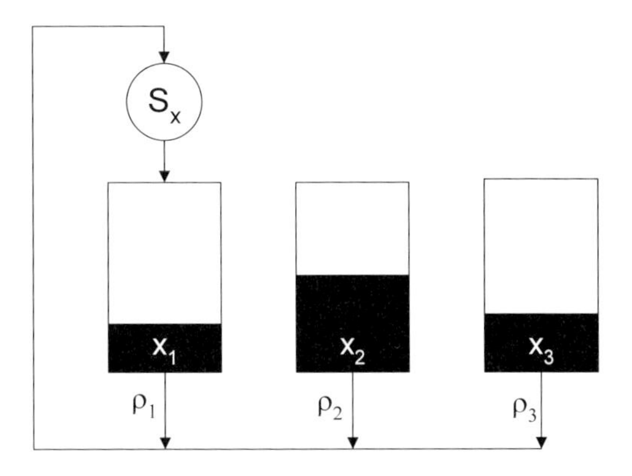

Figure 1.4: A switched arrival system for three machines.

the server continues to fill a buffer until another buffer empties (see Fig. 1.4). At that moment the server switches instantaneously to the empty buffer. The rate at which the server fills a buffer is equal to the sum of all the work coming out of all the buffers. This models a policy known in industrial engineering as the CONWIP policy (constant work in progress) which releases a new product into a machine or factory once a finished product has been released. Hence the system is closed and the total amount of work in the system is constant. Let $x_i(t)$ denote the amount of work in buffer i at time t. We choose the total amount of work in all of the buffers to be 1, i.e.

$$\sum_{i=1}^{N} x_i = 1. \tag{1.6}$$

We also choose the time to produce the total amount of work as the characteristic time and set it to 1 also. Hence the production rates ρ_i satisfy

$$\sum_{i=1}^{N} \rho_i = 1. \tag{1.7}$$

Let j be the position of the server. The server switches to the first buffer that empties. This happens after a time

$$\tau = \min_{i \neq j} \{x_i(t)/\rho_i\}. \tag{1.8}$$

For $t_0 \leq t \leq t_0 + \tau$ the buffer state is determined by the following linear equations:

$$x_i(t) = \begin{cases} x_i(t_0) - \rho_i(t - t_0) & \text{for } i \neq j, \\ x_i(t_0) + (1 - \rho_i)(t - t_0) & \text{for } i = j. \end{cases} \tag{1.9}$$

When this system is sampled at the times when a buffer empties the continuous model becomes a discrete-event model of the form

$$G(x) = x + \min_{k \neq j} \left(\frac{x_k}{\rho_k} \right) (\mathbf{1}_j - \rho), \tag{1.10}$$

where $\mathbf{1}_j$ is a vector with all zeros except for a 1 in the jth position and ρ is a vector containing the work rates ρ_i. The buffer state evolves on the simplex $\sum x_i = 1$. For $N = 3$ machines this simplex becomes an equilateral triangle and the hybrid system $G(x)$ maps the boundaries of the triangle onto each other. Obviously each individual map in $G(x)$ is expanding on the average with a rate of 2. Hence, no matter what the production rates of the different machines are, there is always at least one region in phase space that is chaotic. Whether that region is a chaotic attractor or just leads to transient chaos depends on the details of the production rates. For instance, for the completely symmetric case when all the machines are identical and hence the production rate is $\rho = 1/3$, the expansion is constant everywhere and the associated Lyapunov exponent is given as $\ln 2$. Peters et al. [33] have recently analyzed the bifurcations associated with varying the individual production rates of the machines. Katzorke et al. [18, 36] have studied a discretized version of this problem representing a discrete order flow. They show that the dynamics becomes periodic for rational production rates. Control schemes employing finite buffer sizes [15], setup times switching from one buffer to another [38, 39] or timed idling of one machine [18, 35] have been discussed.

One major drawback of the switched arrival system is the fact that the chaotic dynamics is strictly internal to the production – the total throughput through the set of machines is always constant and does not reflect the chaotic dynamics. Rem and Armbruster [35] introduced switching or maintenance time into the model. The basic idea is that the server cannot instantaneously switch from one machine to the other but needs a certain setup time during which the machine will not produce. In that way, the system will lose production whenever a queue is empty. The outflux values will typically not be chaotic but switch between a few discrete levels. However, the time between switches will become chaotic. Coupling switched arrival systems horizontally and vertically now leads to a much more realistic model of a layered production system. The analysis in [35] shows that the average throughput for the different layers will self-adjust to a steady state, although the dynamics of each individual layer is chaotic. Figure 1.5 shows the aggregate work levels in a three-layer, three-machine network working under the switched arrival protocol and with a CONWIP policy.

Coupling the switched arrival systems to a network opens up many new and exciting research avenues:

- The consequences of self-organized production networks are intriguing. So far, a typical production network is set up to have a constant flux through the network. This is typically done by expensive studies that determine the actual throughput at any particular station. Observed bottlenecks are identified and additional resources are allocated to them if possible. If production networks are able to organize themselves, that would reduce the need for difficult measurements and adjustments. It is interesting to note that the theme of "self-organization", which is the major rationale for the linear bucket brigade production system in the last section, is picked up in these production networks too. Further studies on the performance of self-organized networks and their stability towards systematic and stochastic parameter changes will be very important.

- The production networks are natural examples for a merging of the concepts of control of chaos [32] and chaotic synchronization [34]. So far, individual layers of machines have been controlled to optimal trajectories. On the other hand, while the network does not seem to synchronize completely, it nevertheless exchanges information between layers that

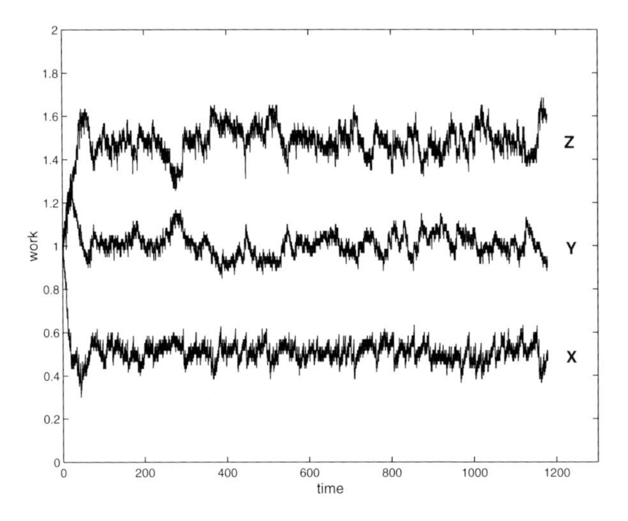

Figure 1.5: Work levels as a function of time for the three-layer system.

leads to the described self-organization. An obvious extension of the concepts of control and synchronization in chaotic systems is to determine whether there exist parameter regimes and production policies such that by controlling one layer in the network to an optimal trajectory, the other layers will become synchronized to that optimal dynamics. It seems obvious that by controlling *one* layer you will not be able to control a very large network but it is unclear whether there exists a neighborhood of layers that can be controlled. Issues of global vs. local control will be studied. Work along these lines is in progress.

1.4 Dynamics of Supply Chains

In recent years the focus of research into production dynamics has switched from modeling and optimizing individual factories to the study of a whole supply chain from the supplier's supplier to the customer's customer. One of the major open problem in that field is the question of how to scale up the simulation and analysis that worked well for an individual factory to the much larger problem of a whole supply chain. This fact was acknowledged by NSF when they started an RFP entitled "Scalable Enterprise Initiative" [31] that called for new ideas to simulate production flows in large systems. One typical response to that problem is to use discrete event simulators which have been quite successful as models for individual factories to model a whole supply network. The major problem with such an approach is that typical simulations for reasonably complicated production (e.g. in the semiconductor industry) take much too long. As a result, DES are not a useful tool for fast evaluations of management and policy decisions. Asynchronous and parallel simulations [22], together with the usual increases in raw computer power, may over time solve that problem.

An alternative approach has been developed recently [1, 2] extending concepts from gas dynamics and traffic theory to production flows (see also [20, 14]). This approach not only will allow very fast simulations, it also allows for hierarchical models that, at least on their lowest levels, can be understood qualitatively, leading to *insight* into the dynamics of supply chains. While fluid systems (see Sect. 1.3) consider the work in the factory as a continuous variable, they still describe the flow through the factory by a discrete graph: work moves from one machine to another, moving from one node in a graph to the next. In order to generate a simple universal model for flows in a factory we assume that a product going through our factory undergoes many steps and that a limiting description of a continuous production line is a reasonable model. In that way we actually merge features of the underlying models of the bucket brigade and the fluid models. As a result, the relevant quantity moving through a factory is the work density (WIP density) $\rho(\xi, t)$ describing the amount of parts at time t at completion stage ξ. Again, $\xi = 0$ denotes the beginning of the factory while a part at $\xi = 1$ is completed. A *work-conserving* production hence naturally leads to a conservation law for the WIP density:

$$\rho_t + (v(\rho)\rho)_\xi = 0, \tag{1.11}$$

where $v(\rho)$ describes the velocity of the product moving in the factory. The exact nature of the transport velocity $v(\rho)$ is the major modeling issue. Following adiabatic gas theory or Lighthill and Whitham's [25] traffic model we may describe the functional dependence of the velocity via a state equation

$$v = f(\rho). \tag{1.12}$$

We note that the units of ρ are [parts][stage] and the units of $v(\rho)\rho$ are [parts][time]. This suggests that in conventional nomenclature of process control and performance simulation $\rho(\xi, t)$ represents local WIP density and the flux $\rho(\xi, t)v(\xi, t)$ is the local throughput at stage ξ at time t, respectively. As a starting point for the state equation it has to respect the fundamental law of factory physics, Little's law [26]. For a single factory Little's law may be written as

$$N = \tau\lambda, \tag{1.13}$$

where N is the time-averaged load of the factory (WIP), τ is the mean cycle or throughput time over all outputs (TPT) and λ is the start rate. Little's law is fundamentally a deterministic law and results from mass conservation. However, by amending it with a description of the stochastic processes in a factory we can generate a state equation characterizing the factory. In its simplest case the stochastic process is represented by its means. For instance, modeling a factory as a single linear queue with Markov arrival and processing rates (an $M/M/1$ queue) the mean cycle time τ can be determined as a function of the start rate λ and processing rate μ [13] to be

$$\tau = \frac{1}{\mu - \lambda}. \tag{1.14}$$

Therefore, the relationship between WIP and cycle time becomes

$$\tau = \frac{1}{\mu}(1 + N). \tag{1.15}$$

Equation (1.14) shows that μ becomes the maximal or critical start rate. With $v = 1/\tau$ we have the desired state equation. Equation (1.15) shows that WIP is a linear function of the cycle time with a slope μ. The linear relationship between TPT and WIP is intuitively obvious: a part entering the queue will have to wait until the queue is served (N/μ time units) plus its own processing time, $1/\mu$ time units.

A linear relationship as in Eq. (1.15) is true for a queueing network that has product form [29, 6], i.e. the whole network can be replaced by an effective queue. However there are several key factors that will lead to a queueing network that cannot be approximated by a product network with constant production rates. Specifically, semiconductor production lines are re-entrant with highly complex topologies. As a result they show moving bottlenecks and other nonlinear behavior. In addition, there is anecdotal evidence from real re-entrant factories [17] that show a stronger than linear increase in the average throughput time τ as the loading of the factory is increased. In addition, Lu et al. [27] show that through dispatch and scheduling rules one can reduce the variance of cycle times inside semiconductor manufacturing plants while Dai and Weiss [11] show that some scheduling rules lead to unstable networks, i.e. WIP grows to infinity. Furthermore, operator fatigue will lead to increased error rates, as the load in the factory is increased. As a result, the network cannot be approximated by a product network with constant production rates any more but rather through queues whose production rate depends on the length of the queue. Unfortunately, the true state equation for a re-entrant factory is not known, as large factories are rarely (if ever) run in equilibrium and controlled experiments are obviously impossible. It is worth noting here that discrete event simulators are not a good substitute for real experiments specifically if human interaction is part of the cause for the increase in variance.

In order to develop the type of results that can be achieved with a continuum-model approach based on a conservation-law model and to show the feasibility of fast simulations for large production and supply networks we choose a specific model for a re-entrant factory. We therefore *assume* a state equation analogous to the Lighthill–Whitham traffic model of the form

$$v(\rho) = v_0 \left(1 - \frac{\bar{\rho}(t)}{L}\right), \tag{1.16}$$

with

$$\bar{\rho}(t) = \int_0^1 \rho(\xi, t) d\xi.$$

Here v_0 is the speed for the empty factory and L is the maximal load (capacity of the factory). All the following analysis can easily be adapted to fit any other state equation that has been validated through real data or justified through heuristic arguments. As a result of Eq. (1.16) the throughput time, instead of showing a singularity with respect to the start rate, has a singularity with respect to WIP, i.e.

$$\tau = \frac{\tau_0}{L - \bar{\rho}}, \tag{1.17}$$

where L is called the critical load. Clearly the TPT increases without bound as $\bar{\rho} \to L$. This implies that the factory can only run in equilibrium for WIP less than L. It is instructive to

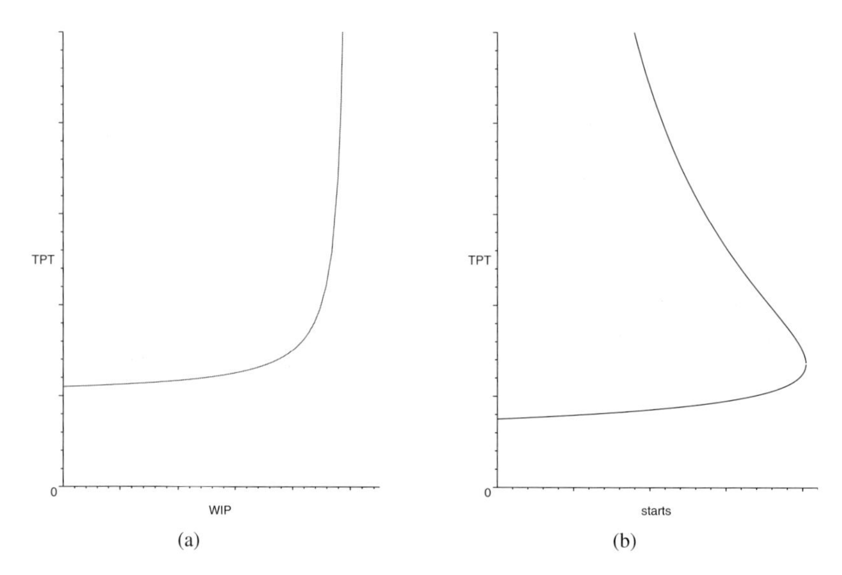

Figure 1.6: (a) Schematic of TPT as a function of WIP, Eq. (1.17), (b) TPT as a function of start rate.

input Eq. (1.17) into Little's law: Fig. 1.6 shows TPT as a function of WIP (Eq. (1.17)) and TPT as a function of start rate. Notice that there is still a critical start rate λ_c: For $\lambda < \lambda_c$ there exist two equilibria, for $\lambda > \lambda_c$ there are none, a case of a saddle-node bifurcation.

There is a crucial difference between the state equation for our factory models, Eq. (1.17), and the Lighthill–Whitham model for traffic flow. Our factory models represent re-entrant factories. Therefore, as a first-order model, the velocity is uniform in the whole factory and a local density fluctuation influences the velocity everywhere. In contrast, traffic velocity depends on the local density only. A more accurate model could use a state equation of the form

$$v(\xi, t) = v_0 \left(1 - \frac{\int_0^1 w(\xi, s)\rho(s, t)ds}{L} \right).$$

Here appropriate weight functions $w(\xi, s)$ describe the influence of WIP at stage s on the velocity of WIP at stage ξ and can be used to model in more detail the complicated topologies of re-entrant product flow.

To conclude, we re-state the full model, with $g(\xi)$ an initial density distribution, and taking into account that v as in Eq. (1.17) does not depend on ξ:

$$\begin{aligned} &\rho_t + v(\rho)\rho_\xi = 0, \\ &v(\rho) = v_0 \left(1 - \frac{\bar{\rho}(t)}{L} \right), \\ &\rho(\xi, 0) = g(\xi), \\ &\rho(0, t)v(t) = \lambda(t). \end{aligned} \tag{1.18}$$

Notice that the start rate $\lambda(t)$ into the factory enters as the boundary condition for the local throughput at $\xi = 0$.

This model shares many features with thermodynamic transport equations. In particular, modeling the relationship between density and velocity through a state equation is typically called an adiabatic approximation: factory flow is modeled as if it were always in equilibrium, following adiabatically the state equation (1.17). As a result, we can expect that fast transients will be modeled poorly but that slow transients and averages will be modeled very well.

A more elaborate model including a *dynamic* equation for the evolution of the velocity is presented in [2]. Here we derive the Boltzmann or Euler equations for the transport along a factory and generate an additional conservation law for the time evolution of the velocity. Modeling a factory or part of a factory as a single queue (linear flow, not re-entrant), we derive

$$\begin{aligned} &\rho_t + (v(\rho)\rho)_\xi = 0, \\ &v_t + v v_\xi = 0, \end{aligned} \tag{1.19}$$

with boundary conditions

$$\begin{aligned} &v(0,t) = \frac{v_0}{1+\bar{\rho}(t)}, \\ &\bar{\rho}(t) = \int_0^1 \rho(\xi,t)d\xi, \\ &v(0,t)\rho(0,t) = \lambda(t), \end{aligned}$$

where $\lambda(t)$ is the prescribed influx into the factory and $1/v_0$ is the production rate of the queue.

1.4.1 Simulation and Control

Modeling production flow through conservation laws of hydrodynamic type opens up the vast literature of scientific computing for these equations (e.g. [23, 24]). As a result, fast and accurate simulations of Eq. (1.18) or Eq. (1.19) are straightforward. In addition, theoretical analysis of Eq. (1.18) can be done. For instance, linear stability analysis of the two equilibria of Fig. 1.6 leads to a linear delay equation with the high-speed–low-WIP equilibrium being stable and the low-speed–high-WIP equilibrium being unstable.

The method of characteristics can be used to solve the following control problem: change the production flow of the factory from one steady state, corresponding to an outflux that meets a specific constant demand, to another one. We assume that the demand $d(t)$ changes like

$$d(t) = \begin{cases} d_1 & t < 0, \\ d_2 & t > 0. \end{cases}$$

Due to the production lag, any change in the start of the factory will leave the factory at some later time. Hence there will be a mismatch of the desired output d_2 and the actually produced output at least until the change in influx has moved through the factory. We call that mismatch *backlog*. The control problem is to design an influx $\lambda(t)$ that would move the system from the equilibrium ρ_1 corresponding to a production rate of d_1 to a new equilibrium ρ_2 corresponding to a production rate d_2, generating an output such that the backlog becomes zero in the shortest possible time. We show in [1] how to arrive at the following heuristic:

If the jump in demand is small (leading to a small enough backlog) then the optimal strategy will be to get the factory to its maximal density instantly and keep it there for the right amount of time, such that when this extra product leaves the factory, the backlog is zero.

It is an open problem to design an optimal control strategy for *all* jumps in demand. Simulations and control experiments can be viewed on the webpage `http://math.duke.edu/~daniel/supply_chain.html`.

Acknowledgment

This paper reviewed collaborative work with Esma Gel, Karl Kempf, Dan Marthaler, Junko Murakami, Bart Rem and Christian Ringhofer. Support through grants from NSF (DMS-0204543) and from Intel Corporation is gratefully acknowledged.

Bibliography

[1] D. Armbruster, D. Marthaler, and C. Ringhofer, *A continuum model for a re-entrant factory*, preprint, Arizona State University (2002).

[2] D. Armbruster, D. Marthaler, C. Ringhofer, *Kinetic and fluid model hierarchies for supply chains*, to appear in SIAM MMS (2003).

[3] D. Armbruster, E.S. Gel, *Bucket Brigades Revisited: Are they always effective?* preprint, Arizona State University (2003).

[4] D. Armbruster, E.S. Gel, and J. Murakami, *Bucket brigades and learning presented at INFORMS 2002*, San Jose, paper in preparation (2003).

[5] J.J. Bartholdi and D.D. Eisenstein, *A production line that balances itself*, Operations Research **44**(1) (1996), 21–34.

[6] F. Baskett, K.M. Chandy, R.R. Muntz, and F.G. Palacios, *Open, closed and mixed networks of queues with different classes of customers*, Journal of the Association of Computing Machinery **22** (1975), 248–260.

[7] D.P. Bischak, *Performance of a manufacturing module with moving workers*, IIE Transactions **28** (1996), 723–733.

[8] L.A. Bunimovich, *Method of stabilization of a target regime in manufacturing and logistics*, Chapter 2, this book.

[9] C.J. Chase, J. Serrano, and P.J. Ramadge, *Periodicity and chaos from switched flow systems: contrasting examples of discretely controlled continuous systems*, IEEE Transactions on Automatic Control **38**(1) (1993), 70–83.

[10] H. Chen, J.M. Harrison, A.Mandelbaum, A. Van Ackere, and L.M. Wein, *Empirical evaluation of a queueing network model for semiconductor wafer fabrication*, Operations Research **36** (1988), 202–215.

[11] J.G. Dai and G. Weiss, *Stability and isntability of fluid models for certain re-entrant lines*, Mathematics of Operations Research **21** (1996), 115–134.

[12] See e.g. Factory Explorer, WWK products (1996).

[13] D. Gross and C.M. Harris, *Fundamentals of Queueing Theory*, Wiley, New York, 1985.

[14] D. Helbing, *Modeling and optimization of supply networks: Lessons from traffic dynamics*, Chapter 6, this book.

[15] C. Horn and P.J. Ramadge, *A topological analysis of a family of dynamical systems with non-standard chaotic and periodic behavior*, International Journal of Control **67**(6) 1997, 979–996.

[16] F.M. Bevis and D.R. Towill, *Continued learning and the work study observer*, Work Study and Management Services **18** (1974), 420–427.

[17] K. Kempf, Intel, personal communications (2001).

[18] I. Katzorke and A. Pikovsky, *Chaos and complexity in simple models of production dynamics*, Discrete Dynamics in Nature and Society **5** (2000), 179–187.

[19] P.R. Kumar, *Re-entrant lines*, Queueing Systems **13** (1993), 87–110.

[20] E. Lefeber, *Nonlinear models for control of manufacturing systems*, Chapter 5, this book.

[21] R.I. Leine, D.H. Van Campen, and B.L. Van der Vrande, *Bifurcations in nonlinear discontinuous systems*, Nonlinear Dynamics **23** (2000), 105–164.

[22] P. Lendermann, B.P. Gan, and L.F. McGinnis, *Distributed simulation with incorporated APS procedures for high-fidelity supply chain optimization*, in: Proceedings of the 2001 Winter Simulation Conference, Washington DC (2001), pp. 1138–1145.

[23] R.J. LeVeque, *Finite Difference Methods for Differential Equations*. Draft version for use in AMath 585-6, University of Washington (1998).

[24] R.J. LeVeque, *Numerical Methods for Conservation Laws*, Birkhäuser-Verlag, Basel, 1992.

[25] M.J. Lighthill and G.B. Whitham, *On kinematic waves II. A theory of traffic flow on long crowded roads*, Proceedings of the Royal Society, Series A, **229** (1955), 317–345.

[26] J.D.C. Little, *A proof for the queuing formula* $L = \lambda W$, Operations Research **9** (1961), 383–387.

[27] S.C.H. Lu, D. Ramaswamy, and P.R. Kumar, *Efficient scheduling policies to reduce mean and variance of cycle-time in semiconductor manufacturing plants*, IEEE Transactions on Semiconductor Manufacturing **7** (1994), 374–385.

[28] G.F. Newell, *Approximation methods for queues with application to the fixed-cycle traffic light*, SIAM Review **7**(2) (1965), 223–240.

[29] R. Nelson, *Probability, Stochastic Processes, and Queueing Theory*, Springer Verlag, New York, 1995.

[30] G.F. Newell, *Scheduling, location, transportation and continuum mechanics; some simple approximations to optimization problems*, SIAM Journal of Applied Mathematics **25**(3) (1973), 346–360.

[31] Scalable Enterprise Initiative, NSF 2000.

[32] E. Ott, C. Grebogi, and J.A. Yorke, *Controlling Chaos*, Phys. Rev. Lett. **64** (1990), 3215.

[33] K. Peters, J. Worbs, U. Parlitz, and H.-P.Wiendahl, *Manufacturing systems with restricted buffer sizes*, Chapter 3, this book.

[34] A. Pikovsky, M. Rosenblum, and J. Kurths, *Synchronization*, Cambridge University Press, 2001.

[35] B. Rem and D. Armbruster, *Control and synchronization in switched arrival systems*, Chaos **13**(1) (2003), 128–137.

[36] B. Scholz-Reiter, M. Freitag, A. Schmieder, A, Pikovsky, and I. Katzorke, *Modelling and analysis of a re-entrant manufacturing system*, Chapter 4, this book.

[37] S.N. Simic, K.H. Johansson, S. Sastry, and J. Lygeros, *Towards a geometric theory of hybrid systems*, in: *Hybrid Systems: Computation and Control*, edited by N. Lynch and B.H. Krogh, LNCS **1790**, pp. 421–436, Springer, 2000.

[38] T. Ushio, H. Ueda, and K. Hirai, *Controlling chaos in a switched arrival system*, Systems & Control Letters **26** (1995), 335–339.

[39] T. Ushio, H. Ueda, and K. Hirai, *Control of Chaos in Switched Arrival Systems with N Buffers*, Electronics and Communications in Japan, Part 3, **83**(8) (2000), 81–86.

2 Method of Stabilization of a Target Regime in Manufacturing and Logistics

L.A. Bunimovich

We discuss a hierarchy of models of production lines in work-sharing manufacturing. This hierarchy is based on the constraints which appear in concrete applications, e.g. in the apparel industry, supply chains, condition-based maintenance, etc. The dynamics of the simplest (basic) models in the hierarchy is completely described. These results allow us to suggest a general approach to the optimization of many systems in operations research and logistics, stabilization of a target regime, and an algorithm for its implementation, which proved to be efficient in applications.

2.1 Introduction

Evolution of systems in manufacturing and logistics consists of alternating productive and decision-making stages. At the productive stage the system (e.g. an assembly line) operates continuously, and all the productive elements (workers, robots, etc.) at the corresponding production lines work at this stage (in their intrinisic regimes). In an ideal situation this stage of production would last forever. However, in reality it will unavoidably stop because there is no more unfinished work (work in progress, WIP) on the line, some worker got bumped by an upstream worker (collision), etc. At such moments the system cannot operate in the same regime as it used before, and some actions must be made to "tell" the system what it should do next. It is in these stages that the system should be managed, i.e. the continuous production gets interrupted by some decision-making process.

A standard goal in managing any production line is to increase the throughput. The main difficulty in the management of a flow line is to balance it, i.e. to divide the work equally among the productive elements (workers). The corresponding rules for decision making are called policies.

A typical logistics system consists of a set of servers and of a set of customers. The examples include production lines where the workers (or unfinished items which they carry) are customers and workstations (machines) are servers, the fleets of aircraft where servers are the flights from some fixed schedule or gates in airports and aircraft are the customers, etc. Customers form queues which wait for the servers (facilities).

A policy says which customer from the queue the server should pick after serving another customer. For instance, the most popular (and in a sense, the most natural) policy is first-in-first-out (FIFO). The problem is to find such a policy which maximizes throughput, i.e. which minimizes an average time spent by the customers in the system (an average service time).

Nonlinear Dynamics of Production Systems. Edited by G. Radons and R. Neugebauer

ISBN 3-527-40430-9

In operations research and logistics all such systems were traditionally considered as intrinsically stochastic ones. This approach seems to be very natural because of the complexity of such systems and the presence of many elements (servers and customers) as well as of many uncontrollable factors. Actually some branches of probability theory, such as queueing theory, inventory control, etc., appeared because of these applications.

However, rather recently another approach emerged, where the systems were modeled in logistics and operations research was modeled as dynamical systems. Such dynamical systems are inherently nonlinear. Indeed, as we already mentioned, the evolution of such systems consists of continuous stages (of "production") interrupted by the stages of decision making. Therefore these models should be piecewise continuous, i.e. nonlinear ones.

The advantage of the (nonlinear) dynamics approach is that it is aimed at understanding global dynamics as well as local dynamics of the system, while the stochastic modeling deals with statistical analysis, which essentially focuses only on the average characteristics of the dynamics. Both approaches are useful though and have their advantages. For instance, stochastic modeling usually is easier to perform because it does not require a very careful analytic approximation of the corresponding (empirical) distributions. Everybody knows that a reasonable approximation can be found among one out of 10–15 standard distributions. In fact statistical (average) characteristics are much more robust than the characteristics of individual orbits. Dynamical (deterministic) systems, though, even with a few degrees of freedom, can exhibit a very complicated behavior which is difficult to study analytically.

It is clear, however, that the best approach to study the evolution of any system would be to develop a relevant dynamical model of this system, to study it analytically and then to investigate how the always present random noise influences the dynamics. However, this approach is not very practical, especially for such a rich variety of complex systems which one encounters in industry and manufacturing, where room for experiments is also rather limited. Therefore some new general approaches should be developed to tackle these systems. One such approach has been recently suggested [1]. It is built upon the algorithm of optimization [2] developed for production lines in work-sharing manufacturing. In work-sharing manufacturing each productive element (worker) has the skills to work at any place (on any workstation) on the production line.

Three major components of this approach are the method of stabilization of the target regime, a hierarchy of relevant models and the optimization algorithm for work-sharing manufacturing.

2.1.1 Stabilization of a Target Regime (STR Method)

Although exact relevant models of manufacturing processes are usually not known, it is quite clear what are the goals in managing any concrete system. Therefore any manager (and often a worker) would tell you what is the optimal (target) regime for this system.

From the point of view of dynamics a regime (of evolution) is nothing else than an orbit (trajectory) of the system. Therefore the first task is to organize the decision-making (managing) process so that the orbits with this optimal regime are realized in the operation of this system.

However, it is not enough. Such regimes should be stable, i.e. the system should be able to re-instate such a regime under the action of the (always present) perturbations. Therefore, it is another task for the decision making to *stabilize* this target regime (orbit). In principle, the system can have several optimal orbits (regimes). To make the system operate in the target regime the decision-making process must ensure that all (or almost all, i.e. with probability 1, where probability is just the volume in the phase space of our system, formed by all possible initial conditions) orbits of the system will converge (be attracted) to one of the target orbits (i.e. orbits which correspond to the target regime). As the result the target regime becomes globally stable.

Therefore, the implementation of the STR method consists of the following stages of the decision-making process.

(i) Make the target regime realized in the set of all orbits of the system.

(ii) Stabilize the orbits with the target regime, i.e. make them the only stable asymptotic orbits in the system.

STR would remain just an abstract approach unless there are situations where it could be effectively applied. We will show that there is a large class of models, which naturally appear in applications, where this method works. Moreover, it is possible for the models in this class to ensure that there exists only one orbit which carries a target regime, and this orbit is a global attractor in the corresponding system. In particular, the applications discussed in [2–5] can be viewed as special examples of the general STR method. Another specific example is provided by the chaos-control approach.

2.1.2 Constraints-based Hierarchy of Models

Several classes of models relevant to many processes in operations research and logistics were introduced recently [1]. These classes form a hierarchy. Such hierarchy (or sometimes hierarchies) of basic models, from the simplest ones to more complicated, exists virtually in all branches of science and engineering. They are usually based on a number of parameters, equations, etc. The hierarchy of dynamical models for systems in logistics, manufacturing and operations research which we propose [1] is rather based on the constraints imposed on the corresponding system. Naturally for more sophisticated (complex) models not only the analysis becomes more complicated but also some new questions arise.

The simplest models in this hierarchy can be completely studied analytically [1]. Therefore they can serve as the basic models for the theory of dynamical systems in manufacturing, which currently cannot even be considered as being in an infancy, but hopefully will be born soon. Such completely solvable models proved to be especially important in nonlinear dynamics where the behavior of the systems is intrinsically complex.

2.1.3 The Algorithm of the Optimal Management of the Systems in Work-sharing Manufacturing

The STR method together with the comprehensive analysis of some models in the hierarchy suggests an algorithm which allows for the optimal management of such systems. This algorithm consists of two steps. In the first all optimal possibilities (decisions) should be determined for an unconstrained model. Then out of these optimal decisions the one should be chosen that is the most consistent with the existing constraints. For instance, out of all possible optimal partitions of the entire work between the workers on a homogeneous production line should be chosen the one which is the closest to the configuration of the real line (i.e. partition of this line into workstations).

The structure of the paper is the following. In Sect. 2.2 we define the hierarchy of models under study. Section 2.3 describes their dynamics. In Sect. 2.4 we discuss a general algorithm of optimization for such systems. Section 2.5 discusses some possible generalizations and other applications of the proposed approach.

2.2 The Hierarchy of Models

Consider a dynamical system generated by the motion of N particles (workers) in a straight segment $[0, L], 0 \leq L < \infty$. Each particle moves with the velocity $v_i(x) > 0, i = 1, 2, \ldots, N$, $x \in [0, L]$. Particles are allowed to pass each other.

This model is called the one-way street (OWS) [1]. It describes the motion of (unfinished) items along a production line (where passing is allowed, e.g. a line with parallel workstations), the motion of (a brigade) of pickers along the aisles of a warehouse, the motion of goods in a simple supply chain, etc.

Certainly the dynamics of the OWS model is not completely defined yet. Indeed, what happens when one of the particles reaches the end ($x = L$) of the street? It is exactly at this moment that the decision-making process comes into play. There are various possibilities here. For instance, should this particle (if it is, for example, a public bus or any other asset in some fleet of assets) go for maintenance or should it be placed at the origin $x = 0$ or some other (which one?) point $0 < x < L$? Or, should this particle (worker) take raw material and start to work with a new item at $x = 0$ or should it take over the unfinished item which is carried by the next worker upstream?

All these are the examples of different rules (decisions) which are actually implemented in some applications. Together with such rule the dynamics of the OWS model becomes completely defined. We will refer to the OWS model as the one where decision rules are not specified. Therefore it is in fact not just one model but a class of models.

One gets the one-way narrow street (OWNS) model if in the OWS model the particles are not allowed to pass each other. This restriction is imposed in many production lines, motion of pickers in warehouses, etc. [2–5]. Therefore the positions $x_i(t)$, $i = 1, 2, \ldots, N$, of the particles at any moment of time t satisfy the relations $0 \leq x_1(t) \leq x_2(t) \leq \cdots \leq x_n(t) \leq L$.

Observe that for the OWNS model decision making comes into play not only when one of the workers reaches the end $x = L$. Indeed, some particles could be faster than their downstream neighbors. Then at some moment $\hat{t}$ the positions of such particles can become equal, i.e. $x_i(\hat{t}) = x_{i+1}(\hat{t}), i \leq N-1$. Such moments of time will be called blocking times. At the blocking times decisions must be made on how the system should proceed further because the passing is not allowed.

One of the possibilities here [4, 5] is that the ith particle starts to move with the velocity $v_{i+1}(x)$ of the downstream particle until a point x', where $v_i(x') < v_{i+1}(x')$. Such a point may not exist though and then another decision problem pops up when both these workers simultaneously reach the end $x = L$.

There are different decision rules which naturally come to mind. These rules characterize different approaches to the manufacturing in production lines. (Recall that we consider here only work-sharing manufacturing, i.e. it is assumed that each worker has the skills to work at any position (workstation) on the line, or any particle is allowed to visit the entire street.)

The oldest, and still the most used approach is zone manufacturing, where each worker is assigned to some zone (usually a collection of several consecutive workstations) on a production line. Therefore in zone manufacturing each worker is allowed to work only in his zone. It is a very hard (and often impossible) task to balance (equally distribute the work in) an assembly line in the framework of zone manufacturing. In particular, it requires a complicated work-content model which is very hard (if ever possible) to develop and, besides, the zones require constant monitoring.

There are two major challenges in manufacturing to ensure an optimal throughput. The first is to keep a high production rate, i.e. optimally all workers should work all the time. It is the case in zone manufacturing. However, another obvious condition of a high throughput is a fast removal of finished items from the line, and minimization of work in progress, i.e. of a number of unfinished items on a line. Clearly, in zone manufacturing (unless the zones are created ideally and are adjusted all the time) there is always a bottleneck (e.g. the least skillful, i.e. the slowest worker, or the one who was assigned to the most time-consuming zone, etc.). Observe, also, that zones can consist only of the (already existing) workstations (in a rather general situation when only one worker is allowed to work on any workstation at a time). Therefore the options to create different zones can also be quite limited.

To overcome these difficulties Toyota introduced the so-called TSS (Toyota Sewing Products Management System) rules, where a concept of fixed zones has been abolished. TSS rules refer to the flow lines which are OWNS systems. The first TSS rule is that the last (downstream) worker after finishing his product sends it off and then walks back upstream to take over the work of his predecessor, who walks back and takes over the work of his predecessor and so on, until after relinquishing his product, the first worker walks back to the start to begin a new product.

OWNS models with the first TSS rule are sometimes called bucket brigade (BB) production lines because motion of workers in these lines resembles that of people trying to fight a fire. (In this case $x = 0$ corresponds to the well, and $x = L$, the end of the "production line", corresponds to the fire.)

Therefore in BB production lines there are no fixed zones. The main problem in managing BB lines is to find an optimal sequencing of workers. Another problem is to compute the production rate of a BB. Indeed, contrary to the production lines with fixed zones, in BBs workers may not work all the time with their intrinsic production rates, but they must sometimes slow down when being blocked by the next worker downstream. Already there is some confusion with the notion of BB lines in the literature devoted to the dynamics of production lines. Sometimes the authors refer to BBs to mean only such BB production lines where the workers are sequencing along (downstream) the line in such a way that $v_1(x) \leq v_2(x) \leq \cdots \leq v_N(x)$, $x \in [0, L]$, i.e. in the increasing order of their velocities. Observe that if $v_i(x) > v_j(x)$ but $v_i(y) < v_j(y)$ for some points $x, y \in [0, L]$ then the sequencing from the slowest to the fastest is just impossible (not well defined). We mean by a BB production line any production line where decision making is done according to the TSS rules at the moments of blocking as well as when the last (downstream) worker finishes his job.

Still, BB production lines are not completely defined yet. Indeed, the first TSS rule says what is the rule of operation of the line only in the case when the last worker finished the job, but it says nothing about a decision that should be made at the moments of blocking. There is a natural reason for that because TSS rules were invented specifically for the apparel industry where production lines consist of workstations. Because in this industry two workers cannot simultaneously work on the same workstation, then, according to the second TSS rule, all blocked workers must wait until the corresponding workstation(s) become available.

However the simplest models of flow lines are not separated into pieces (workstations) [1, 2, 4, 5]. Therefore, at the moments of blocking there are two different "natural" decisions to choose from. The one, exploited in [4, 2, 1, 5], says that the blocked worker should continue to work downstream and acquire the velocity of the worker which blocked him.

Another possible decision is for the blocked worker to leave his unfinished item and walk back upstream and take over the work of his predecessor. Then this predecessor also walks upstream and takes over the work of his predecessor, and so on. So, in the event of blocking the line partially resets itself according to the second TSS rule. The obvious advantage of this rule is that no worker ever slow down. However, the WIP increases because the blocked workers leave unfinished items on the line. (It is worthwhile to mention that in all models of the BB lines it is assumed that the workers walk back with "infinite" velocity, i.e. no time is wasted in the resets of the line. This assumption is, in fact, close to reality, because the velocity of walking is much higher than the velocity with which a worker is progressing along the line when he is working.)

We will show that BB lines with such partial resets often demonstrate chaotic behavior. It should be contrasted with the models where blocked workers continue to move along the line, where the dynamics was found to be always eventually periodic [1, 2, 4].

Finally, we introduce a special subclass of the BB models. This subclass is defined by a simplified technical condition that velocities of all workers on the line are constant, i.e. $v_i(x,t) = v_i > 0$ for any $x \in [0, L]$, $t > 0$ and $i = 1, 2, \ldots, N$. These models will be called basic models (BMs). Some of such models were studied numerically in [4] for $N = 2$ and 3. The assumption that the velocities of workers are constant is rather technical. (It is worthwhile to mention though that in real production lines often the velocities of workers are indeed practically constant [3].)

The basic models are particularly important because they (at least in some cases) are completely solvable [1]. Therefore the analysis of the BMs provides the firm intuition for future analytical, numerical and experimental studies of the systems in the work-sharing industries.

We conclude this section by the following diagram representing the proposed hierarchy of the models of the flow lines.

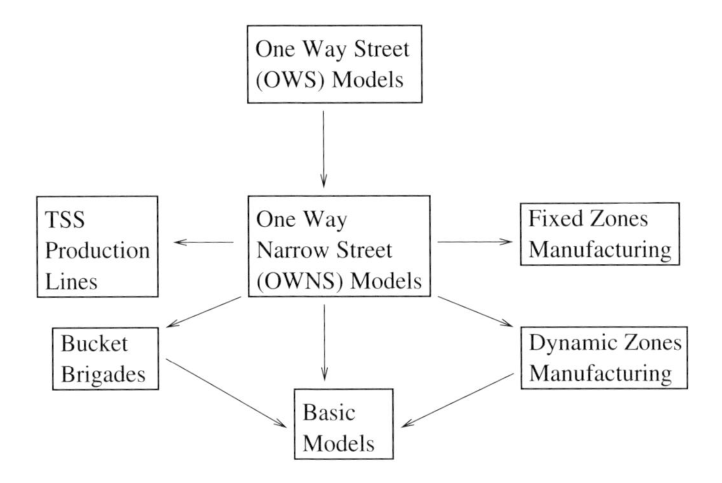

2.3 Dynamics of the Models in the Hierarchy

We start with the obvious observation that dynamics of models in the fixed zones manufacturing is trivial. Indeed, in this case workers do not interact, and they do not even intersect in space, being confined to their zones. Therefore all models in the fixed zones manufacturing deal with non-interacting particles (a kind of an ideal gas). The production rate of brigades of workers in the fixed zones manufacturing is always the maximal possible one, because (it is assumed that) each worker works all the time with his intrinsic rate, and no blocking occurs. WIP though grows with time linearly and therefore (without additional expensive and time consuming to handle buffers) such lines eventually must be reset in some way, which is very difficult (if even possible) to figure out how to properly do it [3, 5–10].

Consider now OWNS models with fixed zones. The phase space of these systems is the N-dimensional simplex $\mathcal{D}_L(N) = \{X = (x_1, x_2, \ldots, x_N) : 0 \leq x_1 \leq x_2 \leq \cdots \leq x_N \leq L\}$, where x_i is the position of the ith worker. These models are dynamical systems with continuous ("real") time. Luckily OWNS models have several global Poincaré sections. (Recall that it is unusual for a generic dynamical system to have even one global Poincaré section.) These global Poincaré sections correspond to the moments of reset. The standard (and arguably the most convenient choice) is $\{x_1 = 0\}$, i.e. we observe the system only at the moments when the first worker is at the beginning of the line (e.g. he takes raw material to start working on a new item). Hence we will follow the evolution of orbits of OWNS models at the moments when all (for BB models) or all partial (for dynamic zones (DZ) models) resets just have been made, i.e. all (or the blocked particle and all particles upstream) jumped into positions of their immediate predecessors, and x_1 becomes equal to zero.

We denote by $f : \tilde{\mathcal{D}}_L(N) \to \tilde{\mathcal{D}}_L(N)$ the corresponding Poincaré map, where $\tilde{\mathcal{D}}_L(N) = \{X \in \mathcal{D}_L(N), x_1 = 0\}$. The map f is continuous provided that the velocities of all particles are bounded from above by some constant. (The proof of the continuity of f [1] does not use the strict positivity of velocities of the workers and can be equally applied, for example, to BBs with workstations.)

As a corollary of the continuity of f one gets [1] that any OWNS model has at least one fixed point. For the dynamical system with continuous time (in the entire simplex $\mathcal{D}_L(N)$) to fixed points of the map f correspond periodic orbits.

The following statement (see [1]) is crucial for understanding the dynamics of the BB models.

Absence of Blocking Lemma. If $X^* = (x_2^*, x_3^*, \dots, x_N^*)$ is a fixed point of the BB model then at this point there is no blocking.

Observe that this statement deals with the discrete-time BB models (or more precisely with their Poincaré sections). Indeed it is easy to construct such examples of BB models (but not of basic models) where some workers were blocked for some time during the periodic motion which corresponds to a fixed point of the Poincaré section. It is easy to see that such blocking can ever occur only if at least for one pair of the particles $v_i(x) > v_{i+k}(x)$ in some subset $A \subset [0, L]$ of the production line and $v_i(x) < v_{i+k}(x)$ in another subset $B \subset [0, L]$.

However the following proposition [1] demonstrates that a (continuous) time-periodic orbit, which corresponds to a fixed point of a BB model with discrete time, has no moments of blocking.

Absence of Blocking at Any Time. Let X^* be a fixed point of the BB model with discrete time. Then the periodic orbit corresponding to X^* of the BB model with continuous time does not have any moments of blocking.

Indeed, let $X^*(t) = (x_1^*(t), x_2^*(t), \dots, x_N^*(t))$ be the orbit of the BB model with continuous time that corresponds to the fixed point X^*. Suppose that there exists such moment of time t' that $x_i^*(t') = x_{i+1}^*(t')$ for some $i = 1, 2, \dots, N-1$, i.e. blocking occurs at $t = t'$. But it means that $x_{i+1}^* < x_i^*(t')$, which contradicts the assumption that X^* is a fixed point in view of the absence of blocking lemma.

It has been already shown that any OWNS model has fixed points. For BB models one fixed point can be explicitly constructed.

In fact, there is always such point in the phase space $\mathcal{D}_L(N)$, where the initial positions of workers are such that the time required for the ith worker to reach the initial position of the $(i+1)$st worker does not depend upon the number i of the worker, $i = 1, 2, \dots, N-1$. In other words, at such point the entire work is equally partitioned between the productive elements. It is easy to compute coordinates of this fixed point $\hat{X} = (\hat{x}_2, \dots, \hat{x}_N)$. Indeed the condition of the balance implies that

$$\frac{\hat{x}_2^2}{\int_0^{\hat{x}_2} v_1(x)dx} = \frac{(\hat{x}_3 - \hat{x}_2)^2}{\int_{\hat{x}_2}^{\hat{x}_3} v_2(x)dx} = \dots = \frac{(\hat{x}_N - \hat{x}_{N-1})^2}{\int_{\hat{x}_{N-1}}^{\hat{x}_N} v_{N-1}(x)dx} = \frac{(L - \hat{x}_N)^2}{\int_{\hat{x}_N}^{L} v_N(x)dx}. \quad (2.1)$$

For BMs the expression (2.1) can be simplified to

$$\hat{x}_k = L \frac{\sum_{i=1}^{k-1} v_i}{\sum_{i=1}^{N} v_i}. \quad (2.2)$$

The next statement [1] is very important for understanding dynamics of BB models. It claims that any BB model has one and only one fixed point.

Uniqueness of Balance Point. The fixed point (2.1) is the only fixed point of a BB model.

This statement immediately follows [1] from the absence of blocking lemma. Indeed if there exist two fixed points then in both of them there must be no blocking. In other words, in both these points the entire work should be equally partitioned between the workers, which is impossible.

The absence of blocking and the uniqueness of a balance point suggest that one should look for a maximal throughput in a BB model at its unique fixed point. However, such stationary states in the real systems are required to be "physically realizable," i.e. stable and sufficiently robust.

We are able to show [1] that these properties hold for a somewhat more narrow class of BB models, namely for the basic models (BMs). There are quite strong reasons (theoretical as well as practical, empirical ones) to believe that the basic models capture the main features of many processes in manufacturing and logistics. (To differentiate the basic models of BBs from the basic models of the dynamic zone manufacturing we will denote the last by BMDZ. For BB-models, and therefore for the BMs as well, the problem of maximization of the throughput is equivalent to finding of the optimal sequencing of particles (workers) on a flow line $[0, L]$. (Here and in what follows we discuss BMs of BBs.)

Clearly the maximal throughput for basic models equals $\sum_{i=1}^{N} v_i$. Indeed, this system has the maximal throughput when no one particle gets blocked by its downstream neighbor. Exactly such situation occurs at the point $\hat{X}$.

However, only stable fixed points are of interest for applications. Moreover, a size of the basin of attraction of a fixed point is of crucial importance. In fact, it defines the maximum amplitude of perturbations which do not destroy this stable regime. Therefore the size of the basin of attraction defines how robust is the corresponding stable regime of motion.

It is easy to see that a basic model can even have infinitely many periodic points. Consider, for instance, a line with two workers ($N = 2$) which have equal velocities ($v_1 = v_2$). Obviously all orbits of this model have period two (besides the one which corresponds to the initial position of the second particle right in the middle, i.e. $x_2 = L/2$). Observe that it is (2.2) for $N = 2$.

In the absence of blocking the dynamics of any basic model is linear, because in this case each particle moves with constant velocity. Another main feature of a basic model is that if any particle gets blocked, then it remains blocked until the next reset.

Because any basic model is also a BB model it has only one fixed point (2.2). This point can be stable, unstable or neutral.

Consider, for example, the case $N = 3$. Then the fixed point (2.2) is stable in the region $\{v_1/v_2 < 1, 0 \leq v_2 < v_1+v_3\}$. The fixed point becomes neutral at the boundary of this region, and it becomes unstable outside this region. (These results were first obtained numerically [4], but exact computations are not hard to perform as well.)

Can the basic models have another regime with the maximal throughput besides the one at the fixed point (2.2)? The first candidate to consider would be the set of periodic points of a basic model. The answer, based on the absence of blocking lemma, is negative though. Namely, if $v_i \neq v_j$ for all pairs i, j, $1 \leq i, j \leq N$, then at any periodic point (with period $p > 1$) there is blocking.

The following statement [1] is the central one in the theory of basic models.

Global Stability Theorem. If the fixed point (2.2) of a basic model is (locally) stable then it is also globally stable.

Indeed, the dynamics of a basic model is linear in the absence of blocking. It is easy to see that any event of blocking occurs at the boundary $\partial\tilde{\mathcal{D}}_L(N)$ of the phase space $\tilde{\mathcal{D}}_L(N)$. Observe that $\tilde{\mathcal{D}}_L(N)$ is a $(N-2)$-dimensional simplex. Therefore $\partial\tilde{\mathcal{D}}_L(N)$ consists of a finite number of convex closed sets (faces of $\tilde{\mathcal{D}}_L(N)$).

It is easy to see that a point $X \in \tilde{\mathcal{D}}_L(N)$ may not be attracted to X^* only if its orbit $\{f^k X\}$, $k > 0$, hits the boundary $\partial\tilde{\mathcal{D}}_L(N)$ infinitely many times. Therefore such orbit must hit some face of $\tilde{\mathcal{D}}_L(N)$ infinitely many times.

Let $f_C : C \to C$ is the first return map induced by f. It is easy to see that f_C is continuous. Then it must have a fixed point $Y \in \bigcap_{i=\infty}^{\infty} f^n C$. Such fixed point must be a periodic point of f (with a finite period). But f is linear in the interior of $\tilde{\mathcal{D}}_L(N)$ and has there the unique fixed point X^*. Therefore such Y cannot exist, and we come to a contradiction.

It immediately follows from this theorem that a sequencing of particles provides the maximal throughput $\sum_{i=1}^{N} v_i$ in any basic model of a BB iff the corresponding fixed point is (locally!) stable. So we have a simple local criterion to verify global stability of the optimal regime for this class of models.

We demonstrated that dynamics of BBs, and particularly of the basic models of BBs, is often rather simple, in a sense that these models have the unique fixed point which in many cases is a global attractor. It is known as well [1, 2, 4] that some models of this class can have globally stable periodic points.

Totally different behavior is demonstrated by the models in the dynamic zone manufacturing and, particularly, the basic models of this type. Namely the dynamics of models of the dynamic zone manufacturing production lines is typically chaotic. We will not discuss here any general results for the dynamic zones manufacturing, but consider instead just one simple but transparent example.

Consider a two-worker flow line. Let the velocities of the workers be constant (v_1 and v_2) everywhere on the line, i.e. it is a basic model. Recall that in dynamic zones manufacturing a worker, when being blocked, leaves his unfinished item at the point of blocking and walks back to take over the unfinished item from his upstream neighbor, or to start the new item at the beginning of the line.

If $v_1 \leq v_2$ then blocking never occurs, and this model is equivalent to a BB model. If, however, $v_1 > v_2$ then the situation changes drastically. The corresponding (Poincaré) map f has the form (we let here $L = 1$ for the sake of simplicity)

$$f(x) = \begin{cases} x(1 - v_2/v_1)^{-1} & \text{if } 0 \leq x \leq (1 - v_2/v_1), \\ -\frac{v_1}{v_2}\,x + \frac{1}{v_2} & \text{if } (1 - v_2/v_1) < x \leq 1. \end{cases} \tag{2.3}$$

This piecewise-linear map is expanding and has a finite Markov partition. It is well known (see e.g. [11]) that such maps are strongly chaotic, i.e. they have the unique stochastically stable absolutely continuous invariant measure, they are ergodic, mixing, and time correlations in such systems decay exponentially.

The same results hold for piecewise-continuous expanding maps. Such maps naturally arise as the models of flow lines in dynamic zones manufacturing when velocities of the workers are non-constant.

2.4 Algorithm of Stabilization of the Target Regime for OWS Models

The results discussed in previous sections allow one to suggest the algorithm of performance optimization for OWNS systems. It has been already mentioned that, different to the BB model, in general OWNS models there are some additional constraints. Recall that the only constraint in BB models is that the order of particles is preserved. (One may think of the additional constraints as of some forces distributed over the segment $[0, L]$.)

OWNS models are much more complicated than the BB models. Consider, for example, a TSS production line. In these models production lines (the segment $[0, L]$) get partitioned into $M \geq N$ non-overlapping intervals I_i, $i = 1, 2, \ldots, M, \bigcup_{i=1}^{M} I_i = [0, L]$. The TSS constraint (the TSS second rule) is that no pair of particles can be at the same moment of time at the one and the same interval I_i. This model is proved to be relevant to the production lines in the apparel industry, where the intervals I_i are referred to as workstations. If a worker finished work at the ith workstation I_i, and the next workstation I_{i+1} is occupied by another worker then the ith worker gets blocked at the left-most point of I_{i+1} until his successor finishes work at the $(i + 1)$st workstation. The rest of workers in TSS lines, at the moment when the last worker finishes work at the last workstation I_M, is the same as in BB models. Another type of constraints appear, for example, in the hoist problem [12]. In these production lines the role of workstations is played by tasks, and the one of the workers by the printed circuit boards (PCBs). The constraints here are imposed by the upper and lower bounds on times that PCBs are allowed to spend on each task. The problem for this system is again to maximize the throughput. It has been shown [3] that TSS models have only one fixed point if the workers are sequenced in the increasing order of their velocities. (In general such sequencing is not possible though because some workers can be faster in some parts of the line but slower in another of its parts [13].)

However, this fixed point may not provide for the highest production rate. Consider, for example, the TSS production line $[0, L]$ with three workstations $[0, L/8)$, $[L/8, 5L/8)$ and $[5L/8, L)$. The "lengths" of workstations are defined by the time that is required for some "ideal" worker to perform the work on a given workstation (work content).

Let the velocities of workers be $v_1 = 1$, $v_2 = 3$, $v_3 = 4$. The corresponding fixed point is $\{x_2^* = L/8, x_3^* = 13L/32\}$. The production rate at this point equals $\frac{16}{19}\, v_1 + \frac{12}{19}\, v_2 + v_3$. Change now the order of workers in the line to v_1, v_3, v_2. Then the fixed point is $\{x_2^{**} = L/8, x_3^{**} = 5L/8\}$. The production rate at (x_2^{**}, x_3^{**}) is $v_1 + v_2 + v_3$, i.e. the maximal possible one.

This example, together with the results in the previous section, suggest the following algorithm of optimization (maximization of the throughput and the balancing of work) for the work-sharing production lines [1, 2].

The algorithm consists of two steps:

(1) Find all sequencings of productive elements (workers) in a production line such that the corresponding dynamical models have a stable stationary state (stable fixed point).

(2) Choose among the sequencings found at the step (1) such a one that is the most consistent with the configuration of the production line.

By the configuration of the production line we mean here its partition into the workstations, into the tanks with the corresponding workstations, etc. A measure of "consistency" in (2) depends upon a concrete model. For instance, for TSS production lines the optimal sequencing of workers corresponds to such a stable point where the positions of the workers are maximally close (with respect to the work time) to the upstream boundaries of the workstations on this line.

This algorithm is the special case of the general approach to optimization in manufacturing, which we call the stabilization of a target regime (STR). The STR method can be considered as the generalization of a control of chaos. Indeed, the control of chaos is aimed at finding a regular (periodic) orbit and at making such orbit stable by some procedure performed on the controlled system. In logistics stabilization of a target regime is required for the systems with chaotic as well as with regular dynamics.

2.5 Concluding Remarks

OWS models are rather general and relevant for many processes in logistics, not just for production lines and supply chains. Consider, for instance, the so-called condition-based maintenance (CBM) problem [14]. In this case one deals with a fleet of assets (trucks, aircraft, buses, exchangeable parts of machines, etc.) which should perform work (according to some fixed schedule of, for example, flights) but must also be sent for maintenance after some time and/or after a long extensive use. The problem is to maintain a sufficient fleet of usable assets at any time.

A production line $[0, L]$ for CBM can be interpreted as a "life axis", i.e. no asset can "survive" in business after reaching the end $x = L$ and should be removed from the fleet for servicing. At this moment there should be a "reset" on the line, i.e. some brand new or (just) returned from maintenance asset should be placed at the origin $x = 0$.

To make the models in our hierarchy more realistic one should assume that the particles on the line are "aging" and, on the other hand, the brand new particle, that substituted the departed one, can be already a different one (new type of trucks, specially trained workers, etc.). To handle this situation we must allow that the particles' velocities depend on time, i.e. $v_i = v_i(x, t)$, where the time coordinate t carries the information about a number of replaced particles, on their types, etc. Hence, the model becomes a non-autonomous one. Non-autonomous dynamical systems are much more difficult to analyze. Therefore the role of completely solvable relevant models becomes even more important.

To illustrate this consider the following two practical consequences of the global stability theorem. One of them has to do with the motion of pickers in warehouses, where there are no constraints on the movements of pickers besides the one that the order of pickers must be preserved. In this case the sequencing of pickers from the slowest to the fastest is always preferable, if the stock-keeping units and orders are uniformly distributed along the aisle. The application of this sequencing resulted in essential increase of pick rates [3, 5]. Another practical consequence of this theorem deals with CBM. It is well known that the probability that any given asset (or part) will fail increases with time. Therefore the usage of the "older" assets should be more intensive than of the "younger" ones. It will increase the rate of usage ("productivity") of the assets in the fleet during their lifetime (from the beginning of their usage till the time of maintenance or replacement).

There are some questions related to OWS which we did not discuss here. For instance, a passing is allowed in the OWS models. Therefore, for such systems one of the important problems is synchronization (of brigades of pickers working in different aisles of a warehouse, of subfleets of aircraft which are based in different airports, etc.).

We had shown here that different policies can result in quite different dynamics of logistics systems from a regular to a strongly chaotic one. (Another such example can be found e.g. in [15].) Therefore it is very important to develop dynamics-based policies for operation of such systems. There is little doubt that such policies will be more effective than traditional too general ("philosophical") policies like, for example, FIFO.

Bibliography

[1] L.A. Bunimovich, *Dynamical systems and operations research*, Discrete and Continuous Dynamical Systems **1B** (2001), 209–218.

[2] L.A. Bunimovich, *Controlling production lines*, in: *Handbook of Chaos Control*, edited by H. Schuster, Wiley-VCH, Berlin, 1999, 324-343.

[3] J.J. Bartholdi and D.D. Eisenstein, *A production line that balances itself*, Operations Research **44** (1996), 21–34.

[4] J.J. Bartholdi, L.A. Bunimovich, and D.D. Eisenstein, *Dynamics of two- and three-worker "bucket brigade" production lines*, Operations Research **47** (1999), 488–491.

[5] J.J. Bartholdi, D.D. Eisenstein, and R.D. Foley, *Performance of bucket brigades when work is stochastic*, Operations Research **49** (2001), 374–385.

[6] J. Ostalaza, L.J. Thomas, and J.O. McClain, *The use of dynamic (state dependent) assembly line balancing to improve throughput*, Journal of Manufacturing and Operations Management **31** (1990), 105–133.

[7] D. Bishak, *Performance of manufacturing model with moving workers*, IIE Transactions **28** (1996), 213–219.

[8] E. Zavadlav, J.O. McClain, and L.J. Thomas, *Self-buffering, self-balancing, self-flushing production lines*, Management Science **42** (1996), 1151–1164.

[9] B.J. Schroer, J. Wang, and M.G. Ziemke, *A look at TSS through simulations*, Bobbin Magazine (July 1991), 114–119.

[10] F.S. Hillier and B.W. Boling, *On the optimal allocation of work in symmetrically unbalanced production line systems with variable operation times*, Management Science **25** (1979), 317–326.

[11] A. Lasota and M.C. Mackey, *Chaos, Fractals and Noise*, Springer-Verlag, New York, 1994.

[12] H. Chen, C. Chu, and J.-M. Proth, *Cyclic scheduling of a hoist with time window constraints*, IEEE Transactions on Robotics and Automation **14** (1998), 144–152.

[13] D. Ambruster and E.S. Gel, *Dynamics of bucket brigades with varying workers speed* Preprint, ASU, 2001.

[14] S.R. Venkatesh, M. Dorobantu, and J.E. Rogan, *On optimal policies for fleet maintenance*, Preprint, UTRC, 2000.

[15] C. Chase, J. Serrano and P.J. Ramage, *Periodicity and chaos from switching flow systems*, IEEE Transactions on Automatic Control **38** (1993), 70–83.

3 Manufacturing Systems with Restricted Buffer Sizes

K. Peters, J. Worbs, U. Parlitz, and H.-P. Wiendahl

Practically, the capacity of buffers in manufacturing is limited; therefore policies have to be established to prevent overfilling or vacancies in connected manufacturing systems. Even without irregular influences like unforeseen breakdowns and disturbances of machines simple manufacturing systems can develop highly complex dynamics just with simple operating rules.

We consider hybrid models of strange billiard type for small connected manufacturing systems and demonstrate that for such models limited buffer capacities can induce a number of unusual bifurcations and deterministic chaos. By means of Poincaré map techniques we discuss different dynamical behaviors. Different dynamical behaviors also affect logistic parameters like throughput time distributions and production rates. On the more practical side similar results are obtained for models with discrete material flow.

3.1 Introduction

Modern manufacturing systems are typically large networks of various production and storage facilities. Their dynamics is governed by complex, system-immanent inter-dependences involving both deterministic and stochastic elements as well as unforeseen external influences. A basic feature of most manufacturing systems is the existence of policies controlling the allocation of scarce resources among competing tasks.

Whereas the classical queueing theory focuses on the influence of stochastic arrival and departure processes on the formation of queues, mostly in the language of equilibrium solutions, a number of recent investigations reveal the fact that manufacturing systems are dynamical systems (see [1] and references therein). In the context of dynamical systems rule (policy)-dependent switchings between different operation modes and parameters are essential nonlinearities, leading to complex behavior in manufacturing systems. To analyze such influences it is reasonable to neglect any stochastic influence and the discreteness of the material flow. This approach leads to dynamical systems consisting of piecewise-defined continuous-time evolution processes interfaced with some logical or decision-making process. Such systems which are described by continuous as well as discrete state variables are called *hybrid systems*.

In the present work we investigate layout structures where one work unit has to load a number of subsequent units or one unit has to serve some previous workstations. Such *m:n* connections are basic topological structures in the layout of manufacturing systems among re-entrant structures [2, 3] and pure linear chains. We consider especially the limited capacity of buffers in these m:n connections. Figure 3.1 gives an illustration of the discussed structure in a manufacturing system. First models of the described type have been analyzed by Chase et

Nonlinear Dynamics of Production Systems. Edited by G. Radons and R. Neugebauer

ISBN 3-527-40430-9

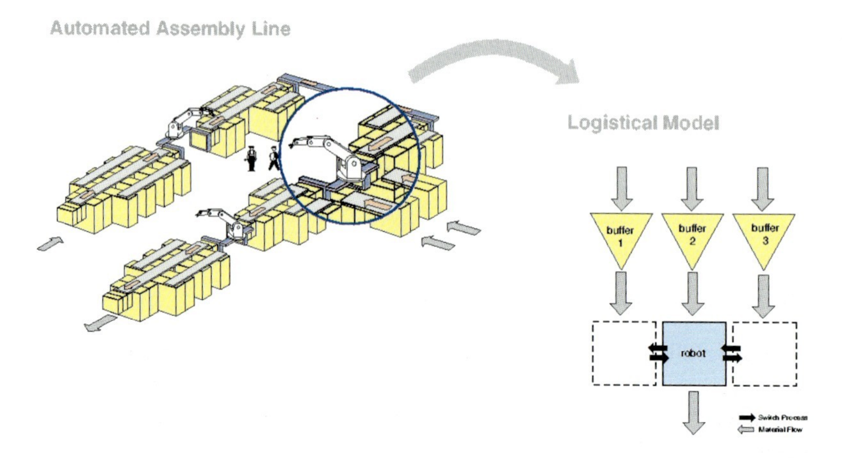

Figure 3.1: A part of a production line for crankshafts as an example of a switched arrival system [9].

al. [4]. The work of Chase et al. was extended by Schürmann and Hoffman [5]. Recently, the switched arrival system was studied in [6] in the context of discrete material flow and used to study the interaction of parallel layers of switched arrival systems, coupled through production losses [8].

In the context of manufacturing systems our investigations lead to the conclusion that for a given layout and policy the dynamics may be affected significantly (i.e. driven from an equilibrium state to chaotic behavior, for instance) by restricted buffer sizes. This feature, on the other hand, gives the ability to optimize production costs by suitable adjustment of buffer capacities or chaos-control methods [7, 6].

The remainder of the paper is organized as follows. In Sect. 3.2 we introduce two standard examples of hybrid models for manufacturing systems, derive the framework of strange billiards and summarize results for the dynamics of the examples. Some related models are briefly discussed. In Sect. 3.3 consequences of the dynamics on the performance of manufacturing systems are considered, and possible optimizations are given. In Sect. 3.4 we present numerical evidence that the dynamics of discrete deterministic queueing models is similar to the dynamics of hybrid models. Finally, in Sect. 3.5 we summarize the paper.

3.2 Hybrid Models

The basic unit of most manufacturing systems is a workstation containing a server (e.g. a machine) to perform some processing and a buffer, queueing the material before it can be processed. In the present work we model connected workstations by a certain class of hybrid systems consisting of basic units, called *tanks* in the following, interacting by discrete events. These tanks $i = 1, ..., n$ have maximum capacities $b_i, i = 1, 2, ..., n$, and can be continuously filled with fluid at rates λ_i and emptied with rates μ_i. The fluid content x_i of each tank is controlled by switching the inflow or outflow according to switching rules. We note that the modeling of buffers and workstations as tanks in a flow-orientated manner incorporates the well-known funnel approach [10] and has some similarities to the fluid models used in queueing theory [11, 12]. The full state in state space for connected systems of the described type contains both continuous variables x_i and a discrete (symbolic) variable q labeling the discrete state of the system (i.e. the on or off state of inflows or outflows). At the discrete event times t_m the full state of the system changes according to $[x(t_m), q(t_m)] \mapsto [x(t_m), q(t_m^+)]$ (with $x = (x_1, x_2, ..., x_n)$).

3.2.1 Switched Arrival and Server Systems

A common problem is the connection of work systems, i.e. one unit has to load several subsequent work units or one work unit has to serve some previous work units. Because the parallel (foregoing or subsequent) work units in general may process different products such connections are generally operated under scheduling policies summarized under the term CAF *clear-a-fraction*. That means the server picks one of the parallel buffers and serves it at its full ability until a switch to another buffer is required by some rule. The question is: which rule to select, and what are the consequences for the efficiency of production processes if some fixed switching rules are applied? Generally empty buffers reduce the (effective) departure rate, while full buffers reduce the (effective) arrival rate. Both cases reduce the production rate. Moreover, switchings themselves can produce inefficiencies and costs themselves. Therefore conventional switching rules try to minimize switching. Suggestive methods are switchings if the just-served buffer has become empty or an overfilling requires a switching of the server.

Let us investigate such strategies by a hybrid modeling approach where we approximate the buffers by tanks. Consider a system consisting of n parallel tanks, and one server as shown in Fig. 3.2. At a time, the server can be attached to one tank, only. This server has either to empty all tanks (Fig. 3.2a), in which case they are assumed to fill themselves continuously, or the tanks, which are all emptied continuously, will be filled by a single switching server (Fig. 3.2b). We call, according to [4], the first situation a *switched server system*, and the second, where the input to parallel tanks is delivered by a single server, a *switched arrival system*.

The discrete (symbolic) variable q labels the discrete state of the system (i.e. the position of the switching server).

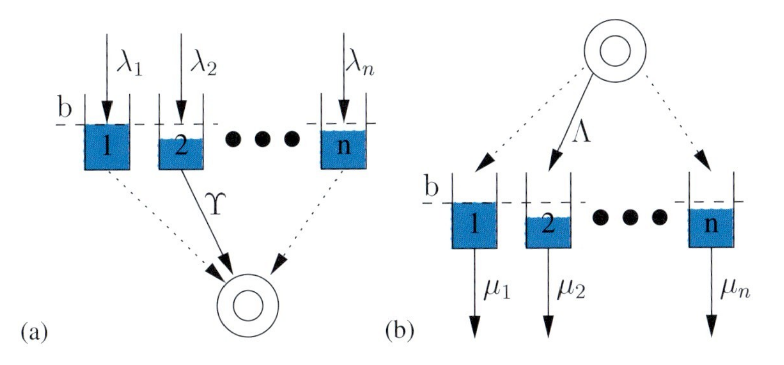

Figure 3.2: (a) n-tank switched server system. (b) n-tank switched arrival system.

The switching policies under consideration are the following. For the switched server system:

- If a tank is filled to its maximum the server instantaneously starts to serve this tank.

Another event that may occur is a tank reaching a lower limit level. The value of this reference level does not influence the dynamics; thus we refer to it as an empty level. If the currently served tank becomes empty before the first rule applies, a deterministic function of the current state should be used to determine the next tank that goes in service. We point out one possible rule:

- Serve the next tank in cyclical order.

For the switched arrival system the policy is inverse: a tank will be instantaneously served if it becomes empty, and the second switching rule has to be applied if a tank is full before another becomes empty.

For the dynamical systems defined here a specific *balanced* regime exists if the total outflow matches the inflow. The balanced state guarantees the normal operation of a manufacturing system and is in industrial engineering assured under CONWIP (constant work in process) policies. Thus we require in the following for the switched server system $\sum_i \lambda_i = \Upsilon$ and for the switched arrival system $\sum_i \mu_i = \Lambda$. This implies that the total content of the system is constant. Without loss of generality we therefore normalize $\Upsilon = 1 = \sum_i \lambda_i$, $\Lambda = 1 = \sum_i \mu_i$ and $\sum_i x_i = 1$, respectively. For simplicity we choose here the same maximum capacity b for all tanks, $b_i = b$. As long as the system stays in the discrete state q (i.e. the server is attached to tank q) the equation of motion for x is simply:

$$\dot{x} = v_q, \tag{3.1}$$

where $v_q = \lambda - e_q$ for the switched server system and $v_q = e_q - \mu$ for the switched arrival system.

Here $x = (x_1, x_2, ..., x_n)$ is the continuous state vector, $\{v_q | q = 1, ..., n\}$ is the set of velocity vectors, $\lambda = (\lambda_1, \lambda_2, ..., \lambda_n), \mu = (\mu_1, \mu_2, ..., \mu_n)$ are constant vectors and e_q is the qth canonical unit vector in $\mathbb{R}^n$. The evolution of the system inside $\mathbb{R}^n$ reveals the structure of a strange billiard. The current state moves uniformly and linearly inside the bounded region $S_n^*(b) = \{x \in \mathbb{R}^n | \sum_i x_i = 1, 0 \le x_i \le b; \text{for } i = 1, ..., n\}$. The evolution changes according to the switching rules if the continuous state x hits the boundary $\delta S_n^*(b)$ of $S_n^*(b)$. At this time instant t_m the full state of the system changes as $[x(t_m), q(t_m)] \mapsto [x(t_m), q(t_m^+)]$ and a new velocity vector is selected from the set $\{(v_q) | q = 1, ..., n\}$. This looks like a strange reflection at the boundary. But in contrast to ordinary physical billiards, which are invertible Hamiltonian systems, these systems are neither Hamiltonian nor invertible.

Consider the Poincaré map $G : \delta S_n^* \to \delta S_n^*$ that describes the dynamics on the boundary δS_n^*. For the switched arrival as well as for the switched server system two successive hits of the boundary (at t_m and t_{m+1}) are determined uniquely by $x(t_m)$ and we can formally write

$$x(t_{m+1}) = G(x(t_m)) = x(t_m) + v_q \Delta t_m, \tag{3.2}$$

where $\Delta t_m = t_{m+1} - t_m$ is $\Delta t_m = \min([x_q(t_m)/(1 - \lambda_q)]; [(b - x_i(t_m))/\lambda_i]_{i \neq q})$ for the switched server system and $\Delta t_m = \min([(b - x_q(t_m))/(1 - \mu_q)]; [x_i(t_m)/\mu_i]_{i \neq q})$ for the switched arrival system.

3.2.2 Limiting Cases

Whereas the following sections are dedicated to a more detailed view on systems with three parallel tanks it will be instructive to consider limiting cases for the dynamics of the general switched server and switched arrival systems.

For the normalized switched systems (with $n > 2$) two limiting cases with respect to the shape of S_n^* exist. For $b = 1$, S_n^* is the usual n-simplex embedded in $\mathbb{R}^n$ given as $\bar{S}_n = S_n^*(b=1) = \{x \in \mathbb{R}^n | \sum_i x_i = 1, 0 \le x_i \le 1; \text{ for } i = 1, ..., n\}$. If $b = 1/(n-1)$, S_n^* is a geometrically similar regular n-simplex which is now smaller and inverted, given as $\tilde{S}_n = S_n^*(b = 1/(n-1)) = \{x \in \mathbb{R}^n | \sum_i x_i = 1, 0 \le x_i \le 1/(n-1); \text{ for } i = 1, ..., n\}$. Figure 3.3 shows an illustration of these two limiting cases for $n = 3$.

Now we shall examine the properties of the Poincaré map G on the boundaries $\delta\bar{S}_n$ and $\delta\tilde{S}_n$. For the switched server system on $\bar{S}_n$ it was shown [4] that G is everywhere contracting. In contrast, for the switched arrival system on $\bar{S}_n$ the map G is chaotic and the invariant measure can be derived by constructing the Frobenius–Perron operator. The piecewise-constant probability measure invariant under G is given by

$$p^*(\bar{s}_{x_i=0,n}) = \frac{1}{d}\mu_i(1-\mu_i), \tag{3.3}$$

with $d = \sum_i \mu_i(1-\mu_i)$ and $\bar{s}_{x_i=0,n} = \{x \in \mathbb{R}^n | x_i = 0, x \in \bar{S}_n\}$ denoting the ith face of $\bar{S}_n$ ($\delta\bar{S}_n = \bigcup_{i=1}^{n} \bar{s}_{x_i=0,n}$) [5].

For $\tilde{S}_n$ choose the transformations $\tilde{e}_q = \frac{1}{n-1}(\mathbb{I} - e_q)$ and $\tilde{\mu} = (n-1)^{-1}(\mathbb{I} - \lambda)$, $\tilde{\lambda} = (n-1)^{-1}(\mathbb{I} - \mu)$ with $\mathbb{I} = \sum_{i=1}^{n} e_i$. This constitutes a complete orthonormal system (with mirrored handedness) where the equation of motion and the Poincaré map G for the switched server system on $\tilde{S}_n$ are (up to the scaling factor $1/(n-1)$) just the ones of the switched arrival system on $\bar{S}_n$ and vice versa.

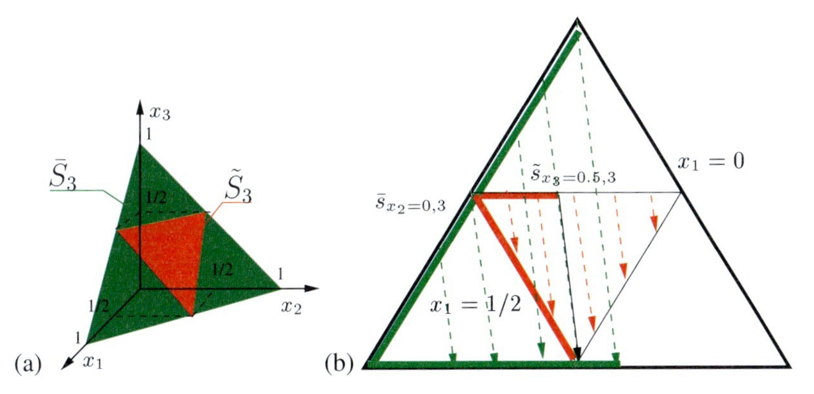

Figure 3.3: Illustration of the continuous part of the state space embedded in $\mathbb{R}^3$. (a) The two 3-simplexes $\bar{S}_3$ (green) and $\tilde{S}_3$ (red) that correspond to the limiting cases $b = 1$ and $b = 1/2$, respectively. (b) Contracting and expanding properties of one and the same velocity vector inside $\bar{S}_3$ and $\tilde{S}_3$, respectively. The green edge of $\bar{S}_3$ (corresponding to $\bar{s}_{x_2=0,3}$) is mapped under v_3 to a shorter segment at the bottom of $\bar{S}_3$, whereas the red segment of the top edge of $\tilde{S}_3$ (subset of $\tilde{s}_{x_3=1/2,3}$) is expanded by the same flow.

For $b < 1/(n-1)$ the system dynamics is the same as for $b = 1/(n-1)$, restricted to a smaller and smaller n-simplex up to $b = 1/n$ where the simplex vanishes.

We conclude that for more than two parallel tanks the switched arrival system behaves chaotically if the tank capacities are large with respect to the total content of the system and may be regularized by decreasing the tank capacity, whereas the switched server systems behave chaotically just for small capacities.

3.2.3 Dynamics and Bifurcations

In the following we discuss the dynamics with respect to b as bifurcation parameter. For more details see [13]. We restrict our analysis to $n = 3$ and consider the switched server system. Applying the results of Sect. 3.2.2 the transfer to the switched arrival system is straightforward.

For three tanks the dynamics is restricted to S_3^* which lies on a two-dimensional manifold, and is generally a 6-simplex. The boundary δS_3^* is one dimensional.

After suitable changes of coordinates ($T : \delta S_3^* \to [0,1]; x \mapsto X$) and re-scaling of the parameters the resulting Poincaré maps are piecewise-linear maps of the unit interval onto itself (Fig. 3.4). For $b = 1$ and $b = 1/2$ we obtain the above-discussed limiting cases. For the switched server systems in between the limiting cases $b = 1$ and $b = 1/2$ we obtain a parameter-dependent morphing between an everywhere-contracting map and a Bernoulli-

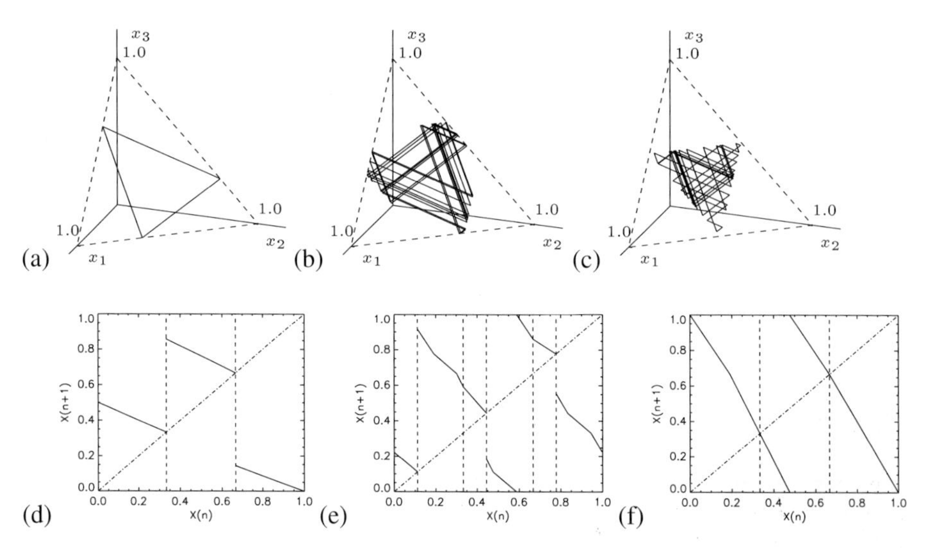

Figure 3.4: Sample trajectories of a three-tank switched server model for different values of the maximum tank capacity b, filling rates $\lambda_1 = 0.3$, $\lambda_2 = 0.4$, $\lambda_3 = 0.3$. Upper thresholds (a) $b = 1.0$, (b) $b = 0.62$, (c) $b = 0.5$. Below are shown Poincaré maps of the switched server system. The main segments (corresponding to faces of S_3^*) are indicated by dashed lines. $X(n) = X(n+1)$ is shown by a dot-dashed line. (d) Everywhere-contracting Poincaré map with two discontinuities for $b = 1.0$ where only switchings at the lower threshold take place. (e) $b = 0.6$. (f) Bernoulli-type Poincaré map for $b = 0.5$ where only switchings at the upper threshold take place.

type map (Fig. 3.4e). The mechanism, typical for strange billiards, is the expansion of linear branches and the creation of new branches of the map with decreasing b.

In general, the maps have six main segments, associated with the six faces of S_3^*, denoted as $s_{x_1=0}, s_{x_3=b}, s_{x_2=0}, ..., s_{x_2=b}$. Each of the six segments possesses a number of subsegments g (up to three subsegments for $s_{x_i=b}$ and up to two for $s_{x_i=0}$), where the slope of the Poincaré map is constant. Whereas the exact slope is defined by $\{\lambda_i | i = 1, 2, 3\}$, some general properties depend only on the main segments (faces of S_3^*) to which the subsegment maps. A segment belonging to $g(s_{x_i=b} \mapsto s_{x_i=0})$ or $g(s_{x_i=0} \mapsto s_{x_i=b})$ provides just interval exchanges. A segment $g(s_{x_i=b} \mapsto s_{x_j=b})$ contributes expanding segments whereas segments $g(s_{x_i=0} \mapsto s_{x_j=0})$ lead to contracting properties. For segments $g(s_{x_i=0} \mapsto s_{x_j=b})$ the slope is given by $a = -(\lambda_i/\lambda_j)$ and the segments are contracting, neutral or expanding depending on $\{\lambda_i\}$. Thus, for the strange billiard expanding properties are closely related to faces where $x_i = b$. The discontinuities of the map are given by the vertices $s_{x_1=0}/s_{x_3=b}$, $s_{x_2=0}/s_{x_1=b}$ and $s_{x_3=0}/s_{x_2=b}$. These discontinuities are crucial for the dynamic behavior.

It is obvious that not all (sub) segments are present for all b. Moreover, the properties of some segments depend on $\{\lambda_i\}$. Therefore *different types of maps*, connected with different admissible orbits, specify the dynamics for different b. This is crucial for the occurrence of chaotic behavior.

Since a chaotic attractor is governed by more local expansion than contraction during its evolution chaotic behavior is only possible if branches with expanding properties exist and are visited sufficiently often. A simple example for the creation of expanding segments and the associated change of the dynamics is depicted in Fig. 3.5a. Here with $\lambda_i = 1/3$ the only expanding segments $g(s_{x_i=b} \mapsto s_{x_j=b})$ appear simultaneously for $b < 2/3$ (which can be derived from simple geometrical arguments) and in this parameter region chaotic behavior is also obtained.

In a parameter region where neither new segments appear nor existing segments vanish, the slope and (dis)continuity of the branches are invariant for changing b but their start point and extension are altered. In this case a *bifurcation* can occur, if a point of a periodic orbit of the map hits the border of a segment. At this *border* the map is *either continuous or has a discontinuity*. In the billiard picture the hit of a segment border corresponds to a reflection point at the boundary, which hits a vertex.

If a periodic point moves from one segment (slope a_L) on a branch to another with slope a_R *border-collision bifurcations* [14] can occur. Because all slopes in the Poincaré map are smaller than zero, either both slopes of the segments (in an n-fold Poincaré map for a periodic orbit) are greater than zero (even period) or smaller than zero (odd period). Therefore, as pointed out in [15], only three possible types of border-collision bifurcations of periodic orbits are possible if the two branches are continuous at the border: either period-doubling bifurcation takes place or we have a unique attractor on both segments, only the path of the fixed point changes. The third possibility is a periodic attractor that vanishes.

Whereas the role of these bifurcations in the entire scenario is limited, the *discontinuities* are in some sense essential for the dynamics of the systems. When due to varying b a periodic point hits a point of discontinuity, it is mapped to a completely distinct site. It is well known that critical points caused by a discontinuity can induce a rich variety of dynamical behavior in maps on the unit interval. This includes *period-adding scenarios* [17], special types of *intermittency* [16] and unusual transitions to chaos.

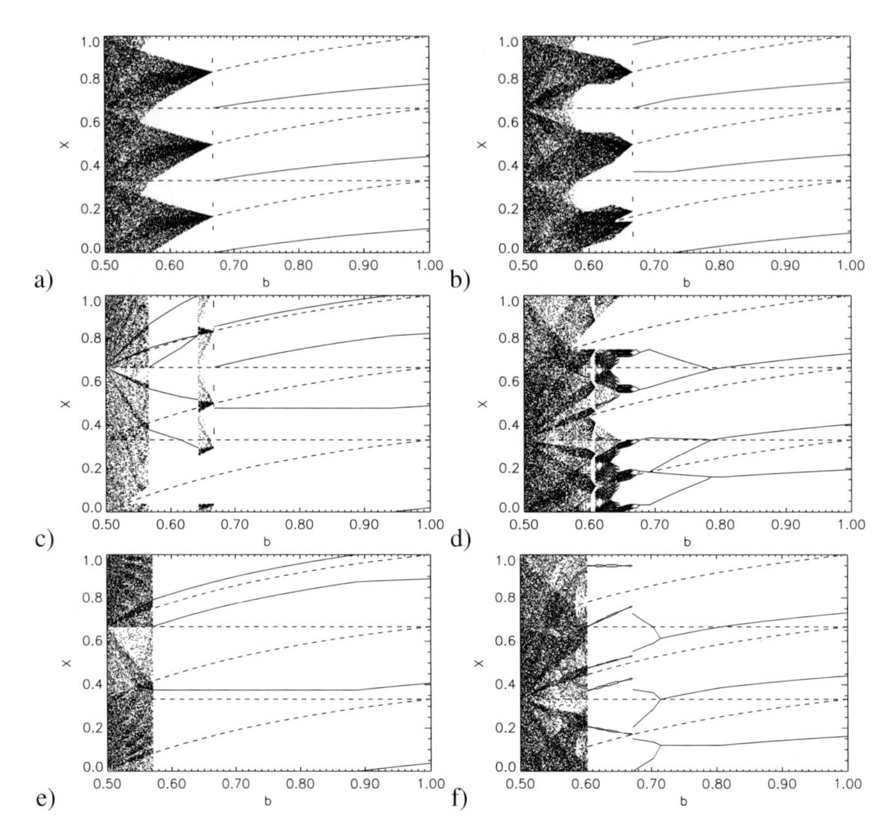

Figure 3.5: Bifurcation diagrams for three-tank switched server models showing the normalized boundary coordinate X vs. maximum tank capacity b. The main segments are indicated by dashed lines. The diagrams have been computed for different filling rates $\lambda_{1,2,3}$: (a) $\lambda_1 = \lambda_2 = \lambda_3 = 1/3$. (b) $\lambda_1 = 0.3$, $\lambda_2 = 0.4$, $\lambda_3 = 0.3$. (c) $\lambda_1 = 0.1$, $\lambda_2 = 0.8$, $\lambda_3 = 0.1$. (d) $\lambda_1 = 0.45$, $\lambda_2 = 0.15$, $\lambda_3 = 0.4$. (e) $\lambda_1 = 0.1$, $\lambda_2 = 0.6$, $\lambda_3 = 0.3$. (f) $\lambda_1 = 0.5$, $\lambda_2 = 0.2$, $\lambda_3 = 0.3$.

In view of the different numbers of segments for different b, the number of critical points in the maps and the fact that the segments are relatively small, it is clear that the entire structure of the dynamics of strange billiards can be quite complicated. Figure 3.5 provides some examples of bifurcation diagrams. Usually *co-existing attractors* touching different subsets of δS_3^* are possible.

To describe the *full dynamics of the hybrid system* not only the filling levels of the tanks (x) or their transformed equivalents X (given by iteration of the Poincaré map) at the switching times are required but also the switching times t_m or the intervals Δt_m. The interval Δt_m is uniquely determined by $x(t_m)$ or $X(t_m)$ and is given by a piecewise-linear function $M : [0, 1] \to \mathbb{R}, X(t_m) \mapsto \Delta t_m$. We remark that $M(X)$ has segments located in the same manner as with $G(X)$. The segments of $M(X)$ are of constant slope, determined by $\{\lambda_i\}$ and for segments $m(s_{x_i=b} \mapsto s_{x_i=0})$ and $m(s_{x_i=0} \mapsto s_{x_i=b})$ $M(X) = \text{const}$. The function $M(X)$ is not invertible.

3.2.4 Modified Switching Rules

As pointed out above, the switching rules for the switched server system and for the switched arrival system (at the lower and upper thresholds, respectively) are not pre-determined by system requirements. Two further possible choices for the switched server rule at the lower threshold are:

- (SR2b) Serve the tank that needs the most time to become empty.
- (SR2c) Serve the tank that needs the shortest time to become empty.

The main impact of such modified switching rules is an increasing number of critical points (discontinuities) in the Poincaré map. This essentially changes the admissible orbits and makes co-existing orbits more likely. On the other hand the fundamental bifurcation mechanisms are not changed. Nonetheless the impact of changed switching rules on the dynamical behavior is remarkable (cf. Fig. 3.6).

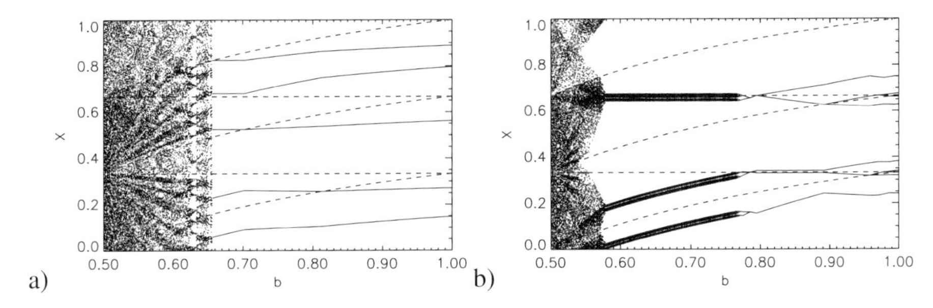

Figure 3.6: Bifurcation diagrams for three-tank switched server models. The main segments are indicated by dashed lines. (a) Filling rates $\lambda_1 = 0.45, \lambda_2 = 0.15, \lambda_3 = 0.40$, switching rule SR2b is applied. (b) Same filling rates as (a) but switching rule SR2c is applied.

3.2.5 Manufacturing Systems with Setup Times

As we have mentioned at the beginning, usually different buffers (i.e. tanks in hybrid models) contain different types of products. Thus in general a server incurs a 'setup time' τ (a bounded delay) when changing over to serve a new tank. During this time interval τ the server is not able to work at its full ability, causing production losses and finally additional costs which are discussed in the following section.

One way to take into account losses caused by setup times is to assume a fixed amount of costs for a switching. This implies in some sense that the whole system is stopped for a time τ and continues after this interval just at the same state it has before the switching took place. On the other hand one may argue that a server can work during a certain setup time, but with a reduced rate. As an example for the dynamics under this assumption we consider here switched server systems with $n = 3$ and the following additional switching rule:

- (LR) If required, switch the server according to the rules given in Sect. 3.2.1. During a time interval τ after a switching maintain the left tank at its previous level, i.e. this tank gets no input for τ. Also the empty rate μ of the server is reduced by the rate that this left tank normally obtains.

For simplicity we assume an identical τ for all switches. The additional rule implies a number of new discrete states for the hybrid system (Fig. 3.7a). For three tanks we obtain three additional states where one tank is out of service and three further states where two tanks are simultaneously out of service. Therefore the introduction of a setup time with reduced production rate also changes the dynamical behavior (cf. Fig. 3.7c). Furthermore, the switches between different discrete states occur not only at the boundary δS_3^* and the following discrete state depends on the previous one.

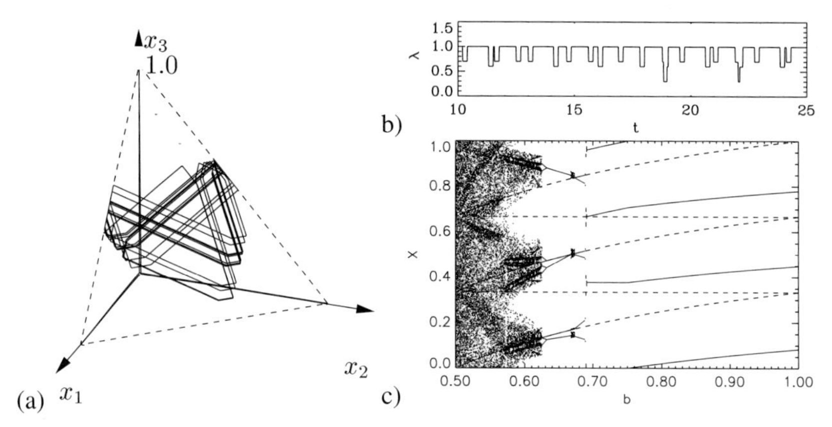

Figure 3.7: Dynamics of a three-tank switched server model ($\lambda_1 = 0.3, \lambda_2 = 0.4, \lambda_3 = 0.3$) with an additional setup time $\tau = 0.2$. The rule LR is applied. (a) Sample trajectory of the tank content at $b = 0.6$. (b) Short time series of the production rate ($\lambda(t) = \sum \lambda_i(t) = \mu(t)$) for the system if the maximal tank content is $b = 0.6$. With every switching made, the productivity is temporarily reduced. (c) Bifurcation diagram of the system showing points (in normalized boundary coordinates) where the trajectory strikes δS_3^* vs. b. Note that not every state change in the hybrid model takes place at δS_3^* . The main segments are indicated by dashed lines.

3.3 Performance of Manufacturing Systems

In the foregoing section the influence of restricted tank content and different switching rules on the dynamics of switched server and switched arrival systems was studied. Now we address the question of how different dynamics may affect the performance of manufacturing systems, modeled as hybrid systems.

In a practical view chaotic behavior, induced by small tank capacities for instance, has two major consequences. On the one hand, the temporal evolution of tank (buffer) contents is not predictable for long times in a chaotic system. This spreads on a statistical level the distributions of inter-switching and throughput times and makes in general the sequence of different products at the departure point irregular. Thus the production planning may be affected. On the other hand the performance of manufacturing itself is influenced by the number of switchings per time interval for instance.

3.3.1 Evaluation of Cost Functions

For models without explicit production losses through setup times we study the performance through cost functions. In the context of manufacturing two types of cost functions have to be considered. Usually the setup of buffer capacities causes costs. For simplicity we assume that these costs are proportional to the maximal buffer capacity and call them buffer cost:

$$k_x = cb, \tag{3.4}$$

where c is a constant. On the other hand consider the switching cost function:

$$k_n = \lim_{T\to\infty} \frac{N_T}{T}, \tag{3.5}$$

where N_T is the number of switches in the time interval T. Here we assume that every switching of the server causes fixed costs. The dependence of the cost functions on the maximal buffer content b for switched server systems is depicted in Fig. 3.8.

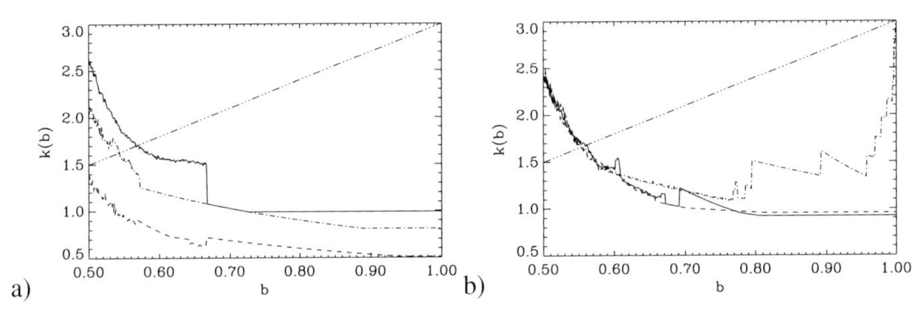

Figure 3.8: Cost functions for three-tank switched server systems vs. maximal buffer content b. The buffer costs for $c = 3$ are given by the three-dot-dashed line. (a) The solid curve gives the switching costs for $\lambda_1 = 0.3, \lambda_1 = 0.4, \lambda_1 = 0.3$. The dot-dashed curve gives the switching costs for $\lambda_1 = 1, \lambda_2 = 0.8, \lambda_3 = 0.1$ and the dashed curve for $\lambda_1 = 0.1, \lambda_2 = 0.6, \lambda_3 = 0.3$. Compare the bifurcation diagrams given in Fig. 3.5. (b) Switching costs for a three-tank switched server system $(\lambda_1 = 0.45, \lambda_1 = 0.15, \lambda_1 = 0.4)$ under different switching rules. Solid line: switching rules as in Sect. 3.2.1, dashed line: SR2b is applied, dot-dashed line: SR2c is applied. Compare the bifurcation diagrams given in Fig. 3.6.

For models designed according to Sect. 3.2.5 the production loss due to switchings can be calculated directly. We consider the production rate:

$$\rho = \lim_{T\to\infty} \frac{\int_0^T dt\mu(t)}{\int_0^T dt\mu_0}, \tag{3.6}$$

where $\mu(t) = \sum_i \lambda_i(t)$ is the actual empty rate while $\mu_0 = 1$ through normalization. Figure 3.9 provides some examples of production rates for switched server systems, depending on b. Every attractor of the system has its own switching costs and production rate respectively. Therefore discontinuities in the switching cost function and production rate are typical if the system changes the attractor. The switching costs become high (and the production rate decreases) if

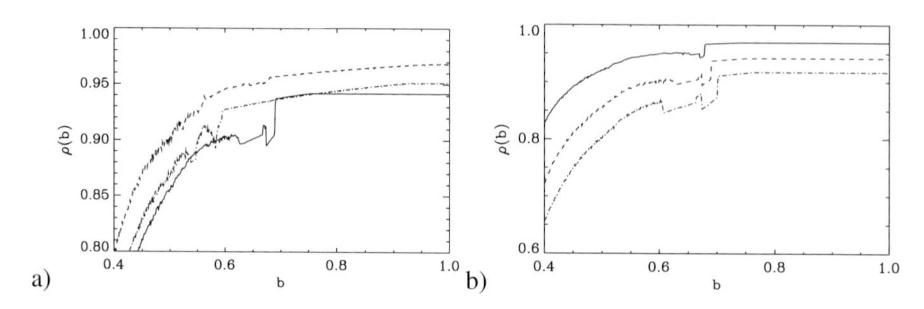

Figure 3.9: Production losses due to setup times for three-tank switched server systems operated with rule LR. The production rate $\rho(b)$ vs. maximal buffer content is shown. Note that a lossless production will result in $\rho = 1.0$. (a) $\rho(b)$ for different parameters and a setup time $\tau = 0.2$. The solid curve gives $\rho(b)$ for $\lambda_1 = 0.3, \lambda_1 = 0.4, \lambda_1 = 0.3$. The dashed curve is for $\lambda_1 = 1, \lambda_2 = 0.8, \lambda_3 = 0.1$ and the dot-dashed curve is for $\lambda_1 = 0.1, \lambda_2 = 0.6, \lambda_3 = 0.3$. (b) Production rate $\rho(b)$ for a three-tank switched server system with $\lambda_1 = \lambda_3 = 0.3, \lambda_2 = 0.4$ if different setup times are applied. Solid curve: $\tau = 0.1$, dashed curve: $\tau = 0.2$ and dot-dashed curve: $\tau = 0.3$.

the attractor comes close to vertices, because there the time interval to the next switching is short. It is obvious in the examples that the chaotic behavior for small b reduces the performance of switched server systems.

3.3.2 Optimization and Chaos Control

It is common to reduce buffer costs by lowering b. In the switched arrival system this can also prevent chaotic behavior but it induces chaos connected with high switching costs or production losses in the switched server system.

Optimization may therefore address organizational tasks to achieve parameter combinations in a desired range. This can include appropriate chosen switching thresholds or filling rates. Optimization may alternatively consider appropriately chosen switching rules for the lower (upper) threshold in the switched server (arrival) system. As we have seen the dynamics is influenced strongly by the switching rule and another rule may lead to stable orbits with lower costs.

Another way may be the active control of the dynamics. To optimize both cost functions for the switched server system the maximal buffer content b can be reduced, and then chaos-control methods can be applied in the chaotic region to reduce the switching costs. To apply this strategy one of the unstable periodic orbits, which minimizes the switching cost and is embedded in the chaotic attractor, has to be stabilized using the OGY control method [7] for instance. With the given Poincaré map it is straightforward to find periodic orbits and to calculate their costs. The control scheme is quite simple [6, 8]. If a trajectory near the target orbit comes close to the boundary δS^* where the switching takes place, we have to evaluate the virtual intersection point of the current trajectory and the reflected target trajectory. If this intersection point lies before the boundary we have to switch a little bit earlier and if the intersection point is behind the boundary switching is delayed to bring the trajectory closer to the target orbit. The earlier or later switching can also be seen as an appropriate small shift of the boundary. This kind of control is only possible if both directions are reachable, i.e. no tank runs completely empty due to the delayed switching. This makes the control scheme well suited for the switched server system with small b.

3.4 Switched Discrete Deterministic Systems

The hybrid modeling of manufacturing systems in connection with the strange-billiard concept is a powerful tool to evaluate the potential dynamical features of a given topology of manufacturing systems under priority rules and to obtain analytical results in the language of nonlinear dynamics. However, in manufacturing reality, one has frequently a discrete rather than a continuous material flow (which was described by a fluid model in the previous sections).

Thus we now dismiss the approximation of continuous material flows and study the behavior of the previously investigated systems under the assumption of deterministic, but discrete flow of parts. We assume fixed inter-arrival times (T_i) and inter-departure times (Θ) for the buffers and servers. The switching rules with respect to the buffer filling levels under consideration should be analogous to the rules introduced in Sect. 3.2.1. The main difference between both types of modeling is that a change in the server position now can occur only at specific times, i.e. when the processing of a part is finished. A normalization as with the hybrid models is not feasible in the discrete case. Rather we have to start with an initial buffer content $x_i \in \mathbb{N}$ and even discrete thresholds for the buffer capacities. Additionally, the initial condition defines the time to the next event inside the intervals, given by the inter-event times. In some sense hybrid models are the limiting case of discrete models with infinite total content or infinitesimal inter-event times. Following the terminology of the (stochastic) queueing theory such models should be described as *deterministic queueing models*. In principle the same system of queues, servers and routers as in queueing theory are considered, but instead of stochastic processes the systems are governed by deterministic point processes.

3.4.1 Dynamics

Due to the doubly discrete nature of the system (in space and time) the dynamics of the discrete model is fundamentally distinct from the continuous hybrid model. The discreteness implies a countable, finite number of possible states (filling levels of buffers) and therewith a periodic behavior of the trajectory. Usually a larger number of co-existing orbits in the discrete state space are obtained. Nonetheless the discrete model shows in some sense a similar dynamics as the hybrid model even if the transition from the continuous to the discrete material flow cannot be done in a mathematical framework. Especially a correspondence of similar coarse-grained dynamics in certain parameter regions is obtained, and, even in deterministic queueing models bifurcation-like phenomena arise (cf. Fig. 3.10a). Small buffer capacities in the switched server system even lead to a complicated behavior which may be dubbed quasi-chaotic. For the discrete model it is possible to obtain cycle times for single parts. The distribution of throughput times is determined by the dynamic behavior, and therefore this logistic parameter also depends on the maximal buffer content (Fig. 3.10d, e).

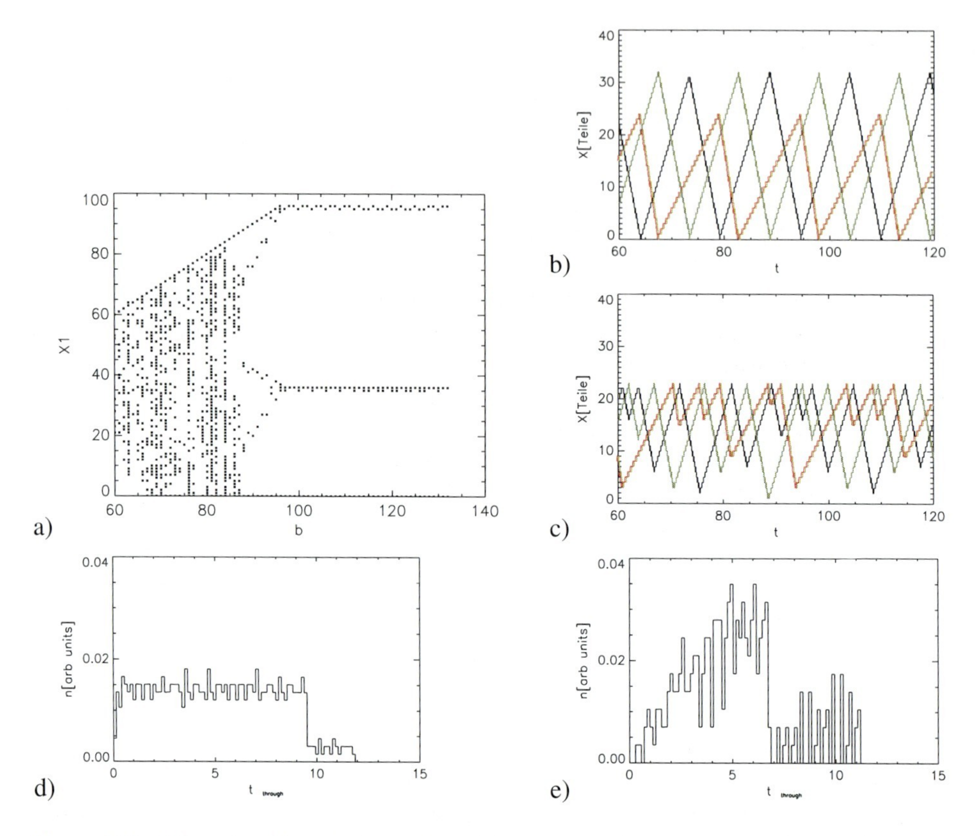

Figure 3.10: Bifurcation-like behavior and transition to quasi-chaotic states in a deterministic queueing model of a switched server type with three buffers. Parameters: $T_1 = 0.3$, $T_2 = 0.5$, $T_3 = 0.3$, $\Theta = 0.1153$. (a) Bifurcation diagram; the content of the first buffer (measured in parts) at switching times of the server vs. the maximal buffer capacity is shown. Total content: 132 ± 1. (b), (c) Time series of buffer content for a total content of 42 ± 1 parts: (b) maximal buffer capacity $b = 40$, (c) maximal buffer capacity $b = 23$. (d), (e) Histograms of throughput times for the whole system with a total content of 42 ± 1 parts for different maximal buffer capacities. (d) $b = 40$, (e) $b = 23$.

3.4.2 Small Stochastic Disturbances

As a further step towards fully realistic modeling of manufacturing we carried out numerical simulations of deterministic queueing models with small stochastic disturbances of inter-event times, as they are typical for highly automated manufacturing systems. The results indicate that such small disturbances mainly lead to slow changes in the total content of the systems. Thus the dynamics is reasonably well approximated by the deterministic queueing models for short time scales. At larger time intervals the system may (because of the varying total content) slowly drift thought different dynamics of the pure deterministic model. This leads, for instance, to unexpected chaotic behavior, if the total content of the systems rises above a certain value.

3.5 Conclusion

We have analyzed the dynamical behavior of idealized manufacturing systems governed by the interaction of scheduling policies and limited buffer capacities in the deterministic limit case.

As a first step dynamical systems belonging to a certain class of switched flow hybrid systems are investigated. Due to the continuity of the trajectory at the switching points in the continuous part of the state space their behavior can be characterized as a strange billiard. In Sect. 3.2.2 we have shown that with the introduction of upper thresholds switched arrival and switched server systems are limiting cases of one and the same class of switched flow systems. For systems with three tanks these limiting cases show chaotic and periodic dynamics both in switched server and switched arrival systems.

By sampling the dynamics at the switching points piecewise-linear Poincaré maps were obtained. When the bifurcation parameter b is varied linear segments of the Poincaré map appear, disappear and change their extension whereas the slope of the segments is constant. We have demonstrated that the change of extensions leads to bifurcations in the strange billiard. But these bifurcation scenarios are different from the usual scenarios, obtained for maps on the unit interval where the slope of linear segments changes. We further explored changes by different switching policies (Sect. 3.2.4). Such changes in general also change the Poincaré maps and therefore the dynamics of the systems.

By estimation of cost functions or calculation of production losses due to setup times, it became obvious that dynamical behavior directly influences the performance of manufacturing systems.

Despite their relative simplicity the systems investigated here develop a complex dynamics, even without irregular influences. Furthermore, not only the policies themselves contribute to this behavior, but also the chosen thresholds are crucial where these policies become active. For larger networks of work units with even more rules, thresholds and – in a hybrid modeling – a huge number of possible discrete states, therefore, a highly complex behavior is expected.

Acknowledgment

The authors gratefully acknowledge financial support from the Volkswagen Foundation (Grant No. I/76 279–280).

Bibliography

[1] L. Bunimovich, *Dynamical systems and operations research: a basic model*, Discrete and Continuous Dynamical Systems B **1**(2) (2001), 209–218.

[2] T. Beaumariage and K. Kempf, *The nature and origin of chaos in manufacturing systems*, Proceedings of the 5th IEEE/SEMI Advanced Semiconductor Manufacturing Conference (1994), 169–174.

[3] I. Diaz-Rivera, D. Armbruster, and T. Taylor, *Periodic orbits in a class of re-entrant manufacturing systems*, Operations Research, to appear.

[4] C. Chase, J. Serrano, and P.J. Ramadge, *Periodicity and chaos from switched flow systems: contrasting examples of discretely controlled continuous flow systems*, IEEE Transactions on Automatic Control **38**(1) (1993), 70–83.

[5] T. Schürmann and I. Hoffman, *The entropy of strange billiards inside n-simplexes*, Journal of Physics A **28** (1995), 5033–5039.

[6] I. Katzorke and A. Pikovsky, *Chaos and complexity in simple models of production dynamics*, Discrete Dynamics in Nature and Society (2001), 179–187.

[7] E. Ott, C. Grebogi, and J.A. Yorke, *Controlling Chaos*, Physical Review Letters **64**(11) (1990), 1196–1199.

[8] B. Rem and D. Armbruster, *Control and Synchronization in switched arrival systems*, Chaos **13**(1) (2003), 128–137.

[9] J. Worbs and K. Peters, *Analysis of the logistical performance of manufacturing processes with methods of nonlinear dynamics*, in: 3rd CIRP International Seminar on Intelligent Computation in Manufacturing Engineering, ICME 2002, Ischia (Italy), ISBN 88-87030-44-8.

[10] P. Nyhuis and H.-P. Wiendahl, *Logistische Kennlinien*, Springer Verlag, Berlin, Heidelberg, 1999 (in German).

[11] L. Kleinrock, *Queueing Systems*, Vols. 1, 2, John Whiley, New York, 1975.

[12] H. Chen and A. Mandelbaum, *Hierarchical modeling of stochastic networks, Part I: Fluid models* 47–100 in: D.D. Yao (ed.) *Stochastic Modeling and Analysis of Manufacturing Systems*, Springer Verlag, Berlin, Heidelberg, 1994, pp. 47–100.

[13] K. Peters and U. Parlitz, *Hybrid systems forming strange billiards*, International Journal of Bifurcation and Chaos **19**(9) (2003), 2575–2588.

[14] H.E. Nusse, E. Ott, and J.A. Yorke, *Border-collision bifurcations: An explanation for observed bifurcation phenomena*, Physical Review E **49**(2) (1994), 1073–1076.

[15] S. Banerjee and C. Grebogi, *Border collision bifurcations in two-dimensional piecewise smooth maps*, Physical Review E **59**(4) (1999), 4052–4061.

[16] S.-X. Qu, S. Wu and D.-R. He, *Multiple devil's staircase and type V intermittency*, Physical Review E **57**(1) (1998), 402–411.

[17] S. Coombes and A.H. Osbaldestin, *Period adding bifurcations and chaos in a periodically stimulated excitable neural relaxation oscillator*, Physical Review E **62** (3) (2000), 4057–4066.

4 Modeling and Analysis of a Re-entrant Manufacturing System

B. Scholz-Reiter, M. Freitag, A. Schmieder, A. Pikovsky, and I. Katzorke

Re-entrant manufacturing systems can frequently be found in semiconductor industries, where the same work systems are used repeatedly for different stages of processing. Due to feedback loops in the material flow, re-entrant manufacturing systems show complex dynamic behavior making it difficult to predict or control the performance of the systems.

The first part of the paper presents a dynamical model of a re-entrant manufacturing system. This model shows different qualities of dynamic behavior, depending on the ratio between work load and system capacity. The under-loaded system has the properties of a quasi-periodic driven dissipative dynamical system. The over-loaded system shows dynamics between quasi-periodicity and chaos.

The second part of the paper presents a simulation model based on a dynamical control concept. The simulation model consists of different modules containing the system parameters. The system variables – buffer levels and processing phases – indicate the state of the manufacturing process. This model will be used to develop, test and evaluate dynamic control methods and to investigate the influence of different control policies on dynamics and performance of this re-entrant manufacturing system.

4.1 Introduction

4.1.1 Re-entrant Manufacturing Systems and Models

Re-entrant manufacturing systems are production systems that use the same work systems repeatedly for different stages of processing. This can be found in job shops as well as in flow shops, where the route of a certain type of part through the system is pre-determined, but forms one or more feedback loops. Such re-entrant lines can frequently be found in the semiconductor industry, where wafers undergo a high number of processing stages at a lower number of work systems.

The re-entrant nature of material flow, disparate processing times for different processing stages at a machine and machine down times make scheduling and control of re-entrant manufacturing systems quite difficult. The situation becomes even more complicated if the manufacturing system has to process two or more different types of parts. Then the problem of setups and lot sizing becomes relevant.

Investigations of re-entrant manufacturing systems aim at stability of scheduling and release policies, optimal control for maximum performance as well as understanding of dynamic

Nonlinear Dynamics of Production Systems. Edited by G. Radons and R. Neugebauer

ISBN 3-527-40430-9

system behavior. For these purposes, different types of models were developed in the past. They can roughly be classified into *flow models*, mostly realized by differential equations, and *discrete models*, realized by a queueing network or event-discrete simulation. At both types of models, events within the system can be modeled as *stochastic* or *deterministic*. A processing time, for example, can vary randomly or be constant.

The most common models of re-entrant manufacturing systems are *queueing networks*, where a machine and its input buffer are represented by a server with a queue in front of it. The discrete part flow through the network is described by formulae of stochastic queueing theory. Such queueing models are good tools for analyzing the overall system performance under different release and scheduling policies. This has been done intensively in the past by various authors, especially by Kumar and his colleagues, e.g. in [5].

The analytic investigation of queueing models is appropriate for small systems with only a few components. But larger numbers of machines, buffers and processed parts lead to complex queueing networks that are hard to analyze using the formulae of queueing theory. In that case, it is appropriate to use *discrete-event simulation* for analyzing the system. The simulation provides results about stability and performance of the system, plus the chance to analyze its dynamic behavior. Beaumariage and Kempf [1] analyzed a re-entrant manufacturing system via simulation and showed the sensitive dependence of throughput times and output patterns on initial conditions as well as scheduling and release policies. They postulated that chaotic dynamics was the reason for this unstable behavior. But for a proper investigation of the dynamics of a re-entrant manufacturing system, a deterministic model based on nonlinear dynamics theory may be essential.

Such a *dynamical model* focuses more on the temporal evolution of the re-entrant manufacturing system than on its performance. The *system state* can be described by *variables* such as "buffer levels" and "processing phases". These variables span the *state space* of the system. The temporal evolution of the system is represented by *trajectories* in this state space. The state space and the run of the system trajectory can be used for analyzing the dynamics of the system. Such dynamical models were used by Hanson et al. [3] and Diaz-Rivera et al. [2]. Hanson et al. investigated the stability of their model and showed that – contrary to [1] – no chaotic behavior occurs. Diaz-Rivera et al. used a Poincaré map of a re-entrant manufacturing system and proved that only periodic orbits appear. Another dynamical model of a re-entrant manufacturing system, developed by Katzorke [4], will be presented in Part I of this paper.

4.1.2 Control Policies and Their Analysis

Control of re-entrant manufacturing systems means an optimal choice and combination of release and scheduling policies to realize high throughput and low work in process. The control policies should be robust against uncertainties and must provide good system performance.

A release policy regulates the part input into the system. It determines the points in time at which parts arrive as well as the quantity of any release. Release policies are applicable if the re-entrant manufacturing system is more or less autonomous. If it is organized in a pull-type production line or supply chain, the arrival of parts into the system is determined exogenously and cannot be controlled by the system.

A scheduling policy regulates the withdrawal of parts from the buffer for processing at the machine. The buffer can contain (i) different types of parts and (ii) parts at different stages of processing. The scheduling policy determines which type of part and which processing stage has the highest priority for processing next. The most common scheduling policies are the *first-come-first-served* policy (FCFS), also known as *first-in-first-out* (FIFO), and the *last-buffer-first-served* policy (LBFS), an implementation of the pull principle. A good overview of scheduling policies for re-entrant manufacturing systems can be found in [5].

From a systems perspective, the control policies represent the information subsystem, which generates the dynamics of the physical subsystem (machines, buffers, parts). The release policy generates the material input into the system; the scheduling policy controls the material flow through the system, thus linking the control policies to the system dynamics. This motivates the use of dynamical models, mentioned above, to analyze the system dynamics and to develop control policies based on nonlinear dynamics theory.

Due to the specific features of re-entrant manufacturing systems, the development of appropriate control policies is not trivial. It is well known that, for example, scheduling problems are NP-hard. Therefore, control policies are mostly verified by simulation. This provides results about the *stability* of the chosen control policies as well as *performance* and *dynamics* of the system.

Stability of a manufacturing system is simply a matter of work load versus processing capacity. That means that the arrival rate of parts may not exceed the processing rate of the overall system. Unfortunately, in re-entrant manufacturing systems, the total system capacity cannot easily be derived from single-machine capacities. Here, the re-entrant structure of the system and the chosen scheduling policy create additional capacity constraints, also called virtual bottlenecks [7]. This can lead to an accumulation of parts to infinity despite a low work load – the system turns unstable. This phenomenon has led to the extensive investigation of various combinations of re-entrant flow structures and scheduling policies, especially with respect to their stability properties. The main result was that the LBFS policy was proven stable for any kind of re-entrant manufacturing system [6].

The goal of controlling re-entrant manufacturing systems is to reach good performance. Common performance measures are the total work in process and its costs, the throughput volume and the mean throughput time and its variance. The total work in process and the throughput time are related to each other by Little's law: work in process = arrival rate of parts $\times$ throughput time [5]. With that, low work in process minimizes not only costs but also shortens the throughput time of a part. A small variance of the mean throughput time ensures meeting the due dates, allows better planning of releases into the system and leads to better coordination of further operations, such as assembly. Maximizing throughput volume seems to be a matter of maximizing capacity utilization. But a high-capacity utilization requires high work in process, which leads to high costs and long throughput times. If the throughput times increase, the throughput volume decreases temporarily. So the throughput volume cannot be controlled directly, but is a result of minimizing work in process and throughput time. All performance measures result from the chosen scheduling and release policies. Some combinations of policies optimize one performance measure but worsen the others. But, in general, the LBFS policy is proven to be almost always the best for re-entrant manufacturing systems [6].

Despite extensive investigations of re-entrant manufacturing systems, there is little research on developing an understanding about their dynamics. The re-entrant structure and a high work load lead to complex dynamic behavior, while small changes in release or scheduling policies lead to large effects on system performance [1]. These complex dynamics (and possibly deterministic chaos) makes it difficult to predict or control the performance of re-entrant manufacturing systems. This motivates an analysis of the system dynamics (i) to decide whether the system is periodic, quasi-periodic or chaotic and (ii) to find a mapping of system attractors to performance measures, which will serve as a basis for application of dynamic control methods, derived from nonlinear dynamics theory.

This paper is structured as follows: Part I introduces a specific re-entrant manufacturing system, develops a dynamical model and analyzes its dynamics. Part II introduces a dynamical concept for manufacturing control and presents a simulation platform for an analysis of different scheduling and control policies.

Part I

4.2 "Two Products – Two Stages" Re-entrant Manufacturing System

The re-entrant manufacturing system considered here consists of a single work system, where two varying types of parts have to pass the system twice for processing. Such a manufacturing system is depicted in Fig. 4.1. The two varying parts – labeled A and B – arrive at this work system and enter the input buffer. After the first processing stage, the parts are denoted as A′ and B′ and return to the input buffer. After the second processing stage, the parts are denoted as A″ and B″ and leave the work system.

The parts in the input buffer are organized as a queue according to the FIFO policy. No advanced scheduling rule is applied. In the case of simultaneous arrival of two or more parts, the priority for entering the input buffer is $A > B > A' > B'$. This means that at the input of the machine a series of four symbols A, B, A', B' is encountered, which is transformed to the corresponding series of A', B', A'', B'' at the output. It is assumed that parts A and B arrive

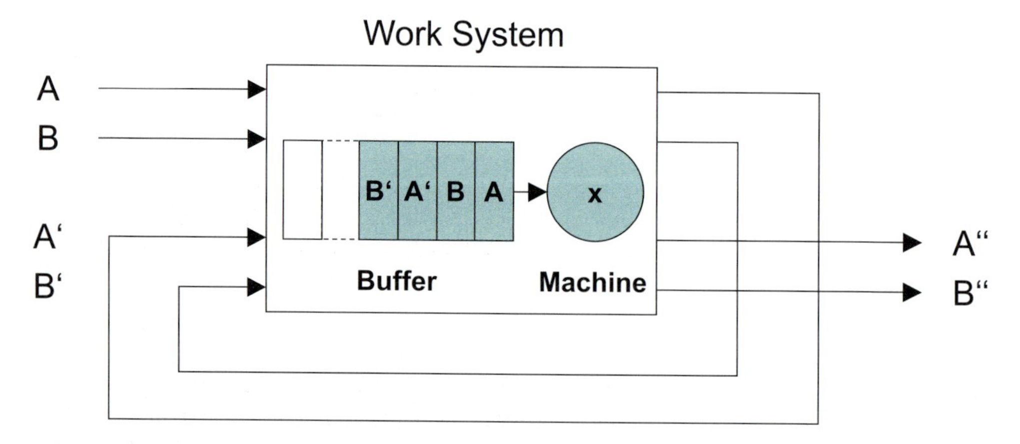

Figure 4.1: Schematic part flow for the "two products – two stages" re-entrant manufacturing system. Two varying parts A and B have to pass the work system twice until they leave it as A″ and B″.

at deterministic periodic time intervals t_A and t_B, respectively. Other relevant parameters are the processing times for parts A, B, A', B'. These times are denoted as $T_A, T_B, T_{A'}, T_{B'}$,

respectively. It is assumed that all these times are constants. For the dynamics of the system it is essential that the times $t_{A,B}$, $T_{A,B,A',B'}$ are arbitrary real numbers. This makes the whole process at least as complex as quasi-periodic. Here, the symbolic characterization of the process will be used to investigate this system.

Depending on its capacity, the machine will be able to process all incoming parts, or not. In terms of the parameters of the system, i.e. of the times $t_{A,B}$, $T_{A,B,A',B'}$, this can be formulated as follows: let us consider a time interval T, which is large enough for statistical investigations. During this time $N_A = T/t_A$ parts of type A and $N_B = T/t_B$ parts of type B are delivered. The total processing time for these parts is $T_{\text{process}} = N_A(T_A + T_{A'}) + N_B(T_B + T_{B'})$.

Thus, the critical processing capacity is described by the parameter τ:

$$\tau = \frac{T_{\text{process}}}{T}, \tag{4.1}$$

$$\tau = \frac{T_A + T_{A'}}{t_A} + \frac{T_B + T_{B'}}{t_B}. \tag{4.2}$$

The following types of systems can be distinguished according to the value of τ:

- $\tau < 1$: under-loaded system: the incoming part flow does not exhaust the system capacity;
- $\tau = 1$: balanced system: the incoming part flow fits exactly the processing capacity;
- $\tau > 1$: over-loaded system: the processing capacity is insufficient to process all incoming parts.

Because of this classification, the queue will disappear, remain or grow with time depending on the value of τ.

4.3 Dynamical Model

The work system modeled here consists of the two subsystems "buffer" and "machine" (see Fig. 4.1). The overall system state thus consists of the current states of the buffer and the machine. The current state of the buffer is described by a sequence of symbols in the queue, namely A, B, A', B'. The current state of the machine is described by a continuous variable x ($0 \le x \le 1$) which can be viewed as the processing phase. The phase x grows linearly in time:

$$\frac{dx}{dt} = \frac{1}{T_S}, \tag{4.3}$$

where S ($S \in A, B, A', B'$) stands for the part which is in process. As x reaches the value 1, the processing of the part ends and the queue is re-arranged: the first-in part enters the machine and – in the case $S = A$ or $S = B$ – the corresponding part (A' or B') is added to the end of the queue as soon as the parts leave the machine. Additionally, every time step t_A and t_B, parts A and B are added to the end of the queue. In the general case of incommensurate arrival times t_A and t_B, this latter operation can be considered as a quasi-periodic driving of the system.

The parameters governing the dynamics are the times t_{A}, t_{B}, $T_{\mathrm{A}}, T_{\mathrm{B}}, T_{\mathrm{A}'}, T_{\mathrm{B}'}$. The following parameters are used: $t_{\mathrm{A}} = 1, t_{\mathrm{B}} = 1 + \sqrt{5}, T_{\mathrm{A}} = T_{\mathrm{A}'} = (2\sqrt{2})^{-1}, T_{\mathrm{B}} = T_{\mathrm{B}'} = c^{-1}(2 - \sqrt{2})(\sqrt{5} - 1)^{-1}$. Here the parameter c governs the balance condition. Using the parameters, one obtains from Eq. (4.2): $\tau = c^{-1}(1 + (c - 1)/\sqrt{2})$.

4.4 Analysis of Dynamics

4.4.1 Sensitivity to Initial Conditions

The sensitivity of the model to initial conditions is analyzed because a sensitive dependence of the system dynamics on small changes in initial conditions is an indicator for a chaotic dynamics. For this purpose, the model is slightly perturbed by a small shift of the arrival times t_{A} and t_{B}. Then, changes in the symbolic sequence in the queue are observed. In the case of a balanced or over-loaded system, the change in the symbolic sequence never disappears, because the queue is never empty and the undisturbed arrival times cannot be restored (see Fig. 4.2). In the under-loaded case, the queue is empty from time to time. In these periods, the perturbation of the shifted arrival times disappears, and the unperturbed symbolic sequence is restored (see Fig. 4.2). Thus, the effect of a perturbation on the under-loaded dynamics is only temporary: it exists only during the time interval $\delta t_1 < t < \delta t_2$ after the perturbation is imposed. Here δt_2 is the time at which the sum of all empty queue states exceeds the perturbation and is inversely proportional to $(1 - \tau)$.

4.4.2 Ergodicity and Stationarity

To test the system for ergodicity, one has to determine how the observed transition probabilities in a time series (i.e. probabilities to find pairs of symbols like AB) depend on the initial conditions. For $c \neq 1$, it is found that the transition probabilities do not depend on the initial conditions while, for $c = 1$, such dependence is observed (see Fig. 4.3).

Qualitatively, such a dependence on the balance condition can be understood as follows: in the under-loaded case, the dynamics as shown above is not sensitive to the initial conditions and is fully determined by the quasi-periodic drive (input flow of parts). Therefore, the system is ergodic. In the over-loaded case, the queue grows; thus any perturbation in the dynamics of the system increases because it affects more and more parts in the queue. As a result, effective "averaging" over all initial perturbations occurs, resulting in ergodic behavior. In the balanced case, a perturbation in the initial conditions remains roughly constant over time because the queue length is constant. Therefore, different initial conditions do not "mix" and the statistics can depend on them, thus breaking the ergodicity.

Stationarity of the observed symbolic time series is ensured in the under-loaded case, where the dynamics is essentially driven by the quasi-periodic input, and thus follows the stationarity of the quasi-periodic process. The stationarity is tested numerically by the χ^2-criterion

$$\chi^2 = \sum_{i}^{n_c} \frac{(s_i^- q_i)^2}{s_i + q_i}, \tag{4.4}$$

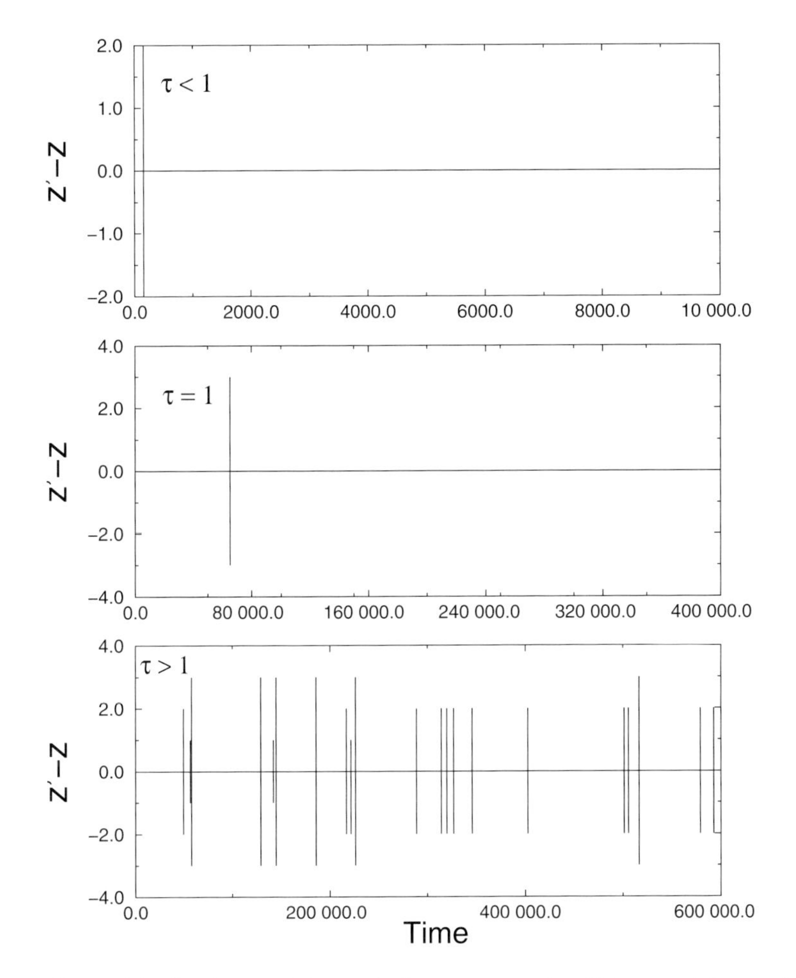

Figure 4.2: Sensitivity of the symbolic sequences: the difference of the symbols of the perturbed (at time step 4) and the unperturbed sequences. Upper panel: under-loaded system $c = 1.1$, middle panel: balanced system $c = 1$, lower panel: over-loaded system $c = 0.9$. The perturbation is $c_{\mathrm{p}} = c \times 1.00001$ in all cases.

where s_i is the number of the observed events of a combination, for instance (A, B), from one section of the time series and q_i the number of observed events of the identical combination in another section of the time series. Further, n_{c} is the number of all possible combinations.

For values of the parameter τ not too close to 1, the stationarity hypothesis could always be confirmed. If τ is close to 1 (e.g. $\tau \approx 0.995$), there are difficulties in applying the test because the statistical properties indicate very slow modulations, which would require averaging over extremely long time intervals.

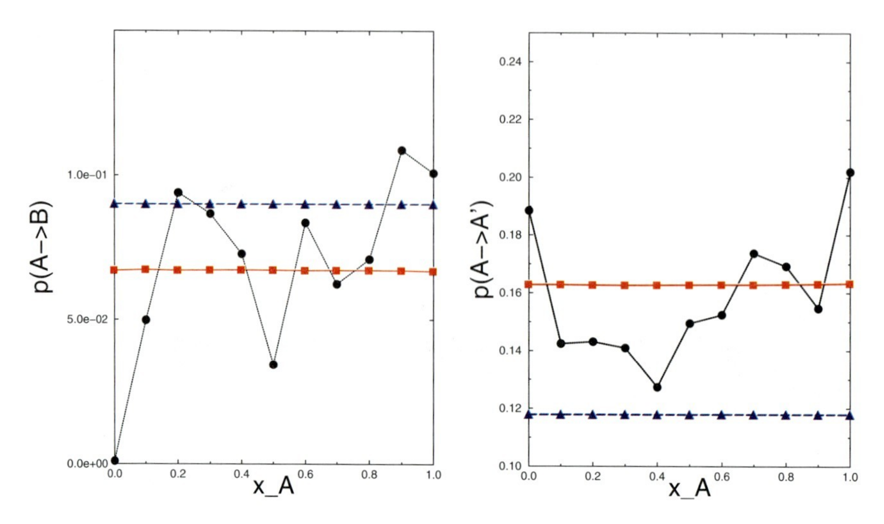

Figure 4.3: The probabilities of finding a pair (A, B) (left) and a pair $(\mathrm{A}, \mathrm{A}')$ (right) in the queue are shown for the balanced (solid line, filled circles), the over-loaded (dotted line, filled squares) and for the under-loaded (dashed line, filled triangles) systems depending on the initial conditions x_{A}. Note that only the balanced system exhibits a dependence of p on the initial conditions.

4.4.3 Correlations

For the calculation of the normalized autocorrelation function, a time series z_i is used:

$$C(k) = \frac{\langle z_{i+k} z_i \rangle - \langle z \rangle^2}{\langle z^2 \rangle - \langle z \rangle^2} \,. \tag{4.5}$$

In the under-loaded case, the autocorrelation function demonstrates a pattern typical of quasi-periodic dynamics, returning to a value near 1 at a relatively regular rate. This supports the above conclusion that, in this case, the behavior of the system is completely determined by the quasi-periodic driving. In the case of perfect balance ($\tau = c = 1$), the dynamic behavior depends on initial conditions, so the autocorrelation function depends on them as well. Two examples are shown in Fig. 4.4. In all observed situations, the correlations are close to 1 for some large time shifts k, thus indicating quasi-periodicity. In some cases, the correlations are not close to 1 for small k. This means that the process is more complex than a quasi-periodic one.

The largest complexity is achieved in the over-loaded case. Here the value of the auto-correlation function does not return to 1 even for large time shifts k. An example of the autocorrelation function for $0.98 \leq c \leq 0.999$ is presented in Fig. 4.5. This function looks like a quasi-periodic one on a small scale, but with a slowly varying envelope with numbers less than 1. This envelope is shown in Fig. 4.5 for different values of parameter c, i.e. for different levels of violation of the balance condition. One can see that, for larger deviations of τ from 1, the correlations decay faster and to a lower level.

From the shape of the correlation function it follows that the spectrum is neither discrete nor absolutely continuous. Indeed, it neither decreases to zero nor returns to 1. Because

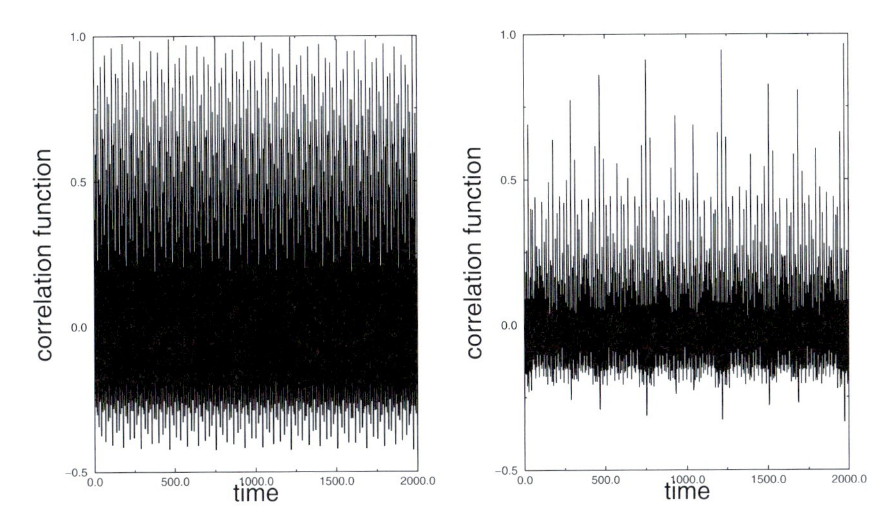

Figure 4.4: Balanced system: dependence of the correlation functions on the initial conditions, i.e. the start position of the parts (left-hand side: $x_{\mathrm{A}} = 0$, $x_{\mathrm{B}} = 0$, $x_{\mathrm{A}'} = 1$ and $x_{\mathrm{B}'} = 0$, right-hand side: $x_{\mathrm{A}} = 0.5$, $x_{\mathrm{B}} = 0.5$, $x_{\mathrm{A}'} = 0$ and $x_{\mathrm{B}'} = 0$). Sequences (2^{19} symbols) are measured at the output of the work system and consist of all four symbols A, B, A′ and B′. The values of these correlation functions return to the value 1 after some time steps. From this, one can assume that the system will be quasi-periodic in all cases.

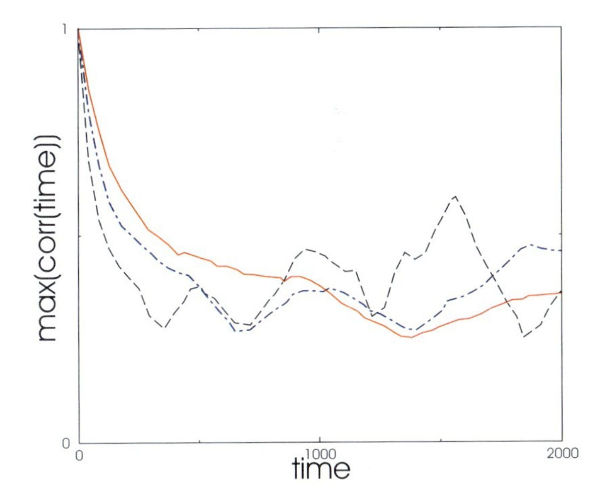

Figure 4.5: The envelope drawn through the peaks of the correlation function. Parameter values used: $c = 0.98$ (dashed line), $c = 0.99$ (dot-dashed line) and $c = 0.995$ (solid line).

the integrated correlation function does not decrease to a value close to zero, the spectrum is apparently not singular continuous. Thus, the situation is quite complicated and it is hardly possible to extract the singular continuous component from the spectrum, because it also has very strong discrete components. Thus, the results described in the last section are of purely conjectural nature. At present, no mathematical descriptions of the spectral measure and the properties of the integrated correlation function exist for this type of system.

Part II

4.5 Dynamical Concept for Manufacturing Control

Modern manufacturing systems are highly complex in their structures and dynamics. Even quite simple systems as investigated in Part I of this paper can show complex dynamic behavior. To control such manufacturing systems, concepts from nonlinear dynamics theory could help to manage their dynamic complexity and possibly provide strategies and methods to control the manufacturing process.

This motivates the development of a dynamical concept for manufacturing control. Within this concept, the manufacturing system is considered to be a dynamical system. Such a system has a set of parameters and system variables. The behavior of the manufacturing system as a whole follows its intrinsic dynamics, which is influenced by the system parameters. As a result of the

dynamics, the system variables change over time.

The system variables span the state space of the system. The temporal evolution of the system is represented by trajectories in this state space. The state space and the run of the system trajectories are used for analyzing the dynamics of the system and provide possibilities to control the system.

The state of the manufacturing system at any point in time is defined by a point on the trajectory in the state space. It is thus possible to influence and control the state and the evolution of a manufacturing system by manipulation of the system trajectory. For this, several methods for the control of dynamical systems have been developed.

A dynamical system can be controlled by variation of the system parameters to force the system variables onto a desired trajectory. The variables that are usually controlled in a manufacturing system are inventory levels or work in process. The temporal evolution of these variables depends on the system dynamics. The idea behind the dynamical concept is to control these dynamics by variation of the system parameters, which are considered flexible and capable of being influenced.

Because of the high number of different parameters of a manufacturing system, they have to be combined into the following functional groups: *structure, capacity, operational rules, release policies, scheduling/queueing policies* [8]. These functional groups generate the dynamics of a manufacturing system, and enable and influence the part flow through the system. They are at first a framework for modeling the manufacturing system and thereafter provide possibilities to control the production process.

Within the dynamical control concept, the variables of the system such as *buffer levels* and *processing phases* are monitored continuously. These variables represent the current system state, which is the basis for controlling the system. Due to the reference to the *current* state of the system, the control takes place continuously and dynamically by an adjustment of the functional groups mentioned above to force the buffer contents and therewith the overall work in process to reach a desired level. The goal is a close-to-real-time control of the production process to meet the current requirements of the manufacturing system.

To verify the dynamical control concept, the "two products – two stages" re-entrant manufacturing system – introduced in Part I of this paper – is used for developing dynamic control methods. Here, the functional group "structure" is considered constant – the other functional groups can be used for control.

4.6 Simulation Model

A simulation platform that provides an event-discrete simulation of the manufacturing process was developed with the objective of developing, testing and evaluating dynamic control methods as well as investigating the influence of different control policies on the dynamics of the considered re-entrant manufacturing system. An important feature of this simulation platform is the possibility to monitor and record all processes and events so as to capture the dynamics of the system.

The simulation platform is based on the dynamical concept introduced in the last section. The functional groups are implemented in modules. The main module contains the "structure", here the structure of the "two products – two stages" re-entrant manufacturing system (see Fig. 4.1). This main module consists of further modules (subsystems) which contain the other four functional groups. The subsystem "capacity" contains the capacity of the input buffer (minimum and maximum buffer levels) and the capacity of the machine (processing times for every type of part at every processing stage). Thereby, the processing times can randomly vary or be constant. The subsystem "operational rules" controls the buffer input and output. It contains rules such as "stop input if buffer full" or "stop output if buffer empty". The subsystem "release policy" defines the part input into the system. Parts can arrive at deterministic points in time (constant release periods or inter-arrival times respectively) or randomly.

The subsystem "scheduling/queueing policy" determines the priority of the different types of parts at the different stages of processing (i) for entering the input buffer in the case of simultaneous arriving of two or more parts and (ii) for the withdrawal of parts from the buffer for processing at the machine. The first task of this subsystem is due to the existence of only one main buffer for the different types of parts and processing stages. Here, the type of part or the stage with the lower priority will be shifted from simulation step t to $t+1$. The second task realizes the application of a specific scheduling policy such as LBFS or simply FIFO. But the most interesting feature of this subsystem is the possibility to implement dynamic scheduling policies. In doing so, the policy picks out a type of part and a stage depending on the current system state.

The current state of the system is defined by the system variables. To grasp this system state, the main module "structure" contains monitors for observing the system variables. These are: the buffer levels of the different types of parts and processing stages ($\mathrm{A}, \mathrm{B}, \mathrm{A}', \mathrm{B}'$) as well as the total buffer content, the buffer input and output patterns and the progress of processing of a part in the machine (processing phase). The temporal evolutions of these system variables represent the dynamics of the system. They will be used for analysis of dynamic behavior caused by different static or dynamic control policies.

To verify the performance of a chosen control policy, there are routines for calculating the throughput times for every part and the total throughput as well. It should be noted that no mean values are used. The whole simulation model is deterministic. Nevertheless, stochastic influences can be implemented additionally.

4.7 Analysis of Scheduling Policies

To demonstrate the functioning of the simulation model, the influence of different scheduling policies on buffer levels, part processing and throughput is investigated. The processing time of every part at every stage is set to 5. The total cycle time of every part including transit times is 13. The inter-arrival times of both types of parts A and B are set to 20. Due to the successive processing of every part and every stage, the system is overloaded by factor 1.3. This overload is used to point out the effects of the different scheduling policies.

The scheduling policy applied first is first-buffer-first-served (FBFS). This policy prioritizes the buffer that contains parts waiting for the lowest stage of processing. That means in this case that the parts A and B, which are waiting for processing at its first stage, have a higher priority than the parts A′ and B′, which are waiting for processing at its second stage.

The second scheduling policy applied is last-buffer-first-served (LBFS). This policy prioritizes the buffer that contains parts waiting for the highest stage of processing. That means that the parts A′ and B′, which are waiting for processing at its second stage, have a higher priority than the parts A and B, which are waiting for processing at its first stage.

In both cases, part type A always has a higher priority for processing than part type B. The following Figs. 4.6 and 4.7 show (i) the buffer levels of the different types of parts and processing stages and the total buffer content, and (ii) the part processing in the machine.

Due to the FBFS policy and the general A-over-B priority, parts A have the highest priority and B′ the lowest. This leads to an increasing amount of B′ in the buffer up to a value of 13 (see Fig. 4.6, upper panel). All other buffer levels oscillate only between 0 and 1. So the increasing B′ level is responsible for the increasing of the total buffer content up to 14. The direct effect of the FBFS policy can be seen in the lower panel of Fig. 4.6. All of the 20 released parts B were processed at its first stage (B) but only seven parts were processed at its second stage (B′). That means that only seven of 20 released parts could be finished – the throughput of part B is only 35% of its maximum.

In the case of the LBFS policy and still the general A-over-B priority, parts A′ have the highest priority and B the lowest. This leads to an increasing amount of B in the buffer up to a value of 7 (see Fig. 4.7, upper panel). All other buffer levels oscillate only between 0 and 1. So the increasing B level is responsible for the increasing of the total buffer content up to 8. It should be noted that the maximum buffer level is only about 57% of the maximum buffer level in the FBFS case. The reason for this fact can be seen in the lower panel of Fig. 4.7. Only 14 of 20 released parts B were processed at its first stage (B) but 13 of these 14 parts were processed at its second stage (B′). That means that 13 of 20 released parts could be finished – the throughput of part B is 65% of its maximum. That means that in comparison with the FBFS policy, the LBFS policy leads almost to a doubling of the overall throughput and almost to a halving of the maximum buffer level or the overall work in process respectively.

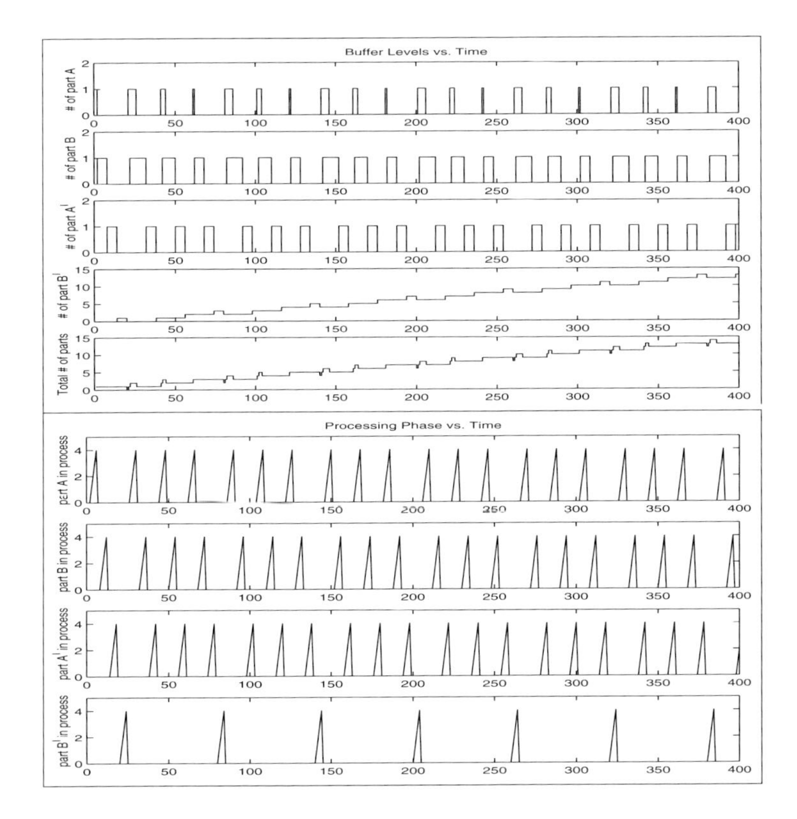

Figure 4.6: Buffer levels and part processing under the FBFS policy.

The scheduling policies FBFS and LBFS, which have been analyzed here, are static priority rules. But, in spite of their static nature, such advanced scheduling policies lead to a better performance than the widely used FIFO policy in the majority of cases. Furthermore, the development of dynamic scheduling policies, which change their priority scheme depending on the current system state, is in process. This will provide a dynamic control of the processing and completion of parts.

4.8 Conclusion and Outlook

This paper (i) introduced a dynamical model of a re-entrant manufacturing system and analyzed its dynamics and (ii) presented a simulation model based on a dynamical concept for manufacturing control.

Part I of this paper described a "two products – two stages" re-entrant manufacturing system and developed a dynamical model of this system. The dynamics is governed by the ratio of work load to system capacity. If the system capacity is larger than the work load, the system is under-loaded and has the properties of a quasi-periodically driven dissipative dynamical

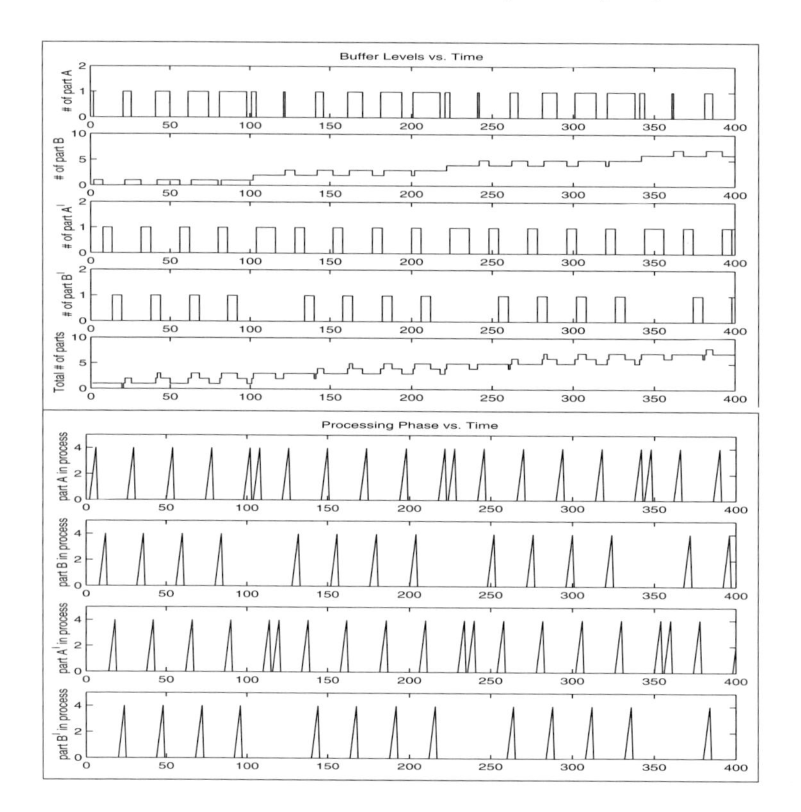

Figure 4.7: Buffer levels and part processing under the LBFS policy.

system. One can relate this to the fact that for large production rates, the queue disappears and the idle time intervals damp out perturbations.

When the system is balanced (work load = system capacity), the dissipativity is no longer given. In this case, the queue is never empty and the perturbations do not decay. Moreover, the dynamics depends on the initial state. This can be related to the fact that the perturbations do not grow either.

In the third case, when the system capacity is less than the work load, the queue grows continuously and the dynamics becomes more complex than in the quasi-periodic case. One can understand this as an effect of the growing queue which leads to an effective growth of the perturbations. This growth is, however, not fast enough to yield chaos. The correlations in the over-loaded case neither return to one nor decay to zero.

Further investigations need to be devoted to the classification of the particular kind of system that seems to be neither periodic nor chaotic. In addition, suitable controlling methods have to be developed for an optimization of such systems.

Part II of this paper introduced a dynamical concept for manufacturing control. Within this concept, the manufacturing system is considered to be a dynamical system. The *buffer levels* and the *processing phases* were defined as system variables. The general parameters of a manufacturing system were combined into the functional groups *structure, capacity, operational rules, release policies, scheduling/queueing policies.*

This dynamical concept was realized in a simulation platform and a simulation model of the "two products – two stages" re-entrant manufacturing system. All functional groups were implemented in specific modules that can be used to control the system variables. The functioning of the simulation model was exemplified using the functional group "scheduling/queueing policies" for control of the processing and completion of parts. It was shown that the LBFS scheduling policy leads to a much better system performance than the FBFS scheduling policy.

Moreover, the dynamical concept and its implementation in the simulation platform provide a dynamic control of the manufacturing process using all functional groups. For example, the module "release policy" can be used for a part release depending on the current system state, the module "capacity" can be used for a dynamic adjustment of the buffer or machine capacity and so on. Consequently, the developed simulation platform is a useful tool for further work regarding dynamic control of manufacturing systems.

Bibliography

[1] T. Beaumariage and K. Kempf, *The nature and origin of chaos in manufacturing systems*, in: IEEE/SEMI Advanced Semiconductor Manufacturing Conference (1994), pp. 169–174.

[2] I. Diaz-Rivera, D. Armbruster, and T. Taylor, *Periodic orbits in a class of re-entrant manufacturing systems*, Mathematics of Operations Research **25**(4) (2000), 708–725.

[3] D. Hanson, D. Armbruster, and T. Taylor, *On the stability of re-entrant manufacturing systems*, in: A. Beghi et al. (eds.): *Mathematical Theory of Networks and Systems*, Proceedings of the 31 MTNS, Il Poligrafo, Padova, 1998, pp. 937–940.

[4] I. Katzorke, *Modeling and Analysis of Production Systems*, Ph.D. thesis, Faculty of Mathematics and Science, University of Potsdam (2002), pp. 27–43.

[5] P.R. Kumar, *Scheduling manufacturing systems of re-entrant lines*, in: D.D. Yao (ed.): *Stochastic Modeling and Analysis of Manufacturing Systems*, Springer, New York, 1994, pp. 325–360.

[6] S.H. Lu and P.R. Kumar, *Distributed scheduling based on due dates and buffer priorities*, IEEE Transactions on Automatic Control **36** (1991), 1406–1416.

[7] E.H. Nielsen, *Performance spread of re-entrant system structures*, in: D. Hanus and J. Talácko (eds.): Proceedings of the 16th International Conference on Production Research, Czech Association of Scientific and Technical Societies, Prague, 2001, CD-ROM.

[8] B. Scholz-Reiter, M. Freitag, and A. Schmieder, *A dynamical approach for modeling and control of production systems*, in: S. Boccaletti, B.J. Gluckman, J. Kurths, L.M. Pecora, and M.L. Spano (eds.): *Experimental Chaos*, Proceedings of the 6th Experimental Chaos Conference, American Institute of Physics, Melville, New York, 2002, pp. 199–210.

5 Nonlinear Models for Control of Manufacturing Systems

E. Lefeber

Current literature on modeling and control of manufacturing systems can roughly be divided into three groups: fluid models, queueing theory and discrete-event models. Most fluid models describe linear time-invariant controllable systems without any dynamics. These models mainly focus on throughput and are not concerned with cycle time. Queueing theory deals with relationships between throughput and cycle time, but is mainly concerned with steady-state analysis. In addition, queueing models are not suitable for control theory. Discrete-event models suffer from state explosion. Simple models of manufacturing systems can be studied and analyzed, but for larger problems the dimension of the state grows exponentially. In addition, most control problems studied are supervisory control problems: the avoidance of undesired states. An important class of interesting manufacturing control problems asks for proper balancing of both throughput and cycle time for a large nonlinear dynamical system that never is in steady state. None of the mentioned models is able to deal with these kinds of control problems. In this paper, models are presented which are suitable for addressing this important class of interesting manufacturing control problems.

5.1 Introduction

In this paper we are interested in the problem of how to ramp up a manufacturing line, or to be more precise: we are interested in models that are suitable for obtaining a proper solution to this problem. For that reason we propose a *computationally feasible*, *dynamic* model that incorporates both *throughput* and *cycle time*.

The model we propose is a flow model, based on the theory of modeling traffic flow. The idea is to consider the flow of products as a compressible fluid flow. The flow model we propose is not to be confused with the flow model as initiated by Kimemia and Gershwin [21] for modeling failure-prone manufacturing systems, nor with the fluid models or fluid queues as proposed by queueing theorists [16, 33], nor with the stochastic fluid model as introduced by Cassandras et al. [10].

All of the three mentioned fluid models from the literature are throughput-oriented. These models do not explicitly contain information about cycle time. Also, the processing times of machines are assumed to be deterministic. As a result, a property of these models is that any feasible throughput can be achieved by means of zero inventory.

A class of models available in the literature are models based on relations from queueing theory [22, 23], see e.g. [9, 32]. Although these results give valuable insight into steady-state behavior of manufacturing lines, a disadvantage is that only the steady state is concerned.

Nonlinear Dynamics of Production Systems. Edited by G. Radons and R. Neugebauer

ISBN 3-527-40430-9

No dynamic relations are available. Therefore, these models cannot be used for studying the problem of how to ramp up a manufacturing line.

A third class of models are discrete-event models like for instance the class of discrete-event systems as studied by Ramadge and Wonham [28]. These models do include dynamics, and both throughput and cycle time are incorporated. Unfortunately, as all states in which a manufacturing system can be have to be considered, these models are almost unsuitable for practical use. However, a promising modeling approach consists of the so-called max-plus-linear discrete-event systems with variability expansion, as studied in [8].

To summarize: roughly three classes of models for manufacturing lines have been studied in the literature so far: discrete-event models that suffer from state explosion, queueing theory that contains only steady-state results and fluid models that do not incorporate cycle times. In this paper we propose a new class of flow models, by considering the flow of products to be a compressible fluid flow. However, first we recall the fluid models as currently available in the literature and illustrate some of their shortcomings. We also prepare ways to (partially) overcome these shortcomings.

5.2 Extensions to the Standard Fluid Model

As mentioned in the introduction, one of the advantages of fluid models is that these models incorporate the dynamical behavior of manufacturing systems. Unfortunately, these models do not take into account cycle times. In this section we present extensions to the fluid model that (partially) overcome this disadvantage. However, before we can present this extension we first have to present the fluid model as currently used in the literature.

5.2.1 A Common Fluid Model

The current standard way of deriving fluid models is most easily explained by means of an example. Therefore, consider a simple manufacturing system consisting of two machines in series, as displayed in Fig. 5.1. Let $u_0(t)$ denote the rate at which jobs arrive to the system at time t, let $u_i(t)$ denote the rate at which machine M_i produces lots at time t, let $y_i(t)$ denote the number of lots in buffer B_i at time t ($i \in \{1, 2\}$) and let $y_3(t)$ denote the number of lots produced by the manufacturing system at time t. Assume that machines M_1 and M_2 have a maximum capacity of respectively μ_1 and μ_2 lots per time unit. This provides us with all information for deriving a fluid model.

Clearly the rate of change of the buffer contents is given by the difference between the rates at which lots enter and leave the buffer. Under the assumption that the number of lots can be

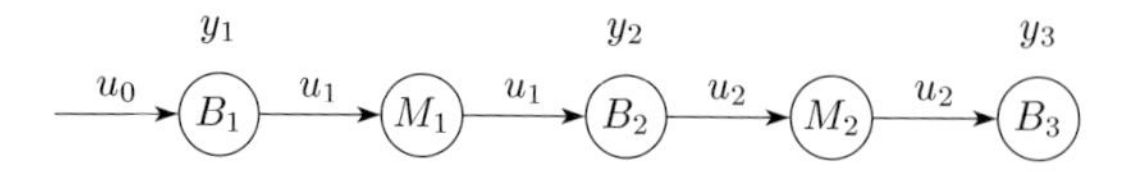

Figure 5.1: A simple manufacturing system.

considered continuous, this observation leads to the following fluid model:

$$\begin{aligned}\dot{y}_1(t) &= u_0(t) - u_1(t),\\ \dot{y}_2(t) &= u_1(t) - u_2(t),\\ \dot{y}_3(t) &= u_2(t),\end{aligned} \tag{5.1}$$

which can also be expressed as follows:

$$\dot{x}(t) = \begin{bmatrix} 0 & 0 & 0 \\ 0 & 0 & 0 \\ 0 & 0 & 0 \end{bmatrix} x(t) + \begin{bmatrix} 1 & -1 & 0 \\ 0 & 1 & -1 \\ 0 & 0 & 1 \end{bmatrix} u(t), \tag{5.2a}$$

$$y(t) = \begin{bmatrix} 1 & 0 & 0 \\ 0 & 1 & 0 \\ 0 & 0 & 1 \end{bmatrix} x(t) + \begin{bmatrix} 0 & 0 & 0 \\ 0 & 0 & 0 \\ 0 & 0 & 0 \end{bmatrix} u(t), \tag{5.2b}$$

where $u = [u_0, u_1, u_2]^{\mathrm{T}}$ and $y = [y_1, y_2, y_3]^{\mathrm{T}}$. We also have capacity constraints on the input, as well as the constraint that the buffer contents should remain positive. These constraints can be expressed by means of the following equations:

$$0 \leq u_1(t) \leq \mu_1, 0 \leq u_2(t) \leq \mu_2 \quad \text{and} \quad y_1(t) \geq 0, y_2(t) \geq 0, y_3(t) \geq 0. \tag{5.3}$$

System (5.2) is a controllable linear system of the form $\dot{x} = Ax + Bu, y = Cx + Du$ as extensively studied in control theory. Note that the description (5.2) is not the only possible input/output/state model which yields the input/output behavior (5.1). To illustrate this, consider the change of coordinates

$$x(t) = \begin{bmatrix} 1 & -1 & 0 \\ 0 & 1 & -1 \\ 0 & 0 & 1 \end{bmatrix} \bar{x}(t), \tag{5.4}$$

which results in the following input/output/state model:

$$\dot{\bar{x}}(t) = \begin{bmatrix} 0 & 0 & 0 \\ 0 & 0 & 0 \\ 0 & 0 & 0 \end{bmatrix} \bar{x}(t) + \begin{bmatrix} 1 & 0 & 0 \\ 0 & 1 & 0 \\ 0 & 0 & 1 \end{bmatrix} u(t), \tag{5.5a}$$

$$y(t) = \begin{bmatrix} 1 & -1 & 0 \\ 0 & 1 & -1 \\ 0 & 0 & 1 \end{bmatrix} \bar{x}(t) + \begin{bmatrix} 0 & 0 & 0 \\ 0 & 0 & 0 \\ 0 & 0 & 0 \end{bmatrix} u(t). \tag{5.5b}$$

We would like to study the response of the output of the system (5.2), or equivalently (5.5). Assume that initially we start with an empty production line (i.e. $x(0) = 0$), that both machines have a capacity of 1 lot per unit time (i.e. $\mu_1 = \mu_2 = 1$) and that we feed the line at a rate of 1 lot per time unit (i.e. $u_0 = 1$). Furthermore, assume that machines produce at full capacity, but only in case something is in the buffer in front of it, i.e.

$$u_i(t) = \begin{cases} \mu_i & \text{if } y_i(t) > 0 \\ 0 & \text{otherwise} \end{cases} \qquad i \in \{1, 2\}. \tag{5.6}$$

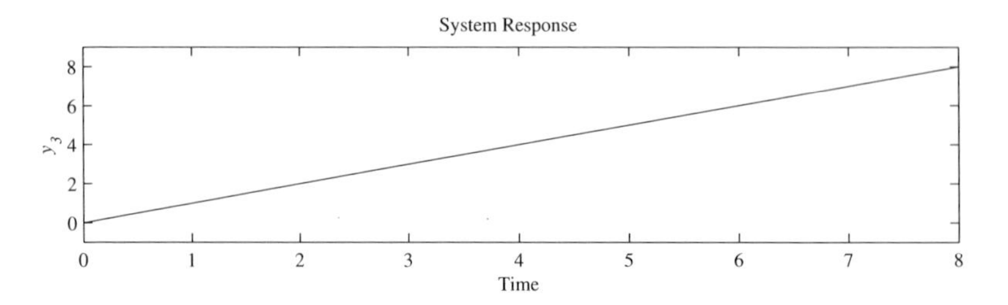

Figure 5.2: Output of the manufacturing system using model (5.1).

Under these assumptions, the resulting contents of buffer B_3 are as displayed in Fig. 5.2. Notice that immediately lots start coming out of the system. Clearly this is not what happens in practice. Since both machines M_1 and M_2 need to process the first lot, it should take the system at least $\frac{1}{\mu_1} + \frac{1}{\mu_2}$ time units before lots can come out. This illustrates our statement that cycle times are not incorporated in fluid models as currently available in the literature. Now we are ready for formulating an extension to the standard fluid model as presented.

5.2.2 An Extension

In the previous subsection we noticed that in the standard fluid model lots immediately come out of the system, once we start producing. A way to overcome this problem is to explicitly take into account the required delay. Whenever we decide to change the production rate of machine M_1, buffer B_2 notices this $1/\mu_1$ time units later. As a result the rate at which lots arrive to buffer B_2 at time t is equal to the rate at which machine M_1 was processing at time $t - 1/\mu_1$. This observation results in the following model (see also Fig. 5.3):

$$\begin{aligned} \dot{y}_1(t) &= u_0(t) - u_1(t), \\ \dot{y}_2(t) &= u_1\left(t - \frac{1}{\mu_1}\right) - u_2(t), \\ \dot{y}_3(t) &= u_2\left(t - \frac{1}{\mu_2}\right). \end{aligned} \tag{5.7}$$

Clearly the constraints (5.3) also apply to the model (5.7).

We expect that this model shows a response which is closer to reality. Assume that for the system (5.7) we also have $\mu_1 = \mu_2 = 1$ lot per time unit, and that we perform the same experiments as in the previous subsection, i.e. start from $x(0) = 0$, apply $u_0 = 1$ and Eq. (5.6). The resulting response of buffer B_3 is displayed in Fig. 5.4. If we compare the results from Fig. 5.4 to that of Fig. 5.2 we see that no products enter buffer B_3 during the first 2.0 time

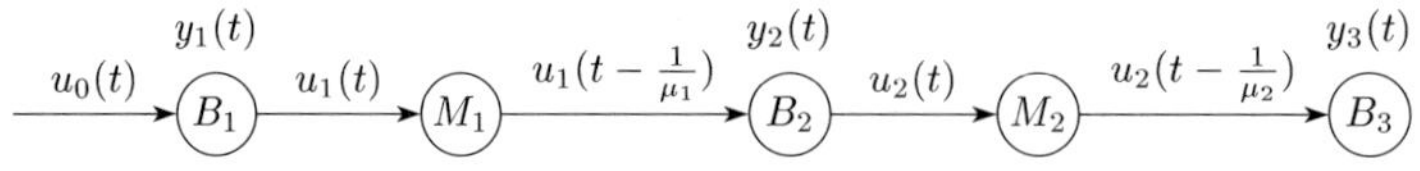

$y_1(t)$ $y_2(t)$ $y_3(t)$
$u_0(t)$ → B_1 → $u_1(t)$ → M_1 → $u_1(t - \frac{1}{\mu_1})$ → B_2 → $u_2(t)$ → M_2 → $u_2(t - \frac{1}{\mu_2})$ → B_3

Figure 5.3: A simple manufacturing system revisited.

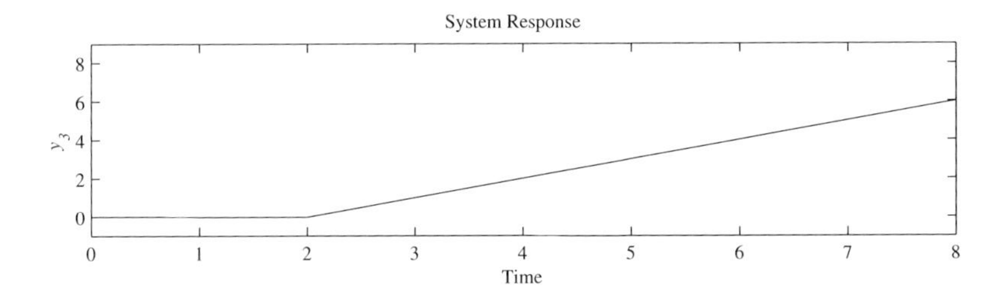

Figure 5.4: Output of the manufacturing system using model (5.7).

units in case we use the extended fluid model. Clearly the extended fluid model produces more realistic results than the standard fluid model.

5.2.3 An Approximation to the Extended Fluid Model

In the previous subsection we proposed an extended version of the standard fluid model. Although the model (5.7) still is a linear model, standard linear control theory is not able to deal with this model, due to the time delay. For controlling the model (5.7) we have to rely on control theory of infinite-dimensional linear systems. For a good introduction to infinite-dimensional linear systems, see e.g. [11].

Instead of using infinite-dimensional linear systems theory, another possibility would be to approximate the time delays by means of a Padé approximation. When we use second-order Padé approximations, the model (5.7) can be approximated as:

$$\dot{x} = \begin{bmatrix} \mathbf{0} & \mathbf{0} & 0 & 0 & \mathbf{0} & 0 & 0 \\ \mathbf{0} & \mathbf{0} & 0 & 0 & \mathbf{0} & 0 & 0 \\ 0 & 0 & 4 & 6 & -3 & 0 & 0 \\ 0 & 0 & 0 & 4 & 0 & 0 & 0 \\ \mathbf{0} & \mathbf{0} & 0 & 0 & \mathbf{0} & 0 & 0 \\ 0 & 0 & 0 & 0 & 4 & 6 & -3 \\ 0 & 0 & 0 & 0 & 0 & 4 & 0 \end{bmatrix} x + \begin{bmatrix} \mathbf{1} & \mathbf{0} & \mathbf{0} \\ \mathbf{0} & \mathbf{1} & \mathbf{0} \\ 0 & 0 & 0 \\ 0 & 0 & 0 \\ \mathbf{0} & \mathbf{0} & \mathbf{1} \\ 0 & 0 & 0 \\ 0 & 0 & 0 \end{bmatrix} u, \tag{5.8a}$$

$$y = \begin{bmatrix} \mathbf{1} & -\mathbf{1} & 0 & 0 & \mathbf{0} & 0 & 0 \\ \mathbf{0} & \mathbf{1} & -3 & 0 & -\mathbf{1} & 0 & 0 \\ \mathbf{0} & \mathbf{0} & 0 & 0 & \mathbf{1} & -3 & 0 \end{bmatrix} x + \begin{bmatrix} \mathbf{0} & \mathbf{0} & \mathbf{0} \\ \mathbf{0} & \mathbf{0} & \mathbf{0} \\ \mathbf{0} & \mathbf{0} & \mathbf{0} \end{bmatrix} u. \tag{5.8b}$$

Notice the structure in (5.8). In bold face we can easily recognize the dynamics (5.5). The additional dynamics is needed for approximating the time delays.

If we initiate the system (5.8) from $x(0) = 0$ and feed it at a rate $u_0 = 1$ while using Eq. (5.6), we obtain the system response as depicted in Fig. 5.5. It is clear that we do not get the same response as in Fig. 5.4, but the result is rather acceptable from a practical point of view. At least it is closer to reality than the response as displayed in Fig. 5.2.

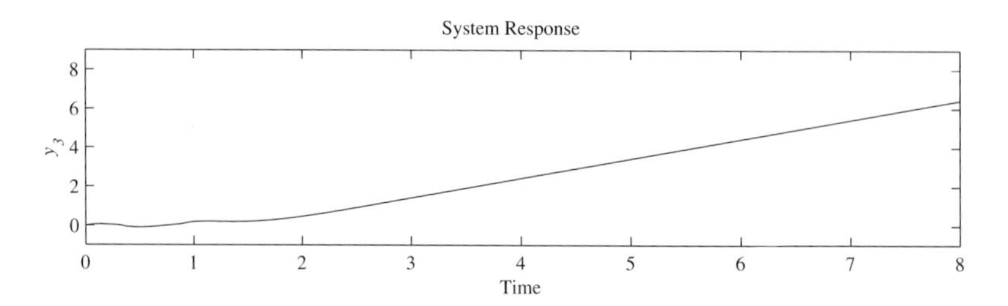

Figure 5.5: Output of the manufacturing system using model (5.8).

5.2.4 A Hybrid Model

In the previous subsections, we provided some extensions to the standard fluid model by taking into account the time delay lots encounter due to the processing of machines. We also mentioned the constraints (5.3) that have to be obeyed. These are constraints that we have to take into account when designing a controller for our manufacturing system. The way we dealt with these constraints in the previous subsections was by requiring the machines to produce only in case the buffer contents in front of that machine were positive, cf. Eq. (5.6).

A way to extend the standard fluid model (5.2) is to think of these constraints in a different way. As illustrated in Subsect. 5.2.1, when we turn on both machines, immediately lots start coming out of the system. This is an undesirable feature that we would like to avoid. In practice, the second machine can only start producing when the first machine has finished a lot. Keeping this in mind, why do we allow machine M_2 to start producing as soon as the buffer contents of the buffer in front of it are positive? Actually, machine M_2 should only start producing as soon as a whole product has been finished by the machine M_1. In words: machine M_2 should only start producing as soon as the buffer contents of the buffer in front of it becomes 1. Therefore, we should not allow for a positive u_2 as soon as $y_2 > 0$, but only in case $y_2 \geq 1$.

When we consider the initially empty system (5.2), i.e. $x(0) = 0$, and assume

$$u_i(t) = \begin{cases} \mu_i & \text{if } y_i(t) \geq 1 \\ 0 & \text{otherwise} \end{cases} \qquad i \in \{1, 2\}, \tag{5.9}$$

the resulting system response to an input of $u_0 = 1$ is shown in Fig. 5.6. Notice that we obtain exactly the same response as in Fig. 5.4.

Unfortunately, this is not all. The change in the constraints as proposed is not sufficient. It is in case we ramp up our manufacturing systems, but in case we ramp down it is not. Suppose that after a while we do not feed the manufacturing line any more, i.e. after a while we have $u_0 = 0$. In that case machine M_1 builds off the contents of the buffer B_1, until exactly one product remains. As soon as $y_1 = 1$, the machine is not allowed to produce any more due to the constraint we imposed. This is not what we would like to have. Therefore, in case $u_1 = 0$, machine M_1 should be allowed to produce until $y_1 = 0$.

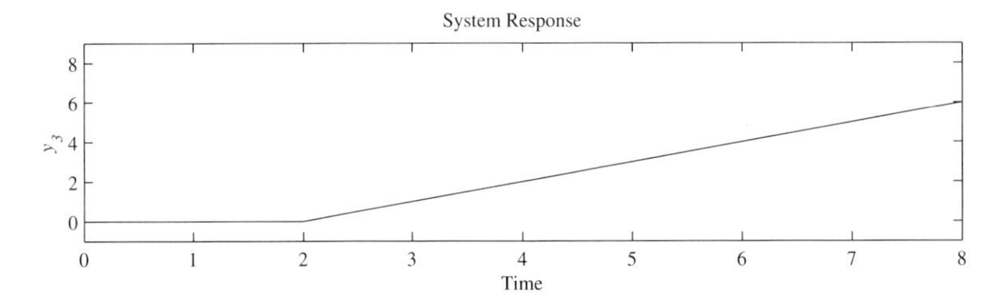

Figure 5.6: Output of the manufacturing system.

Under these conditions, we could also think of our model operating in different modes. For the manufacturing system under consideration we can distinguish the following modes:

$$\begin{array}{llllll}
\text{mode 1:} & 0 \le y_1 \le 1, & 0 \le y_2 \le 1, & u_0 = 0, & u_1 \ge 0, & u_2 = 0. \\
\text{mode 2:} & 0 \le y_1 \le 1, & 0 \le y_2 \le 1, & u_0 \ge 0, & u_1 = 0, & u_2 \ge 0. \\
\text{mode 3:} & 1 \le y_1, & 0 \le y_2 \le 1, & & u_1 = 0, & u_2 \ge 0. \\
\text{mode 4:} & 1 \le y_1, & 0 \le y_2 \le 1, & & u_1 \ge 0, & u_2 = 0. \\
\text{mode 5:} & 0 \le y_1 \le 1, & 1 \le y_2, & u_0 = 0, & u_1 \ge 0. & \\
\text{mode 6:} & 0 \le y_1 \le 1, & 1 \le y_2, & u_0 \ge 0, & u_1 = 0. & \\
\text{mode 7:} & 1 \le y_1, & 1 \le y_2. & & &
\end{array}$$

In all of these modes, the system dynamics is described by Eq. (5.2).

In fact, what we just presented is a hybrid systems model of the manufacturing system under consideration. The description as just presented is also known as that of piecewise affine (PWA) systems [31]. Other well-known descriptions are linear complementarity (LC) systems [17, 30] and mixed logical dynamical (MLD) systems [7]. In [5, 18] it was shown that (under certain assumptions like well-posedness) these three descriptions are equivalent. This knowledge is useful, as each modeling class has its own advantages (cf. [3]). Stability criteria for PWA systems were proposed in [15, 20], and control and state-estimation techniques for MLD hybrid models have been presented in [4, 6, 7]. These results can now be applied for controlling the hybrid systems model of our manufacturing system.

5.3 A New Flow Model

In the previous section we proposed to replace the standard fluid model (5.1) with the model (5.7) which contains a time delay. In that way we could overcome the shortcoming of the standard fluid model that once we start producing, immediately lots come out of the system. We also presented a Padé approximation of this time-delayed model, as well as a hybrid systems model that produced the desired delays.

Although the proposed models do not suffer from the problem that lots come out of the system as soon as we start producing, cycle times are not truly present in these models. It is not possible to determine the time it takes lots to leave once they have entered the system. As mentioned in the introduction, we are interested in dynamic models that incorporate both throughput *and* cycle time.

Therefore, the models presented in the previous section are (still) not satisfactory. Furthermore, according to these models any feasible throughput can be achieved by means of zero inventory. In this section we present a dynamic model that does incorporate both throughput and cycle time. This dynamic model is inspired by the continuum theory of highway traffic. Therefore, before presenting this dynamic model we first present some results from traffic theory.

5.3.1 Introduction to Traffic Flow Theory: the LWR Model

In the mid 1950s Lighthill and Whitham [25] and Richards [29] proposed a first-order fluid approximation of traffic flow dynamics. This model nowadays is known in traffic flow theory as the LWR model.

Traffic behavior for a single one-way road can be described using three variables that vary in time t and space x: flow $u(x,t)$, density $\rho(x,t)$ and speed $v(x,t)$. The first observation is that flow is the product of speed and density:

$$u(x,t) = \rho(x,t)v(x,t) \qquad \forall x,t. \tag{5.10}$$

Second, for a highway without entrances or exits, the number of cars between any two locations x_1 and x_2 ($x_1 < x_2$) needs to be conserved at any time t, i.e. the change in the number of cars between x_1 and x_2 is equal to the flow entering via x_1 minus the flow leaving via x_2:

$$\frac{\partial}{\partial t}\int_{x_1}^{x_2} \rho(x,t)dx = u(x_1,t) - u(x_2,t), \tag{5.11a}$$

or in differential form:

$$\frac{\partial \rho}{\partial t}(x,t) + \frac{\partial u}{\partial x}(x,t) = 0. \tag{5.11b}$$

The two relations (5.10) and (5.11) are basic relations that any model must satisfy. As we have three variables of interest, a third relation will be needed. For this third relation, several choices can be made. The LWR model assumes in addition to the relations (5.10) and (5.11) that the relation between flow and density observed under steady-state conditions also holds when flow and density vary with x and/or t; i.e. for a homogeneous highway:

$$u(x,t) = S(\rho(x,t)). \tag{5.12}$$

The model (5.10), (5.11), (5.12) can predict some things encountered in traffic rather well. In order to overcome some of the deficiencies of the LWR model, in the early 1970s higher-order theories have been proposed where the relation (5.12) has been replaced by another partial differential equation, containing diffusion or viscosity terms. Unfortunately, these extended models experience some undesirable properties, as made clear in [13]. The most annoying of these properties is the fact that in these second-order models cars can travel backwards. Second-order models that do not suffer from this deficiency have been presented in [19, 34]. However, for our modeling purposes the first-order LWR model (5.10), (5.11), (5.12) is sufficient.

5.3.2 A Traffic Flow Model for Manufacturing Flow

In the previous subsection we introduced the LWR model from traffic flow theory. This model describes the dynamic behavior of cars along the highway at a macroscopic level and contains information both about the number of cars passing a certain point and about the time it takes cars to go from one point to the next. The observation we make in this paper is that we can not only use this model for describing the flow of cars along the highway, but also for describing the flow of products through a manufacturing line.

Consider, instead of a homogeneous highway, a homogeneous manufacturing line, i.e. a manufacturing line that consists of a lot of identical machines. Let t denote the time and let x the position in the manufacturing line. The behavior of lots flowing through the manufacturing line can also be described by three variables that vary with time and position: flow $u(x,t)$ measured in unit lots per unit time, density $\rho(x,t)$ measured in unit lots per unit machine and speed $v(x,t)$ measured in unit machines per unit time. Now we can relate these three variables by means of Eqs. (5.10), (5.11) and (5.12), where in Eq. (5.12) the function S describes the relation between flow and density observed under steady-state conditions.

To make this last statement more explicit, consider a manufacturing line where all machines have exponentially distributed processing times and an average capacity of μ lots per unit time. Furthermore, consider a Poisson arrival process where lots arrive to the first machine with a rate of λ lots per unit time ($\lambda < \mu$), and assume that buffers have infinite capacity. Then we know from queueing theory [22] that the average number of lots in each workstation (consisting of a buffer and a machine) in steady state is given by

$$N = \frac{\frac{\lambda}{\mu}}{1 - \frac{\lambda}{\mu}} = \frac{\lambda}{\mu - \lambda}. \tag{5.13}$$

In words: in steady state we have

$$\rho(x,t) = \frac{u(x,t)}{\mu - u(x,t)}, \tag{5.14}$$

from which we can conclude that in steady state:

$$u(x,t) = \frac{\mu\rho(x,t)}{1 + \rho(x,t)}. \tag{5.15}$$

For this example, this is the mentioned function $S(\rho)$.

With this information we can conclude that the dynamics of this manufacturing line might be described by means of the partial differential equation

$$\frac{\partial \rho}{\partial t} + \mu \frac{\partial}{\partial x}\left(\frac{\rho}{1+\rho}\right) = 0. \tag{5.16a}$$

Together with the relations

$$u = \frac{\mu\rho}{1+\rho} \text{ and } v = \frac{u}{\rho} \text{ or } v = \frac{\mu}{1+\rho} \tag{5.16b}$$

this completes our model.

Notice that contrary to the fluid models presented in the previous sections, the dynamic model (5.16) is able to incorporate the stochasticity as experienced in manufacturing lines. If the manufacturing line would be in steady state, the throughput and cycle time as predicted by the model (5.16) will be exactly the same as those predicted by queueing theory. However, contrary to queueing theory, the model (5.16) is not a steady-state model, but also incorporates dynamics. Therefore, the model (5.16) is a dynamic model that incorporates both throughput and cycle time. Furthermore, given the experience in the field of fluid dynamics, the model is computationally feasible as well.

5.4 The Manufacturing Flow Model Revisited

In Sect. 5.3 we noticed that for the standard fluid model (5.1) it is possible to achieve any feasible throughput by means of zero inventory. Even when we are not interested in cycle times, this is still a major shortcoming of the standard fluid models. Using insight from the flow model as derived in the previous section, this shortcoming of standard fluid models can be overcome.

Consider the fluid model (5.16). Discretization of this model (with respect to x only, see also [12]) yields

$$\begin{aligned} \dot{x}_1 &= u_0 - \frac{\mu x_1}{1+x_1}, \\ \dot{x}_2 &= \frac{\mu x_1}{1+x_1} - \frac{\mu x_2}{1+x_2}, \\ \dot{x}_3 &= \frac{\mu x_2}{1+x_2}. \end{aligned} \tag{5.17}$$

Notice that the discretized model (5.17) can also be seen as a system of the form (5.1) where instead of (5.6) we use

$$u_i(t) = \frac{\mu_i y_i}{1+y_i}, \qquad i \in \{1,2\}. \tag{5.18}$$

What we can learn from this observation is that in case we move from deterministic processing times to stochastic processing times, apparently we should replace the inputs (5.6) with (5.18). In that case, to each throughput rate corresponds a non-zero steady-state work-in-progress level which is equal to the one predicted by queueing theory. Furthermore, notice that whenever we start from a feasible initial condition, i.e. the buffer contents initially are non-negative, the conditions (5.3) are always met.

More can be said about the model (5.17). In Sect. 5.2 we mainly were considered with the output of the manufacturing line, i.e. we were mainly concerned with the signal $y_3(t) = x_3(t)$. Even though the model (5.17) clearly is a nonlinear model, it has a nice structure: the model is feedback-linearizable [26, 27]. To make this statement more explicit, consider the following change of coordinates:

$$\begin{aligned} z_1 &= \frac{\mu^2(x_1 - x_2)}{(1+x_1)(1+x_2)^3}, \\ z_2 &= \frac{\mu x_2}{1+x_2}, \\ z_3 &= x_3, \end{aligned} \tag{5.19a}$$

together with the input

$$u_0 = \frac{(1+x_1)^2(1+x_2)^2}{\mu^2} v - \frac{2\mu(x_1 - x_2)}{1+x_2} + \frac{3\mu(x_1 - x_2)(1+x_1)}{(1+x_2)^3} + \frac{\mu x_1}{1+x_1}, \tag{5.19b}$$

where v can be an arbitrary signal. If we apply (5.19) to the system (5.17) we obtain the system

$$\dot{z}(t) = \begin{bmatrix} 0 & 0 & 0 \\ 1 & 0 & 0 \\ 0 & 1 & 0 \end{bmatrix} z(t) + \begin{bmatrix} 1 \\ 0 \\ 0 \end{bmatrix} v(t), \tag{5.20a}$$

$$y_3(t) = \begin{bmatrix} 0 & 0 & 1 \end{bmatrix} z(t) + \begin{bmatrix} 0 \end{bmatrix} v(t), \tag{5.20b}$$

which is a *linear* system. After applying the nonlinear change of coordinates and feed forward (5.19) we can control the output of the manufacturing line by means of standard linear control theory, as made clear by the system (5.20).

Another standard nonlinear control technique that can be used for controlling the system (5.17) is backstepping, cf. [24, 26].

5.5 Concluding Remarks

In the literature roughly three classes of models for manufacturing lines have been studied so far: fluid models that do not incorporate cycle times, queueing theory that contains only steady-state results and discrete-event models that suffer from state explosion.

In this paper we presented a flow model for modeling manufacturing lines, based on the theory of modeling traffic flow. The presented model is the first computationally feasible dynamic model that incorporates both throughput and cycle time. This model is a suitable model for addressing dynamic control questions like how to ramp up a given manufacturing line.

We also illustrated that the presented flow model can give valuable insights on how to modify the standard fluid models from the literature in case we would like to deal with non-deterministic processing times of machines.

The idea to use traffic flow models for modeling the dynamics of manufacturing systems emerged only recently. Related work can be found in [1, 2]. Also, the book [14] provides a good introduction to the subject.

Issues like the relation between variability of manufacturing systems and turbulence, the influence of scheduling policies on the relation (5.15), extensions to higher-order models (like [19, 34], while keeping in mind the observations in [13]), correct discretization schemes (cf. [12]), control of these flow models and last but not least the validity of these models will be the subject of future study.

Bibliography

[1] D. Armbruster, D. Marthaler, and C. Ringhofer, *Kinetic and fluid model hierarchies for supply chains*, accepted by SIAM Journal on Multiscale Modeling and Simulation (2003).

[2] D. Armbruster, D. Marthaler, and C. Ringhofer, *Modeling a re-entrant factory*, submitted to Operations Research (2003).

[3] A. Bemporad, *An efficient technique for translating mixed logical dynamical systems into piecewise affine systems*, in: Proceedings of the 41st Conference on Decision and Control, Las Vegas, NV, USA (December 2002), 1970–1975.

[4] A. Bemporad, F. Borrelli, and M. Morari, *Piecewise linear optimal controllers for hybrid systems*, in: Proceedings of the 2000 American Control Conference, Chicago, IL, USA (June 2000), 1190–1194.

[5] A. Bemporad, G. Ferrari-Trecate, and M. Morari, *Observability and controllability of piecewise affine and hybrid systems*, IEEE Transactions on Automatic Control **45**(10) (2000) 1864–1876.

[6] A. Bemporad, D. Mignone, and M. Morari, *Moving horizon estimation for hybrid systems and fault detection*, in: Proceedings of the 1999 American Control Conference, San Diego, California, USA (June 1999), 2471-2475.

[7] A. Bemporad and M. Morari, *Control of systems integrating logic, dynamics, and constraints*, Automatica **35** (1999), 407–427.

[8] T.J.J. van der Boom, B. de Schutter, and B. Heidergott, *Stochastic reduction in MPC for stochastic max-plus-linear systems by variability expansion*, in: Proceedings of the 41st Conference on Decision and Control, Las Vegas, NV, USA (December 2002), 3567–3572.

[9] J.A. Buzacott and J.G. Shantikumar, *Stochastic Models of Manufacturing Systems*, Prentice Hall, Englewood Cliffs, NJ, USA, 1993.

[10] C.G. Cassandras, Y. Wardi, B. Melamed, G. Sun, and C. Panayiotou, *Perturbation analysis for on-line control and optimization of stochastic fluid models*, IEEE Transactions on Automatic Control **47**(8) (2002), 1234–1248.

[11] R.F. Curtain and H. Zwart, *An Introduction to Infinite-Dimensional Linear Systems Theory*, Number 21 in *Texts in Applied Mathematics*, Springer-Verlag, Berlin, Germany, 1995.

[12] C.F. Daganzo, *A finite difference approximation of the kinematic wave model of traffic flow*, Transportation Research Part B **29**(4) (1995), 261–276.

[13] C.F. Daganzo, *Requiem for second-order fluid approximations of traffic flow*, Transportation Research Part B **29**(4) (1995), 277–286.

[14] C.F. Daganzo, *A Theory of Supply Chains*, Springer-Verlag, Heidelberg, Germany, 2003.

[15] R.A. DeCarlo, M. Branicky, S. Petterson, and B. Lennartson, *Perspectives and results on the stability and stabilizability of hybrid systems*, Proceedings of the IEEE **88**(7) (2000), 1069–1082.

[16] J.M. Harrison, *Brownian Motion and Stochastic Flow Systems*, John Wiley, New York, 1995.

[17] W.P.M.H. Heemels, J.M. Schumacher, and S. Weiland, *Linear complementarity systems*, SIAM Journal on Applied Mathematics **60**(4) (2000), 1234–1269.

[18] W.P.M.H. Heemels, B. de Schutter, and A. Bemporad, *Equivalence of hybrid dynamical models*, Automatica **37**(7) (2001), 1085–1091.

[19] R. Jiang, Q.-S. Wu, and Z.-J. Zhu, *A new continuum model for traffic flow and numerical tests*, Transportation Research Part B **36** (2002), 405–419.

[20] M. Johansson and A. Rantzer, *Computation of piece-wise quadratic Lyapunov functions for hybrid systems*, IEEE Transactions on Automatic Control **43**(4) (1998), 555–559.

[21] J. Kimemia and S.B. Gershwin, *An algorithm for the computer control of a flexible manufacturing system*, IIE Transactions **15**(4) (1983), 353–362.

[22] L. Kleinrock, *Queueing Systems, Vol. I: Theory*, John Wiley, New York, 1975.

[23] L. Kleinrock,. *Queueing Systems, Vol. II: Computer Applications*, John Wiley, New York, 1976.

[24] M. Krstić, I. Kanellakopoulos, and P. Kokotović, *Nonlinear and Adaptive Control Design*, Series on *Adaptive and Learning Systems for Signal Processing, Communications, and Control*, John Wiley, New York, 1995.

[25] M.J. Lighthill and J.B. Whitham, *On kinematic waves. I: Flow movement in long rivers. II: A theory of traffic flow on long crowded roads*, Proceedings of the Royal Society A **229** (1955), 281–345.

[26] R. Marino and P. Tomei, *Nonlinear Control Design*, Prentice-Hall, Hemel Hempstead, Hertfordshire, England, 1995.

[27] H. Nijmeijer and A.J. van der Schaft, *Nonlinear Dynamical Control Systems*, Springer-Verlag, New York, 1990.

[28] P.J. Ramadge and W.M. Wonham, *Supervisory control of a class of discrete-event systems*, SIAM Journal on Control and Optimization **25** (1987), 206–230.

[29] P.I. Richards, *Shockwaves on the highway*, Operations Research **4** (1956), 42–51.

[30] A.J. van der Schaft and J.M. Schumacher, *Complementarity modelling of hybrid systems*, IEEE Transactions on Automatic Control **43** (1998), 483–490.

[31] E.D. Sontag, *Nonlinear regulation: The piecewise linear approach*, IEEE Transactions on Automatic Control
26(2) (1981), 346–358.

[32] R. Suri, *Quick Response Manufacturing: A Companywide Approach to Reducing Lead Times*, Productivity Press, Portland, Oregon, 1998.

[33] R.J. Williams, *Reflecting diffusions and queueing networks*, in: Proceedings of the International Congress of Mathematicians, Berlin, Germany, 1998 (invited paper).

[34] H.M. Zhang, *A non-equilibrium traffic model devoid of gas-like behavior*, Transportation Research Part B **36** (2002), 275–290.

6 Modeling and Optimization of Production Processes: Lessons from Traffic Dynamics

D. Helbing[1]

We will develop and study models of supply networks and how they relate to vehicular traffic. These models allow us to take into account the nonlinear, dynamical interactions of different production units and to test alternative management strategies with respect to their potential impacts. In this way, one can understand the pre-conditions of the so-called bull-whip effect (i.e. the fact that small variations in the consumption rate can cause large variations in the production rate of companies generating the requested product). Moreover, we will show how the nonlinear dynamics of a particular supply chain in semiconductor production has been optimized by means of the "slower-is-faster effect" known from panicking pedestrian crowds. Driven many-particle models of pedestrian motion also offer solutions for other typical problems of nonlinear production processes such as the coordination of robots the efficient segregation of different kinds of objects or the frictionless merging of object flows at bottlenecks. Finally, from the simulation of pedestrian behavior one can learn how fluctuations could be used to increase the order in the system, how to speed up certain production processes or how to compensate for delays in a series of production steps.

6.1 Modeling the Dynamics of Supply Networks

Concepts from statistical physics and nonlinear dynamics have been very successful in discovering and explaining dynamical phenomena in traffic flows [1, 2]. Many of these phenomena are based on mechanisms such as delayed adaptation to changing conditions and competition for limited resources, which are relevant for production systems as well. Therefore, economists [3], traffic scientists [4], mathematicians [5] and physicists [6] have recently pointed out that methods used for the investigation of traffic dynamics are also of potential use for the study of supply networks. In this contribution, our primary attention will be directed towards the *nonlinear interaction* between different production units or production processes and the resulting *dynamics,* while classical queueing theory mainly focusses on stochastic fluctuations of production processes in a stationary state.

[1]The author wants to thank for the partial support by SCA Packaging Ltd. and for invitations by Prof. Radons and Prof. West. He acknowledges inspiring discussions with Dieter Armbruster, Carlos Daganzo, Illés Farkas, Arne Kesting, Megan Khoshyaran, Christian Kühnert, Péter Molnár, Takashi Nagatani, Tadeusz Płatkowski, Dick Sanders, Pétr Šeba, Thomas Seidel, Tamás Vicsek, Torsten Werner, and Ulrich Witt. The author is also grateful to Dominique Fasold and Tilo Grigat for preparing the schematic illustrations. Last, but not least, he has appreciated the warm hospitality and the challenging intellectual atmosphere at the Santa Fe Institute.

Nonlinear Dynamics of Production Systems. Edited by G. Radons and R. Neugebauer

ISBN 3-527-40430-9

6.1.1 Modeling One-dimensional Supply Chains

For simplicity, let us start with a model of one-dimensional supply chains. The assumed model consists of a series of u suppliers b, which receive products from the next "upstream" supplier $b-1$ and generate products for the next "downstream" supplier $b+1$ [7, 8]. The final products are delivered to the consumers $u+1$ (see Fig. 6.1). The consumption and delivery rates are typically subject to perturbations, which may cause variations in the stock levels and deliveries of upstream suppliers. This is due to delays in the adaptation of their delivery rates.

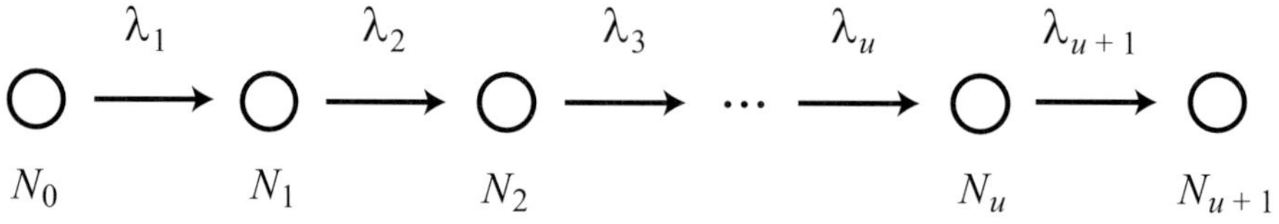

λ_1 λ_2 λ_3 λ_u λ_{u+1}

N_0 N_1 N_2 ... N_u N_{u+1}

Figure 6.1: Illustration of the one-dimensional supply chains treated in this paper, including the key variables of the model. Circles represent different suppliers b, N_b their respective stock levels and λ_b the delivery rate to supplier b or the production speed of this supplier. $b = 0$ corresponds to the resource sector generating the basic products and $b = u+1$ to the consumer sector.

To study the resulting dynamics, let us denote the stock level ("inventory") at supplier b by N_b. It changes in time t according to the equation

$$\frac{dN_b}{dt} = \lambda_b(t) - \lambda_{b+1}(t)\,. \tag{6.1}$$

Here, λ_b has the meaning of the rate at which supplier b receives ordered products from supplier $b-1$, while λ_{b+1} is the rate at which he delivers products to the next downstream supplier $b+1$. Therefore, Eq. (6.1) is just a continuity equation which reflects the conservation of the quantity of products. (It is easy to generalize this equation to cases where products are lost. One would just have to add a term of the form $-\gamma_b N_b(t)$.) Boundary conditions must be formulated for $b = 0$, which corresponds to the supplier of the raw materials (fundamental resources), and for $b = u+1$, which corresponds to the consumers. That is, λ_0 is the supply or production rate of the basic product, while λ_{u+1} is the consumption rate.

The question remains of how the delivery rates λ_b evolve in time. It is reasonable to assume that the temporal change of the delivery rate is proportional to the deviation of the actual delivery rate from the desired one W_b (the order rate) and its adaptation takes on average some time interval τ. According to this, we have the equation

$$\frac{d\lambda_b}{dt} = \frac{1}{\tau}\left[W_b(t) - \lambda_b(t)\right]\,. \tag{6.2}$$

The order rate W_b will usually be reduced with increasing stock levels N_a, but their temporal changes dN_a/dt may be taken into account as well, e.g. when the stock levels are forecasted. Therefore, it is natural to assume a general dependence of the form

$$W_b(t) = W_b\left(\{N_a(t)\}\,, \{dN_a(t)/dt\}\right)\,. \tag{6.3}$$

The function W_b reflects the management strategy, i.e. the order policy regarding the desired delivery rate as a function of the actual stock levels $N_a(t)$ or anticipated stock levels $N_a(t) +$

$\Delta t\, dN_a(t)/dt \approx N_a(t+\Delta t)$ (in first-order Taylor approximation). The simplest strategy of supplier b would be to react to the own stock level $N_b(t)$. However, it may be useful to consider also the stock levels of the next downstream suppliers $a > b$, as these determine the future demand, and the stock levels of the next upstream suppliers $a < b$, as they determine future deliveries or shortages of the product ("out of stock" situations). We will, therefore, assume that

$$W_b(t) = W_b\left(\{N_a(t)\}, \{dN_a(t)/dt\}\right) = W(N_{(b)}(t)), \tag{6.4}$$

where W denotes a supplier-independent management strategy and

$$N_{(b)}(t) = \sum_{c=-n}^{n} w_c \left(N_{b+c} + \Delta t\, \frac{dN_{b+c}}{dt} \right) \tag{6.5}$$

is a weighted mean value of the own stock level and the ones of the next n upstream and n downstream suppliers (with $2n+1 \le u$). For $a = b+c < 0$ and $a = b+c > u$, the weights w_c are always set to zero, and they are normalized to one:

$$\sum_{c=-n}^{n} w_c = 1\,. \tag{6.6}$$

For $\Delta t = 0$, the management adapts the delivery rate λ_b to the *actual* weighted stock level $N_{(b)}(t)$ while, for $\Delta t > 0$, the management orients at the *anticipated* weighted stock level. The parameter Δt has the meaning of the forecast time horizon, and the specification of the parameters w_c and Δt reflects the management strategy.

6.1.2 "Bull-whip Effect" and Stop-and-Go Traffic

It is possible to study the stability of the steady-state solution of Eqs. (6.1) and (6.2) analytically. This stationary solution is given by $N_b = N_0$ and $\lambda_b = \lambda_0 = W(N_0)$. Small deviations from this stationary solution will fade away, if the stability condition

$$\tau < \Delta t + \frac{1}{|W'(N_0)|}\left(\frac{1}{2} + \sum_{c=-n}^{n} c w_c\right) \tag{6.7}$$

is fulfilled, where W' means the derivative of the management function W [9]. That is, the supply chain behaves stably if the adaptation time τ is small, the forecast time horizon Δt is large or the change of the management function W with changes in the stock levels N_b is small. However, if this condition is invalid, perturbation will grow and generate oscillations in the inventories $N_b(t)$. This so-called "bull-whip effect" has, for example, been reported for beer distribution [10]. Similar dynamical effects are known for other distribution or transportation chains with significant adaptation times. Series of production processes have similar features as well. In this case, the index b represents the different successive production steps or machines, λ_b describes the corresponding production rate and the management function W_b reflects the desired production rate as a function of the stock levels N_b in the respective output buffers.

Note that, in the case $N_{(b)} = N_b$ (i.e. $w_0 = 1$ and $w_c = 0$ for $c \neq 0$), the stability condition (6.7) agrees exactly with that of the optimal velocity model [11], which is a particular microscopic traffic model. This car-following model assumes an acceleration equation of the form

$$\frac{dv_b(t)}{dt} = \frac{V_{\text{opt}}(d_b(t)) - v_b(t)}{\tau} \tag{6.8}$$

and the complementary equation

$$\frac{dd_b(t)}{dt} = -[v_b(t) - v_{b+1}(t)]\,. \tag{6.9}$$

In contrast to the above supply chain model, the index b represents single vehicles, $v_b(t)$ is their actual velocity of motion, V_{opt} the so-called optimal (safe) velocity, which depends on the distance $d_b(t)$ to the next vehicle ahead, and τ is an adaptation time. Comparing this equation with Eq. (6.2), the velocities v_b would correspond to the delivery rates λ_b, the optimal velocity V_{opt} to the desired delivery rate W_b and the inverse vehicle distance $1/d_b$ would approximately correspond to the stock level N_b (apart from a proportionality factor). This shows that the analogy between supply chain and traffic models concerns only their mathematical structure, but not their interpretation, although both relate to transport processes. Nevertheless, this mathematical relationship can give us hints how methods, which have been successfully applied to the investigation of traffic models before, can be generalized for the study of supply networks.

Compared to traffic dynamics, supply networks and production systems have some interesting new features: Instead of a continuous space, we have discrete production units b, and the control function $W_b(\cdots)$ is different from the empirical velocity–density relation in traffic. While in traffic flow, the velocity–density relation is empirically given, the new feature of supply chains is that the management has a large degree of freedom how to specify $W_b(\cdots)$, for example as a function of the own stock level and of the stock levels of other suppliers, if this information is available. With suitable strategies, the oscillations can be mitigated or even suppressed (see Sect. 6.1.4). Moreover, production systems may operate in different regimes, and small changes of parameters may have tremendous effects (compare Fig. 6.3d to 6.3e, and this with Figs. 6.3f, g). Finally, production systems are frequently supply networks with complex topologies rather than one-dimensional supply chains, i.e. they have additional features compared to (more or less) one-dimensional freeway traffic (see Sect. 6.1.8). They are more comparable to street networks of cities [12].

6.1.3 Dynamical Solution and Resonance Effects

In the vicinity of the stationary state, it is possible to calculate the dynamical solution of the one-dimensional supply chain model with $w_0 = 1$ and $w_c = 0$ for $c \neq 0$ [13]. For this, let $\delta N_b(t) = N_b(t) - N_0$ be the deviation of the inventory from the stationary one, and $\delta\lambda_b(t) = \lambda_b(t) - W(N_0)$ the deviation of the delivery rate. The linearized model equations read

$$\frac{d\delta N_b}{dt} = \delta\lambda_b(t) - \delta\lambda_{b+1}(t) \tag{6.10}$$

and

$$\frac{d\delta\lambda_b}{dt} = \frac{1}{\tau}\left[W'(N_0)\left(\delta N_b + \Delta t\,\frac{d\delta N_b}{dt}\right) - \delta\lambda_b(t)\right]. \tag{6.11}$$

Deriving Eq. (6.11) with respect to the time t and inserting Eq. (6.10) results, with $B = -W'(N_0) = |W'(N_0)| > 0$, in the following set of second-order differential equations:

$$\frac{d^2\delta\lambda_b}{dt^2} + \underbrace{\frac{1 + B\,\Delta t}{\tau}}_{=2\gamma}\frac{d\delta\lambda_b}{dt} + \underbrace{\frac{B}{\tau}}_{=\omega_0{}^2}\delta\lambda_b(t) = \underbrace{\frac{B}{\tau}\left(\delta\lambda_{b+1}(t) + \Delta t\,\frac{d\delta\lambda_{b+1}}{dt}\right)}_{=f_b(t)}. \tag{6.12}$$

This corresponds to the differential equation for the damped harmonic oscillator with damping constant γ, eigenfrequency ω_0 and driving term $f_b(t)$. The eigenvalues of this (system of) equation(s) are

$$\omega_{1,2} = -\gamma \pm \sqrt{\gamma^2 - \omega_0^2} = \frac{-(1 + B\,\Delta t) \pm \sqrt{(1 + B\,\Delta t)^2 - 4B\tau}}{2\tau}. \tag{6.13}$$

The set of equations (6.12) can be solved successively, starting with $b = u$ and progressing to lower values of b. For example, assuming periodic oscillations of the form $f_u(t) = f_u^0\cos(\alpha t)$, after a transient time of about $3/\gamma$ we find

$$\delta\lambda_u(t) = f_u^0 F\cos(\alpha t - \varphi), \tag{6.14}$$

with

$$\tan\varphi = \frac{-2\gamma\alpha}{\alpha^2 - \omega_0^2} = \frac{-(1 + B\,\Delta t)\alpha}{\alpha^2\tau - B} \tag{6.15}$$

and

$$F = \frac{1}{\sqrt{(\alpha^2 - \omega_0^2)^2 + 4\gamma^2\alpha^2}} = \frac{1}{\sqrt{(\alpha^2 - B/\tau)^2 + (1 + B\,\Delta t)^2\alpha^2/\tau^2}}, \tag{6.16}$$

where the dependence on the frequency ω_0 is important to undestand the resonance effect mentioned in the next section. Equations (6.12) and (6.14) imply that

$$f_{u-1}(t) = \frac{B}{\tau}\left(\delta\lambda_u(t) + \Delta t\,\frac{d\delta\lambda_u}{dt}\right) = f_{u-1}^0\cos(\alpha t - \varphi - \delta_{u-1}), \tag{6.17}$$

with

$$\tan\delta_{u-1} = -\alpha\,\Delta t \qquad \text{and} \qquad f_{u-1}^0 = \frac{B}{\tau}f_u^0 F\sqrt{1 + (\alpha\,\Delta t)^2}. \tag{6.18}$$

The oscillation amplitude increases if $f_{u-1}^0/f_u^0 > 1$. One can show that this can happen for $B > 1/(2\tau)$, which corresponds to Eq. (6.7) for $w_0 = 1$ [13]. That is, supply chains behave unstably if the adaptation time τ is too large or if the management reacts too strongly to changes in the stock level (corresponding to a large value of $B = |W'(N_0)|$).

6.1.4 Discussion of Some Control Strategies

According to the stability condition (6.7), a supply chain can be stabilized (i.e. oscillations in the delivery rates and stock levels can be reduced) by several strategies: (1) by reduction of the adaptation time τ, (2) by anticipation of the temporal evolution of the inventories ($\Delta t > 0$), (3) by taking into account the inventories N_a of other suppliers $a = b + c$ with $w_c > 0$ for $c > 0$ and (4) by modification of the functional form of the management function $W(\cdots)$. In the case of perturbations in the consumption rate, numerical simulation results are as follows [9]:

- Anticipation of the own future inventory is an efficient means to stabilize the production system. Even anticipation time horizons Δt considerably smaller than the adaptation time τ are sufficient to reach complete stability.

- The adaptation to a variation in the consumption rate tends to be better, if not only the own inventory, but also the inventories of downstream suppliers or the consumer sector itself are taken into account by so-called *"pull strategies"*. In contrast, considering the inventories of upstream suppliers corresponding to *"push strategies"* tends to destabilize the system (cf. [14]). It is the direction of the information flow in the system which is responsible for this: the oscillations in the consumption rate travel upstream, as in stop-and-go traffic [1].

- Although the linear stability analysis gives a good idea under which conditions the oscillation amplitude in the system becomes zero, further implications are limited, because nonlinear effects dominate when the evolving oscillation amplitudes become large. For example, the emerging oscillation frequency ω in the system does often neither correspond to the frequency α of the external perturbation nor to the frequency which is most unstable according to the linear stability analysis. Instead, it is often much smaller than expected [6] and must be determined by simulations [9]. When in the weighted stock level $N_{(b)}(t) = w_0 N_b(t) + (1 - w_0) N_a(t)$, the weight $(1 - w_0)$ of another inventory $N_a(t)$ is increased in the management strategy, there is a surprise: as expected, the oscillation amplitudes are significantly reduced, when the second next downstream supplier is taken into account with $N_a(t) = N_{b+2}(t)$ instead of the next downstream one with $N_a(t) = N_{b+1}(t)$. However, considering the variation in the consumption rate itself with $N_a(t) = N_{u+1}(t)$ has a very weak stabilization effect, although the consumer sector is located even further downstream [9]. This point is related with resonance effects: according to Sect. 6.1.3, there is a frequency dependence of the emerging oscillation amplitudes.

In conclusion, there are non-trivial and unexpected effects in the behavior of one-dimensional supply chains. Therefore, simulation models describing the nonlinear interactions and dynamics of supply chains and production processes are relevant for their optimization. From the practical point of view it is, for example, useful that Eq. (6.7) allows one to estimate the maximum adaptation time τ or the minimum forecast time horizon Δt supporting a stable supply chain. Moreover, the stabilizing effect of a reaction to inventories of downstream suppliers suggests exchanging these data online. Note that our conclusions regarding the stabilization by forecasts and the consideration of downstream stock levels are expected to be transferable to more complex systems than the one-dimensional supply chains treated here. They should also be applicable to cases where suppliers are characterized by different parameters, to situations with limited buffers and transport capacities [15] or to supply networks [13]. Some of these aspects will be included in the more general model of production networks discussed in the following.

6.1.5 Production Units in Terms of Queueing Theoretical Quantities

We will investigate a system with u production units (machines or factories) $b \in \{1, 2, \ldots, u\}$ producing or using p different products $i, j \in \{1, 2, \ldots, p\}$. The respective production process is characterized by parameters c_b^j and p_b^i: in each production step, production unit b requires c_b^j products ("educts") $j \in \{1, \ldots, p\}$ and produces p_b^i products $i \in \{1, \ldots, p\}$.[2] The number of production steps of production unit b per unit time is a measure of the throughput and shall be represented by $Q_b^{\text{out}}(t)$. It can be related to the variables used in queueing theory [16]: let λ_b be the feeding (arrival) rate, C_b the number of parallel channels, μ_b the overall processing (departure) rate (i.e. C_b times the processing rate of a *single* channel),

$$\rho_b(t) = \lambda_b(t)/\mu_b(t) \tag{6.19}$$

the utilization and S_b the storage capacity of production unit b (see Fig. 6.2).

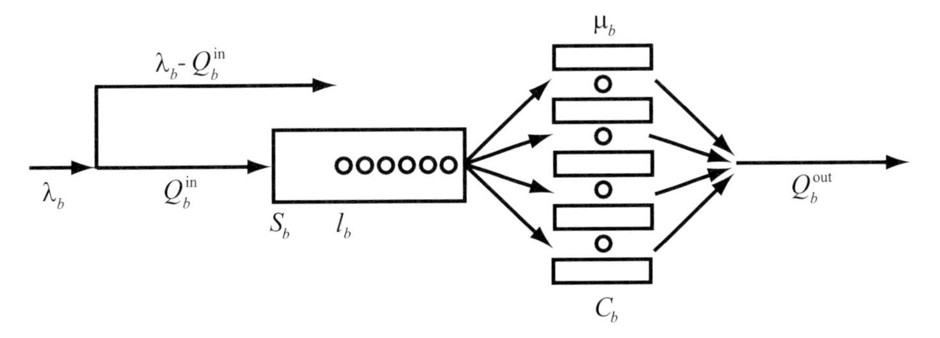

Figure 6.2: Schematic illustration of a production unit b as a queueing system with a limited storage capacity S_b and C_b parallel production channels (after [6]). The arrival rate λ_b, the departure rate μ_b as well as the inflow Q_b^{in} and the outflow Q_b^{out} are indicated.

The inflow Q_b^{in} into the buffer is dependent on the queue length l_b. If this is smaller than the storage capacity S_b, it is given by the arrival rate λ_b, otherwise it is zero due to a lack of buffer space:

$$Q_b^{\text{in}}(l_b) = \begin{cases} \lambda_b & \text{if } l_b < S_b, \\ 0 & \text{otherwise.} \end{cases} \tag{6.20}$$

[2] Note that the indices i and j are formally running over all p possible products, and c_b^j or p_b^i are typically non-zero only for a few products i or educts j, which depends on the production unit b. Moreover, the products i are normally different from the educts j, i.e. $c_b^j > 0$ normally implies $p_b^i = 0$ and $p_b^i > 0$ normally implies $c_b^j = 0$.

The outflow Q_b^{out} agrees with the processing rate μ_b, if all C_b channels are occupied, i.e. $l_b \geq C_b$. Otherwise, only a proportion l_b/C_b of the channels is active, i.e.

$$Q_b^{\text{out}}(l_b) = \begin{cases} \mu_b & \text{if } l_b \geq C_b, \\ \mu_b \, l_b/C_b & \text{if } l_b < C_b. \end{cases} \tag{6.21}$$

If $P_b(l_b)$ denotes the probability of finding a queue of length l_b, the average inflow is given by

$$\langle Q_b^{\text{in}} \rangle = \sum_{l_b} Q_b^{\text{in}}(l_b) P_b(l_b) = \lambda_b \Bigg(1 - \underbrace{\sum_{l_b=S_b}^{\infty} P_b(l_b)}_{=p_b^{\lambda}} \Bigg) , \tag{6.22}$$

while the average outflow is given by

$$\langle Q_b^{\text{out}} \rangle = \sum_{l_b} Q_b^{\text{out}}(l_b) P_b(l_b) = \mu_b \Bigg(1 - \underbrace{\sum_{l_b=0}^{C_b-1} \frac{C_b - l_b}{C_b} P_b(l_b)}_{=p_b^{\mu}} \Bigg) . \tag{6.23}$$

Generally speaking, the fractions p_b^{λ} and p_b^{μ} are measures of inefficiency in production due to full or empty buffers (or other reasons). In the stationary case, the average in- and outflow agree with each other and determine the (average) throughput $\langle Q_b \rangle$:

$$\langle Q_b \rangle = [1 - p_b^{\lambda}(\rho_b, C_b, S_b)]\lambda_b = [1 - p_b^{\mu}(\rho_b, C_b, S_b)]\mu_b = \frac{\langle l_b \rangle}{\langle T_b \rangle} . \tag{6.24}$$

As indicated, it is also given as the quotient of the average queue length $\langle l_b \rangle$ and the average waiting time $\langle T_b \rangle$ (which is known as Little's law). Both are functions of the utilization ρ_b, the number C_b of parallel channels, the storage capacity S_b and possibly other variables as well. For example, for a $M/M/1 : (S_b/\text{FIFO})$ process (one channel with first-in-first-out serving, storage capacity S_b, Poisson-distributed arrival times and exponentially distributed service intervals), one finds for $\lambda_b \leq \mu_b$

$$\langle Q_b \rangle = \lambda_b \frac{1 - {\rho_b}^{S_b}}{1 - {\rho_b}^{S_b+1}} \overset{\rho_b \to 1}{\longrightarrow} \mu_b \frac{S_b}{S_b + 1} . \tag{6.25}$$

Note that not only the expected value, but also the standard deviation of the queue length and the waiting time diverge for $\rho_b \to 1$ [16]. Therefore, efficient production is often related to a utilization $\rho_b \leq 0.7$.

6.1.6 Calculation of the Cycle Times

Apart from the productivity or throughput Q_b of a production unit, production managers are highly interested in the cycle time, i.e. the time interval between the beginning of the generation of a product and its completion. The problem is similar to determining the travel times of vehicles entering a traffic jam [12]. Let T_b denote the process cycle time between entering the queue of production unit b and leaving it, assuming that all c_b^i required educts i for one production cycle are transported together and located at the same place in the queue. The change of the queue length l_b in time is then given by the difference between the inflow and the outflow at time t:

$$\frac{dl_b}{dt} = Q_b^{\text{in}}(l_b(t)) - Q_b^{\text{out}}(l_b(t)) \,. \tag{6.26}$$

On the other hand, the waiting educts move forward Q_b^{out} steps per unit time. For this reason, the waiting time $t_b(t)$ until one of the channels is reached is given by the implicit equation

$$l_b(t) - C_b = \int\limits_t^{t+t_b(t)} dt'\, Q_b^{\text{out}}(l_b(t')) = \int\limits_{-\infty}^{t+t_b(t)} dt'\, Q_b^{\text{out}}(l_b(t')) - \int\limits_{-\infty}^{t} dt'\, Q_b^{\text{out}}(l_b(t')) \,, \tag{6.27}$$

and the overall time T_b required for the processing of the product corresponds to the sum of the waiting time t_b and the treatment time by one of the channels:

$$T_b(t) = t_b(t) + \frac{C_b}{\mu_b(t + t_b(t))} \,. \tag{6.28}$$

From Eqs. (6.26) and (6.27), one can finally derive a delay-differential equation for the waiting time under varying production conditions [6]:

$$\frac{dt_b}{dt} = \frac{Q_b^{\text{in}}(l_b(t))}{Q_b^{\text{out}}(l_b(t + t_b(t)))} - 1 \,. \tag{6.29}$$

As the production initially starts with a waiting time of $t_b(0) = 0$ (when the factory or production unit b is opened), this equation can be solved numerically as a function of the outflow $Q_b^{\text{out}}(t')$. In this way, it is possible to determine the waiting time t_b and process cycle time T_b.

6.1.7 Feeding Rates, Production Speeds and Inventories

In the absence of capacity constraints, we just have the relation $\lambda_b = \rho_b^0 \mu_b$ for the feeding rate, where the actual utilization ρ_b agrees with the desired utilization ρ_b^0, e.g. $\rho_b^0 = 0.7$. However, the production of a product requires the presence of all required parts (educts). Therefore, the actual feeding rate λ_b is determined by the *minimum* of the desired production speed $\rho_b^0 \mu_b$ and the delivery rates λ_b^j of the required educts j:

$$\lambda_b(t) = \min_j(\rho_b^0 \mu_b, \{\lambda_b^j\}) \,. \tag{6.30}$$

In the following, we will assume that the delivery rates λ_b^j of educts j are proportional to the desired production speed $\rho_b^0 \mu_b$ and to the number N_j of available educts j, divided by the quantity c_b^j of educts needed for one production step: $\lambda_b^j(t) = V_b^j \rho_b^0 \mu_b N_b^j(t)/c_b^j$. Due to transport constraints V_b^j, $V_b^j \rho_b^0 \mu_b$ is the maximum transport rate for getting educt j into the production unit b.

As in Eq. (6.2), we will assume that the adaptation of the production speed $\rho_b^0 \mu_b$ to changing demand is again delayed by some adaptation time τ_b:

$$\frac{d(\rho_b^0 \mu_b)}{dt} = \frac{1}{\tau_b} \left[W_b(N_b, \dots) - \rho_b^0 \mu_b \right] . \tag{6.31}$$

In the case $\lambda_b = \rho_b^0 \mu_b$ when transport constraints do not matter, this equation exactly agrees with our previous formula (6.2), while deviations may occur when required educts are not delivered at the desired rate. This makes production networks considerably more sensitive and complex than one-dimensional supply chains. It is reasonable to assume that the management function W_b increases with decreasing stock levels N_i of the products i production unit b produces, but that it saturates due to financial, spatial or technological limitations and inefficiencies in the processing of high-order flows. In the following, we will therefore use a function of the form

$$W_b(N_b, \dots) = \max\left(A_b \frac{1 + B_b N_b}{1 + B_b N_b + D_b {N_b}^2}, 0 \right) \tag{6.32}$$

with $1/N_b(t) = \sum_i p_b^i / N_i(t)$ and suitably chosen parameters A_b, B_b and D_b. N_b is something like a weighted inventory of the produced products. Moreover, if no products are lost, the stock level (inventory) N_i of product i changes according to the conservation equation

$$\frac{dN_i}{dt} = \sum_b [p_b^i Q_b^{\text{in}}(l_b(t)) - c_b^i Q_b^{\text{out}}(l_b(t))] , \tag{6.33}$$

as $p_b^i Q_b^{\text{in}}(l_b(t))$ is the number of products i finished by production unit b per unit time, while $c_b^i Q_b^{\text{out}}(l_b(t))$ is the number of educts entering its queue per unit time.

If the storage capacity S_b is appropriately chosen and the buffer is sufficiently filled, we may assume that $C_b < l_b < S_b$ and $Q_b = Q_b^{\text{in}} = Q_b^{\text{out}} = \lambda_b$. In the following, we will focus on this particular case for simplicity, but other cases can be numerically treated as well. Moreover, we will assume something like a conservation of materials or value: first of all, the quantity of product i consumed by the production units b should be generated somewhere, i.e.

$$\sum_{b=1}^{u+1} c_b^i = \sum_{b=1}^{u+1} p_b^i . \tag{6.34}$$

Second, the quantity of educts consumed by some production unit b corresponds to the quantity of its generated products, i.e.

$$\sum_{i=0}^{p} c_b^i = \sum_{i=0}^{p} p_b^i . \tag{6.35}$$

In the following, we will discuss the case $p_b^i = 1$, if $b = i$, otherwise 0. This defines the u production units b through their $p = u$ respective main products i and implies Leontief's classical input–output model from macroeconomics [17] as stationary solution [6]. Moreover, we can define

$$c_b^0 = 1 - \sum_{i=1}^{u} c_b^i \quad \text{and} \quad c_{u+1}^i = 1 - \sum_{b=1}^{u} c_b^i . \tag{6.36}$$

Values $c_b^0 > 0$ allow us to describe the inflow of basic resources $i = 0$, while $c_{u+1}^i > 0$ allows one to describe the depletion of products by an additional consumer sector $b = u + 1$. The boundary conditions are completely defined by specifying $N_0(t)$ and $Q_{u+1}(t)$ (see below).

6.1.8 Impact of the Supply Network's Topology

Assuming the case $Q_b(t) = Q_b^{\text{in}}(t) = Q_b^{\text{out}}(t) = \lambda_b(t)$, we will now discuss simulations based on the equations specified in Sect. 6.1.7 for three different supply networks sketched in Fig. 6.3a–c, each with five levels: (a) a one-dimensional supply chain with five production units, (b) a "supply ladder" with 10 production units and (c) a hierarchical supply tree with 31 production units. By introducing random variables ξ_b^l, which were assumed to be equally distributed in the interval $[-\eta, \eta]$, we can take into account a heterogeneity η in the individual parameters characterizing the different production units. Here, we have chosen $N_0(t) = N_0 = 20$, $N_i(0) = 20(1 + \xi_i)$, $\tau_b = 180(1 + \xi_b)$, $V_b^0 = V$, $V_b^j = Vc_b^j(1 + \xi_b^j)$ for $j > 0$, $\rho_b^0(0)\mu_b = W_b(N_b(0))$, $Q_{u+1}(t) = W_{u+1}(N_0)\min_i(1, N_i(t)V/c_{u+1}^i)[1 + 0.1\sin(0.04t)]$ and $A_b = 100/V$, $B_b = 0.01$, $D_b = 0.02$. For the one-dimensional supply chain, we have $c_b^i = 1$, if i delivers to b, otherwise $c_b^i = 0$. For the supply ladder and the hierarchical supply tree, we have $c_b^i = 0.5$, if an arrow points from i to b (see Figs 6.3b, c), otherwise $c_b^i = 0$. c_b^0 and c_{u+1}^i are defined in accordance with Eq. (6.36). These specifications guarantee that, *for* $\eta = 0$, i.e. if the production units are characterized by identical parameters, *the dynamics of the inventories is the same for all three discussed network topologies.* However, the topology matters a lot if we have a heterogeneity $\eta > 0$ in the model parameters (see Fig. 6.3e,f). Moreover, as our dynamical model of supply networks assumes nonlinear interactions, small changes of the stationary inventory N_0 or the relaxation time τ_b can have large effects (see Figs. 3 and 6 in [6]). We may also have a transition from small oscillations of relatively high frequency to large oscillations of low frequency, when we change the transport capacity a little (cf. Fig. 6.3d, e).

6.1.9 Advantages and Extensions

As the variables in the model are operational and measurable, the model can be tested and calibrated with empirical data. The above model of supply networks is flexible and easy to generalize. For this reason, it can be adapted to various applications. Our approach can be related to microscopic considerations such as queueing theory or event-driven (Monte Carlo) simulations of production processes, but, as it focusses on the average dynamics, it is numerically much more efficient and, therefore, suitable for online control. Nevertheless, the formulae can be extended by noise terms to reflect stochastic effects. Our system of coupled differential equations would then become a coupled system of stochastic differential equations (Langevin equations), where the noise amplitudes would be determined via relationships from queueing theory.

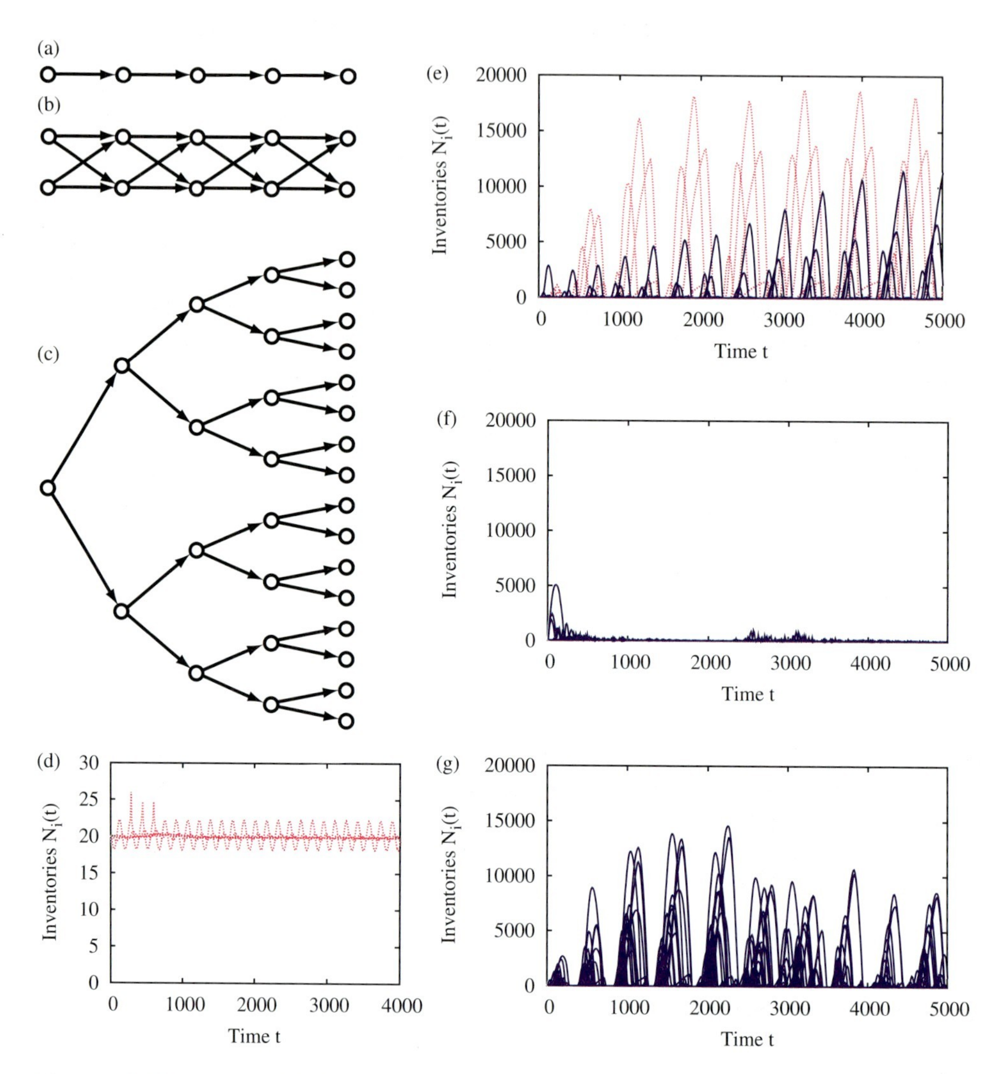

Figure 6.3: Illustration of different supply networks and their dynamics: (a) one-dimensional supply chain with five production units, (b) "supply ladder" with five levels and (c) hierarchical supply tree. For identical parameters and strategies ($\eta = 0$), the dynamics of the inventories is the same for all three network topologies (a)–(c): for the transport capacity $V = 0.045$, the dynamics is shown in (d), while (e) corresponds to $V = 0.047$; see the light dotted lines. The drastic change from small oscillations of high frequency to huge oscillations of low frequency above a critical threshold of the transport capacity V indicates a phase transition. The topology matters a lot when the individual parameters vary. The dark solid lines correspond to a heterogeneity of $\eta = 0.25$ in the case of (e) a one-dimensional supply chain, (f) a supply ladder and (g) a hierarchical supply network. One can conclude that heterogeneity in supply networks can considerably decrease the undesired oscillation amplitudes in the inventories. The strongest effect is found for supply ladders, which is relevant for the design of robust supply networks.

6.2 Many-particle Models of Production Processes

Instead of modeling product flows by mean-value equations, as above, we may also simulate manufacturing processes as driven many-particle systems, where the discrete products and transport devices play the role of particles. Their interactions can be specified in a way which reproduces the observed arrival and departure rates. In a recent study, for example, we have been able to increase the actual throughput of a chain of production processes in a major semiconductor factory by up to 39% [18]. In this particular supply chain, silicon wafers require a series of chemical processes similar to the development of a photographic film. Between the chemical treatments, the wafers must be washed in water basins, and in the end they need to be dried (see top of Fig. 6.4). The treatment times can be varied within certain time intervals. If the treatment takes longer or shorter, the wafers will be of poor quality and cannot be sold. The main problem in this system is the limited transport capacity of the handler, i.e. the device which has to move the wafers around. If several sets of wafers had to be moved at the same time, the treatment times for one of these would probably exceed the critical time threshold. Therefore, the challenge is to find a schedule that resolves conflicts in desired handler usage (as one channel, namely the handler, has to serve several products in parallel). The problem to achieve coordination among several elements is similar to the coordination of pedestrians in the merging area in front of a door (see Sect. 6.2.1).

In the first step, we have analyzed the time requirements of the different treatment and transport processes in detail. Generally speaking, these time requirements are (more or less) deterministic, and queueing effects in the supply chain result from dynamical interactions of the different units, namely conflicts in desired handler usage. These conflicts imply a delayed service (cf. Sects. 6.1.1 and 6.1.7). Based on a variable Gantt diagram (with variable treatment times), we have resolved these conflicts and reduced the related delays by harmonization, i.e. coordination of the different treatment times. This has usually been reached by *increasing* the treatment times, as is indicated in the middle of Fig. 6.4 by longer bars in the optimized schedule (shown on the right) compared to the original schedule (on the left). Different sets of wafers are distinguished by different shades of gray, while the treatment times belonging to the same "run" (series of treatments) are represented by the same shade of gray. One can clearly see that, in the optimized schedule with longer treatment times, the waiting times between successive runs are significantly decreased, resulting in much higher throughputs (see bottom of Fig. 6.4). That is, instead of stop-and-go patterns (waiting and usage periods of the chemical or water basins), we have reached a more or less continuous usage pattern (cf. Sects. 6.1.2 and 6.1.4).

6.2.1 Learning from Pedestrians

Note that the above-described optimization of some processes in semiconductor manufacturing is an example for the application of the "slower-is-faster effect", which had been discovered for panicking pedestrian crowds [19]. However, a closer investigation shows that one could learn many more strategies from pedestrian behavior to improve production processes with nonlinear dynamics, as the basic features of pedestrian streams and many production systems are the same: (1) the system consists of a large number of similar *entities* (individuals, particles, products, boxes, ...). (2) The entities are externally or internally *driven*, i.e. there is some energy

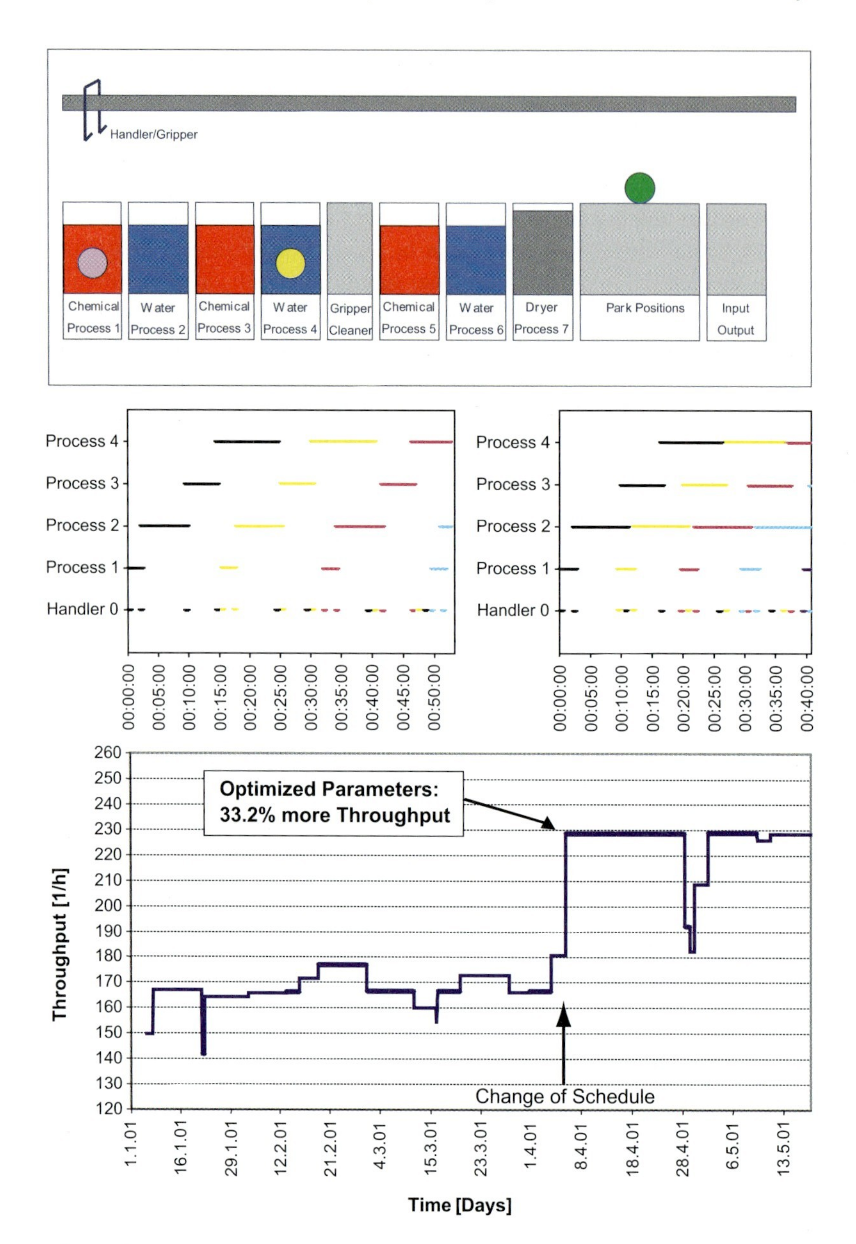

Figure 6.4: Top: schematic representation of the successive processes of a wet bench, i.e. a particular supply chain in semiconductor production. Middle: the Gantt diagrams illustrate the treatment times of the first four processes, where we have used the same shades of gray for processes belonging to the same run, i.e. the same set of wafers. The left-hand diagram shows the original schedule, while the right-hand one shows an optimized schedule based on the "slower-is-faster effect" (see Sect. 6.2.3). Bottom: the increase in the throughput of a wet bench by switching from the original production schedule to the optimized one was found to be 33%, in some cases even higher. (After [18].)

input, e.g., they can move. (3) The entities *interact nonlinearly,* i.e. under certain conditions small variations can have large effects. In other words, the system behavior is dominated by the interactions rather than the boundary conditions (the external control). (4) There is a *competition for resources* such as time (slots), space, energy, etc. (5) Each entity has a certain *extension* in space or time, or a certain demand. (6) When entities come too close to each other, *frictional and obstruction effects* occur. Therefore, we are trying to transfer our knowledge of traffic dynamics to the optimization of real production processes.

Pedestrian models have, for example, been successfully applied to the coordination of robots [20]. In the following, we will give a short introduction to this model, while details can be found in the available reviews [1, 21, 22]. The so-called *social-force model* of pedestrian dynamics describes the different competing motivations of pedestrians by separate force terms similar to granular flows, and it has the following advantages: (1) the social-force model takes into account the flexible usage of space (i.e. compressibility), but also the excluded-volume and friction effects which play a role at extreme densities. (2) The model assumptions are simple and plausible. (3) There are only a few model parameters to calibrate. (4) The model is robust and naturally reproduces many different observations without modifications of the model. (5) Nevertheless, it is easy to consider individual differences in the dynamic behavior, and extensions for more complex problems are possible.

The basic version of the social-force model assumes that the change of the location $\vec{x}_\alpha(t)$ of some pedestrian α in the course of time t is given by the actual velocity $\vec{v}_\alpha(t)$, i.e. $d\vec{x}_\alpha/dt = \vec{v}_\alpha(t)$. Moreover, the acceleration $d\vec{v}_\alpha/dt$ is specified by a sum of "social forces", e.g. the driving force $(v_\alpha^0\vec{e}_\alpha - \vec{v}_\alpha)/\tau_\alpha$, which describes the adaptation of the actual velocity $\vec{v}_\alpha$ to the desired velocity v_α^0 and the desired walking direction $\vec{e}_\alpha(t)$ within a certain acceleration time τ_α, the repulsive forces $\vec{f}_{\alpha\beta}(\vec{x}_\alpha, \vec{v}_\alpha, \vec{x}_\beta, \vec{v}_\beta)$ with respect to other pedestrians β, the repulsive forces $\vec{f}_{\alpha k}(\vec{x}_\alpha, \vec{v}_\alpha, t)$ with respect to obstacles k and fluctuation forces $\vec{\xi}_\alpha(t)$ reflecting individual variations in behavior:

$$\frac{d\vec{v}_\alpha}{dt} = \frac{v_\alpha^0\vec{e}_\alpha - \vec{v}_\alpha(t)}{\tau_\alpha} + \sum_{\beta(\neq\alpha)} \vec{f}_{\alpha\beta}(t) + \sum_k \vec{f}_{\alpha k}(t) + \vec{\xi}_\alpha(t)\,. \tag{6.37}$$

For reasonable specifications of the interaction forces see [1, 19, 23, 24].

This model describes many self-organization phenomena in pedestrian crowds in a natural and realistic way. For example, for "relaxed" pedestrians in normal situations with small fluctuation amplitudes, our microsimulations of pedestrian counterflows in corridors reproduce the empirically observed *segregation* of opposite flow directions into lanes [1, 23, 24, 25], see Fig. 6.5a. However, for large fluctuation amplitudes corresponding to "nervous" pedestrians, we find a "freezing by heating effect" characterized by a breakdown of "fluid" lanes and the emergence of "solid" blockages [23], see Fig. 6.5b. The same model also reproduces the observed oscillations of the flow direction at bottlenecks [1, 24], see Fig. 6.5c. Cellular automaton Java applets visualizing these phenomena are available on the Internet (see `www.helbing.org/Pedestrians/Corridor.html, /Door.html`). These findings are obviously relevant for production processes involving granular flows or requiring the segregation of different kinds of particles or objects.

Due to mutual interactions and environmental impacts, pedestrians suffer delays and experience different travel times, even if their desired velocities v_α^0 are the same. In order to

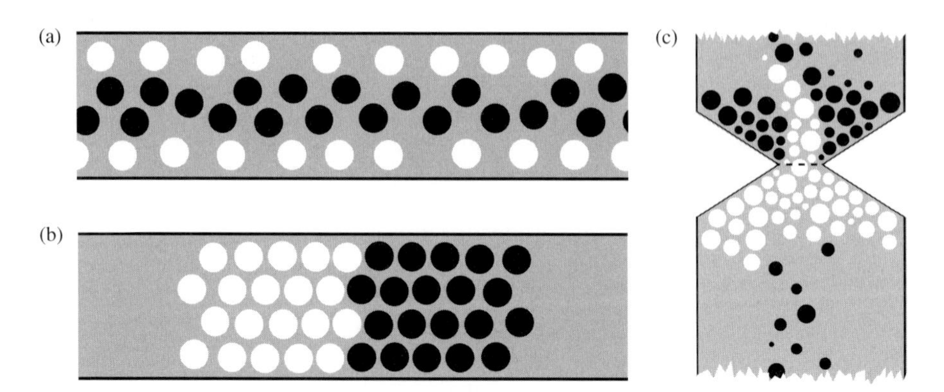

Figure 6.5: (a) Segregation of opposite flow directions into lanes for the case of small noise amplitudes (after [1, 21, 23]; cf. also [24, 25]). White disks represent entities (pedestrians or objects) moving from left to right, black ones move the other way round. (b) For sufficiently high densities and large fluctuations, we observe the noise-induced formation of a crystallized, "frozen" state (after [1, 23]). (c) Bottlenecks are passed by clusters of entities in alternating directions, giving rise to oscillatory flows. The phenomena (a) to (c) are relevant for heterogeneous object flows, segregation processes, the coordination of robots, and production processes with competing goals.

compensate for this and arrive in time, pedestrians adapt their desired velocities. If l_α is the total length of the way and $s_\alpha(t) = \int_0^t dt'\ \vec{v}_\alpha(t') \cdot \vec{e}_\alpha(t')$ the portion of the way they have completed until time t, a reasonable adaptation strategy would be

$$v^0_\alpha(t) = \min\left(\frac{l_\alpha - s_\alpha(t)}{T_\alpha - t}, v_\alpha^{\max}\right), \tag{6.38}$$

where T_α is the required arrival time and $v_\alpha^{\max}$ the maximum speed. This strategy will eventually lead to different actual speeds, which will also segregate into lanes: fast lanes and slow lanes. A similar strategy may be used to compensate for delays in production processes.

6.2.2 Optimal Self-organization and Noise-induced Ordering

Lane formation (segregation) is an *optimal self-organization* phenomenon [25] resulting from the combined action of driving and repulsive forces: under certain conditions, one can show that lane formation maximizes the average speed, and the resulting pattern is "fair" for both directions of motion [25]. That is, the efficiency in the overall system is optimized based on local interactions. For this reason, one can speak of a distributed control mechanism or even distributed intelligence. Such mechanisms are usually cost effective and robust with respect to local failures, which is interesting for applications. When the amplitude of the fluctuations $\vec{\xi}_\alpha(t)$ is increased, we expect a reduced degree of order in the system. For large noise amplitudes, we find indeed a breakdown of self-organized patterns such as lanes. In this case, we face a disordered system with homogeneous distributions in space. However, for medium noise amplitudes, one can observe an increase in the level or order. For example, instead of many narrow lanes, one can find a few wide ones (see Fig. 6.6). This phenomenon

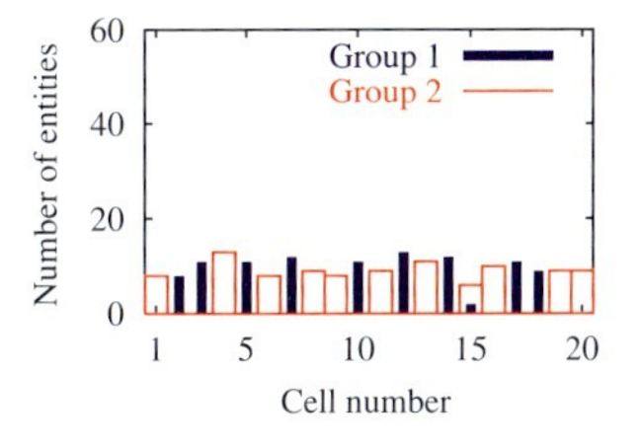

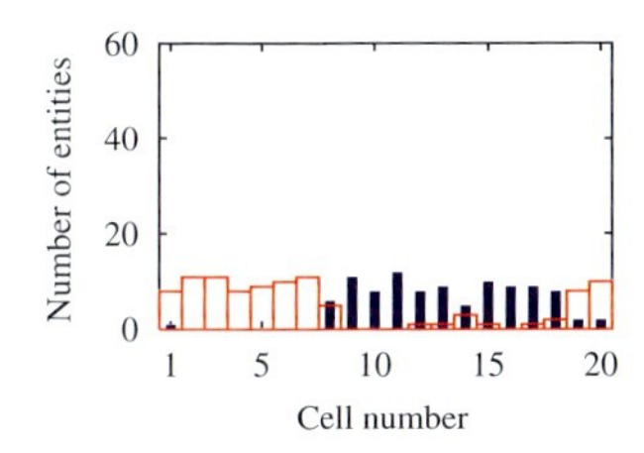

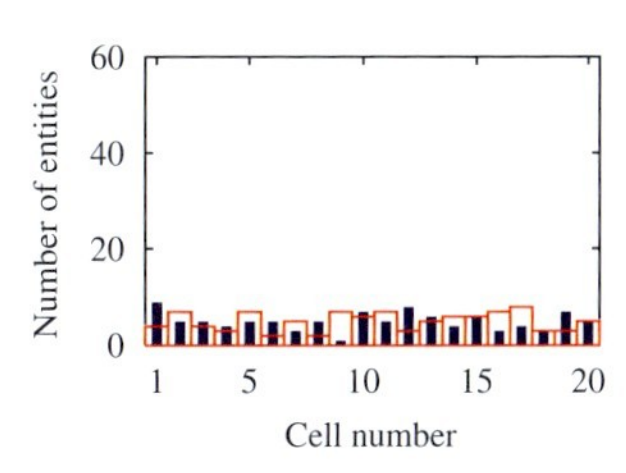

Figure 6.6: Resulting distribution (at a cross section of a corridor) for two different groups of entities (walking directions) in a segregation (lane-formation) process at small (left), medium (middle) and large noise strengths (right). The lower number of lanes at medium fluctuation amplitudes indicates *noise-induced ordering,* while the homogeneous distribution at large noise amplitudes corresponds to disorder or "freezing by heating". (After [26].)

is called *noise-induced ordering* [26, 27]. We have learned that a careful choice of the noise strength can speed up the time-dependent increase of the order very much. Moreover, after a given, large enough time period, the system has reached a typical level of order, which depends significantly on the fluctuation strength. In conclusion, a variation of the "applied" fluctuation strength together with a proper choice of the "treatment times" would allow one to control pattern formation in several respects: (1) the speed of ordering, (2) the typical length scale in the system and (3) the level of ordering. A time-dependent variation of the control parameters should even facilitate switching between the support and suppression of structure formation, e.g. between demixing and homogenization. These points are, for example, relevant for the production, properties, handling and transport of heterogeneous materials, for flow control and efficient separation techniques for different kinds of particles or objects.

6.2.3 "Slower-is-Faster Effect" in Merging Flows

"Freezing by heating" is one of the phenomena observed in panic stampedes. Another one is the *"faster-is-slower effect" or "slower-is-faster effect"* [19]. It is caused by arching and clogging at bottlenecks like exits, which implies irregular outflows (see Fig. 6.7a). The reason is frictional interactions, when the entities touch each other. This happens when the driving forces exceed a certain critical threshold and the diameter of the bottleneck is small. Below the critical threshold, outflows are regular and efficient (see Fig. 6.7b), i.e. entities arrange perfectly among each other in the merging area in front of a bottleneck. This mutual coordination is thanks to the nonlinear repulsive interactions. It requires neither communication nor fluctuations. For simulations see `http://angel.elte.hu/~panic/` or `http://www.panics.org`. The relevance for production processes involving merging flows (packing processes) or granular flows through hoppers (filling processes) is obvious.

6.2.4 Optimization of Multi-object Flows

Due to the nonlinear dynamics, driven multi-object flows decisively depend on the geometry of the boundaries. This calls for innovative solutions, which utilize the self-organization in

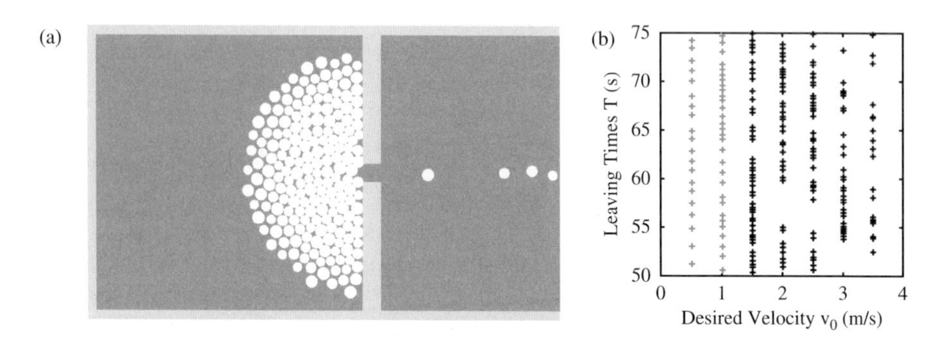

Figure 6.7: (a) When the desired velocities v_α^0 are too high (e.g. in panic situations), pedestrians come so close to each other that their physical contacts cause the build up of pressure and obstructing friction effects, which results in temporary arching and clogging. (b) This is related with an irregular and reduced outflow, while the outflow is regular for small enough desired velocities ($v_\alpha^0 \leq 1.5$ m/s) [1, 19, 21]. It is interesting that suitable obstacles can improve the outflow by reducing the pressure (see `http://angel.elte.hu/~panic/` for online Java simulations.) This has also implications for filling and packing processes with merging flows.

nonlinearly interacting many-particle systems. We will exemplify this for the improvement of standard elements of pedestrian facilities [22] (see Fig. 6.8):

1. At high pedestrian densities, the lanes of uniform walking direction tend to disturb each other, as pedestrians expand into areas of low density and try to overtake each other. This often leads to mutual obstructions of the opposite walking directions. The lanes can be stabilized by a series of small obstacles in the middle of the walkway, see Fig. 6.8 (left) which, in the direction of motion, has a similar effect as a wall.

2. The flow at bottlenecks can be improved by a funnel-shaped construction (see center of Fig. 6.8). Interestingly, the optimal funnel shape resulting from an evolutionary optimization is convex [22].

3. Oscillatory changes of the walking direction and periods of standstill in between do not only occur for counterflows at doors, but also when different flows *cross* each other. The loss of efficiency caused by this can be reduced by railings inducing roundabout traffic, see Fig. 6.8 (right). Roundabout traffic can already be triggered and stabilized by an obstacle in the middle of a crossing, because it suppresses the phases of "vertical" or "horizontal" motion in the intersection area. In our simulations, this increased efficiency up to 13%.

It is natural to use similar solutions for the improvement of multi-object flows in production processes as well.

6.3 Summary and Outlook

In this contribution, we have shown that various concepts from traffic theory can be transferred to production processes with several implications for their optimization. We just mention

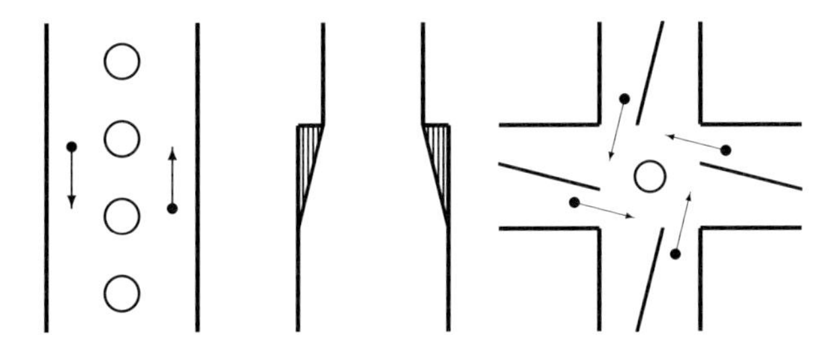

Figure 6.8: Improved standard elements of pedestrians facilities: corridors (left), bottlenecks (middle) and intersections (right). Empty circles represent obstacles such as columns or trees, while full circles with arrows symbolize pedestrians and their walking directions (after [22]). Similar solutions may also be applied to other multi-object flows, e.g. filling processes.

optimal self-organization (regarding the segregation or merging of multi-object flows) or noise-induced ordering. Moreover, the "slower-is-faster effect" has already been successfully applied to semiconductor production. The reason for the similarities of traffic and production systems is the presence of moving entities (persons or objects), which interact in a nonlinear way with obstructive and frictional effects. Therefore, a competition for limited resources (such as capacities, time or space) takes place. The question is how to distribute them in an efficient and fair way. Answers have, for example, been developed in the areas of game theory and of intelligent transportation systems. The theory of complex driven many-particle systems has made significant contributions to this.

We have also sketched a novel theory of supply networks. This theory is developed to help understand the dynamical phenomena, breakdowns, instabilities and inefficiencies of production and supply networks. It is well suited for an efficient simulation and online control of production systems, and effects of fluctuations can be studied as well. Oscillations in the inventories ("the bull-whip effect"), resulting from nonlinear interactions of production units, may occur on considerably slower time scales than the variation of the consumption rate, which may explain the existence of business cycles. Moreover, we have discovered that the supply network's topology has a significant impact on the resulting production dynamics, and that a heterogeneity in the parameters characterizing the different production units can stabilize the production considerably. Production can also be stabilized by forecasting temporal changes in the inventories, by reduction of the time required to adapt the production speed, by consideration of the stock levels of downstream suppliers and by modification of the management function.

Bibliography

[1] D. Helbing, *Traffic and related self-driven many-particle systems*, Reviews of Modern Physics **73** (2001), 1067.

[2] T. Nagatani, *Statistical physics of vehicular traffic and some related systems*, Reports on Progress in Physics **65** (2002), 1331;
D. Chowdhury, L. Santen, and A. Schadschneider, *The physics of traffic jams*, Physics Reports **329** (2000), 199.

[3] U. Witt and G.-Z. Sun, *Myopic behavior and cycles in aggregate output*, in: Jahrbücher f. Nationalökonomie u. Statistik, Vol. 222–3, Lucius & Lucius, Stuttgart, 2002, p. 366.

[4] C. Daganzo, *A Theory of Supply Chains*, Springer, New York, 2003;
C. Daganzo, *On the stability of supply chains*, Operations Research, submitted (2002).

[5] D. Marthaler, D. Armbruster, and C. Ringhofer, *A mesoscopic approach to the simulation of semiconductor supply chains*, in: Proceedings of the International Conference on Modeling and Analysis of Semiconductor Manufacturing, edited by G. Mackulak et al., 2002, p. 365;
B. Rem and D. Armbruster, *Control and synchronization in switched arrival systems*, Chaos **13** (2003), 128;
I. Diaz-Rivera, D. Armbruster, and T. Taylor, *Periodic orbits in a class of re-entrant manufacturing systems*, Mathematics and Operations Research **25** (2000), 708.

[6] D. Helbing, *Modeling supply networks and business cycles as unstable transport phenomena*, New Journal of Physics **5** (2003), 90.1.

[7] E. Lefeber, *Nonlinear models for control of manufacturing systems*, Chapter 5, this book.

[8] D. Armbruster, *Dynamical systems and production systems*, Chapter 1, this book.

[9] T. Nagatani and D. Helbing, *Stability analysis and stabilization strategies for linear supply chains*, submitted to Physics A, see preprint http://arXiv.org/abs/cond-mat/0304476 (2003).

[10] E. Mosekilde and E.R. Larsen, *Deterministic chaos in the beer production-distribution model*, System Dynamics Review **4**(1–2) (1988), 131–147;
J.D. Sterman, *Business Dynamics*, McGraw-Hill, Boston, 2000.

[11] M. Bando, K. Hasebe, A. Nakayama, A. Shibata, and Y. Sugiyama, *Dynamical model of traffic congestion and numerical simulation*, Physical Review E **51** (1995), 1035.

[12] D. Helbing, *A section-based queueing-theoretical traffic model for congestion and travel time analsis in networks*, Journal of Physics A: Mathematical and General, **36** (2003), L593–L598.

[13] D. Helbing, S. Lämmer, P. Šeba, and T. Platkowski, *Stability of regular and random supply networks*, preprint (2003).

[14] W.J. Hopp and M.L. Spearman, *Factory Physics*, McGraw-Hill, Boston, 2000.

[15] K. Peters, J. Worbs, U. Parlitz, and H.-P. Wiendahl, *Manufacturing systems with restricted buffer sizes*, Chapter 3, this book.

[16] R. Hall, *Queueing Methods for Service and Manufacturing*, Prentice Hall, Upper Saddle River, NJ, 1991;
T. Saaty, *Elements of Queueing Theory with Applications*, Dover, New York, 1983.

[17] W.W. Leontief, *Input-Output Economics*, Oxford University Press, New York, 1966.

[18] D. Fasold, Master's thesis, TU Dresden (2001).

[19] D. Helbing, I. Farkas, and T. Vicsek, *Simulating dynamical features of escape panic*, Nature **407** (2000), 487.

[20] P. Molnár and J. Starke, *Control of distributed autonomous robotic systems using principles of pattern formation in nature and pedestrian behavior*, IEEE Transactions on Systems, Man, and Cybernetics B **31** (2001), 433;
M. Brecht et al., *Three index assignment of robots to targets: An experimental verification*, in: Proceedings of the International Conference on Intelligent Autonomous Systems, edited by E. Pagallo et al., 2000, p. 156.

[21] D. Helbing, I.J. Farkas, P. Molnár, and T. Vicsek, *Simulation of pedestrian crowds in normal and evacuation situations*, in: Pedestrian and Evacuation Dynamics, edited by M. Schreckenberg and S.D. Sharma, Springer, Berlin, 2002, p. 21.

[22] D. Helbing, P. Molnár, I. Farkas, and K. Bolay, *Self-organizing pedestrian movement*, Environment and Planning B **28** (2001), 361.

[23] D. Helbing, I. Farkas, and T. Vicsek, *Freezing by heating in a mesoscopic system*, Physical Reviews Letters **84** (2000), 1240.

[24] D. Helbing and P. Molnár, *Social force model of pedestrian dynamics*, Physical Review E **51** (1995), 4282.

[25] D. Helbing and T. Vicsek, *Optimal self-organization*, New Journal of Physics **1** (1999), 13.1.

[26] D. Helbing and T. Platkowski, *Self-organization in space and induced by fluctuations*, International Journal of Chaos Theory and Applications **5** (2000), 47.

[27] D. Helbing and T. Platkowski, *Drift- or fluctuation-induced ordering and self-organization in driven many-particla systems*, Europhysics Letters **60** (2002), 227.

Part II

Machine Tools and Manufacturing Processes

This part is devoted to problems arising in connection with mechanical machining and manufacturing processes. A classical dynamical problem that was considered scientifically already a century ago in the work of F. Taylor and which also nowadays is not fully understood is that of machine chatter. An understanding of machine chatter and the means of its compensation or avoidance is of great importance since the appearance of this phenomenon enhances tool wear and strongly degrades the quality of the machined surface. Today it is still a major obstacle for reaching high production rates e.g. in high-speed milling, the nowadays most preferred and efficient cutting process. In the following part five papers present the very recent developments mainly in the field of regenerative (secondary) chatter, i.e. the regeneration of waviness at the workpiece surface. The paper by G. Stepan et al. emphasizes the differences between turning and milling, where chatter corresponds to self-interrupted and parametrically interrupted cutting, respectively. Mathematically this difference is reflected in the appearance of autonomous vs. non-autonomous differential delay equations. T. Kalmar-Nagy and F.C. Moon in their contribution show that the combined effect of mode-coupled tool vibrations and the regenerative effect can lead to chatter and a stability analysis is performed. The paper by R. Rusinek et al. provides an advanced model, which by including dry friction between tool and chip with a nonlinear Rayleigh term also explains self-excited vibrations, a primary source of chatter. Secondary, regenerative chatter is discussed as a resonance phenomenon leading to synchronization. In contrast to the previous papers the contribution of N. van de Wouw et al. aims in addition at an experimental verification of their model of regenerative chatter for high-speed milling. Their approach leads to non-autonomous differential delay equations as in the paper of G. Stepan et al., finally, however, an autonomous approximation appears to be satisfactory. The last paper in this series by H.K. Tönshoff et al. deals with chatter occurring with cylindrical grinding. The important new point here is the development of realistic chatter-compensation strategies and their practical implementation.

Machine chatter is not the only process that influences the cutting quality. Improvements are nowadays achieved by the use of coated tools, and a successful control of the thermal behavior of tool systems enhances the precision of the cutting process. The chip forming process itself is therefore of special interest. A. Behrens et al. present simulations of this process and the thermal effects on the tool by use of the Finite Element Method (FEM) and compare it with experimental results. The effect of coated tools is analyzed also with FEM computations in the contribution of G. Schmidt et al. and the paper of R. Rentsch et al. aims in the same direction applying Molecular-Dynamics (MD) simulations. In the work of J. Konvicka et al. the thermal behavior of a modular tool system of a working milling machine and the means of compensating the resulting displacements are investigated numerically with the FEM and subsequently compared with experimental results.

The last two contributions within this part deal with mechanical forming technologies. The one by M. Kleiner et al. investigates sheet-metal spinning. Possible failure modes of this method due to wrinkling were identified as dynamical instability in contrast to similar effects in the deep-drawing technique. The use of time-series analysis and application of the FEM yields an understanding of the dynamical nature of the wrinkling process. As a result feedback control may be used to suppress wrinkle formation. The work of H.J. Holl et al. considers nonlinear vibrations during a strip-rolling and coiling process in a Steckel Mill facility. For the latter the variable-mass problem is solved and implemented with corresponding FEM techniques.

7 Nonlinear Dynamics of High-speed Milling Subjected to Regenerative Effect

G. Stépán, R. Szalai, and T. Insperger

The regenerative effect is a widely accepted cause of self-excited vibrations in machine tools. Their prediction has always been one of the most difficult research areas. For conventional turning, the subcritical nature of Hopf bifurcations at the stability limits of stationary cutting has been proved recently. For thread cutting, subcritical co-dimension 2 Hopf bifurcations were also observed experimentally. These investigations were based on the analysis of autonomous delay-differential equations. The system behavior 'outside' the unstable periodic motion is strongly affected by the loss of contact between the workpiece and the tool. Milling is a kind of cutting where loss of contact between teeth and workpiece occurs typically in a periodic way. This leads to non-autonomous governing equations similar to the damped, delayed Mathieu equation. The parametric excitation in the delay-differential equation leads to secondary Hopf bifurcations, and also to period-doubling bifurcations if the speed of cutting is high enough. A low number of milling teeth and high speed together lead to highly interrupted cutting when the Poincaré mapping of the governing equation can be constructed in closed approximate form. The subcritical nature of both period-doubling and secondary Hopf bifurcations is shown, and the development of chaotic oscillations is also explained. The global nonlinear dynamics of turning and high-speed milling are compared as self-interrupted and parametrically interrupted cutting processes.

7.1 Introduction

Machine tool vibrations have a negative effect on the quality of the machined surface of the workpiece. One of the most important causes of instability in the cutting process is the so-called regenerative effect. The physical basis of this phenomenon is well known in the literature (see [1, 2]); still, it is one of the most difficult dynamical problems of engineering. Referring to its infinite-dimensional nature, it is often compared to the problem of turbulence in fluid mechanics. Because of some external perturbations, the tool starts a damped oscillation relative to the workpiece, and the surface of the workpiece becomes wavy (see the simplest planar mechanical model of orthogonal cutting in Fig. 7.1). After a revolution of the workpiece (or tool) the chip thickness will vary at the tool because of this wavy surface. As a consequence, the cutting force depends on the actual and delayed values of the relative displacement of the tool and the workpiece. The length of the delay is equal to the time period τ of the revolution of the workpiece (or tool). This delay is the central idea of the regenerative effect (see [3]).

Nonlinear Dynamics of Production Systems. Edited by G. Radons and R. Neugebauer

ISBN 3-527-40430-9

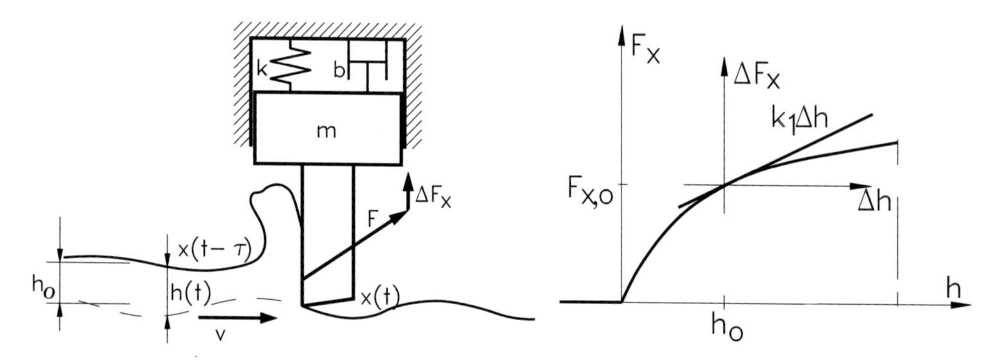

Figure 7.1: Mechanical model and cutting force characteristic.

In case of conventional turning, the linear stability analysis of stationary cutting provides a complicated, sometimes fractal-like stability chart in the plane of the technological parameters (cutting speed, chip width, chip thickness). The existence of unstable periodic motions around the stable stationary cutting was already predicted in the early 1980s by the experiments in [4]. The corresponding subcritical Hopf bifurcation at the stability limit of turning has been proven analytically only recently [5–9]. The experiments and introductory analytical study in [10] prove the existence of saddle-like unstable quasi-periodic motions, too, in certain finite regions of the system parameters. The description of the global dynamics of the cutting process 'outside' the unstable limit cycle, or unstable torus, requires the analysis of the so-called self-interrupted cutting. This means that a stable quasi-periodic or even chaotic motion may exist when the tool leaves the workpiece for certain short time periods during cutting. The time intervals of these no-contact cases are regulated by the system itself. This is referred to as self-interrupted cutting. The next section summarizes the corresponding results for turning in order to characterize the major differences between the nonlinear dynamics of turning and high-speed milling.

High-speed milling is one of the most preferred and efficient cutting processes nowadays. It is a challenging task for researchers to explore its special dynamical properties, including the stability conditions of the cutting process and the nonlinear vibrations that may occur near to the stability boundaries. These dynamical properties are mainly related to the underlying regenerative effect in the same way as is in the case of the classical turning process. Still, some new phenomena appear for low-immersion milling as predicted in [11, 12]. These phenomena were also reported in [13–15] in the case of milling, independently from the immersion or speed characteristics of the milling processes.

High-speed milling usually means also low immersion (very small chip thickness) with a relatively small number (two to four) of cutting edges on the tool. There is either no contact between the tool and the workpiece, or there is only one edge in contact for a (relatively short) time. In these cases, highly interrupted machining can well approximate the whole machining process. In the simplest models of highly interrupted machining, the ratio of time spent cutting to not cutting is a small parameter. This leads to a mechanical model where the free vibration of the tool is perturbed periodically by an impact to the workpiece. This impact results in a sudden change in the tool oscillation velocity and it depends on the actual chip thickness, that is, on the difference of the present and the previous tool positions. In this respect, the model includes the classical regenerative effect described for turning. There is one major difference, though: the periods of no-contact are regulated by the cutting speed parameter. Accordingly, high-speed milling is a kind of *parametrically interrupted* cutting as opposed to the *self-interrupted* cutting arising in unstable turning processes.

The simplest possible, but still nonlinear highly interrupted cutting model leads to a two-dimensional discrete mathematical model. Bifurcation analysis can be carried out along stability limits related to period-doubling bifurcations and (secondary) Hopf bifurcations. These require center manifold reduction and normal-form transformation. The tedious algebraic work can be carried out in closed form and leads to nonlinear phenomena similar to the one experienced in the case of the Hopf bifurcation in the turning process. The resulting subcritical bifurcations are presented in analytical form in the subsequent sections. The existence of stable period-two vibrations is also shown 'outside' the unstable period-two vibrations. The stable ones can, however, quickly bifurcate to chaotic oscillations with increasing chip width. This is also shown by numerical investigation.

7.2 Nonlinear Dynamics of Turning

In the following subsections, the nonlinear mechanical model of turning is introduced, and the corresponding time-delayed mathematical model is derived. Then the results of the local bifurcation analyses are summarized, and the global dynamics is explained in order to compare it to the nonlinear dynamics of high-speed milling.

7.2.1 Modeling of Turning

The simplest mechanical model of orthogonal cutting is presented in Fig. 7.1. There is one cutting edge only, and it is continuously in contact with the workpiece material. The elastic tool is characterized by modal parameters like the angular natural frequency $\omega_{\mathrm{n}} = \sqrt{k/m}$, the relative damping factor $\varsigma = b/(2m\omega_{\mathrm{n}})$ and the angular frequency $\omega_{\mathrm{d}} = \omega_{\mathrm{n}}\sqrt{1-\zeta^2}$ of the damped free tool oscillation. The theoretical chip thickness is h_0 and the constant chip width is w. The actual chip thickness is

$$h(t) = h_0 + x(t-\tau) - x(t)\,. \tag{7.1}$$

The x component of the nonlinear cutting force F can be calculated in accordance with the experimentally verified three-quarter rule [23]:

$$F_x = Kwh^{3/4} = Kw(h_0 + x(t-\tau) - x(t))^{3/4}, \tag{7.2}$$

where K is an experimentally identified constant parameter. This formula is valid in a certain region around the theoretical chip thickness but, clearly, $F_x = 0$ for negative chip thickness in case of possible large x oscillations (see the cutting force characteristics in Fig. 7.1). The third-degree Taylor-series approximation of the cutting force variation

$$\Delta F_x = F_x - F_{x0} \tag{7.3}$$

with respect to the chip thickness variation

$$\Delta h = h(t) - h_0 = x(t-\tau) - x(t) \tag{7.4}$$

assumes the form

$$\Delta F_x \approx k_1(\Delta h) - \frac{1}{8h_0}k_1(\Delta h)^2 + \frac{5}{96h_0^2}k_1(\Delta h)^3. \tag{7.5}$$

The stationary cutting force F_{x0} and the so-called cutting coefficient k_1 assume the forms

$$F_{x0} = Kwh_0^{3/4}\,, \quad k_1 = \left.\frac{\partial F_x}{\partial h}\right|_{h_0} = \frac{3}{4}\frac{Kw}{\sqrt[4]{h_0}}\,, \tag{7.6}$$

respectively. The equation of motion has the simple form

$$\ddot{x}(t) + 2\varsigma\omega_{\mathrm{n}}\dot{x}(t) + \omega_{\mathrm{n}}^2 x(t) = \frac{1}{m}\Delta F_x. \tag{7.7}$$

If the cutting force variation is substituted here using Eqs. (7.5) and (7.4), and the dimensionless time $\tilde{t} = \omega_{\mathrm{n}} t$ is introduced, then the nonlinear delay-differential equation

$$\begin{aligned} &x''(\tilde{t}) + 2\varsigma x'(\tilde{t}) + (1+\tilde{w})x(\tilde{t}) - \tilde{w}x(\tilde{t}-\tilde{\tau}) \\ &= -\tfrac{1}{8h_0}\tilde{w}(x(\tilde{t}) - x(\tilde{t}-\tilde{\tau}))^2 + \tfrac{5}{96h_0^2}\tilde{w}(x(\tilde{t}) - x(\tilde{t}-\tilde{\tau}))^3 \end{aligned} \tag{7.8}$$

is obtained with its linear part on the left-hand-side. The dimensionless chip width $\tilde{w}$ can be calculated as the ratio of the cutting coefficient and the modal stiffness of the machine tool:

$$\tilde{w} = \frac{k_1}{m\omega_{\mathrm{n}}^2} = \frac{k_1}{k} = \frac{3}{4}\frac{K}{k\sqrt[4]{h_0}}w, \tag{7.9}$$

while the dimensionless time delay $\tilde{\tau} = \omega_{\mathrm{n}}\tau$ and the dimensionless angular velocity $\tilde{\Omega} = \Omega/\omega_{\mathrm{n}}$ are related to the angular velocity Ω of the cylindrical workpiece in the following way:

$$\tilde{\tau} = \frac{2\pi}{\tilde{\Omega}} = \frac{2\pi}{\Omega}\omega_{\mathrm{n}} = \tau\omega_{\mathrm{n}}. \tag{7.10}$$

If v denotes the cutting speed, τv is the circumference of the cylindrical workpiece.

7.2.2 Bifurcation Analysis of Turning

The stability analysis of the stationary turning means the investigation of the trivial solution of the delay-differential equation (7.8) that can be based on the characteristic function

$$\lambda^2 + 2\varsigma\lambda + (1+\tilde{w}) - \tilde{w}\exp(-2\pi\lambda/\tilde{\Omega}) = 0. \tag{7.11}$$

The locations of the infinitely many characteristic roots λ depend on three dimensionless parameters only. The so-called stability chart is presented in the plane of the parameters $\tilde{\Omega}$ and $\tilde{w}$ related to the chosen technology, while the modal damping ζ of the machine tool is a

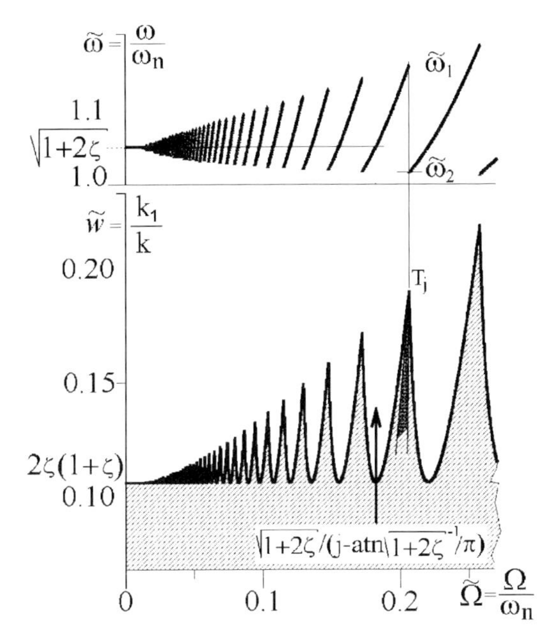

Figure 7.2: Stability chart of turning.

fixed parameter chosen to be 0.05 here. The shaded region in the chart of Fig. 7.2 shows the parameter domain where the characteristic roots are in the left half of the complex plane, that is, where the stationary cutting is stable.

There are two sets of points where closed-form bifurcation analysis can be carried out. The details of these long algebraic calculations are presented in [5] and [10]. We briefly summarize these results here to make them comparable to the special nonlinear phenomena of high-speed milling.

At the minimum points

$$\tilde{\Omega}_{\mathrm{cr},j} = \frac{\sqrt{1+2\varsigma}}{j - \frac{1}{\pi}\mathrm{atn}\frac{1}{\sqrt{1+2\varsigma}}}, \qquad j = 1, 2, 3, \ldots, \tilde{w}_{\mathrm{cr}} = 2\varsigma(1+\varsigma) \tag{7.12}$$

of the stability lobes, Hopf bifurcation occurs referring to self-excited vibrations with the dimensionless vibration frequency $\tilde{\omega} = \sqrt{1+2\varsigma}$ also shown above the chart of Fig. 7.2. The infinite-dimensional centre manifold reduction and the normal-form calculation prove the subcritical sense of the bifurcation, that is, unstable periodic vibration exists around the stable stationary cutting approximated in the form

$$x(t) = \frac{8}{\sqrt{5}} h_0 \left(1 + \frac{11}{30}\varsigma\right) \sqrt{1 - \frac{\tilde{w}}{2\varsigma(1+\varsigma)}} \cos(\sqrt{1+2\varsigma}\,\omega_{\mathrm{n}} t). \tag{7.13}$$

At the peak points T_j, $j = 1, 2, 3, \ldots$ of the stability lobes, co-dimension 2 Hopf bifurcations occur. This means that a sector can be identified in the stable parameter domain (see the gray-shaded region below T_j in Fig. 7.2) where two unstable periodic motions exist together with an unstable quasi-periodic oscillation 'around' the stable stationary cutting. The corresponding two dimensionless vibration frequencies $\tilde{\omega}_{1,2}$ of these oscillations are also presented above the stability chart in Fig. 7.2. The structure of the corresponding phase space is explained when the global dynamics of the system is discussed in the next subsection.

7.2.3 Global Dynamics of Self-interrupted Cutting

Since there are unstable periodic motions, the existence of a global attractor is not explained by the local bifurcation analysis presented in the previous section. Experiments and numerical simulations show that there is no attractor in the delayed system – the large-amplitude stable motion exists only when the tool leaves the workpiece for certain time periods. As Fig. 7.3 shows, the vibration amplitudes of the tool may become so large that the tool is not in contact with the workpiece between points Q_1 and Q_2 and then between Q_3 and Q_4.

In these intervals, there is zero cutting force, no regenerative effect and the system is described by the equation of the damped oscillator:

$$x''(\tilde{t}) + 2\varsigma x'(\tilde{t}) + (1 + \tilde{w})x(\tilde{t}) = 0. \tag{7.14}$$

Since the trivial solution in this system is globally asymptotically stable, the tool quickly enters the material of the workpiece again, and the de-stabilising delay effect is switched on again. The system will have switches between these two dynamics: the damped free flight of the tool described by Eq. (7.14) and the regenerative cutting described by Eq. (7.8). This may be either periodic, quasi-periodic, chaotic or transient chaotic motion. The bifurcation diagram of Fig. 7.4 represents the global dynamics of turning in the vicinity of the stability limit at the minimum point (7.12) of the stability lobes. Transient chaotic motion may occur when the tool entering the workpiece arrives back to the infinite-dimensional phase space of regenerative chatter inside the unstable limit cycle, and the chaotic switches between the two different dynamics disappear suddenly, and stable stationary cutting is approached.

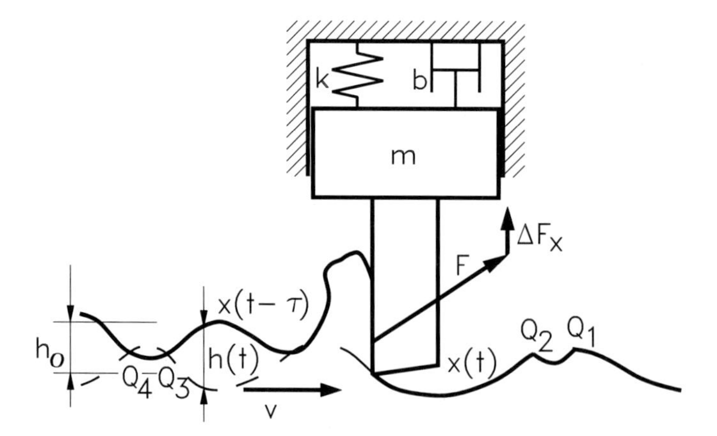

Figure 7.3: Self-interrupted cutting process.

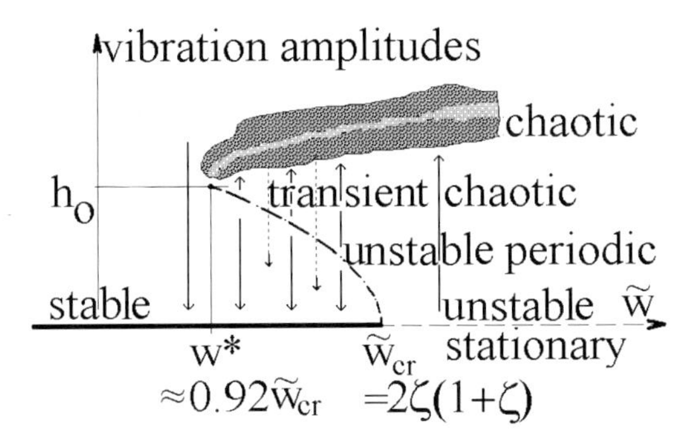

Figure 7.4: Bifurcation diagram and global dynamics of turning.

It is still unresolved, however, how the unstable periodic motion is connected to the branch of the (possibly chaotic) attractor 'outside'. The location of this turning point can be estimated where the vibration amplitude of the unstable limit cycle reaches the theoretical chip thickness h_0. Then for small damping values ($\varsigma < 0.05$), Eqs. (7.12) and (7.13) yield

$$h_0 = \frac{8}{\sqrt{5}} h_0 \left(1 + \frac{11}{30}\varsigma\right) \sqrt{1 - \frac{\tilde{w}}{\tilde{w}_{\mathrm{cr}}}} \quad \Rightarrow \quad \tilde{w}^* \approx \left(1 - \frac{5}{64} - \frac{22}{384}\varsigma\right) \tilde{w}_{\mathrm{cr}} \approx 0.92\, \tilde{w}_{\mathrm{cr}}. \tag{7.15}$$

This means that globally stable stationary cutting is likely to be guaranteed only at about 8% below that critical chip width calculated from linear stability theory. The experimental results in [4] give even smaller values for the limit of global stability. Near a notch of the stability chart, for example, they measured $w^* = 2.7$ [mm] and $w_{\mathrm{cr}} = 3.1$ [mm], that is, the global limit is about $1 - 2.7/3.1 = 13\%$ below the linear one.

The bifurcation scenario is even more complicated at the peak point T_j of the stability lobes in Fig. 7.2. Related experiments with thread cutting around a co-dimension 2 bifurcation point are described in detail in [10]. When the system parameters are taken from the dark-shaded sector below the point T_j in Fig. 7.2, the surface plot of the machined workpiece in Fig. 7.5 clearly presents the shadow of an unstable quasi-periodic oscillation with the two nearby frequencies $\omega_{1,2}$ also shown in Fig. 7.2. This explains the clear beating effect in the first part of the signal.

As Fig. 7.5 also shows, the tool starts leaving the workpiece during its fourth round between the points $Q_1 - Q_2$ and $Q_3 - Q_4$ in the same way as explained in Fig. 7.3. From that point, the possibly chaotic oscillation leaves a fractal-like material surface behind in a similar way as explained for the co-dimension 1 Hopf bifurcation case at the notches of the stability lobes.

The measured surface of a workpiece in Fig. 7.5 hides an intricate infinite-dimensional phase-space structure for the co-dimension 2 Hopf bifurcation case at a certain peak T_j of the stability lobes. The governing equations (7.8) and (7.14) with the switching condition $h(t) = 0$ in Eq. (7.1) contain the time delay both in Eq. (7.8) and in Eq. (7.1). This time delay is responsible for the infinite-dimensional nature of the phase space (see details in [16, 17]).

In Fig. 7.6, the plane $(x, \dot{x})$, together with the axis x_∞ imitating all the other coordinates of infinite number, represents this infinite-dimensional phase space. It shows two saddle-like

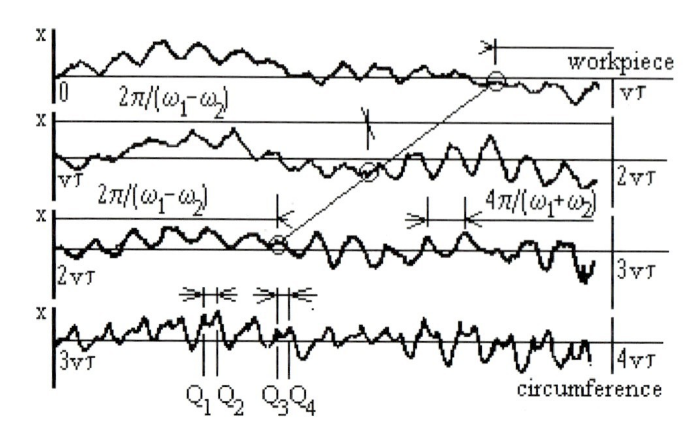

Figure 7.5: Measured surface of a machined workpiece.

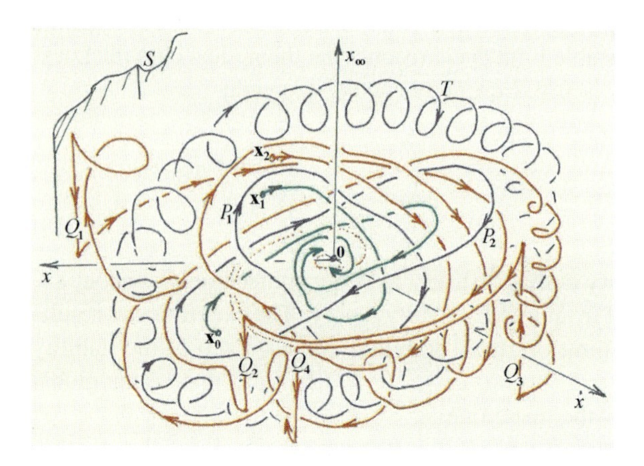

Figure 7.6: Self-interrupted cutting – global dynamics in infinite-dimensional phase space.

unstable periodic solutions P_1 and P_2. Their stable manifolds – embedded in the infinite-dimensional space – cross each other at the origin O only. It also shows a saddle-like unstable torus T that combines the frequencies of the two periodic solutions. The torus is covered by a quasi-periodic solution.

Solutions starting 'inside' the periodic solutions, like the ones with initial data $\mathbf{x}_0$ or $\mathbf{x}_1$, converge to O representing the stable stationary cutting. However, a trajectory starting 'outside' the periodic solutions and the torus, like the one starting at $\mathbf{x}_2$ close to P_1, gets involved in the dynamics of the limit cycle, spiralling 'outwards', then approaches the torus, gets involved in the quasi-periodic motion, then it leaves this torus, too, and spirals further 'outwards' while it hits the (non-stationary!) hyper-surface S. This surface represents the wavy workpiece surface here, and crossing this surface means that the tool leaves the workpiece.

The loss of contact results in a switch to the damped free flight of the tool starting at Q_1. Then the trajectory will continue to spiral towards the origin O in the phase plane $(x, \dot{x})$ in accordance with the damped, one degree-of-freedom oscillator (7.14). Before the trajectory could reach the origin, it will hit the (moving!) hyper-surface S at the point Q_2 – the tool enters the workpiece again. The trajectory is switched back to the infinite-dimensional phase space of regenerative chatter (7.8). Then the process is continued close to the unstable periodic motion P_1, then around the unstable torus T, then free flight between Q_3 and Q_4, and so on. The resulting motion is likely to be chaotic, with a chance of jumping back to the regenerative chatter 'inside' the unstable periodic and quasi-periodic motion. This could happen after a long chaotic transient, if it can happen at all.

The above-described phase-space structure represents a cutting process where the free flights of the tool are regulated by the cutting process itself. This is called self-interrupted cutting. High-speed milling has a qualitatively different global dynamics in this respect.

7.3 Nonlinear Vibrations of High-speed Milling

In case of milling, the number z of cutting edges is more than 1, and they are not in contact with the workpiece continuously, even when the stationary cutting is stable. The dimensionless governing equation of the mechanical model in Fig. 7.7 assumes the form

$$x''(\tilde{t}) + 2\varsigma x'(\tilde{t}) + (1 + \tilde{w}(\tilde{t}))x(\tilde{t}) - \tilde{w}(\tilde{t})x(\tilde{t} - \tilde{\tau}) = \text{higher-order terms.} \tag{7.16}$$

The dimensionless chip width varies periodically with a period equal to the dimensionless time delay $\tilde{\tau} = \tau\omega_{\mathrm{n}}$, that is

$$\tilde{w}(\tilde{t} + \tilde{\tau}) = \tilde{w}(\tilde{t}), \tag{7.17}$$

due to the periodically varying number (1 or 2 in Fig. 7.7) of cutting edges being in contact with the workpiece. The time delay τ is the tooth-pass period here, that is,

$$\tau = 2\pi/(z\Omega). \tag{7.18}$$

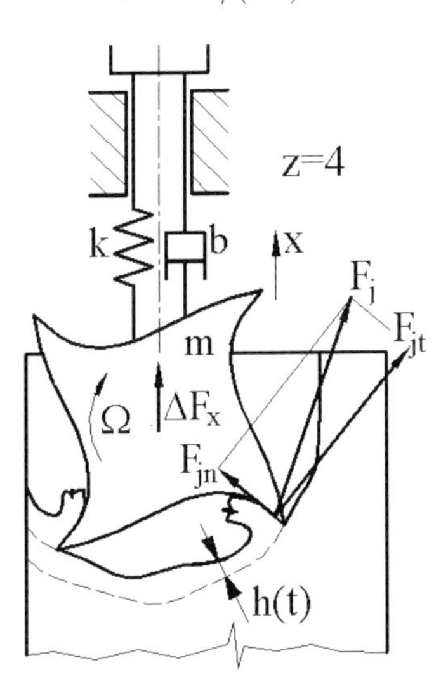

Figure 7.7: Mechanical model of milling.

The derivation of Eq. (7.16) can be found in [18]. The stability analysis of this delayed Mathieu-type equation is difficult due to the combination of the delay and the parametric excitation [19]. The model and analysis of high-speed milling is somewhat easier since the number of cutting edges is small ($z = 2$), the immersion is low, and consequently the ratio of the times spent cutting to not cutting is very small for each cutting edge. The corresponding mechanical model and its analysis are presented in the subsequent subsections.

7.3.1 Modeling of High-speed Milling

The mechanical model of low-immersion high-speed milling is presented in Fig. 7.8. The modal parameters of the machine tool are the same here as they are for turning (see Sect. 7.2.1, Fig. 7.1):

$$\omega_{\rm n} = \sqrt{k/m}, \quad \varsigma = b/(2m\omega_{\rm n}), \quad \omega_{\rm d} = \omega_{\rm n}\sqrt{1-\zeta^2}. \tag{7.19}$$

The cutting force characteristics are the same as in Fig. 7.1. The only difference between the models of turning and high-speed milling is the contact time between the tool and workpiece. In case of high-speed milling, the cutting is highly interrupted, that is, the contact periods are very short compared to the tooth-pass period τ in Eq. (7.18). Consequently, the tool edge spends most of the time in free flight.

The theoretical chip thickness is denoted by h_0, again. The actual chip thickness is either zero for no contact, or

$$h(t_j) = h_0 + x(t_j - \tau) - x(t_j), \tag{7.20}$$

where $t_j - \rho\tau$ is the initial time instant of the jth contact period between the tool and the workpiece ($j = 1, 2, \ldots$). We consider that the contact time $\rho\tau$ is so short ($\rho \ll 1$) that the position of the tool, and also the chip thickness, do not change during this period of time. This

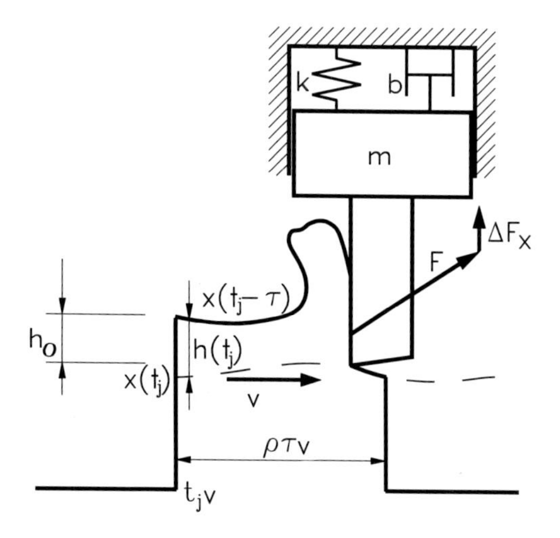

Figure 7.8: Mechanical model of high-speed milling.

approximation is deeply analyzed and justified in [11], and also confirmed experimentally in [12].

Thus, the equations of motion can be constructed for the two parts of the tool motion in the following way. For the free flight of the tool, we have

$$\ddot{x}(t) + 2\zeta\omega_{\mathrm{n}}\dot{x}(t) + \omega_{\mathrm{n}}^2 x(t) = 0\,, \quad t \in [t_j - \tau\,,\, t_j - \rho\tau) \tag{7.21}$$

with initial conditions

$$x_{j-1} = x(t_{j-1}) \approx x(t_{j-1} - \rho\tau), \quad v_{j-1} = \dot{x}(t_{j-1}), \tag{7.22}$$

where, again, we consider that the position of the tool does not change much during the short contact period. For the contact period, we have

$$m\ddot{x}(t) = -c\dot{x}(t) - kx(t) + F_x \approx F_x, \quad t \in [t_j - \rho\tau, t_j), \tag{7.23}$$

where the usual condition of the classical impact theory is applied: all the forces except the contact ones (actually, except the cutting force) are negligible. The initial conditions are as follows:

$$x_j = x(t_j - \rho\tau) \approx x(t_j), \quad v_j^- = \dot{x}(t_j - \rho\tau). \tag{7.24}$$

The nonlinear cutting force can be calculated in accordance with Eq. (7.2):

$$F_x = Kwh^{3/4} = Kw(h_0 + x(t-\tau) - x(t))^{3/4} \approx Kw(h_0 + x_{j-1} - x_j)^{3/4}, \tag{7.25}$$

where w denotes the chip width, again. The chip-thickness variation (7.4) simplifies to

$$\Delta h = h(t_j) - h_0 = x_{j-1} - x_j, \tag{7.26}$$

and the Taylor series (7.5) of the cutting force with respect to the chip-thickness assumes the form

$$F_x \approx F_{x0} + k_1(x_{j-1} - x_j) - \frac{1}{8h_0}k_1(x_{j-1} - x_j)^2 + \frac{5}{96h_0^2}k_1(x_{j-1} - x_j)^3, \tag{7.27}$$

where the stationary cutting force F_0 and the cutting coefficient k_1 are defined in Eq. (7.6).

The solution of the impact equation (7.23) for the time interval $\rho\tau$ assumes the form

$$m(v_j - v_j^-) = \rho\tau F_x, \tag{7.28}$$

that is

$$v_j = \frac{\rho\tau}{m}F_0 + \frac{\rho\tau}{m}k_1(x_{j-1} - x_j) + v_j^- - \frac{\rho\tau}{8mh_0}k_1(x_{j-1} - x_j)^2 + \frac{5\rho\tau}{96mh_0^2}k_1(x_{j-1} - x_j)^3, \tag{7.29}$$

where x_j and v_j^- can be calculated as a linear function of x_{j-1} and v_{j-1} by means of the well-known solution of the equation of motion (7.21) of the free damped oscillation of the tool with the initial conditions (7.22).

With the above-determined coefficients we can construct the nonlinear discrete model

$$\begin{bmatrix} x_j \\ v_j \end{bmatrix} = \mathbf{A} \begin{bmatrix} x_{j-1} \\ v_{j-1} \end{bmatrix} + \begin{bmatrix} 0 \\ \sum\limits_{h+k=2,3;\ h,k\geq 0} b_{hk} x_{j-1}^h v_{j-1}^k \end{bmatrix} + \begin{bmatrix} 0 \\ \frac{\rho\tau}{m} F_{x0} \end{bmatrix}, \quad (7.30)$$

where the coefficients in $\mathbf{A}$, and b_{hk} in the nonlinear part, are determined from the corresponding solutions of the series of free flights and impacts of the tool, that is, from Eqs. (7.21), (7.22) and (7.29). The linear part assumes the actual form

$$\mathbf{A} = \begin{bmatrix} \frac{e^{-\zeta\omega_n\tau}}{\sqrt{1-\zeta^2}} \cos(\omega_d\tau - \varepsilon) & \frac{e^{-\zeta\omega_n\tau}}{\sqrt{1-\zeta^2}} \sin(\omega_d\tau) \\ -\frac{\omega_n e^{-\zeta\omega_n\tau}}{\sqrt{1-\zeta^2}} \sin(\omega_d\tau) + \frac{\rho\tau k_1}{m}\left(1 - \frac{e^{-\zeta\omega_n\tau}}{\sqrt{1-\zeta^2}}\cos(\omega_d\tau - \varepsilon)\right) & \frac{e^{-\zeta\omega_n\tau}}{\sqrt{1-\zeta^2}}\left(\cos(\omega_d\tau + \varepsilon) - \frac{\rho\tau k_1}{m\omega_n}\sin(\omega_d\tau)\right) \end{bmatrix}. \quad (7.31)$$

In these formulae, the phase angle ε satisfies $\tan\varepsilon = \zeta/\sqrt{1-\zeta^2}$.

7.3.2 Bifurcation Analysis of High-speed Milling

In case of high-speed milling, the linear stability analysis of stationary cutting is based on the characteristic equation of the linear part of the difference equation (7.30):

$$\det(\lambda\mathbf{I} - \mathbf{A}) = 0. \quad (7.32)$$

In stable cases, the characteristic multipliers $\lambda_{1,2}$ are located in the open unit disk of the complex plane. The stability boundaries in the stability chart of Fig. 7.9 are calculated from the condition

$$|\lambda_{1,2}| = 1. \quad (7.33)$$

The analysis of Eq. (7.32) with Eq. (7.31) shows that there are two kinds of loss of stability: period-doubling (or flip) bifurcation occurs when $\lambda_1 = -1$ at

$$\tilde{w}\big|_{\text{cr}} = \frac{\rho\, k_1|_{\text{cr}}}{m\omega_n^2} = \frac{\sqrt{1-\varsigma^2}}{\omega_n\tau}\frac{\text{ch}(\zeta\omega_n\tau) + \cos(\omega_d\tau)}{\sin(\omega_d\tau)}, \quad \omega_n\tau = \frac{2\pi}{z\tilde{\Omega}}, \quad (7.34)$$

and Neimark–Sacker (or secondary Hopf) bifurcation occurs when $\lambda_{1,2} = \exp(\pm i\varphi)$ at

$$\tilde{w}\big|_{\text{cr}} = \frac{\rho\, k_1|_{\text{cr}}}{m\omega_n^2} = -2\frac{\sqrt{1-\varsigma^2}}{\omega_n\tau}\frac{\text{sh}(\zeta\omega_n\tau)}{\sin(\omega_d\tau)}, \quad \omega_n\tau = \frac{2\pi}{z\tilde{\Omega}}. \quad (7.35)$$

The bifurcations along the stability limits can be distinguished with the help of the vibration frequencies of the self-excited vibrations above the stability chart of Fig. 7.9 (the full structure of these frequencies is presented in [20] experimentally, too). The chart is constructed in the

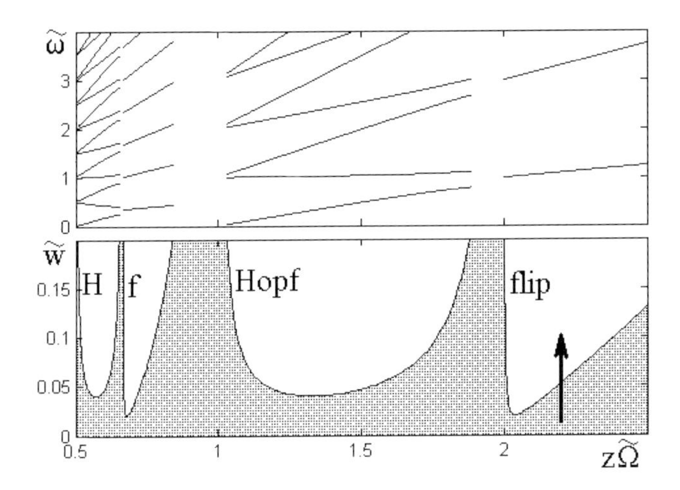

Figure 7.9: Stability chart of high-speed milling.

plane of the same parameters as in Fig. 7.2 with somewhat modified definitions related to the new parameters ρ and z:

$$z\tilde{\Omega} = z\frac{\Omega}{\omega_{\mathrm{n}}} = \frac{2\pi}{\tau\omega_{\mathrm{n}}}, \qquad \tilde{w} = \frac{\rho k_1}{k} = \frac{\rho k_1}{m\omega_{\mathrm{n}}^2} = \frac{3}{4}\frac{K\rho}{k\sqrt[4]{h_0}}w\,. \tag{7.36}$$

These parameters coincide with the ones in (7.9) and (7.10) defined for turning with $z = 1$ and $\rho = 1$.

Period-doubling Bifurcation. At the stability boundary (7.34) of period-doubling vibrations, the eigenvalues of the coefficient matrix $\mathbf{A}$ are

$$\lambda_1 = -1, \quad \lambda_2 = e^{-\zeta\omega_{\mathrm{n}}\tau}\left(\mathrm{sh}(\zeta\omega_{\mathrm{n}}\tau) + \cos(\omega_{\mathrm{d}}\tau)\right), \tag{7.37}$$

where $|\lambda_2| < 1$. With the help of the corresponding critical eigenvectors $\mathbf{s}_1$ and $\mathbf{s}_2$, the Jordan transformation matrix can be constructed in the form $\mathbf{T} = \begin{bmatrix} \mathbf{s}_1 & \mathbf{s}_2 \end{bmatrix}$ and the new variables ξ and η are introduced in accordance with col $[x\ y] = \mathbf{T}\,\mathrm{col}\,[\xi\ \eta]$:

$$\begin{bmatrix} x \\ y \end{bmatrix} = \begin{bmatrix} \sin(\omega_{\mathrm{d}}\tau) & \sin(\omega_{\mathrm{d}}\tau)/\omega_{\mathrm{n}} \\ -\omega_{\mathrm{n}}\cos(\omega_{\mathrm{d}}\tau - \varepsilon) - \omega_{\mathrm{d}}e^{\zeta\omega_{\mathrm{n}}\tau} & -\zeta\sin(\omega_{\mathrm{d}}\tau) + \sqrt{1-\zeta^2}\mathrm{sh}(\zeta\omega_{\mathrm{n}}\tau) \end{bmatrix} \begin{bmatrix} \xi \\ \eta \end{bmatrix}. \tag{7.38}$$

In this way, and also by using a constant shift transformation, system (7.30) can be transformed into

$$\begin{bmatrix} \xi_j \\ \eta_j \end{bmatrix} = \begin{bmatrix} -1 & 0 \\ 0 & \lambda_2 \end{bmatrix} \begin{bmatrix} \xi_{j-1} \\ \eta_{j-1} \end{bmatrix} + \begin{bmatrix} \sum\limits_{h+k=2,3;\ h,k\geq 0} c_{hk} \xi_{j-1}^h \eta_{j-1}^k \\ \sum\limits_{h+k=2,3;\ h,k\geq 0} d_{hk} \xi_{j-1}^h \eta_{j-1}^k \end{bmatrix}, \tag{7.39}$$

where λ_2 is given in Eq. (7.37), and the coefficients c_{hk}, d_{hk} of the nonlinear terms are obtained from the coefficients b_{hk} in Eq. (7.30) with the same linear transformation (7.38).

In Eq. (7.39), one can calculate the second-degree approximation of the center manifold in a much simpler way as in the case of the infinite-dimensional delayed model of turning. In this case, the center manifold is a curve in the (x, y) plane tangent to the subspace spanned by the eigenvector $\mathbf{s}_1$ corresponding to the critical eigenvalue $\lambda_1 = -1$, and tangent to the ξ axis in the (ξ, η) plane. The solutions of Eq. (7.39) converge to this manifold, since $|\lambda_2| < 1$ is true always for the other eigenvalue of the linear coefficient matrix $\mathbf{A}$ (see [21]). The second-order approximation of the center manifold simply reads

$$\eta = \frac{d_{20}}{1-\lambda_2}\xi^2 + \ldots, \tag{7.40}$$

where the parameter

$$d_{20} = -\frac{\omega_{\mathrm{n}}}{2h_0}\sin(\omega_{\mathrm{d}}\tau)\frac{\cos(\omega_{\mathrm{d}}\tau) + \mathrm{ch}(\zeta\omega_{\mathrm{n}}\tau)}{\cos(\omega_{\mathrm{d}}\tau) + \mathrm{ch}(\zeta\omega_{\mathrm{n}}\tau) + 2\mathrm{sh}(\zeta\omega_{\mathrm{n}}\tau)} \tag{7.41}$$

is the result of a lengthy algebraic calculation. The dynamics restricted to this center manifold is described by the scalar one-dimensional discrete dynamical system

$$\xi_j = -\xi_{j-1} + c_{20}\xi_{j-1}^2 + \left(c_{30} + c_{11}\frac{d_{20}}{1-\lambda_2}\right)\xi_{j-1}^3. \tag{7.42}$$

The second-degree term of Eq. (7.42) can be eliminated by means of a near-identity nonlinear transformation

$$\xi = u + \frac{c_{20}}{2}u^2 + \ldots \tag{7.43}$$

to obtain the normal form

$$u_{j+1} = (-1+\mu)\,u_j + \delta u_j^3, \tag{7.44}$$

with $\mu = 0$ for the unperturbed case. In the perturbed normal form (7.44), μ is a small positive perturbation of the critical characteristic multiplier $\lambda_1 = -1$. Then the sub- or supercritical nature of the period-two solution is determined by the parameter δ, since the amplitude of the emerging period-two vibration is estimated by

$$r = \sqrt{-\mu/\delta}. \tag{7.45}$$

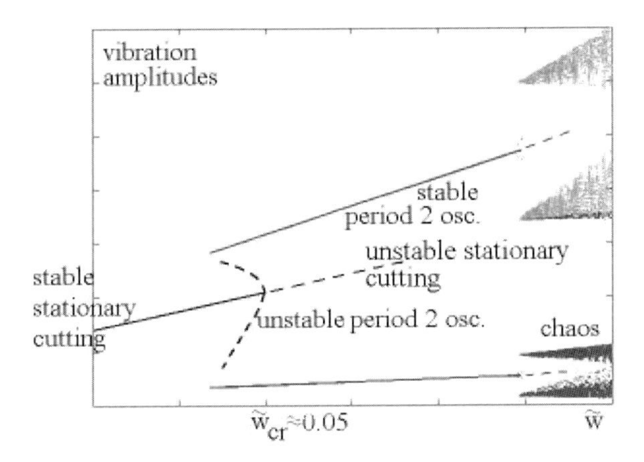

Figure 7.10: Bifurcation diagram and global dynamics of high-speed milling.

If $\delta > 0$ then the flip bifurcation is supercritical, if $\delta < 0$ then it is subcritical. The parameter δ (similar to the Poincaré–Ljapunov constant of the Hopf bifurcation) is the result of a long algebraic calculation that was also checked by computer algebraic software:

$$\delta = c_{20}^2 + c_{30} + \frac{c_{11}d_{20}}{1-\lambda_2} = -\frac{5}{12h_0^2}\frac{\sin^2(\omega_{\mathrm{d}}\tau)\,(\cos(\omega_{\mathrm{d}}\tau) + \mathrm{ch}(\varsigma\omega_{\mathrm{n}}\tau))}{\cos(\omega_{\mathrm{d}}\tau) + \mathrm{ch}(\varsigma\omega_{\mathrm{n}}\tau) + 2\mathrm{sh}(\varsigma\omega_{\mathrm{n}}\tau)} < 0. \quad (7.46)$$

Since the sign of δ is always negative, the flip bifurcation is *subcritical*, just as it is for turning. This means that an unstable period-two motion (limit cycle) exists around the stable period-one motion (stable fixed point of the iteration). The corresponding bifurcation diagram is shown in Fig. 7.10.

Neimark–Sacker Bifurcation. At the stability boundary (7.35) of quasi-periodic self-excited vibrations, the eigenvalues of the coefficient matrix **A** are complex conjugate:

$$\lambda_{1,2} = \exp(\pm i\varphi), \quad (7.47)$$

where $|\lambda_{1,2}| = 1$. With the help of the corresponding critical eigenvectors and the new variables ξ and η, system (7.30) can be transformed into

$$\begin{bmatrix} \xi_j \\ \eta_j \end{bmatrix} = \begin{bmatrix} \cos\varphi & \sin\varphi \\ -\sin\varphi & \cos\varphi \end{bmatrix} \begin{bmatrix} \xi_{j-1} \\ \eta_{j-1} \end{bmatrix} + \begin{bmatrix} \sum\limits_{h+k=2,3;\ h,k\geq 0} c_{hk}\xi_{j-1}^h \eta_{j-1}^k \\ \sum\limits_{h+k=2,3;\ h,k\geq 0} d_{hk}\xi_{j-1}^h \eta_{j-1}^k \end{bmatrix}. \quad (7.48)$$

The bifurcation calculation has the same steps as in the case of the flip bifurcation. There is no need for center manifold reduction in this case; the normal-form calculation is even more complicated, though. However, the final result is the same: the bifurcation is *subcritical*, again. The details of the lengthy algebraic work can be found in [22]. This means that unstable quasi-periodic motion exists around the stable stationary cutting in the neighborhood of the Neimark–Sacker-type stability limit.

7.3.3 Global Dynamics of Parametrically Interrupted Cutting

Since the bifurcations are subcritical either for the period-doubling bifurcations or for the secondary Hopf bifurcations, unstable period-two or unstable quasi-periodic oscillations exist around the otherwise stable stationary cutting in case of high-speed milling in a similar way as in the case of simple turning. The study of the self-interrupted cutting 'outside' the unstable motions in Sect. 7.2.3 helps to understand the global behavior of high-speed milling, too. A scenario, similar to the bifurcation diagram in Fig. 7.4, is described below in the neighborhood of a period-doubling bifurcation denoted by the thick arrow in Fig. 7.9.

It is likely that, 'outside' the unstable period-two oscillation, there is a region where the tool 'leaves' the workpiece, more exactly, it does not enter the workpiece at each revolution. For example, the existence of a stable period-two oscillation is expected where the tool enters the workpiece only at every second round, and it flies over the possible contact region in between. This means, for example, that a milling tool with two cutting edges will operate in a peculiar way: only one of the edges will enter the workpiece after each revolution of the tool, while the other edge will always fly without any contact with the workpiece.

Analytical calculations confirm this hypothesis at certain parameter regions. If a discrete mapping is constructed in a way that two free-flight periods described by Eqs. (7.21) and (7.22) are connected to one impact described by Eqs. (7.28) and (7.29), another nonlinear discrete mapping can be constructed, similar to Eq. (7.30). The analysis of this even more complicated mapping proves the existence of stable period-two oscillations outside the unstable period-two oscillations (see details in [22]). Further experimental study and observations related to these special vibrations are presented in [24].

Numerical simulations clearly confirm the analytical predictions. The simulation results also show that the tool does not reach the workpiece in each tooth-pass period when it operates in an unstable parameter regime. In Fig. 7.10, the bifurcation diagram is calculated numerically in the neighborhood of the flip bifurcation at $z\tilde{\Omega}_{\mathrm{cr}} \approx 2.2$, $\tilde{w}_{\mathrm{cr}} \approx 0.05$. The unstable period-two vibration is presented by a broken line as a result of the analytical predictions. They are also identified numerically by a simple kind of shooting method. The connection (the existence of a likely turning point) of the stable and unstable period-two oscillations has not been explored yet.

The path following of the stable period-two oscillation that can also be considered as a kind of parametrically interrupted oscillation shows that this motion can quickly bifurcate to *chaotic oscillations* at increasing values of the dimensionless chip thickness $\tilde{w}$ (see Fig. 7.10).

7.4 Conclusions

High-speed milling is usually combined with low immersion and a low number of cutting edges. Since the dynamics of milling contains parametric excitation due to the periodically varying number of active teeth, high-speed milling is very sensitive to this due to the small ratio of time spent cutting to not cutting. This parametric excitation is combined with the regenerative effect related to a time delay equal to the time period of the parametric excitation itself. While the dynamics of conventional turning contains the regenerative effect only, the combination with parametric excitation results in new ways of loss of stability for milling processes.

In case of high-speed milling, the mathematical model describes the new ways of loss of stability via period-two vibrations, and the model also describes the (secondary) Hopf bifurcations similar to the ones of turning. The local nonlinear analysis proves that all these bifurcations are subcritical due to the nonlinear cutting force characteristics either for turning or milling, either for the flip or for the Hopf case. The subcritical nature of the bifurcations explains for engineers why the local stability analysis provides limited results in practice, and why the otherwise linearly stable stationary cutting may be sensitive to slight perturbations.

The study of the global dynamics of cutting 'outside' the unstable motions shows that chaos can easily appear either for turning or high-speed milling. These chaotic oscillations are related to the loss of contact between the tool and the workpiece. While the loss of contact is self-regulated in case of turning, it is determined parametrically in case of high-speed milling. Both global vibrations might be chaotic, but there is a major difference that can be expressed with an extended terminology used for self-excited vibrations and parametrically excited vibrations. In case of unstable stationary cutting, or in case of large perturbations of slightly stable cutting processes, the global attractor of turning describes a kind of *self-interrupted cutting* process, while the global attractor of high-speed milling is either a period 2 or a chaotic oscillation representing a *parametrically interrupted cutting* process.

Bibliography

[1] J. Tlusty, A. Polacek, C. Danek, and J. Spacek, *Selbsterregte Schwingungen an Werkzeugmaschinen*, VEB Verlag Technik, Berlin, 1962.

[2] S.A. Tobias, *Machine Tool Vibration*, Blackie, London, 1965.

[3] G. Stépán, *Delay-differential equation models for machine tool chatter*, in: *Dynamics and Chaos in Manufacturing Processes*, edited by F.C. Moon, Wiley, New York, 1998.

[4] H.M. Shi and S.A. Tobias, International Journal of Machine Tool Design and Research **24** (1984), 45–69.

[5] G. Stépán and T. Kalmár-Nagy, in: Proceedings of the 1997 ASME Design Engineering Technical Conferences, Sacramento, California, paper no. DETC97/VIB-4021 (CD-ROM).

[6] T. Kalmár-Nagy, G. Stépán, and F.C. Moon, Nonlinear Dynamics **26** (2001), 121–142.

[7] J.R. Pratt and A.H. Nayfeh, Philosophical Transactions of the Royal Society **359** (2001), 759–792.

[8] D.E. Gilsinn, Nonlinear Dynamics **30** (2002), 103–154.

[9] A.H. Nayfeh, C.M. Chin, and J. Pratt, Journal of Manufacturing Science and Engineering **119** (1997), 485–493.

[10] G. Stépán, Philosophical Transactions of the Royal Society **359** (2001), 739–757.

[11] M.A. Davies, J.R. Pratt, B. Dutterer, and T.J. Burns, Journal of Manufacturing Science and Engineering **124**(2) (2002), 217–225.

[12] P.V. Bayly, J.E. Halley, B.P. Mann, and M.A. Davies, in: Proceedings of the ASME 2001 Design Engineering Technical Conferences, Pittsburgh, Pennsylvania, paper no. DETC2001/VIB-21581, 2001 (CD-ROM).

[13] T. Insperger and G. Stépán, Periodica Polytechnica, Mechanical Engineering **44**(1) (2000), 47–57.

[14] T. Insperger and G. Stépán, in: Proceedings of the Symposium on Nonlinear Dynamics and Stochastic Mechanics, Orlando, Florida, AMD-241 (2000), 119–123.

[15] B. Balachandran, Philosophical Transactions of the Royal Society **359** (2001), 793–820.

[16] J.K. Hale and S.M.V. Lunel, *Introduction to Functional Differential Equations*, Springer-Verlag, New York, 1993.

[17] G. Stépán, *Retarded Dynamical Systems*, Longman, Harlow, 1989.

[18] T. Insperger, B.P. Mann, G. Stépán and P.V. Bayly, International Journal of Machine Tools and Manufacture **43**(1) (2003), 25–34.

[19] T. Insperger and G. Stépán, Proceedings of The Royal Society, Mathematical, Physical and Engineering Sciences **458**(2024) (2002), 1989–1998.

[20] T. Insperger, G. Stépán, P.V. Bayly, and B.P. Mann, Journal of Sound and Vibration **262** (2003), 333–345.

[21] J. Guckenheimer and P. Holmes, *Nonlinear Oscillations, Dynamical Systems, and Bifurcations of Vector Fields*, Springer-Verlag, New York, 1983.

[22] R. Szalai, *Nonlinear Vibrations Of Interrupted Cutting Processes*, M.Sc. thesis, Budapest University of Technolgy and Economics, 2002.

[23] J. Tlusty, *Manufacturing Processes and Equipment*, Prentice Hall, New Jersey, 2000.

[24] G. Stépán, R. Szalai, B. Mann, P. Bayly, T. Insperger, J. Gradisek, and E. Govekar, in: Proceedings of the 2003 ASME Design Engineering Technical Conferences, Chicago, Illinois, paper no. DETC03/VIB-48572, 2003 (CD-ROM).

8 Mode-coupled Regenerative Machine Tool Vibrations

T. Kalmár-Nagy and F.C. Moon

In this paper a new three-degree-of-freedom lumped-parameter model for machine tool vibrations is developed and analyzed. One mode is shown to be stable and decoupled from the other two, and thus the stability of the system can be determined by analyzing these two modes. It is shown that this mode-coupled non-conservative cutting tool model including the regenerative effect (time delay) can produce an instability criterion that admits low-level or zero chip thickness chatter.

8.1 Introduction

One of the unsolved problems of metal cutting is the existence of low-level, random-looking (maybe chaotic) vibrations (or pre-chatter dynamics, see [17]). Some possible sources of this vibration are the elasto-plastic separation of the chip from the workpiece and the stick–slip friction of the chip over the tool. Recent papers of Davies and Burns [9], Wiercigroch and Krivtsov [43], Wiercigroch and Budak [41] and Moon and Kalmár-Nagy [27] have addressed some of these issues. Numerous researchers investigated single degree-of-freedom regenerative tool models [39, 13, 34, 11, 18, 28, 20, 36, 21, 38, 37]. Even though the classical model [39] with nonlinear cutting force is quite successful in predicting the onset of chatter [19], it cannot possibly account for all phenomena displayed in real cutting experiments. Single-degree-of-freedom deterministic time-delay models have been insufficient so far to explain low-amplitude dynamics below the stability boundary. Also, real tools have multiple degrees of freedom. In addition to horizontal and vertical displacements, tools can twist and bend. Higher-degree-of-freedom models have also been studied in turning, as well as in boring, milling and drilling [32, 2, 1, 44]. In this paper we will examine the coupling between multiple-degree-of-freedom tool dynamics and the regenerative effect in order to see if this chatter instability criterion will permit low-level instabilities.

Coupled-mode models in aeroelasticity or vehicle dynamics may exhibit so-called ‘flutter’ or dynamics instabilities (see e.g. [8]) when there exists a non-conservative force in the problem. One example is the follower force torsion-beam problem as in [15]. In the present work we assume that the chip-removal forces rotate with the tool, thereby introducing an unsymmetric stiffness matrix which can lead to flutter and chatter. Tobias [39] called this mode-coupled chatter. Often this model of chatter is analyzed without the regenerative effect. In this paper we will show that the combination of a mode-coupling non-conservative model and a time delay can produce an instability criterion that admits low-level or zero chip thickness chatter. There is no claim in this paper to having solved the random or chaotic low-level dynamics since only

Nonlinear Dynamics of Production Systems. Edited by G. Radons and R. Neugebauer

ISBN 3-527-40430-9

linear stability analysis is presented here. But the results shown below provide an incentive to extend this model into the nonlinear regime. A dynamic model with the combination of two-degree-of-freedom flutter model with time delay may also be applicable to aeroelastic problems in rotating machinery, where the fluid forces in the current cycle depend on eddies generated in the previous cycle. However, the focus of this paper is on the physics of cutting dynamics.

The structure of the paper is as follows. In Sect. 8.2 an overview of the turning operation is given, together with the description of chatter and the regenerative effect. The equations of motion are developed in Sect. 8.3. The model parameters are estimated in Sect. 8.4. Analysis of the model is performed in Sect. 8.5 and conclusions are drawn in Sect. 8.6.

8.2 Metal Cutting

The most common feature of machining operations (such as turning, milling and drilling) is the removal of a thin layer of material (the chip) from the workpiece using a wedge-shaped tool. They also involve relative motion between the workpiece and the tool. In turning the material is removed from a rotating workpiece, as shown in Fig. 8.1.

The cylindrical workpiece rotates with constant angular velocity Ω [rad/s] and the tool is moving along the axis of the workpiece at a constant rate. The feed f is the longitudinal displacement of the tool per revolution of the workpiece, and thus it is also the nominal chip thickness. The translational speed of the tool is then given by

$$v_{\text{tool}} = \frac{\Omega}{2\pi} f. \tag{8.1}$$

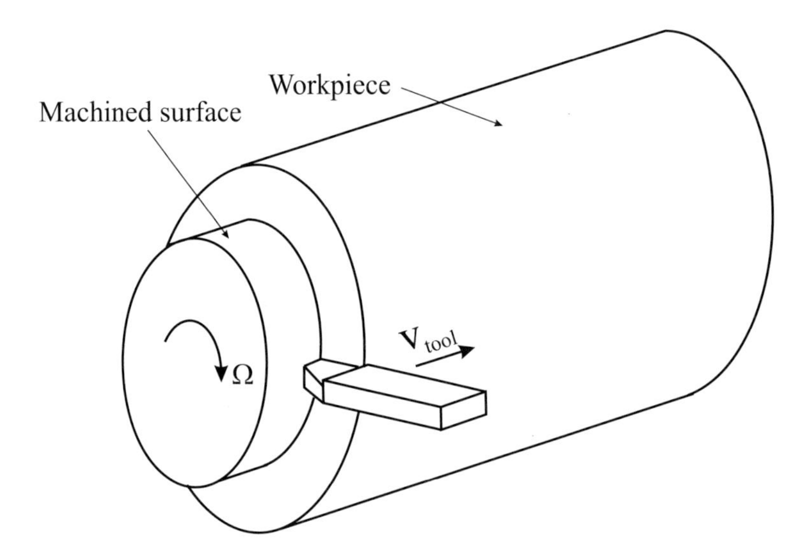

Figure 8.1: Turning.

The interaction between the workpiece and the tool gives rise to vibrations. One of the most important sources of vibrations in a cutting process is the *regenerative effect*. The present cut and the one made one revolution earlier might overlap, causing chip-thickness (and thus cutting force) variations. The associated time delay is the time period τ of one revolution of the workpiece:

$$\tau = \frac{2\pi}{\Omega}. \qquad (8.2)$$

The phenomenon of the large-amplitude vibration of the tool is known as *chatter*. A good description of chatter is given by Tobias [39], one of the pioneers of modern machine tool vibration research: "The machining of metal is often accompanied by a violent relative vibration between work and tool which is called the chatter. Chatter is undesirable because of its adverse effects on surface finish, machining accuracy, and tool life. Furthermore, chatter is also responsible for reducing output because, if no remedy can be found, metal removal rates have to be lowered until vibration-free performance is obtained."

Johnson [18] summarizes several qualitative features of tool vibration:

- The tool always appears to vibrate while cutting. The amplitude of the vibration distinguishes chatter from small-amplitude vibrations.
- The tool vibration typically has a strong periodic component which approximately coincides with a natural frequency of the tool.
- The amplitude of the oscillation is typically modulated and often in a random way. The amplitude modulation is present in both the chattering and non-chattering cases.

Tool vibrations can be categorized as self-excited vibrations [24, 25] or vibrations due to external sources of excitation (such as resonances of the machine structure) and can be periodic, quasiperiodic, chaotic or stochastic (or combinations thereof). A great deal of experimental work has been carried out in machining to characterize and quantify the dynamics of metal cutting. Recently, a number of researchers have provided experimental evidence that tool vibrations in turning may be chaotic [26, 5, 18, 3]. Other groups however now disavow the chaos theory for cutting and claim that the vibrations are random noise [42, 12].

8.2.1 Oblique Cutting

Although many practical machining processes can adequately be modeled as single-degree-of-freedom and orthogonal, more accurate models demand a chip-formation model in which the cutting velocity is not normal to the cutting edge. Figure 8.2 shows the usual oblique chip-formation model, where the inclination angle i (measured between the cutting edge and the normal to the cutting velocity in the plane of the machined surface) is not zero, as in orthogonal cutting. The cutting velocity is denoted by v_{C}, the chip flow angle is η_{c}, the thickness of the undeformed chip is f, the deformed chip thickness is f_2 and the chip width is w. The three-dimensional cutting force acting on the tool insert is decomposed into three mutually orthogonal forces: F_{C}, F_{T}, F_{R}. The cutting force F_{C} is the force in the cutting direction, the thrust force F_{T} is the force normal to the cutting direction and machined surface, while the radial force F_{R} is normal to both F_{C} and F_{T}. While orthogonal cutting can be modeled as a two-dimensional process, oblique cutting is a true three-dimensional plastic flow problem [30].

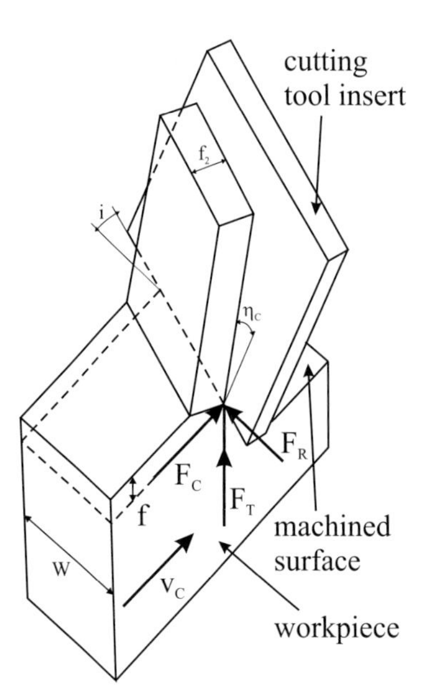

Figure 8.2: Oblique chip-formation model.

8.3 Three-degree-of-freedom Model of Metal Cutting

Figure 8.3 shows a tool with a cutting chip (insert) both in undeformed and deformed states of the tool. The three degrees of freedom are horizontal position (x), vertical position (z) and twist (ϕ). In the lumped-parameter model (Fig. 8.4) all the mass m of the beam is placed at its end (this effective mass is equivalent to modal mass for a distributed beam).

The equations of motion are the following:

$$m\ddot{z} + c_z\dot{z} + k_z z = F_z, \tag{8.3}$$

$$m\ddot{x} + c_x\dot{x} + k_x x = F_x, \tag{8.4}$$

$$I\ddot{\phi} + c_\phi\dot{\phi} + k_\phi\phi = M_y. \tag{8.5}$$

Figure 8.5 shows the forces acting on the tool tip.

As the tool bends about the x axis, the direction of the cutting velocity (and main cutting force) changes, as shown in Fig. 8.6. In order to derive the equations of motion, two coordinate systems are defined: an inertial frame $(\mathbf{I}, \mathbf{J}, \mathbf{K})$ fixed to the tool and a moving frame $(\mathbf{i}, \mathbf{j}, \mathbf{k})$ fixed to the cutting velocity. The force acting on the insert can then be written as

$$\mathbf{F} = -F_\mathrm{T}\mathbf{I} + F_\mathrm{R}\mathbf{J} - F_\mathrm{C}\mathbf{K} \tag{8.6}$$

or

$$\mathbf{F} = F_x\mathbf{i} + F_y\mathbf{j} + F_z\mathbf{k}, \tag{8.7}$$

where $\mathbf{i}$, $\mathbf{j}$, $\mathbf{k}$ are unit vectors in the x, y, z directions, respectively.

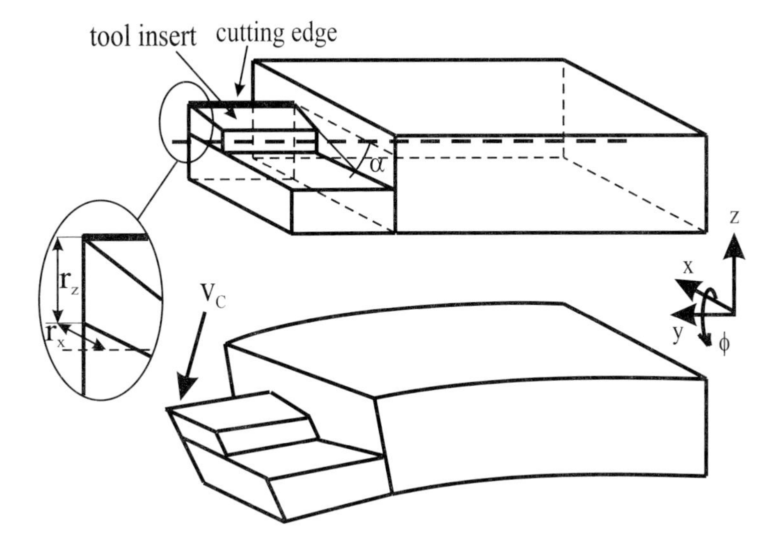

Figure 8.3: 3 DOF metal-cutting model.

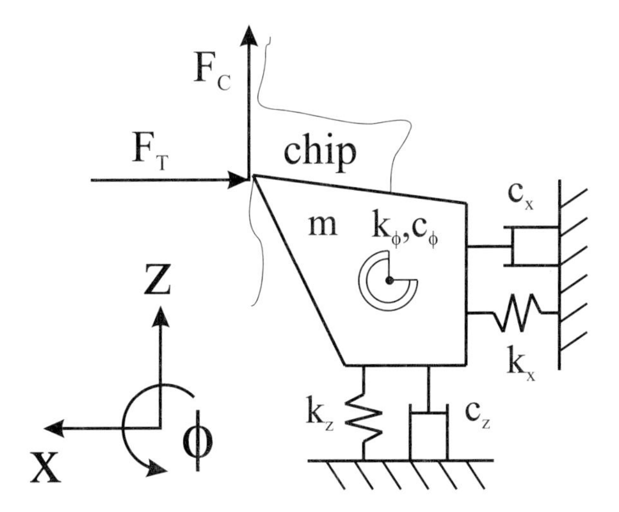

Figure 8.4: 3 DOF lumped-parameter model.

Here

$$F_x = -F_{\mathrm{T}}, \tag{8.8}$$
$$F_y = F_{\mathrm{C}} \sin\beta + F_{\mathrm{R}} \cos\beta, \tag{8.9}$$
$$F_z = F_{\mathrm{R}} \sin\beta - F_{\mathrm{C}} \cos\beta. \tag{8.10}$$

The bending also results in a pitch ψ (shown in Fig. 8.6). This is not a separate degree of freedom, but nonetheless it will influence the inclination angle.

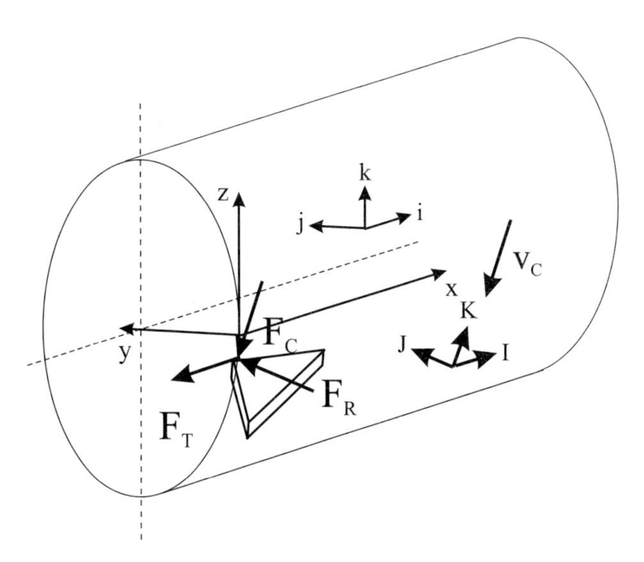

Figure 8.5: Forces on the tool tip.

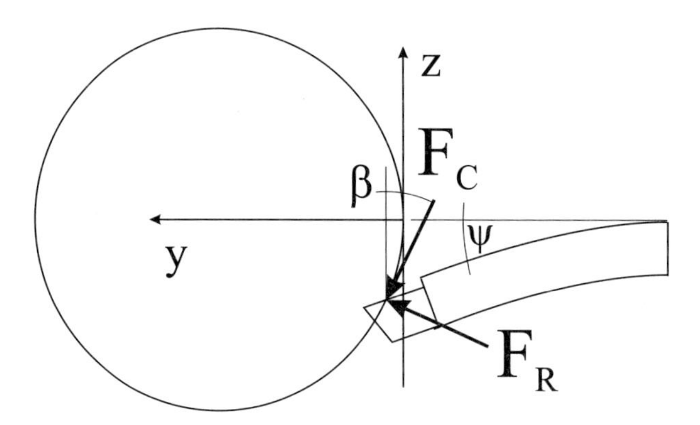

Figure 8.6: Direction of cutting velocity.

The following assumptions are used in deriving the equations of motion:

- The forces that act on the insert are steady-state forces.
- The width of cut w (y position) is constant.
- All displacements are small.
- Yaw is negligible.

Steady-state forces refer to time-averaged quantities. The effect of rate-dependent cutting forces was studied by Saravanja-Fabris and D'Souza [33], Chiriacescu [7] and Moon and Kalmár-Nagy [27]. Next we find the position of the tool tip in the fixed system of the platform.

To do so we have to find the rotation matrix $\mathbf{R}$ that describes the relationship between the moving frame $(\mathbf{i}, \mathbf{j}, \mathbf{k})$ and the fixed frame $(\mathbf{I}, \mathbf{J}, \mathbf{K})$:

$$\begin{pmatrix} \mathbf{i} & \mathbf{j} & \mathbf{k} \end{pmatrix} = \mathbf{R} \begin{pmatrix} \mathbf{I} & \mathbf{J} & \mathbf{K} \end{pmatrix}. \tag{8.11}$$

Using the Tait–Bryant angles $\{\psi, \phi\}$ we express $\mathbf{R}$ as a product of two consecutive planar rotations (pitch–roll system):

$$\mathbf{R} = \mathbf{R}_2 \mathbf{R}_1. \tag{8.12}$$

The cross section is first rotated about $\mathbf{I}$ by the pitch angle ψ. The corresponding rotation matrix is

$$\mathbf{R}_1 = \begin{pmatrix} 1 & 0 & 0 \\ 0 & \mathrm{c}\psi & \mathrm{s}\psi \\ 0 & -\mathrm{s}\psi & \mathrm{c}\psi \end{pmatrix}, \tag{8.13}$$

where the abbreviations $\mathrm{c} = \cos$, $\mathrm{s} = \sin$ were used. The second rotation is about the $\mathbf{J}_2$ (the rotated $\mathbf{J}$) axis through the roll angle ϕ (with respect to the toolholder):

$$\mathbf{R}_2 = \begin{pmatrix} \mathrm{c}\phi & 0 & \mathrm{s}\phi \\ 0 & 1 & 0 \\ -\mathrm{s}\phi & 0 & \mathrm{c}\phi \end{pmatrix}. \tag{8.14}$$

$\mathbf{R}$ can then be calculated by Eq. (8.12):

$$\mathbf{R} = \begin{pmatrix} \mathrm{c}\phi & -\mathrm{s}\phi\mathrm{s}\psi & \mathrm{c}\psi\mathrm{s}\phi \\ 0 & \mathrm{c}\psi & \mathrm{s}\psi \\ -\mathrm{s}\phi & -\mathrm{c}\phi\mathrm{s}\psi & \mathrm{c}\psi\mathrm{c}\phi \end{pmatrix}. \tag{8.15}$$

The position of the tool tip can be expressed in the fixed frame as

$$\mathbf{r}^* = \mathbf{R} \begin{pmatrix} r_x \\ 0 \\ r_z \end{pmatrix} = \begin{pmatrix} r_x\mathrm{c}\phi + r_z\mathrm{c}\psi\mathrm{s}\phi \\ r_z\mathrm{s}\psi \\ r_z\mathrm{c}\psi\mathrm{c}\phi - r_x\mathrm{s}\phi \end{pmatrix}. \tag{8.16}$$

The roll-producing moment can then be calculated as

$$\begin{aligned} M_y &= (\mathbf{r}^* \times \mathbf{F}) \cdot \mathbf{j} \\ &= F_{\mathrm{T}}\left(r_x\mathrm{s}\phi - r_z\mathrm{c}\phi\mathrm{c}\psi\right) + F_{\mathrm{C}}\mathrm{c}\beta\left(r_x\mathrm{c}\phi + r_z\mathrm{c}\psi\mathrm{s}\phi\right) - F_{\mathrm{R}}\mathrm{s}\beta\left(r_x\mathrm{c}\phi + r_z\mathrm{c}\psi\mathrm{s}\phi\right). \end{aligned} \tag{8.17}$$

In the following we assume small displacements and small angles and neglect nonlinear terms. The angle β is taken to be proportional to the vertical displacement, i.e. $\beta = -nz$ $(n > 0)$ and so is the pitch, i.e. $\psi = kz$ $(k > 0)$.

$$m\ddot{x} + c_x\dot{x} + k_x x = -F_{\mathrm{T}}, \tag{8.18}$$

$$m\ddot{z} + c_z\dot{z} + k_z z = -\left(F_{\mathrm{C}} + nz\bar{F}_{\mathrm{R}}\right), \tag{8.19}$$

$$I\ddot{\phi} + c_\phi\dot{\phi} + k_\phi\phi = M_y = \phi\left(r_x\bar{F}_{\mathrm{T}} + r_z\bar{F}_{\mathrm{C}}\right) - r_z F_{\mathrm{T}} + r_x F_{\mathrm{C}} + nzr_x\bar{F}_{\mathrm{R}}, \tag{8.20}$$

where $\bar{F}_C$, $\bar{F}_R$ and $\bar{F}_T$ denote the constant terms in F_{C}, F_{R} and F_{T}, respectively.

8.3.1 Cutting Forces

Generally we assume that the cutting forces F_{C}, F_{T}, F_{R} depend only on the inclination angle i and chip thickness f (see Fig. 8.2) and the rake angle α (see Fig. 8.3). We again emphasize that the chip width w is considered constant in the present analysis. Our hypothesis here is that F_{C} and F_{T} depend linearly on both the rake angle and chip thickness (see Sect. 8.4.2) in the following manner:

$$F_{\mathrm{C}} = -l_{\mathrm{C}}\alpha + m_{\mathrm{C}}t_1 + F_{\mathrm{C0}}, \tag{8.21}$$

$$F_{\mathrm{T}} = -l_{\mathrm{T}}\alpha + m_{\mathrm{T}}t_1 + F_{\mathrm{T0}}, \tag{8.22}$$

where m_{C} and m_{T} are cutting force coefficients, while l_{C} and l_{T} are angular cutting force coefficients (they show how strong the force dependence is on rake angle). The variable t_1 is the chip thickness variation (the deviation from the nominal chip thickness). The constant forces F_{C0} and F_{T0} arise from cutting at a nominal chip thickness. The radial cutting force can be expressed as [30]

$$F_{\mathrm{R}} = \sin i \frac{F_{\mathrm{C}} \cos i \,(i - \sin\alpha) - F_{\mathrm{T}}}{\sin^2 i \sin\alpha + \cos^2 i}, \tag{8.23}$$

where Stabler's flow rule [35] $\eta_{\mathrm{C}} = i$ was used. The effective rake angle depends on the initial rake angle and the roll:

$$\alpha = \alpha_0 - \phi, \tag{8.24}$$

while the inclination angle will depend on the initial inclination angle (i_0) as well as the pitch:

$$i = i_0 - \psi. \tag{8.25}$$

The chip thickness depends on the nominal feed and the position of the tool tip (both the present and the delayed ones). The displacement of the tool tip is due to translational and rotational motion as shown in Fig. 8.7. Here the dashed line corresponds to the position vector of the tool tip in the undeformed configuration, while the solid line depicts how this vector rotates (ϕ) and translates (due to the displacements x and z). The chip thickness is then given by

$$t_1 = t_{10} + x - x_\tau + r_z \sin(\phi - \phi_\tau) \approx t_{10} + x - x_\tau + r_z (\phi - \phi_\tau), \tag{8.26}$$

where x_τ and ϕ_τ denote the delayed values $x(t-\tau)$ and $\phi(t-\tau)$, respectively. Then the cutting forces can be written as

$$F_{\mathrm{C}} = m_{\mathrm{C}}(x - x_\tau) + (l_{\mathrm{C}} + r_z m_{\mathrm{C}})\phi - r_z m_{\mathrm{C}}\phi_\tau + \overbrace{m_{\mathrm{C}}t_{10} + F_{\mathrm{C0}} - \alpha_0 l_{\mathrm{C}}}^{\bar{F}_{\mathrm{C}}}, \tag{8.27}$$

$$F_{\mathrm{T}} = m_{\mathrm{T}}(x - x_\tau) + (l_{\mathrm{T}} + r_z m_{\mathrm{T}})\phi - r_z m_{\mathrm{T}}\phi_\tau + m_{\mathrm{T}}t_{10} + F_{\mathrm{T0}} - \alpha_0 l_{\mathrm{T}}. \tag{8.28}$$

If the initial inclination angle is assumed to be zero, the expression for F_{R} will simplify to

$$F_{\mathrm{R}} = k\left(\bar{F}_{\mathrm{T}} + (\sin\alpha_0 - 1)\,\bar{F}_{\mathrm{C}} + t_{10}\,(m_{\mathrm{C}}(1 - \sin\alpha_0) - m_{\mathrm{T}})\right) z. \tag{8.29}$$

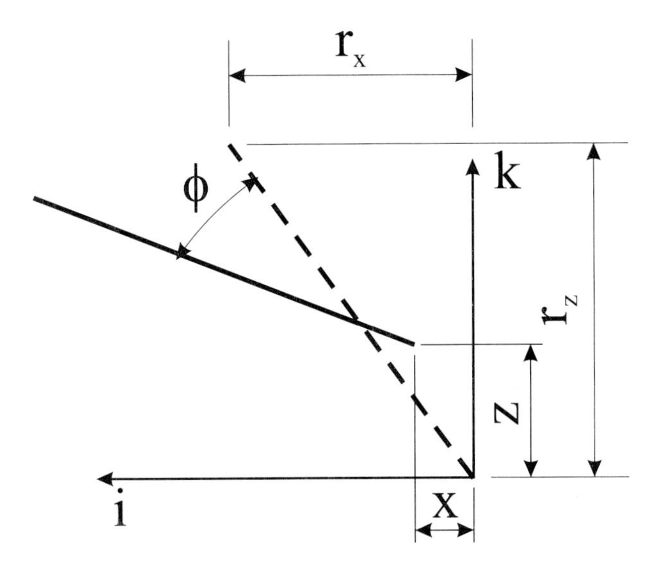

Figure 8.7: Motion of the tool tip.

8.3.2 The Equations of Motion

Substituting Eqs. (8.27) and (8.28) into Eqs. (8.18)–(8.20) and eliminating the constant (by translation of the variables) results in

$$m\ddot{z} + c_z\dot{z} + k_z z = -n\bar{F}_{\mathrm{R}} z - m_{\mathrm{C}}\left(x - x_\tau\right) - \left(l_{\mathrm{C}} + r_z m_{\mathrm{C}}\right)\phi + r_z m_{\mathrm{C}}\phi_\tau, \tag{8.30}$$

$$m\ddot{x} + c_x\dot{x} + k_x x = -m_{\mathrm{T}}\left(x - x_\tau\right) - \left(l_{\mathrm{T}} + r_z m_{\mathrm{T}}\right)\phi + r_z m_{\mathrm{T}}\phi_\tau, \tag{8.31}$$

$$\begin{aligned} I\ddot{\phi} + c_\phi\dot{\phi} + k_\phi\phi &= -r_z m_{\mathrm{T}}\left(x - x_\tau\right) \\ &- r_z\left(l_{\mathrm{T}} + r_z m_{\mathrm{T}} - m_{\mathrm{C}} t_{10} - F_{\mathrm{C}0} + \alpha_0 l_{\mathrm{C}}\right)\phi \\ &+ r_z^2 m_{\mathrm{T}}\phi_\tau, \end{aligned} \tag{8.32}$$

where now $(x,\, z,\, \phi)$ represent deviations from the steady values of the original displacements. As we can see, the x and ϕ equations are uncoupled from the z equation, so the stability of the system is determined by Eqs. (8.31), (8.32). Equations (8.31), (8.32) can also be written as

$$\ddot{x} + 2\zeta_x\omega_x\dot{x} + \left(\omega_x^2 + \frac{m_{\mathrm{T}}}{m}\right)x + \frac{1}{m}\left(l_{\mathrm{T}} + r_z m_{\mathrm{T}}\right)\phi = \frac{m_{\mathrm{T}}}{m}x_\tau + r_z\frac{m_{\mathrm{T}}}{m}\phi_\tau, \tag{8.33}$$

$$\begin{aligned} &\ddot{\phi} + 2\zeta_\phi\omega_\phi\dot{\phi} + \frac{r_z m_{\mathrm{T}}}{I}x + \left(\omega_\phi^2 + \frac{r_z}{I}\left(l_{\mathrm{T}} + r_z m_{\mathrm{T}} - m_{\mathrm{C}} t_{10} - F_{\mathrm{C}0} + \alpha_0 l_{\mathrm{C}}\right)\right)\phi \\ &= \frac{r_z m_{\mathrm{T}}}{I}x_\tau + r_z^2\frac{m_{\mathrm{T}}}{I}\phi_\tau, \end{aligned} \tag{8.34}$$

where

$$\omega_x = \sqrt{\frac{k_x}{m}}, \qquad \omega_\phi = \sqrt{\frac{k_\phi}{I}}. \tag{8.35}$$

By introducing the non-dimensional time and displacement

$$\hat{t} = t/T, \qquad \hat{x} = x/X, \tag{8.36}$$

we obtain

$$\begin{aligned}
&\hat{x}'' + 2\zeta_x\omega_x T\hat{x}' + \left(\omega_x^2 + \frac{m_{\mathrm{T}}}{m}\right)T^2\hat{x} + \frac{1}{m}\left(l_{\mathrm{T}} + r_z m_{\mathrm{T}}\right)\frac{T^2}{X}\phi \\
&= \frac{m_{\mathrm{T}}}{m}T^2 x_\tau + r_z\frac{m_{\mathrm{T}}}{m}\frac{T^2}{X}\phi_\tau,
\end{aligned} \tag{8.37}$$

$$\begin{aligned}
&\phi'' + 2\zeta_\phi\omega_\phi T\phi' + \frac{r m_{\mathrm{T}}}{I}T^2 X\hat{x} \\
&+ \left[\omega_\phi^2 + \frac{r_z}{I}\left(l_{\mathrm{T}} + r_z m_{\mathrm{T}} - F_{\mathrm{C}0} - m_{\mathrm{C}}t_{10} + \alpha_0 l_{\mathrm{C}}\right)\right]T^2\phi \\
&= \frac{r_z m_{\mathrm{T}}}{I}T^2 X x_\tau + r_z^2\frac{m_{\mathrm{T}}}{I}T^2\phi_\tau.
\end{aligned} \tag{8.38}$$

With the choice of the following scales

$$T = \frac{1}{\omega_x}, \qquad X = \sqrt{\frac{I}{m}}, \tag{8.39}$$

the equations assume the form ($\hat{\tau} = \omega_x\tau$)

$$\hat{x}'' + 2\zeta_x\hat{x}' + k_{11}\hat{x} + k_{12}\phi = r_{11}\hat{x}_{\hat{\tau}} + r_{12}\phi_{\hat{\tau}}, \tag{8.40}$$

$$\phi'' + 2\hat{\zeta}_\phi\phi' + k_{21}\hat{x} + k_{22}\phi = r_{21}\hat{x}_{\hat{\tau}} + r_{22}\phi_{\hat{\tau}}, \tag{8.41}$$

where

$$k_{11} = 1 + \frac{m_{\mathrm{T}}}{\omega_x^2 m}, \qquad k_{12} = \frac{l_{\mathrm{T}} + r_z m_{\mathrm{T}}}{\omega_x^2\sqrt{Im}}, \tag{8.42}$$

$$k_{21} = \frac{r_z m_{\mathrm{T}}}{\omega_x^2\sqrt{Im}}, \qquad \hat{\zeta}_\phi = \zeta_\phi\frac{\omega_\phi}{\omega_x}, \tag{8.43}$$

$$k_{22} = \left(\frac{\omega_\phi}{\omega_x}\right)^2 + \frac{r_z}{\omega_x^2 I}\left(l_{\mathrm{T}} + r_z m_{\mathrm{T}} - m_{\mathrm{C}}t_{10} - F_{\mathrm{C}0} + \alpha_0 l_{\mathrm{C}}\right), \tag{8.44}$$

$$r_{11} = \frac{m_{\mathrm{T}}}{\omega_x^2 m}, \qquad r_{12} = r_{21} = \frac{r_z m_{\mathrm{T}}}{\omega_x^2\sqrt{Im}}, \tag{8.45}$$

$$r_{22} = \frac{r_z^2 m_{\mathrm{T}}}{\omega_x^2\sqrt{Im}}. \tag{8.46}$$

Note that the stiffnesses k_{12} and k_{21} are different. This is characteristic of non-conservative systems [4, 31]. In many mechanical systems this non-conservativeness is due to the presence of following forces.

8.4 Estimation of Model Parameters

In the following we estimate different terms in Eqs. (8.42)–(8.46) to establish their relative strengths in order to simplify the model.

8.4.1 Structural Parameters

The toolholder is assumed to be a rectangular steel beam. The length of the toolholder is relatively short for normal cutting, while it can be longer for boring operations [22]. So we assume l to be between 0.05 m and 0.3 m. The width and height are usually of order of a centimeter. The stiffnesses for such a cantilevered beam can be in the following ranges:

$$k_x \simeq 10^4\text{–}10^7 \ \frac{\text{N}}{\text{m}}, \tag{8.47}$$

$$k_z \simeq 10^5\text{–}10^7 \ \frac{\text{N}}{\text{m}}, \tag{8.48}$$

$$k_\phi \simeq 1000\text{–}10\,000 \ \frac{\text{N}}{\text{rad}}. \tag{8.49}$$

Since a lumped-parameter approximation is used, the mass at the end of the massless beam is assumed to be the modal mass. The vibration frequencies are then

$$\omega_x \simeq 100\text{–}5000 \ \frac{\text{rad}}{\text{s}}, \tag{8.50}$$

$$\omega_z \simeq 100\text{–}10\,000 \ \frac{\text{rad}}{\text{s}}, \tag{8.51}$$

$$\omega_\phi \simeq 1000\text{–}10\,000 \ \frac{\text{rad}}{\text{s}}. \tag{8.52}$$

The ratio ω_ϕ/ω_x varies between 2 and 10 (the shorter the tool the higher the ratio).

8.4.2 Cutting Force Parameters

Experimental cutting force data during machining of 0.2% carbon steel is shown in Fig. 8.8 [30]. The graph shows the forces F_{C} and F_{T} for different rake angles ($\alpha = -5^\circ$ and 5° for top and bottom figures, respectively). The width of cut and chip thickness were 4 mm and 0.25 mm, respectively. Since our model assumes constant cutting speed, forces were taken from these graphs at the value 200 m/s of the cutting speed and plotted against rake angle (Fig. 8.9). The constants l_{C} and l_{T} were found as the slopes of the lines corresponding to $t_1 = 0.25$ mm:

$$l_{\text{C}} = 1580\frac{\text{N}}{\text{rad}}, \qquad l_{\text{T}} = 3150\frac{\text{N}}{\text{rad}}. \tag{8.53}$$

A linear relationship is assumed between forces at zero rake angle and chip thickness, i.e.

$$F_{\text{C}} = F_{\text{C0}} + m_{\text{C}} t_1, \tag{8.54}$$

$$F_{\text{T}} = F_{\text{T0}} + m_{\text{T}} t_1, \tag{8.55}$$

where these coefficients were determined to be

$$m_{\text{C}} = 6 \times 10^6 \frac{\text{N}}{\text{m}}, \quad F_{\text{C0}} = 458 \text{ N}, \tag{8.56}$$

$$m_{\text{T}} = 1.65 \times 10^6 \frac{\text{N}}{\text{m}}, \quad F_{\text{T0}} = 784 \text{ N}. \tag{8.57}$$

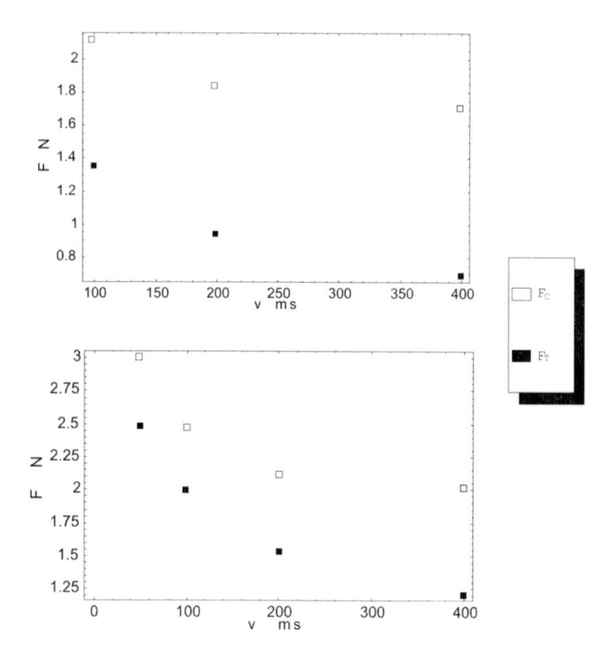

Figure 8.8: Forces in oblique cutting of 0.2% carbon steel. $\alpha = -5^\circ$ (top) and $\alpha = 5^\circ$ (bottom). $f = 0.125$ mm. After [30].

8.4.3 Model Parameters

Since

$$\frac{r_z}{\omega_x^2 I}\left(l_\mathrm{T} + r_z m_\mathrm{T} - m_\mathrm{C} t_{10} - F_{\mathrm{C}0} + \alpha_0 l_\mathrm{C}\right) \ll \left(\frac{\omega_\phi}{\omega_x}\right)^2 \tag{8.58}$$

this term will be neglected, i.e.

$$k_{22} = \left(\frac{\omega_\phi}{\omega_x}\right)^2. \tag{8.59}$$

Also, the term r_{22} is very small, so it is neglected:

$$r_{22} = 0. \tag{8.60}$$

8.5 Analysis of the Model

With the approximations (8.59), (8.60) the model (8.40), (8.41) can be written as the matrix equation

$$\ddot{\mathbf{x}} + \mathbf{C}\dot{\mathbf{x}} + \mathbf{K}\mathbf{x} = \mathbf{R}\mathbf{x}_\tau, \tag{8.61}$$

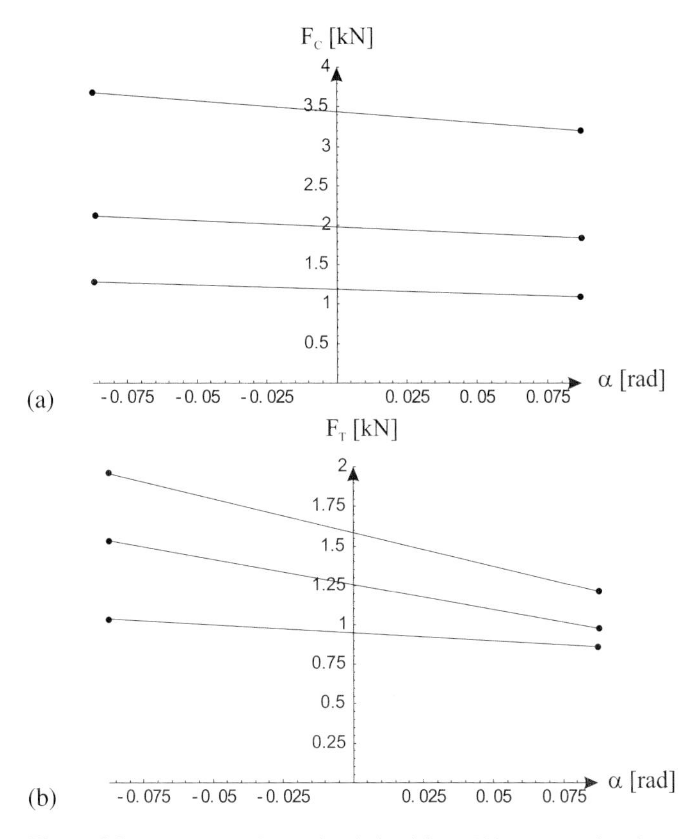

Figure 8.9: Forces vs. rake angle (derived from [30]). (a) cutting force, (b) thrust force.

where

$$\mathbf{x} = \begin{pmatrix} \hat{x} \\ \phi \end{pmatrix}$$

and the matrices are given by

$$\mathbf{C} = \begin{bmatrix} 2\zeta_x & 0 \\ 0 & 2\zeta_\phi \end{bmatrix}, \qquad \mathbf{K} = \begin{bmatrix} 1+p & a+pq \\ pq & k_{22} \end{bmatrix}, \tag{8.62}$$

$$\mathbf{R} = \begin{bmatrix} p & q \\ q & 0 \end{bmatrix}. \tag{8.63}$$

Here we introduced the parameters

$$p = \frac{m_\mathrm{T}}{\omega_x^2 m}, \qquad q = \frac{r_z}{X} = r_z\sqrt{\frac{m}{I}}, \tag{8.64}$$

where the constants a and k_{22} are

$$a = \frac{l_\mathrm{T}}{\omega_x^2\sqrt{Im}}, \qquad k_{22} = \left(\frac{\omega_\phi}{\omega_x}\right)^2. \tag{8.65}$$

It is characteristic of systems with non-symmetric stiffness matrix that they can lose stability either by divergence (buckling) or by flutter. Chu and Moon [8] examined divergence and flutter instabilities in magnetically levitated models. Kiusalaas and Davis [23] studied stability of elastic systems under retarded follower forces. Recently, several numerical methods were proposed to investigate stability of linear time-delay systems (see [6, 10, 16, 29]).

8.5.1 Classical Limit

If $q = 0$ the equations reduce to

$$x'' + 2\zeta_x x' + (1+p)\,x + a\phi = px_\tau, \tag{8.66}$$

$$\phi'' + 2\zeta_\phi \phi' + c\phi = 0. \tag{8.67}$$

The ϕ equation is uncoupled from the x equation and reduces to that of a damped oscillator. Its equilibrium $\phi = 0$ is asymptotically stable and thus it does not affect the stability of the x equation. In this case we recover the one-degree-of-freedom classical model [40].

8.5.2 Stability Analysis of the Undamped System Without Delay

First we perform linear stability analysis of the system

$$\ddot{\mathbf{x}} + \mathbf{K}\mathbf{x} = \mathbf{0}, \tag{8.68}$$

where the matrix $\mathbf{K}$ is non-symmetric and of the form ($k_{22} > 0$)

$$\mathbf{K} = \begin{bmatrix} k_{11} & k_{12} \\ k_{21} & k_{22} \end{bmatrix}. \tag{8.69}$$

Assuming the solutions in the form

$$\mathbf{x} = \mathbf{d}e^{i\omega t}, \tag{8.70}$$

we obtain the characteristic polynomials

$$\left(\mathbf{K} - \omega^2 \mathbf{I}\right)\mathbf{d} = 0, \tag{8.71}$$

which have non-trivial solution if the determinant of $\mathbf{K} - \omega^2\mathbf{I}$ is zero:

$$\begin{vmatrix} k_{11} - \omega^2 & k_{12} \\ k_{21} & k_{22} - \omega^2 \end{vmatrix} = \left(k_{11} - \omega^2\right)\left(k_{22} - \omega^2\right) - k_{21}k_{12} = 0. \tag{8.72}$$

The characteristic equation for the coupled system becomes

$$\omega^4 - (k_{11} + k_{22})\,\omega^2 + k_{11}k_{22} - k_{21}k_{12} = 0. \tag{8.73}$$

Divergence (static deflection, buckling) occurs when $\omega = 0$ (or $\det \mathbf{K} = 0$), that is when

$$k_{11}k_{22} - k_{21}k_{12} = 0. \tag{8.74}$$

If $\omega \neq 0$, then the characteristic equation (8.73) can be solved for ω^2 as

$$\omega^2 = \frac{1}{2}\left(k_{11} + k_{22} \pm \sqrt{(k_{11} + k_{22})^2 - 4\left(k_{11}k_{22} - k_{21}k_{12}\right)}\right). \tag{8.75}$$

For stable solutions, both solutions should be positive. Since $k_{22} > 0$, this is the case if

$$0 \leq k_{11}k_{22} - k_{21}k_{12} \leq \left(\frac{k_{11} + k_{22}}{2}\right)^2. \tag{8.76}$$

The two bounds correspond to divergence and flutter boundaries, respectively. With the stiffness matrix in Eq. (8.62),

$$k_{11} = 1 + p, \qquad k_{12} = a + pq, \tag{8.77}$$

$$k_{21} = pq. \tag{8.78}$$

In the plane of the bifurcation parameters q, p the divergence boundaries are given by

$$p = \frac{1}{2q^2}\left(k_{22} - aq \pm \sqrt{4k_{22}q^2 + (k_{22} - aq)^2}\right) \tag{8.79}$$

and the flutter boundary is characterized by

$$p = \frac{1}{1 + 4q^2} \times \left(k_{22} - 2aq - 1 \pm 2\sqrt{q\left(a\left(1 - k_{22}\right) + a^2 q - q\left(k_{22} - 1\right)^2\right)}\right). \tag{8.80}$$

Figure 8.10 shows these boundaries in the (q, p) parameter plane for $a = 1$, $k_{22} = 2$. The different stability regions are indicated by the root location plots.

8.5.3 Stability Analysis of the Two-degree-of-freedom Model with Delay

In this section we include the delay terms in the analysis. In order to be able to study how these terms influence the stability of the system, we introduce a new parameter, similar to the overlap factor [39].

First we analyze the system with no damping:

$$\ddot{\mathbf{x}} + \mathbf{K}\mathbf{x} = \mu \mathbf{R}\mathbf{x}_\tau. \tag{8.81}$$

When $\mu = 0$ we recover the previously studied Eq. (8.68), while $\mu = 1$ corresponds to Eq. (8.61) without damping. The characteristic equation is

$$\det\left(-\lambda^2 \mathbf{I} + \mathbf{K} - \mu e^{-\lambda\tau}\mathbf{R}\right) = 0, \tag{8.82}$$

or

$$\lambda^4 + \left(k_{11} + k_{22} - \mu p e^{-\lambda\tau}\right)\lambda^2 + k_{11}k_{22} - k_{12}k_{21} + \mu e^{-\lambda\tau}\left(q\left(k_{12} + k_{21}\right) - pk_{22}\right) - \mu^2 q^2 e^{-2\lambda\tau} = 0. \tag{8.83}$$

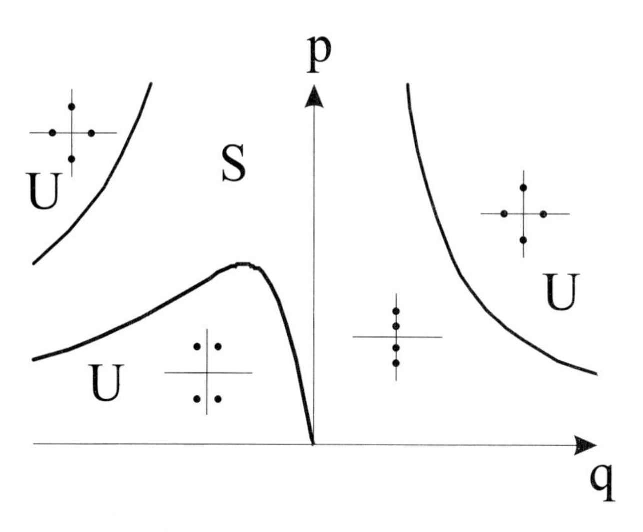

Figure 8.10: Stability boundaries of the undamped 2 DOF model without delay.

Substituting $\lambda = i\omega$, $\omega \geq 0$ yields a complex equation that can be separated into two real ones (the second equation was divided by $\mu \sin(\tau\omega) \neq 0$):

$$\omega^4 - (k_{11} + k_{12})\,\omega^2 + k_{11}k_{22} - k_{12}k_{21}$$
$$+ \mu \cos(\tau\omega) \left(p\omega^2 + q\,(k_{12} + k_{21}) - pk_{22}\right) - \mu^2 q^2 \cos(2\tau\omega) = 0, \tag{8.84}$$

$$p\omega^2 + q\,(k_{12} + k_{21}) - pk_{22} + 2\mu q^2 \cos(\tau\omega) = 0. \tag{8.85}$$

We solve the second equation for $\cos(\tau\omega)$:

$$\cos(\tau\omega) = \frac{p\omega^2 + q\,(k_{12} + k_{21}) - pk_{22}}{-2\mu q^2}. \tag{8.86}$$

Using this relation and the identity $\cos(2\tau\omega) = 2\cos(\tau\omega)^2 - 1$ in the real part, Eq. (8.84) results in

$$\omega^4 - (k_{11} + k_{22})\,\omega^2 + k_{11}k_{22} - k_{12}k_{21} + \mu^2 q^2 = 0. \tag{8.87}$$

Divergence occurs where $\omega = 0$, that is where

$$k_{11}k_{22} - k_{12}k_{21} + \mu^2 q^2 = 0.$$

Substituting the elements of the stiffness matrix as given in Eq. (8.62) yields

$$-\left(q\ (a + q\ (p - \mu))\ (p - \mu)\right) + k_{22}\ (1 + p - p\,\mu) = 0, \tag{8.88}$$

which can be solved for p as

$$\frac{1}{2\,q^2}\Bigg(k_{22}\ (1 - \mu) + q\ (2\,q\,\mu - a)$$
$$\pm \sqrt{(k_{22}\,(\mu - 1) + q\,(a - 2\mu q))^2 + 4\,q^2\ (k_{22} + q\,\mu\ (a - q\,\mu))}\Bigg). \tag{8.89}$$

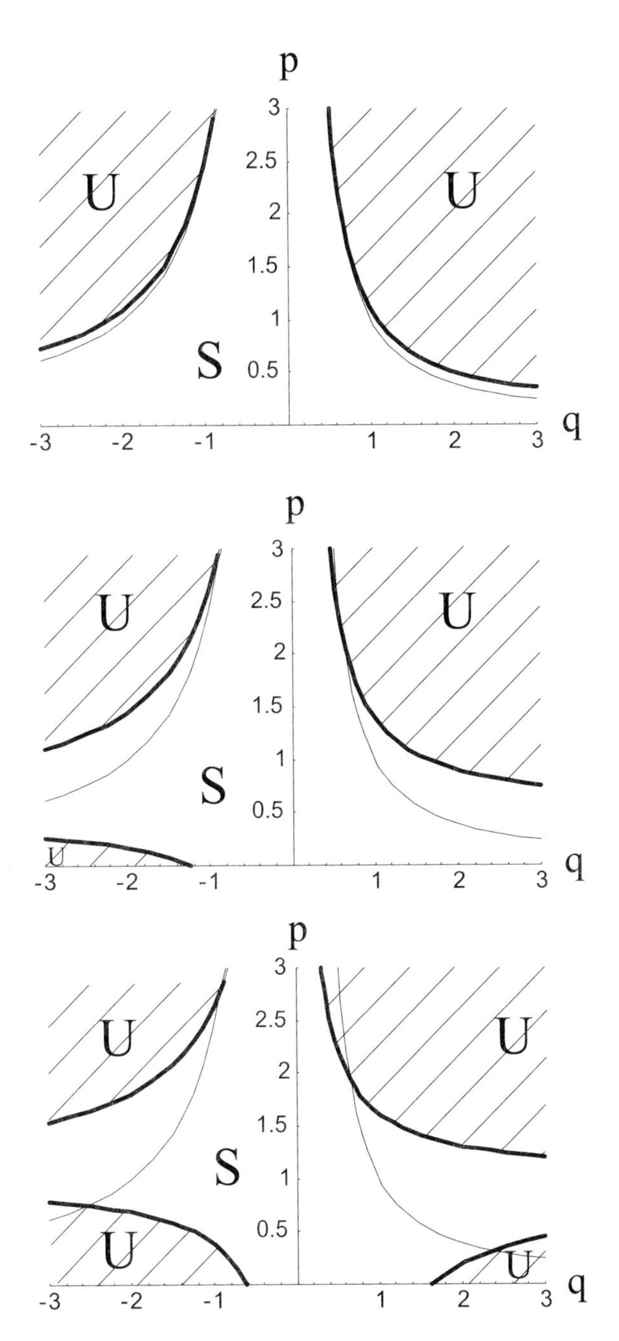

Figure 8.11: The change of the divergence boundary for system (8.5.2), $\tau = 1$. $\mu = 0.1, 0.5$ and 1 for top, middle and bottom figures, respectively.

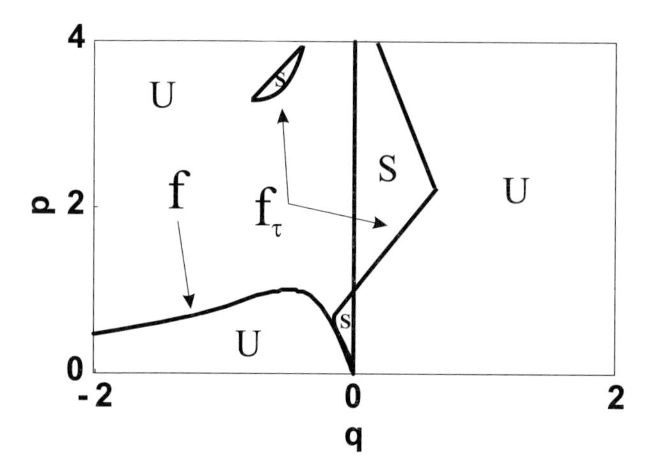

Figure 8.12: Flutter boundary of (8.5.2) with $\mu = 0.01$. S and U denote stable and unstable regions.

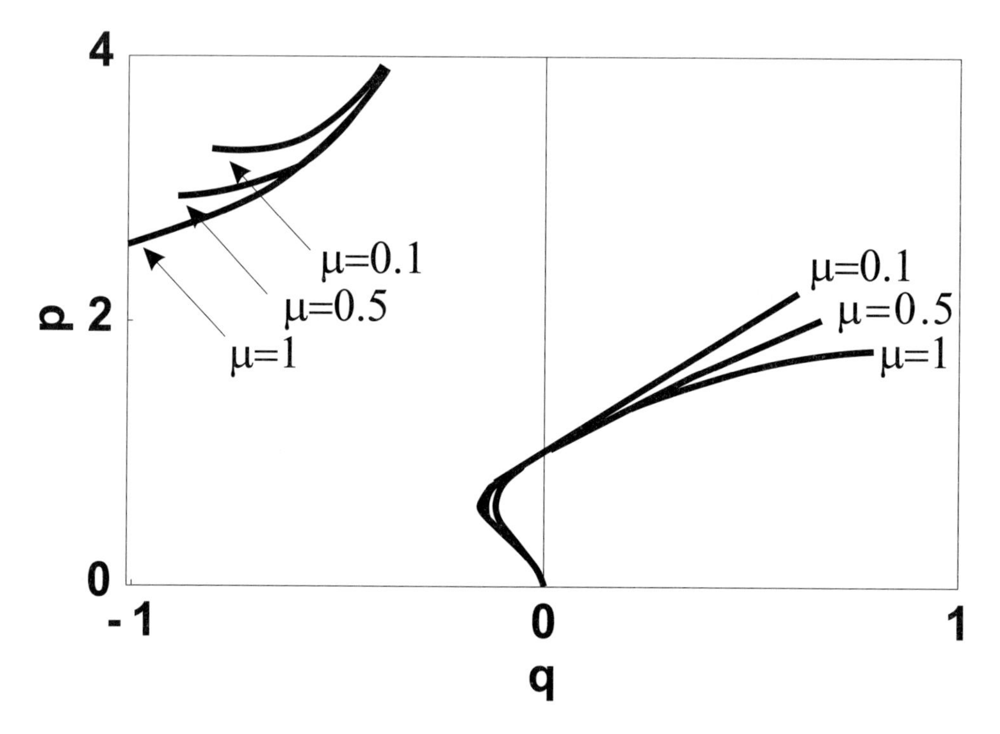

Figure 8.13: Flutter boundary as a function of μ ($\mu = 0.1, 0.5, 1$).

The change of the divergence boundary is shown in Fig. 8.11 (top, middle and bottom) for $\mu = 0.1, 0.5$ and 1 while the delay was set to 1. Flutter occurs for $\omega > 0$, and the boundary can be found by numerically solving Eqs. (8.85), (8.87) for p and q for a given μ. Figure 8.12 shows the flutter boundary for small μ (0.01) together with the flutter boundary (8.80). Figure 8.13

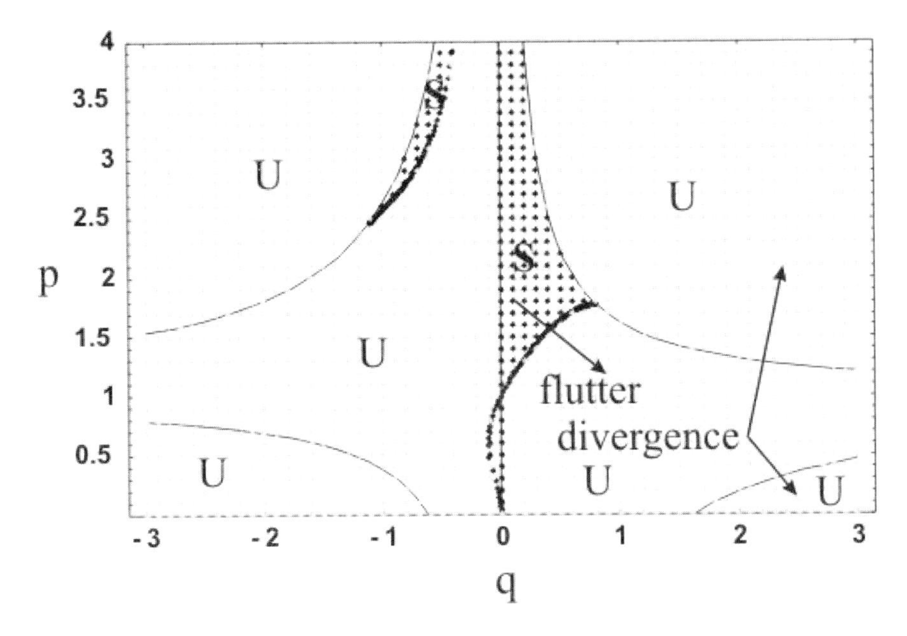

Figure 8.14: Stability chart for the undamped system (8.81), $\mu = 1$, $\tau = 1$.

shows how this boundary changes with increasing μ ($\mu = 0.1, 0.5, 1$). Figure 8.14 shows the full stability chart, complete with both the divergence and flutter boundaries, for $\mu = 1$. To validate this stability chart the parameter space (p, q) was gridded and the delay-differential equation (8.61), (8.62) was integrated with constant initial function (note that the amplitude does not matter for linear stability) at the grid points. The integration was carried out for 15τ intervals of which the first 5τ intervals were discarded. Stability was determined by whether the amplitude of the solution grew or decayed. Dark dots correspond to stable numerical solutions. This figure can also explain a practical trick used in machine shops: sometimes, to avoid chatter, the tool is placed slightly *above* the centerline. We note that increasing q moves the system into the stable region of the chart.

Now we examine the effect of damping on the size of stability regions. It is an important step, as it is known [14] that damping can have a destabilizing effect in non-conservative systems. The damping coefficients ζ_x and ζ_ϕ are taken to be 0.01, while the ratio of frequencies ω_ϕ/ω_x was changed in Fig. 8.15 (this is the same as keeping this ratio fixed and increasing ζ_ϕ). As the figure shows, the size of the stability regions increases with added damping. And, finally, we show how the lobes of the conventional stability chart deform with the added parameter q ($0 \leq q \leq 1$). Figure 8.16 shows that increasing q results in the 'birth' of unstable regions. These upside-down lobes are actually lobes of the classical model for $p < 0$ (p is the non-dimensional cutting force coefficient, which is positive for a cutting force that is an increasing function of chip thickness). While linear instability at zero chip thickness is not physically meaningful, the parameter p also depends on how the cutting force depends on the cutting conditions. However, even for small values of q the stability boundary is lowered and this reduces the strength of forcing necessary to produce small-amplitude nonlinear vibrations.

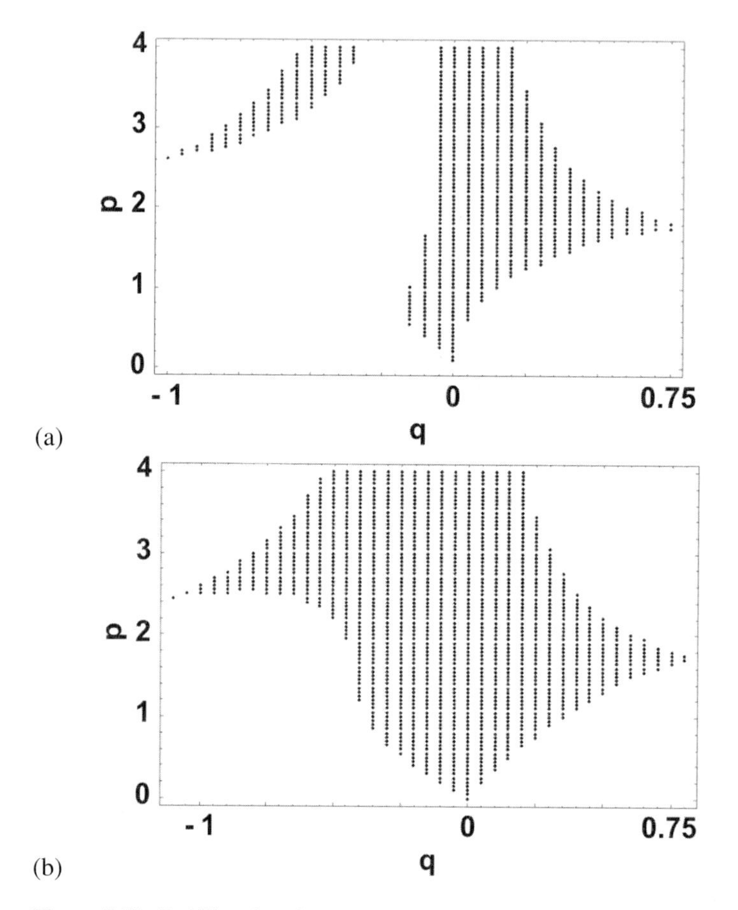

Figure 8.15: Stability chart for the 3-DOF model. (a) $\omega_\phi/\omega_x = 2$, (b) $\omega_\phi/\omega_x = 10$.

8.6 Conclusions

A new 3-DOF model derived may help explain at least two phenomena in metal cutting. The first is that off-centering the tool might help avoiding chatter. The second phenomenon is the observation that small-amplitude tool vibrations can arise below the classical stability boundary. As shown, the added degrees of freedom result in unstable regions below the one predicted by the one-DOF classical model. To summarize the important observations:

- The 3-DOF model results in coupling between twist and lateral bending.
- The model can exhibit both divergence and flutter instabilities.
- Damping seems to increase the size of stability regions.
- The tool offset produces new regions of instability (the upside-down lobes).

This model is based on the assumption of rate-independent cutting forces, i.e. forces that do not exhibit hysteresis [27]. It does not include temperature effects either [9].

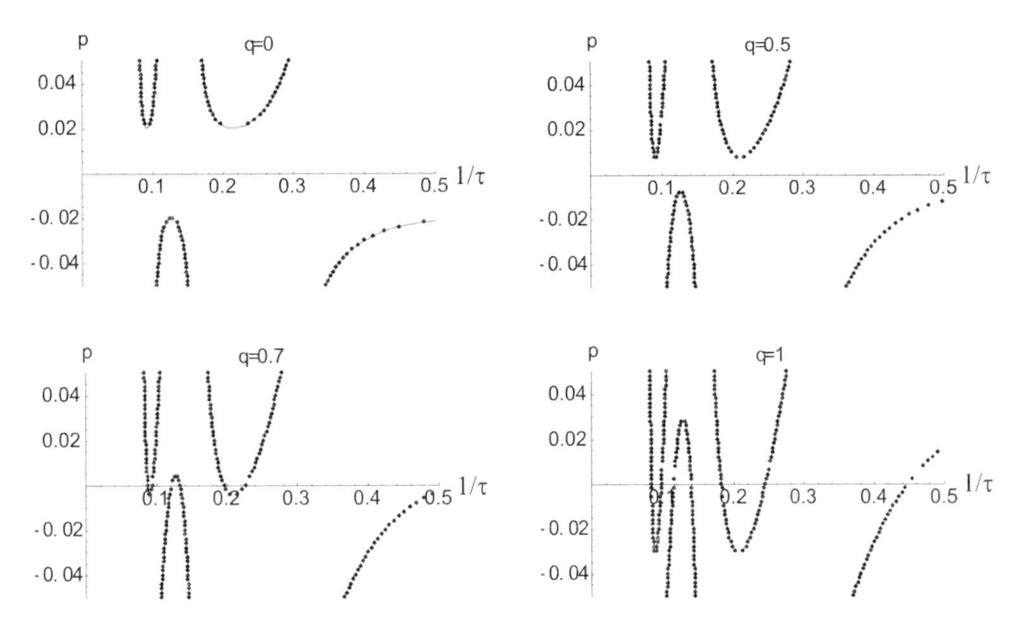

Figure 8.16: Stability charts for the 3-DOF model with increasing q.

Finally, only the analysis of a full nonlinear model could characterize the nature of vibrations and provide estimates of vibration amplitudes for the low chip thickness unstable regions.

Bibliography

[1] B. Balachandran, *Nonlinear dynamics of milling processes*, Philosophical Transactions of the Royal Society **359** (2001), 793–819.

[2] S.A. Batzer, A.M. Gouskov, and S.A. Voronov, *Modeling the vibratory drilling process*, Proceedings of the 17th ASME Biennial Conference on Vibration and Noise 1999, pp. 1-8.

[3] B.S. Berger, I. Minis, Y.H. Chen, A. Chavali, and M. Rokni, *Attractor embedding in metal cutting*, Journal of Sound and Vibration **184**(5) (1995), 936–942.

[4] V.V. Bolotin, *Nonconservative Problems of the Theory of Elastic Stability*, The MacMillan Company, New York, 1963.

[5] S.T.S. Bukkapatnam, A. Lakhtakia, and S.R.T. Kumara, *Analysis of sensor signals shows turning on a lathe exhibits low-dimensional chaos*, Physical Review E **52**(3) (1995), 2375–2387.

[6] S.-G. Chen, A.G. Ulsoy, and Y. Koren, *Computational stability analysis of chatter in turning*, Journal of Manufacturing Science and Engineering **119** (1997), 457–460.

[7] S.T. Chiriacescu, *Stability in the Dynamics of Metal Cutting*, Elsevier, New York, 1990.

[8] D. Chu, and F.C. Moon, *Dynamic instabilities in magnetically levitated models*, Journal of Applied Physics **54**(3) (1983), 1619–1625.

[9] M.A. Davies and T.J. Burns, *Thermomechanical oscillations in material flow during high-speed machining*, Philosophical Transactions of the Royal Society **359** (2001), 821–846.

[10] K. Engelborghs and D. Roose, *Numerical computation of stability and detection of Hopf bifurcations of steady state solutions of delay differential equations*, Advances in Computational Mathematics **10** (1999), 271–289.

[11] M.S. Fofana, Ph.D. thesis, University of Waterloo (1993).

[12] J. Gradišek, E. Govekar, and I. Grabec, *Time series analysis in metal cutting: chatter versus chatter-free cutting*, Mechanical Systems and Signal Processing **12**(6) (1998), 839–854.

[13] N.H. Hanna and S.A. Tobias, *A theory of nonlinear regenerative chatter*, Transactions of the American Society of Mechanical Engineers – Journal of Engineering for Industry **96**(1) (1974), 247–255.

[14] G. Herrmann and I.-C. Jong, *On the destabilizing effect of damping in nonconservative elastic systems*, Transactions of the ASME **32** (1965), 592–597.

[15] C.S. Hsu, *Application of the tau-decomposition method to dynamical systems subjected to retarded follower forces*, Transactions of the American Society of Mechanical Engineers – Journal of Applied Mechanics **37**(2) (1970), 259–266.

[16] T. Insperger and G. Stépán, *Semi-discretization method for delayed systems*, International Journal for Numerical Methods in Engineering **55**(5) (2002), 503–518.

[17] M. Johnson and F.C. Moon, *Nonlinear techniques to characterize prechatter and chatter vibrations in the machining of metals*, International Journal of Bifurcation and Chaos **11**(2) (2001), 449–467.

[18] M.A. Johnson, Ph.D. thesis, Cornell University (1996).

[19] T. Kalmár-Nagy, J.R. Pratt, M.A. Davies, and M.D. Kennedy, *Experimental and analytical investigation of the subcritical instability in turning*, in: Proceedings of the 1999 ASME Design Engineering Technical Conferences, paper no. DETC99/VIB-8060, 1999.

[20] T. Kalmár-Nagy, G. Stépán, and F.C. Moon, *Subcritical Hopf bifurcation in the delay equation model for machine tool vibrations*, Nonlinear Dynamics **26** (2001), 121–142.

[21] T. Kalmár-Nagy, Ph.D. thesis, Cornell University (2002).

[22] S. Kato, E. Marui, and H. Kurita, *Some considerations on prevention of chatter vibration in boring operations*, Journal of Engineering for Industry **91** (1969), 717–730.

[23] J. Kiusalaas and H.E. Davis, *On the stability of elastic systems under retarded follower forces*, International Journal of Solids and Structures **6**(4) (1970), 399–409.

[24] G. Litak, J. Warmiński, and J. Lipski, *Self-excited vibrations in cutting processes*, in: 4th Conference on Dynamical Systems – Theory and Applications (1997), 193–198.

[25] B. Milisavljevich, M. Zeljkovic, and R. Gatalo, *The nonlinear model of a chatter and bifurcation*, in: Proceedings of the 5th Engineering Systems Design and Analysis Conference, ASME, 2000, 391–398.

[26] F.C. Moon and H.D.I. Abarbanel, *Evidence for chaotic dynamics in metal cutting, and classification of chatter in lathe operations*, F.C. Moon (ed), Summary Report of a Workshop on Nonlinear Dynamics and Material Processing and Manufacturing, organized by the Institute for Mechanics and Materials, held at the University of California, San Diego, 1995, 11–12, 28–29.

[27] F.C. Moon and T. Kalmár-Nagy, *Nonlinear models for complex dynamics in cutting materials*, Philosophical Transactions of the Royal Society **359** (2001), 695–711.

[28] A. Nayfeh, C. Chin, and J. Pratt, *Applications of Perturbation methods to tool chatter dynamics*, in: F.C. Moon (ed.), Dynamics and Chaos in Manufacturing Processes, John Wiley, New York, 1998.

[29] N. Olgac and R. Sipahi, *An exact method for the stability analysis of time-delayed LTI systems*, IEEE Transactions on Automatic Control **47**(5) (2002), 793–797.

[30] P.L.B. Oxley, *Mechanics of Machining: An Analytical Approach to Assessing Machinability*, John Wiley, New York, 1989.

[31] Y.G. Panovko and I.I. Gubanova, *Stability and Oscillations of Elastic Systems*, Consultants Bureau, New York, 1965.

[32] J. Pratt, Ph.D. thesis, Virginia Polytechnic Institute and State University (1996).

[33] N. Saravanja-Fabris and A.F. D'Souza, *Nonlinear stability analysis of chatter in metal cutting*, Transactions of the American Society of Mechanical Engineers – Journal of Engineering for Industry **96**(2) (1974), 670–675.

[34] H.M. Shi and S.A. Tobias, *Theory of finite amplitude machine tool instability*, International Journal of Machine Tool Design and Research **24**(1) (1984), 45–69.

[35] G.V. Stabler, *Advances in Machine Tool Design and Research*, Pergamon Press, Oxford, 1964, Chapter *The Chip Flow Law and its Consequences*, p. 243.

[36] G. Stépán, *Modelling nonlinear regenerative effects in metal cutting*, Philosophical Transactions of the Royal Society **359** (2001), 739–757.

[37] G. Stépán, R. Szalai, and T. Insperger, *Nonlinear dynamics of high-speed milling subjected to regenerative effect*, Chapter 7, this book.

[38] E. Stone and S. A. Campbell, *Stability and bifurcation analysis of a nonlinear DDE model for drilling*, Submitted to Journal of Nonlinear Science (2002), to appear 2004.

[39] S.A. Tobias, *Machine Tool Vibration*, Blackie, London, 1965.

[40] S.A. Tobias and W. Fishwick, *The chatter of lathe tools under orthogonal cutting conditions*, Transactions of the American Society of Mechanical Engineers **80**(5) (1958), 1079–1088.

[41] M. Wiercigroch and E. Budak, *Sources of nonlinearities, chatter generation and suppression in metal cutting*, Philosophical Transactions of the Royal Society **359** (2001), 663–693.

[42] M. Wiercigroch and A. H-D. Cheng, *Chaotic and stochastic dynamics of orthogonal metal cutting*, Chaos, Solitons and Fractals **8**(4) (1997), 715–726.

[43] M. Wiercigroch and A.M. Krivtsov, *Frictional chatter in orthogonal metal cutting*, Philosophical Transactions of the Royal Society **359** (2001), 713–738.

[44] N. van de Wouw, R.P.H. Faassen, J.A.J. Oosterling, and H. Nijmeijer, *Modelling of High-Speed Milling for Prediction of Regenerative Chatter*, Chapter 10, this book.

9 Influence of the Workpiece Profile on the Self-excited Vibrations in a Metal Turning Process

R. Rusinek, K. Szabelski, and J. Warminski

The metal turning process described by a one degree of freedom model, including a nonlinear cutting force as a function of cutting depth and cutting velocity, has been analyzed in this paper. Self-excitation in the model caused by a dry friction between a tool and a chip has been modeled by a nonlinear Rayleigh term. Dynamics of the system has been considered in two variants, the first pass of the tool, when the profile of the workpiece is ideally smooth, and the second pass of the tool, when the profile was shaped in the previous pass. The results for two cases have been compared and the resonance regions and bifurcation points for this process have been presented. The analysis for the data taken from the real experimental process has been conducted.

9.1 Introduction

The material removal process plays an important role in the manufacture of products. Modern manufacturing requires using hardware integrated with CAD/CAM systems. However, the responsibility for the selection of technological parameters still belongs to an engineer or a designer. In this paper, we look at a specific aspect of the cutting process, namely the influence of rough cutting parameters on vibrations in the finish turning.

In most cases researchers consider a cutting process restricted to orthogonal cutting. The first considerable paper treating the above matter was Taylor's article [11]. Next the works of Pisspanen [10] and Merchant contributed to create an orthogonal cutting model [7, 8]. From that time a lot of scientists have tried to explain the dynamical behavior of a turning system. In the 1960s the first model of self-excited vibrations had been developed based on orthogonal metal cutting [12]. In such systems, unwanted excessive vibrations between the tool and the workpiece are called chatter [15, 16]. This phenomenon introduces some difficulties during machining; therefore, the acquaintance of parameters for which chatter is present has great meaning. A very important moment which inspired researchers to penetrating analysis was the time after publication of articles by Grabec and cooperators [2, 3]. They have proved the possibility of occurrence of chaotic vibrations. Moreover, the appearance of chaotic oscillations has been confirmed by Wiercigroch and co-workers [14–16]. They have used the nonlinear model, which allows for the change of cutting force vs. cutting depth and velocity. Nayfeh et al. have shown an interesting approach [9]. They have applied the analytical method of to analyze a regenerative cutting process. Chaotic vibrations in a rough cutting process have also been considered by Litak [5]. The dynamic problems connected with an orthogonal turning

Nonlinear Dynamics of Production Systems. Edited by G. Radons and R. Neugebauer

ISBN 3-527-40430-9

process have been raised by Marghitu et al. [6]. The vibrations of a tool–workpiece system induced by random disturbances in a straight turning process, and their effect on a product surface, have been considered in [4]. Therefore, the detailed analysis of a profile got from rough cutting and its influence on finish cutting has been carried out in [13].

In this paper we have applied a simple one degree of freedom model reduced to the radial direction. The motion of a workpiece in this direction is the most important for the quality of a product. The proposed model allows for nonlinearities of a thrust force and a profile from a previous tool pass. This work is also an attempt of analysis of nonlinear and non-orthogonal turning.

9.2 Modeling of Turning Process

According to Merchant's model, two phenomena are responsible for the cutting force: a shear of material, and a friction between a tool and a chip. A cutting force has been distributed on two components: the shear force and the perpendicular one or the friction force and the perpendicular one. The forces create a so-called "Merchant's wheel". The model of the force assumes simultaneous occurrence of the friction (F_{t}) and the shear force (F_{s}). The forces mentioned above are presented in the tool working back plane in Fig. 9.1, where R is the component of the cutting force in the plane perpendicular to the symmetry axis of the workpiece; ϕ is the shear angle; β is the angle between forces F_z and R; γ_p is the tool rake; F_z is the cutting force; F_y is the thrust force.

The analysis of forces and velocities leads to the relationship which describes the thrust force (F_y):

$$F_y = K_y a_{\mathrm{p}}^{\rho} \left[\mathrm{Sgn}\left(v_{\mathrm{w}y}\right) - \alpha_y v_{\mathrm{w}y} + \beta_y v_{\mathrm{w}y}^3 \right] H(a_{\mathrm{p}}), \tag{9.1}$$

in which a_{p} is an actual depth of cut, $v_{\mathrm{w}y}$ is a relative velocity between a tool and a chip in direction y, $H()$ means Heaviside's function and K_y is cutting resistance. The expression

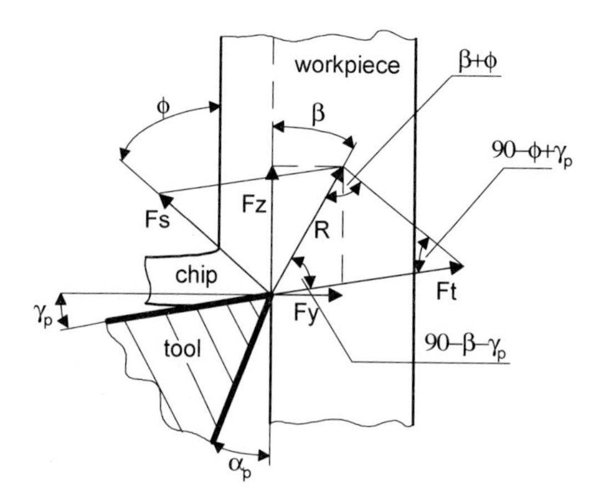

Figure 9.1: Cutting forces in tool working back plane.

given in the square brackets of Eq. (9.1) presents the self-excited Rayleigh term. The depth and the velocity are defined as

$$a_{\mathrm{p}} = a_{\mathrm{po}} + a_{\mathrm{pp}} + a_{\mathrm{pd}} - y,$$
$$v_{\mathrm{w}y} = v'_{\mathrm{w}y} - \dot{y}, \quad v'_{\mathrm{w}y} = \frac{v_c}{\eta k_{\mathrm{h}}}. \tag{9.2}$$

Here ρ, α_y, β_y, η, k_{h} are technological constants, $y, \dot{y}$ are respectively displacement and velocity of a workpiece, a_{po} is an assumed depth of cut, v_{c} is a cutting velocity, a_{pp} models a kinematic profile which maps a shape of the tool from rough cutting and a_{pd} determines a dynamic profile connected with vibrations in rough cutting. When $a_{\mathrm{pp}} = a_{\mathrm{pd}} = 0$ then we have a smooth surface cutting.

A physical model of the turning process reduced to the y direction is shown in Fig. 9.2, where k_y is a system stiffness, c_y is a system damping and ω_2 is an angular velocity of a workpiece during a finishing cutting. The thrust force F_y acting on the workpiece is also presented in Fig. 9.1. Such defined force has nonlinear character for the sake of an actual cutting depth (a_{p}) and a relative velocity ($v_{\mathrm{w}y}$) (Fig. 9.3).

The profile of the tool from the previous pass influences the actual cutting depth in connection with it; in Eq. (9.2) a_{pp} describes a profile obtained during roughing [13]. The profile obtained after the first pass maps the shape of the tool (Fig. 9.4).

That profile has been approximated by three terms of a Fourier series:

$$a_{\mathrm{pp}} = a_0 - a_1 \cos \omega_{\mathrm{p}} t - b_1 \sin \omega_{\mathrm{p}} t. \tag{9.3}$$

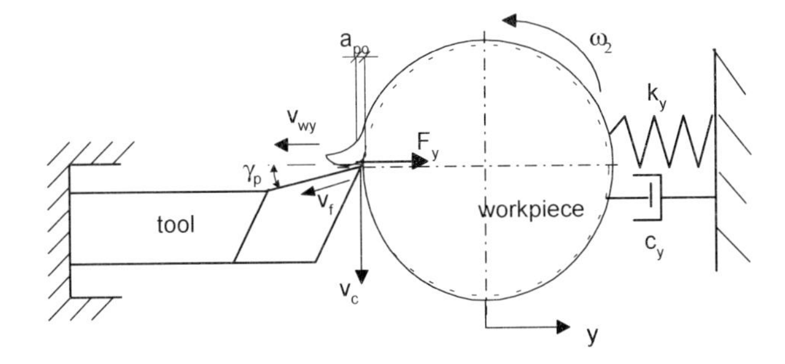

Figure 9.2: Physical model of turning process.

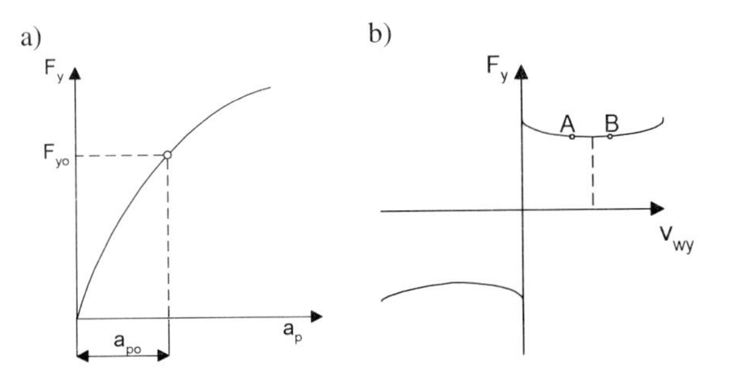

Figure 9.3: Thrust force vs. (a) cutting depth, (b) relative velocity between a tool and a chip.

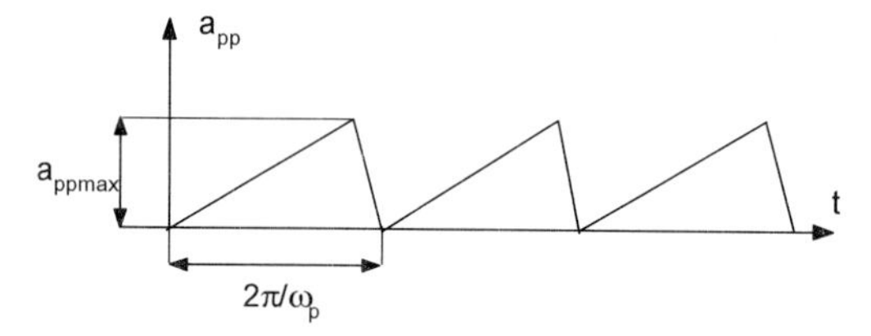

Figure 9.4: Profile mapping the tool shape.

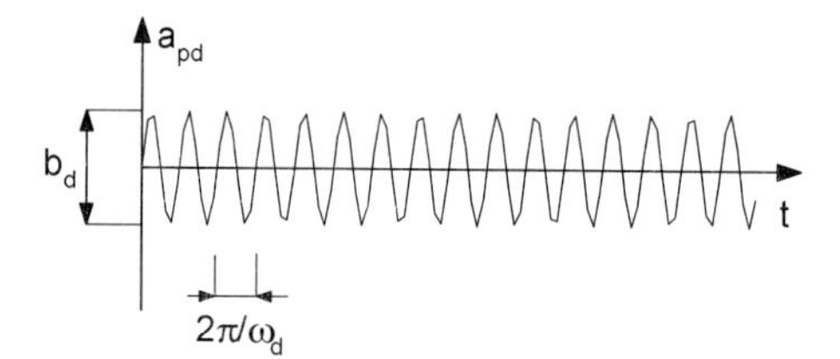

Figure 9.5: Profile comes from vibrations of the tool first pass.

Frequency ω_p is defined as

$$\omega_\mathrm{p} = \omega_2 \left(1 - \frac{f_2}{f_1}\right), \tag{9.4}$$

where f_1, f_2 are the feeds of the rough and finish turning. The frequency ω_p depends on the angular velocity of the workpiece and ratio of the feed in the second and the first tool passes (see details in [13]).

The effect connected with the kinematics of the rough cutting and geometry of the tool has been shown above. The second effect connected with the dynamics of the rough cutting has been presented below. Vibrations in a rough cutting also produce some workpiece profile. Therefore, this workpiece has a surface roughness, which additionally affects the cutting thickness. In the first approximation it can be presented by the equation (Fig. 9.5)

$$a_\mathrm{pd} = b_\mathrm{d} \sin \omega_\mathrm{d} t. \tag{9.5}$$

ω_d is near the natural frequency of the system and b_d denotes an amplitude of vibrations in a first tool pass – an amplitude of the dynamic profile.

Based on this model, the differential equation of motion is given by

$$m\ddot{y} + k_y y + c_y \dot{y} = F_y - F_{y\mathrm{o}}, \tag{9.6}$$

where

$$F_{y\mathrm{o}} = K_y a_{\mathrm{po}}^{\rho} \left(1 - \alpha_y \frac{v_\mathrm{c}}{\eta k_\mathrm{h}} + \beta_y \left(\frac{v_\mathrm{c}}{\eta k_\mathrm{h}}\right)^3\right). \tag{9.7}$$

$F_{y\mathrm{o}}$ is a static force occurring in the case of cutting without vibrations ($y = 0$).

9.3 Analytical Investigations of Primary Cutting

At the beginning let us consider a primary cutting (roughing) and assume that the surface is ideally smooth. Then $a_{\text{pp}} = a_{\text{pd}} = 0$. Moreover, let us assume that the tool and the workpiece does not lose contact, thus $a_{\text{p}} > 0$ and $H(a_{\text{p}}) = 1$. By introducing a dimensionless time τ and a coordinate y_1 we obtain:

$$a_{\text{po}}p_{y\text{o}}^2\ddot{y}_1 + 2n_y a_{\text{po}}p_{y\text{o}}\dot{y}_1 + a_{\text{po}}p_{y\text{o}}y_1 = \frac{1}{m}K_y a_{\text{po}}^{\rho}((1-y_1)^{\rho}(\text{Sgn}[v_{wy}]$$
$$-\alpha_y(v'_{wy} - a_{\text{po}}p_{y\text{o}}\dot{y}_1) + \beta_y(v'_{wy} - a_{\text{po}}p_{y\text{o}}\dot{y}_1)^3) - (1 - \alpha_y v'_{wy} + \beta_y {v'_{wy}}^3)), \quad (9.8)$$

where

$v_{wy} = v'_{wy} - a_{\text{po}}p_{y\text{o}}\dot{y}_1$,

$\tau = p_{y\text{o}}t$ is a dimensionless time,

$y_1 = \dfrac{y}{a_{\text{po}}}$ is a dimensionless coordinate,

$2n_y = \dfrac{c_y}{m_y}$ is a viscous damping coefficient

and

$p_{y\text{o}} = \sqrt{\dfrac{k_y}{m_y}}$ is the natural frequency of the linear system.

After the expansion of the function $(1-y_1)^{\rho}$ in a Taylor series we get the differential equation in the following form:

$$\ddot{y}_1 + p^2 y_1 = K\dot{y}_1 + Dy_1^2 + G\dot{y}_1 y_1 + Cy_1^2\dot{y} + J\dot{y}_1^2$$
$$+ Fy_1\dot{y}_1^2 + By_1^2\dot{y}_1^2 + I\dot{y}_1^3 + Ey_1\dot{y}_1^3 + Ly_1^2\dot{y}_1^3 + W, \quad (9.9)$$

where

$$p^2 = 1 + \frac{a_{\text{po}}^{\rho-1}K_y\rho}{m_y p_{y\text{o}}^2}(\text{Sgn}[v_{wy}] - v'_{wy}\alpha_y + {v'_{wy}}^3\beta_y), \quad (9.10)$$

$$W = \frac{a_{\text{po}}^{\rho-1}K_y}{m_y p_{y\text{o}}^2}(\text{Sgn}[v_{wy}] - 1), \quad L = \frac{\beta_y a_{\text{po}}^{\rho+2}K_y\rho p_{y\text{o}}}{2m_y}(1-\rho), \quad (9.11)$$

$$D = \frac{a_{\text{po}}^{\rho-1}K_y\rho}{2m_y p_{y\text{o}}^2}(\text{Sgn}[v_{wy}] - v'_{wy}\alpha_y + {v'_{wy}}^3\beta_y)(\rho-1), \quad (9.12)$$

$$K = \frac{a_{\text{po}}^{\rho}K_y}{m_y p_{y\text{o}}}(\alpha_y - 3\beta_y {v'_{wy}}^2) - \frac{2n_y}{p_{y\text{o}}}, \quad G = \frac{a_{\text{po}}^{\rho}K_y\rho}{m_y p_{y\text{o}}}(3\beta {v'_{wy}}^2 - \alpha_y), \quad (9.13)$$

$$C = \left(\frac{a_{\text{po}}^{\rho}K_y\rho\alpha_y}{2m_y p_{y\text{o}}} - \frac{3a_{\text{po}}^{\rho}K_y\rho {v'_{wy}}^2\beta_y}{2m_y p_{y\text{o}}}\right)(\rho-1), \quad J = \frac{3\beta a_{\text{po}}^{\rho+1}K_y v'_{wy}}{m_y}, \quad (9.14)$$

$$F = \frac{-3\beta_y a_{\text{po}}^{\rho+1}K_y\rho v'_{wy}}{m_y}, \quad B = \frac{3\beta_y a_{\text{po}}^{\rho+1}K_y\rho v'_{wy}}{2m_y}(\rho-1), \quad (9.15)$$

$$I = \frac{-\beta_y a_{\mathrm{po}}^{\rho+2} K_y p_{yo}}{m_y}, \quad E = \frac{\beta_y a_{\mathrm{po}}^{\rho+2} K_y \rho p_{yo}}{m_y}. \tag{9.16}$$

The differential equation (9.9) has been solved employing the method of multiple scales. The solution of Eq. (9.9) is expressed by the series of a small parameter ε:

$$y_1(\tau, \varepsilon) = y_{1,0}(T_0, T_1, T_2) + \varepsilon y_{1,1}(T_0, T_1, T_2) + ...,$$
$$\tau = T_0 + \varepsilon T_1 + ..., \tag{9.17}$$

where T_o is a fast scale, $T_1, T_2...$ are slow scales and ε is a formal small parameter.

Calculating derivatives we obtain:

$$\frac{dy_1}{d\tau} = \frac{\partial y_{1,0}}{\partial T_0} + \varepsilon \frac{\partial y_{1,1}}{\partial T_0} + \varepsilon \frac{\partial y_{1,0}}{\partial T_1} + \varepsilon^2 \frac{\partial y_{1,1}}{\partial T_1},$$
$$\frac{d^2 y_1}{d\tau^2} = \frac{\partial^2 y_{1,0}}{\partial T_0^2} + \varepsilon \frac{\partial^2 y_{1,1}}{\partial T_0^2} + 2\varepsilon \frac{\partial^2 y_{1,0}}{\partial T_1 \partial T_0} + 2\varepsilon^2 \frac{\partial^2 y_{1,1}}{\partial T_1 \partial T_0} + \varepsilon^2 \frac{\partial^2 y_{1,0}}{\partial T_1^2} + \varepsilon^3 \frac{\partial^2 y_{1,1}}{\partial T_1^2}. \tag{9.18}$$

Putting Eq. (9.17) into Eq. (9.9) and grouping powers of ε we get the system of equations:

$$\frac{\partial^2 y_{1,0}}{\partial T_0^2} + p^2 y_{1,0} = 0, \tag{9.19}$$

$$\frac{\partial^2 y_{1,1}}{\partial T_0^2} + p^2 y_{1,1} = -2\frac{\partial^2 y_{1,0}}{\partial T_0 \partial T_1} + L_1 y_{1,0}^2 \left(\frac{\partial y_{1,0}}{\partial T_0}\right)^3 + E_1 y_{1,0} \left(\frac{\partial y_{1,0}}{\partial T_0}\right)^3$$
$$+ I_1 \left(\frac{\partial y_{1,0}}{\partial T_0}\right)^3 + B_1 y_{1,0}^2 \left(\frac{\partial y_{1,0}}{\partial T_0}\right)^2 + F_1 y_{1,0} \left(\frac{\partial y_{1,0}}{\partial T_0}\right)^2 + J_1 \left(\frac{\partial y_{1,0}}{\partial T_0}\right)^2$$
$$+ C_1 y_{1,0}^2 \frac{\partial y_{1,0}}{\partial T_0} + G_1 y_{1,0} \frac{\partial y_{1,0}}{\partial T_0} + K_1 \frac{\partial y_{1,0}}{\partial T_0} + D_1 y_{1,0}^2 + W_1. \tag{9.20}$$

A solution of Eq. (9.19) has the complex form:

$$y_{1,0}(T_1) = A_1(T_1) e^{ipT_0} + \overline{A_1}(T_1) e^{-ipT_0}, \tag{9.21}$$

where

$$A_1 = \frac{1}{2} a_1 e^{i\varphi}, \quad \overline{A_1} = \frac{1}{2} a_1 e^{-i\varphi}. \tag{9.22}$$

a_1 and φ are dimensionless amplitude and phase respectively. The dimensionless amplitude is defined analogously to the dimensionless coordinate y_1:

$$a_1 = \frac{a}{a_{\mathrm{po}}}, \tag{9.23}$$

where a is a vibration amplitude.

Putting the solution (9.21) into Eq. (9.20) and eliminating secular terms yields

$$\frac{1}{2} i K_1 p a_1 e^{i\varphi} + \frac{1}{8} i C_1 p a_1^3 e^{i\varphi} + \frac{1}{8} F_1 p^2 a_1^3 e^{i\varphi} + \frac{3}{8} i I_1 p^3 a_1^3 e^{i\varphi} + \frac{1}{16} i L_1 p^3 a_1^5 e^{i\varphi}$$
$$- 2ip \left(\frac{1}{2} \dot{a}_1 e^{i\varphi} + \frac{1}{2} i a_1 \dot{\varphi} e^{i\varphi}\right) = 0, \tag{9.24}$$

where $K = \varepsilon K_1, C = \varepsilon C_1, F = \varepsilon F_1, I = \varepsilon I_1, L = \varepsilon L_1$.

Separating the real and the imaginary parts, we finally obtain the modulation equations representing the phase and the amplitude:

$$\dot{\varphi} = -\frac{1}{8}F_1 p a_1^2, \tag{9.25}$$

$$\dot{a}_1 = \frac{1}{16}(8K_1 a_1 + 2C_1 a_1^3 + 6I_1 p^2 a_1^3 + L_1 p^2 a_1^5). \tag{9.26}$$

Then the amplitude of vibrations in the steady state, $\dot{a}_1 = 0$, has the form:

$$a_1 = \sqrt{\frac{-3I_1}{L_1} - \frac{C_1}{L_1 p^2} - \frac{\sqrt{-32L_1 K_1 p^2 + (2C_1 + 6I_1 p^2)^2}}{2L_1 p^2}} \tag{9.27}$$

or

$$a_1 = \sqrt{\frac{-3I_1}{L_1} - \frac{C_1}{L_1 p^2} + \frac{\sqrt{-32L_1 K_1 p^2 + (2C_1 + 6I_1 p^2)^2}}{2L_1 p^2}} \tag{9.28}$$

and the trivial solution $a_1 = 0$.

From a practical point of view stability of the solutions is very significant, especially the trivial one. Disturbing the equations $\dot{a}_1 = f_1(a_1, \varphi)$, $\dot{\varphi} = f_2(a_1, \varphi)$ near the steady state we obtain

$$\tilde{a}_1(\tau) = a_1(\tau) + \delta_1(\tau), \quad \tilde{\varphi}(\tau) = \varphi(\tau) + \delta_2(\tau). \tag{9.29}$$

Subtracting non-disturbed solutions from disturbed ones, and substituting $\delta_1 = C_1 e^{\mu\tau}$, $\delta_2 = C_2 e^{\mu\tau}$ we obtain the characteristic equation:

$$\begin{vmatrix} \left(\frac{\partial f_1}{\partial a}\right)_0 - \mu & \left(\frac{\partial f_1}{\partial \varphi}\right)_0 \\ \left(\frac{\partial f_2}{\partial a}\right)_0 & \left(\frac{\partial f_2}{\partial \varphi}\right)_0 - \mu \end{vmatrix} = 0, \tag{9.30}$$

which after expansion is given by

$$\mu\left(\mu - \frac{1}{16}(8K_1 + 6C_1 a_1^2 + 18I_1 p^2 a_1^2 + 5L_1 p^2 a_1^4)\right) = 0, \tag{9.31}$$

where μ is the root of Eq. (9.31) which determines the solution stability.

9.4 Numerical Analysis of Primary Cutting

Exemplary numerical calculations of primary (rough) cutting have been made for the parameters presented in Table 9.1. Figure 9.6 presents the stability of a trivial solution. AR and NS denote analytical results and numerical simulations respectively. The trivial solutions for the parameters lying above the curve AR are stable; therefore vibrations do not appear – they are damped. Such a situation is presented in a phase portrait (Fig. 9.7a), where a trajectory tends to zero. The opposite case takes place when the parameter values lie below the curve AR. Then the trivial solutions are unstable and in the system chatter appears. It is visible in the phase portrait (Fig. 9.7b), where the trajectory tends to a limit cycle. The horizontal part in Fig. 9.7b points to the case when $y' = v'_{wy}$.

Table 9.1: Applied parameters.

Parameter	Value	Unit
a_{po}	0.003	m
$v'_{\mathrm{w}y}$	1.5	m/s
k_y	992.472	kN/m
c_y	5	kg/s
m	0.51	kg
K_y	44	kN/m
ρ	0.95	
α_y	0.4	
β_y	0.05	

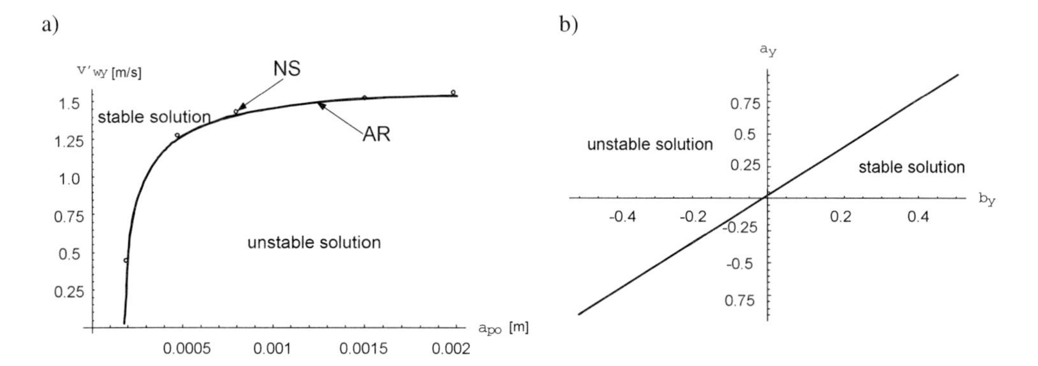

Figure 9.6: Stability of trivial solution vs. (a) cutting depth and velocity, (b) parameters α_y, β_y.

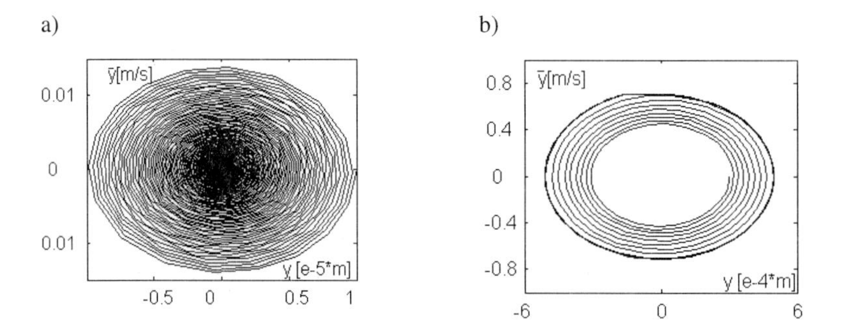

Figure 9.7: Phase portrait for $a_{\mathrm{po}} = 0.001$ m; (a) $v'_{\mathrm{w}y} = 1.7$ m/s, (b) $v'_{\mathrm{w}y} = 0.7$ m/s.

The numerical simulations have been conducted to verify the analytical results. The points NS describe a border of the stability (Fig. 9.6a). We can notice a good agreement between both methods. Parameters α_y, β_y are responsible for the self-exciting phenomenon. Their influence on stability of the trivial solution is shown in Fig. 9.6b. Under the straight line we have stable solutions, whereas over the line there are unstable solutions.

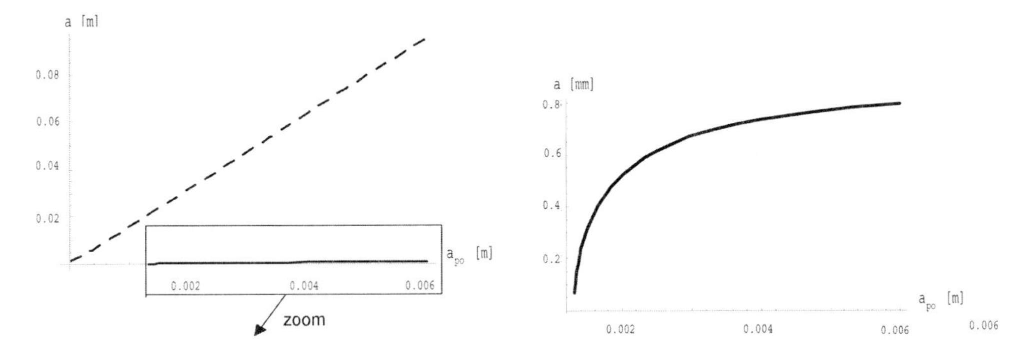

Figure 9.8: Non-trivial amplitude a vs. cutting depth a_{po}.

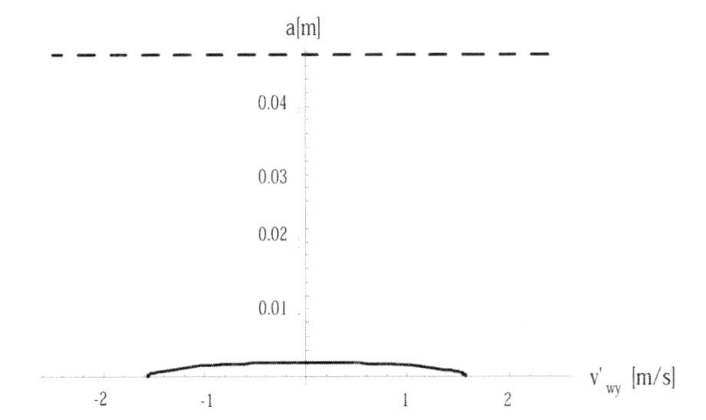

Figure 9.9: Non-trivial amplitude a vs. relative velocity $v'_{\mathrm{w}y}$.

Stability of non-trivial solutions represented by (9.27) and (9.28) is plotted in Figs. 9.8 and 9.9. The dashed line represents the unstable solution and the solid line the stable one. The amplitude presented here is connected with the dimensionless amplitude a_1 by Eq. (9.23). From a practical point of view only the stable amplitude is attainable. The stable non-trivial solution does not exist for a cutting depth less than about 0.0013 m. For bigger values, the amplitude of the stable solutions (solid line) increases, as presented in Fig. 9.8.

The non-trivial stable solution appears within the range of velocity from -1.6 to 1.6 m/s (Fig. 9.9). A minus sign denotes the reverse direction of velocity and thus the solutions on both sides of the axis are equivalent.

9.5 Numerical Investigation of Finishing Cutting Dynamics

In this part of the paper we have analyzed the model which additionally takes into account a profile obtained from the first tool pass. Then a_{pp} (9.3) and a_{pd} (9.5) are not equal to zero. The rough surface profile has been obtained for the following parameters: tool–cutting-edge angle $\chi_{r1} = \chi'_{r1} = 60$ deg, $f_1 = 2$ mm/rev. Finishing cutting has been performed for the parameters which are presented in Table 9.2 and in Table 9.1.

Table 9.2: Parameters used in the simulation.

χ_{r2} deg	n_2 rad/s	f_2 mm/rev	ω_{p} rad/s	a_{o} mm	a_1 mm	b_1 mm	ω_{d} rad/s	b_{d}
60	104.9	0.09	100.0	0.866	0	0.5514	p_{yo} =1395	$0.1a_{\mathrm{po}}$

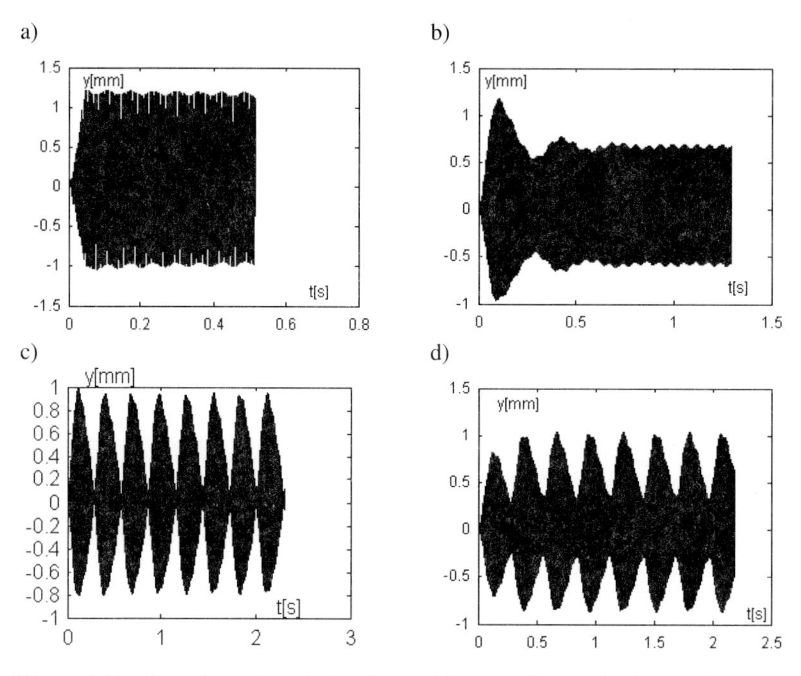

Figure 9.10: Time histories of vibrations (a) $b_{\mathrm{d}} = 1/3a_{\mathrm{po}}$, (b) $b_{\mathrm{d}} = 1/5a_{\mathrm{po}}$, (c) $b_{\mathrm{d}} = 1/7a_{\mathrm{po}}$, (d) $b_{\mathrm{d}} = 1/10a_{\mathrm{po}}$.

A numerical analysis has been conducted using the software Matlab-Simulink® and Dynamic Solver [1]. The time histories of vibrations occurring during the finishing cutting for different b_{d} are presented in Fig. 9.10. Big modulations of amplitude occur for a small value of b_{d}. However, if b_d has a value near a_{po} the modulations are smaller.

To compare the second pass of the tool with the first one, oscillations during the rough cutting have been presented in Fig. 9.11. In this case, the modulations of the amplitude do not occur.

When the parameters $\alpha_y = 0.4$, $\beta_y = 0.05$ correspond to point A, which is situated on the decreasing branch of the curve in Fig. 9.3, we obtain the amplitude–frequency characteristic shown in Fig. 9.12. Figure 9.13 illustrates the case in which $\alpha_y = 0.08$, $\beta_y = 0.07$ and we have point B on the increasing branch of the curve. In the former case, in the neighborhood of the first resonance ($\omega_{\mathrm{d}} = p_{yo}$) the local decrease and increase of the vibration amplitude is visible. It is worth noting that in the latter case the amplitudes are smaller and the local decrease of the amplitude does not occur. The second resonance ($\omega_{\mathrm{d}} = 2p_{yo}$) is visible only for the first case.

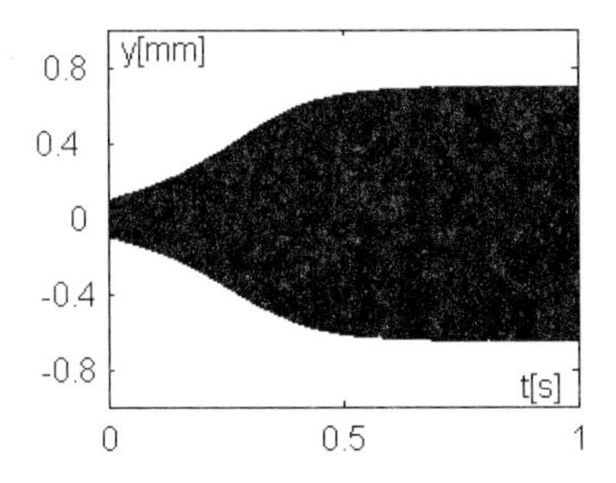

Figure 9.11: Vibrations during rough cutting.

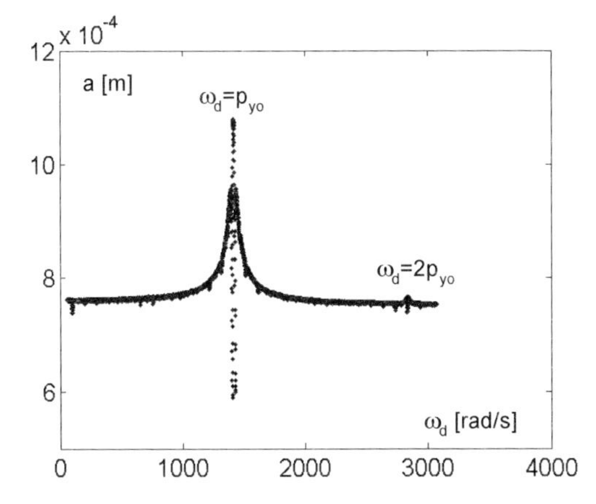

Figure 9.12: Amplitude–frequency characteristic for $\alpha_y = 0.4$, $\beta_y = 0.05$.

Near $\omega_\mathrm{d} = \omega_\mathrm{p} = 100$ rad/s we can see a reduction of the amplitude. Then the excitation ω_d is in antiphase with ω_p. In both cases a breakdown of the amplitude is present too.

The time histories for different values of ω_d are plotted below (Figs. 9.14–9.17). Column (a) represents the results for $\alpha_y = 0.4$, $\beta_y = 0.05$ whereas column (b) corresponds to $\alpha_y = 0.08$, $\beta_y = 0.07$. When $\omega_\mathrm{d} \cong p_{y\mathrm{o}}$ we have oscillations near the first resonance (Fig. 9.16) and while $\omega_\mathrm{d} \cong 2p_{y\mathrm{o}}$ near the second resonance (Fig. 9.17).

Vibrations for $\alpha_y = 0.08$, $\beta_y = 0.07$ have an amplitude significantly smaller than for $\alpha_y = 0.4$, $\beta_y = 0.05$. Besides, a displacement of the vibration center is visible.

The bifurcation diagrams reveal a phenomenon of synchronization where the vibration frequency corresponds to the excitation frequency. The dark area in Fig. 9.18 represents quasi-periodic oscillations while the solid line denotes periodic vibrations.

This effect is the clearest for $\omega_\mathrm{d} \cong p_{y\mathrm{o}}$; then the widest region of synchronization exists. When $\omega_\mathrm{d} \cong 2p_{y\mathrm{o}}$ and $\omega_\mathrm{d} \cong 0.5p_{y\mathrm{o}}$ the synchronization is less visible.

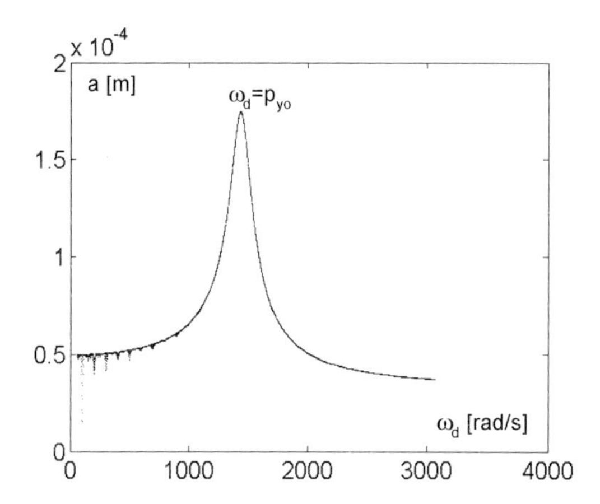

Figure 9.13: Amplitude–frequency characteristic for $\alpha_y = 0.08$, $\beta_y = 0.07$.

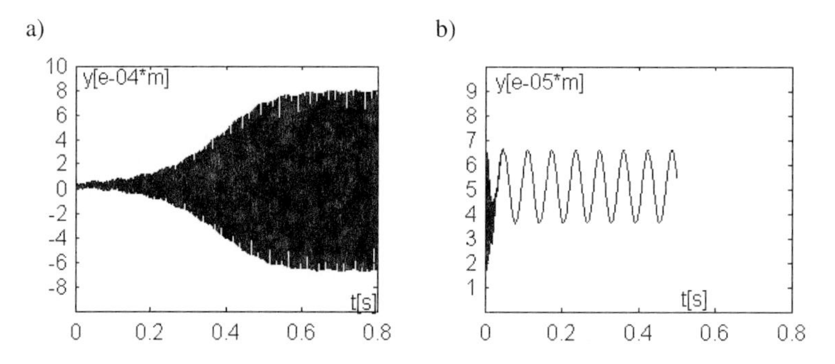

Figure 9.14: Time histories $\omega_\mathrm{d} = \omega_\mathrm{p} = 100$ rad/s, (a) $\alpha_y = 0.4$, $\beta_y = 0.05$, (b) $\alpha_y = 0.08$, $\beta_y = 0.07$.

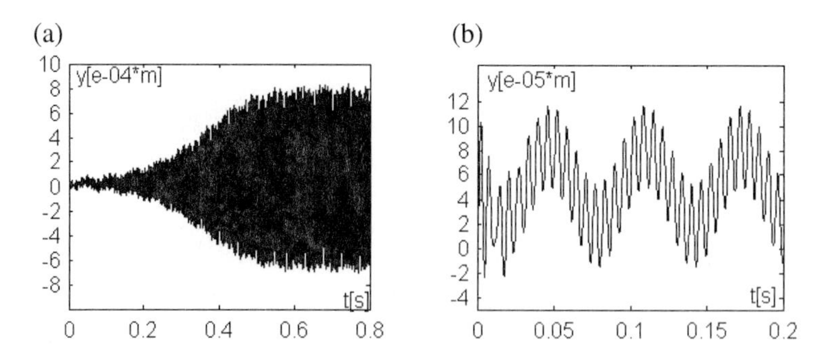

Figure 9.15: Time histories $\omega_\mathrm{d} = \omega_\mathrm{p} = 1000$ rad/s, (a) $\alpha_y = 0.4$, $\beta_y = 0.05$, (b) $\alpha_y = 0.08$, $\beta_y = 0.07$.

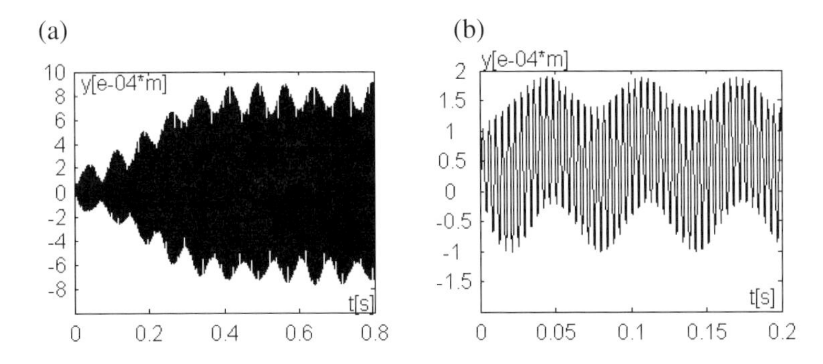

Figure 9.16: Time histories $\omega_d \cong p_{yo}$, (a) $\alpha_y = 0.4$, $\beta_y = 0.05$, (b) $\alpha_y = 0.08$, $\beta_y = 0.07$.

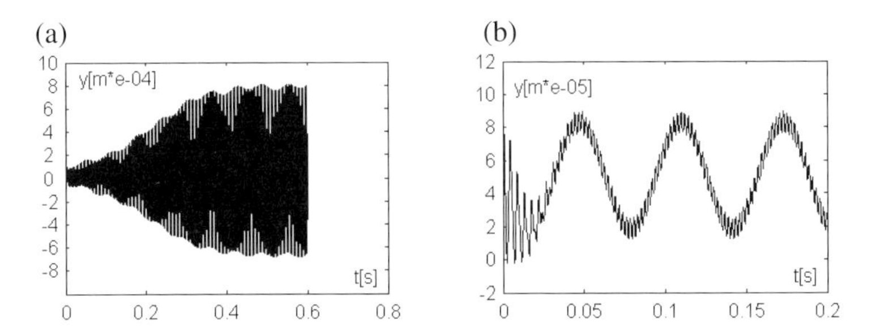

Figure 9.17: Time histories $\omega_d \cong 2p_{yo}$, (a) $\alpha_y = 0.4$, $\beta_y = 0.05$, (b) $\alpha_y = 0.08$, $\beta_y = 0.07$.

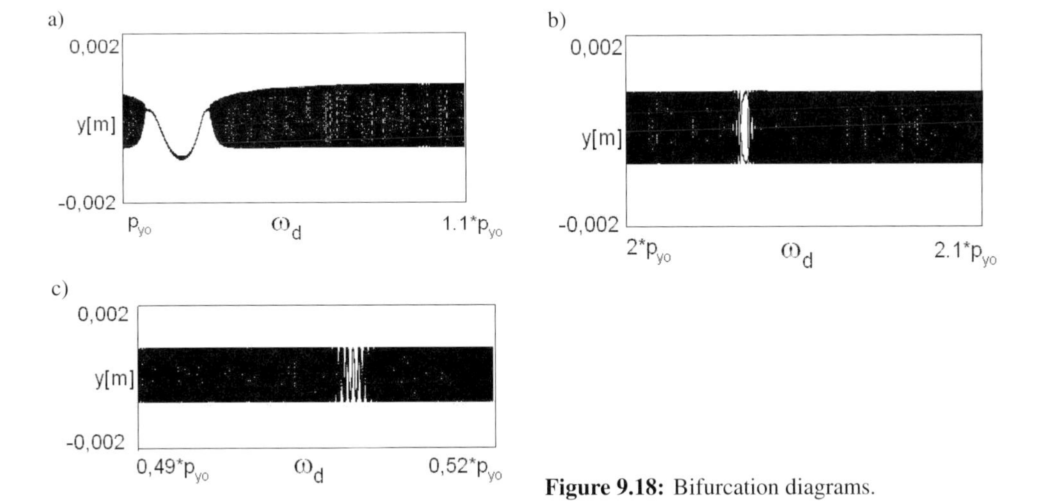

Figure 9.18: Bifurcation diagrams.

9.6 Conclusions

In the paper a study of the distribution of cutting forces has been conducted. On the grounds of the study a nonlinear model of the thrust force has been built. The new approach takes into consideration simultaneous acting of a shear and a friction force as well as using the Rayleigh term for modeling of self-excitation (9.1). Such a defined nonlinear force is the main cause of the chatter.

In a real cutting process we tend to avoid chatter. The self-excited vibrations deteriorate the final surface quality. It has been proved that there are parameters for which the amplitude of vibrations is stable and equals zero. It is worth pointing out that both the technological and the physical parameters of the model are responsible for the stability. Generally cutting is the most effective when cutting depth is small and cutting velocity is big enough. As far as the finishing cutting is concerned, a profile after the first tool pass has a big influence on the vibrations character during the regenerative cutting. In the finishing cutting, the modulations of the amplitude and the phenomenon of synchronization appear. Near the fundamental resonance the region of synchronization is the widest and the clearest. Outside of the region the quasi-periodic oscillations are observed. The interesting effect of the local decreasing of the amplitude is visible near the fundamental resonance when velocity $v_{\mathrm{w}y}$ is on the decreasing branch of the characteristic of the force vs. velocity. Then a small change of ω_{d} produces a sudden and big variation of the amplitude.

Bibliography

[1] J.M. Aquirregabiria, *Dynamics Solver*, The University of the Basque Country, Bilbao, 2001.

[2] I. Grabec, *Chaotic dynamics of the cutting process*, International Journal of Machine Tools and Manufacture **28**(1) (1998), 19–32.

[3] J. Gradisek, E. Govekar, and I. Grabec, *A chaotic cutting process and determining optimal cutting parameter values using neural networks*, International Journal of Machine Tools and Manufacture **36**(10) (1996), 1161–1172.

[4] J. Lipski, G. Litak, R. Rusinek, K. Szabelski, A. Teter, J. Warmiñski, and K. Zaleski, *Surface quality of a work material's influence on the vibrations of the cutting process*, Journal of Sound and Vibration **252**(4) (2002), 729–737.

[5] G. Litak, *Chaotic vibrations in a regenerative cutting process*, Chaos Solitons and Fractals **13** (2002), 1531–1535.

[6] D.B. Marghitu, B.O. Ciocirlan, and N. Craciunoiu, *Nonlinear dynamics in orthogonal turning process*, Chaos Solitons and Fractals **12** (2001), 2343–2352.

[7] M. Merchant, *Mechanics of metal cutting process. Part I. Orthogonal cutting and a type-2 chip*, Journal of Applied Physics **16** (1945), 267–275.

[8] M. Merchant, *Mechanics of metal cutting process. Part II. Plasticity conditions in orthogonal cutting*, Transactions of the ASME Journal of Applied Mechanics **A11** (1945), 318–324.

[9] A.H. Nayfeh, Chin Char-Ming, and J. Pratt, *Perturbation mthods in nonlinear dynamics - applications to machining dynamics*, Journal of Manufacturing Science and Engineering **119** (1997), 485–493.

[10] V. Pisspanen, *Theory of chip formation*, Teknillinen Aikaauslenti **27** (1937), 315-322 (in Finnish).

[11] F. Taylor, *On the arto of cutting metal*, Transactions of the ASME **28** (1907), 31–248.

[12] J. Tlusty, *Dynamics of high speed milling*, Transactions of the ASME Journal of the Engineering Industry **108** (1986), 59–67.

[13] J. Warminski, J. Lipski, K. Zaleski, K. Szabelski, G. Litak, A. Teter, and R. Rusinek, *Modelling of nonlinear vibrations in secondary turning process*, in: Proceedings of the Dynamics and Control of Mechanical Processing, Budapest, Publishing Company of Technical University of Budapest (1999).

[14] M. Wiercigroch, *Chaotic vibration of a simple model of the machine tool - cutting process system*, Journal of Vibration and Acoustics **119** (1997), 468–475.

[15] M. Wiercigroch and E. Budak, *Sources of nonlinearities, chatter generation and suppression in metal cutting*, Philosophical Transactions of the Royal Society **359** (2001), 663–693.

[16] M. Wiercigroch and A.M. Krivtsov, *Frictional chatter in orthogonal metal cutting*, Philosophical Transactions of the Royal Society **359** (2001), 713–738.

10 Modeling of High-speed Milling for Prediction of Regenerative Chatter

N. van de Wouw, R.P.H. Faassen, J.A.J. Oosterling, and H. Nijmeijer

High-speed milling is widely used to manufacture products. Examples of application areas are the aerospace industry and the mould industry. Cost-related considerations place high demands on the material removal rate and surface generation rate. However, in this respect restrictions on the process parameters, determining these two rates, are imposed by the occurrence of regenerative chatter. Chatter is an undesired instability phenomenon, which causes both reduced product surface quality and tool wear. In this paper, the milling process is modeled, based on dedicated experiments on both the behavior of the workpiece material and the machine dynamics. Moreover, an efficient method for determining the chatter boundaries in the model is proposed and applied to the model in order to predict chatter boundaries in the process parameters, such as the spindle speed and depth of cut, which both influence the material removal rate and surface generation rate. Finally, experiments are performed to estimate these chatter boundaries in practice. Comparison of the modeled chatter boundaries with these experimental results confirms the validity of the model and the effectiveness of the stability analysis proposed.

10.1 Introduction

The milling process is used widely in many sectors of industry. Some examples are the fabrication of moulds and the aeroplane building industry, where large amounts of material are removed from a large structure. The milling process is most efficient if the material removal rate is as large as possible, while maintaining a high quality level. For a certain machine–tool–workpiece combination, the main factors that have influence on this removal rate are the spindle speed, the depth of cut (axial and radial) and the feed rate.

During the milling process, chatter can occur at certain combinations of axial depth of cut and spindle speed. This is an undesired phenomenon, since the surface of the workpiece becomes non-smooth as a result of heavy vibrations of the cutter. Moreover, the cutting tool and machine wear out rapidly and a lot of noise is produced when chatter occurs. Several physical mechanisms causing chatter can be distinguished [1]. Wierchigroch [2], Wierchigroch and Krivtsov [3] and Grabec [4] showed that friction between the tool and the workpiece can cause chatter. Another cause for chatter is due to the thermodynamics of the cutting process [1, 5]. In [6], the phenomenon of mode coupling is discussed as a cause of chatter. Chatter due to these physical mechanisms is often called primary chatter. Secondary chatter is caused by the regeneration of waviness of the surface of the workpiece. This regenerative chatter is

Nonlinear Dynamics of Production Systems. Edited by G. Radons and R. Neugebauer

ISBN 3-527-40430-9

considered to be one of the most important causes of instability in the cutting process. This type of chatter will be considered in this paper.

Several studies have been made since the late 1950s regarding regenerative chatter, by e.g. Tobias and Fishwick [7], Tobias [8], Tlusty and Polacek [6], Merrit [9] and Altintas [10]. It was shown that the border between a stable cut (i.e. no chatter) and an unstable cut (i.e. with chatter) can be visualized in terms of the axial depth of cut as a function of the spindle speed. This results in a stability lobes diagram (SLD). Using these diagrams it is possible to find the specific combination of machining parameters which results in the maximum chatter-free material removal rate.

In order to predict the stability boundaries related to chatter, an accurate dynamic model for the milling process is needed. Therefore, in Sect. 10.2, the modeling of the milling process is described. This model will be used to find the stability limit of the milling process. The model consists of a part which describes the machine–tool interaction and a part which describes the machine dynamics. For both model parts, dedicated experiments are performed to support such modeling. The resulting model is described by a set of linear delay-differential equations. In Sect. 10.3, the D-partitioning method [11, 12] is used to analyze the stability boundary (in terms of spindle speed and depth of cut) of the equilibrium point of those differential equations, which represents a stable cut. The results of such a stability analysis for the model, constructed in Sect. 10.2, are compared to experimentally determined stability boundaries in Sect. 10.4 for validation purposes. Finally in Sect. 10.5, conclusions are presented.

10.2 Modeling

In the milling process, material is removed from a workpiece by a rotating tool, which has one or more cutting teeth. While the tool rotates, it translates in the feed direction at a certain speed. A schematic representation is shown in Fig. 10.1. The parameters shown are the spindle speed Ω, the feed per tooth f_z, the axial depth of cut a_p and the radial depth of cut a_e. As a result of the feed rate and the rotating cutter, the chip thickness is not constant, but periodic.

The milling process is an interaction between a milling machine and a workpiece. This interaction is shown in a block diagram in Fig. 10.2.

A certain displacement of the cutter, related to the feed, is dictated to the spindle of the machine. The static chip thickness (h_stat) is a result of this displacement. While achieving this displacement the cutter–workpiece interaction results in a force $\underline{F}$ acting on the cutter, if the cutter is in cut. This force on the cutter causes a displacement of the cutter. This fact reflects the main difference between the milling process and e.g. the sawing process. Namely, in the sawing process a certain force is dictated to the tool, whereas in the milling process a certain displacement is dictated to the tool.

The mechanism described above results in vibrations of the tool. This causes a wavy surface of the workpiece. The next cutting tooth encounters this wavy surface and generates its own wavy surface. The chip thickness is, therefore, the sum of the static and dynamic chip thicknesses. The static chip thickness is a function of the feed rate f_z and the rotation angle of the cutter ϕ: $h_\mathrm{stat}(t) = f_z \sin(\phi(t))$, which expresses that the chip thickness is measured in the radial direction, see Fig. 10.1. Hereto, the tooth path is assumed to be a circular arc. The dynamic chip thickness is a result of the vibration of the tooth in cut in x and

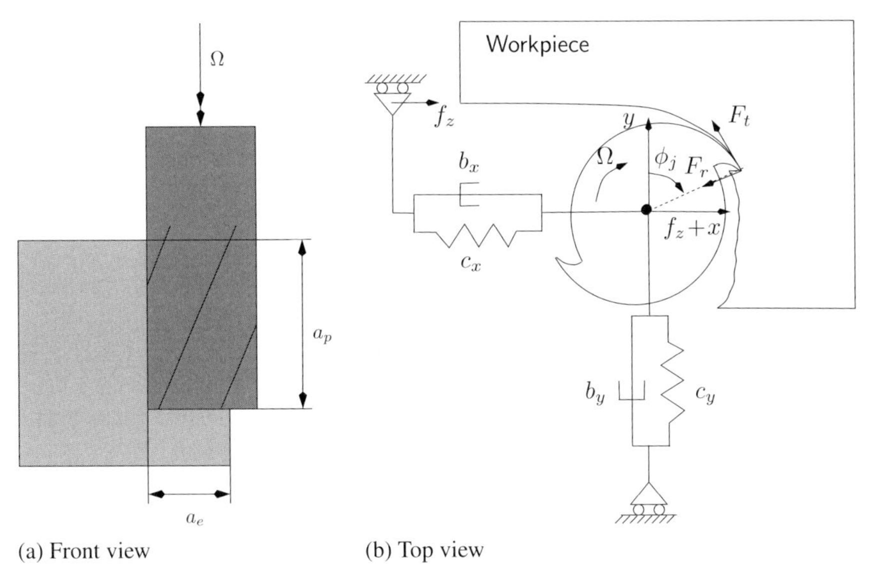

(a) Front view (b) Top view

Figure 10.1: Schematic representation of the milling process.

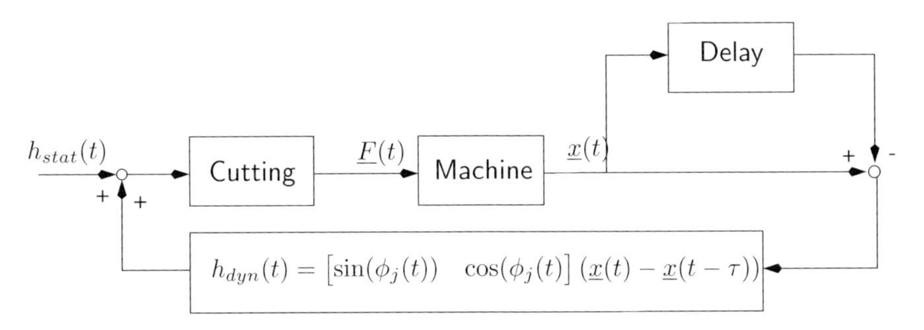

Figure 10.2: Block diagram of the milling process.

y directions (see Fig. 10.1) and the vibration of the previous tooth in cut in these directions. These displacement coordinates x and y are gathered in a displacement column $\underline{x} = [x \ \ y]^{\mathrm{T}}$. The difference between the actual displacement of the cutter and the dictated displacement of the cutter causes a dynamic chip thickness $h_{\text{dyn}}(t)$. This dynamic chip thickness is defined by $h_{\text{dyn}}(t) = [\sin(\phi(t)) \quad \cos(\phi(t))](\underline{x}(t) - \underline{x}(t-\tau))$. So, it can be seen that the force on the cutter is not only dependent on the current cutter displacement, but also on the displacement of the previous tooth. Therefore, the tooth-passing period determines the delay time τ in Fig. 10.2. When the spindle speed Ω is given in rpm and z is the number of teeth of the cutter, the tooth-passing period is defined as

$$\tau = \frac{60}{z\,\Omega}. \tag{10.1}$$

Appropriate modeling of the cutting process and machine dynamics (see Fig. 10.2) is discussed in Sects. 10.2.1 and 10.2.2, respectively. The material model will be constructed on the basis of dedicated experiments in Sect. 10.2.1. Note that, in Fig. 10.1, the dynamics of the spindle system is modeled by a mass–spring–damper system. However, of course the dynamics, in both x and y directions, may very well be more complex and in Sect. 10.2.2 a multi-modal model will be proposed, for both directions, based on experiments.

10.2.1 Material Model

As was outlined in Sect. 10.2, the milling process is an interaction between the cutting process and the machine dynamics. In this section, we focus on the cutting process. In the literature, various models have been proposed to model the radial and tangential cutting forces F_{r} and F_{t} as a function of the cutting parameters, such as the depth of cut and the feed rate. For some of these models, the forces acting on a single cutting tooth are shown in Table 10.1. The parameter $K_{\mathrm{c1.1}}$ in the model by Kienzle is defined as the force that is needed to cut a chip of 1 mm by 1 mm.

As was stated before, the static chip thickness is approximated by $h_{\mathrm{stat}} = f_z \sin\phi_j$, where ϕ_j is the rotation angle of tooth j. The dynamic chip thickness is assumed to be the difference between the vibration x and y of the current tooth at rotation $\phi_j(t)$ and the vibration x and y of the previous tooth at $(t-\tau)$, when $\phi_j(t) = \phi_{j-1}(t-\tau)$. Consequently, the chip thickness h encountered by tooth j can be described as a function of rotation angle ϕ_j by ($h = h_{\mathrm{stat}} + h_{\mathrm{dyn}}$)

$$h_j\left(\phi_j\left(t\right)\right) = f_z \sin\phi_j(t) + (x(t) - x(t-\tau))\sin\phi_j(t) + (y(t) - y(t-\tau))\cos\phi_j(t)\,, \tag{10.2}$$

with

$$\phi_j(t) = \Omega\,t + j\,\vartheta, \quad j = 0, 1, \ldots, z-1, \tag{10.3}$$

where Ω is expressed in rad/s and $\vartheta = \frac{2\pi}{z}$ is the angle between two consecutive teeth. The radial and tangential forces are zero when the tooth is not in cut. This can be modeled by multiplying the equations which describe the force of Table 10.1 by a function $g_j(\phi_j(t))$, which describes whether a tooth is in or out of cut. The tooth is in cut if $\phi_{\mathrm{s}} \le \phi_j \le \phi_{\mathrm{e}}$, where ϕ_{s} and ϕ_{e} are the start and exit angles, respectively. This function is given by

$$g_j(\phi_j(t)) = \frac{1}{2}\left(1 + \mathrm{Sgn}\left(\sin\left(\phi_j(t) - \psi\right) - p\right)\right) = \begin{cases} 1, & \phi_{\mathrm{s}} \le \phi_j(t) \le \phi_{\mathrm{e}}, \\ 0, & \text{otherwise}, \end{cases}\,, \tag{10.4}$$

with

$$\tan\psi = \frac{\sin\phi_{\mathrm{s}} - \sin\phi_{\mathrm{e}}}{\cos\phi_{\mathrm{s}} - \cos\phi_{\mathrm{e}}}, \quad p = \sin\left(\phi_{\mathrm{s}} - \psi\right). \tag{10.5}$$

Table 10.1: Different models for the cutting force.

Author	Model		Year
Kienzle and Victor [13]	$\bar{F}_{\mathrm{t}} = a_{\mathrm{p}}\,K_{\mathrm{c1.1}}\,\bar{h}^{1-m}$		1950s
Altintas and Budak [14, 10]	$F_{\mathrm{t}} = a_{\mathrm{p}}\,K_{\mathrm{tc}}\,h + a_{\mathrm{p}}\,K_{\mathrm{te}}$,	$F_{\mathrm{r}} = a_{\mathrm{p}}\,K_{\mathrm{rc}}\,h + a_{\mathrm{p}}\,K_{\mathrm{re}}$	1995, 2000
Stépán and Insperger [15, 16]	$F_{\mathrm{t}} = a_{\mathrm{p}}\,K_{\mathrm{t}}\,h^{x_{\mathrm{F}}}$,	$F_{\mathrm{r}} = a_{\mathrm{p}}\,K_{\mathrm{r}}\,h^{x_{\mathrm{F}}}$	2000, 2001

In Table 10.1, the material model used by Insperger and Stépán [15] is shown. From now on, we will use this exponential model, since experiments (such as described later on in this section) showed that for a fixed spindle speed Ω chatter may occur for one specific feed rate f_z where it may not for another feed rate. At low feed rates (0.08 or 0.12 mm/tooth) chatter occurred at some spindle speeds while, for higher feed rates, chatter did not occur at the same spindle speed. Such feed-rate dependency is not modeled by the linear model of Altintas and Budak [14, 10], whereas the exponential model takes a feed-rate dependency into account. As will be shown the estimated stability limit indeed increases if the feed rate increases when the exponential model is used. This behavior was also found empirically when performing the cutting tests described in the next section. Multiplying this model with Eq. (10.4) gives for a single tooth:

$$\begin{aligned} F_{\mathrm{t}_j}(t) &= K_{\mathrm{t}}\, a_{\mathrm{p}}\, h_j(t)^{x_{\mathrm{F}}}\, g_j(\phi_j(t)), \\ F_{\mathrm{r}_j}(t) &= K_{\mathrm{r}}\, a_{\mathrm{p}}\, h_j(t)^{x_{\mathrm{F}}}\, g_j(\phi_j(t)), \end{aligned} \tag{10.6}$$

where $0 < x_{\mathrm{F}} < 1$. Using Eqs. (10.2), (10.4) and (10.6) an expression for the cutting forces can be derived:

$$\begin{aligned} \underline{F}(t) &= \begin{bmatrix} F_x(t) \\ F_y(t) \end{bmatrix} \\ &= a_{\mathrm{p}} \sum_{j=0}^{z-1} g_j(\phi_j(t))\left(f_z \sin\phi_j(t) + x(t,t-\tau)\sin\phi_j(t) + y(t,t-\tau)\cos\phi_j(t)\right)^{x_{\mathrm{F}}} \\ &\quad \times \begin{bmatrix} -K_{\mathrm{t}}\cos\phi_j(t) - K_{\mathrm{r}}\sin\phi_j(t) \\ K_{\mathrm{t}}\sin\phi_j(t) - K_{\mathrm{r}}\cos\phi_j(t) \end{bmatrix}, \end{aligned} \tag{10.7}$$

with $\underline{x}(t,t-\tau) = \underline{x}(t) - \underline{x}(t-\tau)$. This equation can also be linearized around $\underline{x} = \underline{0}$, which corresponds to a stable cut without chatter, resulting in $\underline{F} = \underline{F}(\underline{x} = \underline{0}) + \Delta\underline{F}$. The linearized force $\Delta\underline{F}$ can then be written as:

$$\Delta\underline{F} = a_{\mathrm{p}}\, \underline{k}(t)\underline{x}(t,t-\tau), \tag{10.8}$$

with the matrix $\underline{k}(t)$ defined by

$$\begin{aligned} \underline{k}(t) = \begin{bmatrix} k_{xx} & k_{xy} \\ k_{yx} & k_{yy} \end{bmatrix} &= \sum_{j=0}^{z-1} f_z^{x_{\mathrm{F}}-1}\, x_{\mathrm{F}}\, g_j\left(\phi_j(t)\right) \sin^{x_{\mathrm{F}}}\phi_j(t) \\ \times \begin{bmatrix} -\left(K_{\mathrm{t}}\cos\phi_j(t) + K_{\mathrm{r}}\sin\phi_j(t)\right) & -\sin^{-1}\phi_j(t)\cos\phi_j(t)\left(K_{\mathrm{t}}\cos\phi_j(t) + K_{\mathrm{r}}\sin\phi_j(t)\right) \\ \left(K_{\mathrm{t}}\sin\phi_j(t) - K_{\mathrm{r}}\cos\phi_j(t)\right) & \sin^{-1}\phi_j(t)\cos\phi_j(t)\left(K_{\mathrm{t}}\sin\phi_j(t) - K_{\mathrm{r}}\cos\phi_j(t)\right) \end{bmatrix}&. \end{aligned} \tag{10.9}$$

If the spindle is modeled as a two-dimensional linear mass–spring–damper system, the milling process is described as

$$\underline{M}\,\ddot{\underline{x}}(t) + \underline{B}\,\dot{\underline{x}}(t) + \underline{C}\,\underline{x}(t) = a_{\mathrm{p}}\,\underline{k}(t)\,\underline{x}(t,t-\tau), \tag{10.10}$$

with $\underline{M}$, $\underline{B}$ and $\underline{C}$ the mass, damping and stiffness matrices, respectively. It should be noted that, in Sect. 10.2.2, a higher-order model for the machine dynamics is constructed.

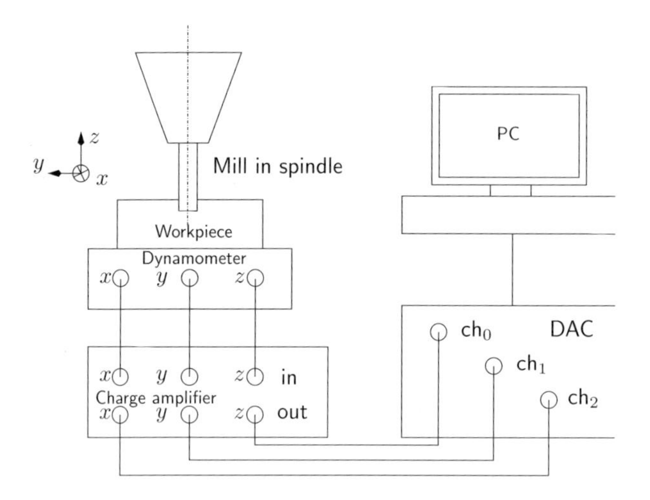

Figure 10.3: Schematic representation of the setup for the cutting experiments.

In order to validate the model described above and to find the values of the material parameters K_t, K_r and x_F of the model, experiments are performed. A schematic representation of the experimental setup is shown in Fig. 10.3. An aluminum 6082 (DIN 3.2315.71) workpiece is mounted on a dynamometer (Kistler 9255). Using this dynamometer and a charge amplifier, the forces in x, y and z directions can be measured. The dynamometer is placed on the machine bed in such a way that x is the feed direction, y is the normal direction and z is the axial direction. At full immersion, a cut is made and the forces are measured. The cuts have been made at TNO Industrial Technology using a Mikron HSM 700 milling machine and a Kelch shrink-fit toolholder. The tool used is a 2-tooth, 10-mm-diameter Jabro Tools JH420 cutter. The spindle speed has been varied from 5000 rpm to 40 000 rpm with increments of 5000 rpm. At each spindle speed the feed rate has been varied from 0.08 mm/tooth to 0.24 mm/tooth with increments of 0.04 mm/tooth. Two cuts have been made at each combination of spindle speed and feed rate. The sampling frequency of the measurements is 30 kHz for spindle speeds below 35 000 rpm and 50 kHz at spindle speeds of 35 000 and 40 000 rpm.

For an experiment performed at 20 000 rpm, a fit of the exponential model using the mean cutting forces, $\bar{F}_x = \int_0^T F_x(t)dt$, $\bar{F}_y = \int_0^T F_y(t)dt$, is performed taking all measurements at that spindle speed into account. The result is shown in Fig. 10.4. The material parameters of this exponential model can be found in Table 10.2. Corresponding measured forces and the modeled forces at a feed rate of $f_z = 0.16$ mm/tooth are shown in Fig. 10.5. If the spindle speed is changed, the values for the material parameters also change to a certain extent. For every spindle speed, the parameters of the exponential model are obtained by fitting this model

Table 10.2: Material parameters for cutting at 20 000 rpm.

K_t	473.7	N/mm$^{1+x_\mathrm{F}}$
K_r	127.5	N/mm$^{1+x_\mathrm{F}}$
x_F	0.78	–

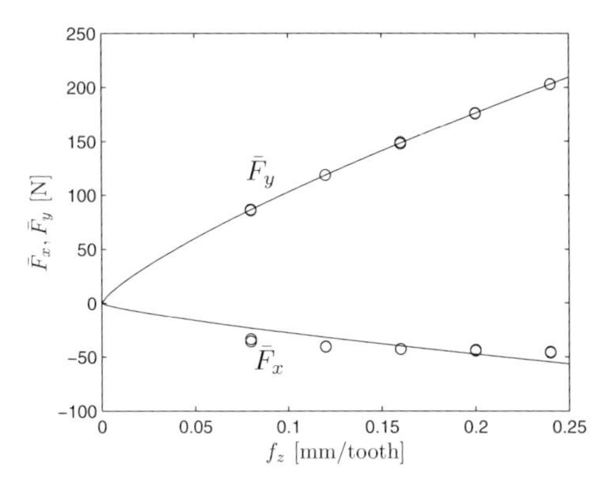

Figure 10.4: Fitting the material parameters on the mean measured forces at 20 000 rpm. Material parameters can be found in Table 10.2.

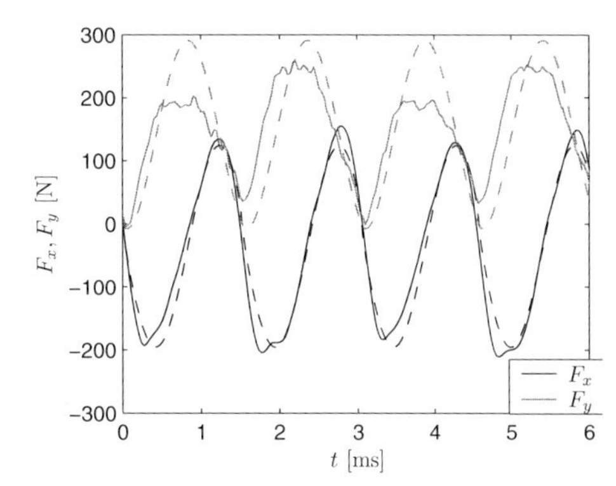

Figure 10.5: Forces in x and y directions as a function of time at 20 000 rpm and a feed rate of $f_z = 0.16$ mm/tooth (solid: experiments, dashed: model).

to the experimental results at these spindle speeds. In Fig. 10.6, the material parameters of the exponential model are shown as a function of spindle speed.

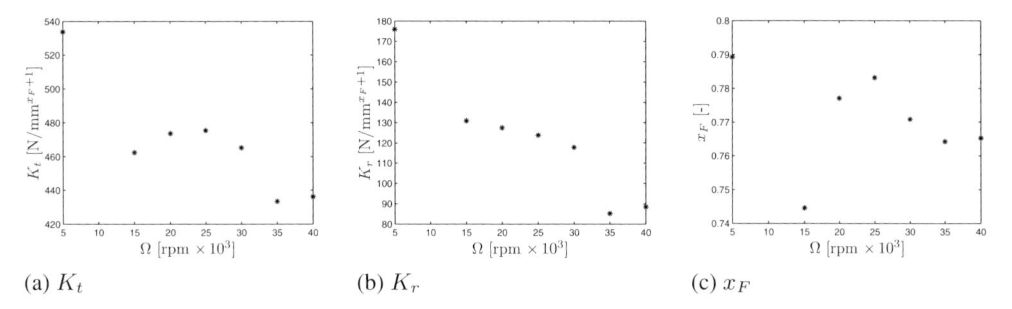

(a) K_t (b) K_r (c) x_F

Figure 10.6: Material parameters of the exponential model as a function of spindle speed.

10.2.2 Machine Model

The second part of the block diagram of Fig. 10.2 represents the modeling of the machine dynamics, i.e. the tool, toolholder and spindle. In this section, the machine dynamics will be modeled based on dedicated experiments.

In the literature [17–19] the machine system is assumed to be a 1 DOF or 2 DOF linear second-order (mass–spring–damper) system. However, experiments will show that a higher-order model is necessary to describe the machine dynamics. The dynamics can be described by the transfer function matrix between cutting forces $\underline{F} = [F_x \;\; F_y]^{\mathrm{T}}$ and displacements of the cutter $\underline{x} = [x \;\; y]^{\mathrm{T}}$:

$$\underline{X}(s) = \underline{H}_{\underline{x}\,\underline{F}}(s)\underline{F}(s) = \begin{bmatrix} H_{11}(s) & H_{12}(s) \\ H_{21}(s) & H_{22}(s) \end{bmatrix} \underline{F}(s), \tag{10.11}$$

where $\underline{X}(s) = \mathcal{L}(\underline{x}(t))$ and $\underline{F}(s) = \mathcal{L}(\underline{F}(t))$ with $\mathcal{L}(\cdot)$ the Laplace operator. Each entry of this transfer function matrix can be modeled using the following model form:

$$H_{ij}(s) = \frac{b_m\, s^m + b_{m-1}\, s^{m-1} + \ldots + b_1\, s + b_0}{a_n\, s^n + a_{n-1}\, s^{n-1} + \ldots + a_1\, s + a_0}, \tag{10.12}$$

where the coefficients a and b differ for different i and j. It will be assumed that $H_{ij} = 0$ for $i \neq j$. In other words, the dynamics in x and y directions are decoupled. These transfer functions describe the dynamics of the cutter (mill), due to flexibility of the mill, and the dynamics of the spindle and toolholder, from now on called the spindle dynamics. Firstly, experiments are performed to measure the spindle dynamics at different spindle speeds. Secondly, experiments are performed, at $\Omega = 0$, to measure the dynamics of the mill. Let us now discuss these experiments performed to retrieve H_{11} and H_{22}.

In order to measure the dynamic behavior of the spindle system, consisting of the tool, toolholder and spindle, and the influence of the spindle speed on this dynamic behavior, impulse tests are performed. Using these tests it is possible to find a frequency response function between the force applied to the tool and the displacement, which provides information on the dynamical behavior of the spindle system. These impulse tests are performed at various spindle speeds, in order to study the spindle-speed dependency of the machine dynamics. A schematic representation of the setup is shown in Fig. 10.7. A 50-mm-long, 10-mm-diameter carbide cylinder is mounted in a Kelch shrink-fit toolholder and used on a Mikron HSM 700 milling machine. Since the spindle is rotating while being hit by the impulse hammer, a mill cannot be used. The first natural frequency of the cylinder is approximately 4 kHz and the dynamics of toolholder and spindle is related to a lower-frequency range. Therefore, the dynamics of the toolholder and spindle can be distinguished from the dynamics of the cylinder and identified separately.

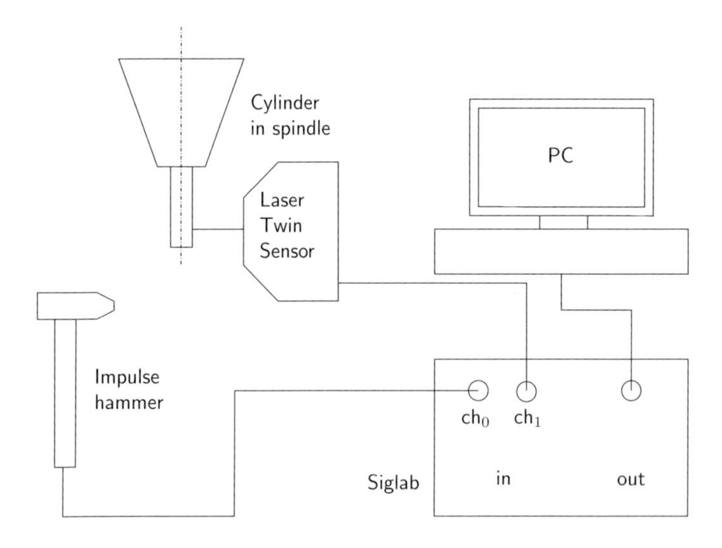

Figure 10.7: Schematic representation of the setup for the impulse tests.

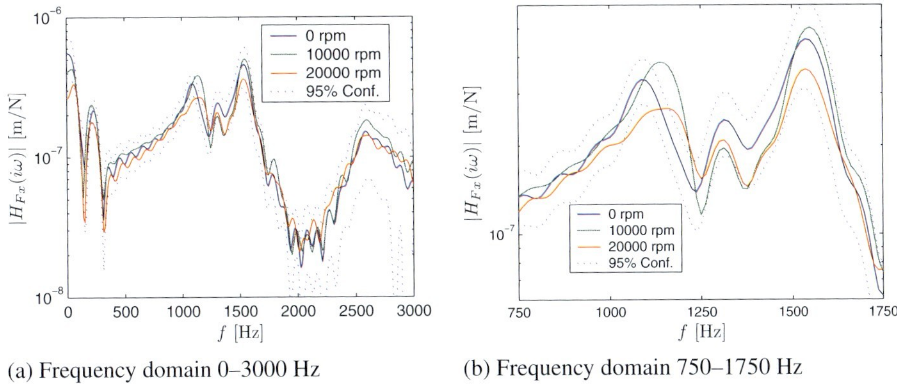

(a) Frequency domain 0–3000 Hz (b) Frequency domain 750–1750 Hz

Figure 10.8: $|H_{11}|$ at different spindle speeds. Dotted line: 95% confidence interval at 0 rpm.

The displacement of the cylinder in x or y direction is measured with an LMI laser twin sensor (LTS 15/3). An impulse force hammer is used to hit the cylinder and to measure the force applied to it. A Siglab dynamic signal analyzer is used for data-acquisition purposes. The spindle speed has been varied from 0 to 25 000 rpm, with increments of 5000 rpm. At each spindle speed, 20 impacts have been performed in both x and y directions, while the laser has been placed opposite to the place of impact.

In Fig. 10.8, the absolute value of the measured frequency response function $H_{11}(i\omega)$ is depicted at different spindle speeds (mean of 20 measurements). Statistical significance tests showed that the differences between these frequency response functions at different spindle speeds are significant, especially in the frequency range 750–1750 Hz, in which the most important resonances related to the spindle dynamics are situated. Since the natural frequency of the cylinder lies in the order of 4 kHz and the measured natural frequencies are much lower, the latter frequencies are not the natural frequencies of the cylinder but are related to the toolholder and spindle. Identical experiments are performed to measure $H_{22}(i\omega)$.

Next, the same type of experiments is performed on a non-rotating mill in order to capture the dynamics of the real cutter tool. Since the flexibility modes of the tool do not depend on the spindle speed, only measurements at 0 rpm are performed. The total dynamics of the toolholder and spindle and the mill can now be constructed by superposition of the individual dynamics of the toolholder and spindle on the one hand and the dynamics of the mill on the other hand, see Fig. 10.9. In doing so, up to 2530 Hz the spindle dynamics is taken into account and above 2530 Hz the mill dynamics is dominant.

The total machine dynamics, see Fig. 10.9, can now be modeled using Eqs. (10.11) and (10.12). In order to obtain the parameters of this model, the distance between the measured and modeled frequency response functions in the complex plane is minimized using an optimization routine and a least-squares-type objective function. In Fig. 10.9, the resulting modeled frequency response function, formulated by an 18th-order model, is compared to the measured frequency response function for $\Omega = 0$. When performing this identification step, the highest level of importance was assigned to the highest resonance peaks in the absolute value of the experimental frequency response function, because these resonances are dominant in the stability analysis, as will be shown in Sect. 10.4. The same parametric estimation

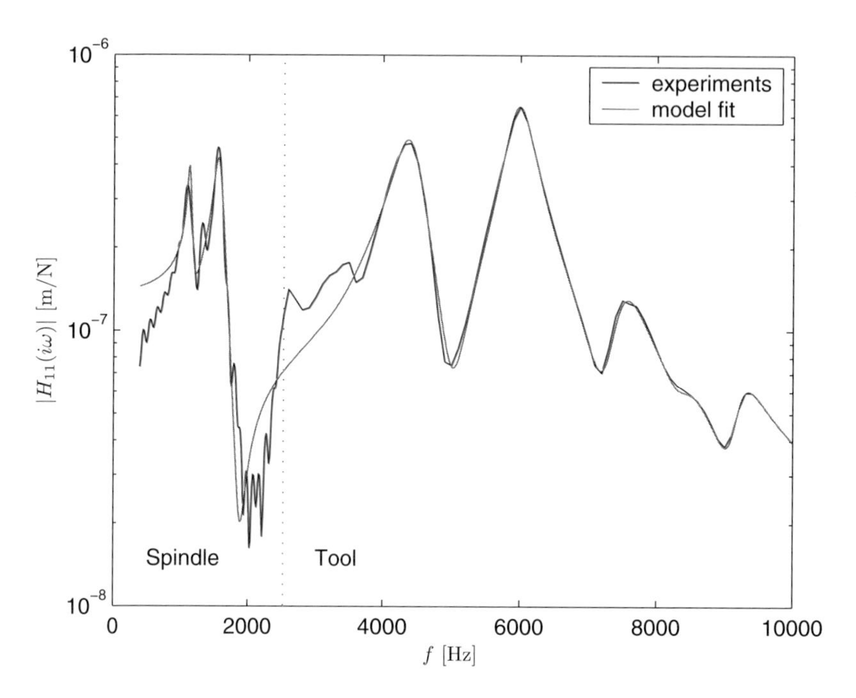

Figure 10.9: Measured $|H_{11}|$ for $\Omega = 0$ rpm incorporating both tool dynamics and toolholder and spindle dynamics vs. corresponding modeled tool and spindle dynamics.

procedure is applied to $H_{22}(i\omega)$, which closely resembles $H_{11}(i\omega)$. Since it is assumed that $H_{12}(i\omega) = H_{21}(i\omega) = 0$, all information on $\underline{H}_{\underline{x}\,\underline{F}}(s)$ is available for $\Omega = 0$. Of course, the same parametric identification procedure was performed for other spindle speeds, resulting in different dynamic models for different spindle speeds.

10.2.3 The Total Milling Model

The models for the material behavior and the dynamics of tool, toolholder and spindle can be combined to form a model for the milling process using the block diagram of the milling process as depicted in Fig. 10.2. This leads to the following description of the milling process in the Laplace domain:

$$\underline{H}_{\underline{x}\,\underline{F}}^{-1}(s)\underline{X}(s) = a_{\mathrm{p}}\underline{K}(s) * \left(1 - e^{-s\tau}\right)\underline{X}(s), \tag{10.13}$$

where $\underline{H}_{\underline{x}\,\underline{F}}(s)$ is defined by Eqs. (10.11) and (10.12), the matrix $\underline{K}(s) = \mathcal{L}(\underline{k}(t))$ is the Laplace transform of the matrix $\underline{k}(t)$ related to the material properties and '$*$' denotes the convolution operator.

It should be noted that Eq. (10.13) represents the Laplace transform of a set of linear non-autonomous delay-differential equations, where the non-autonomous nature is due to the explicit time dependency of $\underline{k}(t)$. For the remainder of this paper, $\underline{k}(t)$ will be approximated by its zeroth-order Fourier approximation:

$$\underline{k} = \frac{1}{\tau} \int_0^{\tau} \underline{k}(t)\, dt, \tag{10.14}$$

which, since $\underline{k}(t)$ is only non-zero if a tooth is actually in cut, equals

$$\underline{k} = \frac{1}{\vartheta} \int_{\phi_s}^{\phi_e} \underline{k}(\phi)\, d\phi, \tag{10.15}$$

where ϑ is the angle between two consecutive teeth, $\phi = \Omega t$ and $\underline{k}(t)$ is defined by Eq. (10.9) for the linearized exponential material model. This approximation is quite common in literature and dramatically simplifies the stability analysis since it transforms the model into an autonomous model [10]. Consequently, the convolution operation in Eq. (10.13) changes to a normal multiplication:

$$\underline{\underline{H}}_{x\,F}^{-1}(s)\underline{X}(s) = a_p \underline{\underline{K}} \left(1 - e^{-s\tau}\right) \underline{X}(s), \tag{10.16}$$

where $\underline{\underline{K}}$ is a constant matrix.

10.3 Stability Analysis of the Milling System

In this section, the model of the milling system proposed in the previous section (see Eq. (10.13)) will be used for the purpose of stability analysis. The linear autonomous delay-differential equation describing the dynamics of Eq. (10.16) in the time domain exhibits one unique equilibrium point: $\underline{x} = \underline{0}$, which corresponds to the desired no-chatter situation. Therefore, stability of this equilibrium point corresponds to the stability of the milling process and instability of the equilibrium point corresponds to a response with chatter. It should be noted that the change in stability properties of the equilibrium point in the linearized, autonomous system corresponds to a Hopf bifurcation of the equilibrium point of the corresponding nonlinear, autonomous system. When the time averaging of $\underline{k}(t)$ is omitted, the milling process is described by a non-autonomous delay-differential equation. In that case, in general, the no-chatter situation corresponds to a periodic solution with a period time corresponding to the delay time τ. Bifurcations of this solution, a secondary Hopf or a flip bifurcation, lead to chatter [20].

Here, the method of D-partition [12] will be used to assess the stability of this equilibrium point. This method was used by e.g. Stépán [11, 21] to investigate the stability of the milling process using a single-degree-of-freedom, single-mode milling model. Note that, for a given spindle speed Ω, the stability of the equilibrium point depends on the axial depth of cut a_p. So, the stability analysis will aim at finding the critical value for a_p, at a given spindle speed Ω, which forms the stability boundary, allowing for the construction of so-called stability lobes diagrams.

10.3.1 Method of D-partition

The method of D-partition uses the criterion that an equilibrium point of a system, described by a linear, autonomous delay-differential equation, is asymptotically stable if and only if all the roots of its characteristic equation lie in the open left-half complex plane. It should be noted that a delay-differential equation has an infinite number of poles. The characteristic equation corresponding to Eq. (10.16) is given by

$$\det\left(\underline{H}_{xF}^{-1}(s) - a_{\mathrm{p}}\underline{K}\left(1 - e^{-s\tau}\right)\right) = 0. \tag{10.17}$$

A certain choice for the system's parameters (for example a_{p}) determines the number of poles in the open left-half complex plane. The parameter space can be divided into domains $D(k, n-k)$, $0 \le k \le n$ which contain all the points with poles with k negative real parts and $n-k$ positive real parts. This is called D-partitioning. The domain of asymptotic stability is the domain $D(n, 0)$. An increase of the number of roots with positive real parts can only occur if a certain pole crosses the imaginary axis from the left to the right. This corresponds with the situation that a certain point in parameter space moves from the domain $D(k, n-k)$ to $D(k-1, n-(k-1))$. Therefore, the borders of the D-partitions are the map of the imaginary axis $s = i\,\omega$, with $-\infty < \omega < +\infty$ on the parameter space.

Let us introduce a new complex variable $S = s\tau$ and use this to re-write the characteristic equation (10.17) as

$$\det\left(\underline{H}_{xF}^{-1}\left(\frac{S}{\tau}\right) - a_{\mathrm{p}}\underline{K}\left(1 - e^{-S}\right)\right) = 0. \tag{10.18}$$

Now, we will not use this equation to determine the poles of the system for given parameters, but we will determine the values for the parameter a_{p} for which at least one pole lies on the imaginary axis ($s = i\omega$). Using the fact that $\underline{H}_{xF}(s)$ is given by Eq. (10.11), with $H_{12} = H_{21} = 0$, and choosing $S = i\omega^*$, with $\omega^* = \omega\tau$, which corresponds to $s = i\omega$, Eq. (10.18) transforms to

$$a_0\, a_{\mathrm{p}}^2 + a_1\, a_{\mathrm{p}} + 1 = 0, \tag{10.19}$$

with

$$a_0 = (1 - \cos\omega^* + i\sin\omega^*)^2\, H_{11}(i\omega^*/\tau)\, H_{22}(i\omega^*/\tau)(k_{xx}k_{yy} - k_{xy}k_{yx}), \tag{10.20}$$

$$a_1 = -(1 - \cos\omega^* + i\sin\omega^*)(k_{xx}H_{11}(i\omega^*/\tau) + k_{yy}H_{22}(i\omega^*/\tau)). \tag{10.21}$$

The axial depth of cut as a function of ω^*, $a_{\mathrm{p}}(\omega^*)$, can then be found by

$$a_{\mathrm{p}}(\omega^*) = \frac{-a_1 \pm \sqrt{a_1^2 - 4\,a_0}}{2\,a_0}. \tag{10.22}$$

The critical axial depth of cut (with respect to stability) is defined as the depth of cut a_{p} in the parameter set defined by $\{a_{\mathrm{p}}(\omega^*) : \mathrm{Im}\, a_{\mathrm{p}}(\omega^*) = 0 \wedge \mathrm{Re}\, a_{\mathrm{p}}(\omega^*) > 0\}$, for which $|\mathrm{Re}\; a_{\mathrm{p}}(\omega^*)|$ has its minimum value, since for $a_{\mathrm{p}} = 0$ all poles are in the open left-half complex plane when all the modes of the machine dynamics are damped, which is always the case in practice. The value for ω^* for which this occurs is the dimensionless chatter frequency $\omega^* = \omega_{\mathrm{c}}^*$. The real chatter frequency that corresponds to the dimensionless chatter frequency is $\omega_{\mathrm{c}} = \omega_{\mathrm{c}}^*/\tau$. In summary, the following steps need to be taken in order to use the method of D-partition to find the chatter boundary in terms of a_{p} (for a specific value of the spindle speed):

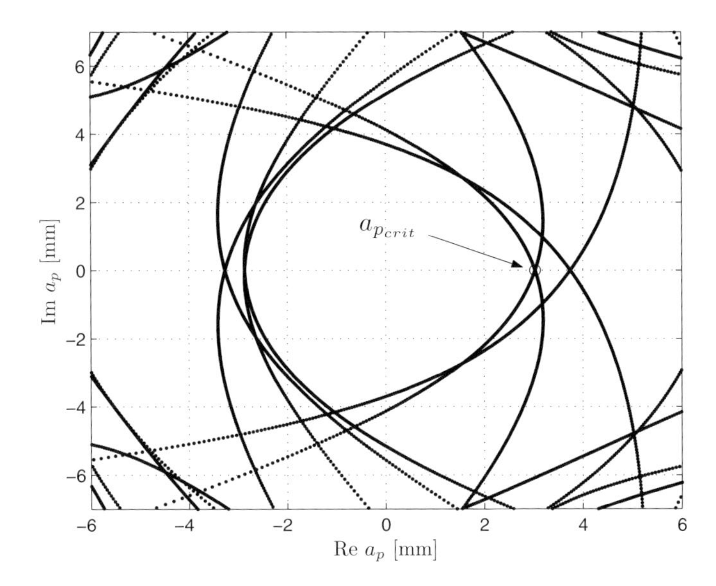

Figure 10.10: D-partition diagram for a spindle speed of 25 000 rpm.

1. Choose a certain spindle speed Ω, and calculate the delay factor τ by Eq. (10.1);
2. Choose a domain for ω^*;
3. In the characteristic equation (10.18), substitute $S = i\omega^*$;
4. Solve Eq. (10.19) for a_p. Now, $a_\mathrm{p}(\omega^*)$ is known, but $a_{\mathrm{p}_\mathrm{crit}}$ still has to be found;
5. $\omega^* = \omega_\mathrm{c}^*$ if Im $a_\mathrm{p}(\omega_\mathrm{c}^*) = 0$, $Re\, a_\mathrm{p}(\omega_\mathrm{c}^*) > 0$ and $|Re\, a_\mathrm{p}(\omega_\mathrm{c}^*)|$ has its minimum value. By scanning the positive real axis, it is the point where a D-curve crosses the real axis for the first time. This is shown in Fig. 10.10. In this figure, the dotted lines represent the boundaries between the domains $D(k, n-k)$;
6. Calculate the chatter frequency $\omega_\mathrm{c} = \omega_\mathrm{c}^*/\tau$;
7. Repeat all steps for different spindle speeds.

In the second step of this procedure, a choice for the domain of ω^* should be made, such that ω_c^* lies in this domain. For a milling model, in which the lowest angular eigenfrequency of the machine dynamics is denoted by ω_l and the highest eigenfrequency by ω_h, a suitable choice for this domain is: $0.5\omega_\mathrm{l}\tau < \omega^* < 1.5\omega_\mathrm{h}\tau$.

10.4 Results

The model of the milling process, constructed in the previous section, will be used to pursue a stability analysis. Stability lobe diagrams can be computed using the analysis method illuminated in Sect. 10.3. In order to study the accuracy of the model, the stability boundaries

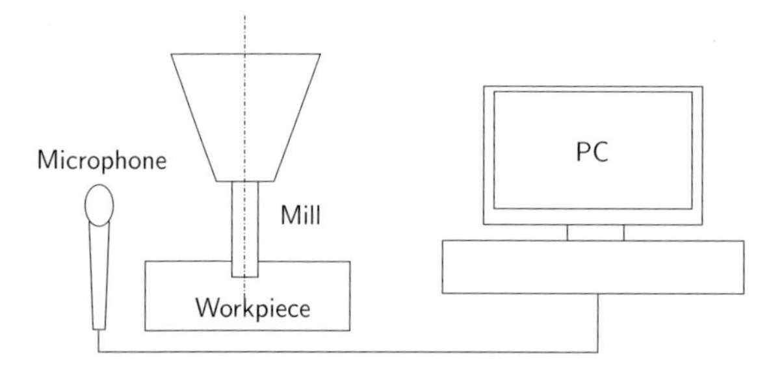

Figure 10.11: Experimental setup for the determination of the chatter boundary.

obtained in this way need to be validated with the stability border determined in practice. The model used in this analysis uses the material parameters given in Table 10.2. Moreover, the spindle-speed dependency of the machine dynamics is taken into account in the following way. At discrete values for the spindle speed Ω, $\Omega = 0, 5000, 10\,000, 15\,000, 20\,000, 25\,000$ rpm, models for the machine dynamics were constructed. When the stability analysis is performed for an arbitrary spindle speed (not at one of the discrete values for which the models were constructed) in order to find the critical value for the depth of cut a_{p} at that spindle speed, the dynamic model corresponding to the closest spindle speed is used. In this way, the spindle-speed dependency of the machine dynamics is accounted for in the stability analysis.

For validation purposes, the stability boundary, in terms of a_{p} as a function of Ω, has to be determined experimentally. Hereto, a series of cuts have been made at the Mikron HSM 700 milling machine using an aluminum 6082 (DIN 3.2315.71) workpiece, a Kelch shrink-fit toolholder and a 2-tooth, 10-mm-diameter Jabro Tools JH420 cutter. For every spindle speed investigated, an initial depth of cut is chosen such that no chatter occurs. Whether chatter is occurring or not is detected using the experimental setup depicted in Fig. 10.11 and the software program Harmonizer. This program scans the sound of the cutting process measured using the microphone. If the energy of the measured sound signal at a certain frequency exceeds a certain threshold, the cut is marked as exhibiting chatter. This threshold can be set automatically by Harmonizer, but it can also be set manually. Using Harmonizer, the chatter frequency is also measured. The frequency at which the energy level exceeds the threshold level is marked as the chatter frequency.

In Fig. 10.12, the experimental results are compared with the modeled stability lobes. Clearly, the prediction of the stability lobes diagram is good. The spindle-speed dependency of the model can be recognized in the stability lobe diagram by the fact that the minima of the stability boundary, in Fig. 10.12, vary (increase) with the spindle speed Ω. In Fig. 10.13, the measured chatter frequencies are compared with the modeled chatter frequencies and the natural frequencies of the spindle and tool. Clearly, the modeled chatter frequencies also resemble the measured chatter frequencies very well. Moreover, it can be concluded that the dynamics of the non-rotating mill highly influences the stability lobes, since the chatter frequency is always close to a resonance frequency of the mill. The natural frequencies of the spindle are much lower. This indicates that the dynamics of the spindle, which is spindle-speed dependent, is less important than the dynamics of the mill, which is spindle-speed independent. It can be concluded that for this specific, rather slender, tool this speed dependency of the spindle

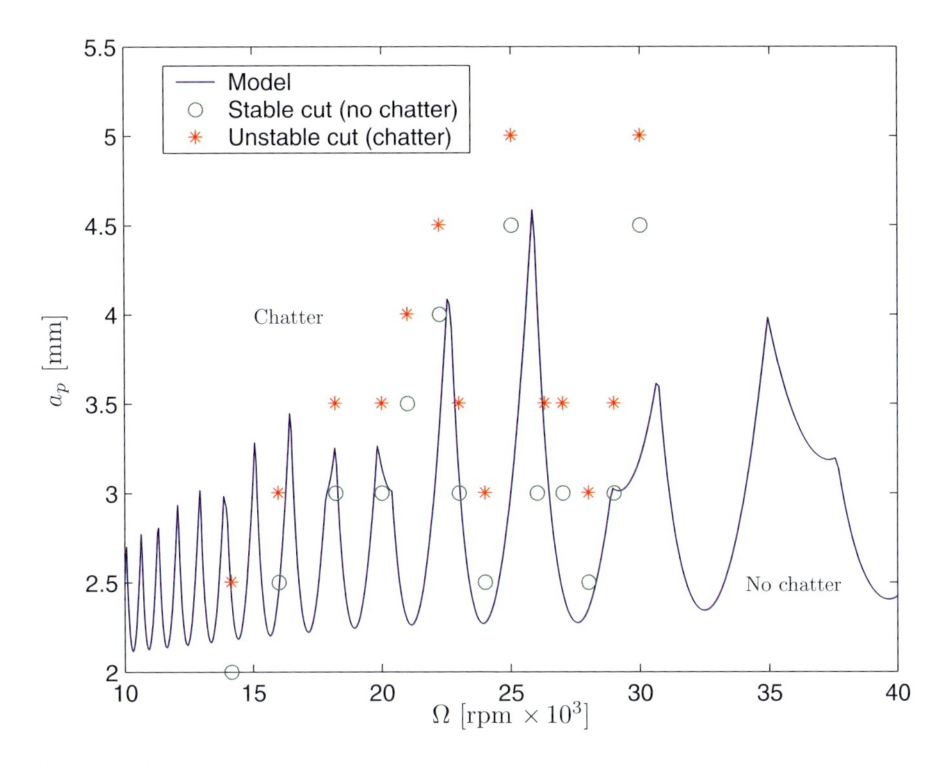

Figure 10.12: Modeled stability lobes, compared with experimentally performed cuts.

dynamics does not influence the stability lobes diagram to a great extent. However, when a shorter, thicker mill would have been used (resulting in extremely high resonance frequencies for the mill dynamics), the spindle dynamics may become dominant. In such a case, it is important to include the spindle-speed dependency of the spindle dynamics in the stability analysis. So, the fact that the minima of the stability boundary in Fig. 10.12 vary with Ω is due to the spindle-speed dependency in the material model. This can be understood by realizing that the material parameters K_t and K_r show a tendency to decrease with increasing Ω, see Fig. 10.6, which results in lower cutting forces and a higher stability boundary (in terms of a_p).

10.5 Conclusions

In this paper, a dynamic model for the milling process has been constructed based on dedicated experiments. The model comprises a material model and a model for the machine dynamics. For the material model, an exponential model (of the cutter forces in terms of the chip thickness) is used to account for feed-rate dependencies in the stability lobes diagram. A linear, higher-order model for the machine dynamics has been constructed, including dynamic modes of the tool and the spindle and toolholder. The spindle dynamics appeared to be spindle-speed dependent. The total model is described by a linear, higher-order delay-differential equation.

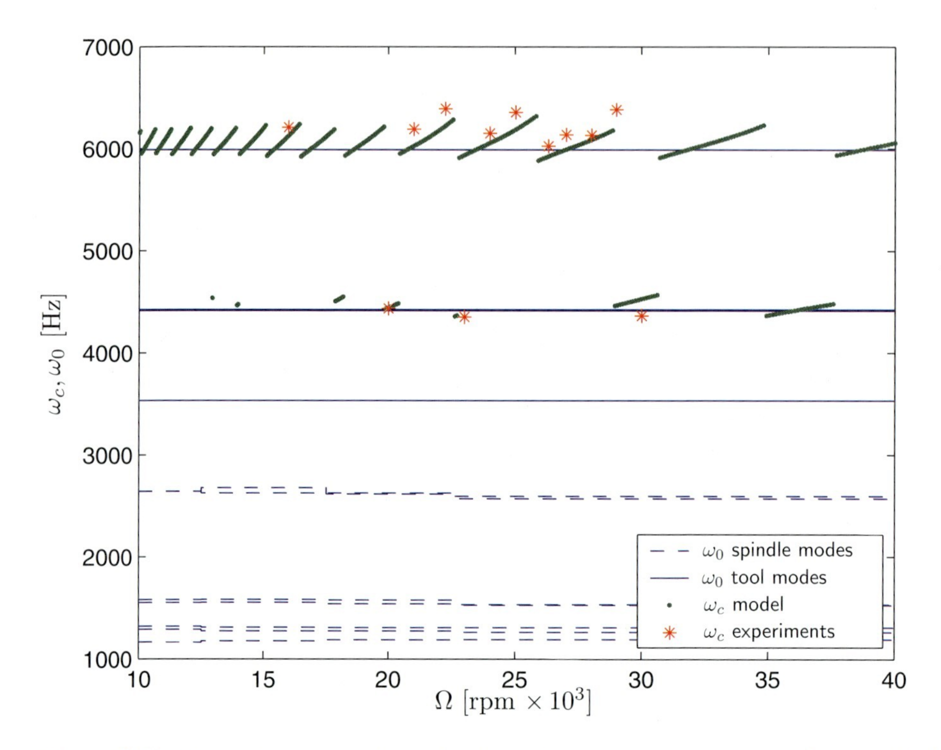

Figure 10.13: Chatter frequencies. Asterisks: chatter frequencies in experiments, dots: modeled chatter frequencies, solid lines: natural frequencies of the tool, dashed lines: speed-dependent natural frequencies of the spindle in x and y directions.

The method of D-partition is used to analyze the stability of the unique equilibrium point of this differential equation, corresponding to a process situation without chatter. In this way, the stability lobes diagram is constructed. An advantage of this method over the method used in [10] is that while using the method of D-partition the critical depth of cut can be found for a specific spindle speed, whereas while using the method of Altintas the critical depth of cut for a specific chatter frequency, which is an unknown response variable, is found. This allows us, in case of the method of D-partition, to account for spindle-speed dependencies of the milling model, whereas this is not possible in the method of [10].

It was shown that the constructed model provides an accurate prediction for the stability lobes diagram when it is compared to an experimentally determined stability lobes diagram. Moreover, the spindle-speed dependency of the model clearly influences that stability lobes diagram. For additional information on this work we refer the reader to [22].

Further research will focus on the extension of the model (and the analysis techniques needed to analyze the stability properties of such a model). One important extension relates to the inclusion of the (periodically) time-dependent nature of the milling process as was presented in [20] for a one-dimensional model.

Bibliography

[1] M. Wiercigroch and E. Budak, *Sources of nonlinearities, chatter generation and suppression in metal cutting*, Philosophical Transactions of the Royal Society of London **359**(A) (2001), 663–693.

[2] M. Wiercigroch, *Nonlinear dynamics and chaos of orthogonal metal cutting*, in: M. Wiercigroch, A. de Kraker (eds.), *Applied Nonlinear Dynamics and Chaos of Mechanical Systems with Discontinuities*, World Scientific, London, 2000, pp. 361–401.

[3] M. Wiercigroch and A.M. Krivtsov, *Frictional chatter in orthogonal metal cutting*, Philosophical Transactions of the Royal Society of London **359**(A) (2001), 713–738.

[4] I. Grabec, *Chaotic dynamics of the cutting process*, International Journal of Machine Tools and Manufacture **28** (1988), 19–32.

[5] M.A. Davies, T.J. Burns, and C.J. Evans, *On the dynamics of chip formation in machinig hard materials*, Annals of the CIRP **46** (1997), 25–30.

[6] J. Tlusty and M. Polacek, *The stability of machine tools against self-excited vibrations in machining*. in: Proceedings of the ASME International Research in Production Engineering Conference, Pittburgh, USA, 1963, 465–474.

[7] S.A. Tobias and W. Fishwick, *The chatter of lathe tools under orthogonal cutting conditions*, Transactions of the ASME **80** (1958), 1079–1088.

[8] S.A. Tobias, *Machine Tool Vibration*, Blackie, London, 1965.

[9] H.E. Merrit, *Theory of self-excited machine tool chatter*. Journal of Engineering for Industry **87** (1965), 447–454.

[10] Y. Altintas, *Manufacturing Automation*, Cambridge University Press, 2000.

[11] G. Stépán, *Delay-differential equation models for machine tool chatter*, in: F.C. Moon (ed.), *Dynamics and Chaos in Manufacturing Processes*, Wiley, New York, 1998, pp. 162–192.

[12] B. Porter, *Stability Criteria for Linear Dynamical Systems*, Oliver & Boyd, London, 1967.

[13] O. Kienzle and H. Victor, *Spezifische schnittkrafte bei der metallbearbeitung*, Werkstofftechnik und Machinenbau **45**(7) (1957), 224–225.

[14] Y. Altintas and E. Budak, *Analytical prediction of stability lobes in milling*, Annals of the CIRP **44**(1) (1995), 357–362.

[15] T. Insperger and G. Stépán, *Stability of the milling process*, Periodica Polytechnica Mechanical Engineering **44**(1) (2000), 47–57.

[16] G. Stépán, *Modelling nonlinear regenerative effects in metal cutting*, Philosophical Transactions of the Royal Society London **359**(A), (2001), 739–757.

[17] B. Balachandran and M.X. Zhao, *A mechanics based model for study of dynamics of milling operations*, Meccanica **35** (2000), 89–109.

[18] H.Z. Li and X.P. Li, *Modeling and simulation of chatter in milling using a predictive force model*, International Journal of Machine Tools and Manufacture **40** (2000), 2047–2071.

[19] E.P. Nosyreva and A. Molinari, *Analysis of nonlinear vibrations in metal cutting*, International Journal of Mechanical Sciences **40**(8) (1998), 735–748.

[20] T. Insperger, G. Stépán, P.V. Bayly, and B.P. Mann, *Multiple chatter frequencies in milling processes*, Journal of Sound and Vibration **262** (2003), 333–345.

[21] G. Stépán, *Retarded Dynamical Systems: Stability and Characteristic Functions*, Longman, Harlow, 1989.

[22] R. Faassen, N. van de Wouw, J. Oosterling, and H. Nijmeijer, *Prediction of regenerative chatter by modelling and analysis of high-speed milling*, International Journal of Machine Tools and Manufacture **43**(14) (2003), 1437–1446.

11 Nonlinear Dynamics of an External Cylindrical Grinding System and a Strategy for Chatter Compensation

H.K. Tönshoff, B. Denkena, J. Jacobsen, B. Heimann, O. Schütte, K. Grudzinski, and A. Bodnar

Nonlinear effects play an important role in high-precision manufacturing processes. Especially in cylindrical grinding operations, many problems occur related to nonlinearities like stick–slip effects in the guideways, kinematic transmission errors in the feed drive, progressive stiffness of the grinding wheel or impact-like oscillations of the workpiece. In the cooperative project *Analysis of Process and Structure of Dynamic Nonlinear Precise Machining Systems* that is financed by the German VolkswagenStiftung extended nonlinear models have been worked out for the parts *grinding contact, machine structure, guideways* and *feed drive* of a grinding system. With these models strategies were developed that allow the usage of the machine's infeed drive for chatter compensation. The main aim is to overlay the slow infeed motion with oscillations synchronous to the wheel's rotational frequency, to compensate the part of the wheel waviness that is caused by eccentricity and unbalance. To correctly move the heavy tool head with a frequency of about 35 Hz and only a few micrometers amplitude, online information about the contact situation and the specific waviness on the wheel surface is necessary as well as advanced knowledge about the nonlinear behavior of the kinematic chain of infeed drive, machine structure and guideways.

11.1 Introduction

Chatter vibrations are one of the main process-limiting effects in grinding. They reduce the productivity and can affect the machining accuracy of the workpiece. There are two different types of vibrations. On the one hand forced chatter is mostly caused by an unbalanced or eccentric grinding wheel, or by oscillations of the machine tool not related to the process, provoking a varying grinding force. On the other hand, self-excited chatter originates in the dynamic process caused mainly by the formation of a regenerative waviness.

To reduce chatter, various approaches are possible like changing the grinding wheel speed, passive or active damping. The basic idea of passive damping is to lower the peaks of the amplitudes in the frequency response of the system [7]. As the natural frequencies usually depend on the ever-changing workpiece and vary due to a nonlinear stiffness of the grinding wheel surface, it is time consuming to adjust passive systems.

Active systems like the piezo-actuators of Gosebruch [8] and Michels [9] have to cope with the same nonlinearities. Therefore, it is difficult to extinguish or reduce chatter just by an inverse-signal feedback. Advanced strategies have to be applied. However, piezo-translators are also problematic for their high costs and high-voltage power supply.

Nonlinear Dynamics of Production Systems. Edited by G. Radons and R. Neugebauer

ISBN 3-527-40430-9

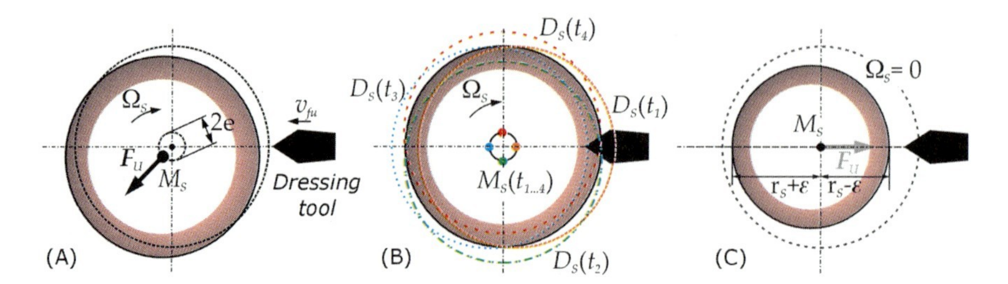

Figure 11.1: Eccentricity of a grinding wheel dressed under unbalance.

In this paper another idea is presented which does not afford any special and expensive actuators and promises to be a good solution for chatter problems caused by unbalance and eccentricity.

As is shown in Fig. 11.1 a small unbalance of the grinding wheel, which cannot be avoided in practical application, leads to an eccentricity ε of the wheel following its center of mass M_S. Viewed from the dressing tool, ε is zero, but as soon as the workpiece comes in contact with the wheel, the rotational speed, stiffness and position of the spindle changes so that viewed from the workpiece ε becomes greater than zero and a part of the waviness acts on the contact.

If this low section of the wheel, that is visible for example in records of [10] or [11], passes the workpiece, the contact force decreases. The workpiece, that underlies a permanent broadband excitation by the geometrically undefined cutting edges of the grains, gains more freedom for oscillations. In addition, the low section initially has the same phase angle as the unbalance force, which will also act partially on the workpiece.

Chatter Compensation Using the Infeed Drive

Conventional approaches of compensation take place at the natural frequency of the wheel–workpiece system of some 100 Hz. Instead of external actuators, the new idea is to use the existing infeed drive, to avoid the stimulation of vibrations by an eccentric wheel and its ever-present unbalance forces. The grinding wheel rotational frequency, f_S = 30–50Hz in conventional grinding processes, is one order below the chatter frequency.

The successful usage of the infeed drive implies an advanced knowledge of the complete grinding system with evident and hidden nonlinearities. The controlled motion of the infeed drive must create an oscillation of the wheel head that has an amplitude of only a few micrometers and a rather high frequency regarding the inertia of the wheel head.

Therefore, an extended modeling of nonlinear effects inside the guideways and the machine body was done, to exactly predict the wheel motion during compensation, see Sect. 11.3. The infeed drive controller was modified elaborately, to cope with nonlinearities inside the power train, see Sect. 11.4. To develop identification strategies for chatter and the waviness of the grinding wheel the dynamic behavior of the workpiece system in contact with the grinding wheel was intensely studied, see Sect. 11.2.

Figure 11.2 shows the structure and the information flow of the combined project from the University of Hannover, Germany, and the Politechnika Szczecinska, Poland.

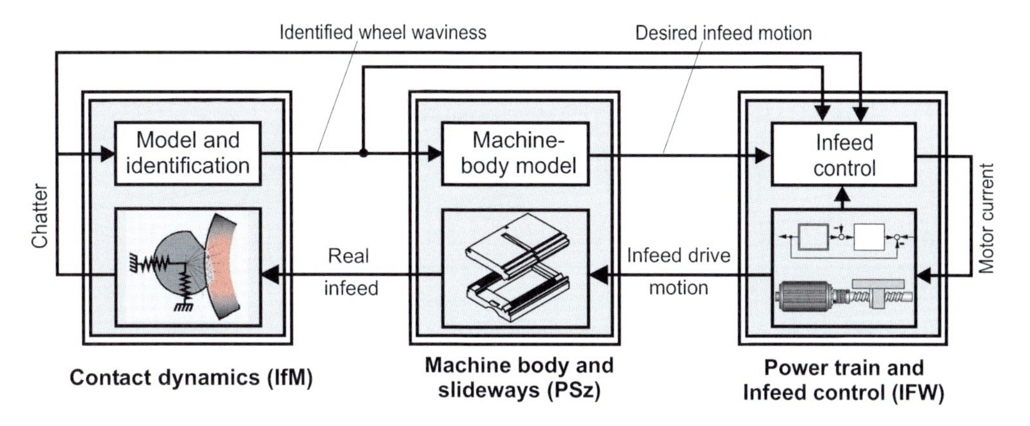

Figure 11.2: Structure and information flow of the combined project for chatter compensation.

11.2 Wheel–Workpiece Dynamics

Besides the undefinedness of grain engagement the grinding system behaves highly nonlinearly. The contact stiffness and the wheel stiffness are progressive. The contact situation between the workpiece and the non-rotating centers changes permanently causing variations of the workpiece eigenfrequencies and friction damping of large amplitudes.

Assisted by an unbalance of the wheel, Fig. 11.1, and stick–slip effects of the feed mechanism, changes of the contact force produce a regenerative waviness on the wheel and lead to chatter vibrations.

11.2.1 Chatter Vibrations

With a freshly dressed wheel, the grinding process runs optimally. The workpiece performs random oscillations of small amplitude producing a surface of high quality.

In the case of chatter vibrations the oscillation mode changes to a kind of impact motion as is shown in Fig. 11.3, region (i). This behavior can be described by the small piecewise-defined model of a one-mass oscillator of Fig. 11.4. The small dots on the curves in Fig. 11.3 mark the points when the system stiffness changes.

Two cases must be distinguished. First, when the grinding wheel is in contact with the workpiece, Fig. 11.4a. Secondly, when the workpiece oscillates at its eigenfrequency, Fig. 11.4b. In regions (ii) in Fig. 11.3 the workpiece is closely coupled to the grinding wheel. Both oscillate with a frequency of 650 Hz. Due to the combined stiffness the oscillation frequency is higher than the 485 Hz of region (i), where the fraction of the free oscillation is bigger than the fraction of the impact leading to a lower mean stiffness.

The chatter frequency that has the highest peak in a spectrum is the 485 Hz of the impact oscillation. With rising amplitude this chatter frequency decreases because the fraction of the type b oscillations increases.

The waviness shown by the dotted curves of the y_{w}^{α} vs. time plot in Fig. 11.3 has its origin in the dressing procedure as was shown in Sect. 11.1, see Fig. 11.1.

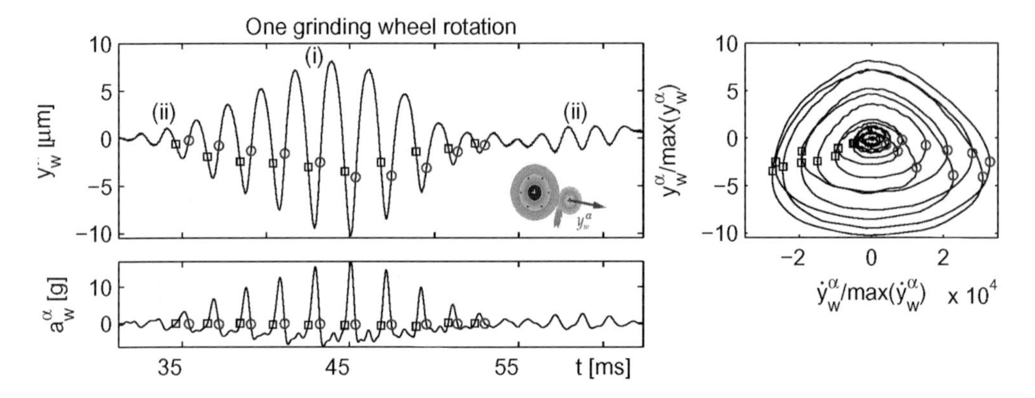

Figure 11.3: Workpiece oscillations during chatter vibrations.

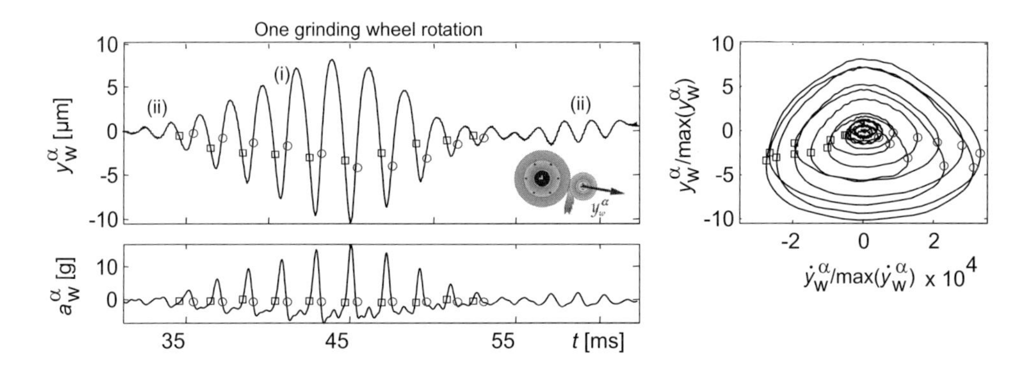

Figure 11.4: Principal model for workpiece dynamics in the case of chatter vibrations.

11.2.2 Compliance

To determine the dynamic compliance of the grinding system different excitation methods and actuators have been studied. In general, it is said that nonlinear systems should be investigated using a stepped-sinus excitation with a controlled force amplitude. A piezo-electric actuator inside a test workpiece was used to apply forces similarly to contact forces. Also, magnetic actuators were developed to investigate the compliance during the process, Fig. 11.5.

However, to examine the natural behavior of the grinding contact and to detect nonlinearities as they appear during the process, impulse excitation has shown the best test results. The changing frequency shown in Fig. 11.6 due to nonlinear stiffness of the grinding wheel surface cannot be found by monofrequency excitation, but only by free vibration analysis. If a stepped-sinus sweep is applied to the grinding system, large vibration amplitudes are needed to find nonlinear behavior in the form of the typical resonance jumps in the frequency response function (FRF).

In Fig. 11.6, the contact force $F_{\mathrm{S}y}$ measured with a sensor in the contact zone between wheel and workpiece and the workpiece displacement y_{w} calculated from the workpiece acceleration

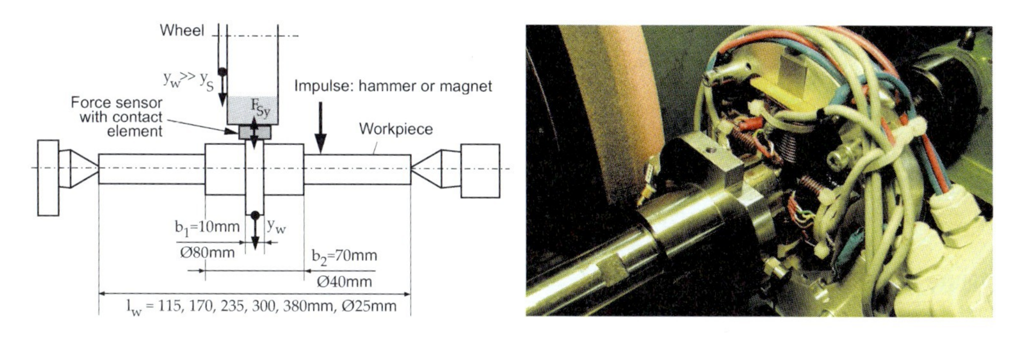

Figure 11.5: Workpieces for grinding and compliance tests.

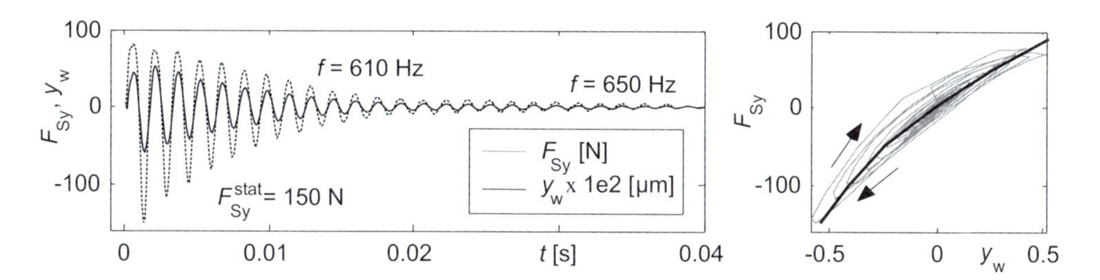

Figure 11.6: Impulse test results showing the nonlinear force–displacement relation.

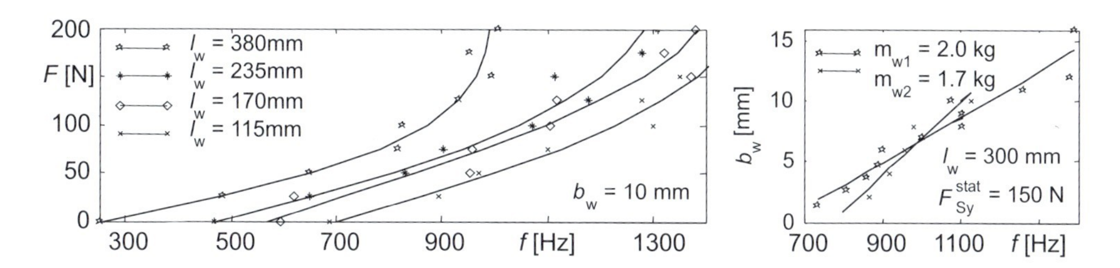

Figure 11.7: Dependency of workpiece eigenfrequencies upon contact force and contact width.

are displayed. Beneath a hysteresis, the curve shows a progressive stiffness, which can be approximated by a polynomial of second order.

Due to the nonlinear stiffness, the natural frequency of the system in total decreases with higher amplitude, as can be observed during the chatter burst shown in Fig. 11.3 and in general during chatter development in the process. The hysteresis is due to relaxation effects of the bonding, where the grains were pressed in during strain. When the strain relieves, the bonding must recover and relieves with a delay, causing a higher system stiffness during the relief.

During sets of 10 impulses for each contact situation the system behaves almost similarly from the first to the last impulse. Thus, a history of the contact surface can be neglected.

Furthermore, the natural frequency increases with higher contact forces F_{Sy} and decreases with the contact width b_w, as shown in Fig. 11.7. The frequency–force relation was least-square

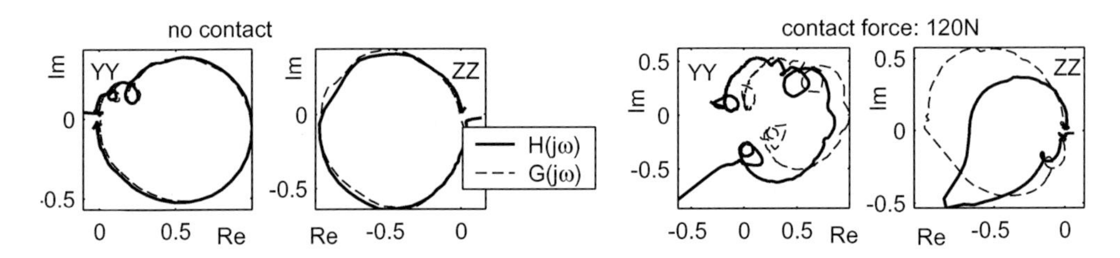

Figure 11.8: FRF of the workpiece and the corresponding Hilbert transforms with and without contact.

approximated by a polynomial of order three. The frequency–width relation is almost linear. The deviation of the data is high, due to the undefined contact conditions.

11.2.3 Hilbert Transform

The Hilbert transform that was computed from stepped-sinus excitation data normal to the contact (Y) and tangentially to the contact (Z) and the corresponding workpiece responses show another nonlinearity, Fig. 11.8 [12]. Without contact the workpiece behaves like an almost linear system. In the case of contact in the normal direction (YY) the results show no sign of large nonlinearity, either. But, in the tangential direction (ZZ) the workpiece's Hilbert transform looks similar to the Hilbert transform of a soft spring: The higher the vibration amplitudes grow, the fewer grains are able to maintain contact with the workpiece causing a lower stiffness.

11.2.4 Chatter Detection

The human ear is by far the best chatter detector known in industry. Unfortunately, it is available only in non-automated production and small workshops. Therefore, automated chatter detection is needed which is done usually by threshold-value methods [13]. But, signal levels in general are valid for only one combination of machine parameters, workpiece, cooling fluid and wheel type as Popp [14] showed for example in the case of acoustic emission signals.

Therefore, industrial applications of chatter-detection strategies would work best without a user-defined threshold value. Using the wavelet transformation on workpiece oscillation signals the following two methods are proposed.

Wavelet Detector

In Fig. 11.9 detail coefficients (Di) of a wavelet transformation are plotted for different grinding times. A Daubechies-type wavelet (scaling 5 in MATLAB notation) was used for this time-domain filtering. It looks like the chatter burst in Fig. 11.3 and fits it much better than sinusoidal waves that are used for example in a fast Fourier transform.

At an early grinding time $t_S \approx 20$–50 s the workpiece performs a kind of random motion of low amplitude. Wavelets of different scaling fit the signal with low amplitudes of almost the same height. When chatter arises at a grinding time of $t_S \approx 110$ s the typical burst patterns in the signal develop. The burst-like wavelet of the chatter frequency band (D5 in Fig. 11.9) now

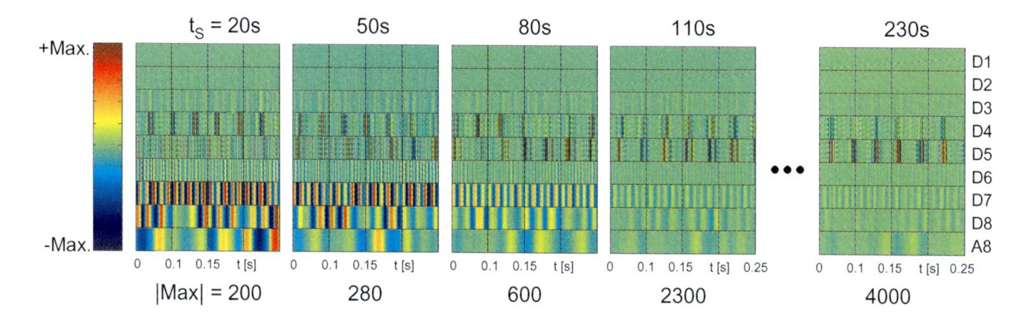

Figure 11.9: Wavelet coefficients during the development of chatter.

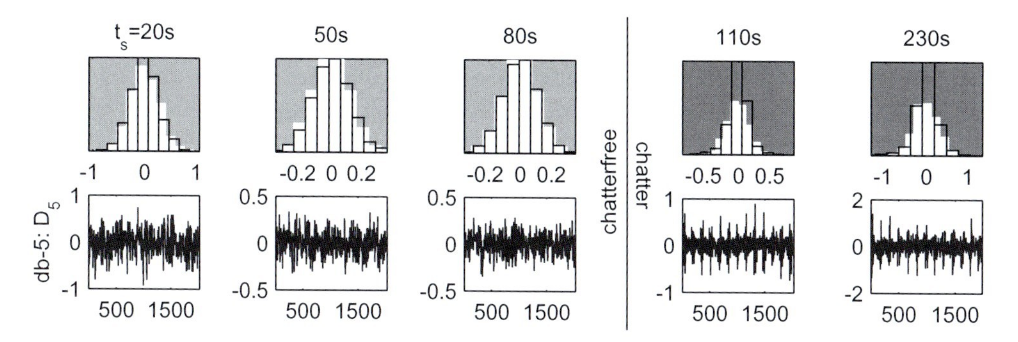

Figure 11.10: Lilliefors test applied on wavelet coefficients in the chatter-frequency band.

fits much better than the others. Its detail coefficient has an amplitude many times higher. In general, the proposed chatter-detection strategy is: check for *active* wavelet frequency *bands*, where the amplitude reaches more than p percent of the maximum value of all amplitudes, with $p \leq 100$. If the number of such *active bands* decreases chatter is present. The value of p does not depend on the process parameters. For chatter detection just to stop the process $p \geq 20$ can be chosen. For early chatter detection $p \geq 50$ produced suitable results.

Gauss Detector

A strategy that also evaluates the change of workpiece motion from random oscillation to directed chatter vibrations is the Lilliefors test [20].

At first, the standard deviation σ_x and the mean value $\bar{x}$ of a time series $x(t)$ are computed. Thereafter, the actual distribution of the signal (rectangles in Fig. 11.10) is compared with the nominal distribution calculated from σ_x and $\bar{x}$ (white bars in Fig. 11.10). If the maximum height difference of neighboring bars in the distribution plot is above a critical value, the hypothesis "$x(t)$ is distributed normally" is rejected. The significance level of the test, which is used to calculate the critical value, is independent of the process parameters and was chosen as the default value $\alpha = 0.05$.

If displacement or grinding force data are used, results of the algorithm are poor. However, if the data is a wavelet coefficient of the chatter-frequency band, an efficient chatter detector can be built up, because the wavelet reacts very subtly on the first formation of chatter patterns, as already shown in Fig. 11.9.

11.3 Modeling of Mechanical Structure Dynamics

The work on modeling the mechanical structure of the grinding machine was carried out in the Technical University of Szczecin (Poland). Owing to the necessity of the integration of the elaborated system representation into one common machine–workpiece–cutting process dynamic model, its structure had to be simplified to the degree enabling straightforward application of the model in the control procedure of the grinding wheel–workpiece vibration compensation.

When modeling the mechanical structure of the grinding machine used in the investigations, only the grinding head and its drive were taken into consideration. Other subassemblies of the machine were treated as stiff enough to neglect their influence on the grinding process dynamics. This decision was supported by the results of experimental investigations that were conducted on the machine [16].

The modernized version of the grinding wheel head is shown in Fig. 11.11. It comprises a DC motor M, lead screw L with hydraulic thrust/radial bearing B and hydraulic nut, table T moving on rolling guides (rolls shown as spring connections), rotary table R fastened with screws to the table. The upper support S and enables the pivoting of the grinding head H, carrying the grinding spindle motor D and the head and the grinding wheel G.

11.3.1 Model of Guideway Connection

It was shown [1] that screwed connections which are pre-stressed with high forces reveal almost linear static stress–strain characteristics, and their stiffness is relatively high compared to the stiffness of lightly loaded or movable connections. Therefore, it was decided to neglect the compliances of connections "a" and "b" marked in Fig. 11.11 during the modeling of the grinding head. This decision was taken after numerical assessment of their stiffness [17].

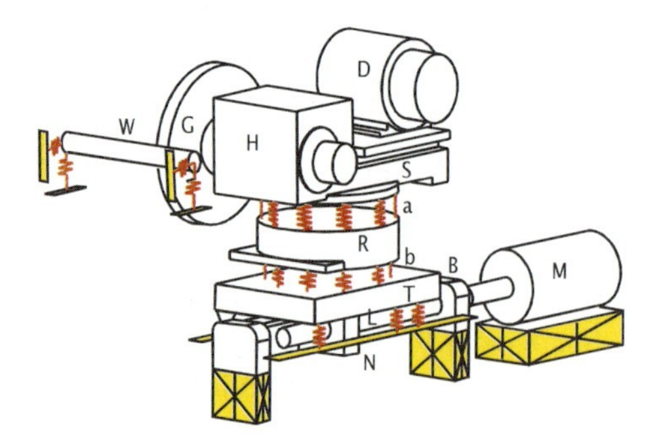

Figure 11.11: Sketch of grinding wheel head dynamic system.

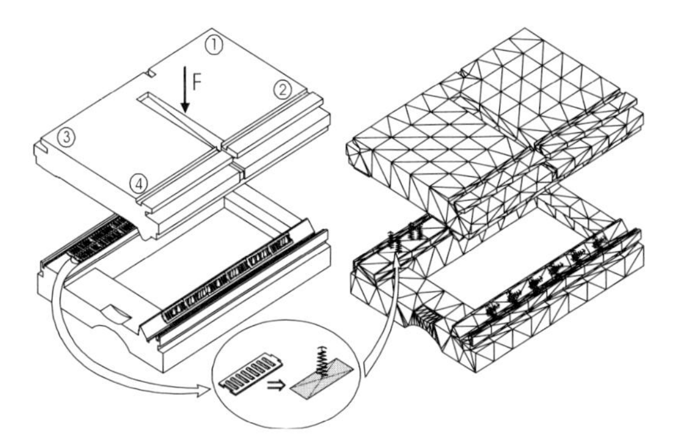

Figure 11.12: Table of the grinding wheel head and its rolling guides. Numbers show the places of displacement measurements under the action of static force F.

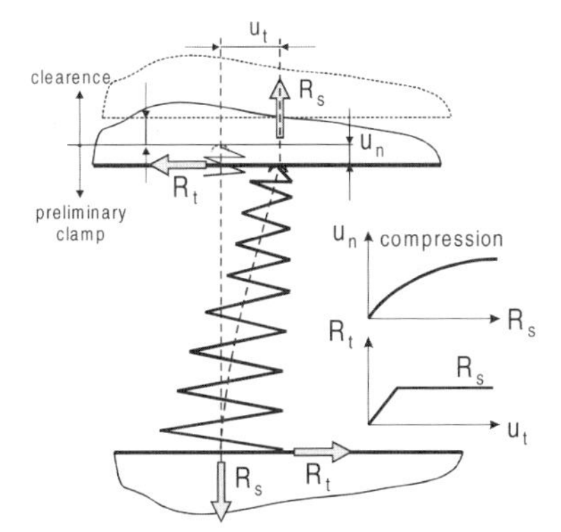

Figure 11.13: Spring-frictional contact element designed for modeling joints of machine bodies.

Numerical calculations were also carried out for finding the stiffness of the hydraulic bearing and the screw–nut connection. Both exhibit sufficiently high stiffness to be ignored in the final model. The most compliant connections that were left in the model of the grinding head were rolling guides. These connections are also the sources of nonlinearity in the system [15]. The guideway system of the head is shown in Fig. 11.12. It contains two guides – a plain one and a V-shaped one.

A special spring-frictional contact element, shown in Fig. 11.13, was developed [17] for modeling such connections. In rolling guides it models a single rolling element or a single cage containing rolls or needles. This element takes into account such features of real contact joints and phenomena occurring in them, as: unilaterality of characteristics of the contact elastic bond (subject to compression only), deformability of surface layers in contacting bodies, physical nonlinearities of elastic characteristics, static contact friction, clearances Δ, initial clamps u_{n}, as well as shape errors on mating surfaces. Hence, the nonlinearities of physical and geometrical nature are considered.

The experimental static examinations of two cages [1] containing rolling elements, used in the analyzed guideway system of the grinding machine, enabled the estimation of the stiffness of a single contact element applied in the model. Its deformations can be approximated with the power function

$$\delta = cQ^m, \tag{11.1}$$

where Q is the loading force and c and m are constants determined experimentally. Calculation of static stiffness of the whole connection was carried out twice. In the first calculation, the FEM model of the table was applied; in the second the solids which cooperate through guideways were treated as perfectly rigid bodies and idealized according to the assumptions of the rigid finite-element method. The obtained results showed that the second, much simpler method can be applied without loss of accuracy of computations [16].

The following procedure was used in static calculations. For finding generalized static displacements $\mathbf{q}$ of solids being in contact the so-called "external loads correction method" [17] was applied. In this method, the displacements are found in an iterative process from the equation describing the balance of forces operating in the connection:

$$\mathbf{K} \cdot \mathbf{q}(i) = \mathbf{Q} + \mathbf{Q}_{\mathrm{c}}(i) + \mathbf{Q}_{\mathrm{r}}(i), \tag{11.2}$$

where $\mathbf{K}$ is the system contact stiffness matrix (constant), $\mathbf{Q}$ is the external load, $\mathbf{Q}_{\mathrm{c}}$ and $\mathbf{Q}_{\mathrm{r}}$ are variable corrective matrices, and i is the preciding iteration number.

Within the ith iteration, the forces existing in contact elements, resulting from the action of external load $\mathbf{Q}$ and correcting load $\mathbf{Q}_{\mathrm{c}}(i)$, are calculated. The movable solid of the joint (the slide) is subject to reactive forces: $\mathbf{R}_{\mathrm{c}}(i)$ existing in elements being compressed and $\mathbf{R}_{\mathrm{t}}(i)$ which exist in elements being tensioned. For the next iteration $(i + 1)$, it is assumed that the external load corrective component $\mathbf{Q}_{\mathrm{c}}$ is equal to oppositely directed reactive forces of tensioned elements calculated in the preceding iteration $\mathbf{R}_{\mathrm{t}}(i)$:

$$\mathbf{Q}_{\mathrm{c}}(i+1) = -\mathbf{R}_{\mathrm{t}}(i). \tag{11.3}$$

In slideway connections, when the loading force is not perpendicular to the guide, an additional corrective component $\mathbf{Q}_{\mathrm{r}}$ is needed. It adds to the load adequate friction forces that occur in contact elements under compression. This component can be ignored in rolling guides owing to a very small rolling resistance.

Wishing to obtain a good approximation of static properties of rolling guides, displacements of the table corners were measured under increasing and decreasing static load F. The results of measurements were compared with those obtained from calculations, and an identification procedure was applied for minimization of errors between four experimentally determined and four computed hysteresis loops. In consequence the identified static stiffnesses were employed in dynamic calculations.

The dynamic response of the grinding head to the harmonic force acting on the grinding wheel in direction α (see Fig. 11.4) was calculated. The direction and amplitude of the force were taken close to possible values that appear in chatter vibrations in the system. This time, the computational effort was much greater, as on every time step one had to prove which contact elements are in operation, and which are loose. The steady-state motion of the grinding wheel–workpiece contact point and its decomposition into harmonics are shown in Fig. 11.14. It is apparently nonlinear when the frequency reaches the first eigenfrequency of the system

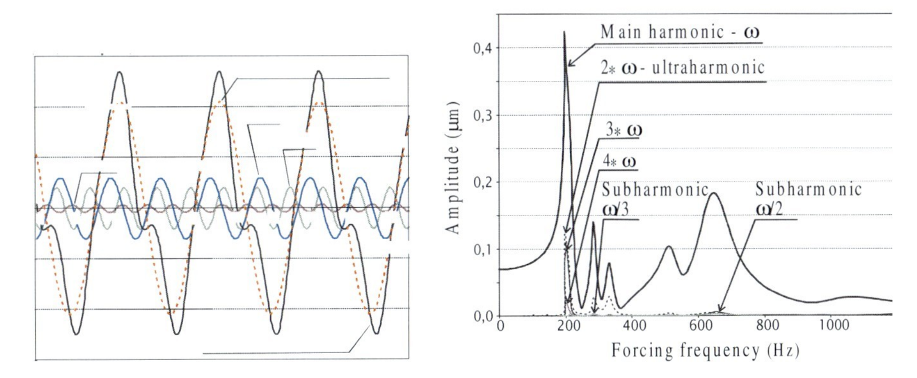

Figure 11.14: Time course and frequency content of displacements of the grinding wheel – workpiece contact point.

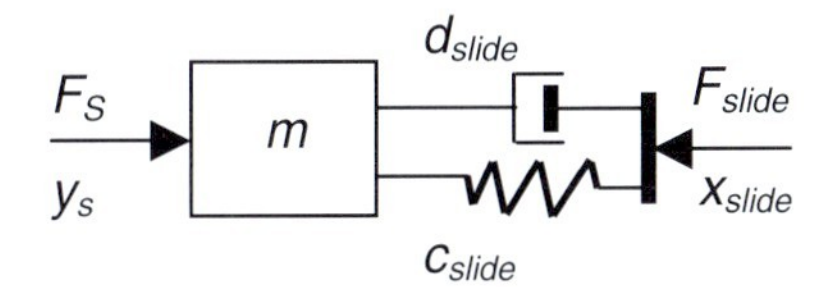

Figure 11.15: Reduced nonlinear model of the grinding wheel head dynamic system.

(approx. 200 Hz). At the first resonance almost 6% of rolling elements enclosed in both guide connections lose periodically their contact with the opposite surface. At higher and lower frequencies the system behaves almost like a linear one.

The obtained characteristics helped to synthesize a reduced model of the system. It is shown in Fig. 11.15. While, in this research work, special interest was assigned to the transmission properties of the grinding wheel head, the final model is a two input – two output system. It receives from one side the excitation F_s from the grinding process and transmits signals to the drive and, on the other side, it obtains the excitation F_{slide} from the drive, originating from the drive-control signals or caused by the drive-motion kinematic errors. These excitations are transmitted in the opposite direction – to the grinding wheel. The nonlinear characteristic of the spring was approximated with a polynomial having unknown coefficients. These coefficients, as well as the mass of the system, were then estimated through an optimization process. The objective function was the minimization of the sum of square errors between frequency characteristics of the original and reduced models at selected frequencies. The model may undergo further simplifications if they appear to be necessary for speeding up the vibration compensation control algorithms; e.g. it can be linearized.

11.3.2 Resonances in Guideway System

In this work, for modeling phenomena taking place on surfaces of contacting bodies a special two-dimensional contact element is used. It models nonlinear normal stiffness of the contact, damping and tangential friction forces which exist in real joints.

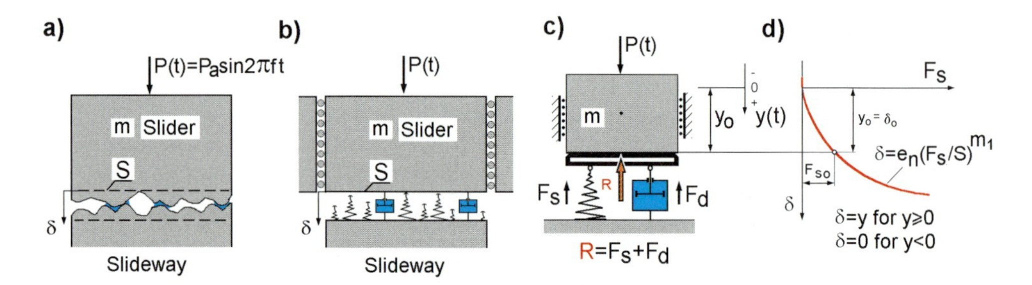

Figure 11.16: Considered dynamic system (a, b), its model (c) and static characteristic (d).

In the previous works of the authors [1] the influence of normal micro-vibration of contacting machine tool slideway elements on the value of friction forces and on the kind of motion was analyzed. It was shown that the contact vibrations can strongly reduce friction forces opposing the motion. They can as well stabilize as destabilize the motion.

The problem analyzed in this section is the question of resonance conditions that may appear in contact vibrations. Investigations connected with this problem are usually limited to the analysis of the main resonance [18]. However, as was shown in the previous subsection, in nonlinear systems beside the main resonance a number of super- and sub-harmonic resonances may exist [19]. In the case of the guideway connection of the machine tool, especially the resonances that have lower frequencies than the main resonance are of special interest.

In Fig. 11.16 the model of the analyzed joint is shown. The slide and its supporting body are treated as infinitely stiff. Rolling elements in the connection (or unevenness of its surface) are modeled with the help of contact elements. The equation of motion for the model shown in Fig. 11.16c can be written in the following form:

$$m\ddot{y} + Sh\delta^l\dot{\delta} + Sc_\mathrm{n}\delta^{m_1} - mg - P(t) = 0, \tag{11.4}$$

where $y, \dot{y}$ are the normal displacement and normal velocity of the slide, S is the contacting area, c_n, m_2, h and l are constants determined experimentally and $\delta = y$ for $y > 0$ or $\delta = 0$ for $y \leq 0$.

When the excitation force is applied to the slide, resonance conditions may appear in the system. Numerical solutions of Eq. (11.4) for harmonic excitation with a normal force of $P(t) = P_\mathrm{a}\sin 2\pi f t$ could demonstrate a very different nature, depending on amplitude P_a and frequency f of the force and parameters of the analyzed system. Calculation results that were obtained so far are being prepared, together with their detailed analysis, for a separate publication. In the following part of this section only some selected, but most representative, results are presented.

In Fig. 11.17 the main resonance characteristics are demonstrated. They were obtained for a few small amplitudes of excitation.

Figure 11.18 shows characteristics of ultra-harmonic resonances for excitation equal to 80% of the slide weight; P_a = 0.8 mg. From the results presented in Fig. 11.18 one can see that at $f < f_0$ a number of interesting phenomena occur in the system, like high-amplitude vibrations with complex motion, resonance peaks, jump changes of amplitude, bifurcations and chaotic motion.

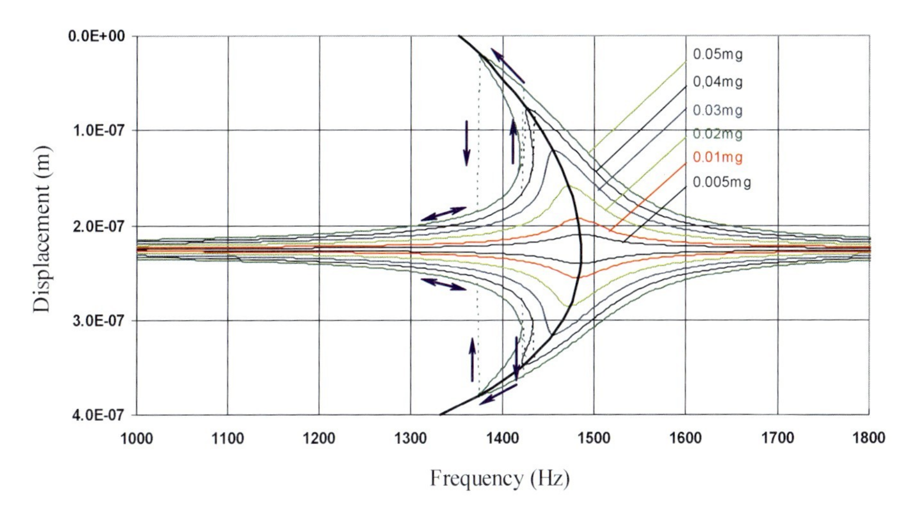

Figure 11.17: Main resonance characteristics obtained at small amplitudes of excitation force – in the range (0.005–0.05) mg.

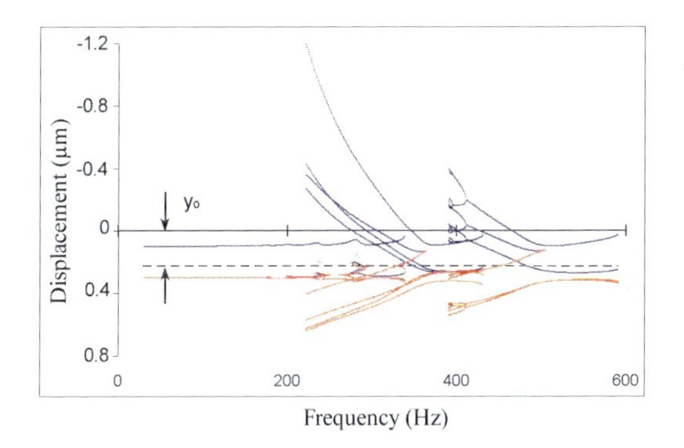

Figure 11.18: Ultra-harmonic resonance response of the system shown in Fig. 11.5 at excitation amplitude $P_a = 0.8\,\text{mg}$ ($f < f_0$).

The analyzed system is very sensitive to the initial conditions and parameter f. The frequency of excitation and the direction of its changes influence the solutions – the system may have one, two, four or more regular solutions, and in particular conditions also chaotic solutions.

This could be seen in Fig. 11.19a which shows the cascade of period-doubling bifurcation leading to chaotic solutions. Figure 11.19b shows a Poincaré map illustrating the evolution of the phase flow section for f being in the range 420–397.3 Hz, and in Fig. 11.20 the attractor (approx. 30 000 points) of chaotic solutions of Eq. (11.4) for $f = 397.3\,\text{Hz}$.

The examples presented in this section demonstrate the high complexity and diversity of kinds of forced vibration responses that may appear in the machine tool contact joints – as well in sliding, as in rolling connections.

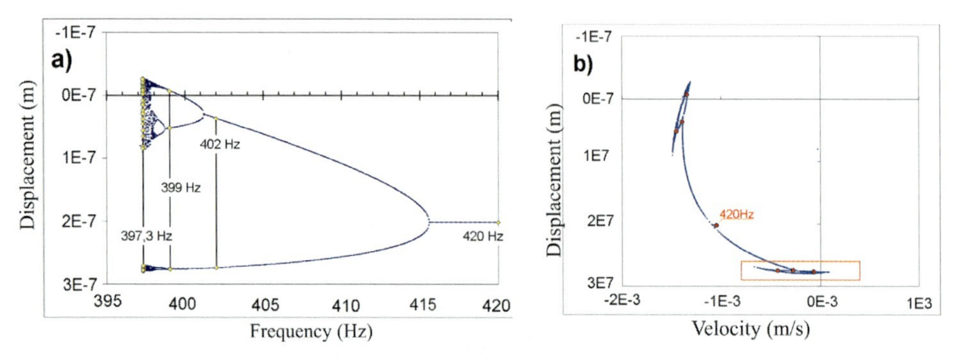

Figure 11.19: Cascade of period-doubling bifurcation plot (a) and Poincaré map (b) showing the flow evolution for f in the range 420–397.3 Hz.

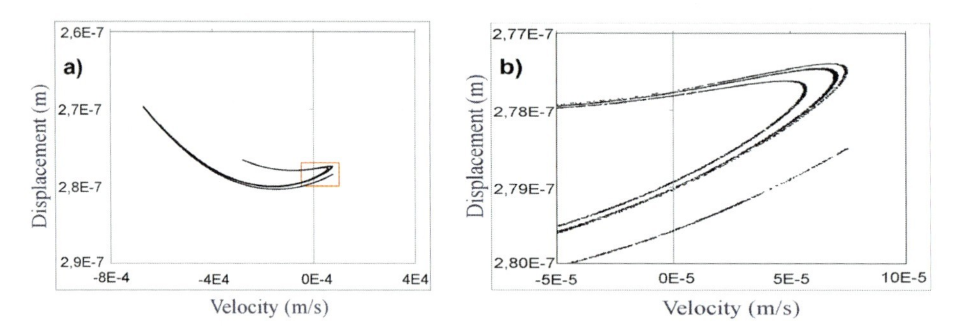

Figure 11.20: Poincaré section (a) of a chaotic attractor of the forced system (Fig. 11.19) and detail (b).

11.4 Feed Drive

For the proposed chatter compensation strategy many detailed properties of the feed drive have to be analyzed that have nonlinear influences on the force flux like non-constant stiffness. They have to be compensated by an adaptive control that also has to apply the oscillating torque synchronous to the grinding wheel rotation that is the basic idea of the compensation strategy, see Sect. 11.1. This method seems to be a good future solution for industrial application because there is no need for further drives or actuators.

The aim of the application is to compensate an eccentricity of up to $5\,\mu\mathrm{m}$ at a rotational speed of $3000\,\mathrm{rpm}$. The nearly constant rotational speed of the wheel allows us to design the controller in the position domain [2].

Due to the resulting stresses caused by the oscillation the mechanical elements of the force flux, especially the system screw–nut and the fixed bearing, have to be very stiff. Additional rotation of the drive would be required to overcome the displacement of the bearing and of the connection between the screw and the nut. The necessary stiffness can be achieved by using hydrostatic components. They are commonly integrated in modern grinding machines, also because of their high damping of undesired vibrations. In conventional ballscrew–nut connections small oscillations with a very low feed rate cause abrasion [3]. Hydrostatic elements do not have this problem due to their missing solid-body contact.

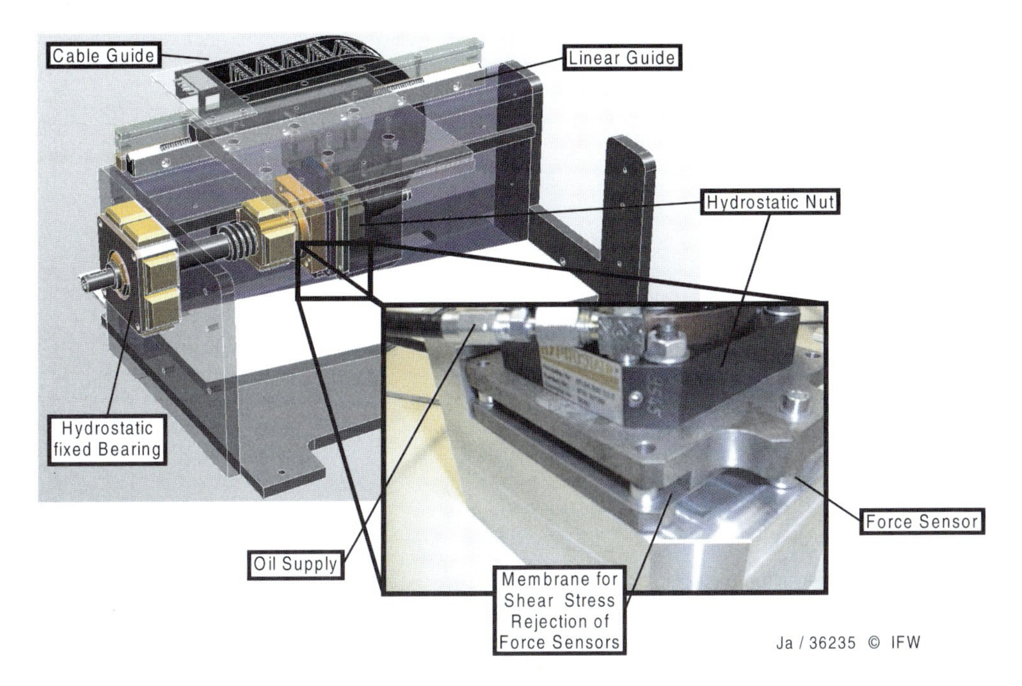

Figure 11.21: Test Bench with Membrane Construction.

Construction of the Test Bench

The test bench for measuring the static and dynamic behavior of the linear guides, the hydrostatic bearing and hydrostatic screw is shown in Fig. 11.21 as an opened CAD model. It is a very massive construction, otherwise the force of the engine would be exhausted by the elastic compliance. The hydrostatic nut is mounted to two piezo-electric force sensors. Driven by a servo drive and using an encoder, a torque sensor and a displacement sensor, the static characteristics can be measured as well as the frequency response of up to much more than the aspired 50 Hz.

The length of the loaded part of the screw depends on the slide's position. So the varying screw's stiffness influences the properties of the feed drive in a nonlinear way.

After measuring the dynamic attributes a controller is to be developed to imply the anti-phase oscillation force on the feed slide. Finally, the project's grinding machine will be equipped with the hydrostatic components of the test bench.

Special care has to be taken for the integration of the force sensor, based on piezo-electric ceramics. A membrane construction is used to prevent excessive shear stresses to the sensor (photograph in Fig. 11.21).

The membrane is mounted parallel to the two sensors, but it only receives a negligible fraction of the total infeed force. The main part of this force is transferred by the sensors due to their high stiffness compared to the high bending compliance of the membrane. The conversion of a torque to a forward motion produces a torque to be held by the hydrostatic nut. This torque is received by the membrane, which is stiff for torque transmission.

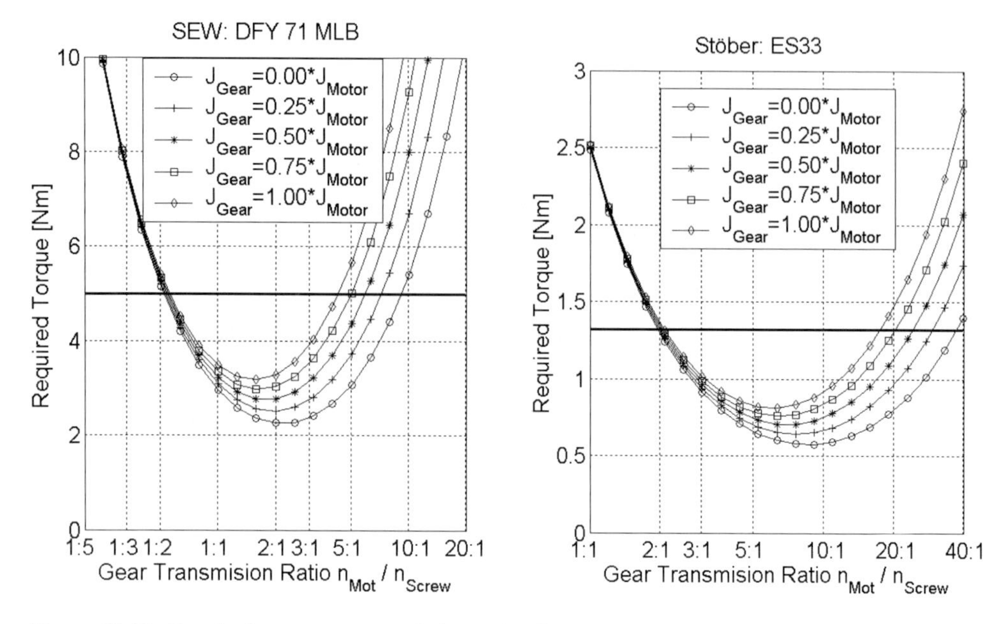

Figure 11.22: Required torque vs. transmission ratio of motors: SEW-DFY 71 MLB and Stöber-ES 33.

11.4.1 Requirements for the Infeed Drive

Due to the required dynamics a high-quality drive has to be used. The friction has to be low to avoid stick–slip effects. During the grinding process the infeed speed is quite low and a slipping slide could cause manufacturing errors. But not even all modern motors are suitable. A certain relationship between the torque and the moment of inertia of the motor is necessary to meet the aim, depending on the moment of inertia of the hydrostatic screw, the bearing, the drive and the slide's mass. Many motors have enough power but cannot be used without a gear. For the described requirements it is difficult to use a gear due to its slackness or compliance. It would inhibit a high-frequency movement of the slide. Also, modern machine tools use direct drives for their better precision, so a gear should not be used here.

Figure 11.22 shows the necessary torque for two servo drives depending on a gear transmission ratio and the gears' inertia. An infinite stiffness is assumed; therefore a safety factor has to be considered. The diagrams are based on the data of the applied hydrostatic components and the motor as well as a slide mass of 200 kg and a screw pitch of 10 mm.

The x axis is assigned with the gear transmission ratio and the ordinate with the required torque, depending on the moment of inertia of the gear. The horizontal line indicates the maximum torque of the motor. If the required torque is below this line (regarding the safety factor), the operating point is suitable. The diagrams indicate that the SEW motor can be used without a gear while the Stöber motor cannot.

11.4.2 Nonlinear Effects and Control Scheme

Nonlinearities

Because of the high dynamics which are necessary for the feed drive in the grinding machine the dynamic characteristics of all components in the force flux have to be analyzed. There is also a big variety of nonlinearities to be considered like occurring slackness and hysteresis effects. It is shown how to model them in [4]. One of the advantages of hydrostatic elements is the missing solid-body contact, so neither a slackness nor a hysteresis can influence the system. The absence of gears also leads to a better force flux of the infeed drive.

Another nonlinear effect is well known as "stick–slip". This effect is caused by falling friction characteristics at low speeds of drives or linear guides and leads to an improper feed of the grinding wheel [5]. Depending on the motor friction characteristic these effects can be neglected or may have a big influence, making that it is not possible to control a slow rotation also with a speed-dependent compensation of the friction.

A control scheme with anticipatory control and disturbance rejection is shown in Fig. 11.23. The nonlinear disturbance rejection compensates the nonlinear effects like cogging and friction of the motor. The screw is divided into several blocks, because only the ideal transmission ratio is insufficient to describe it. Also, the contact stiffness and the force-dependent damping have to be considered. The properties of the slide and the machine structure were already described in Sect. 11.3.

The anti-chatter-control bases on a waviness detection as presented in Sect. 1.5. Alternatively, the waviness can be evaluated by an analysis of the acoustic emission signal.

11.4.3 Compensation of Cogging

The torque of servo drives is not constant during one rotation. An additional periodic torque exists due to permanent magnets inside the motor. This so-called cogging force significantly influences the position control of the motor, especially when it is running at a very low speed.

This periodic torque has to be compensated as well as the speed-dependent friction. Up to a certain degree this can be done by an anticipatory control, releasing the controller from this task. So its dynamics do not affect the uniform motion any more [6]. The cogging torque has small effects at a high motor speed because the alternating torque has a small time to influence the rotation. At very low speeds a PID control can compensate the alternating torque, assuming there are no stick–slip problems. A mid-range of rotational speeds cannot be well controlled. An experimental iterative approach is made to detect the amplitude and phase of the cogging torque. It requires a motor with low friction, which can be controlled at low speed. When the rotational speed of such a motor is controlled to be $1^\circ/\text{s}$ by a PID control, the output of the control is a first approximation of the cogging torque. A frequency analysis results in an amplitude and a phase of the dominant harmonic. In a next step the output of the control is superposed by a sine wave with these data. The idea is to release the control from the main part of the cogging torque. Although the rotation is quite slow, a control deviation remains, containing the cogging periodicity. The compensation still is not complete. The same experiment is repeated in a second experiment with compensation. This time the deviation is much smaller, but it is necessary to repeat these experiments each time with a better approximation of the cogging until the result does not improve any more.

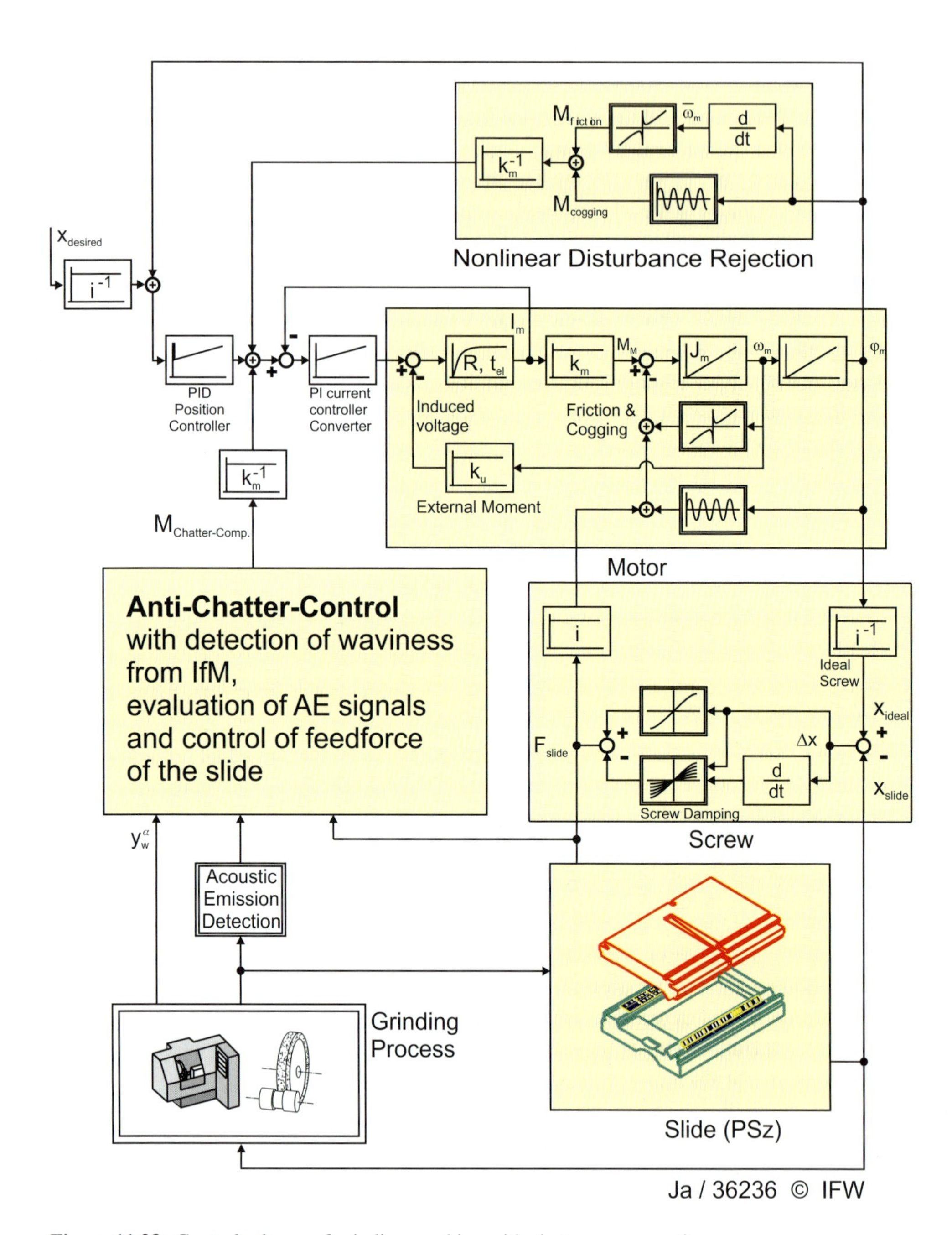

Figure 11.23: Control scheme of grinding machine with chatter compensation.

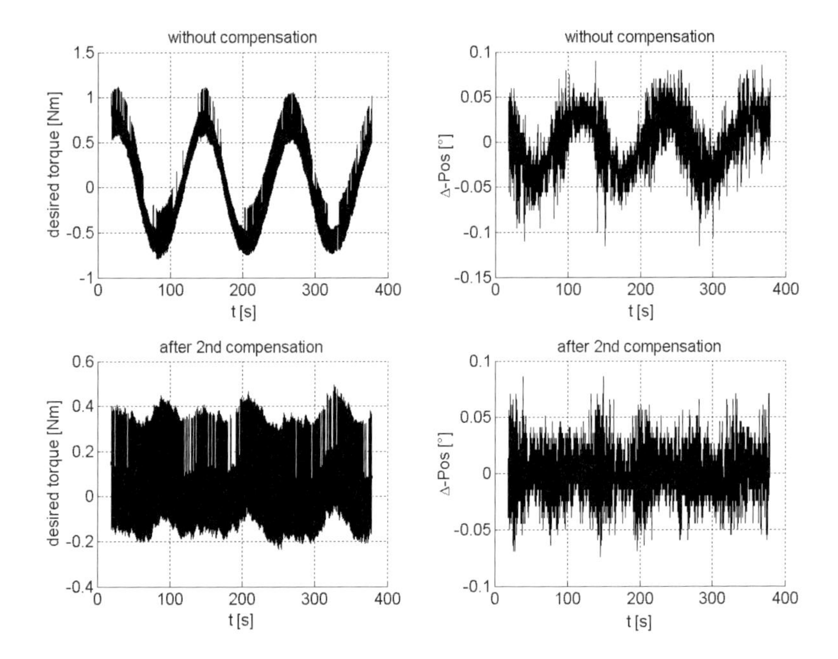

Figure 11.24: Compensation of Cogging.

In Fig. 11.24 the results of the compensation of the cogging effect are shown. After the second pass of the compensation algorithm the remaining cogging torque is about 5% of the original one. The position errors due to the cogging are nearly eliminated.

11.5 Waviness Compensation

The relative waviness between workpiece and wheel can be estimated for the proposed compensation strategy using an identification procedure on the workpiece oscillation signal (see Fig.11.25). As was shown in Fig. 11.3 the lift-off and touch-down points were calculated, see Sect. 11.2.1. The mean value of these two curves is fitted with a sine curve. Depending on the dynamics of the infeed drive the first harmonic of the base frequency may also be controlled.

The amplitude value and the phase angle of the identified sinusoid are used to compute the infeed motor torque. In doing so, the wheel head model will be used to adjust the phase shift between motor, wheel head and waviness correctly and to avoid stick–slip effects during the unavoidable zero crossings of the infeed motion during compensation. In parallel, the information content of acoustic emission signals is analyzed with respect to the estimation of grinding force variations with the wheels' rotational frequency.

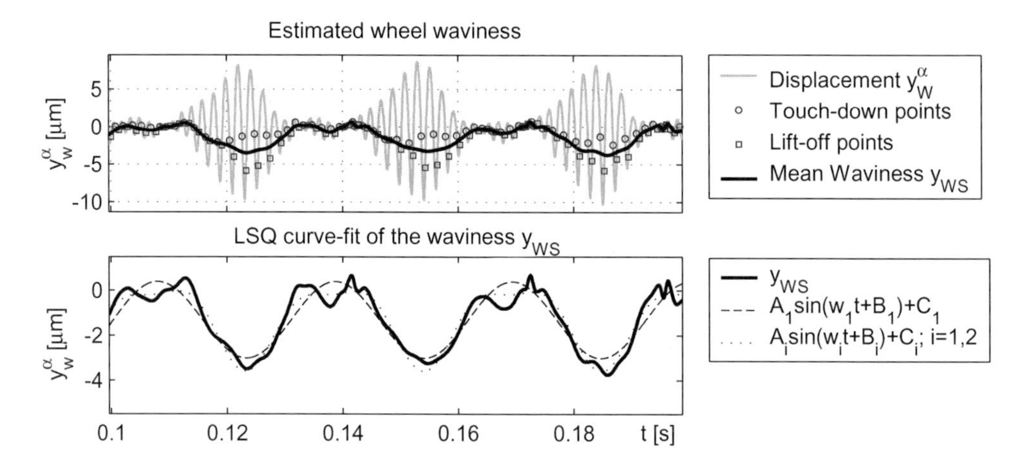

Figure 11.25: Waviness estimation from the workpiece displacement signal.

11.6 Conclusions

The presented new approach to prevent chatter vibrations caused by unbalanced or eccentric grinding wheels uses a wheel-synchronous oscillation of the feed drive. It will assure constant grinding forces without using further expensive drives like piezo-electric actuators. In contrast to previous investigations it avoids the external stimulation of chatter, working at the rotational frequency of the grinding wheel. All the mechanical transfer elements were analyzed regarding nonlinear effects like cogging and friction of the drive, nonlinear stiffness and damping of the applied screw, bearing and machine construction. The detection of chatter and the estimation of a relative waviness between wheel and workpiece were realized by studying the nonlinear workpiece oscillations.

Acknowledgment

The authors gratefully acknowledge the financial support of this work by the Volkswagen-Foundation within the research project *Investigation of Nonlinear Dynamical Effects in Production Systems*.

Bibliography

[1] H.K. Tönshoff, J. Kummetz, B. Heimann, O. Schütte, K. Grudzinski, and A. Bodnar, in: 3rd International Symposium "Investigation of Nonlinear Dynamic Effects in Prod. Systems", TU-Cottbus, 2000.

[2] M. Nakano, J. She, Y. Mastuo, and T. Hino, Control–Engineering-Practice **4**(9) (1996), 1241–1248.

[3] W. Schröder, Ph.D. thesis, TH-Zürich (1996).

[4] C. Kirchmair, Forschungsberichte Künstliche Intelligenz (1999).
[5] S. Webb, Z. Li, C. Cook, in: Proceedings of the Fourth Annual Conference on Mechatronics and Machine Vision in Practice, Toowoomba, Qld., Australia, IEEE Comput. Soc., Los Alamitos CA, USA (1997) pp. 32–37.
[6] J. Kummetz, Ph.D. thesis, University of Hannover (2001).
[7] E. Tellbüscher, Ph.D. thesis, University of Hannover (1986).
[8] H. Gosebruch, Ph.D. thesis, University of Hannover (1990).
[9] F. Michels, Ph.D. thesis, RWTH Aachen (1999).
[10] F. David, Ph.D. thesis, University of Kaiserslautern (1997).
[11] H. Kaliszer, in: Advances in Machine Tool Design and Research, Proceedings of the 11th International MTDR Conference (1970), pp. 615–631.
[12] M. Simon, G.R. Tomlinson, Journal of Sound and Vibration **96**(4) (1984), 421–436.
[13] J. Muckli, *wt Werkstatttechnik*, **91**(5) (2001), 264–268.
[14] C. Popp, Ph.D. thesis, University of Hannover (1992).
[15] N. Back, M. Burdekin, and A. Cowley, in: Proceedings of the 14th International Mach., Tool Des. and Res. Conference, Manchester (London-Basinstoke, 1974), p. 529.
[16] A. Bodnar and O. Schütte, Advances in Manufacturing Science and Technology **24**(4) (2000), 21.
[17] G. Szwengier et al., in: Proceedings of the 9th World Congress on Theory of Machines and Mechanisms, Milan (Politechnico di Milano, Milano, 1995) Vol. 4, p. 2703.
[18] D.P. Hess and A. Soom, Journal of Tribology **113** (1991), 80.
[19] J. Awrejcewicz, Deterministic Vibrations in Discrete Systems, WNT, Warsaw (1996).
[20] H. Lilliefors, Journal of the American Statistical Assoc. **62** (1967).

12 Problems Arising in Finite-Element Simulations of the Chip-Formation Process Under High Speed Cutting Conditions

A. Behrens, K. Kalisch, B. Westhoff, and J. Wulfsberg

The main problem arising in the Finite-Element (FE) simulation of cutting processes under High Speed Cutting conditions (HSC) is due to the chip formation evolving in a very small region. High deviations of stresses, strains and strain rates occur in this area; thus very fine FE-meshes in the simulation are required for simulations.

The first problem is to find suitable models for the separation of the chip material from the remaining workpiece. Furthermore, it is possible to describe the chip formation as a mere plastic deformation assuming ideal ductile material behavior. The different models will be discussed in the paper.

The second problem arises from the material behavior itself. Only a few mathematical models exist for the description of the material behavior influenced by high strain rates. Flow stress curves derived from Split–Hopkinson bar tests are capable of modeling the mathematical flow with satisfactory results.

Another problem occurring in HSC processes is the chip disintegration into segments because of high local shear stress effects in the chip-formation zone. This behavior can also be reproduced in the FE simulations when local refining is applied to the FE mesh.

Real cutting processes need a spatial FE simulation as the chip-formation process is not homogeneous over the cutting edge. Here some 3D simulations that have been performed with a turning process will be discussed.

12.1 Introduction

The conditions and the limits of the cutting process are defined mainly by the characteristics of the chip formation. It also influences forces and temperature distributions as well as the tool wear. The properties of the workpiece surface, formed by the cutting process, are defined by the chip-forming mechanism. These are the main reasons which lead to widespread investigations of the chip-formation process itself.

Necessary experimental values for the understanding of the process, such as the temperature distribution in the chip or in the contact surface of the tool, are very hard to determine due to the extreme conditions in the chip-forming zone. This early led to the development of simple analytical models for the orthogonal cutting process, which were limited by the complex and nonlinear nature of the chip formation. Here the Finite-Element simulation can support the experimental investigations very well.

Nonlinear Dynamics of Production Systems. Edited by G. Radons and R. Neugebauer

ISBN 3-527-40430-9

As the orthogonal cutting process is well suited for basic research, it is used in this article to demonstrate the different material separation algorithms for the simulation of the chip formation. Different material models for the extreme forming conditions in the shearing zone were developed and integrated in the simulations in order to evaluate the quality of these algorithms. The high temperature, stress and strain gradients cause high finite element mesh deformations and efficient re-meshing methods were used to stabilize the simulations. The results from the FE simulations are then compared to experimentally obtained values of cutting forces and chip shapes. Furthermore, we show first results from the verification of the calculated residual stresses in the workpiece subsurface. In the last step the calculations of the planar orthogonal cutting process are extended to a spatial model. Here the spatial outer turning process is demonstrated.

12.2 Orthogonal Cutting Process

12.2.1 Description

First the simple 2D orthogonal cutting process (for process description see Sect. 12.3.1) is explained, which serves for the mere description of the chip contour during the forming process. It is based on [1]. Figure 12.1 shows the different areas in this process. Zone 1 denotes the primary shear zone, where plastic deformation is caused by the material slipping on the shear bands. In the secondary shear zone (zone 2) the chip undergoes an additional shearing due to shear friction stresses at the cutting edge. The material cutting occurs in zone 3, located in front of the tool edge. Zone 4 is characterized by friction at the transition from cutting edge to minor cutting edge. Furthermore, a preliminary deformation zone exists, which influences the workpiece subsurface of the material (zone 5).

12.2.2 Material Laws

In the primary shear zone (zone 1) one has to deal with extremely high localized strains (ε), strain rates ($\dot{\varepsilon}$) and temperature gradients (ϑ) that cannot simply be referred to the uniform state of stresses from the common experimental determination of the material's flow stress behavior

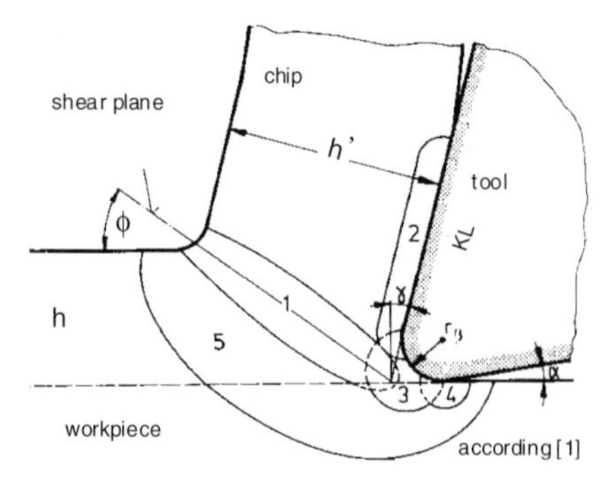

Figure 12.1: Distributed zones in the orthogonal cutting model.

(σ_f). Moreover, the experiments showed a chip segmentation at high cutting velocities in the orthogonal cutting of the carbon steel Ck45 [2]. In order to integrate these conditions into the FE simulations, material laws by Johnson-Cook (Eq. (12.1)) and El-Magd (Eq. (12.2)) were investigated:

$$\sigma_f(\varepsilon, \dot{\varepsilon}, T) = [A + B\varepsilon^n]\,[1 + C\ln(\dot{\varepsilon}/\dot{\varepsilon}_0)]\,[1 - (T*)^m] \quad \text{with } T* = \frac{T - T_0}{T_\mathrm{m} - T_0}, \quad [3] \tag{12.1}$$

$$\sigma_f(\varepsilon, \dot{\varepsilon}, T) = \sigma_f(\varepsilon, \dot{\varepsilon}, T_0)\psi(T) \quad \text{with } \sigma_f(\varepsilon, \dot{\varepsilon}, T_0 = 273\,\mathrm{K}) = K\,(B + \varepsilon^n) + \eta\dot{\varepsilon}. \quad [4] \tag{12.2}$$

The parameter identification in Eq. (12.1) was based on a reference of flow stress curves in [5] with a strain rate of $\dot{\varepsilon}_0$ = 1 s^{-1}. The temperature exponent was assumed to be $m = 1$ in good accordance to [5]. It takes into account a temperature dependency between $\vartheta = 20°\mathrm{C}$ and $\vartheta = 1200°\mathrm{C}$. The melting temperature of the material is T_m = 1808 K and the reference temperature is given as T_0 = 293 K. The second model consists of two parts: for small to medium strain rates, the flow stress behavior is mainly caused by a thermally activated sliding process, whereas at higher strain rates the damping fractions are dominating, leading to a linear flow stress evolution with respect to the strain rate [3]. For limited temperature intervals the parameters n and η can be kept constant. In this way the temperature dependency of the flow stress is integrated into the formula by a suitable temperature function $\psi(T)$ (see [3]). All the necessary material-dependent parameters for the carbon steel Ck45N are determined from the well-known Split-Hopkinson bar tests.

The orthogonal cutting process (for process parameters see Sect. 12.3.1) was simulated using both material models, first varying the cutting velocity between 0 and 600 m/min and the friction factor m between 0, 0.5 and 0.7 (friction behavior modeled by shear friction).

Figures 12.2a–d show the force–velocity behavior on the basis of two temperature functions $\psi(T)$ for Eq. (12.1) and the flow stress–velocity curves resulting from Eq. (12.2). Figure 12.2a and c differ only by a higher temperature coefficient β in the latter diagram. A steeper descent of the flow stress with rising temperature results from this. The calculated cutting forces tend to rise with increasing cutting velocity, in contradiction to the experimental results. This effect is supported by the neglect of friction at lower cutting velocities up to 100 m/min and inaccuracies of the used material descriptions. But the higher temperature coefficient β causes a significant translation of the calculated force curve below the experimentally determined one (cf. Fig. 12.2a and c), whereas in Fig. 12.2a the curve rises more strongly than in Fig. 12.2c.

The significant dependency of the cutting force on the cutting velocity might be ascribed to an over-estimation of the damping influence $\eta\dot{\varphi}$ at low strain rates. It results in a steep ascent of the stresses. The strain rate in Eq. (12.2) goes into the formulation by a (under-proportional) logarithmic function. Figure 12.2d shows the dissatisfying possibility to adapt the calculated values to the experimental ones by reducing the friction factors with increasing cutting velocity. A reduction of the friction factor causes a change of the chip formation by reducing the chip curvature and increasing the chip thickness as well as the contact length. Opposed to this, the simulation forecasts a decrease in chip thickness and contact length and an augmentation of the chip curvature.

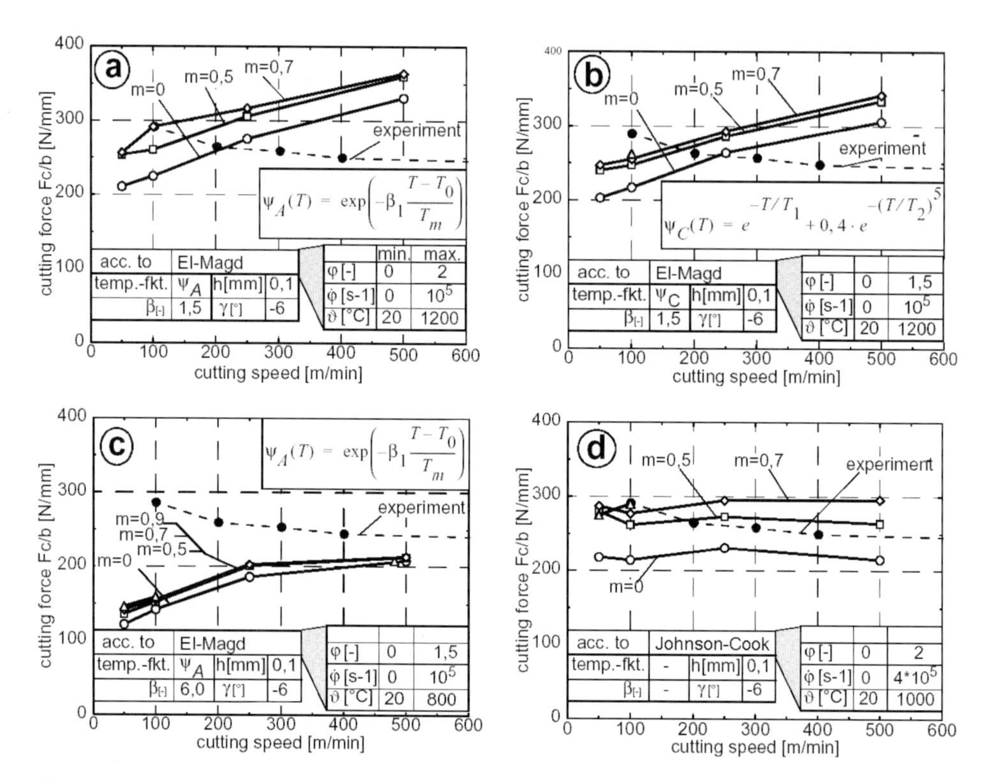

Figure 12.2: Resulting cutting forces due to different flow stress descriptions and temperature functions.

Moreover with this material and under the given cutting conditions ($\gamma = -6°; h = 0.1$ mm) the primary chip formation changes from flowing to segmented chip due to thermal–mechanical instabilities in the primary shearing zone at a cutting speed of approximately $v_c = 750$ m/min [2].

The investigations of the different material laws showed that the models used in Fig. 12.2b and c were only suitable to reproduce the chip-segmentation behavior at the corresponding cutting speed. Eqation (12.1) alone, only used in Fig. 12.2d, cannot meet this demand.

For these reasons Westhoff developed in [6] a modified approach by the numerical combination of the upper two material laws. The cutting force progression and the chip-formation behavior are represented in a better way for cutting velocities up to 3000 m/min. The resulting flow stress behavior is shown in Figs. 12.3 and 12.4.

12.2.3 Remeshing and Chip Separation

There are different possibilities for the description of the activities at the cutting edge (zone 3, see Fig. 12.1) using FEM with the Lagrange formulation. The separation of material can take place at the nodes (node splitting [7], nonlinear spring [8], separation based on contact [6]) or can be achieved by discarding or deactivation of elements [9].

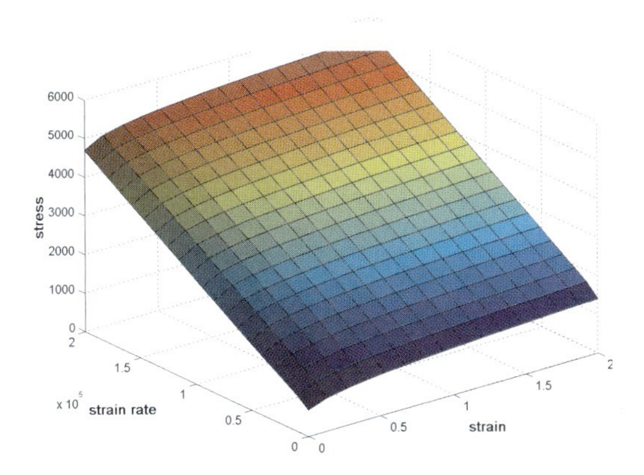

Figure 12.3: Resulting material law at 20°C.

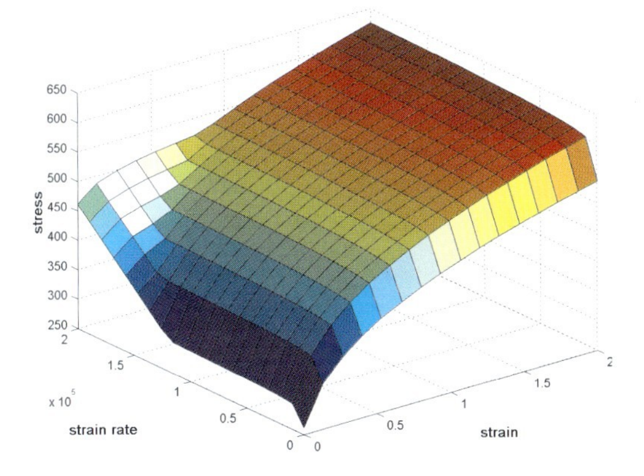

Figure 12.4: Resulting material law at 1000°C.

Furthermore, the activities close to the cutting edge can be described as a pure forming process, which means that there is no material separation. This kind of simulation has an advantage: qualitative information can be obtained concerning the workpiece border zones using various cutting velocities. Unfortunately it was found that this modeling method overestimates the forces and temperatures whereas the simulated chip form corresponds to reality [10].

For the description of chip forming without including the workpiece subsurface the separation based on contact [6] is suitable, if the chip volume is considered exclusively. The nodes are successively separated from a fixed surface by variation of the separation and tangential forces (see Fig. 12.5). Furthermore, the workpiece subsurface is represented by a rigid line segment (separation line in Fig. 12.5). The tool is simplified by being considered to be rigid too. The rigid tool, modeled also as a line segment, has an aiding radius near the line for the workpiece subsurface. This radius partly overlaps this subsurface line (see Fig. 12.5, top) and allows a defined junction between separated nodes and the cutting edge. Having positive rake

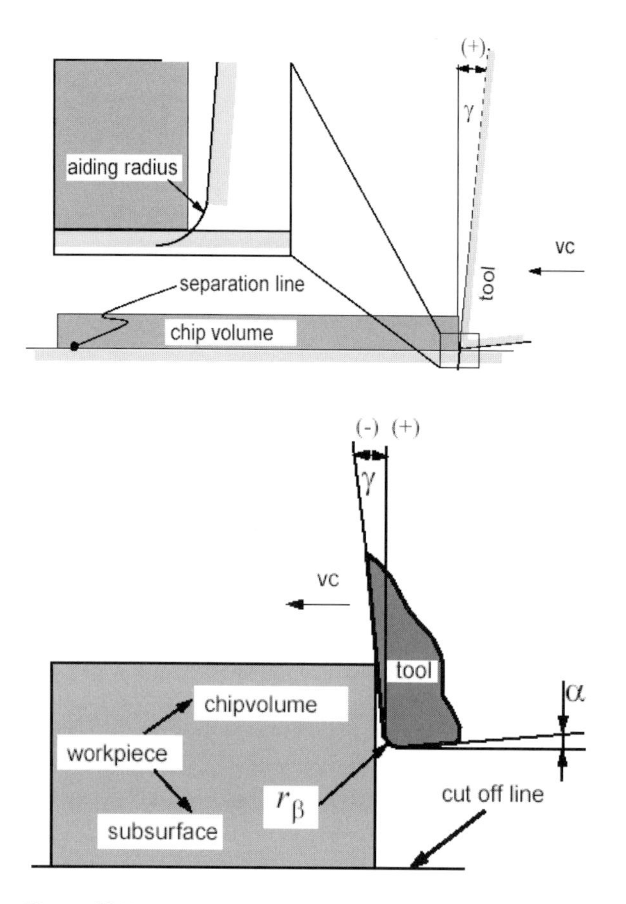

Figure 12.5: Demonstration of the model characteristics using contact-based separation (top) and using the forming model (bottom).

angles, the aiding radius can be neglected, having zero or negative rake angle; the rake friction and the compressive forces in front of the cutting edge inhibit the separation of the nodes and thus a smooth transition to the tool. The used geometric separation criteria were based on the separation condition that a node must have contact with the tool and the separation line (workpiece subsurface) to be separated.

Figure 12.5 shows both methods of process description at the cutting edges which will be examined in this simulation. To be able to compare both approaches, the same parameters for the process characterization specified in Sect. 12.3.1 are used, whereby additionally to the rake angle γ the cutting edge radius r_β and the clearance angle α have to be considered using the forming model.

In the context of the separation algorithm there is also the choice of using a sufficiently high mesh refinement in order to be able to simulate the thermo-mechanical instabilities. These instabilities are caused by high gradients of the forming speeds ($\dot{\varphi}$) at co-existent high temperatures (ϑ) and forming grades (φ) by the material model. For a precise calculation, a very

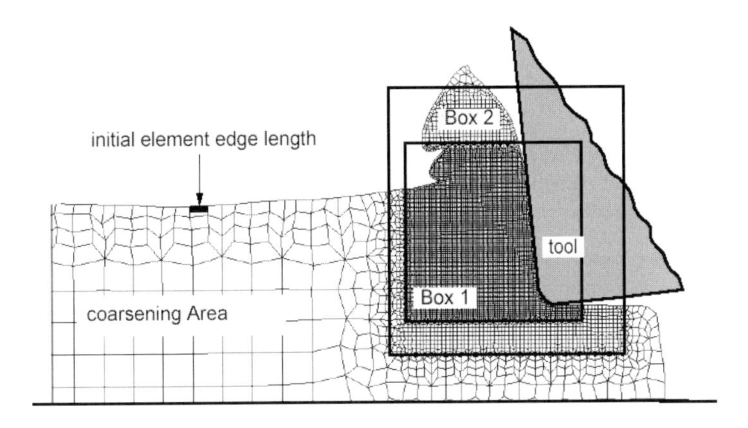

Figure 12.6: Result of a remeshing step in the forging model during the calculation.

high mesh refinement is needed. That is why an element border length of some micrometers is chosen, because the integration over too large elements cannot map the gradients and the discontinuous segment chip forming under high speed cutting conditions cannot be calculated. A lot of FEM programs offer a consistent mesh refinement over the whole model in this context. The enlargement of an extremely fine meshing leads to an enormous increase of computing time, which reduces the possibility of parameter variation. Another point in this discussion is the demand that after at least three increments a new mesh must be calculated. Most of the calculation has the same path steps of third element edge length of the smallest element.

For this reason an interface was implemented which allows us to insert or to change refinement zones (refinement box) before and during the calculation. Furthermore, the re-meshing program obtains miscellaneous information over the interface. This can be information about the starting element edge length, the "movement" speed of the refinement boxes and their size and the coursing level outside the refinement boxes. The result is shown in Fig. 12.6.

12.3 Simulation Results and the Comparison with Experimental Results

12.3.1 Process Parameters

The orthogonal cutting experiments were done with the material Ck45N by [2]. The cutting conditions were realized by a turning off of a 3-mm-wide web with a recessing turning process (small picture in Fig. 12.7). The chip height was 0.1 mm and a small negative tool orthogonal rake of $-6°$ was chosen. The cutting experiments were done without any cooling lubricants and metal carbide was used for the tool. Finally, the blue curve in Fig. 12.8 shows the acquired cutting forces at cutting velocities from 100 m/min up to 3000 m/min, whereas all fundamental process parameters for the simulation model are shown in Table 12.1. In addition the heat transfer between chip and tool was neglected, which results in an over-estimation of the chip temperature. The comparison with experimental results shows the influence of this neglect.

Table 12.1: Parameters of the model construction.

Parameter	Forming model	Separation model
Cutting length	$a = 0.8$ mm at workpiece length of 1mm	
Chip height	$h = 1$mm	
Chip width	$b = 3$ mm	
Friction factor	$m = 0$	
Plastic heat generation	$\chi = 0.9$	
Cutting speed	$V_c = 100–2000$ m/min	
Rake angle	$\gamma = -6°$	
Clearance angle	$\alpha = 7°$	
Radius of cutting edge	$r_\beta = 15$ µm	

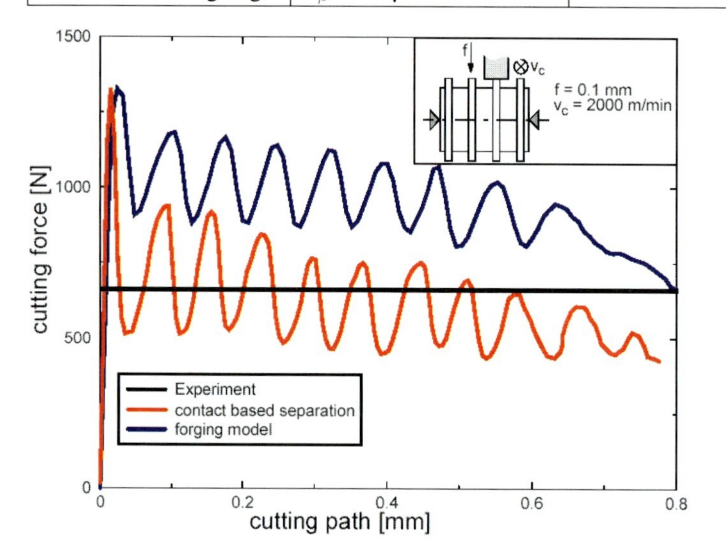

Figure 12.7: Comparison of cutting forces at cutting speed of 2000 m/min.

For example the experiment showed a maximum temperature of 300°C at a chip height of 0.25 mm and at a cutting speed of 1278 m/min in the contact area. The result of the FE simulation indicates a temperature of 500°C at this point. The appointment of the heat-transfer coefficients is a topic of research in further experiments. In addition, the non-influenced part of the workpiece subsurface is neglected in the forming model. This part is represented by a line segment too (see Fig. 12.5, bottom).

Furthermore, a constant friction factor is used for the whole contact area between cutting edge and chip during the computation. At first this factor is chosen to be zero, since the comparison with experimental results (see Fig. 12.7) has shown that mainly in high speed cutting simulation the results only match pretty well with the experimental results if the friction factors are considered to be very small.

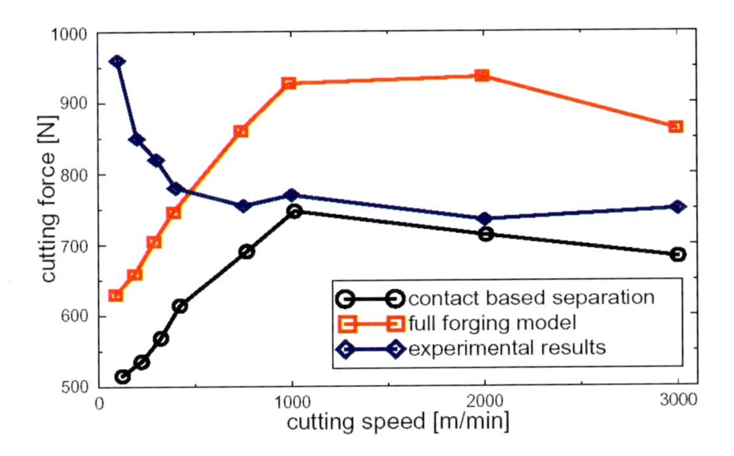

Figure 12.8: Comparison of calculated and measured cutting forces.

12.3.2 Cutting Forces and Chip Geometry

As experimental reference for both simulation models the orthogonal cutting process with the parameters shown in Table 12.1 and a cutting speed between 100 m/min and 3000 m/min was chosen. The verification was done on the basis of the cutting force and the chip form. For this reason the cutting speed curve for a cutting speed of 2000 m/min is shown as an example in Fig. 12.7.

The diagram shows that there is a good conformity between the separation based on the contact model and the experimental results considering the medium cutting force. In contrast, the forming model computation resulted in an over-estimation of the cutting force. While both models result in a force boost at the beginning of cutting, this decreases subsequently due to the slip of the first segment.

Afterwards the cutting force is pulsating (in both models), which is caused by the alternation of the "material compression in front of the tool" and its "slip of a segment". The variation of the friction factors did not result in any significant differences of the computed cutting forces at this cutting speed. The friction factor only enhanced the force maximum before the slip of the first segment.

To identify the quality of the cutting force prediction of both models, the medium cutting forces will be compared (Fig. 12.8). During the analysis in the cutting speed range over 1000 m/min the separation based on the contact model shows a good correspondence with the experimental results, whereas the forming model gives clearly too high cutting force predictions. Nevertheless the qualitative characteristics of both curves are identical, since both are computing a cutting force decrease at the moment of chip segmentation in a cutting speed region over 1000 m/min. The increasing course of the cutting speed curves of both models also shows the discrepancy between modeling and experiment by the neglect of the friction. Out of this it can be reasoned that the friction has a considerable influence on the computed cutting forces by a flow chip forming at cutting speeds lower than 1000 m/min.

An additional verification variable is the calculated chip form, which is exemplarily compared with a real chip produced at the "Institut für Werkstoffkunde Hannover" at cutting speeds

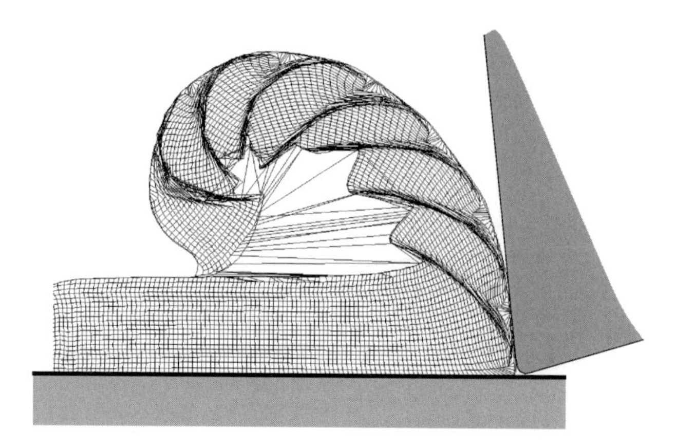

Figure 12.9a: Calculated chip geometries with tool detaching allowed.

of 2000 m/min using the material Ck45N. The chips were produced there by an orthogonal cutting process which is described in Sect. 12.3.1. After the cutting operation the chips were embedded for a micrograph. Due to the neglect of friction, the computed chip forms have a stronger curvature than the real ones (see Fig. 12.9a) because the friction takes effect during the tool–workpiece interaction and during the contact of the chip and the workpiece surface. Both parts of the friction result in a bend up of the real chip. Therefore some more simulations have been made, where the chip separation from the tool was prevented (tool detaching not allowed in Fig. 12.9b) to be able to compare the computed chip forms with the real ones. Figure 12.9b) further shows the temperature distribution inside the chip. The highest temperatures are reached at the tool contact surface and in the shear bands (primary shear zone according to Fig. 12.1).

The comparison between simulation and reality shows few differences in the structure of segmentation (for example the shear angle Φ in Fig. 12.11) and in the distance between two segments. In this case only the third and fourth segments are usable for an accurate comparison because the fifth segment is built now and the first and second segments are influenced by the first cut. With attention to these facts the segmentation ratio is 0.44 in the simulation and 0.46 in the experiment, whereas the segmentation ratio is the defined ratio of the maximum chip height (h_{max}) to the minimum chip height (h_{min}) (see Fig. 12.10). More differences are visible in direct comparison of the chip heights in the calculation, where the delamination of the chip from the tool are not allowed. These effects have to put into perspective in the comparison with the curved chip (Fig. 12.9a), if compared with the real chip height. Then the difference in the chip height amounts to 25% between simulation and experiment, while the difference is 35% according to Fig. 12.10 and Fig. 12.11.

In face of these differences it can be said that the simulation is able to calculate the segmentation behavior of the chip. This result is only possible by using a high mesh refinement (refining boxes) and a modified material law.

The analysis of the calculated contact length shows the differences of the used models (forging model and model with separation based on contact) and is another important part of our interests. In this case the verification with the experimental results makes no sense

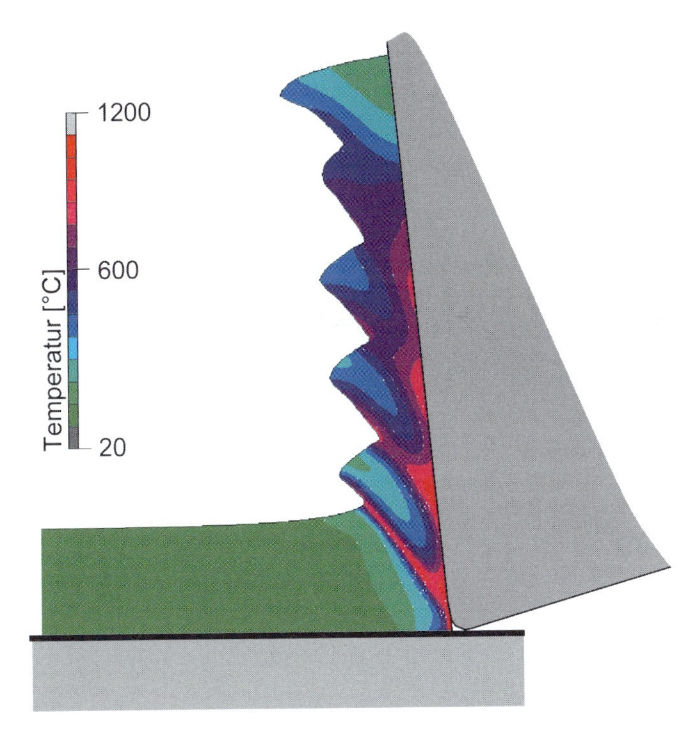

Figure 12.9b: Tool detaching not allowed and temperature distribution in the chip.

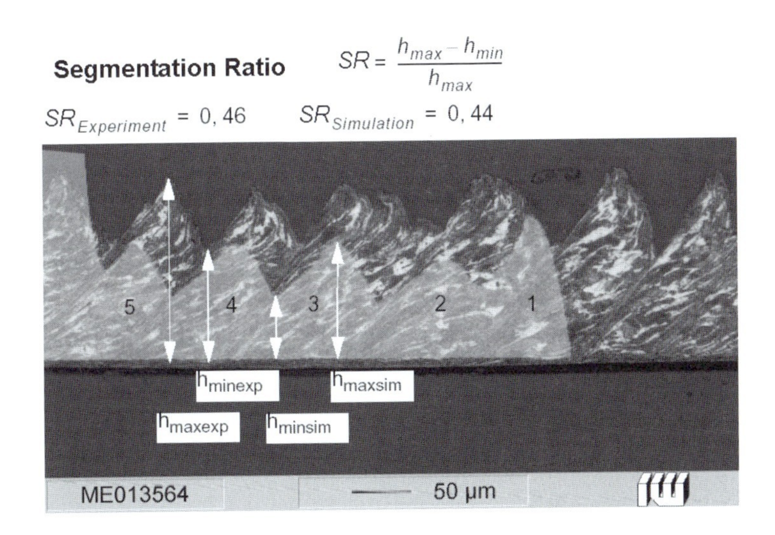

Figure 12.10: Comparison between calculated and experimental chip for segmentation ratio.

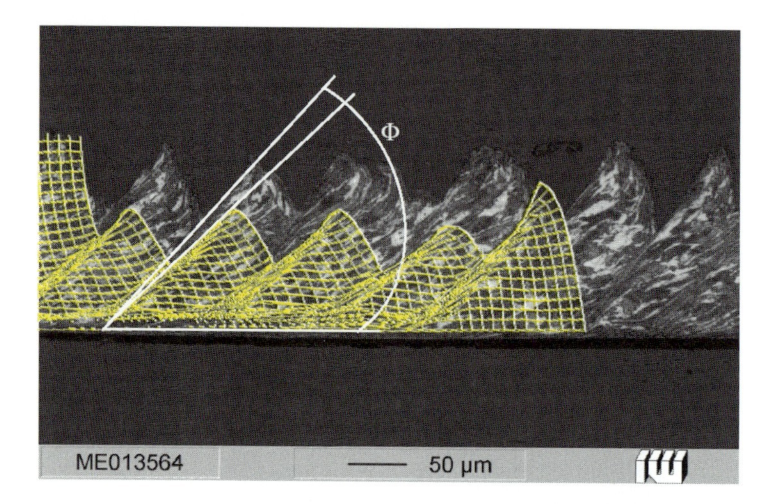

Figure 12.11: Comparison between calculated and experimental chip for shear angle.

because there are no accurate references for this value. The analysis of the contact length should explain if the higher cutting forces in the forming model result from a higher cutting length. The analysis showed a dependence of the cutting length on the cutting speed, but the forming model calculated higher contact lengths than the model where the separation is based on contact. This result concerns the cutting forces because they are an integration value of the boundary forces of the nodes over the whole contact length.

Both models show a decrease of cutting length at an increase of cutting speed. This disagrees with the results in Fig. 12.7, where cutting forces increase with an increase of cutting speed in the lower-speed area up to $v_c = 1000$ m/min (flow chip forming). The probable reason is the unconsidered tribology which has a lot of influence in this cutting speed area. The contact length decreases and oscillates clearly in the high-speed cutting area ($v_c > 1000$ m/min) due to the fact of chip segmentation. This result denotes that the tribology has no influence in the high-speed cutting area on the cutting length and force.

12.3.3 Residual Stresses

The residual stresses were verified in cooperation with the IFW Hannover. The measurement of the residual stresses resulted from samples which were produced by a cycloidal milling process with no overlapping of the cutting paths. In this case the process influences the workpiece subsurface only once. All parameters are nearly identical to Table 12.1, but chip height ($h_{exp} = 0.12$ mm) and rake angle ($\gamma = 18°$) are different in the milling process. That is why the verification of the simulation results is shown as an example.

The residual stresses (shown in Fig. 12.12) were investigated with the same cutting speed as in the experiment although the orthogonal cutting simulation model was used. Thereby a cooling time of 20 s was implemented. In this time the surface temperature of the workpiece has reached room temperature. The residual stresses (σ_{11} in the direction of v_c) show strong variations at a cutting speed of $v_c = 2000$ m/min, which are influenced by the segmentation

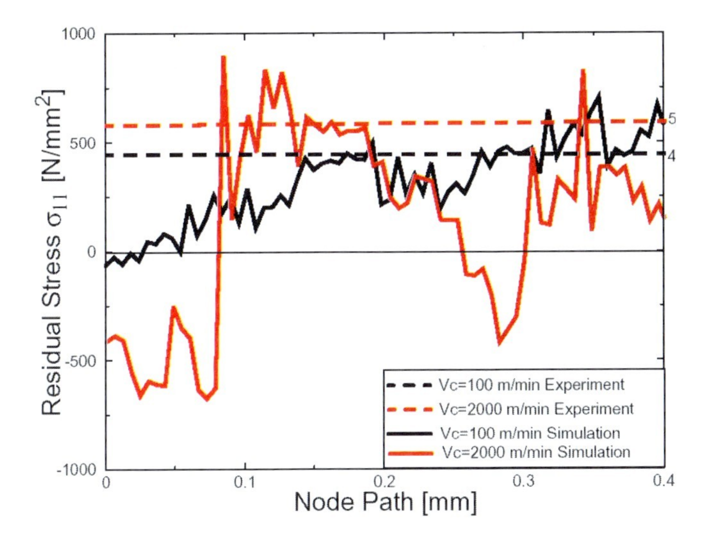

Figure 12.12: Residual stresses on the workpiece subsurface.

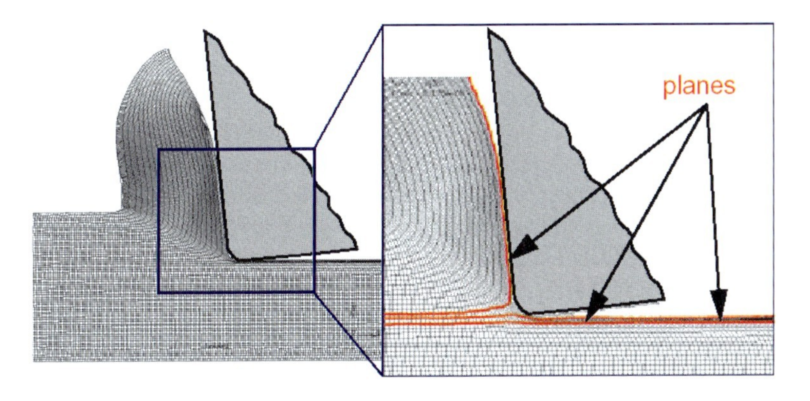

Figure 12.13: Deformed original mesh at a cutting speed of 100 m/min.

chip forming. At the beginning of the first cut one can estimate compressive stresses which are nearly identical to the following tensile stresses. Nevertheless the results show more instability along the cutting path due to the chip segmentation. The model nearly estimates the values of the experiment at a cutting speed of $v_c = 100$ m/min. In conclusion, the forging model can predict the dimension of the residual stresses for both cutting speeds.

12.3.4 Additional Analysis of the Forming Model

Two mechanisms of chip forming (flow and segmented chip) are observed in the area of cutting speed $v_c = 100$–2000 m/min in the analysis of the chip geometry. For this account two examples were chosen in Fig. 12.13 (flow chip at $v_c = 100$ m/min) and in Fig. 12.14 (segmented chip at $v_c = 2000$ m/min, which show the forming of the basic mesh in the forging model.

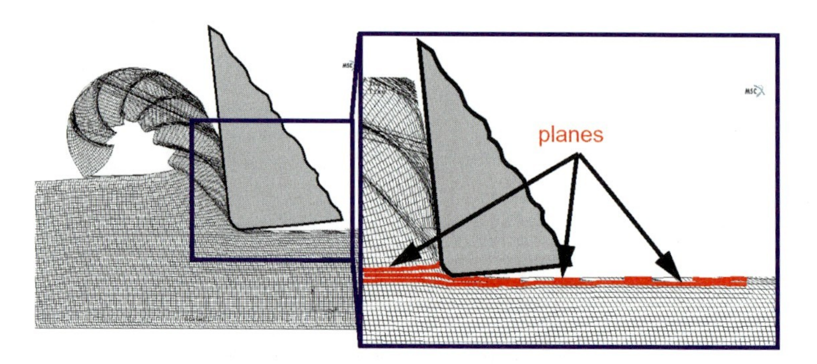

Figure 12.14: Deformed original mesh at a cutting speed of 2000 m/min.

Figure 12.13 shows a definite shear in the basic mesh at a cutting speed of 2000 m/min, so that the segments will be visible clearly. Furthermore Fig. 12.13 and Fig. 12.14 on the right-hand side display the basic mesh analysis of the first cut. In this case a comparison of the deformed basic meshes and the flow lines (red lines) follows at cutting speeds of $v_c = 100$ m/min and $v_c = 2000$ m/min. The flow lines describe a material sharing at the cutting edge, where one part flows in the chip volume while the other part is compressed and causes compressive stresses in the workpiece subsurface. From this it follows that a point (under consideration of the chip width a line) has to exist where the material flow changes direction.

The analysis of the Y component of the local flow velocity in the chip provides this thesis because the Y component changes direction at one certain point. All the material is compressed in the workpiece subsurface below this neutral slip point. Moreover this material behavior is described in the ploughing force model according to [11], but the material separation is not localized only at one point in reality. Rather a material area exists in front of the tool where the material flows in the chip and in the subsurface.

Thereby the ploughing area depends on the chip height h and the radius of the cutting edge r_b [12]. The flowing material from the ploughing area initiates a kind of forging process at the cutting edge which results in a smaller chip height. This means that the whole chip is not removed from the workpiece. The calculation of the neutral slip point shows constant values over all cutting speeds with flow chip forming. This means that the point is not influenced by cutting speed. The coordinate of the neutral slip point oscillates at higher cutting speeds with segmented chip forming, whereas the frequency is the same as the one of the cutting force.

12.4 Analysis of the Thermal Effects on the Tool

From the previously examined model the mechanical loading of the tool can be determined with high accuracy, but not the thermal loading, because this model calculates the first millimeter only.

This would require an extension of the simulation to the point of time where a stable temperature field in the tool (cutting plate) and the tool shank has been reached, which means extending the cutting path to at least 1000 m. Also, a further mesh has to be created at the tool. Both could not be analyzed effectively even on high-performance computers. A more effective

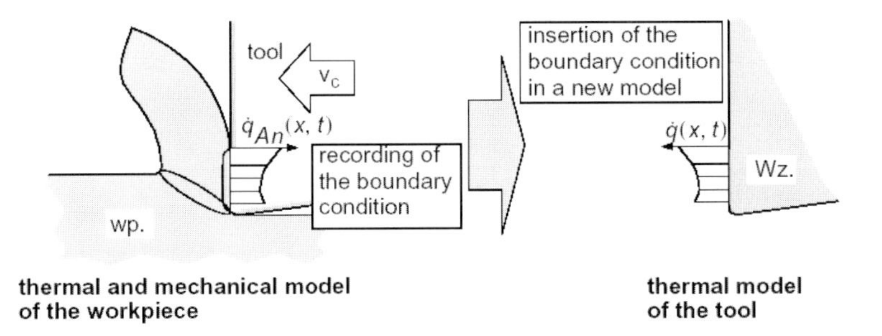

Figure 12.15: Transfer of the results from the first cut to a static analysis of the tool.

way is the transfer of the thermal loading to a separate model which calculates the thermal loading of the tool only. The previous model determines this load as a boundary condition in the new model. As shown in Fig. 12.15, the recorded local heat-flux density $\dot{q}_{An}\,(x,t)$ will be transferred to the new thermal model of the tool.

For this model the thermal simulation results from Sect. 12.3 at a cutting speed of 1000 m/min were used. The properties of the carbide cutting plate are referenced by the producer's specifications in [14] and the thermal and mechanical properties of the tool shank used the data from [13].

A very fine mesh (element edge length = 1.2 µm) was used, in order to reproduce the high temperature gradients in the contact area. Furthermore, a high heat-transfer coefficient ($\alpha_\gamma = 100\,\mathrm{W/m^2\,K}$) was chosen to account for the intensive heat transfer between the cutting plate and the tool shank resulting from the high contact pressure. The heat flow from the tool to the chip was simplified as shown in Eq. (12.3):

$$\dot{q}(x,t) = \dot{q}_{An}(x,t) + \alpha_\gamma[\vartheta_{\text{Chip}} - \vartheta_{\text{Workpiece}}(x,t)]. \tag{12.3}$$

Figure 12.16 shows the temperature field in the tool after a cutting time of 44.3 s. The highest temperature appears almost at the end of the contact area. The temperature analysis at the cutting edge showed a strong increase at the beginning of the simulation. A state of equilibrium is nearly reached at a temperature of 700°C at the end of the cutting time. But it has to be stated also that a stationary state has not been reached yet in the center of the cutting plate.

12.5 3D Model for an Outer Turning Process

The model used for the simulation of an outer turning process is based on the separation model from Fig. 12.5. It should be used for the calculation of the forces and the chip forming and it is a complex spatial model. The knowledge concerning material separation (Sect. 12.2.3) and dynamic material behavior (Sect. 12.2.2) was transferred to this spatial outer turning process shown in Fig. 12.17. The spatial chip forming of this process has a lower complexity than it has in milling processes because there is no change of the chip geometry along the cutting path. In the turning process the kinematics of the tool and thus the geometry of the model are also considerably easier to describe.

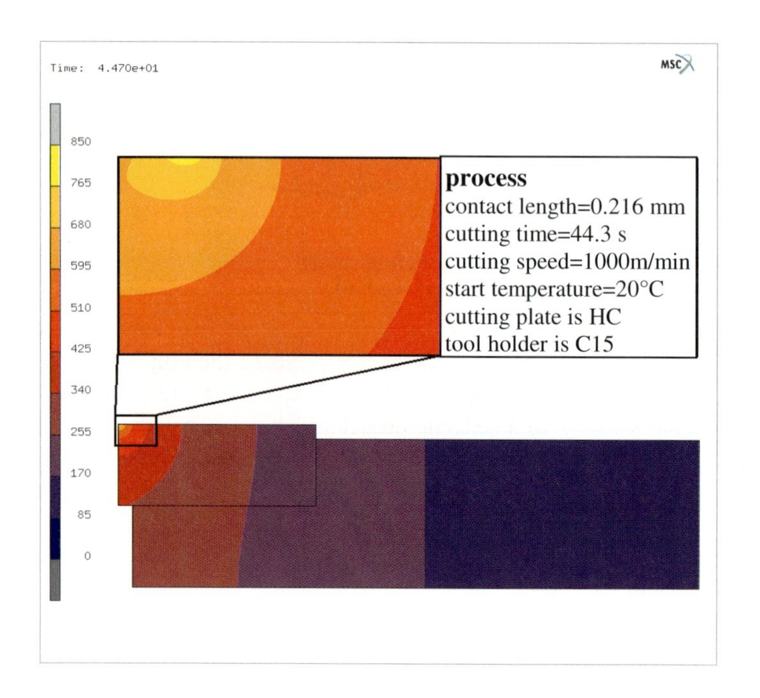

Figure 12.16: Temperature distribution in the tool after 44.3 s.

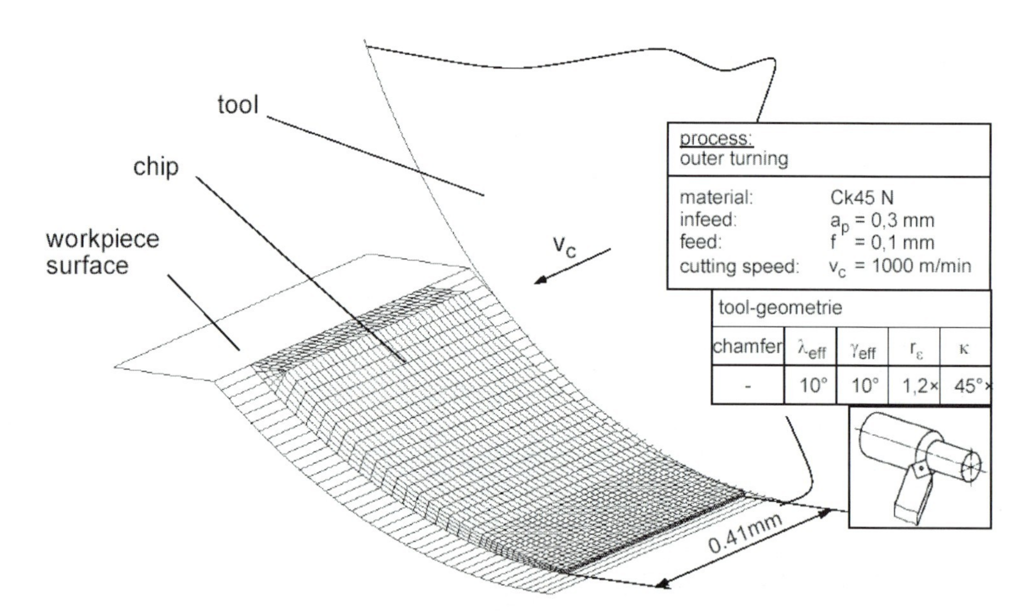

Figure 12.17: Description of the 3D outer turning model.

The chip volume to be calculated was again attached (glued) to the rigid surface of the workpiece and has a length of 0.41mm in the cutting direction. Furthermore, the whole chip volume was modeled with 4836 elements, whereas the transition region to the machined surface was assigned a depth of 7 μm. Although outer turning is a continuous process, only the first cut could be simulated with the available computing capacity. The tool penetrates the workpiece with its rounded cutting edge. Because of the geometry of the tool (effective tool cutting edge inclination and rake angle $\lambda_{\text{eff.}} = 10°$ and $\gamma_{\text{eff.}} = 10°$) a chip which is inclined spatially with respect to the tool plane results. This is in good accordance to the chips observed during the real turning process.

This simple spatial but geometrically complex model illustrates the problems encountered in the simulation of an outer turning process. A very coarse mesh of the chip has to be used compared to planar simulations of an orthogonal cutting process. A re-meshing of the 3D deformable body is not necessary. That is why a smoothing of the steep gradients of temperature, plastic strain and strain rate and thus a loss in accuracy has to be tolerated. Therefore it is not possible to reproduce the thermo-mechanical instabilities (shear bands) in the primary shear zone.

Figure 12.18 shows the deformed chip after the cutting path of $x = 0.2417$ mm. At this point of time the chip is detaching at point B while at the other end (position A) a slight bending upward due to self-contact can be observed. Because of the positive rake angle the cutting edge of the tool contacts the chip first at position A (see Fig. 12.18), where it has the smallest chip thickness. With further penetration of the tool a chip is formed at first here (point A) and the detachment of the chip from the tool begins at point A too. The contact of the elements with the edge is delayed in the feed direction. Thus the chip is already totally separated from

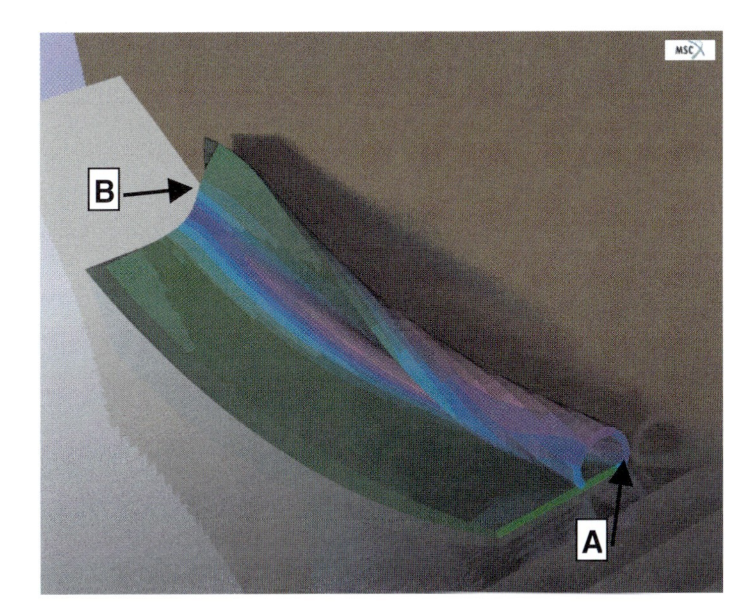

Figure 12.18: Deformed chip after 14.5 μs.

the tool at position A and contacts the surface of the workpiece while at position B the cutting is just beginning.

For the spatial simulation of the chip-formation process a re-meshing is still a requirement. It is needed to avoid the instability of the simulation, because the strong element distortion in the case of chip contact with itself stops the calculation. At the same time the mesh could be refined in order to reproduce the high gradients in the primary shear zone.

12.6 Conclusions and Outlook

In the presented work on the FE simulation of high speed cutting processes the advances in the reproduction of different phenomena of planar chip formation are introduced. This required the formulation of a material law which can reproduce the thermo-mechanical instabilities in the primary shear zone. The potential and the limitations of the description of the processes at the cutting edge were presented also. With the separation based on contact model the cutting forces and the shape of the chip can be predicted well in accordance to experimental data. The drawback of this model is that the properties of the workpiece subsurface cannot be predicted. In this case the forming model has to be used. The model predicts the calculated values of cutting forces and temperature too high, but estimates the residual stresses in the workpiece subsurface well. Further work concerning this topic should combine both models, which means the separation of the workpiece material at the calculated ploughing point. First results show a significant decrease in the cutting forces and the temperature though the chip geometry remains unchanged.

Further, the tribological conditions in the contact area between tool and workpiece were examined. The analysis results for cutting force and cutting length have to result in the conclusion that a significant decrease of the friction coefficient with increasing cutting speed has to be expected. Both simulations and experiments will allow us to estimate the friction coefficients.

Another relevant aspect in the analysis of the cutting process is the loading of the tool. Furthermore, a prediction of the cutting forces is known from the first cut simulations. Section 12.3.2 shows accordance to the experimental data and a rapid achievement of a stationary level. But the first cut simulation is not adequate to predict the thermal loading of the tool. Therefore the thermal loading of the tool was transferred into a new model, where a steady-state temperature distribution at the cutting edge was calculated.

Finally, the models of material separation and dynamic material behavior were transferred to a spatial model of an outer turning process. First simulations dealt with the analysis of the chip geometry. The major problem in spatial models is the description of the complex geometry, especially in milling processes. Also, the geometry of the resulting chip has to be reproduced in three dimensions, leading to a strong distortion of the elements in the mesh. This problem can only be overcome with the insertion of suitable algorithms for the re-meshing into the simulation.

Bibliography

[1] G. Warnecke, *Spanbildung bei metallischen Werkstoffen*, München, Technischer Verlag Resch, 1974.

[2] T. Friemuth, P. Andrae, and R. Ben Amor, *High-speed cutting – potential and demands*, in: First Saudi Technical Conference and Exhibition, Conference Report, Riyadh (2000), 210–223.

[3] S. Hiermaier, Dr.-Ing. Dissertation, Universität der Bundeswehr München, 1996.

[4] E. El-Magd and C. Treppmann, *Stoffgesetze für hohe Dehngeschwindigkeiten. Spanen metallischer Werkstoffe mit hohen Geschwindigkeiten*, Kolloquium des Schwerpunktprogramms der Deutschen Forschungsgemeinschaft, Bonn, 1999.

[5] E. Doege, H. Meyer-Nolkemper, and I. Saeed, *Fließkurven metallischer Werkstoffe*, Hanser Verlag, München, Wien, 1986.

[6] B. Westhoff *Modellierungsgrundlagen zur FE-Analyse von HSC-Prozessen*, Shaker Verlag, Dissertation an der Universität der Bundeswehr Hamburg, 2001.

[7] J. M. Huang and J. T. Black, *An evaluation of chip separation criteria for the FEM simulation of machining*, Journal of Manufacturing Science and Engineering **118** (1996).

[8] B. Zhang and A. Bagchi, *Finite element simulation of chip formation and comparison with machining experiment*, Journal of Engineering for Industry **116** (1994), 289–297.

[9] A. Behrens and B. Westhoff, *Anwendung von MARC auf Zerspanprozesse*, MARC Benutzertreffen, München, 1998.

[10] A. Behrens, K. Kalisch, and B. Westhoff, *Anwendung der FE-Simulation auf den Spanbildungsprozess bei der Hochgeschwindigkeitszerspanung*, Zwischenbericht im Rahmen des DFG Schwerpunktprogrammes "Spanen metallischer Werkstoffe mit hohen Geschwindigkeiten", Hamburg, 2002.

[11] P. Albrecht, *New developments in the theory of the metal- cutting process, Part I, The ploughing process in metal cutting*, Transactions of the ASME **82** (1960), 348–358.

[12] G. Xu, Dr.-Ing. Dissertation, TH Darmstadt, 1996.

[13] N.N., *Stahl-Eisen-Werkstoffblätter (SEW)*, Verein Deutscher Eisenhüttenleute Verlag Stahleisen mbH, Düsseldorf, 1992.

[14] N. N., *Hartmetalle*, Informationsmaterial der Firma WidiaValenite, 2000.

13 Finite-element Simulation of Nonlinear Dynamical Effects in Coating–Substrate Systems

G. Schmidt, R. Leopold, and R. Neugebauer

Coating technology is one of the most important technologies for the higher quality of products and a longer life-cycle time of products and it has a significant impact on the environment [8]. Coated tools have a considerably higher edge life compared to uncoated ones [48], can produce a higher surface quality [24] of the new detail due to less friction and improve the manufacturing process in dry cutting, high-speed cutting and micromachining. Coating–substrate systems are highly influenced by the complexity of the dynamic cutting process. Comprehensive investigations based on finite-element analysis were done to look behind the linear and nonlinear character of this system. The focus lies on the dynamics that appears when cutting conditions are used for non-stationary chip flow. A comprehensive survey of the complex behavior of the cutting process on one hand and the coating–substrate system on the other hand and an outlook for the application of nonlinear methods of time series will be given.

13.1 Introduction

Nowadays, more than three-quarters of all tools are coated with different systems. So far it has been the main problem that most of the technology providers in the field of coating technology have had many years of experience *without* considering any basic information on the substrate–coating physical and mechanical simulation [25]. This means that for new coating systems new mechanical and thermal loading technologies, and especially for new advanced technologies, e.g. dry manufacturing, micromachining and high-speed manufacturing, high-cost experiments have been carried out to find the proper substrate–coating combination. Coating optimization is still mostly achieved by means of 'more and less inspired' trial and error approaches [33]. The stability of coatings at different substrate surfaces depends on the stress–strain situation and must be considered by a closed-loop system [39]. The strength of substrate and coating is one basis for further optimization of the coating technology and for a higher quality of the manufacturing process [4].

Higher precision of manufactured parts and surfaces, reduction of processing time with simultaneous increase of tool life and fully automated production are the major aims of research and development in the field of cutting. In order to fulfill the high demands and in addition to ensure a high process reliability, new methods for theoretical investigations are required. However, the continuous development of new cutting tools with a complex microgeometry, covered with improved coatings and applied to manufacturing of advanced materials, requires high-resolution simulation methods and nonlinear methods.

Nonlinear Dynamics of Production Systems. Edited by G. Radons and R. Neugebauer

ISBN 3-527-40430-9

Coating–substrate systems (CSS) are stressed by the nonlinear dynamical cutting process.

On the other hand, the strength of substrate and coating is the basis for the further optimization of the coating technology and for a higher quality of the manufacturing process. The stability of coatings at different substrate surfaces depends on the stress–strain situation inside the CSS. The main boundary conditions which have to be used in the simulation are the total mechanical tool deformation (bending and torsion), the mechanical and thermal stresses due to the chip flow, the residual stresses due to the coating process, different methods of increasing the interface strength by substrate surface strengthening [46] and finally the clamping mechanism of the cutting insert including nonlinear friction. The finite-element method [3] is widely used in the cutting tool approach. For a real calculation of the coated parts, the dimensions of the coating–substrate model may rise to some million nodes or elements. The clamping boundary conditions are far away from the stress–strain situation in the coating, so that a fine meshing of the layer must be in the range of micrometers or less, and the complete structure including the substrate and the coating is in the range of some mm.

13.2 Mechanics of Chip Formation in Cutting Processes

In order to improve metal cutting processes, i.e. lower part cost, it is necessary to model metal cutting processes at a system level [31]. A necessary requirement of such is the ability to model interactions at the tool–chip interface and, thus, predict cutter performance. Many approaches such as empirical, mechanical, analytical and numerical have been proposed. Some level of testing for model development, either material, machining, or both, is required for all. However, the ability to model cutting tool performance with a minimum amount of testing is of great value, reducing costly process and tooling iterations.

In the CIRP keynote paper of the "Working Group on Modeling of Machining Operations" [25] it is pointed out that new investigations should be directed to chip-forming models with complex directions.

The chip-formation models should be extended to include rounded or chamfered cutting edges, chip-breaking geometry, 3D cutting, interrupted cutting as in milling and various modes of chip formation. Several of these aspects have already been studied by other researchers but could up to now not be integrated in the predictive machining theory.

A large domain of research work completed by various researchers during the last few years primarily deals with the chip forms and the evacuation of chips from the machining zone. An extensive review of this is presented by Jawahir and van Luttervelt in a CIRP keynote paper in 1993 [12]. The study of restricted-contact tools is important in relation to machining with grooved tools having a tool face land. By following the early work involving the slip-line field solutions of Johnson [13] and Usui and Hoshi [47], Jawahir and Oxley showed a quantitative relationship for predicting the chip back-flow angle [11]. The validity of the centered fan slip-line field for a wide range of cutting conditions was also shown through a series of models to take account of the rounded cutting edge and the flank wear [10]. The use of the predictive machining theory for predicting the cutting forces in machining with restricted-contact tools was shown in a recent work by Arseculaeratne and Oxley [1].

Establishing the actual tool–chip contact length for machining with a grooved tool is a very complex task, particularly when the contact within the chip groove is variable and strongly

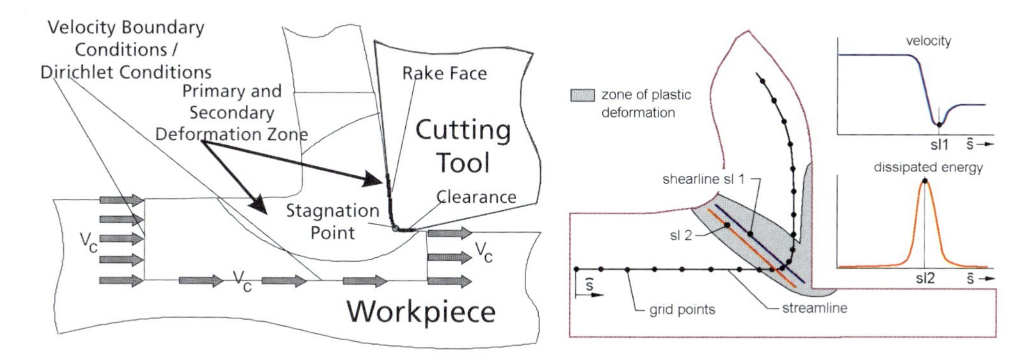

Figure 13.1: Boundary conditions (left) and mechanical-based shear-line definitions (right).

depends on the chip-groove utilization, which in turn is a highly variable factor within a chip-breaking cycle. The challenge is to characterize the levels of contact in metal cutting regions when machining with a complexly grooved tool under 3D cutting conditions [20]. Developing a quick and reliable machine-tool-based testing/measuring method for obtaining the material behavior under large strains and high temperatures on a routine basis will be another challenge, where an analysis of varying work–tool material properties for new tool and work materials could be more readily represented.

The prediction of the regime of chip formation and the resultant chip types is very important. This requires a unified detailed identification of the various chip types, appropriate models for the formation of the various chip types and, perhaps even more important, models to predict the boundaries between the various chip types. Most models are based on consideration of the stress–strain–temperature relation of the work material exposed to intensive shear in the primary shear zone. A recent overview was given by Shaw [44].

After careful examination of movie pictures taken of a cutting processes under a microscope, van Luttervelt was able to construct a dynamic system of four shear planes and a kinematically dead zone to represent the stagnant zone between chip and tool face [25]. This system made it possible to explain and kinematically simulate the formation of segmented chips. These movies also clearly showed that the segmentation started in front of the stagnant zone, close to the cutting edge and not in the primary shear zone, a fact which has been neglected in all subsequent research. Up to now no agreement has been reached on this very basic aspect of chip formation. It is hoped that computational mechanics will be of help here.

The results discussed in the following parts are based on the investigations of the mechanical–physical–chemical and mathematical point of view of cutting operation.

13.2.1 Basic Assumptions of Modeling

We assume the following model (Fig. 13.1) for the mechanical–thermal investigations of the metal cutting process.

There are the following main regions for nonlinear effects:

Ω_1 – Primary deformation zone (shear bands formation – crack initiation)

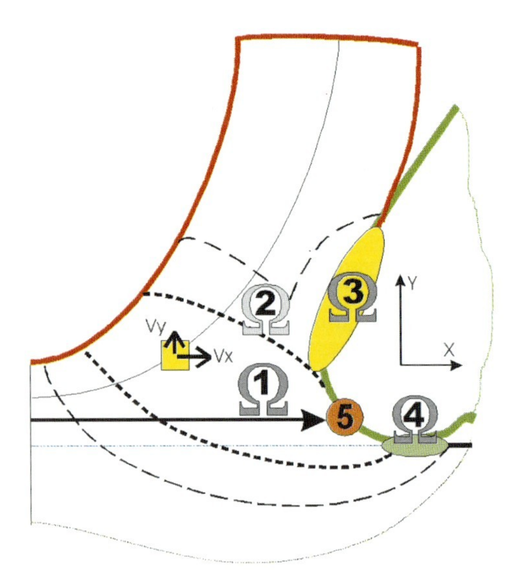

Figure 13.2: Main areas of nonlinear effects in metal cutting.

Ω_2 – Secondary deformation zone to the chip (crack initiation)

Ω_3 – Secondary deformation zone at the rake face – flow region (friction)

Ω_4 – Secondary deformation zone at the flank face (friction)

5 – Stagnation point – built-up edge – nonlinear formation

The lines in the cutting directions are so-called *streamlines* and lines in the orthogonal direction are so-called *potential lines.*

We assume the well-known shear-line definition [28] as shear line 0. This is a geometrical definition of a shear line. In addition to this, mechanical-based definitions of shear lines will be used [18]. The shear line 1 (sl 1) is defined by minimum particle velocities along the stream lines (Fig. 13.2). In conventional machining, the determination of these lines and areas gives enough information for the characterization of the process. For the nonlinear loading of coating–substrate systems additional new aspects have to be taken into consideration.

From the process point of view, there are two main sources for nonlinear external loading of coating–substrate-systems:

a. Process Type

Typically for milling or drilling operations are harmonic vibrations (Fig. 13.3).

b. The Physical Material Removal Process

In addition to this, chip -type effects must be included in a detailed analysis of coating–substrate systems. Figure 13.4 characterizes the three main chip types.

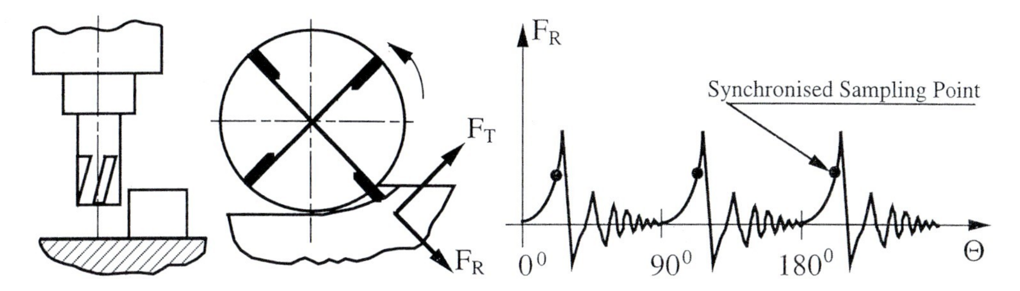

Figure 13.3: Force Signal depending on the cutting tool rotation angle Θ.

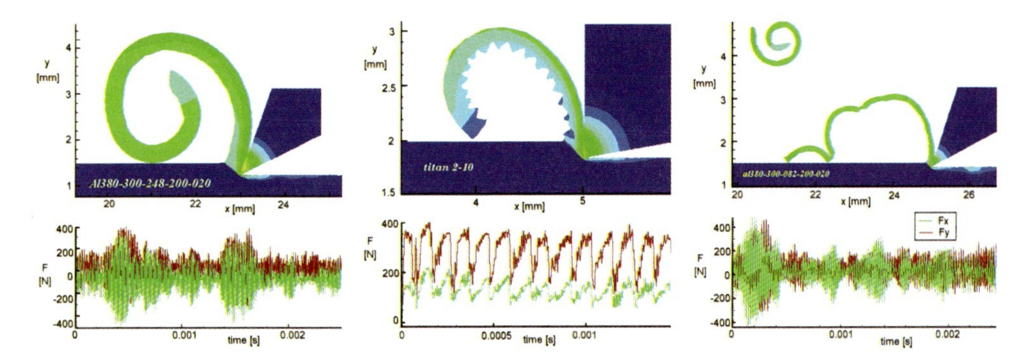

Figure 13.4: Different chip types (left: *flow chip*; middle: *continuous chip*; right: *discontinuous chip*).

13.2.2 Investigations of the Nonlinear Dynamical Cutting Process

There are three main strategies for the simulation of the process.

a. Slip-line–Method-based Investigation

It is a well-known technology with a specialized area of application. The following methods have been widely used by various researchers:

1. The slip-line method, such as the models of Dewhurst [7], Lee and Shaffer [17], Johnson [13], Usui and Hoshi [47], Kudo [15], Shi and Ramalingam [38] and Liu et al. [23].

2. Oxley's [35] analytical method that is based on the analysis of Mohr's stress circle along the tool–chip interface.

3. The minimum-energy approach, such as the models of Merchant [28], Morcos [32], Mittal and Juneja [30] and Mittal et al. [29].

4. The lower-bound method, such as Rubenstein's [41] model.

5. The semi-empirical method, such as Arsecularatne and Oxley's [1] model.

6. The 'dead-metal-zone' method, such as Worthington's [50] method.

7. Jawahir [10] showed that the 'dead-metal zone', which is ahead of the tool face covering the entire restricted land of a cutting tool, is actually a special plastic deformation zone in which the material is not actually stationary but moves at an extremely low speed.

An universal slip-line model follows two most fundamental assumptions of slip-line theory, which are:

- the rigid–plastic material assumption, in which the material shear-flow stress k is constant everywhere in the plastic deformation region. No effects of strains, strain rates and temperatures upon the material shear flow stress are taken into consideration.

- The plane strain deformation assumption, which makes the slip-line model apply only to orthogonal cutting processes. A perfectly sharp cutting edge is also assumed in the universal slip-line model. For the investigation of nonlinear dynamical processes in cutting this method is not applicable.

b. Boundary-element Method (BEM)

The boundary-element method is an important technique in the computational solution of a number of physical or engineering problems. In common with the better-known finite-element method (FEM)and finite-difference method FDM, the boundary-element method is essentially a method for solving partial differential equations (PDEs).

The boundary-element method has the important distinction that only the boundary of the domain of interest requires discretization. For example, if the domain is either the interior or exterior to a sphere then the diagram shows a typical mesh; only the surface is divided into elements. Hence the computational advantages of the BEM over other methods can be considerable. The development of the BEM requires the governing PDE to be re-formulated as a Fredholm integral equation. The main applications of the method are directed to Laplace problems (electrostatics, steady-state heat conduction and potential flow) and Helmholtz problems (acoustics, electromagnetics) and has not been used up to now for nonlinear mechanical–thermal problems.

c. The Finite-element Method (FEM)

There are two fundamental approaches to FEM cutting process simulation: the displacement or Lagrangian formulation and the flow or Eulerian one. In a Lagrangian formulation, the finite-element mesh is attached to the workpiece material, whereas in a Eulerian formulation the workpiece material is assumed to flow through a meshed control volume. Both discussed approaches allow the inclusion of thermal phenomena into the simulation.

The Lagrangian approach seams to be the more natural one: both the chip and the finished workpiece surface arise as the result of an incremental procedure, which reflects the time- dependent process of indenting the cutting tool into the material and forming the chip. Therefore, and since the Eulerian model is applicable to steady-state problems only, most FE investigations for cutting operations are based on the Lagrangian approach, but there are some disadvantages of Lagrange simulation. If the cutting process is simulated as a poor deformation process, the FE mesh suffers the same large deformations as the material. Thus, the large deformation of the finite elements leads to non-regular meshes and re-meshing becomes necessary [2], but only a few FE codes have this facility, especially in case of general 3D meshes.

Opposite to this, the cutting process can be treated as a material separation process. So, the mesh deformation can be substantially reduced, massive deformation of finite elements is restricted to the chip and re-meshing can be avoided in many cases by a good mesh design, but a numerically stable material separation criterion reflecting the real material's behavior must properly be formulated.

The mentioned disadvantages do not occur if the flow formulation in Eulerian coordinates is used, but it must be emphasized that this approach can be used for steady-state cutting processes only. Using the Eulerian formulation, the chip flow is considered as a stationary non-Newtonian fluid in a fixed geometry.

Since the geometry of the chip is *a priori* unknown, it must be calculated iteratively during the solution process by updating the flow domain. The start geometry, i.e. the meshed control volume, can be chosen from experience and must not be very close to the chip's real end geometry. The natural contact length of the chip is determined iteratively too, but by evaluation of calculated contact forces in the contact area. The mathematical and algorithmical details of this Eulerian simulational technique are discussed in detail in [21, 22].

For the investigations of this project the FE code "AdvantEdge™" was used. This is an explicit dynamic, thermo-mechanically coupled finite-element modeling package specialized for metal cutting. Features necessary to model metal cutting accurately include adaptive re-meshing capabilities for resolution of multiple length scales such as cutting edge radius, secondary shear zone and chip load, multiple body deformable contact for tool–workpiece interaction and transient thermal analysis. In order to resolve the critical length scales necessary in the secondary shear zone and the inherent large deformations while maintaining computationally accurate finite-element configurations, adaptive re-meshing techniques are critical. Near the cutting edge radius, the workpiece material is allowed to flow around the edge radius. The finite-deformation kinematic and stress update formulations can be found in Marusich and Ortiz [27]. They are briefly reviewed here.

The balance of linear momentum is written as[1]:

$$\sigma_{ij,j} + \rho b_i = \rho \ddot{u}_i. \tag{13.1}$$

The weak form of the principle of virtual work becomes

$$\int\limits_B \nu_i \sigma_{ij,j} + \nu_i \rho b_i dV = \int\limits_B \rho \nu_i \ddot{u}_i dV. \tag{13.2}$$

Integration by parts and re-arranging terms provides

$$\int\limits_B \rho \nu_i \ddot{u}_i dV + \int\limits_B \nu_{i,j} \sigma_{ij} dV = \int\limits_{\delta B} \nu_i \sigma_{ij} n_j d\Omega + \int\limits_B \nu_i \rho b_i dV, \tag{13.3}$$

[1] We are using the same symbols given in the Theoretical Manual from Third Wave *AdvantEdge*™– version 3.3.

which can be interpreted as a sum of inertial terms plus internal forces, which equals the sum of external forces plus body forces. Finite-element discretization provides

$$\int_B \rho N_a N_b \ddot{u}_{ib} dV + \int_B N_{a,j} \sigma_{ij} dV = \int_{\delta B} N_a \tau_i d\Omega + \int_B \rho N_a b_i dV. \tag{13.4}$$

In matrix form

$$M a_{n+1} + R^{\text{int}}_{n+1} = R^{\text{ext}}_{n+1}, \tag{13.5}$$

where

$$M_{ab} = \int_{B0} \rho_0 N_a N_b dV_{\text{o}} \tag{13.6}$$

is the mass matrix,

$$R^{\text{ext}}_{ia} = \int_{B0} b_i N_a dV_{\text{o}} + \int_{\delta B0\tau} \tau_i N_a d\Omega_0 \tag{13.7}$$

is the external force array and

$$R^{\text{int}}_{ia} = \int_{B0} P_{iJ} N_{a,J} dV_{\text{o}} \tag{13.8}$$

is the internal force array. In the above expressions N_a ($a = 1,\ldots$,numnp) are the shape functions, repeated indices imply summation, a comma (,) represents partial differentiation with respect to the corresponding spatial coordinate and P_{ij} is the first Piola–Kirchhoff stress tensor, analogous to the engineering or nominal stress.

Thermal Equations

Heat generation and transfer are handled via the second law of thermodynamics. A discretized weak form of the first law is given by

$$C\dot{T}_{n+1} + KT_{n+1} = Q_{n+1}, \tag{13.9}$$

where **T** is the array of nodal temperatures,

$$C_{ab} = \int_{Bt} c\rho N_a N_b dV_{\text{o}} \tag{13.10}$$

is the heat-capacity matrix,

$$K_{ab} = \int_{B0} D_{ij} N_{a,i} N_{b,j} dV \tag{13.11}$$

is the conductivity matrix and

$$Q_a = \int_{Bt} s N_a dV + \int_{B\tau q} h N_a dS \tag{13.12}$$

is the heat-source array with h having the appropriate value for the chip or tool. In machining applications, the main sources of heat are plastic deformation in the bulk and frictional sliding at the tool–workpiece interface.

The rate of heat supply due to the first source is estimated as

$$s = \beta \dot{W}^p, \tag{13.13}$$

where $\dot{W}^p$ is the plastic power per unit deformed volume and the Taylor–Quinney coefficient β is of the order of 0.9. The rate at which heat is generated at the frictional contact, on the other hand, is

$$h = -t \left\| v \right\|, \tag{13.14}$$

where **t** is the contact traction and **v** is the jump-in velocity across the contact.

Constitutive Model and Material Characterization

In order to model chip formation, constitutive modeling for metal cutting requires determination of material properties at high strain rates, large strains and short heating times and is essential for prediction of segmented chips due to shear-localization (Sandstrom and Hodowany [43]; Childs [5]). Specific details of the constitutive model used are outlined in Marusich and Ortiz [27]. The model contains deformation hardening, thermal softening and rate sensitivity tightly coupled with a transient heat-conduction analysis appropriate for finite deformations. In a typical high-speed machining event, very high strain rates in excess of 10^5 1/s may be attained within the primary and secondary shear zones. The increase in flow stress, due to strain-rate sensitivity, is accounted for with the relation

$$\left(1 + \frac{\dot{\varepsilon}^p}{\dot{\varepsilon}_0^p}\right) = \left(\frac{\bar{\sigma}}{g\left(\varepsilon^p\right)}\right)^{m_1}, \tag{13.15}$$

where $\bar{\sigma}$ is the effective von Mises stress, g the flow stress, ε^p the accumulated plastic strain, $\dot{\varepsilon}_0^p$ a reference plastic strain rate and m_1 is the strain rate sensitivity exponent. A power hardening law model is adopted with thermal softening. This gives

$$g = \sigma_0 \Theta(T) \left(1 + \frac{\varepsilon^p}{\varepsilon_0^p}\right)^{1/n}, \tag{13.16}$$

where n is the hardening exponent, T is the current temperature, σ_0 is the initial yield stress at the reference temperature T_0 , ε_0^p is the reference plastic strain and $\Theta(T)$ is a thermal softening factor ranging from 1 for a solid at room temperature to 0 for a melt with appropriate variation in between.

Contact

Machining involves contact between the cutting tool and workpiece during chip formation and rubbing on relief surfaces of the workpiece. Additionally, chip–workpiece contact occurs when the chip curls over and touches the workpiece, while chip–chip contact can take place during chip segmentation. A robust and general contact algorithm is necessary to detect and correct all of the scenarios. An explicit predictor–corrector deformable contact algorithm is used. A search to detect node-on-face contact provides mesh inter-penetrations during the time step. During part of the time step one surface acts as the master (rigid) and the other as the dependent slave (deformable). Inter-penetrations of slave nodes are updated via computation of restoring forces during the time step. During the remainder of the time step master and slave surfaces are swapped, restoring forces computed and kinematic compatibility is achieved.

Adaptive Re-meshing

Lagrangian FEM formulations involving finite deformations inherently involve mesh distortion since nodal positions track material points. Mesh distortion can cause deleterious numerical performance such as loss of accuracy, reduction of convergence rates and critical time steps, volumetric locking and element failure via inversion. Additionally, it is highly advantageous to provide mesh gradation where large variations in geometric scales (cutting edge radius and feed) and material instabilities (adiabatic shear localization) occur and need to be resolved. Adaptive meshing techniques are the tool used to overcome such technical barriers in Lagrangian codes. Mesh refinement is effected by element subdivision along the edges of a tetrahedron, creating two smaller tetrahedra. The converse operation of mesh coarsening is performed by collapsing the edge or face of shared elements, creating fewer larger elements. Mesh improvement, i.e. the improvement of an aspect ratio measure, is realized through techniques comprising edge and face swapping and nodal Laplacian smoothing.

13.2.3 Results of Nonlinear Dynamical Loading of the Coating-Substrate System

To investigate the nonlinear dynamic parameters acting on the coating–substrate system, the accuracy and reliability of the numerical system has been checked.

The following main parameters were examined:

Simulation Options

- *Rapid Mode/General Mode*: rapid simulations should normally be done just to get an idea of the cutting situation. The simulation times can be cut by a factor of five, but the accuracy may be off by as much as 20%. It should not be used for force validation or complex analysis where a fine mesh is needed to capture some mechanisms.

- *Residual Stress Analysis*: after the cut is finished, the chip and tool are removed and the workpiece is allowed to thermo-mechanically relax, but it will also drastically increase computation time due to the low coarsening of the mesh and the additional thermo-mechanical calculations. As one result of the investigations it should be pointed out

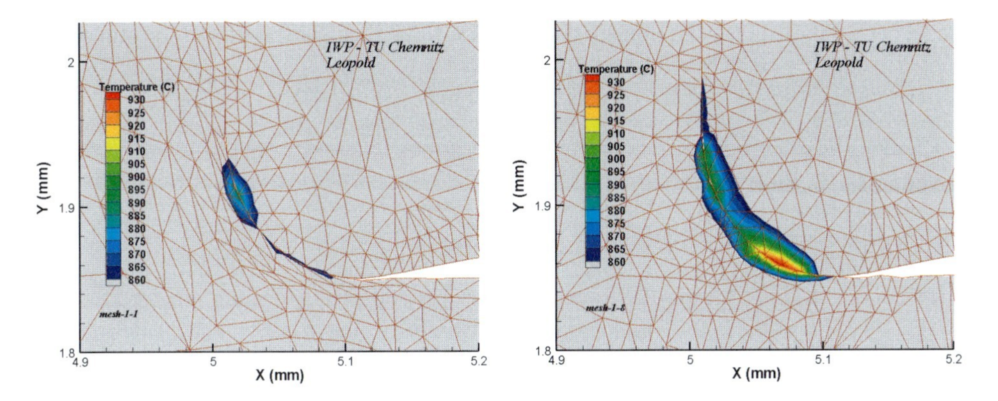

Figure 13.5: The maximum temperature deviation was determined as less than 6.2%.

that calculations of nonlinear dynamical loading of coating–substrate systems need this higher- accuracy but time-consuming mode.

- *Maximum (Minimum) Element Size (200 μm up to 5 μm)* – there is no significant influence on the thermal behavior of the cutting process in the region of coating–substrate systems (see Fig. 13.5).

- *Fraction of Feed* to determine minimum element size (fraction of feed was varied between 0.05 and 0.5) – has no significant influence on the calculated results (0.10 < fraction of feed < 1.2) – deviation of the calculated cutting forces was less than 4%.

- *Fraction of Cutting Edge Radius* to determine minimum element size (Fig. 13.6).

- *Mesh Refinement Factor* and *Mesh Coarsening Factor*: both factors influence the calculation by about 5%.

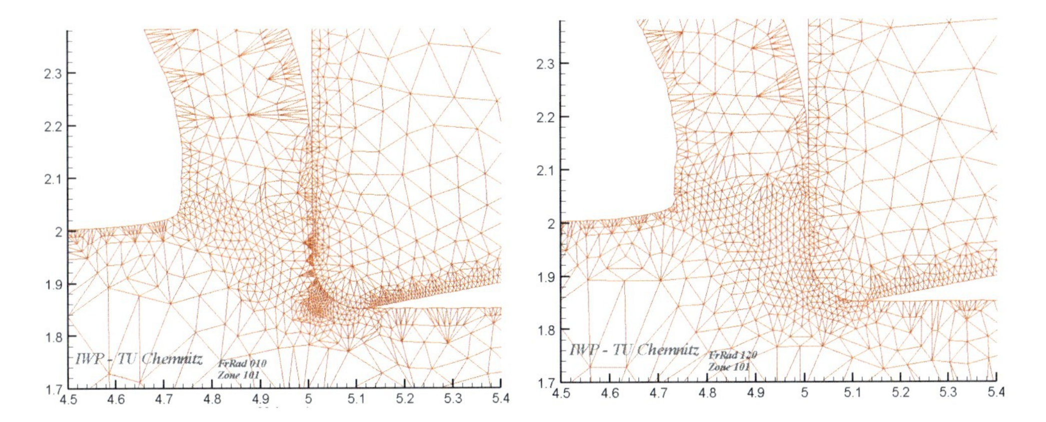

Figure 13.6: Influence of the fraction of radius on the element meshing (left: fine; right: coarse).

Based on these detailed investigations of the accuracy, reliability and stability of the numerical system, investigations with the following parameters were carried out using Ck45 and Ti-6Al-4V as workpiece material, cemented carbide as tool material and TiN as coating.

Shear-band Formation

Several steps of the shear-band-formation process can be distinguished:

a. Starting with a slight plastic deformation in the general metal forming area around the stagnation point of the cutting tool and upward bending of the material's backside.

b. The formation of a deformation zone in front of the cutting tool's stagnation point.

c. A region of the workpiece backside begins a plastic deformation.

d. Furthermore, the two deformation zones join and plastic deformation localizes.

e. In addition, a segment shears begins strongly along the shear band.

f. Finally, a second shear zone may form, which leads to a split shear that is curved downwards.

The same system of split shear band formation has also been observed in other simulations [34] and experimentally [6], so it seems not to be a simulation artifact.

Cutting Forces and Contact Length

One other interesting aspect is related to the cutting forces acting on the contact area of the cutting tool. Figure 13.7 shows the calculated cutting forces and the dynamical contact length. As expected, strong oscillations of the force occur, with a high absolute value of the force, when the deformation is not concentrated, and a lower value during times of shear localization and shearing of the chip along the shear bands.

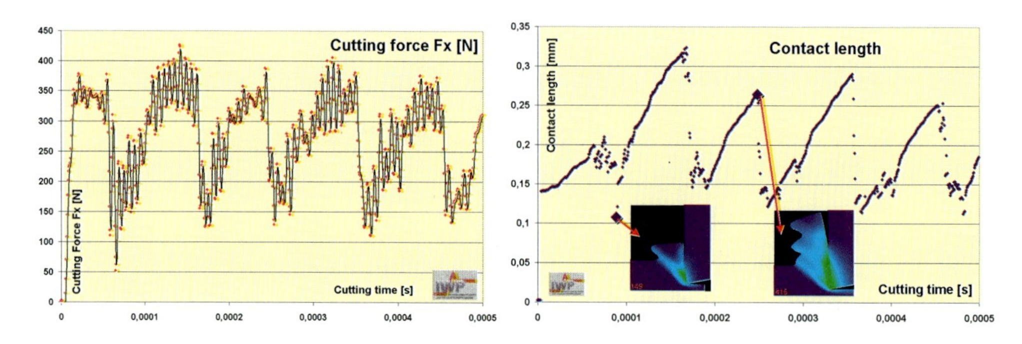

Figure 13.7: Cutting forces and contact length (machining of Ti-6Al-4V).

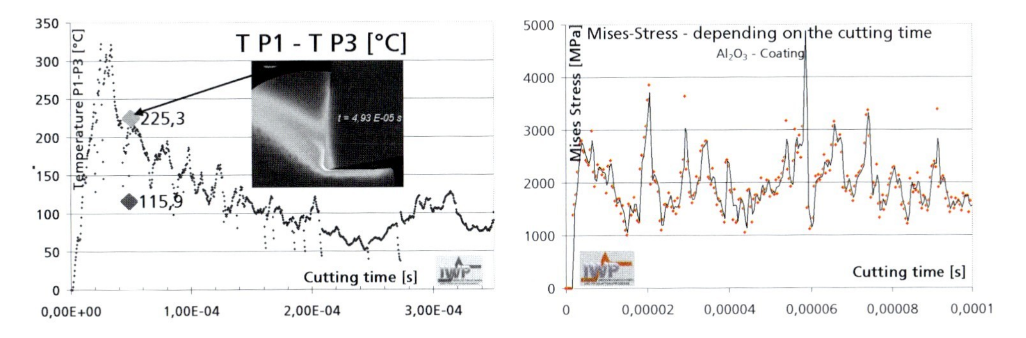

Figure 13.8: Thermo-shock behavior and von Mises stress (machining of Ti-6Al-4V) in coatings.

Nonlinear Dynamically Effects in Coating–Substrate Systems

Based on the above-mentioned simulation process for a TiN-coated cemented carbide cutting tool, there is some nonlinear dynamic behavior of the coating–substrate system itself. Figure 13.8 indicates the thermo-shock behavior and the time-dependent von Mises stress at one point at the coating–substrate interface of the system.

13.3 Modeling and Simulation of Coating–Substrate Systems

Coatings are used for wide industrial application in metal cutting, metal forming and any other wear protecting applications (Fig. 13.9). The majority of carbide cutting tools in use today are coated with chemical vapor deposition (CVD) or physical vapor deposition (PVD) hard coatings. The high hardness, wear resistance and chemical stability of these coatings offer proven benefits in terms of tool life and machining performance [26, 16, 37]. Following their introduction in the late 1960s, CVD technologies have advanced from single layer to various multi-layer and excellent wear-resistant coatings with a total thickness of 4–20 μm [45]. The coatings and coating processes are tailored for specific machining applications and

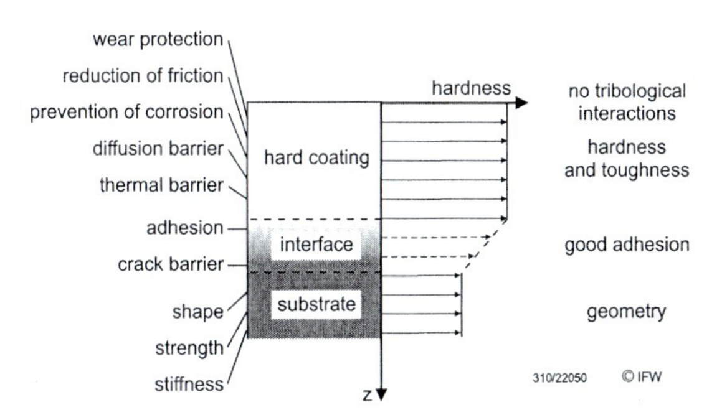

Figure 13.9: Coating–substrate systems.

workpiece materials. In continuous turning operations [9], for example, where the temperature of the insert goes up to 700°C and more, CVD coatings are mainly used. PVD coatings offer performance advantages in applications involving interrupted cuts or those requiring sharp edges, as well as in finishing and other operations [36]. In the industry, high-speed machining (HSM), dry machining (DM) or minimum quantity lubricant (MQL) are recognized as the key manufacturing technologies for higher productivity and lower production costs.

13.3.1 3D Coating–Substrate Simulations Based on Parallel Computing

3D FE computations of stress distributions in layer systems can contribute to the development of new coating–substrate systems. The influence of external loads on the deformation state of coated cutting inserts and on the resulting stress situation can be computed with such a high spatial resolution that critical stress values within and between the coatings can be detected. Combined with the experiences of coating technologists and experimental results, these data can be used to judge the layer system's stability. Since the coatings are very thin compared to the cutting insert's measures, such computations must be performed on highly graded meshes and, due to the nevertheless large computational expense, parallel computers should be used for those investigations. In a previous paper [19] the computational model and results of computations with 2.7×10^6 degrees of freedom for multilayer-coated cutting inserts under real machining conditions, observing the influence of the insert's geometry and of the load's magnitude, were presented (see Fig. 13.10).

The focus of the following investigations was directed to the application of a sequential FE system to investigate coating–substrate systems.

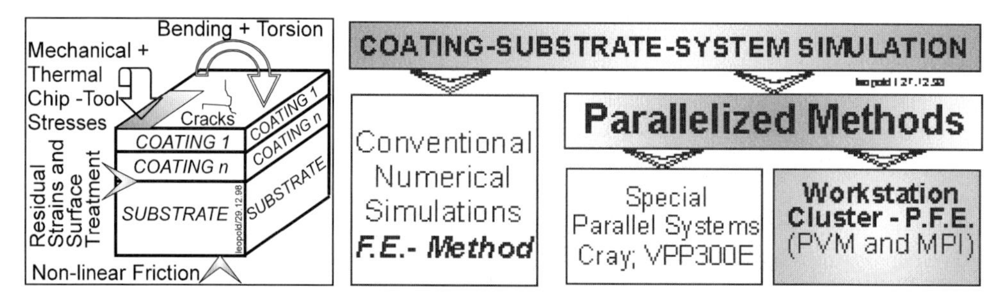

Figure 13.10: Stresses acting on coating–substrate systems and simulation methods.

13.3.2 Indenter Test Simulation

FE Model

The FE simulation of a Vickers indentation test uses a rotationally symmetrical layer–substrate model (Fig. 13.11), where contact areas exist between all elements of the layer vertical to the surface. When the stress component normal to the contact interfaces reaches a critical value, the actual contact node will be released and a crack will open. In the model the distance between two adjacent possible cracks amounts to 0.08 μm. The maximum length of the cracks

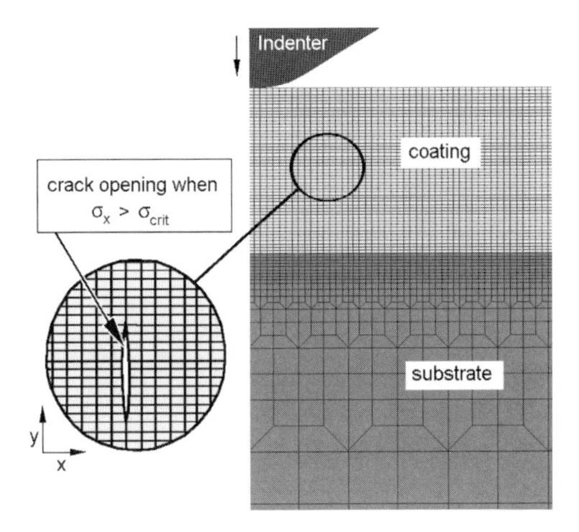

Figure 13.11: FE model for indenter test simulation.

is 4.9 μm and the thickness of the coating is 3.9 μm. So a crack could theoretically propagate into the substrate and the location of arising cracks is not restricted in an unrealistic way.

Interfacial cracks cannot be considered with this model and will be an object of further work.

The model was used to investigate the principal effects of coating, substrate and loading on the crack propagation for three different models:

- Titanium nitride (TiN) on a steel substrate.
- Compact TiN (bulk).
- TiN coating on a tungsten carbide substrate.

As failure criterion for the opening of cracks under tensile stress (mode I) a value given by [49] has been used.

Elastic Behavior of Thin Coatings

The elastic models have been used to analyze stress distributions in dependence on different loadings and for selected types of voids and cracks in the coating and the substrate. Stress concentration between substrate and coating due to cutting forces at the cutting edge is pictured in Fig. 13.12. Figure 13.12 also shows results for a crack in the outer coating area vertical to the surface (rake face) and an interfacial crack between substrate and coating.

Although the substrate and coating materials are considered to be homogeneous for numerical reasons, such a fine FE mesh is useful. The highest practicable number of degrees of freedom was about one million.

Simulation of Failure

The model (Fig. 13.12) enables crack opening between vertical elements. The indenter has a taper angle of 136 degrees with a nose radius of 2 μm.

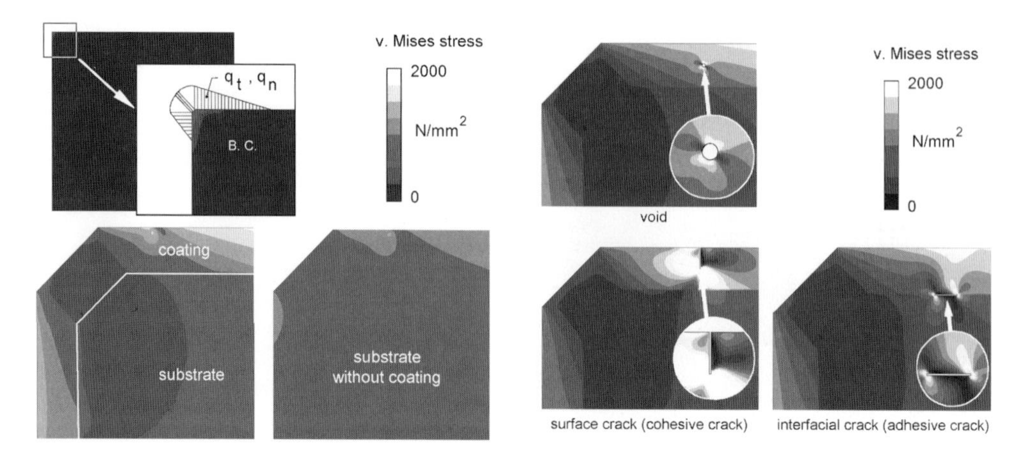

Figure 13.12: Stress concentration at the cutting edge due to cutting forces (left) and stress concentrations around irregularities in the coating (right).

Since the crack opening is based on the supposition of pure mode I behavior no significant normal stresses directed along the crack and shear stresses are acceptable. That assumption is satisfied by the test conditions. Pure bending stresses dominate the analyzed area of the coating. Crack initiation and opening are effected by radial stresses parallel to the surface of the specimen. A radial stress of 2500 N/mm^2 is applied as the critical opening criterion.

Areas of tensile stresses are developing below the indenter in the interface layer and at the surface of the coating to the right of the indenter (see Fig. 13.13).

In accordance with other authors, the characteristic zones with compressive and tensile stresses arise. As soon as the radial stress exceeds the critical value a crack will open and will propagate. Stress and crack distribution depends on substrate and coating properties. Figure 13.13 also shows results for different types of coating and substrate. The stress initiation begins in areas of high radial tensile stress and the propagation is stopped in the compressive stress zone. Only in the case of bulk TiN material do some cracks reach the end of the modeled crack (i.e. contact) zone.

13.4 Time Series Analysis[2]

Time series of calculated cutting forces in x and y direction were analyzed for different tool coatings: no coating, Al_2O_3, TiN, TiN–TiC. The time series were sampled at a frequency f $\cong$ 250 MHz and contained from 90 000 up to 125 000 points, representing approx. 0.4 ms of a cutting process. The median value of the forces increases quickly at the beginning, remains approximately constant for 0.2 ms, and then slowly decreases. For the two coatings (TiN with changed material parameters), the cutting force dependences are significantly different from the ones described above. The increase and decrease of the mean forces are not as distinct as

[2]This investigation has been done by Janez Gradisek, Edvard Govekar and Igor Grabec, Laboratory for Synergetics, Faculty of Mechanical Engineering, University of Ljubljana, Slovenia.

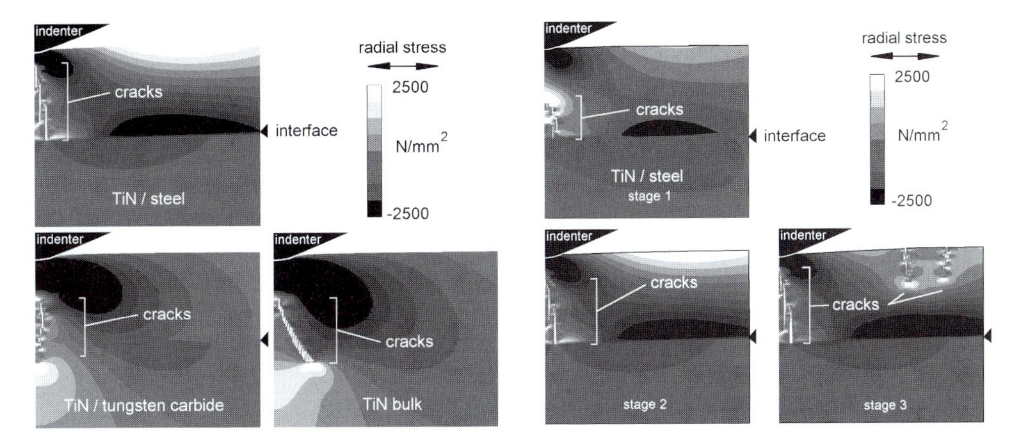

Figure 13.13: Stress distribution and crack propagation during an indenter test (titanium nitride coating on a steel substrate) (left) and characteristic stress distribution and patterns of cracks for various types of substrate and coating (left: titanium nitride/steel; middle: titanium nitride bulk; right: titanium nitride/tungsten carbide) (right).

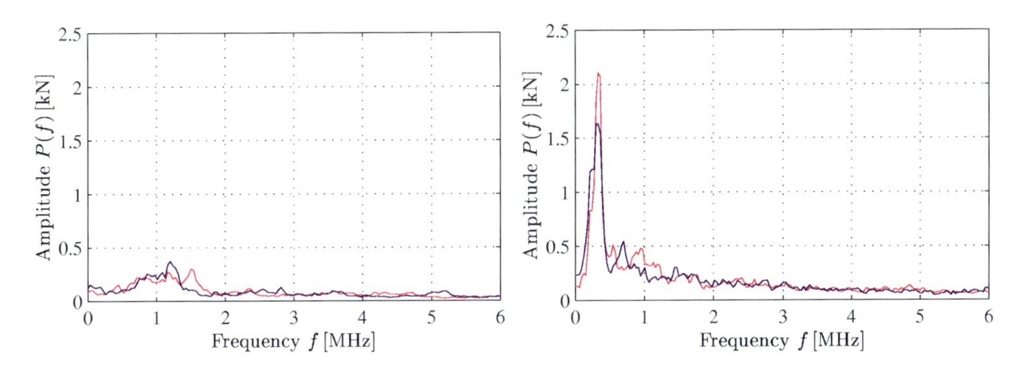

Figure 13.14: Calculated power spectra for cutting Ck45 (1.0503) with two uncoated cutting tools (left) and with two TiN–TIC-coated tools (right).

for the other five coatings. In addition, the mean forces for the last two coatings fluctuate with a larger amplitude.

The amplitude spectra of the segments are shown in Fig. 13.14. The relatively broad spectrum of an uncoated tool with no dominant spectral peaks is in stark contrast with the spectra of coated tools, in which several peaks are observed, dominated by a sharp peak at $f \sim 350$ kHz. Some spectra of coatings are similar to the uncoated tool spectrum. The spectra for special coatings are different from all others; the dominating spectral peak lies at $f \sim 130\,\text{kHz}$ and it is much stronger than the dominant peaks in spectra of other coatings.

13.5 Conclusions

First investigations based on finite-element analysis were done to look behind the linear and nonlinear character of the cutting process. The focus lies on the dynamics that appears when cutting conditions are used for non-stationary chip flow. The material removal process, including the friction behavior, nonlinear material behavior and shear-band formation, is indicated as the main source for the nonlinear character of the process. In addition to this, investigations into the mechanical–thermal character of coating–substrate systems have been done. Further investigations will be concentrated on the physics of the chip-segmentation process based on high-graded FE models and on a much finer discretization of coating–substrate systems to close the gap between the FE simulations in the range of some 100 nm or larger and the molecular–dynamical (MD) simulations in the range of several nanometers [40]. Three workpiece materials will be selected: carbon steel, titanium alloy and aluminum alloy. An additional coating–substrate system will be diamond-like carbon (DLC). Results of this research based on design of experiments will be investigated with different methods of dynamics including specialists in this area.

Bibliography

[1] J.A. Arsecularatne and P.L.B. Oxley, *Prediction of cutting forces in machining with restricted contact tools*, Machining Science and Technology **1** (1997), 95–112.

[2] M. Bäker, J. Rösler, and C. Siemers, *A finite element model of high speed metal cutting with adiabatic shearing*, Computers and Structure **80** (2002), 495–513.

[3] K.-D. Bouzakis et al., *Experimental and FEM analysis of the fatigue behaviour of PVD coatings on HSS substrate in milling*, Annals of the CIRP **47**(1) (1998), 69–72.

[4] J.P. Celis, D. Drees, M.Z. Huq, P.Q. Wu, and M. De Bonte, Surface and Coatings Technology **113** (1999), 165.

[5] T.H.C. Childs, *Material property needs in modeling metal cutting*, in: CIRP Workshop on Modeling Cutting, 1998.

[6] R. Clos, U. Schreppel, and P. Veit, *Temperaturentwicklung bei der Ausbildung adiabatischer Scherbänder*, in: Kolloquium des Schwerpunktprogrammes der DFG, Bonn, 18 November 1999, pp. 123–128.

[7] P. Dewhurst, *On the non-uniqueness of the machining process*, Proceedings of the Royal Society of London Series A **360** (1978), 587–610.

[8] H. Dimigen, *The future role of surface technology in machining*, in: International CIRP/VDI Conference on High-Performance Tools, Düsseldorf, 3–5 November, 1998.

[9] K. Holmberg and A. Matthews, *Coatings Tribology*, Elsevier, Amsterdam, 1994.

[10] I.S. Jawahir, Ph.D. thesis, The University of New South Wales, Australia (1986).

[11] I.S. Jawahir and P.L.B. Oxley, 1988, *New developments in chip control research: moving towards chip breakability predictions for unmanned manufacture*, in: Proceedings of the International Conference of the ASME, MI88, Atlanta, Georgia, USA, 1988, pp. 311–320.

[12] I.S. Jawahir and C.A. van Luttervelt, *Recent developments in chip control research and applications*, Annals of the CIRP, **42**(2) (1993), 659–694.

[13] W. Johnson, *Some slip-line fields for swaging or expanding indenting, extruding and machining for tools with curved dies*, International Journal of Mechanical Sciences **4** (1962), 323–347.

[14] F. Klocke, T. Krieg, K. Gerschwiler, R. Fritsch, V. Zinkann, M. Pöhls, and G. Eisenblätter, *Improved cutting processes with adapted coating systems*, Annals of the CIRP **47**(1) (1998),65–68.

[15] H. Kudo, *Some new slip-line solutions for two-dimensional steady-state machining*, International Journal of Mechanical Sciences **7** (1965), 43–55.

[16] A.A. Layyous, D.M. Frenkel, and R. Israel, Surface and Coating Technology **56** (1992), 89.

[17] E.H. Lee and B.W. Shaffer, *The theory of plasticity applied to a problem of machining*, Transactions of the ASME **73** (1951), 405–413.

[18] J. Leopold, *The application of visioplasticity in predictive modeling the chip flow*, in: Tool loading and surface integrity in turning operations, 3rd Cirp International Workshop on Modelling of Machining Operations", University of New South Wales, Australia, 20 August, 2000.

[19] J. Leopold, J.A. Oosterling, H. van den Berg, N. Renevier, and M.Meisel, *Mechanical and thermal behavior of coating–substrate systems investigated with parallel F.E.*, in: ICMCTF 2002, San Diego, California, USA, 2002.

[20] J. Leopold, H. Weber, A. Beger, M. Fährmann, and B. Schultrich, *Cutting with cermets*, published at TH Karl-Marx-Stadt, 1987.

[21] J. Leopold and G. Schmidt, *Challenge and problems with hybrid systems for the modelling of machining operations*, in: Proceedings of the 2nd CIRP International Workshop on Modelling of machining operations, Nantes, 1999.

[22] J. Leopold (ed.), *Werkzeuge für die Hochgeschwindigkeitsbearbeitung*, Carl-Hanser-Verlag München, Wien, 1999.

[23] X.D. Liu, L.C. Lee, and K.Y. Lam, *A slip-line field model for the determination of chip curl radius*, Journal of Engineering for Industry **117** (1995), 266–271.

[24] D.A. Lucca, E. Brinksmeier, and G. Goch, *Progress in assessing surface and subsurface integrity*, Annals of the CIRP **47**(2) (1998), 669–693.

[25] C.A. van Luttervelt, T.H.C. Childs, I.S. Jawahir, F. Klocke, and P.K. Venuvinod, *Present situation and future trends in modelling of machining operations*, Annals of the CIRP **47**(2) (1998), 587–626.

[26] B. Lux, C. Columbier, H. Atena, and K. Stemberg, Thin Solid Films **138** (1986), 49.

[27] T.D. Marusich, M. Ortiz, *Modeling and simulation of high-speed machining*, Int. J. Num. Meth. Eng. **38** (1995), 3675–3694.

[28] M.E. Merchant, *Basic mechanics of the metal cutting process*, Transactions of the ASME **66** (1944) 168–175.

[29] R.N. Mittal, B.L. Juneja, and G.S. Sekhon, *A solution of the oblique controlled contact continuous cutting problem*, International Journal of Machine Tool Design and Research **20** (1980), 211–221.

[30] R.N. Mittal and B.L. Juneja, *Effect of stress distribution on the shear angle in controlled contact orthogonal cutting*, International Journal of Machine Tool Design and Research **22** (1982), 87–96.

[31] F.C. Moon, *Nonlinear dynamics analysis in production systems; investigation of nonlinear dynamic effects in production systems*, 2nd International Symposium, Aachen, 1999.

[32] W.A. Morcos, *A solution of the free oblique continuous cutting problem in conditions of light friction at chip–tool interface*, Transactions of the ASME Series B **94** (1972), 1124–1130.

[33] R.K. Njiwa, J. Stebut, *Boundary element numerical modelling as a surface engineering tool*, in: Sixth International Conference on Plasma Surface Engineering, Garmich-Partenkirchen, 14–18 September, 1998.

[34] D.R.J. Owen and M. Vaz, Jr., *Computational techniques applied to high-speed machining under adiabatic strain-localisation conditions*, Comput. Meth. Appl. Mech. Eng. **171** (1999), 445.

[35] P.L.B. Oxley, *An analysis for orthogonal cutting with restricted tool–chip contact*, International Journal of Mechanical Sciences **4** (1962), 129–135.

[36] H.G. Prengel, P.C. Jindal, K.H. Wendt, A.T. Santhanam, P.L. Hegde, and R.M. Penich, Surface and Coatings Technology **139** (2001), 25.

[37] H.G. Prengel, W.R. Pfouts, and A.T. Santhanam, Surface and Coatings Technology **102** (1998), 183.

[38] T. Shi and S. Ramalingam, *Modeling chip formation with grooved tools*, International Journal of Mechanical Sciences **35** (1993), 741–756.

[39] M. Reinhard, H.E. Hintermann, H. Penzkofer, and H. Rottmann, *Programmevaluation Oberflächen- und Schichttechnologien*, OSTec, ifo Institut, 1997.

[40] R. Rentsch, F. Brinksmeier, and J. Li, *Investigation of nonlinear dynamic effects in loaded layer-substrate systems through molecular-dynamics simulation*, in: 4th International Symposium on Investigation of Nonlinear Dynamic Effects in Production Systems, Chemnitz, 8–9 April, 2003.

[41] C. Rubenstein, *The mechanism of orthogonal cutting with controlled contact tools*, International Journal of Machine Tool Design and Research **8** (1968), 203–216.

[42] C. Sage, *Cost-effective milling of titanium alloy: high-speed or conventional techniques?*, VDI Berichte No. 1399, 1998, pp. 451–464.

[43] D.R. Sandstrom, J.N. Hodowany, *Modeling the physics of metal cutting in high-speed machining*, Machining Science and Technology, **2** (1998), 343–353.

[44] M.C. Shaw, *Mechanik der Saegezahn Spanbildung*, lecture at the occasion of receiving the Georg-Schlesinger Award 1997 of TU Berlin.

[45] R. Tabersky, H. Van den Berg, and U. König, *Thin solid films*, in: Proceedings of the ICMCTF, 1993.

[46] H.K. Tönshoff, F. Kross, and C. Marzenell, *High-pressure water peening – a new mechanical surface-strengthening process*, Annals of the CIRP **46**(1) (1997), 113–116.

[47] E. Usui and K. Hoshi, *Slip-line fields in metal machining which involve centered fans*, International Research for Production Engineering ASME **61** (1963), 61–67.

[48] V.C. Venkatesh, T.E. Ye, D.T. Quinto, and D.E.P. Hoy, *Performance studies of uncoated, CVD-coated and PVD-coated carbides in turning and milling*, Annals of the CIRP **40**(1) (1991), 545–550.

[49] R. Wiedemann, T. Bertram, and H. Oettel, *Mechanisches Verhalten von Hartstoff-schichten*, Freiberger Forschungshefte, B **297** (1999), 67–79.

[50] B. Worthington, *The effect of rake face configuration on the curvature of the chip in metal cutting*, International Journal of Machine Tool Design and Research **15** (1975), 223–239.

14 Investigation of Nonlinear Dynamic Effects in Loaded Layer–Substrate Systems Through Molecular Dynamics Simulation

R. Rentsch, E. Brinksmeier, and J. Li

Coatings can increase tool life significantly. However, specific failure phenomena, such as cracking and delamination, can be observed during the application of coated tools. As a result, strong variations in tool life and sudden failure can take place. For the strength of a coated tool not only the properties of the coating and the substrate materials alone are important, but also the properties of the generated transition phase in between, the so-called interface. Experimentally, in-situ measurements at thin coatings under loading conditions are difficult, and the insight into the internal processes is limited. In this paper initial results of a molecular dynamics simulation are presented that focuses on the nonlinear dynamic response of tool coatings to cutting forces. The tool coating is modeled as a layer–substrate system considering directly the crystalline structures of coating, substrate and interface. Based on this setup, indentation test simulations are carried out and the nonlinear dynamic response of the layer–substrate system and its possible failure mechanisms are investigated.

14.1 Introduction

Thin hard coatings, such as TiN and other Ti-based alloys, can greatly improve tool life. During the past two decades coated tools were increasingly introduced to cutting processes. While in 1980 the market share for coated cutting tools was only about 2%, it has reached about 90% of all cutting tools sold by the year 2000 [1, 2].

Providing a wide coverage of intermediately ranged mechanical and thermal properties, TiN-coated tools show significantly longer tool life than uncoated cutting tools [1]. However, such tools show typical wear and failure phenomena such as cracking and delaminations. These phenomena result on the one hand from the properties of the coating–substrate system of the tool and, on the other hand, from the load conditions during cutting [2, 3]. Specific machining process parameters in combination with locally changing material properties in the workpiece (grain size and orientation, precipitations etc.) can result in strong variations of the chip-formation process and the cutting force [3–8]. Even when interrupted cutting and process instabilities are not leading to a dynamic tool load because of chatter vibration, the change of the chip formation from a continuous chip to a segmented chip can be a cause of cutting force variations [3, 7]. The dynamics of the cutting forces together with thermal and chemical impacts form the working environment of cutting tools [1, 9, 10].

Nonlinear Dynamics of Production Systems. Edited by G. Radons and R. Neugebauer

ISBN 3-527-40430-9

Usually the properties of hard coatings, i.e. their composite properties as layer–substrate systems, are studied in terms of nanohardness, tribological properties, coating adhesion, etc., applying techniques like impact or indentation tests [11], pin or ball on disk and scratching tests [12–14]. All these tests describe either static properties or allow a post-failure phenomenological material characterization only. In order to determine fatigue and creep properties of hard coatings, dynamic loads are also applied by multiple impact testing [15]. Here static FEM simulations were carried out accompanying the impact tests in order to determine the critical stress on the basis of critical impact loads from the experiments. FEM was further used to simulate the nanoindentation on coated systems [16], covering also TiN coatings with defects [3]. In most cases of FEM modeling the influence and properties of the interface are analyzed and described only in a general and insufficient manner, which does not consider any microstructural properties at all. Because of the lack of experimental data, due to difficulties in measuring thin-film properties, implemented material properties of coatings and interfaces are often modeled by simple approximations. Although the specific failure phenomena of coated tools are well known, their correlation to nonlinear dynamic loads, like in cutting, and the underlying mechanisms are not well understood yet.

Starting from the atomic level configuration, the molecular dynamics (MD) simulation offers the possibility to study the failure mechanisms and the structural influences of coating and interface in detail. The well-defined physical fundamentals of MD allow us to derive macroscopic properties as well [17–20]. For this work an initial layer–substrate system was proposed, which is set up with a Ni layer on a Cu substrate, because of the immediate availability of the necessary potential functions for the MD program. This layer–substrate system represents a combination of a hard coating on a softer substrate, serving as an initial model of TiN coatings. The developed MD program allows us to analyze the coating and interface properties and to carry out indentation simulations. The further focus of this work is on the working stress inside a layer–substrate system and on its nonlinear response to an indenter with respect to the indentation depth.

14.2 Layer–Substrate System Configuration and Material Representation

The employed MD model of the layer–substrate system comprises the layer (coating) and the substrate forming an interface in between. Figure 14.1 shows schematically the initial configuration of the MD model. As coating (in gray color) single-crystalline nickel was chosen and for the substrate (in dark-gray color) copper. In total there are 57 900 atoms in the model: 16 428 atoms belong to the coating material and 41 472 atoms are substrate atoms.

The entire configuration has the dimensions $13.93\,\mathrm{nm} \times 13.93\,\mathrm{nm} \times 4.09\,\mathrm{nm}$ and is made of six layers of Ni atoms on top of 16 layers of Cu atoms. Initially coating and substrate are aligned to the X–Z plane and the Y–Z plane, each on one side. The interactions between atoms, described by the potential functions, are the key elements for MD and similar atomistic level simulations. Their complexity determines directly the efficiency of the computer code. Therefore the complex many-body interactions of atoms and electrons, which are usually described by Schroedinger's equation, were simplified to a potential description between

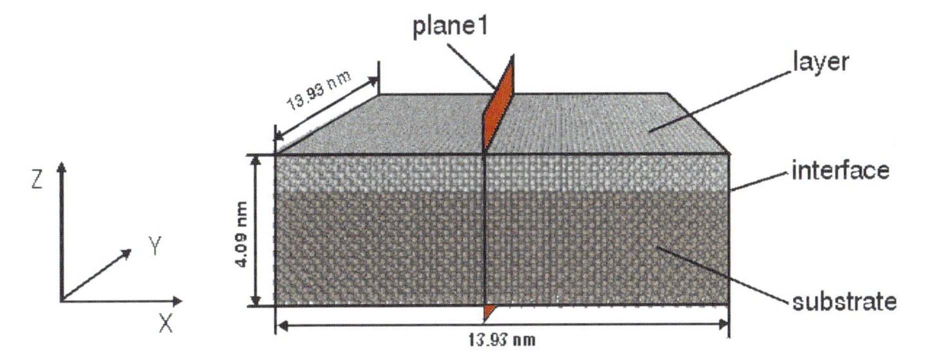

Figure 14.1: Initial configuration of the MD model.

atoms, the potential functions, with respect to atomistic rather than electronic level parameters. Depending on combinations of atomistic level parameters, various potential functions were published in the past. Pair potential functions, which take only separations between atom pairs into consideration, can effectively stabilize lattice structures with equal minimum-energy distances between all neighboring atoms, such as fcc and hcp crystal structures. Their characteristics suit best materials such as rare gases in their solid states, but they have also been applied for other materials. For the initial setup and program development the Lennard–Jones 6-12 pair potential function was employed (Eq. (14.1)). There are only two parameters in the function that describe material properties, the cohesive energy ε and the inter-atomic distance σ. They can be found for many materials in the literature.

$$\text{Lennard–Jones potential function: } \phi(r) = 4\varepsilon\left(\frac{\sigma^{12}}{r^{12}} - \frac{\sigma^6}{r^6}\right) . \tag{14.1}$$

The well depth or minimum of the function is given by the parameter ε for the cohesive energy, while σ defines the position of the energy minimum, the equilibrium inter-atomic distance. The values for the pair potential parameters of all necessary combinations of elements are listed in Table 14.1. Most of them are chosen according to the literature [17, 18], but there were no values found for C-Ni and Cu-Ni interactions. For the indenter material (diamond) interaction with the coating (C-Ni) and the substrate (C-Cu), the same weakly attracting and mostly repulsive potential was chosen. Regarding the Cu-Ni interaction, an intermediate value for the inter-atomic distance was chosen and a rather low cohesive energy. The low cohesive energy enhances delamination of the coating so as to model a layer–substrate system, where the interface represents the weak point of the system, similar to hard coating failure phenomena.

Table 14.1: Values for pair potential parameters.

	Cu-Cu	Ni-Ni	C-C	C-Cu	C-Ni	Cu-Ni
ε [eV]	0.342	0.421	3.68	0.087	0.087	0.114
σ [nm]	0.287	0.279	0.154	0.205	0.216	0.283

14.3 Properties of the Relaxed System

Initially the atoms are placed on ideal lattice positions. At this state, the interface is formed by simply setting up the ideal coating structure at bonding distance above the ideal substrate structure. Besides crystal defects, differences in the material properties between coating and substrate are also responsible for distortions and stresses in the interface. Hence, not only the ideal, but also artificial, structures of coating and substrate need to be relaxed, but the interface region in particular. The relaxation is also necessary in order to establish an equilibrium between layer and substrate structure and, thereby, generating a low stress state. During the relaxation phase free-surface conditions were applied for all directions; therefore the layer–substrate system was able to relax freely in space. The entire initial configuration (cf. Fig. 14.1) was relaxed for 5.0 ps at 300 °K (time step 1.0 fs).

14.3.1 Stress Distributions in the Relaxed System.

The fundamentals of statistical mechanics and thermodynamics provide the basis for linking the micromechanical and instantaneous data in molecular dynamics to macroscopic quantities such as temperature, hydrostatic pressure, etc. Hence, the reliability of MD models can be verified by comparing properties like stress, strains and temperature with e.g. experimental results or continuous mechanics calculations [17, 20]. For this purpose the following stress distributions were calculated as averages over 2500 time steps in order to determine constant values and distributions.

In compound structures, such as layer–substrate systems, often states of high residual stress can be observed due to different material properties or different thermo-mechanical treatment of the structural elements. Common deposition processes, for example, take place at high or elevated temperature (CVD: about 1000 °C; PVD: 400° to 600 °C; Nickel coating at 240 °C) [14]. The stress states of coatings made by these deposition processes result from a continuous growth at elevated temperatures and the subsequent cooling of the compounds. However, the calculated stress distributions presented here may be different from their experimental equivalent as they were generated starting from ideal lattice configurations and relaxing the whole system at 300 °K.

Figures 14.2 and 14.3 show two distributions of normal stress calculated for plane 1 of the relaxed system (cf. Fig. 14.1), σ_{xx} in Fig. 14.2 and σ_{zz} in Fig. 14.3. The entire simulation box is divided into 3703 cells, whose stress tensors were calculated considering all cell member atoms inside. The position of plane 1 was perpendicular to the X axis and crosses the geometrical center of the simulation box. This plane was chosen for symmetry reasons, assuming that σ_{yy} on a perpendicular plane would result in a similar distribution, and for comparison with the indentation effects in Sect. 4 (indentation along the z axis).

The distribution of the normal stress along the x axis σ_{xx} in Fig. 14.2 shows how the stress in the coating is balanced by the stress distribution in the substrate. There are strong gradients of normal stress across the interface on the left- and the right-hand sides of the layer–substrate system, which are related to high shear stress in the interface region. The regions of peak compressive stress in the coating of 1250 MPa are balanced on the substrate side by an equivalent region of tensile stress. The surface of the coating (opposite to the interface side) shows mostly a compressive stress state, except at the outer edges.

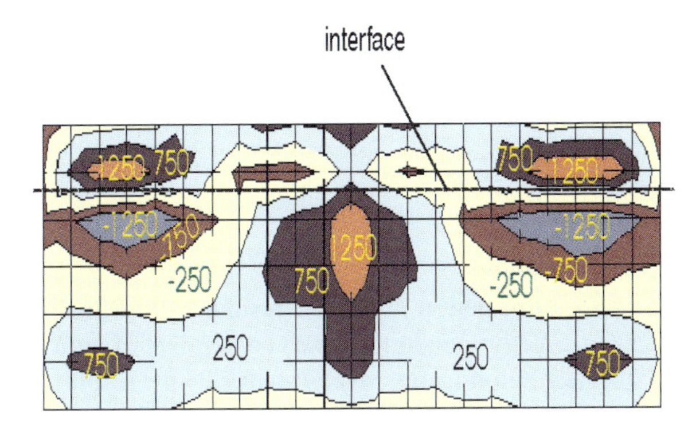

Figure 14.2: Stress distribution σ_{xx} [MPa] in the relaxed system (in plane1).

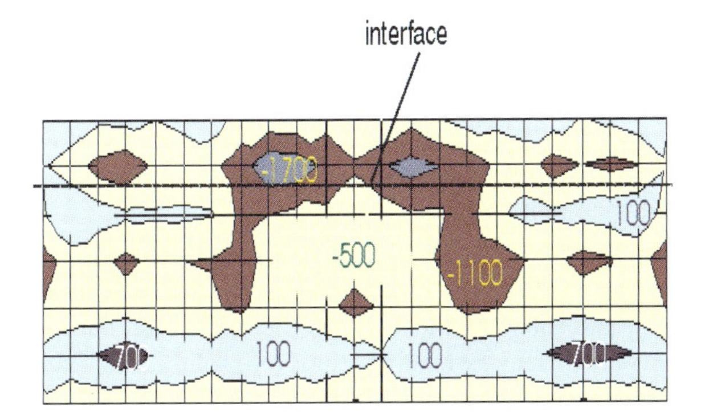

Figure 14.3: Stress distribution σ_{zz} [MPa] in the relaxed system (on plane 1).

Towards the center of the coating the stress level drops fast and changes from compressive to tensile with values up to -750 MPa. The high-stress regions in the substrate are not only balancing the equivalent stress in the coating next to the interface, but also within the substrate itself. The regions of high tensile stress in the substrate next to the interface cause a region of compressive stress in the center of the substrate of equivalent value, right below the interface. The lower region of the substrate experiences a rather moderate level of compressive stress.

In Fig. 14.3 the high tensile stress (-1700 MPa) in the center of the coating reaching into the substrate is the most prevalent feature regarding the normal stress in the z direction σ_{zz}, on plane 1. The other regions show a significantly lower σ_{zz} stress level. The relaxation of the layer–substrate system has led to a slight bending of the coating with a zenith at the center falling off to the outer edges (not visible in the cell representation in Figs. 14.2 and 14.3; cf. the indentation setup below). In order to build the central zenith, the bonds in the region below it need to be elongated and cause, thereby, the high tensile stress values in σ_{zz}, in the coating and the interface in particular.

Coating processes, material properties and substrate topography affect the resulting residual stress states in coatings [2]. In principle, the residual stress in coated systems can be calculated on the basis of microscopic structural measurements using X-ray diffraction or Raman spectroscopy. Typical residual stress values of TiN/HSS tools produced by PVD are about 4 GPa [21]. However, residual stress in coatings can be compressive (+) or tensile (−) with a value of up to 10 GPa depending on material combination and manufacturing process [16, 21].

14.3.2 Interface Properties

The stress distributions in Sect. 14.3.1 provide information about the residual stress in the relaxed layer–substrate system. Since stress is directly related to (local) constraints of lattice structures, it is also possible to identify the strained regions using MD by analyzing the resulting structure regarding atomic positions and displacements. Figure 14.4 shows a top view of four layers of atoms around the relaxed interface, of which each two belong to the coating and two to the substrate structure of the model.

In the very center and at the four corners of the interface, well-matched structures (designated by rectangles in Fig. 14.4) are found with uniform lattice spacing. However, between the two well-matched regions, a region is formed that shows a mismatch between the Ni layers and the Cu substrate (designated by ellipses in Fig. 14.4). An example of a detailed mismatched structure is shown in the left circle and of the well-matched structure in the right circle (solid points are Ni atoms and open circles are Cu atoms).

It is clear to see that in the mismatched region, the coating and the substrate atoms are placed on different positions. Since the Cu-Ni interface potential energy was chosen to be 1/3 of the Cu-Cu potential energy, coating and substrate mostly keep their structure, but the interface yields to the structural constraints first. Here the coating atoms (Ni: solid) are shifted to interstitial positions relative to the lattice structure of the substrate. In the region of the interface where the coating and substrate structure match well (right circle in Fig. 14.4), both lattices coincide closely, but not exactly. The equilibrium inter-atomic distances of copper (0.287 nm) and nickel (0.279 nm) show very close values (mismatch 2.8%).

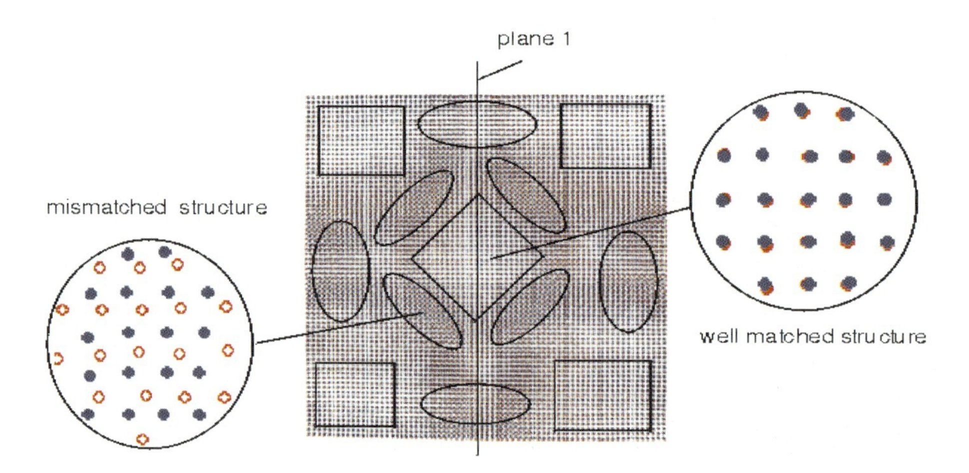

Figure 14.4: Top view of the interface in the relaxed system.

At the interface both lattices have to adjust in order to reduce the local energy as much as possible. Hence, the structures in the so-called 'well-matched' as well as in the 'mismatched' regions of the interface are constrained and are very likely subject to high shear stress, as Figs. 14.2 and 14.3 indicate. The relaxation has not led to any diffusion between layer and substrate. On the one hand, a relaxation at 300 °K does not allow for significant diffusion and, on the other hand, most diffusion constants in metals are beyond practical simulation time.

14.4 Response of the Loaded Layer–Substrate System

The loading of the layer–substrate model was realized by an indentation arrangement. The chosen setup for the indentation simulation was similar to that in nanoindentation experiments. The indenter was designed as a hollow, spherical tip with 7.58 nm in diameter, made of 1518 atoms on fcc lattice sites. Infinite stiffness was assumed, since real indenters used in nanoindentation are much harder than the specimen. The indenter was placed above the relaxed layer–substrate system (Fig. 14.5). Four layers of atoms at the bottom of the relaxed system and three layers on each side form fixed boundaries that balance the indentation load. Further, a thermal layer of four layers of atoms was installed next to the fixed boundaries. Here the thermal layer acts as an energy sink, through which the energy introduced by the indenter can be drained away, simulating heat conduction to the environment. Compared with nanoindentation tests, a significantly higher indentation speed was chosen for computational convenience. A typical speed for nanoindentation tests may be as low as 2.0 nm/s [22]. However 35.0 m/s was chosen for this simulation.

14.4.1 Deformation and Forces

Figure 14.6 shows two force–depth curves of one indentation process. One represents a trend line derived from the other, the instantaneous force–depth curve, which was directly calculated for every time step of the MD simulation. The instantaneous force–depth curve follows closely all processes taking place during indentation, while for the trend line all this information is

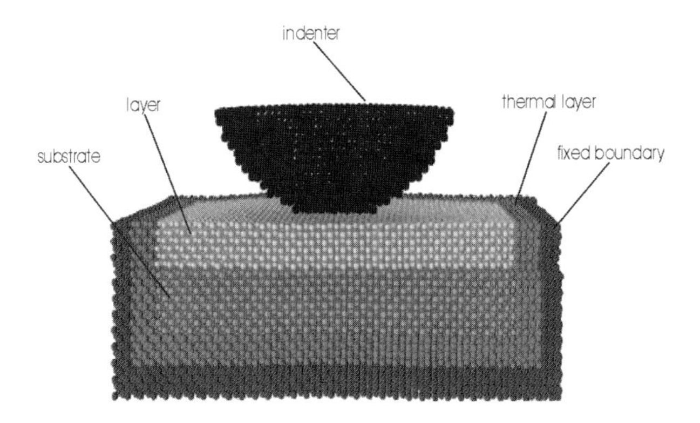

Figure 14.5: Indentation setup.

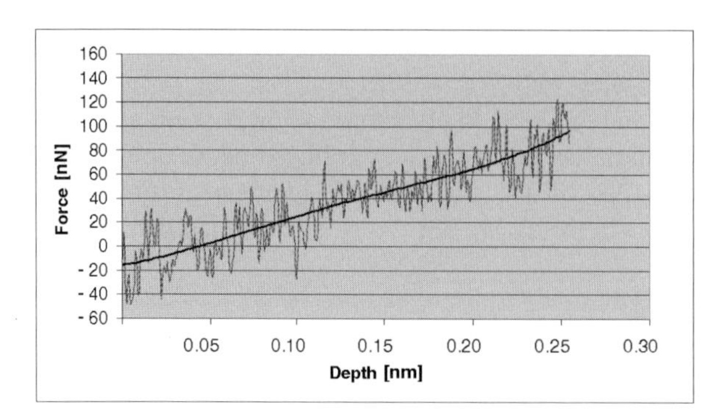

Figure 14.6: Indentation force as function of indentation depth.

filtered away. The indentation forces are calculated as sum of contributions from the layer–substrate system. As the indenter itself is stiff, i.e. its atoms do not vibrate, the force–depth curve describes only the response of the layer–substrate system to the penetration of the tip.

During indentation simulation, the average tool force (trend line) rises continuously from about $-18\,\mathrm{nN}$ at first contact to approximately $97\,\mathrm{nN}$ at 0.25-nm indentation depth. The negative tool force at the beginning of indentation reveals that the somewhat longer-ranged attractive part of the potential function dominates the interaction while the tool approaches the specimen surface. However, with the indenter further moving down, the repulsive contribution of those tool atoms that penetrate the specimen become stronger and the forces change from attractive ('$-$' in Fig. 14.6) to repulsive ('$+$').

The course of the instantaneous force is directly related to the details of the deformation process, the instantaneous stress state at the tip and the formation of dislocations. Initially the indenter causes the formation of an elastic stress field below the surface of the coating until first bonds break and plastic deformation can be observed. The breaking of bonds and the formation as well as the motion of dislocations can directly be related to stress release and, thereby, to the fluctuation of the force–depth curve.

Figures 14.7 and 14.8 show two views of the appearance of the layer–substrate surface at an indentation depth of 0.25 nm. Figure 14.7 shows a cross section of the layer–substrate system with the penetrated indenter on top. The cross section is taken for plane 1, which runs through the center, from the top to the bottom side of the view in Fig. 14.8.

The view in Fig. 14.8 is on top of the surface. For better visibility, the indenter has been deleted from this view and the atoms around the indent with a displacement larger than 35% of the nickel bonding length are colored, so as to identify the area of deformation. The indenter has formed an unsymmetrical indent at the center of the coating surface. Three dislocation lines are visible at the surface, running from the indent towards the boundaries (arrows in Fig. 14.8). All three dislocation lines in Fig. 14.8 appear also in the cross-sectional view in Fig. 14.7. The cross-sectional view clearly shows the deformations caused by the indenter. Right below the tip of the indenter, the coating atoms (bright) were pushed down and formed (three) dislocations, which run from the edges of indenter/coating contact into the layer–substrate system. At the

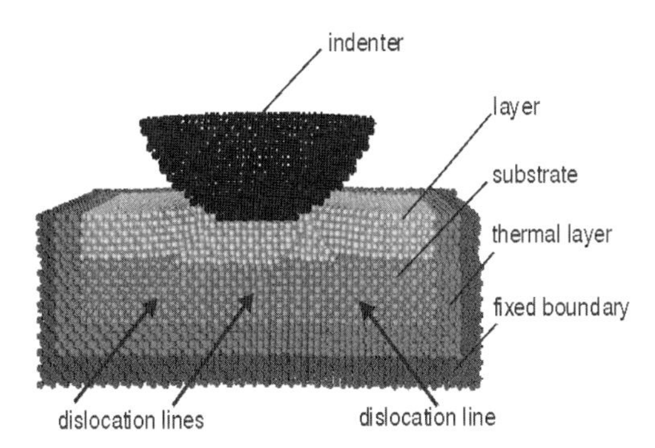

Figure 14.7: Cross section of the layer–substrate system with penetrated indenter (plane 1; indentation depth = $0.25\,\text{nm}$).

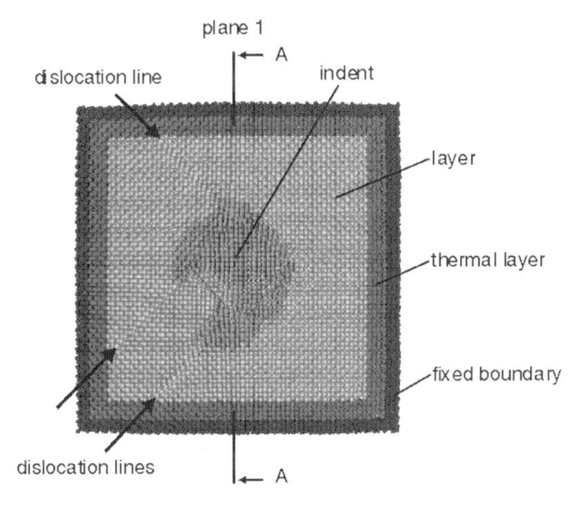

Figure 14.8: Top view of the layer–substrate system with colored indent (indentation depth = $0.25\,\text{nm}$).

left- and the right-hand side of the indenter, the material yielded upward, making room for the atoms that are pushed down by the indenter.

The nonlinear dynamic response of the system appears in the fluctuation of the indenter forces as well as in the formation of dislocations and an unsymmetrical indentation area of the crystalline material. The fluctuations originate in stress release and can directly be related to the formation of specific dislocations or cracks in the model, if observed.

14.4.2 Stress Distribution in the Loaded Layer–Substrate System

Similarly to Figs. 14.2 and 14.3, Figs. 14.9 and 14.10 illustrate the stress distribution in the layer–substrate system at a penetration depth of the indenter of 0.25 nm, Fig. 14.9 for σ_{xx} on plane 1 and Fig. 14.10 for σ_{zz} on plane 1 as well.

Differing from the relaxed system in Fig. 14.2, the distribution of the normal stress σ_{xx} in Fig. 14.9 shows a significantly higher compressive stress level of up to 4000 MPa. The maximum compressive stress appears at the indenter tip, extending into the coating, and at the bottom of the substrate. The center of the substrate shows an average compressive stress of 2500 MPa, which is surrounded by a region of tensile stress of 2000 MPa on its left- and right-hand sides. Right next to the compressive stress maxima below the indenter tip, regions of maximum tensile stress appear on both sides of the indenter, extending on the right-hand side through the coating down to the interface.

The distribution of the normal stress σ_{zz} on plane 1 of the relaxed system (Fig. 14.3) has also changed significantly with the penetrated indenter in Fig. 14.10. The center of the whole layer–substrate system has changed its σ_{zz} stress state from tensile to compressive with its highest compressive stress of 13 000 MPa right below the indenter tip. From here, the stress decreases gradually from 13 000 MPa (compressive) below the indenter to -1250 MPa (tensile) at the sides of the substrate. The σ_{zz} stress distribution also shows tensile stress at the free coating surface reaching to both sides of the indenter.

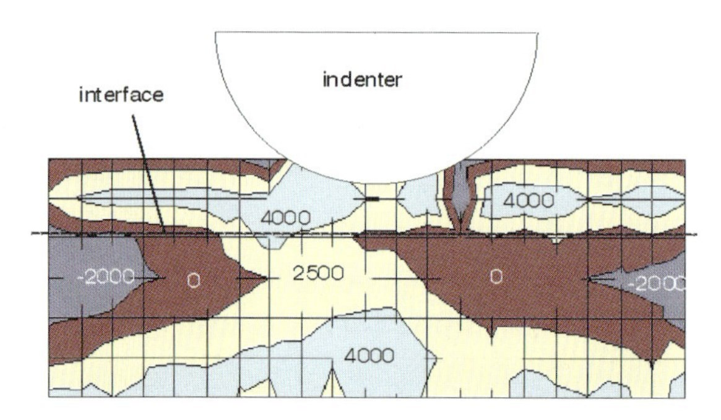

Figure 14.9: Stress distribution σ_{xx} [MPa] in layer–substrate system (on plane 1).

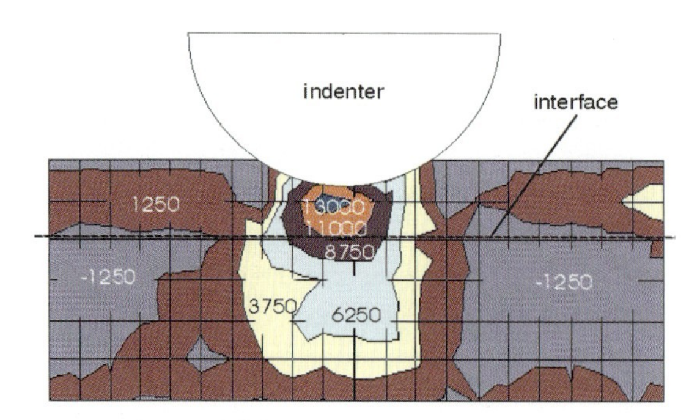

Figure 14.10: Stress distribution σ_{zz} [MPa] in layer–substrate system (on plane 1).

The stress distributions in Figs. 14.9 and 14.10 depend directly on the deformation and dislocations observed in Figs. 14.7 and 14.8. The regions of tensile stress for σ_{xx} on both sides of the indenter in Fig. 14.9 show also for σ_{zz} tensile stress and they are located next to a region of strong compressive stress below the indenter, on the right-hand side in particular. Regions with such strong gradients of normal stress are exposed to high shear stress and can be sources for dislocations and crack formation, particularly at the surface and the interface. When dislocations are formed, stress is released and, relative to each other, both sides of a dislocation move in opposite directions. The dislocations extending from the right-hand side of the indenter (Fig. 14.7) cause such a change in the local stress distribution in the coating, extending down to the interface. The dislocation on the left-hand side of the indenter causes a similar effect.

The two parallel running dislocations in Fig. 14.7 influence obviously also the stress distribution in the substrate. In Fig. 14.10 a region, where the stress state changes from compressive (1250 MPa) to tensile (-1250 MPa), extends from the left-hand side of the indenter down towards the lower left corner of the layer–substrate system. This is exactly along the dislocation line on the left-hand side of the indenter tip.

With the penetrated indenter, the coating shows mostly a compressive state, while the substrate is in compressive state only below the indenter and otherwise, in the surrounding regions, in tensile state. Because of the compressive state on the coating side and the tensile state on the left- and the right-hand sides of the substrate, high shear stress can be anticipated in the interface. Shear stress can similarly be anticipated between the compressive center of the substrate and the surrounding tensile region.

Since there was no load-holding time considered on reaching the indentation depth, a certain part of settling and stress release may lead to lower values than those presented here. During indentation, the structure deformation profile and stress distribution (σ_{zz} on plane 1) are similar to what Landman and his colleagues found in their research, where the maximum compressive stress (8.2 GPa) appears in the contact point [23]. FEM was also employed to study indentation of coated systems. For a DLC coating (3 µm) on tool steel, the maximum working stress calculated by FEM is at the contact area on the coating surface with a maximum compressive stress of 10 GPa [16].

14.5 Conclusions

The developed molecular dynamics model of a layer–substrate system of a Ni coating on a Cu substrate allowed us to study the structure-dependent distribution of residual stress in the system as well as the impact of a stiff indenter and the response of the layer–substrate system to it.

The interface was formed by free relaxation of the Ni layer on the substrate at 300 °K. After relaxation, the residual stress calculation revealed regions of high shear stress in the interface, except for the center of the model, and tensile stress in the center of the coating with compressive stress below in the substrate. The relaxation of the layer–substrate system has led to a slight bending of the coating with a zenith at the center falling off to the outer edges. A structure analysis of the MD model revealed well-matched and mismatched regions at the Ni-Cu interface. The resulting deformation of the layer–substrate system and the local residual stress distribution are due to differences in the lattice constants of the coating and the substrate materials.

The dynamic response to the penetration of the stiff indenter changes the stress state inside the layer–substrate system during indentation. The nonlinear dynamic response of the system appears in the fluctuation of the indenter forces as well as in the formation of dislocations and an unsymmetrical indent. The fluctuations originate in stress release and can directly be related to the formation of specific dislocations or cracks in the model. The indenter formed an unsymmetrical indent at the center of the coating and generated several dislocations as a function of indentation depth, indenter shape and crystalline orientation.

The indentation caused a significant increase of the maximum compressive stress in the layer–substrate system, close to the tip of the indenter and along the direction of indentation in particular. Strong stress gradients were generated at the edge of the contact between indenter and coating, which could serve as sources for the initiation of further dislocations or cracks. A mostly compressive stress state in the coating, in combination with a tensile stress state around the center of the substrate, suggests a strong shear load in the interface as well.

In general, a similar stress distribution as calculated by FEM is observed by MD simulation; dynamic effects inside the layer–substrate system can be seen through the development of the stress distribution in time.

The ongoing investigations focus on the identification of the influence of the specific nonlinear dynamic effects on the fluctuations of the instantaneous indenter force and the related depth of indentation. This will be important for the investigation of the failure mechanisms of the layer–substrate system at nonlinear dynamic indenter load, instead of the here-applied linearly moving indenter, because of the expected complex deformation process. For this purpose, the indenter will be loaded with a nonlinear dynamic force signal from a cutting process. A further goal of the investigations is the combination of a MD indentation model with a FEM model, so as to consider microstructural information and properties in a meso-macroscopic FEM environment.

Bibliography

[1] H.K. Tönshoff, A. Mohlfeld, C. Spengler, and C. Podolsky, in: Proceedings of the Conference on Coatings in Manufacturing Engineering, Thessaloniki, Greece, 1999, pp. 1–20.

[2] H. Holleck, Z. Werkstofftech **17** (1986), 334–341.

[3] G. Schmidt, R. Leopold, and R. Neugebauer, textitFE-simulation of nonlinear dynamical effects in coating–substrate systems, Chapter 13, this book.

[4] G. Stepan, R. Szalai, and T. Insperger, *Nonlinear dynamics of high-speed milling subjected to regenerative effect*, Chapter 7, this book.

[5] N. van de Wouw, R.P.H. Faassen, J.A.J. Oosterling, and H. Nijmeijer, *Modelling of high-speed milling for the prediction of regenerative chatter*, Chapter 10, this book.

[6] R. Rusinek, K. Szabelski, and J. Warminski, *Influence of the workpiece profile on the self-excited vibrations during metal turning process*, Chapter 9, this book.

[7] A. Behrens, J. Wulfsberg, B. Westhoff, and K. Kalisch, *Problems of finite element simulation of chip Formation under high-speed-cutting conditions*, Chapter 12, this book.

[8] E. Govekar, J. Gradisek, I. Grabec, A. Baus, F. Klocke, A. Otto, and M. Geiger, *Dynamics based monitoring of manufacturing processes: detection of transitions between process states*, Chapter 27, this book.

[9] E. Brinksmeier, Precision Engineering **11** (1989), 211–224.

[10] F. Klocke and T. Krieg, Annals of the CIRP **48**(2) (1999), 1–11.

[11] Zhi-Hui Xu and D. Rowcliffe, Surface and Coatings Technology **161**, 44–51 (2002).

[12] Y.L. Su and W.H. Kao, Wear **223** (1998), 119–130.

[13] B.-J. Kim, Y.-C. Kim, D.-K. Lee, and J.-J. Lee, Surface and Coatings Technology **111** (1999), 56–61.

[14] J. Takadoum and H. Houmid Bennani, Surface and Coatings Technology **96** (1997), 272–282.

[15] K.-D. Bouzakis, N. Michailidis, A. Lontos et al., in: Proceedings of the International Conference on Manufacturing Engineering, Halkidiki, Greece, 2002, pp. 491–501.

[16] J. Michler, M. Mermoux, Y. von Kaenel et al., Thin Solid Films **357**(2) (1999), 200–212.

[17] M.I. Baskes, J.E. Angelo, and C.L. Bisson, Modelling and Simululation in Material Science and Engineering **2** (1994), 505–518.

[18] N. Ikawa, Sh. Shimada, H. Tanaka, and G. Ohmori, Annals of the CIRP **40**(1) (1991), 551–554.

[19] R. Rentsch and I. Inasaki, Annals of the CIRP **44**(1) (1995), 295–298.

[20] R. Rentsch and V. Vitek, *Multiscale phenomena in materials – experiments and modeling*, MRS Proceedings USA **578** (1999), 6 pp.

[21] U. Wiklund, J. Gunnars, and S. Hogmark, Wear **232** (1999), 262–269.

[22] D.A. Lucca, M.J. Klopfstein, R. Ghislen, and G. Cantwell, Annals of the CIRP **51**(1) (2002), 483–486.

[23] U. Landman, W.D. Luedtke, N.A. Burnham, and R.J. Colton, Science **248** (1990), 454–461.

15 Simulation, Experimental Investigation and Control of Thermal Behavior in Modular Tool Systems

J. Konvicka, N. Wessel, F. Weidermann, S. Nestmann, R. Neugebauer, U. Schwarz, A. Wessel, and J. Kurths

There is an important interest in compensating thermally induced errors of modular tool systems to improve the manufacturing accuracy. In this study we tested the hypothesis whether we can predict such thermal displacements by using a nonlinear regression analysis, namely the alternating conditional expectation (ACE) algorithm, reliably. First, we analyzed data that were generated by two different finite-element spindle models of modular tool systems. We found that the ACE algorithm is a powerful tool to model the relation between temperatures and displacements for simulated data. Next, we investigated the temperature–displacement relationship in a silent real experimental setup, where the tool system is thermally forced. Again, the ACE algorithm was powerful to estimate the deformation errors. The corresponding errors obtained by using the nonlinear regression approach are 10-fold lower in comparison to multiple linear regression analysis. Finally, we investigated the thermal behavior of a modular tool system in a working milling machine with the aim of controlling the system for a better precision. The thermally induced errors can be estimated with 1–2 μm accuracy using this nonlinear regression analysis. Therefore, this approach seems to be very useful for the development of new modular tool systems.

15.1 Introduction

Accuracy and productivity are very significant criteria in evaluating machine tools and demand a thermally stabilized process with tolerances in the micrometer range [3, 5, 6, 8–12, 14–16, 18–22, 24, 26, 28, 33]. Today the machining should be done without cooling lubricant (dry processing) due to the economic or ecological demands. The tool expands as a result of heating. Distortions around 100 μm appear. Over the past few decades, compensation for defects due to thermal impact has attracted steadily increasing interest. These studies comprise a wide range of methods ranging from computer simulations [1] and internal monitoring in neuronal networks [7, 17, 39] to thermal failure modeling [1, 5, 6, 15, 25, 27, 34]. The last-mentioned is also the approach of our project, which necessarily has engineering and mathematical–statistical aspects.

Nonlinear Dynamics of Production Systems. Edited by G. Radons and R. Neugebauer

ISBN 3-527-40430-9

Engineering Aspects

- Until now, a systematic strategy has not been found to thermally optimize cutting tools with geometrically well-defined tool edges, such as those used in modular tool systems with breaks, and which are linked with the main spindle assembly of the machine tool by an interface. Heat comes from the cutting process and internal heat sources of the machine. Heat entry follows the principles of heat conduction, radiation and convection. Consequently, models have to be developed which represent the time-varying, irregular and nonlinear system behavior as well as possible. Based on this description, we intend to constructively optimize the components, to develop the compensation strategy to minimize thermal influences on the tool and to further increase manufacturing accuracy.

Consequently, essential project targets are

- modeling of the system tool with its boundary conditions, taking into account nonlinear interactions,
- experimental studies of the thermal characteristics of modular tool systems under thermal load,
- abstraction of solution variants to minimize thermally caused deformations and distortions of cutting tools, while being aware of and making use of the nonlinear dynamic system behavior and
- verification of the compensation strategy (with a demonstrator).

Mathematical–Statistical Aspects

The relationship between temperatures and distortions is very complex, even if one considers only a simple bar which is heated on one side [24]. In this relatively simple case the temperature–distortion relationship is no unambiguous function due to the noticeable hysteresis effect observed. A modular tool system, however, consists of numerous individual parts and boundary surfaces which may have a very complicated shape. Thus, it is practically impossible to represent the heat distribution and consequently to describe the associated deformations with classical physical or engineering methods. A statistical approach is used to estimate mappings between appropriate temperature measurements and distortions by means of a nonparametric procedure.

In the following, let $T_i = 1, ..., n_i$ be the temperatures at point i out of n_i measuring points and $S_j = 1, ..., n_j$ the distortions at point j out of n_j points. The numbers of temperature and distortion measuring points may differ, so the same indices do not necessarily stand for equal measuring points. In this investigation, the main target is to determine the distortion at the cutting edge.

Let us first consider the distortions at several measuring points and re-construct them as a function of a set of values that can be measured, such as an appropriate quantity of temperature measurements. Because of well-known hysteresis effects [24] it is necessary to consider a so-called response transformation model, which is of the type

$$\theta(S_j) = \phi(T_1, ..., T_{n_i}), \quad j = 1, ..., n_j, \tag{15.1}$$

where ϕ and θ are nonlinear functions. We use the ACE algorithm (alternating conditional expectations algorithm) [2] in order to estimate the transformations ϕ and θ.

15.2 Investigated Tool Mountings

Three tool mountings with a HSK 63 interface (hollow taper shank; connecting element between the main spindle and the tool) were analyzed. They are distinguished by the type of chucking system. Here, a collet chuck, a hydraulic expansion chuck and a shrink chuck were used.

Collet Chucks

Collet chucks are widely used tool-holder systems. They can be fitted quickly and flexibly to various chucking diameters due to their exchangeability of the collet.

Hydraulic Expansion Chucks

Hydraulic expansion chucks are based on a functional principle differing from that of common chuck systems. In the chuck system, hydraulic internal pressure acting homogeneously is generated upon the so-called chuck application (clamp screw, piston and seal). The internal pressure comprehensively chucks the tool upon an expansion bush. Highest true-running and alternating repeating accuracy values (better than 0.003 mm) are achieved with this chuck system. Furthermore, hydraulic expansion chucks have very good damping characteristics since the tools are embedded into a ring chamber. As a result, not only high surface quality of the workpiece can be achieved, but also significantly enhanced tool-life values, since microchippings due to vibrations on the tool edge, mainly at carbide and PCD tools, are avoided. In the meantime, hydraulic expansion chucks have become standard for powerful machines and operation processes with high-quality requirements.

Power Shrinking Technology

Power shrinking technology is entirely new, and fulfills all the demands of an advanced tool chuck system. Power shrinking technology is characterized by a true-running accuracy of less than 0.003 mm and can transform very high chucking forces. In power shrinking, carbide and HSC tools may be chucked in the same tool holder. The torque values which can be transferred are greatly above the values necessary from the tool. Unlike the collet chuck, in the power shrinking system, there are no moving elements. Thus, the system is insensitive against mechanical impacts and provides a chuck absolutely free of wear. Two sizes of this mounting type were investigated.

Drill

Drills are available in a variety of building types and sizes. We selected one as an example to investigate the distortions due to thermal impacts on the tool's cutting edge in the radial direction. The drill consisted of several components, which are connected with planar supports via cylinder shanks. The cutting edge carrier's spline was adjacent to the tool shank.

15.3 Project Realization

15.3.1 Determination of Replacement Heat-transmission Coefficient for Component Joints in a FE Model

At first we needed a data stock which has not previously been available before completing the test stand. This is required for teaching the model for nonlinear regression (Sect. 15.4). For this reason, data should be generated by FE simulation at first. The model was tuned with measuring values after completion of the test stand to achieve good simulation results. In the FE simulation of tool mountings the definition of the boundary conditions was one of the main requirements. Due to the rotational symmetry of the tool mountings, we had to model only half of the corresponding elements. The joints between tool and tool mounting as well as inside the tool mounting were modeled according to a methodology described in [35, 36]. Here, a replacement material – whose thickness has to be appropriate to the rest of the model due to mesh – is modeled instead of the joint. Within the model, the joint's influence on heat conduction is considered with regard to the heat conductivity of this replacement material. In the model used the joints are modeled with a replacement material of 1-mm thickness. The heat conductivity is 2% of that of the steel. This is a reasonable value under the given circumstances (contact of ground surfaces).

To model joints with FE analysis in a close approximation of reality, we use the relationship in [34] between virtual heat-transmission coefficient in the joint α_{Fu}, surface pressure p and roughness R_z. The related heat-transmission coefficient α_b, related length l_b and reference pressure p_b are designed to represent the equation correctly according to dimensions.

$$\alpha_{Fu} = \left(-0.397 \frac{p l_b}{1000 R_z p_b} + 3.76 \sqrt{\frac{p l_b}{1000 R_z p_b}} + 1.49 \right) \alpha_b,$$

with (15.2)

$$l_b = 1\ \text{mm}, P_b = 10^6 \text{N/m}^2, \alpha_b = \frac{\text{kW}}{\text{m}^2\text{K}}.$$

In general, one has to consider only joints between materials of high heat conductivity, such as aluminum, copper and steel, since they have a great influence on the entire system's heat conductivity in terms of the workpiece dimensions.

Since properties such as specific heat conductivity λ of the means in the joint clearance also act on the joints' virtual heat-transfer coefficient α_{Fu}, we carried out practical tests. The goal of these studies was to determine only significant influencing factors with sufficient accuracy in terms of FE modeling rather than to define all influencing aspects very precisely.

Here, joining of ground surfaces, which is of particular interest for applications in mechanical engineering, was investigated in detail under the influence of low surface pressure (Fig. 15.1).

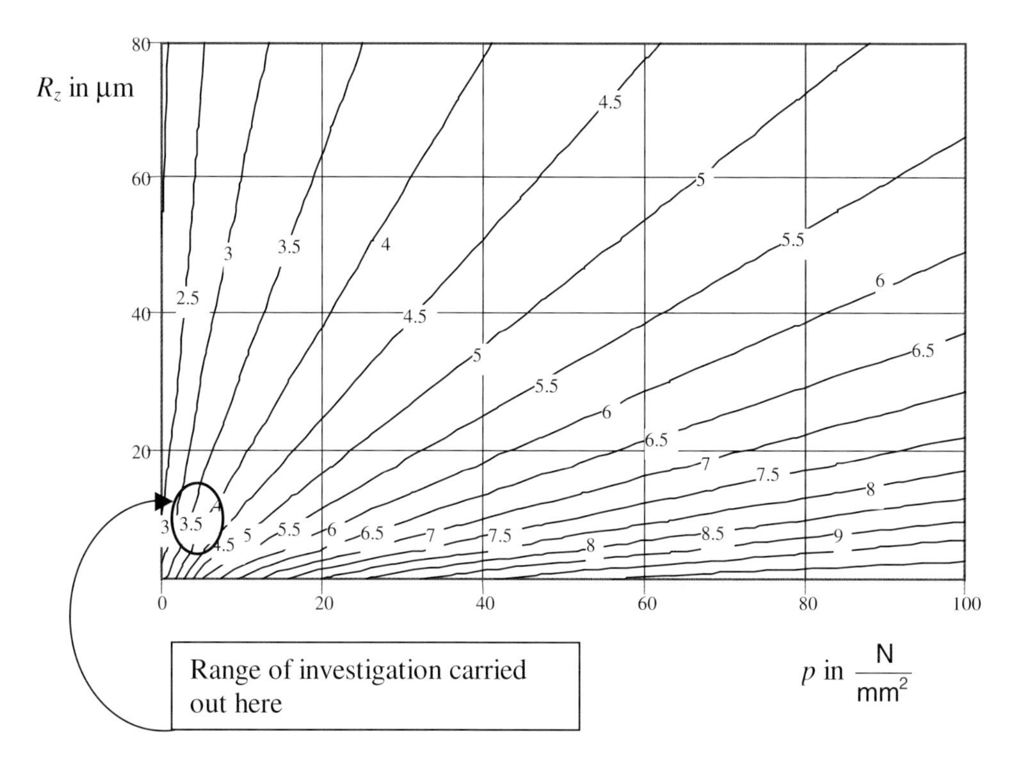

Figure 15.1: Virtual heat-transfer coefficient α_b of a joint as a function of surface pressure p and roughness R_z of contacting surfaces.

Test Setup

The specimen of steel with four joints consisting of five plates ($100 \times 100 \times 20$ mm (Surface area A 100×100 mm)) which are linked by thermally insulated thread bolts is heated in a lateral area parallel to the joints with a heating foil of constant heating power. Heat flow is avoided through the use of heat insulation at the four lateral areas which are perpendicular to the joints. Thus, all of the heat has to flow through the specimen. Surface temperature is measured by temperature probes at both lateral areas parallel to the joints. Heat flow passing through the specimen with joints is expressed by the following equation:

$$\dot{Q} = \frac{A\Delta T}{\frac{\delta_{Fe}}{\lambda} + \frac{4}{\alpha_{Fu}} + \frac{1}{\alpha_{me\beta}}}. \tag{15.3}$$

Another test series was executed with a specimen of the same structure, but without joints, in order to determine the variable $\alpha_{me\beta}$, by which the heat-transmission coefficient at the probe joints is to be described (δ_{Fe} – thickness of the steel plate). Here, heat flow is obtained from the following equation, so that the parameter $\alpha_{me\beta}$ can be calculated:

$$\dot{Q} = \frac{A\Delta T}{\frac{\delta_{Fe}}{\lambda} + \frac{1}{\alpha_{me\beta}}}. \tag{15.4}$$

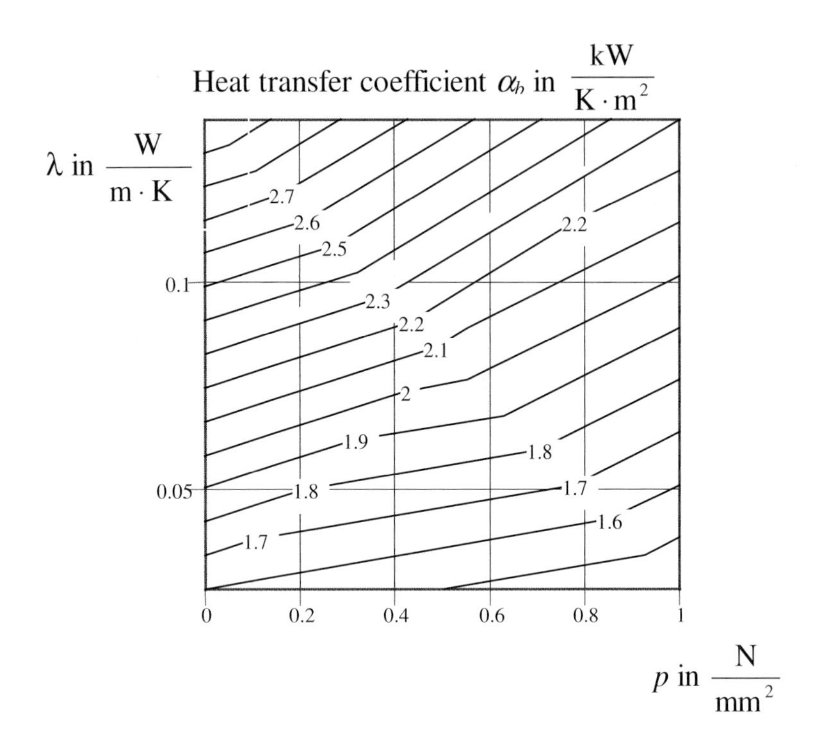

Figure 15.2: Virtual heat transfer coefficient α_b of a joint as a function of surface pressure p and heat conductivity λ of the intermediate means.

In contrast to the studies carried out until now, the virtual heat-transmission coefficient of the joints was studied as a function of surface pressure and heat conductivity of the intermediate means. Dependency on roughness of the contacting areas is assumed as sufficiently known.

15.3.2 Building of the Test Stand for Tool Investigations with Fixed Shaft

Execution of tests is based on the thermal cell established at the Fraunhofer IWU. With the thermal cell, defined reproducible room and foundation temperatures, including local temperature gradients, can be generated. The built-up test stand in the thermal cell (see Fig. 15.3) consists of a body. There, a clamped original main spindle shaft with the hydraulically actuated tool chucking system HSK 63 (in which selected modular tool systems are chucked as tool mounting) is fixed on this body. Electrically driven heating elements enable the intentional introduction of dimensioned heat flows into the tools at the tool chuck via the roller bearing fit of the main spindle shaft.

Heating of the cutting edges, which originally results from the cutting procedure, is generated with a shuttered hot-air blower in an approximately point-wise manner or at the tool tip with a soldering copper.

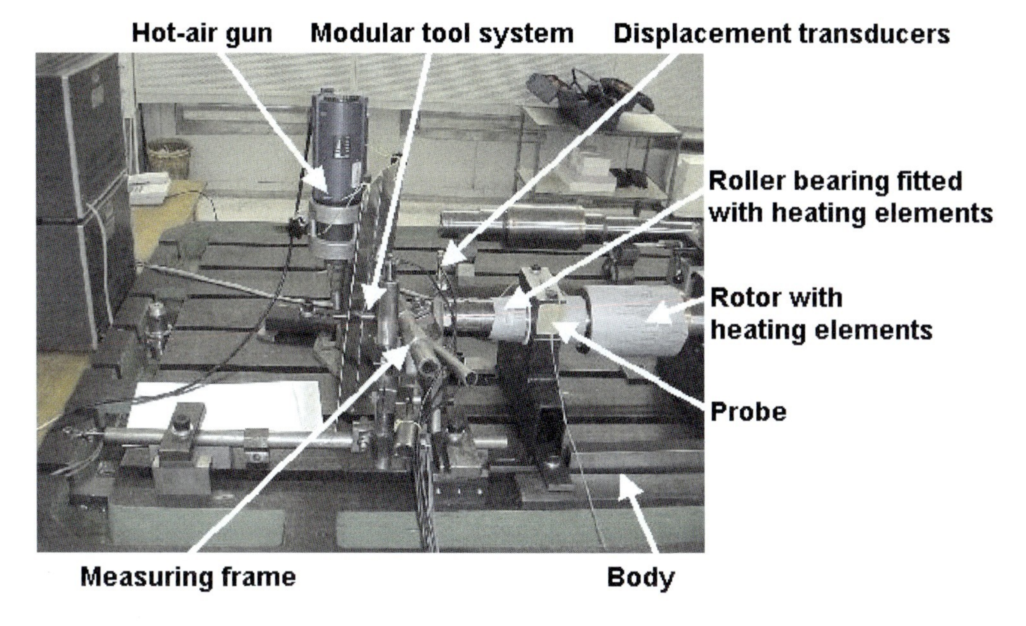

Figure 15.3: Test stand for studies on a fixed shaft.

Main Investigation

In order to obtain an exact possible data series for mathematical modeling of the individual tool mountings, the test stand was refined and the tool mountings were analyzed in further exact test series.

According to this the number of displacement transducers was increased and a fixture was built which is able to measure displacements on the tool's cutting edge. In special tests measuring inaccuracies and failure influences were deliberately removed.

The hot-air blower was replaced by a soldering copper since the latter makes it possible to introduce heat more intentionally and to reduce thermal influence of the measuring structure.

The hydraulic expansion chuck and two building sizes of the power shrinking chuck were studied. A drill was integrated into the test program in order to analyze the thermal influence of the cutting heat on the radial growth of the cutting edges (Fig.15.4).

The tests were performed in a manner analogous to the test series described before. Various heating cycles alternated with breaks of different duration.

15.3.3 Thermographic Investigations

Even the studies carried out at the motionless spindle were to tune the measuring equipment which has to be used for temperature measurement of the rotating spindle. There the measurement of temperatures and distortions was easier. Then the temperature of the rotating spindle was measured with a calibrated thermocamera. The following diagram (see Fig. 15.5) compares the measured and the calculated temperature values at measuring point S_1. A manufacturing cycle was simulated as the load case. In this cycle, 5 min of manufacturing and 3-min

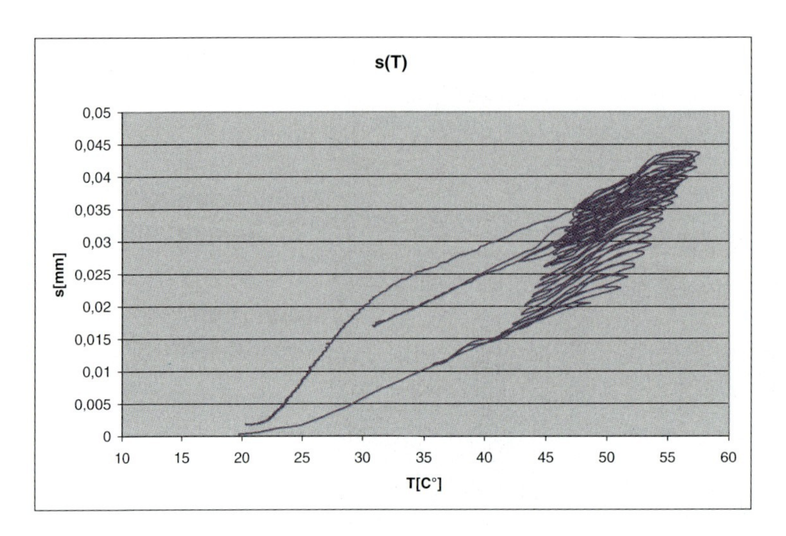

Figure 15.4: Hydraulic expansion chuck spindle + tool heating 5:3; 2×120 min with a break of 30 min.

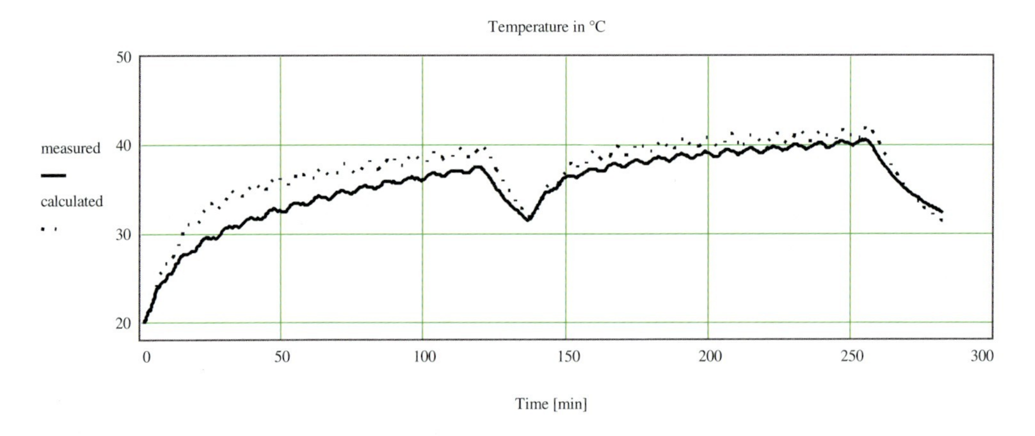

Figure 15.5: Temperature curve at the tool tip on a still standing hydro expansion chuck.

breaks alternate. This cycle is interrupted by a break of about 30 min. A heat-transmission coefficient was used for the calculation (see Eq. (15.5)), which includes the radiation and convection percentages.

$$\alpha_{\text{ges}} = \alpha_{\text{Str.}} + \alpha_{\text{Kon}}, \tag{15.5}$$

$$\alpha_{\text{Str.}} = \sigma \varepsilon_i \frac{\left(T_i^4 - T_j^4\right)}{T_i - T_j} \tag{15.6}$$

where

$$\sigma = 5.67 \times 10^{-8} \frac{\text{W}}{\text{m}^2\text{K}^4} \quad \text{is the Stefan–Boltzmann constant.}$$

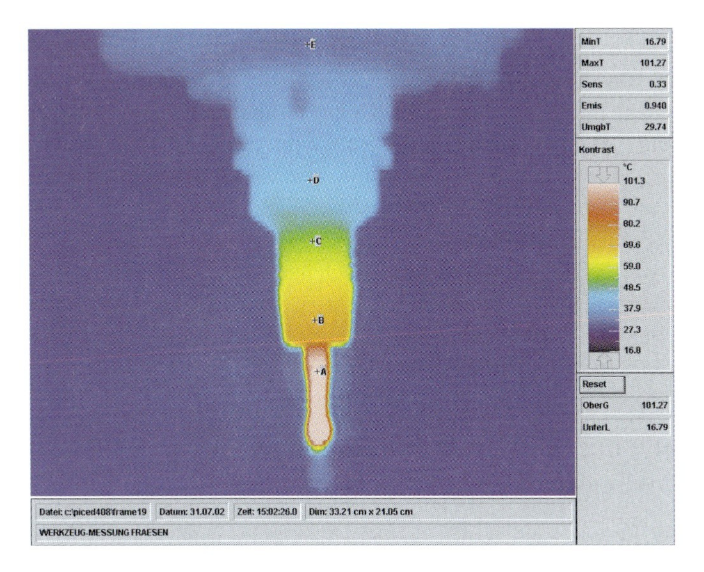

Figure 15.6: Moved tool mounting, recorded with the thermocamera.

The radiation ratio was calculated by Eq. (15.6). The emission ratio ε_i was assumed as $\varepsilon_i = 0.94$ and the tool mountings were painted with a photo varnish which is characterized by this ratio. The convective ratio was calculated with the equation for rotating cylindrical parts. Since the total heat-transmission coefficient depends on absolute temperature, after each calculation step the calculated temperature at each location was used to re-determine the total heat-transmission coefficient.

The calculated and measured temperature curves coincide well. A lower temperature was measured presumably in the first part of the test because the test and measuring equipment was heated too. This heating is relatively large comparing to the intrinsic tool mounting. The heat capacity of the measuring and test equipment was not considered in the calculation.

15.4 Maximal Correlation and Optimal Transformations

The concept of maximal correlation is a very powerful criterion to measure the dependence of two (especially nonlinear) related variables [23]. The main idea of this approach is to measure the maximized correlation of properly transformed variables.

Given a real variable Y and an n-dimensional vector $X = (X_1, ..., X_n)$ in the additive model

$$\Theta(Y) = \Phi(X_1, ..., X_n) = \sum_{i=1}^{n} \Phi_i(X_i). \tag{15.7}$$

Then, the maximal correlation is defined by

$$\Psi(Y, X) := \rho\left(\Theta^*(Y), \Phi^*(X)\right) = \max_{\Theta,\Phi} \left|\rho(\Theta(Y), \Phi(X))\right|, \tag{15.8}$$

where ρ denotes the correlation coefficient. The functions Θ^* and Φ^*, which fulfill the maximal condition, are called the optimal transformation and represent an estimation of the additive model. For the estimation of them an alternating conditional expectation (ACE) algorithm [2] was used. This iterative procedure is non-parametric because the optimal transformations are estimated by local smoothing of the data using kernel estimators. We use a modified algorithm in which the data are rank-ordered before the optimal transformations are estimated. This makes the result less sensitive to the data distribution.

The maximal correlation and optimal transformation approach has been recently applied to nonlinear dynamical systems. It has been used especially to identify delay in lasers and partial differential equations in fluid dynamics [30, 31, 32]. The ACE algorithm turned out to be a very efficient tool for nonlinear data analysis [13, 29, 37].

15.4.1 Reconstruction of Thermally Induced Displacements in Finite-element Models

To investigate whether the nonlinear regression approach described above is appropriate also for modular tool systems, which are used in milling and drilling machines, we firstly analyze the simulated data from finite-element models (FEMs). The tool system is the connecting part between main spindle and milling tool. The main target is the production of clamping forces for accurate machining. There are different types of clamping systems – here we focus on power shrinking and hydraulic chuck tools. The comparison between data sets and the results of the FEM model are used to find an optimal number and optimal locations of measurement points and to study the influence of the controlling parameters. A given regime of rotations leads to the temperatures T_i and the displacements S_j at several measurement points in the FEM.

In the following paragraph, we investigate a rather simply designed tool (main spindle (Fig. 15.7)) with two different measurements. We use the first measurement as a training set $\{S_{t,j}\}_{t=1,\dots,n}, j \in J$, $\{T_{t,i}\}_{t=1,\dots,n}, i \in I$, to compute the optimal transformations and the second as a test series $\{S'_{t,j}\}_{t=1,\dots,n}, j \in J, \{T'_{t,i}\}_{t=1,\dots,n}, i \in I$, to check whether the optimal transformations obtained by the reference series sufficiently describe the temperature–displacement relation. The result of different manufacturing schemes is the only difference between both series, especially with different time scales and regimes of rotations.

Our first aim is the appointment of the temperature–displacement relationship using the same time scales, i.e. estimating the displacements with the knowledge of the temperatures at the time t. A quantitative description of this relation is possible with this approach. We apply a two-step strategy:

(i) Computation of the optimal transformations Θ_j^* and Φ_i^* in the model

$$\Theta_j(S_{t,j}) = \Phi_j(T_{t,i_{i\in I}}) = \sum_{i\in I} \Phi_{j,i}(T_{t,i}) \tag{15.9}$$

from the training series $\{S_{t,j}\}_{t=1,\dots,n}, j \in J, \quad \{T_{t,i}\}_{t=1,\dots,n}, i \in I$.

(ii) Re-construction of the displacements $\{S'_{t,j}\}_{t=1,\dots,n}, j \in J$, of the test series using the temperatures $\{T'_{t,i}\}_{t=1,\dots,n}, i \in I$, of the test series and the optimal transformations Θ_j^* and

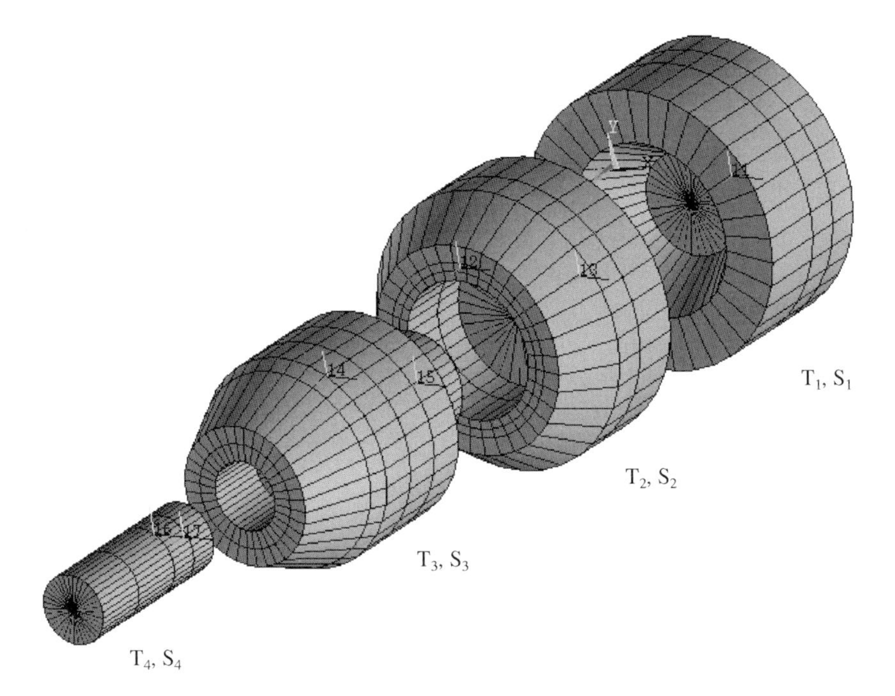

Figure 15.7: Design of the main-spindle model (output of the commercial FE modeling program ANSYS.)

Φ_i^* of the training series computed in (i)

$$\hat{S}'_{t,j} = \hat{\Theta}_j^{*-1}\left(\sum_{i \in I} \hat{\Phi}^*_{j,i}(T'_{t,i})\right), \quad j \in J \tag{15.10}$$

and comparison of the real $S'_{t,j}$ with the estimated $\hat{S}'_{t,j}$ displacements.

The symbol over Θ and Φ denotes a non-parametric estimation for these functions. Practically, we used a nearest-neighbor estimation with $k = 2$ nearest neighbors to estimate these functions [4]. We have tested different numbers of nearest neighbors $k = 1, .., 10$. However there are only slight differences in the estimates.

For two different regimes of rotations we have a simulated data set with values of temperatures $T_{t,i}, i = 2, 3, 4$ and displacements $S_{t,j}, j = 3, 4$ belonging to different parts of the tool. Each series consists of $n = 199$ points with a sampling time of 100 s. The reference series used (training set) is shown in Fig. 15.8a.

In this first investigation, the optimal transformations are additively calculated from temperatures and displacements:

$$\Theta_j(S_{t,j}) = \Phi_{j,1}(T_{t,2}) + \Phi_{j,2}(T_{t,3}) + \Phi_{j,3}(T_{t,4}), \quad j = 3, 4. \tag{15.11}$$

We will limit our discussion on the estimations for S_4 at the front tip of the tool (Fig. 15.6) because the displacements along the cutting edge are the most interesting ones. The results are

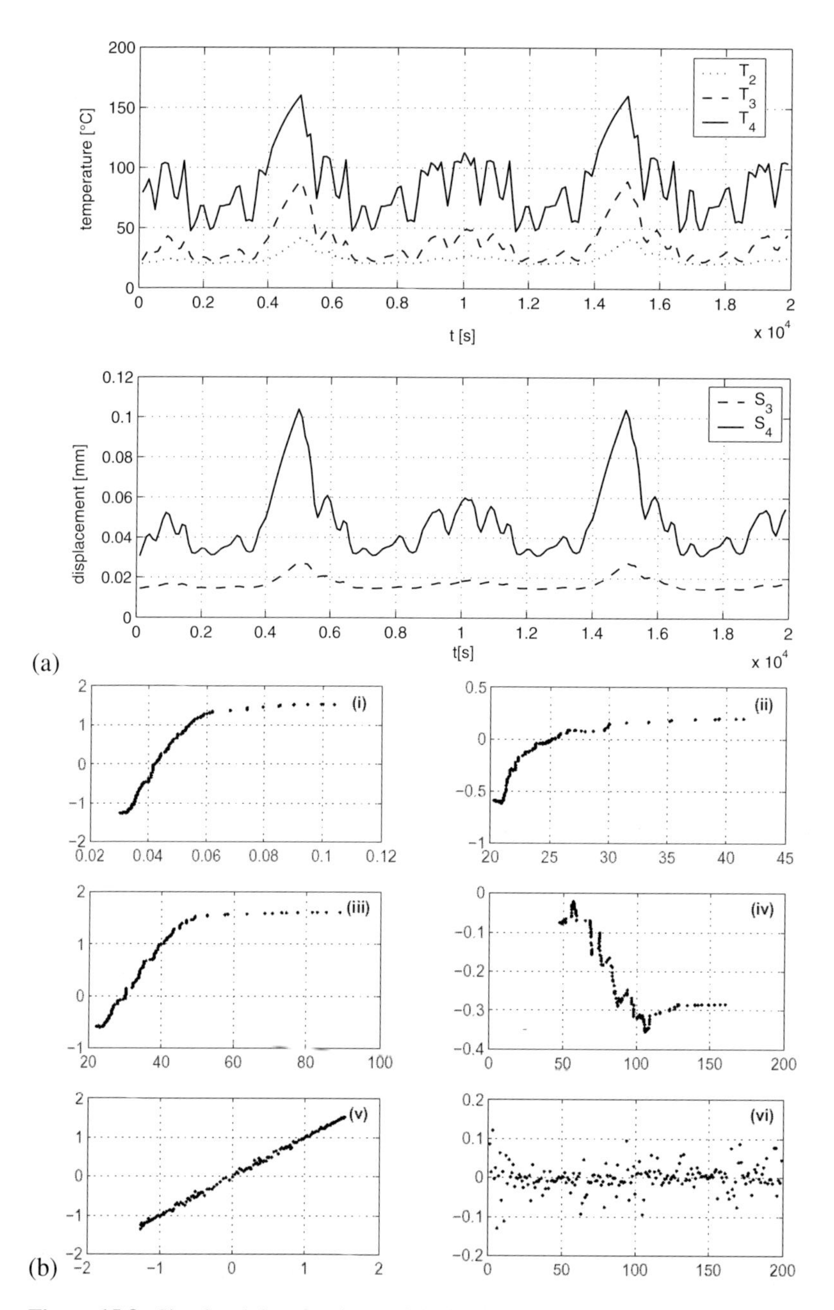

Figure 15.8: Simulated data for the model of Fig. 15.7: (a) training time series (top: temperatures T_2, T_3, T_4, bottom: displacements S_3, S_4), (b) optimal transformations (arbitrary units) for S_4 with respect to the model equation (15.8): (i) $\Theta_4^*(S_{t,4})$, (ii) $\Phi_{4,2}^*(T_{t,2})$, (iii) $\Phi_{4,3}^*(T_{t,3})$, (iv) $\Phi_{4,4}^*(T_{t,4})$, (v) plotting Θ_4^* vs. $\sum \Phi_{4,i}$, (vi) plotting the residuals $\Theta_4^* - \sum \Phi_{4,i}$.

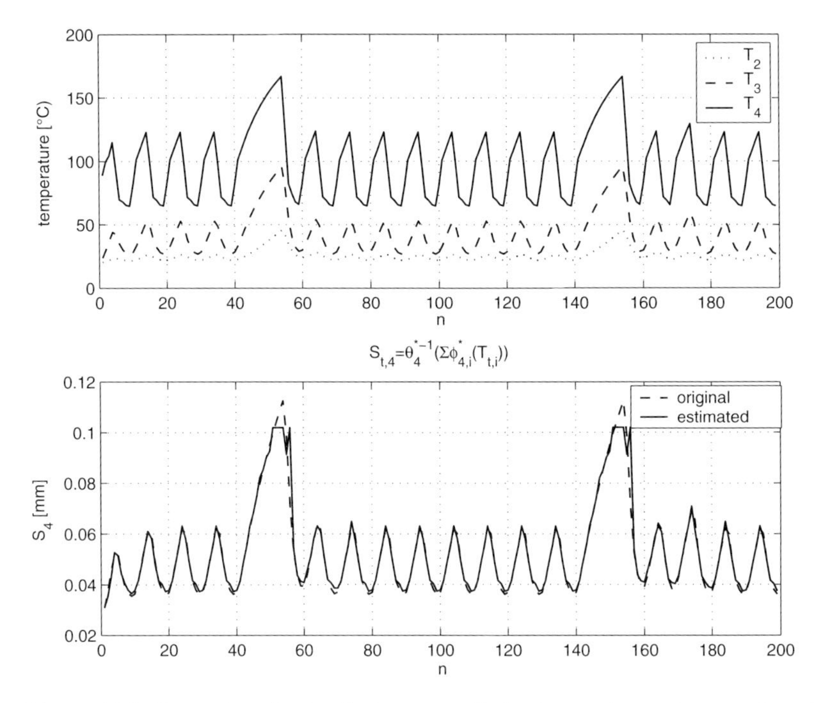

Figure 15.9: Estimations for the displacements S_4 for the test time series, top: the original temperatures, bottom: the original and the estimated displacements of the main-spindle model.

shown in Fig. 15.8b, i.e. the estimations for the optimal transformations Θ_4^* and $\Phi_{4,2}^*, ..., \Phi_{4,4}^*$. The nonlinearity of these functions is a clear indication that linear regression is not sufficient to model the relation between temperatures and displacements – otherwise the optimal transformations should be close to linear. The graph on the lower left corner Fig. 15.8b (v) shows $\Theta_4^*(S_{t,4})$ vs. $\sum_{i=2}^{4} \Phi_{4,i}^*(T_{t,i})$ which should be the identity in the optimal case. In the picture in Fig. 15.6b (vi) on the lower right side we plotted the residuals $\Theta_4^*(S_{t,4}) - \sum_{i=2}^{4} \Phi_{4,i}^*(T_{t,i})$. Assuming the existence of functions Θ and Φ in the model the residuals should vanish and, in fact, the high value obtained for the maximal correlation $\Psi = 0.9994$ indicates a very good estimation. Furthermore, this value is quite high in comparison with the maximal correlation in other applications of the ACE algorithm.

Next, the estimation of displacements was computed by the test data set using the optimal transformations estimated above and the temperatures from the test data set in Eq. (15.10). The estimated curves are quite close to the original ones (Fig. 15.9) except for the cropped peaks. Looking at the original time series and the optimal transformations in Fig. 15.8, we see immediately that the support in the displacement series of the training set is not as wide as the range of the test series. Therefore we have to consider this problem for data from real modular tool systems. Moreover, we have to find minimal measurement points with optimal predictability of the displacements to reduce the dimensionality of the task.

15.4.2 Reconstruction of Thermally Induced Displacements in Real Data

Let S_i, $i = 1, .., 3$ be the measured distortions and T_i, $i = 1, .., 9$ the temperatures measured at the various measuring positions, so that we can assume the following additive model:

$$\Theta_i(S_i) = \sum_{j=1}^{9} \Phi_{j,i}(T_j), \quad i = 1, \ldots, 3. \tag{15.12}$$

In 1985, Breiman and Friedman introduced the alternating conditional expectations (ACE) algorithm where estimation of the optimal transformations Θ_i and $\Phi_{j,i}$ from Eq. (15.12) is possible. In the last section "Reconstruction of Thermally Induced Displacements in Finite Element Models" was demonstrated this approach. Moreover, this approach is able to predict distortions which were generated with a FEM model [38]. Now, the applicability to real data have to be validated.

Two representative measurements from 29/04/2002 and 30/04/2002 are represented by Figs. 15.10 and 15.11. A break in the neighborhood of the measuring point no. 400 during the measurement taken on 29/04/2002 is the only difference between the two measurements. In these measurements the modular tool system was rested, i.e. not active. Load on the machine was simulated by two heatings in order to avoid possible disturbing variables such as convection (see Fig. 15.3).

At first, for reasons of an unambiguous structure, only the temperature-measuring points T_1, T_8 and T_9 were taken into account for modeling (i.e. $\Phi_{j,i} \equiv 0$ for $j \neq 1, 8, 9, i = 1, \ldots, 4$) – this combination of measuring points has proven to be very efficient in predicting

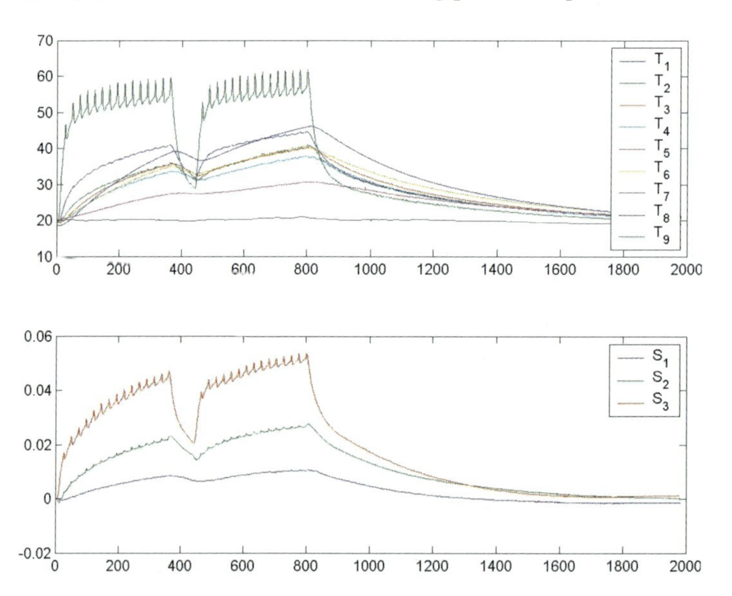

Figure 15.10: Measurement dated 29/04/2002 – T_i, $i = 1, .., 9$ are the temperature values, S_i, $i = 1, .., 3$ the measured distortions (see also Fig. 15.3).

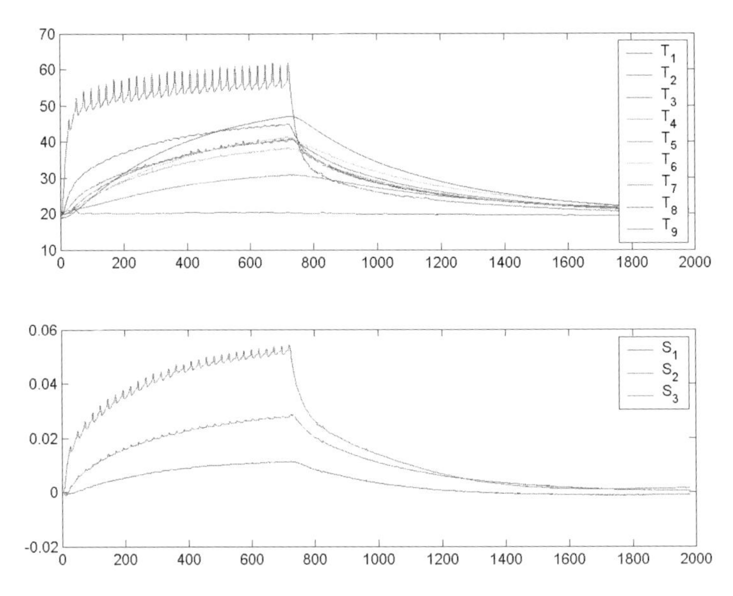

Figure 15.11: Measurement dated 30/04/2002 – $T_i, i = 1, .., 9$ are the temperatures, $S_i, i = 1, .., 3$ the measured distortions (see also Fig. 15.3).

distortions (the algorithm for selection of optimal measuring points is outside the scope of this paper). Figure 15.8 shows the estimated optimal transformations out of the data record from 29/04/2002. Since this temperature–deformation relationship is nonlinear – as can be seen, a linear approach of the type

$$S_i = \sum_{j=1}^{9} a_{j,i} T_j, i = 1, \ldots, 4 \qquad (15.13)$$

can describe this relationship only insufficiently. However, multiple linear regression based on the least-square method was implemented for comparison purposes.

The graph at the bottom right in Fig. 15.12 is characterized by wide deviations between the original values (red) and those estimated upon linear regression (green). However, the distortions estimated upon nonlinear regression are very close to the original. The RMS (root mean square) failure is four to five times less in comparison with linear regression.

15.5 The Thermal Behavior of a Modular Tool System in a Working Milling Machine

Finally, after successfully applying the nonlinear regression approach to the measured data from a silent experimental setup, we investigate now the thermal behavior of a modular tool system in a working milling machine. In this final section, the thermally caused displacements within the modular tool system were compensated as a demonstrator. On a laptop near the machine,

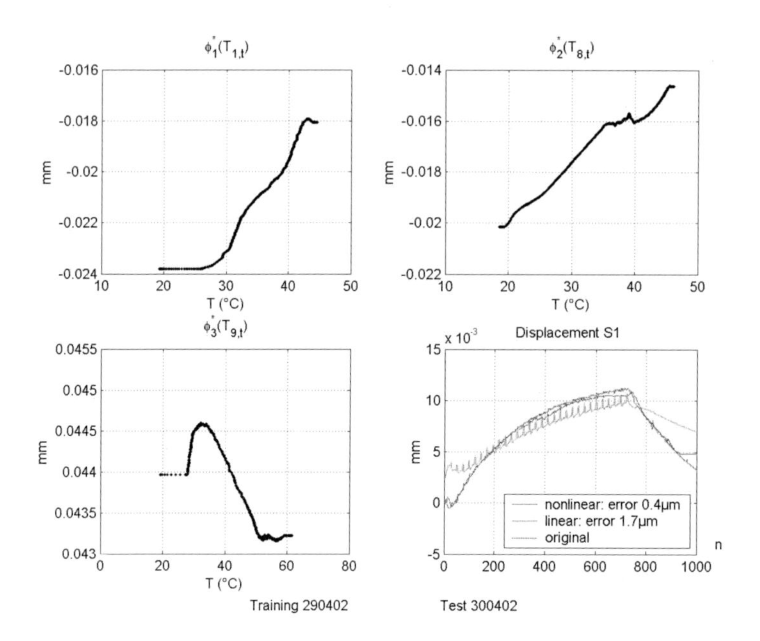

Figure 15.12: Estimation of optimal transformations $\Phi_1^*(T_{1,t})$, $\Phi_2^*(T_{8,t})$ and $\Phi_3^*(T_{9,t})$ with the training data taken on 29/04/2002. The figure at the bottom right illustrates the original displacement S_1 of the measurement dated 30/04/2002 (red curve), the estimation of this distortion with multiple linear regression (green) as well as the estimation of S_1 with the use of optimal transformations (blue) given.

the temperatures measured in the process were entered. Now, using these temperatures, the computer calculated online the displacements based on the temperature–displacement learning set given in Fig. 15.13. Here, four successive measurements from different days were taken as the training set to include several manufacturing schemes and finally to have a sufficiently large range of values. It was possible to compare simultaneously the calculated displacements with the measured ones in the cutting procedure. During this demonstration test, the cutting speed values were repeatedly changed in order to generate deviations from the training measurements and, thus, to reinforce the compensating calculations. The cutting speed values varied from $v_c = 120$ m/min, 200 m/min and 370 m/min. For a comparison of the measured displacements and the estimated ones, see Fig. 15.14.

As can be seen, the nonlinear estimation of the displacement and the original one vary only slightly from each other - the total RMS error is 2.2 µm. Avoiding displacement determination errors such as those around measurement point 17 leads to estimations with 1–2 µm accuracy. In contrast, the estimation using multiple linear regression is significantly worse – the RMS error amounts to 5.0 µm.

15.6 Conclusion

There is an important interest in compensating thermally induced errors of modular tool systems to improve the manufacturing accuracy. In this study, we tested the hypothesis whether

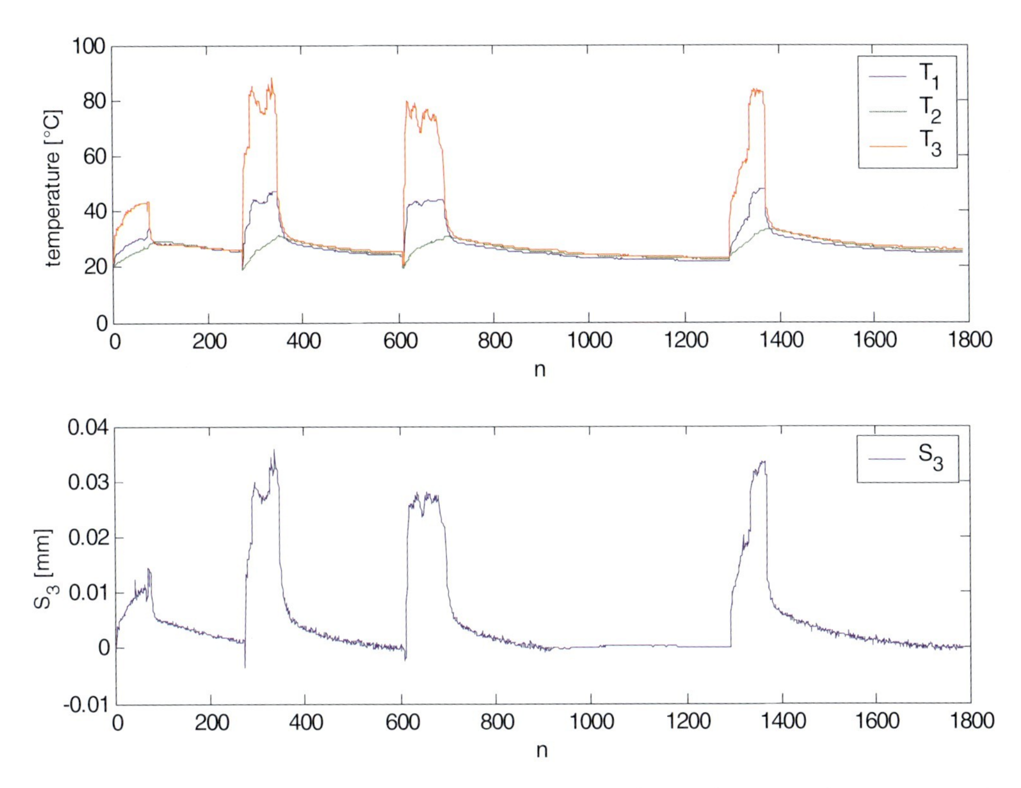

Figure 15.13: Measurement for the 'working tool system' (training set, four successive measurements) – $T_i, i = 1, \ldots, 3$ are the temperatures, S_3 the measured displacement on the tip of the tool system.

we can predict such thermal displacements by using a nonlinear regression analysis, namely the ACE algorithm, reliably. Therefore, we firstly analyzed data that were generated by two different finite-element spindle models of modular tool systems. The paper shows that the ACE algorithm is a powerful tool to model the relation between temperatures and displacements for simulated data. The maximal correlation is larger than 0.999 in both cases, which demonstrates the suitability of the ACE algorithm. The next step was the investigation of the temperature–displacement relationship in a silent real experimental setup where the tool system was heated artificially. Again, the ACE algorithm was powerful to estimate the deformation errors. The latter ones were 10-fold lower using the nonlinear regression approach in comparison to multiple linear regression.

On the tool test stand the heat-transmission coefficient α at the rotating tool mounting was determined. This coefficient is significant for calculations of the thermal behavior. The same tool mountings, which were used already for the motionless shaft, were taken in order to achieve an immediate comparison for the tests. In parallel, tests are executed on a machine tool of the Fraunhofer IWU. These practice tests should make the foundation for a finally valid mathematical model. The distortions are measured immediately or indirectly in milling tests. In fact these distortions appear as a result of the cutting heat due to the interaction between

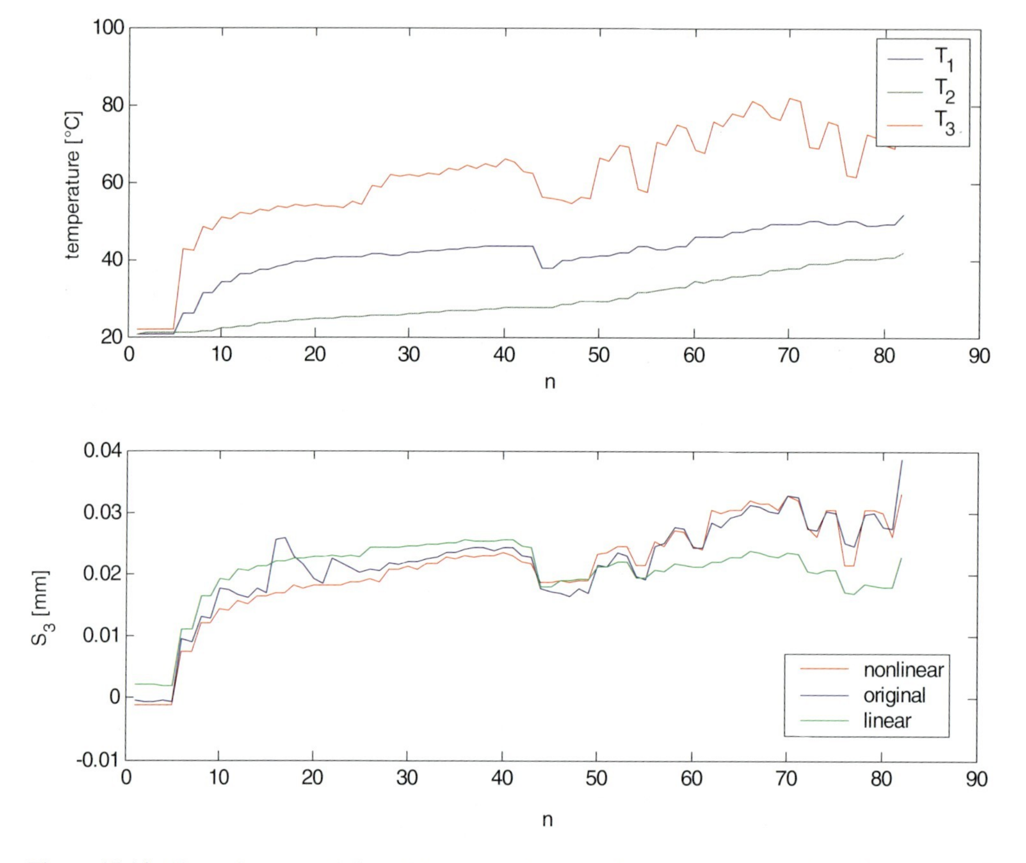

Figure 15.14: Thermal error modeling of the test set based on the estimated optimal transformations of the training set (see Fig. 15.13) and the temperatures from the test set (top). On the bottom the nonlinear and multiple linear estimations of the displacement S_3 as well as the original measured series are plotted.

tool and workpiece. The correct functioning of the compensation algorithm was evaluated in selected tests for PC use.

The nonlinear temperature–displacement functions, which are the basis for thermal error modeling, were estimated firstly from learning measurements using the ACE algorithm and then tested on independent data sets. As it turned out to be very important in this investigation, the support of the training measurements, i.e. the temperature range passed, must be at least as wide as that to be estimated later. Outside the range of values of the training measurements the calculation accuracy is decreasing rapidly. However, in principle, the compensation algorithm described above can be applied to all modular tool systems whose temperature–deformation ratio has been previously determined in a learning trial. Moreover, the algorithm may be immediately transferred to tool systems of similar geometry and consistency. Summarizing the results of this paper, we find that the ACE algorithm is a powerful tool to estimate the optimal transformations in the analyzed models and thus enables a reliable prediction of thermal displacements.

Bibliography

[1] M. Attia and L. Kops, *Computer simulation of nonlinear thermoelastic behavior of a joint in machine tool structure and its effect on thermal deformation*, Journal of Engineering for Industry **101** (1979), 355–361.

[2] L. Breiman and J. Friedman *Estimating optimal transformations for multiple regression and correlation*, Journal of The American Statistical Association **80** (1985), 580–619.

[3] J. Bryan, *International status of thermal error research*, Annals of the CIRP **39**(2) (1990), 645–655.

[4] H. Büning and G. Trenkler, *Nichtparametrische statistische Methoden*, de Gruyter, 1994.

[5] J. Chen and G. Chiou, *Quick testing and modeling of thermally induced errors of cnc machine tools*, International Journal of Machine Tools and Manufacture **35**(5) (1995), 1063–1074.

[6] J. Chen, J. Yuan, and J. Ni, *Thermal error modelling for realtime error compensation*, International Journal of Advanceed Manufacturing Technology **12**(4) (1996), 266–275.

[7] J.-S. Chen, *Computer-aided accuracy enhancement for multiaxis cnc machine tool*, International Journal of Machine Tools and Manufacture **35**(4) (1995), 593–605.

[8] H.S. Damaritürk, Dissertation, TH Darmstadt, Carl Hanser Verlag, Wien, 1990.

[9] P. de Haas, Dissertation, Technische Universität Berlin, (1975).

[10] H. Gartung, Dissertation, TU Braunschweig, (1985).

[11] K. Gebert, *Ein Beitrag zur thermischen Modellbildung von schnell-drehenden Motorspindeln*, Dissertation, Shaker-Verlag, 1997.

[12] B. Hardwick, *Improving the accuracy of cnc machine tools using software compensation for thermally induced errors*, in: Proceedings of the 29th International MATADOR Conference., 1992, pp. 269–276.

[13] T. Hastie and R. Tibshirani, *Generalized Additive Models*. Chapman and Hall, London, New York 1990.

[14] U. Heisel and F. Richter, *Thermal behaviour of industrial robots – measurement, reduction and compensation ofthermal errors*, in: Proceedings of the VII Workshop on Supervising and Diagnostics of Machining Systems, 1996, pp. 111–124.

[15] S. Huang, *Analysis of a model to forecast thermal deformation of ball screw feed drive systems*, International Journal of Machine Tools and Manufacture **35**(6) (1995), 1099–1104.

[16] J. Jedrzejewski and W. Modrzycki, *Intelligent supervision of thermal deformations in high precision machine tools*. in: Proceedings of the 32. International MATADOR Conference., 1997, pp. 457–462.

[17] M. Mitsuishi, T. Okumura, T. Nagao, and Y. Hatamura, *Active thermal deformation compensation based on internal monitoring and a neural network*, in: 7th International Conference on Production/Precision Engineering (7th ICPE) and 4th International Conference on High Technology (4th ICHT), 1994, pp. 215–220.

[18] J.P. Ortmann, *Maschinenentwicklung zum Hochgeschwindigkeitsschleifen von Zahnrädern*, Shaker Verlag, 1995.

[19] Z. Paluncic, *Minimierung thermisch bedingter axialer Spindelverformungen*, Industrie-Anzeiger 106/75, Carl Hanser Verlag, München, Wien, 1984.

[20] G. Popov, *Einfluss der Konvektion auf das thermische Verhalten von Werkzeugmaschinen*, Industrie-Anzeiger 23, 1988, pp. 38f.

[21] G. Popov, *Berechnung, Simulation und Optimierung des thermischen Verhaltens von Werkzeugmaschinen in der Entwicklungsphase* TU Sofia-Verlag, 1996.

[22] J. Rauer, Dissertation, TU Magdeburg, (1991).

[23] A. Renyi, *Probability Theory*, Akadémiai Kiadó Budapest, 1970.

[24] K. Schäfer, *Steuerungstechnische Korrektur thermoelastischer Verformungen an Werkzeugmaschinen*, Berichte aus der Produktionstechnik **94**(1) (1994).

[25] M. Schmid, F. Rieg, *Modelling the temperature of a hydrodynamic coupling*, in: NAFEMS, Numerical Simulation of Heat Transfer, Wiesbaden, 2001.

[26] G. Spur, E. Hoffmann, Z. Paluncic, K. Benzinger, and H.Nymoen, *Thermal behavior optimization of machine tools*, Annals of the CIRP **37**(1) (1988), 401–405.

[27] A. Srivasta, S. Veldhuis, and M. Elbestawit, *Modelling geometric and thermal errors in a five-axis cnc machine tool*, International Journal of Machine Tools and Manufacture **35**(7) (1995), 1321–1337.

[28] P. Tseng, *A real-time thermal inaccuracy compensation method on a machining centre*, International Journal of Advanced Manufacturing Technology **13** (1997), 182–190.

[29] H. Voss, *Analysing nonlinear dynamical systems with nonparametric regression*, in: A. Mees (ed.): Nonlinear Dynamics and Statistics, Birkhäuser, Boston, 2000, pp. 413–434.

[30] H. Voss and J. Kurths, *Reconstruction of nonlinear time delay models from data by the use of optimal transformations*, Physics Letters A **234**(5) (1997), 336–344.

[31] H. Voss and J. Kurths, *Reconstruction of nonlinear time delayed feedback models from optical data*, Chaos Solitons Fractals **10**(4–5) (1999), 805–809.

[32] H.U. Voss, P. Kolodner, M. Abel, and J. Kurths, *Amplitude equations from spatiotemporal binary-fluid convection data*, Physical Review Letters **83**(17) (1999), 3422–3425.

[33] M. Weck, P. McKeown, R.U.H. Bonse, et al., *Reduction and compensation of thermal errors in machine tools*, Annals of the CIRP **44**(2) (1995), 589–598.

[34] M. Weck and W. Schäfer, *Verbesserte Modellbildung für FE – Temperaturfeld- and Verformungsberechnung*, Konstruktion **44**(8) (1992), 333–337.

[35] F. Weidermann and S. Nestmann, *Simulation und Optimierung des thermischen Verhaltens von Werkzeugmaschinenkomponenten*; 1. Dresdner Werkzeugmaschinen-Fachseminar "Thermik an Werkzeugmaschinen", Dresden, 30 November to 1 December 2000.

[36] F. Weidermann, *Praxisnahe thermische Simulation von Lagern und Führungen in Werkzeugmaschinen*, CAD-FEM User's Meeting 2001, Internationale FEM- Technologietage, Potsdam, 17–19 October 2001, point 1.6.6, pp. 1–8.

[37] N. Wessel, A. Voss, H. Malberg, C. Ziehmann, H. Voss, A. Schirdewan, U. Meyerfeldt, and J. Kurths, *Nonlinear analysis of complex phenomena in cardiological data*. Herzschrittmachertherapie und Elektrophysiologie **11** (2000), 159–173.

[38] N. Wessel, J. Aßmus, F. Weidermann. J. Konvicka, S. Nestmann, R. Neugebauer, U. Schwarz, and J. Kurths. *Modeling thermal displacements in modular tool systems*, International Journal of Bifurcation and Chaos (2004) in press.

[39] S. Yang, J. Yuan, and J. Ni, *The improvement of thermal error modeling and compensation on machine tools by cmac neural network*, International Journal of Machine Tools and Manufacture **36**(4) (1996), 527–537.

16 Wrinkling in Sheet Metal Spinning

M. Kleiner, R. Göbel, Ch. Klimmek, B. Heller, V. Reitmann, and H. Kantz

Spinning is a very flexible incremental forming process. Under unsuitable conditions, failure modes such as wrinkles or cracks may occur. In this contribution we report on a series of spinning experiments which demonstrate that wrinkling is a dynamical instability. We present the mathematical description of the spinning process as a non-autonomous dynamical system, being capable of various bifurcations. For the first time, finite-element-method simulations of a rotating blank for three spinning paths have been performed, which are in good agreement with experimental observations, and in particular can show the onset of wrinkling. Time-series analysis of force measurements is used to detect the onset of wrinkling, to study the evolution of wrinkles in a time-resolved manner and to investigate their bifurcation scenario. Altogether, our results suggest that wrinkling could be controlled during the spinning process after its onset by a suitable feedback loop.

16.1 Introduction

Sheet metal spinning is a very flexible manufacturing process which is used to form various metals into axially symmetric hollow bodies in a wide variety of shapes in small and medium volume production. In combination with shear spinning and flowforming it is used to manufacture parts for the chemical, lamp, automotive or aerospace industry.

Figure 16.1 shows the principle of the spinning process. The roller tool operates locally on a so-called 'partial deformation zone' and progressively forms the rotating workpiece in multiple passes, until it conforms to the geometry of the mandrel. The thickness of the workpiece should remain constant throughout the process. A more detailed description of the process is given for example in [1, 2].

The setup of appropriate CNC programs is the central aspect in the manufacturing of parts by means of spinning processes. But the process design is difficult, because a large number of process parameters have to be adjusted accurately. Interactions of parameters and the influence of noise factors must be considered to achieve processes that are cost effective and surely avoid instabilities like wrinkles and cracks. Especially, wrinkling limits the efficency of the process and therefore is in the focus of process optimization [3]. As we will argue in the following, wrinkling in sheet metal spinning is not simply evoked by compressive stresses exceeding the critical (static) buckling load. Instead, it has to be understood also as a dynamical instability, a bifurcation in the sense of dynamical systems' theory. Enriching the conventional approaches of mechanical engineering by theoretical considerations and data-analysis concepts

Nonlinear Dynamics of Production Systems. Edited by G. Radons and R. Neugebauer

ISBN 3-527-40430-9

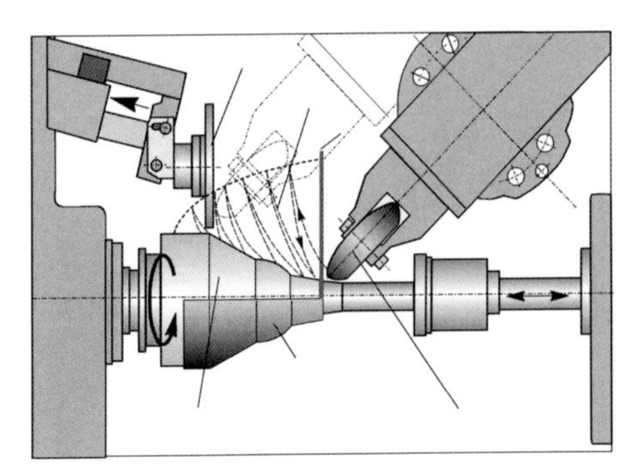

Figure 16.1: Principle of the spinning process.

from nonlinear dynamics therefore opens new perspectives for improvements of the spinning process.

In this paper, we describe how four different approaches are combined with the goal to achieve an as complete as possible understanding of the spinning process and its instabilities leading to wrinkling. Series of experiments making use of the concepts of optimal design have been performed in order to identify the relevant control parameters. Time series data recorded during these experiments are additionally analyzed in order to gain insight into the dynamics of the spinning process. An appropriate mathematical description of the dynamics has been derived, from which the principal dynamical instabilities can be deduced. Finally, a FEM simulation has been set up for a more detailed analysis. Work in progress aims at a full cross-linking of these approaches for their mutual verification. Each of these will be described in the following.

16.2 Wrinkling in Sheet Metal Spinning

Wrinkling is a well-known problem of many sheet metal forming processes. It occurs within the forming of hollow bodies from a blank, when areas of the workpiece material from outer regions are moved to inner diameters. To achieve this, tangential compressive stresses in the plane of the sheet are necessary and result in tangential compression of the material. If this compression exceeds the buckling stability, wrinkling occurs. In sheet metal spinning for this reason the workpiece is formed in multiple passes and supported by the blank-supporting tool [2]. While for deep-drawing a lot of different approaches have been developed to describe and predict wrinkling, this has not likewise happened for sheet metal spinning. For this reason commonly the knowledge of deep-drawing operations is transferred to sheet metal spinning in order to describe wrinkling. At a first glance this seems to be justifiable due to several similarities of the two forming processes. But a deeper analysis of the wrinkling effect shows that there are significant differences that give enough reason for the development of enhanced approaches, considering dynamic effects to describe and predict wrinkling in sheet metal spinning.

For deep-drawing processes Doege et al. [4] describe different types of wrinkles: wrinkles in the flange – called wrinkles of first order – and wrinkles in the free-forming zone between the punch radius and the die radius – called wrinkles of second order. While wrinkles in the flange can be suppressed by the blank-holder force in deep-drawing processes, this is not possible for wrinkles of second order [4]. In this case a change of the tool and workpiece geometry is successful. Opposite to this in sheet metal spinning until now only wrinkles of first order have been observed, or wrinkles that extend over the complete surface. However, further investigations have to be carried out to make sure that this observation is not related to the difficulties in observing the development of the wrinkles within the process.

An overview about different analytical methods for the prediction of flange wrinkling in deep-drawing is given for example in Wang and Cao [5]. Most of the approaches are based on the aspect of plastic instability as described above and use Hill's bifurcation and uniqueness theory published in 1958 [6]. With this theory, wrinkling at double-curved shell elements that are free of any contact can be described. Another common approach is the use of the energy conservation method. Referring to this, wrinkling criteria have been developed for a two-dimensional buckling model of an elastic–plastic annular plate. Additionally, numerical simulations using the finite-element method (FEM) have been developed for deep-drawing and have become the prime tool to predict wrinkling and buckling even for complex geometries.

Using these approaches for sheet metal spinning neglects the differences between metal spinning and deep-drawing – namely that spinning is, in difference to deep-drawing, an incremental forming process and intensively influenced by dynamic aspects. Setting up a finite-element model to describe wrinkling in sheet metal spinning is very difficult and until now realization is only basic. This aspect will be described in detail later on.

The transfer of methods from deep-drawing to describe wrinkling in conventional spinning has been done since the early 1960s. Kobayashi [7] investigated shear spinning and conventional spinning of cones by modifying different approaches used in deep drawing, which were not specified any more closely. For wrinkling in shear spinning he obtained a formula for critical conditions based on the description of geometrical influences. Further, he treats the conventional metal spinning as a special case of shear spinning with a cone angle of zero degrees. But this neglects the reduction of the wall thickness in shear spinning and the different state of stress associated with this [7].

The conventional spinning of cylindrical cups has been investigated by Barkaja and Ruzanov [8]. They tried to calculate the critical flange width in single-pass conventional spinning theoretically. Their theory is based on a 'static' model describing the state of wrinkling at the end of the process for a shell element with a uniform, axial-symmetric load. For this, they were able to calculate the number of wrinkles and to find a criterion for the onset of wrinkling depending on geometrical parameters [8].

Satho and Yanagimoto [9] compared wrinkling in conventional and shear spinning with deep-drawing and have shown that the mechanism of wrinkling in shear spinning is different from that of conventional spinning, but this in turn is more similar to deep-drawing. They investigated the influence of the metallurgical and mechanical influences on wrinkling, both in spinning and shear forming, to compare the differences. Their investigations are based on experimental results for mild steel. As a result they obtained a formula describing the major geometrical influences but also the influence of the feed in shear forming. Additionally, they found a similarity of the wrinkling effect between conventional spinning and deep-drawing that is different from the effects in shear forming and fixed this on the effect of the strain hardening exponent (n-value) of the material [9].

All these approaches neglect dynamic aspects of the wrinkling effect, although Hayama et al. [10] already have found experimentally for shear spinning that there is a large influence of the feed of the roller tool and the revolution of the blank. Additionally, they pointed out the sudden occurrence of wrinkling [10].

Köhne [11] has investigated the wrinkling effect in conventional spinning in his experiments. For some selected experiments he measured amplitude and frequency of wrinkles at different layers on the surface of the workpiece. He basically explained the wrinkling effect by plastic instability and refers to Satho and Yanagimoto [9]. Additionally, he pointed out that an increase of the feed rate (κ = feed/revolution) has a major influence on wrinkling. To prevent wrinkling he suggested the use of the blank-supporting tool with relatively high pressure, a reduction of feed rate, an increase of the number of spinning passes and an adapted control of the spinning-path geometry, which he realized by measuring the forces during the process [11].

In summary, it has been shown experimentally that there is a difference in the wrinkling mechanisms of deep-drawing, shear spinning and conventional spinning. In contrast to deep drawing, metal spinning is significantly affected by the dynamic conditions of the process that are dominated by the feed of the roller tool and the revolution of the circular blank. These dynamic effects seem to amplify the sensibility towards wrinkling significantly. But until the present day no theory exists that takes these aspects into account and explains the mechanism of wrinkling satisfactorily for conventional sheet metal spinning. Because of this, a closer look at the occurrence of wrinkling in metal spinning is necessary.

16.3 Influence of Nonlinear Dynamic Effects on Wrinkling

To investigate the characteristics of wrinkling, first of all the conditions for the occurrence of wrinkling have to be identified. Based on the results from the literature discussed before, different statistically planned experiments have been carried out to find the boundaries for wrinkling and to model the coherences between the process parameters and wrinkling mathematically. With this, proper parameter settings to avoid wrinkling for sure and gain optimized results in the process have been calculated. A detailed description of the experimental setup and the results is given in [2, 3]. All experiments described in the following have been carried out by looking only at the first stages of the process, because there wrinkling appears predominantly. As a result, in addition to several parameters describing the geometry of the spinning path and the tools, especially the feed and revolution of the circular blank have been identified to mainly influence wrinkling. This approves the observation of Hayama et al. [10], that within spinning the effect of parameters influencing dynamic aspects of the process cannot be neglected. But these results, despite their good practical applicability for process optimization, yield no information that helps us to understand the mechanism of wrinkling.

Consequently, to gain deeper knowledge about the wrinkling effect the focus of the investigations in this project, kindly supported by the Volkswagen Foundation, was set on the kind of onset and propagation of wrinkles during the forming process, see Fig. 16.2. For this, the first stage of the first spinning pass has been subdivided into five individual stages that were carried out separately. It was observed that wrinkling spontaneously starts at a single point of the circular blank and propagates with the feed of the roller tool until a complete cycle of wrin-

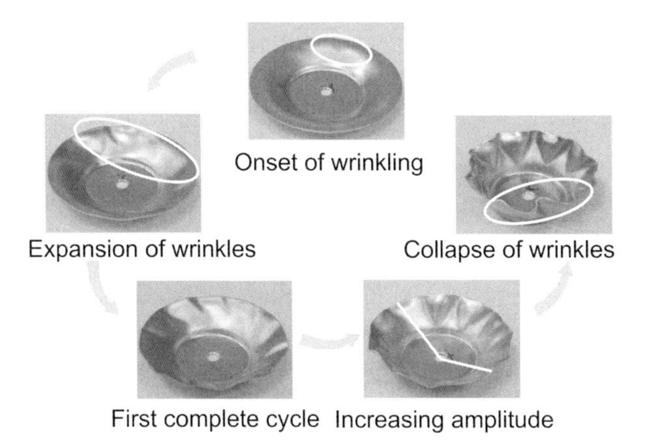

Figure 16.2: Wrinkling in the first spinning pass.

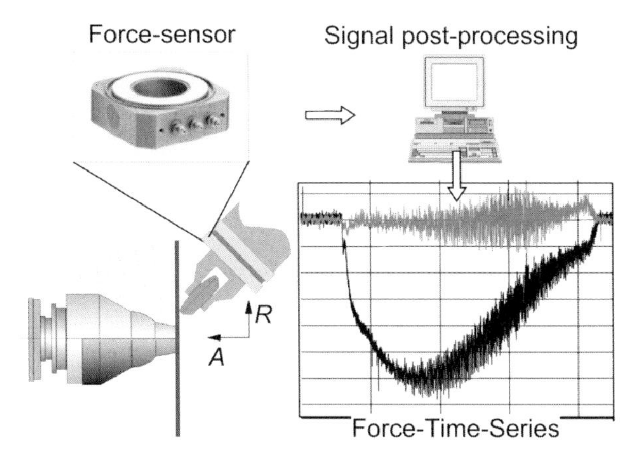

Figure 16.3: Recording of the force–time series.

kles has been passed through. In the following cycles, the amplitude and number of wrinkles increase, until finally the wrinkles collapse. New wrinkles develop by forming substates of existing wrinkles. This happens in a very short period of time. It could be observed that small wrinkles are reversible and can be flattened in following passes. Above a specific amplitude, wrinkling will be irreversible [3].

However, this confirms that in addition to the static causes, wrinkling in sheet metal spinning is influenced or even triggered by dynamic effects which themselves are affected by the revolutions of the circular blank and the feed of the roller tool. These dynamic effects seem to amplify the sensitivity of the process towards wrinkling. Due to the spontaneous occurrence of wrinkling and the typical highly structured patterns of the wrinkles it is assumed that nonlinear effects influence the process severely. To confirm this hypothesis, time-series analysis of forces measured during the process has been used in order to extract information about the dynamics [3].

To obtain data for the time-series analysis the force needed to form the material has been measured. In case of wrinkling, the process is accompanied by intense vibrations that manifest themselves in the force signals. It would be best to detect the corresponding signals as close to the forming zone as possible. But this is made difficult by the fact that the workpiece rotates and it is therefore impossible to apply a sensor to the workpiece. This problem has been solved by detecting the signals that are transmitted to the roller tool fastening device. This is realized by a three-dimensional force-measurement unit that is integrated into the support of the spinning lathe. The wrinkles interact with the roller tool and cause a force peak, which can be used as a measure for the intensity of the wrinkles, see Fig. 16.3. After post-processing these signals are transformed into axial, tangential and radial components of the workpiece coordinate system [2, 3].

A first analysis of the force–time signals has shown a good reproducibility of the process and further on of the amplitude and distribution of the wrinkles in the oscillation of the force–time curve. Moreover, the replication of force–time curves, measured under the same conditions, shows very similar results. The wrinkling effect can be observed especially in the axial and radial components of the forces. At the first spinning pass especially the axial component has proved to be more useful for the analysis [3]. Since the wrinkle formation inside the blank is a phenomenon which is related to its spatial extension and is clearly related to stress and strain tensors at every point on the disk, the measurement of just three scalar components of a force is a strong reduction. In order to interpret the findings in the time-series analysis, it is hence indispensable to consider an appropriate mathematical description of the process as a whole.

16.4 The Spinning Process as a Frictional Contact Problem

In this section, we discuss a mathematical description of the spinning process. We describe the equations of motion, the boundary and initial conditions, and we give an interpretation of the resulting nonlinear PDE problem as an evolutionary variational inequality. We make use of standard notions of continuum mechanics such as stress and strain tensors and their relations. The rotation of the blank leads to the inclusion of gyroscopic forces, whereas the contact problem thereby and by the deformation of the blank during spinning creates a moving boundary problem. Plasticity leads to additional complications.

Suppose that at time t the rotating blank $\Omega(t), t \in [0, \mathcal{T}]$, is in contact with the roller, the dynamics of which is prescribed. Assume also that the time-dependent boundary of $\Omega(t)$ has the representation

$$\Gamma(t) = \partial\Omega(t) = \Gamma_U(t) \cup \Gamma_F(t) \cup \Gamma_{\mathcal{F}}(t) \cup \Gamma_0(t)\,,$$

where $\Gamma_U(t)$ is the surface subjected to displacement boundary conditions, $\Gamma_F(t)$ is the surface part subjected to stress boundary conditions, $\Gamma_{\mathcal{F}}(t)$ is the contact surface between blank and roller and $\Gamma_0(t)$ is the free surface without boundary conditions.

We denote contravariant components of a tensor by superscripts and covariant components by subscripts in the usual way. Repeated indices indicate summation. With u_i and u^i as the components of the displacement, the Lagrangian strain tensor is $\varepsilon_{ij} = \frac{1}{2}\left(u_{j,i} + u_{i,j} + u^k_{\ ,i} u_{k,j}\right)$, where the comma denotes covariant differentiation with respect to the metric of the undeformed body.

Using the second Piola–Kirchhoff stress tensor σ^{ik}, the displacement u_i and including gyroscopic and inertial forces, the equation of motion of the blank in a C^2-setting can be written as [12, 17, 18]

$$(\sigma^{kl}\delta^i_l + \sigma^{kl}u^i_{,l})_{,k} + \rho f^i_A = \rho f^i_I + \rho f^i_C + \rho f^i_Z \,. \tag{16.1}$$

Here δ^i_l is the Kronecker symbol, ρ is the density of the material, f^i_A are body forces, $f^i_I = \ddot{u}^i$ are inertial forces, $f^i_C = 2e^{inm}\omega_n \dot{u}_m$ are the Coriolis forces and $f^i_Z = e^{inl}e_{mnk}\,\omega_l \omega^k u^m$ are the centripetal forces. The numbers ω_n and ω^k are the covariantly and contravariantly written components of the rotation vector, respectively, and e_{mnk} and e^{inl} are the components of the Levi–Civita symbol. Note that in Cartesian coordinates the Coriolis forces and centripetal forces correspond to $2\boldsymbol{\omega} \times \dot{\mathbf{u}}$ and $\boldsymbol{\omega} \times [\boldsymbol{\omega} \times (\mathbf{r} \times \mathbf{u})]$, respectively, where $\boldsymbol{\omega}$, $\mathbf{u}$ and $\mathbf{r}$ are the vectors with components ω_n, u_i and x^i , respectively. In [18, 19], using the existence of running waves for the linearized equation (1), it is shown that for certain values of the frequency $\|\boldsymbol{\omega}\|$ the critical eigenvalue of this linearization is zero and the rotating body becomes unstable. This effect of rotation is the most important at lower frequencies.

The material of the blank is assumed to be elasto-plastic. At any point of the blank the total strain is the sum of an elastic part and a plastic part, i.e. $\varepsilon_{ij} = \varepsilon^{\mathrm{e}}_{ij} + \varepsilon^{\mathrm{p}}_{ij}$. The elastic part $\varepsilon^{\mathrm{e}}_{ij} = L^{\mathrm{e}}_{ijkl}\sigma^{kl}$ is computed by the generalized Hooke's law, where L^{e}_{ijkl} is the elastic deformation tensor. For the plastic part we have $\varepsilon^{\mathrm{p}}_{ij} \neq 0$ if and only if $\mathcal{H}(\sigma^{ij}) \leq 0$, where $\mathcal{H}$ is a yield function, defined (for example) for von Mises material as

$$\mathcal{H}(\sigma^{ij}) = \frac{1}{2}s^{ij}s_{ij} - k^2 \text{ with } s^{ij} = s_{ij} = \sigma^{ij} - \frac{1}{3}\delta^{ij}\sigma^{kk} \,,$$

and $k \neq 0$ as a constant. In the flow theory the plastic strain rate is defined by $\dot{\varepsilon}^{\mathrm{p}}_{ij} = \lambda(\partial Y/\partial\sigma^{ij})$, where λ is a factor of proportionality and Y is a given potential. The total strain rate is $\dot{\varepsilon}_{ij} = \dot{\varepsilon}^{\mathrm{e}}_{ij} + \dot{\varepsilon}^{\mathrm{p}}_{ij}$.

The occurrence of nonzero plastic deformation components in the spinning process is concentrated in two parts of $\Omega(t)$ which are located in an area where the blank loses contact with the mandrel and in the contact area under the roller. The surface $\Gamma_P(t)$ of the union of these areas is a *moving boundary*. For the determination of $\Gamma_P(t)$ it is necessary to use continuity properties of the displacements u_i and the stresses σ^{ij} on this boundary, i.e.,

$$^{(\mathrm{e})}u_i(x,t) = ^{(\mathrm{p})}u_i(x,t)\,,\ ^{(\mathrm{e})}\sigma^{ij}(x,t) = ^{(\mathrm{p})}\sigma^{ij}(x,t), \quad \forall x \in \Gamma_P(t),\ \forall t \in (0,\mathcal{T})\,. \tag{16.2}$$

(The superscripts e and p indicate solutions from the elastic and plastic parts respectively.)

For the blank $\Omega(t)$ and the roller in contact we define on the boundary $\Gamma(t) = \partial\Omega(t)$ the unit normal $\mathbf{n} = (n_i)$. If we assume that $\Gamma(t)$ is Lipschitz, $\mathbf{n}$ exists at almost every point of $\Gamma(t)$. The displacement $\mathbf{u}$ is separated into the tangential and the normal parts as $\mathbf{u} = \mathbf{u}_T + u_N\mathbf{n}$, where $u_N = u_i n^i$. Similarly, for the external surface pressure $\mathbf{p}$ we write $\mathbf{p} = \mathbf{p}_T + p_N\mathbf{n}$. The tangential relative velocity is decomposed into a sticking part and a slipping part, i.e. $\dot{\mathbf{u}}_T = \dot{\mathbf{u}}^{\mathrm{st}}_T + \dot{\mathbf{u}}^{\mathrm{sl}}_T$. The sticking part is included if $\Phi < 0$, where Φ is a yield function, defined for Coulomb's law of friction as $\Phi(\mathbf{p}_T, p_N) := \|\mathbf{p}_T\| - \mu|p_N|$ with a parameter $\mu > 0$ [17]. The sticking part is defined through $\mathbf{p}_T = -k\mathbf{u}^{\mathrm{st}}_T$; the slipping part rate is computed by $\dot{\mathbf{u}}^{\mathrm{sl}}_T = -\dot{\gamma}(\partial\psi/\partial\mathbf{p}_T)$ with a parameter k, a slip potential ψ and a real function γ.

On the surface $\Gamma_U(t)$ the blank is attached to the mandrel. Formally, this is expressed with some given functions U_i as

$$u_i(x,t) = U_i(x,t), \quad \forall x \in \Gamma_U(t)\,, \quad \forall t \in (0,\mathcal{T})\,. \tag{16.3}$$

The blank is rotated with prescribed force F^i, which is applied over the surface $\Gamma_F(t)$ by [12, 17]

$$(\sigma^{kl}\delta^i_l + \sigma^{kl}u^i_{,l})n_k = F^i, \quad \forall x \in \Gamma_F(t)\,, \quad \forall t \in (0,\mathcal{T})\,. \tag{16.4}$$

At the contact surface $\Gamma_{\mathcal{F}}$ between blank and roller the tangential stress is described by

$$\sigma^{ij}n_j - \sigma^{jk}n_j n_k n^i = \mathcal{F}^i, \quad \forall x \in \Gamma_{\mathcal{F}}(t)\,, \quad \forall t \in (0,\mathcal{T})\,. \tag{16.5}$$

In Eqs. (16.4) and (16.5) we again denote by n_i and n^j the covariant and contravariant components of $\mathbf{n}$, respectively.

The initial conditions for the blank are

$$u_i(x,0) = 0\,, \; \dot{u}_i(x,0) = 0\,, \quad \forall x \in \Omega(0)\,. \tag{16.6}$$

Thus the spinning process in the *classical* formulation can be considered as a frictional elastic–plastic contact problem with normal damping defined by the equation of motion given by Eq. (16.1) and the boundary and initial conditions given by Eqs. (16.2)–(16.6). It is well known that, in general, there are no classical solutions from C^2 to the problem posed by Eqs. (16.1)–(16.6). Therefore, we turn to the *weak* or *variational* formulation of the problem. To this end we introduce for $t \in (0,\mathcal{T})$ a space of test functions

$$V_1(t) := \{v \in W^{1,2}(\Omega(t)) : v = 0 \quad \text{on } \Gamma_U(t)\}\,. \tag{16.7}$$

If we take now the scalar product of the terms in Eq. (16.1) with $v_i - \dot{u}_i$, where $v_i \in V_1$ is an arbitrary test function and u_i is the component of the unknown solution, use the integration by parts formula and properties of $v_i - \dot{u}_i$ on the boundary $\Gamma_U(t)$, we get for almost all $t \in (0,\mathcal{T})$ and all $v^i \in V_1$ the identity

$$\begin{aligned}
&\int_{\Omega(t)} (\sigma^{kl}\delta^i_l + \sigma^{kl}u^i_{,l})(v_{i,k} - \dot{u}_{i,k})\, dV \\
&- \int_{\Gamma_F(t)} F^i(v_i - \dot{u}_i)\, dS - \int_{\Gamma_{\mathcal{F}}(t)} [F_N(v_N - \dot{v}_N) + \mathcal{F}_T(v_T - \dot{u}_T)]\, dS \\
&+ \int_{\Omega(t)} \rho(f^i_A - f^i_I - f^i_C - f^i_Z)(v_i - \dot{u}_i)\, dV = 0\,.
\end{aligned} \tag{16.8}$$

Here F_N, v_N and $\mathcal{F}_T, v_T$ denote the normal and tangential components, respectively. If we add a dissipation term for the frictional dissipation to both sides of the last variational equation, we get from Eq. (16.8) for almost all $t \in (0, \mathcal{T})$ the *evolutionary variational inequality*

$$\int_{\Omega(t)} \rho \ddot{u}^i (v_i - \dot{u}_i)\, dV + \int_{\Omega(t)} (\sigma^{kl} \delta_l^j + \sigma^{kl} u^i_{,l})\, (v_{i,k} - \dot{u}_{i,k})\, dV$$

$$+ \int_{\Omega(t)} \rho\, 2\, e^{inm} \omega_n \dot{u}_m (v_i - \dot{u}_i)\, dV + \int_{\Omega(t)} \rho\, e^{inl} e_{mnk}\, \omega_l\, \omega^k u^m (v_i - \dot{u}_i)\, dV$$

$$+ j(v_T) - j(u_T) \geq \int_{\Omega(t)} \rho f_A^i (v_i - \dot{u}_i)\, dV + \int_{\Gamma_F(t)} F^i (v_i - \dot{u}_i)\, dS\,, \qquad (16.9)$$

where $j(v_T) - j(u_T) := \int_{\Gamma_{\mathcal{F}}(t)} |\mathcal{F}_T| (|v_T| - |\dot{u}_T|)\, dS$ satisfies the inequality [17]

$$j(v_T) - j(u_T) + \int_{\Gamma_{\mathcal{F}}(t)} \mathcal{F}_T (v_T - \dot{u}_T)\, dS \geq 0\,, \text{ for a.a. } t \in (0, \mathcal{T})\,.$$

Thus the spinning process as an elastic–plastic deformation and contact problem is described by a second-order time-dependent variational inequality (16.9). The existence of weak solutions for the general contact problem (16.9) is unknown. For special types of material and dynamical contact, theoretical and numerical results are developed in [12–14, 17]. In many papers the idea is to consider families of densely embedded Lebesgue and Sobolev spaces $V_1 \subset V_0 \subset V_{-1}$ of the type

$$V_1 = \{v \in W^{1,2}(\Omega) : v = 0 \text{ on } \Gamma_U\}\,, \quad V_0 = L^2(\Omega) \text{ and } V_{-1} = V' \text{ (the dual)}\,,$$

in order to show that (for fixed σ^{ij}) the weak solution $(u_i, \dot{u}_i)$ of (16.9) belongs to a space of the type $L^2(0, \mathcal{T}; V_0) \times L^2(0, \mathcal{T}; V_{-1})$. Any second-order dynamic variational inequality (16.9) can be written as a first-order dynamic variational inequality. For a class of such inequalities, stability and instability properties of the solutions and some measurement operators are considered in [15, 16]. In summary, the mathematical description of the spinning process poses several fundamental problems and possesses several unsolved mathematical issues, which are partly the subject of current research.

16.5 Time-series Analysis

Having in mind the complexity of the (already idealized) mathematical model as outlined in Sect. 16.4, which, as a partial differential equation, lives in an infinite-dimensional phase space, it might be quite surprising that time-series analysis should yield any reasonable insight into the dynamics of the spinning problem. However, since the stable solutions of the model equations are, in the regime of parameters of interest, most surely fixed points and limit cycles, the solutions are supported by low-dimensional subspaces. Their properties are therefore well represented by time-series data.

Time-series analysis of the spinning process has essentially three goals. Since imaging techniques are currently not available, time-series data should tell us at which point of the spinning path wrinkling sets in, and how it develops over time, thus comparing different settings of the spinning process parameters. Secondly, the temporal evolution of the wrinkle formation is to be compared to what is known from the mathematical model about possible types of instability and bifurcations, and it should be compared to the results of FEM simulations in order to verify the accuracy of the FEM. Lastly, the detection of the onset of wrinkling could be used in an online feedback loop to counteract the wrinkle formation by small parameter changes and hence yield improved performance of the spinning process in manufacturing.

As described in Sect. 16.3, the current experimental equipment supplies time-series data of three components of the force acting between the roller tool and its support, recorded with a sampling rate of 8 kHz. These data are assumed to reflect essentially the forces in between roller tool and workpiece. Wrinkling in the first pass of the roller is most clearly expressed by the axial force component.

The force signals as they are recorded digitally require pre-processing. The raw signals possess a smooth and slow component which is caused by the overall properties of the spinning process, which are functions of the chosen geometry of the spinning path and other parameters. This "baseline signal" is removed from the recordings by subtraction of the suitably smoothed signal (see Fig. 16.3 for a sketch). The remainder fluctuates around zero on a fast time scale (related to the number of revolutions of the workpiece) and contains, among other signatures, the information about the wrinkling process. Since noises and vibrations of the machinery partly conceal the signal of interest, in a next step, such perturbations have to be removed. We employ a combination of band-pass filters and of pattern matching in the time domain in order to eliminate these noise components. All further analysis is performed on signals which are treated in this way.

From all parameters in the spinning process which might influence its dynamics, we choose the two most evident for a more detailed analysis: the rotational speed of the disk and the speed of feed. For 10 different combinations of these two parameters, spinning experiments have been performed. In Fig. 16.4 the parameter values and the wrinkling properties at the end of the experiments are depicted. The number of wrinkles and their amplitude after finishing the forming process differ quite strongly depending on the parameter values chosen. Under the hypothesis that fluctuations of the measured forces (after pre-processing) are direct signatures of wrinkles in the blank, their analysis enables us to study the temporal evolution of the wrinkling process during the spinning process in a much more complete and faster way than is shown in Fig. 16.2. In Fig. 16.5 we present a particular visualization technique based on these recordings: we convert the time t (measured in s) at which the force $f(t)$ is recorded into an angle by $\phi = t\omega$, where $\omega = 2\pi/60 \times$ rpm is the angular frequency. The force value $f(t)$ is transformed into a radius $r(t) = sf(t) + r_0 t$, where s and r_0 are suitable scaling parameters. This representation hence generates a spiral line covering the circular disk, where at each position the modulation of this line represents the actual fluctuation of the force around zero. It hence looks like an image of the wrinkled disk, where, however, the position and amplitude of the wrinkles at each point (ϕ, r) represent the situation at the moment when the roller tool passes over this point. Therefore, this is not a snapshot of the disk at any particular time, but instead it shows the time evolution of the wrinkling process.

In all experiments represented by Fig. 16.4 the spinning path was identical. Then, a quantitative analysis of the features expressed by figures such as Fig. 16.5 comprises a counting of the number of wrinkles as a function of feed and the monitoring of the wrinkle amplitude, since apart from wrinkling the geometry of the blank in different experiments is close to identical for identical feed. Counting wrinkles contains the problem which one encounters already

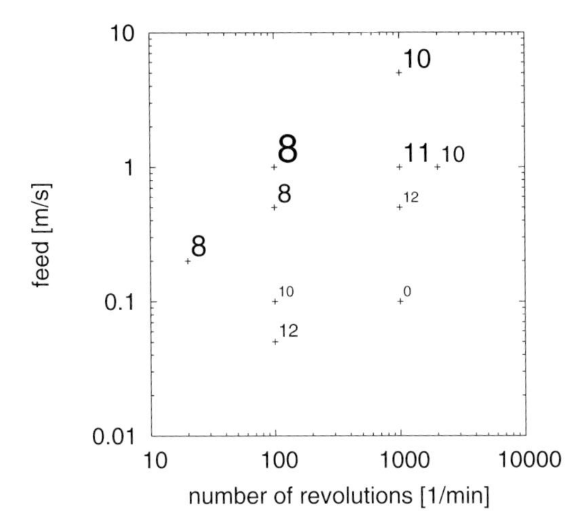

Figure 16.4: The parameter values for which spinning experiments were performed, together with the number of wrinkles in the workpiece. The size of the numbers represents the amplitudes of the wrinkles qualitatively (visual inspection of the final workpiece).

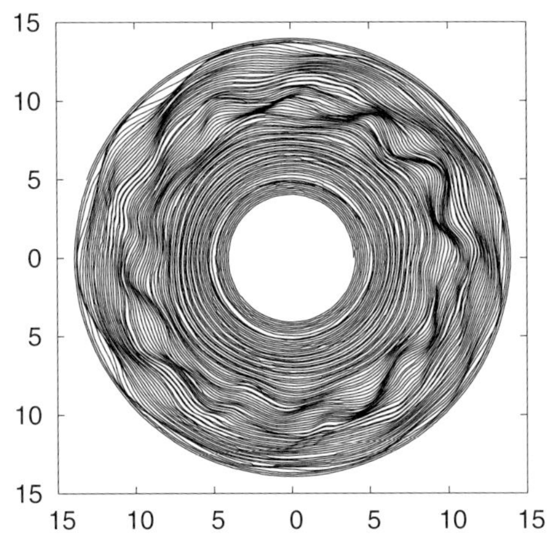

Figure 16.5: The polar representation of time-series data as described in the text, visualizing the time evolution of the wrinkling process.

when inspecting the final workpieces visually: sometimes, it is hard to decide whether a given structure should be interpreted as a wrinkle or not. However, in most cases the regularity of the deformation of the blank enables us to use a suitably smoothed signal (combination of band-pass filters and matching techniques) and to count the number of maxima of the smoothed signal within the time span corresponding to one revolution. In addition, we define the area between this smoothed signal and the baseline as a measure for the intensity of wrinkling. The number of wrinkles per revolution contains as an uncertainty potential fluctuations of the rotational speed of the blank; it is, however, yet unclear whether additionally wrinkles might move.

For most of the parameter combinations where wrinkling occurs, right after its onset four wrinkles appear. Their number and their amplitude increase during the process. However, the number of wrinkles and their intensity in the final state may be smaller than at some intermediate stage. This reflects the dynamical nature of the wrinkling: as long as the process is not stopped, wrinkles represent a kind of plastic wave, whose properties can change, and only when the roller tool is detached from the blank, this wave "freezes" and is turned into a static deformation of the blank. Also, in some of those cases where we cannot detect any wrinkles in the final workpiece, intermediate states do show some wrinkling in our analysis. Right now it has not been verified whether the fluctuations of the force which we interpret as wrinkles in such cases are really due to plastic deformations or whether they represent elastic oscillations of the blank, which would act as inertial forces onto the roller tool.

Previous mathematical studies [20, 21] of dynamical loads acting on elasto-plastic two-dimensional structures have shown that workpieces typically possess a sequence of nonlinear dynamical plastic eigenmodes, which are either stable or unstable, depending on control parameters. It is plausible that these results qualitatively apply also to our elasto-plastic rotating shell. The geometry of these modes, i.e., essentially the number of the wrinkles, should not depend on the rotational speed or the speed of feed, whereas these two parameters will determine which of these modes will be excited and which amplitude they will have. This concept is in agreement with results of the time-series analysis: quite frequently we observe that the number of wrinkles does not change during some time interval. The numbers which then occur are typically 8, 10 or 12, whereas both smaller and larger numbers occur only transiently. If this is a correct interpretation of the dynamics, one can also understand why the intensity and the number of wrinkles can diminish again: during the spinning process, the blank is deformed. This deformation causes a change of the stability properties of the different modes, such that an already excited mode can become a damped mode and may disappear.

The observation that both the amplitude and the number of wrinkles can diminish brings us to the following conclusion: when, in a production line, choosing spinning parameters which are close to the limit of stability, single workpieces will exhibit wrinkling because of small irregularities of, for example, the thickness of the flat blank. This wrinkling could be detected by time-series analysis in an online monitoring system. Knowing that wrinkling beyond its onset can be controlled by a change of parameters, such a change can be performed by a feedback loop. The parameter which appears to be most suitable right now is the feeding speed of the roller tool, which in such a situation should be slightly reduced. A test for this hypothesis will be to design an experiment without feedback, where the first part of the spinning is done at high feed, leading to the onset of wrinkling, and the second part is done at slower feed, where hopefully the wrinkling disappears. If this turns out to be true, then the mathematical theory of nonlinear dynamical systems together with time-series analysis have supplied valuable information for the optimization of the spinning process.

16.6 Finite-element Model

As described before, wrinkling is basically caused by the state of stress exceeding a specific buckling load in the circular blank. To understand wrinkling from this mechanical point of view, it is necessary to investigate the stresses and strains. Due to the predominant dynamics of the process, it is further important to understand in which way the evolution and the changes of the state of stress affect wrinkling. This can be done by analyzing the process using a finite-element model. Until now there have been very few scientific publications regarding the simulation of the spinning process. Recent publications covering this issue were [22–24]. The challenge here is to set up a model that is able to take dynamic effects like vibrations, dynamic forces and wave propagation into account. Nonlinear material behavior, contact and buckling instability of the circular blank make the setup of the model difficult. For this investigation an explicit code has been chosen to solve the problems of contact efficiently [3, 25]. Another advantage of the explicit code is that the necessary dynamic formulations are firmly implemented. An explicit code on the other hand means that a fine mesh will increase computation time enormously. An acceleration of the simulation, as it is being done in simulations of deep-drawing processes by mass scaling or increase of punch velocity, is not possible [2, 3, 25, 27].

A first model was set up using the finite element package PamStamp [11, 15–17]. In Fig. 16.6 the model is illustrated. A detailed definition of the design and the basic parameters can be found in [3, 25–27]. All process parameters (i.e. axial feed, rotary speed and design of the forming pass) were taken from a reference process carried out at the spinning machine. This reference process was designed to manufacture a cup of 100-mm diameter from a 2-mm-thick, mild steel alloy (DC 04), circular blank (200-mm diameter) without any wrinkles using the maximum possible axial feed and angular velocity of the machine. A marginal change of the sensitive process parameters, especially of the forming passes, would have created wrinkles and part damage. The validation of a reference process without wrinkles is more reliable and easier to carry out. The use of this model showed that the shell-element formulation implemented in PamStamp (element type 107) is not capable of mapping the derived values for more than five rotations. An improvement has been made possible by rotating the tool around the workpiece instead of rotating the blank. A detailed comparison of the initial and the improved model-can be found in [3, 25].

One possibility to verify the simulations by experiments is to compare the reference workpieces manufactured in the experiments with the computed model geometrically. This was done by digitizing the manufactured workpieces after every forming pass and importing the data plots into a specially developed evaluation program, using the best-fit algorithm [3]. A second strategy to verify the results was the setup of an additional simulation using the general-purpose FE system LS-Dyna. The advantage of this model is the ability to map the rotation of

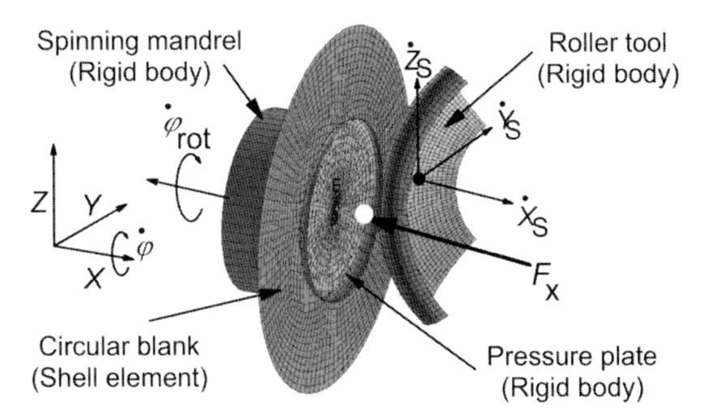

Figure 16.6: PamStamp finite-element model.

the circular blank due to the enhanced shell formulations. To set up the model, the geometry used in PamStamp has been implemented in LS-Dyna. The use of a utilized mesh for the circular blank with rectangular elements and an element side length of 5 mm has shown the best compromise between accuracy and calculation time. The process parameters are modeled according to the reference process as has been already done in PamStamp. The model was set up using a fully integrated shell element of the type "Belytschko-Tsay". The material was expected to be isotropic.

A comparison of both calculated models and the measured data is shown in [3]. In general a very good agreement between both models and the real workpiece with a maximum deviation of the surface lines of 2% was observed. For further model refinement especially further optimization of the numerical settings is necessary, in order to eliminate their effect on the computed results.

Looking at the general development of the stress distribution during the process computed with LS-Dyna, it can be observed that the stresses increase with the progress of the process. Driving the roller tool to the rim of the circular blank, tensile stresses in radial and tangential directions can be observed in the forming zone. On the reverse motion the direction of the stresses changes. To investigate the development of the stresses more detailed time-series data of the principle stresses have been created. These data are to be used for an analysis with methods of nonlinear dynamics in the future to find coherences between the development of the stresses and the onset of wrinkling.

In Fig. 16.7 a time-series plot of the first principal stress of an element located at the outer diameter of the disk calculated with PamStamp is compared with the computation in LS-Dyna. As shown in Fig. 16.7, in the PamStamp model there is a steadily growing amount of first principal stress (tensile stress). This increase of compressive stresses has a dominant effect on the onset of wrinkling once a critical limit is exceeded. Significant is especially that the principal stress changes its values abruptly every 60 ms once it approaches the forming zone,

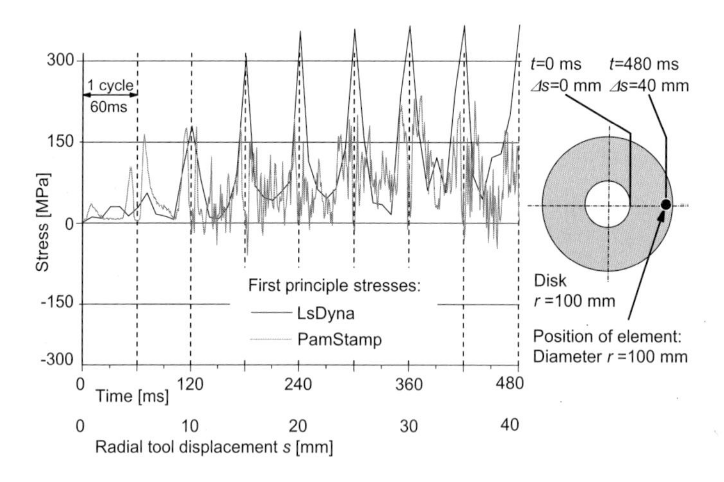

Figure 16.7: Comparison of the first principal stress time series for one element, calculated with LsDyna and with PamStamp.

according to the rotary speed. Within the simulation computed by LS-Dyna, a similar but more smooth run compared to PamStamp can be observed. The mean is shifted to more positive values. The peaks of the stress curve can be observed in this calculation too and are more pronounced. The differences are maybe caused by a more exact formulation of the elements.

A more detailed discussion of the time-series curves can be found in [3]. In further investigations it is important to determine the critical limit for which the onset of wrinkling is initiated. This critical limit will depend upon multiple factors, especially the design of the forming passes, the axial feed of the roller tool and the rotation velocity.

16.7 Conclusions

Within the presented research work the development of an approach to describe and understand the wrinkling effect in conventional sheet metal spinning has been presented. For this the wrinkling effect in conventional sheet metal spinning was investigated in detail. It has been shown and confirmed that in sheet metal spinning wrinkling is not only caused by static effects, namely plastic buckling, but additionally influenced and triggered by dynamic effects significantly. These dynamic effects themselves are dominated by the feed of the roller tool and the revolutions of the circular blank and amplify the sensitivity of the process towards wrinkling. This is a major difference in wrinkling mechanisms of deep-drawing and conventional spinning. Just before wrinkling begins, nonlinear dynamic aspects of wrinkling are predominant. An exact prediction of wrinkling is difficult, since the correct but still idealized mathematical description poses severe problems, such that solutions of the equations of motion are currently not available. In order to overcome this problem, methods of nonlinear dynamics were used. Time-series analysis of the process allows us to gain insights into the time evolution of the wrinkle formation, and hence supplies a much more complete analysis of the experiments than just the study of the final shapes of the blanks would do. The potentially most relevant result is the observation that wrinkles can disappear and hence wrinkle formation might be controllable by feedback loops.

To include the changes of the stresses and strains in dependence on the time into the nonlinear model two FE simulations, one in PamStamp, the other in LS-Dyna, have been developed. This knowledge is essential for the understanding of the wrinkling effect. Both models yield good results for the qualitative distribution of the stresses, whereas the more complex LS-Dyna model seems to be more accurate. By comparing both models the reliability of the simulation results has been increased significantly. Anyhow, the verification of the models has to be developed further. It is necessary to point out that up to now the simulation only shows one specific driven process and by this cannot be generalized for spinning processes driven in a different way. But by increasing the reliability of the verification, the stress time series will provide a good basis for further time-series analysis.

Bibliography

[1] E. v. Finckenstein and H. Dierig, *CNC-drücken*, Annals of the CIRP **39**(1), 267–270 (1990).

[2] M. Kleiner, B. Heller, Ch. Klimmek, R. Göbel and H. Kantz, *Investigations of dynamic instabilities in sheet metal spinning*, in: Proceedings of the 3rd International Symposium on Investigation of Nonlinear Dynamic Effects in Production Systems, Cottbus, 26–27 September 2000.

[3] M. Kleiner, R. Göbel, H. Kantz, Ch. Klimmek and W. Homberg, *Combined methods for the prediction of dynamic instabilities in sheet metal spinning*, Annals of the CIRP **51**(1), 209–214 (2002).

[4] E. Doege, T. El-Dsoki and D. Seibert, *Prediction of necking and wrinkling in sheet-metal forming*, Journal of Materials Processing Technology **50**, 197–206 (1995).

[5] X. Wang and J. Cao, *An analytical prediction of flange wrinkling in sheet metal forming*, Journal of Manufacturing Processes **2**(2), 100–107 (2000).

[6] P. Hill, *A general theory of uniqueness and stability in elastic/plastic solids*, Journal of Mechanics and Physics of Solids **6**, 236–249 (1958).

[7] S. Kobayashi, *instability in conventional Spinning of cones*, Journal of Engineering for Industry, Transactions of the ASME, February, 44–48 (1965).

[8] V. Barkaja and F. Ruzanov, *Bestimmung der kritischen Flanschbreite beim Drücken zylindrischer Näpfe*, Translation, Kuznecno-stampovocnoe Proizvodstvo **10**, 15-17 (1973).

[9] T. Satho and S. Yanagimoto, *On the spinnability of steel sheets*, in: Proceedings of the International Conference on Rotary Metal Working Processes (ROMP), Vol. II, Stratford upon Aven, Great Britain, 1982, pp. 277–286.

[10] M. Hayama, T. Murota and H. Kudo, *Deformation modes and wrinkling of flange on shear spinning*, Bulletin of the ISME **9**(34), 423–433 (1966).

[11] R. Köhne, Dissertation Thesis, University of Dortmund (1984). Published in: Fortschritts-Berichte der VDI Zeitschriften, Reihe 2: Betriebstechnik, No. 77, VDI, ISSN 0341-1656 (1984).

[12] G. Duvant and J.-L. Lions, *Inequalities in Mechanics and Physics*, Springer, Berlin, 1976.

[13] J. Jarušek and Ch. Eck, *Dynamic contact problem with friction in linear viscoelasticity*, Comptes Rendus de l'Academic des Sciences – Series I – Mathematics, Paris **322**, 497–502 (1996).

[14] W. Han and M. Sofonea, *Evolutionary variational inequalities arising in viscoelastic contact problems*, SIAM Journal on Numerical Analysis **38**(2), 556–579 (2000).

[15] V. Reitmann, *Observation stability of controlled evolutionary variational inequalities*, Preprint (2003), `http://www.math.uni-bremen.de/zetem/DFG-Schwerpunkt/`.

[16] V. Reitmann and h. Kantz, *Plastic wrinkling and flutter in sheet metal spinning*, in: EQUADIFF 10 Czechoslovak International Conference on Differential Equations and Their Applications, Prague, 27–31 August 2001, Abstracts, p. 77.

[17] J. Ronda and K.W. Colville, *Comparison of friction models for deep-drawing*, GAMM-Mitteilungen **1**, 39–59 (1995).

[18] J. Padovan, *On gyroscopic problems in elasticity*, International Journal of Engineering Science **16**, 1061–1073 (1978).

[19] A.L. Smirnov, *Vibrations of the rotating shells of revolution.* Prikladnaya mekhanika **5**, 176–186 (1981) (in Russian).

[20] N. Jones and C.S. Ahn, *Dynamic elastic and plastic buckling of complete spherical shells*, International Journal of Solids and Structures **10**, 1357 (1974).

[21] L.H.N. Lee, *Bifurcation and uniqueness in dynamics of elastic-plastic continua*, International Journal of Engineering Science **13**, 69 (1975).

[22] K. Dai, X.C. Gao, D.C. Kang and Z.R. Wang, *Numerical simulation of sheet spinning process*, in: Proceedings of the 6th International Conference in Advanced Technology of Plasticity (ICTF), Vol. 2, Springer, Erlangen, 1999, pp. 1001–1006.

[23] E. Quigley and J. Monaghan, *An Analysis of conventional spinning of light sheet metal*, in: International Conference on Sheet Metal, 1999, pp. 547–554.

[24] E. Quigley and J. Monaghan, *Using a finite element model to study plastic strains in metal spinning*, in: International Conference on Sheet Metal, 2001, pp. 255–262.

[25] Ch. Klimmek, R. Göbel, W. Homberg, H. Kantz and M. Kleiner, *Finite element analysis of sheet metal forming by spinning*, in: Proceedings of the 7th International Conference on Advanced Technology of Plasticity (ICTP), 28–31 October 2002, Yokohama, Japan, pp. 1411–1416.

[26] M. Kleiner, W. Homberg, R. Göbel and Ch. Klimmek, *Process optimsation in sheet metal spinning*, in: Production Engineering 96 Research and Development, Annals of the German Academic Society for Production Engineering WGP, Volume VIII, Issue 2, 2001, pp. 41–44.

[27] Ch. Klimmek, B. Heller, H. Kantz, R. Göbel and M. Kleiner, *Investigation of wrinkling as a dynamic instability in spinning processes*, 18th CAD-FEM Users' Meeting, International Congress on FEM Technology, Vol. 2, Friedrichshafen, 20–22 September 2000.

17 Nonlinear Vibrations During the Pass in a Steckel Mill Strip Coiling Process

H.J. Holl, G. Finstermann, K. Mayrhofer, and H. Irschik

In a Steckel mill facility the strip material is transported, rolled, coiled and manipulated during the production process of hot rolling steel strips. For the computation of the strip vibration during the Steckel rolling process the effects of strip bending and strip elongation must be taken into account and it is necessary to define a non-material control volume with some mass flow across the boundaries. A suitable vibration model of the mechanical parts, namely the rolls, the coiling drum and the rolled strip, is developed first. The mechanical model of the axially moving elastic strip requires Ritz approximations in the region between the boundaries and the rotating rolls. For systems with time-variable masses an extension of the Lagrange equation is necessary. As a characteristic application of the proposed computational strategy, we consider a configuration in which the moving strip leaves the Steckel mill stand and is guided by two pinch rolls and is coiled afterwards on a rotating coiling drum. The rotating coiling drum and the rolls are modeled as rotors with an elastic shaft. The oscillations of the coiling drum and the moving strip are computed using a mechanical model involving proper Ritz approximations. The Newmark method is used to compute the vibrations during the given process time. The simulations of different process conditions are performed for a chosen pass with a certain strip thickness. The results of the time integration of the nonlinear equations of motion visualize the transient behavior of the chosen kinematic and dynamic model variables.

17.1 Introduction

During a production process in a Steckel mill facility, the steel strips are transported, rolled and coiled on drums. The coil is taken out of the Steckel mill as soon as the coiled steel strip has reached the desired thickness. Subsequently it is transported to a storage unit. The process speed of the strip essentially depends on the parameters of the rolling process. The need of higher production output and high quality requires a more accurate mechanical model in order to be able to perform a reliable simulation of the oscillations occurring as a result of different and varying process conditions. A refined simulation model of the coiling process allows a better understanding of the whole process conditions and a better insight with respect to critical parameters, which result in a more efficient production process. When considering the resulting quality of the product, frequently the tension force of the band is an important measure.

Usually it is not possible to consider the motion of all the moving parts of the strip in the simulation under consideration, e.g. because the strip is conveyed from a region which

Nonlinear Dynamics of Production Systems. Edited by G. Radons and R. Neugebauer

ISBN 3-527-40430-9

is very long and not specified in detail. A part of the moving strip is not located within the control volume at the beginning of the computation; thus the mass is not conserved in the described problem. At least the coiling drum has to be included in a non-material control volume with a mass flow across its boundaries, so the present problem is characterized by the fact that the mass is not constant. The mass of the strip moves through the surface into the control volume and is coiled on the drum. Due to the coiled strip the winding radius of the drum and its mass change according to the accumulated strip length as time proceeds. Therefore the vibration theory for dynamic systems with variable mass has to be considered. A suitable dynamical formulation has to be used to study the oscillations of the elastic rotor and the strip during the coiling process, which allows the use of Ritz approximations in the form of the method of finite elements to describe the elastic deformations. For the derivation of the governing equations of motion the Lagrange equations have been used. The commonly used formulation of Lagrange equations is only valid for a constant mass. In the present paper a recently developed version of the Lagrange equations for non-material control volumes is applied to describe the coiling process; see Irschik and Holl [1]. For the considered Steckel mill the dynamic behavior of the rotating drum, the rolls, the elastic bearings and the coiled strip are described by a mechanical model with an elastically moving and a rigid coiled strip. The resulting vibrations due to varying mass and process conditions can disturb the whole production process. The corresponding dynamic effects prevent that the tension force in the strip from being kept constant. The coiling process of a strip has been modeled and analyzed by Cveticanin [2, 3] using different models with some simplifying pre-conditions. In the present paper the mechanical model of the coiling drum and the rotating rolls consists of rotating finite beam elements. The effect of the time-varying radius of the coiling drum is considered. It is assumed that the radius changes linearly with the angle of rotation of the rotor. Furthermore, it is assumed that there are no gaps within the driving unit.

The computation of the solution is performed by numerical time integration of the nonlinear equations of motion and gives the oscillations and the resulting (tension) forces for different process conditions. Typically such calculations need a long computational time and a very high number of time steps, as long strips are coiled and the parameters of the system vary in a wide range. The unconditionally stable version of the Newmark method, see [4] and [5], has been extended to treat the present case. The computational effort of the time-integration method for the high numbers of time steps used and the corresponding long computational time is documented. The influence of certain variations of the speed of the moving strip on the oscillations of the rotating shafts and the tension force in the moving strip is computed and analyzed for the coiling process. The results are presented for one typical pass.

17.2 Mechanical Model of the Coiling Process

The mechanical model of the Steckel mill with the rotating and moving parts is used to derive the equations of motion for the chosen and characteristic degrees of freedom. Rotating elastic finite beam elements and a suitable approximation for the motion of the steel strip are used for the modeling process. A time-varying mass is introduced due to the coiled strip. The system boundaries are defined by a space-fixed control volume, which is shown in Fig. 17.1. The strip

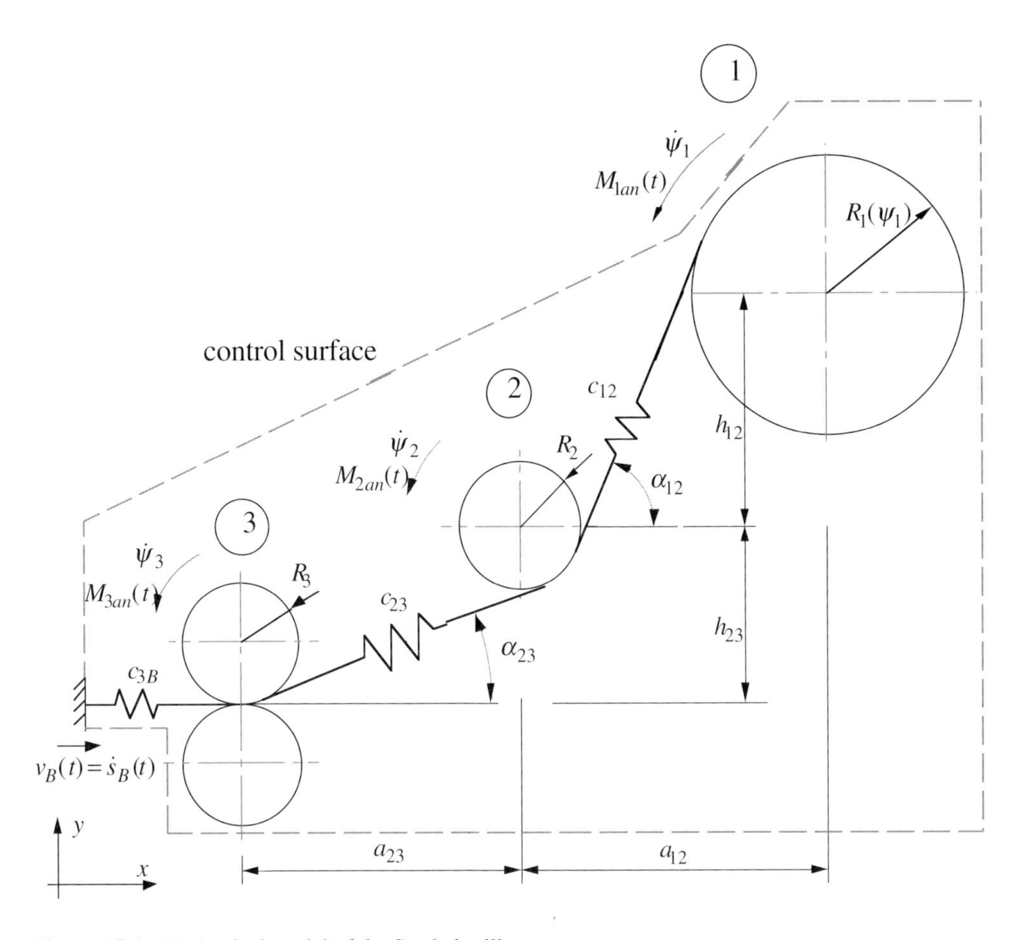

Figure 17.1: Mechanical model of the Steckel mill.

is moving into the control volume, where at the boundaries there is either a given speed or a given tension force of the strip.

To describe the transversal and longitudinal displacement behaviour of the elastic strip between the boundary at the entrance and the coiling drum Ritz approximations are applied, as is usual for the transverse oscillations of axially moving strips. Based on the chosen degrees of freedom (DOFs) and the resulting equations of motion it is possible to compute the transversal and longitudinal vibrations of the steel strip. The complete mechanical model of the Steckel mill of Fig. 17.1 is divided into submodels in order to derive the equations of motion and the forces which are present in the system. In the presented mechanical model it is assumed that the entrance speed $v_B(t)$ of the moving strip is given as a function of time. The three torques $M_{ian}(t)$ in Fig. 17.1 are defined as functions of the time as well.

The index 1 in Fig. 17.1 refers to the coiling drum, 2 refers to the deflector roll and 3 refers to the pinch rolls. The stiffness of the rolls and the bearings, the tension force in the strip and the acting forces and moments are defined in the corresponding submodels. In order to get a suitable mechanical model it is assumed that all motions of the parts under consideration are symmetric with respect to the drawing plane of Fig. 17.1. The rotating elastic shaft of the drum 1 and the deflector roll are elastically supported and are free to move in the plane, while the pinch roll 3 can only move vertically.

The resulting mechanical model of the Steckel mill considers the following effects: due to the used formulation of the Lagrange equations a consistent mass matrix results for the equations of motion. The distributed mass of the strip is considered using a Ritz approximation and a time-variable mass and stiffness of the coiling drum is taken into account. Axial distribution of the mass and transversal vibrations of the rolls are considered and result by application of Ritz approximations. For a non-steady-state speed of the strip the inertia of the steel strip is taken into account. The temperature-dependent stiffness of the strip and the resulting nonlinear transversal oscillations are involved in the mechanical system. It is assumed that the average temperature of a cross section of the strip is taken for the computations of the oscillations. For the transversal oscillations Ritz approximations are used:

$$w^*(\xi,t) = \sum_{i=1}^{m} \varphi_i(\xi) q_i(t). \tag{17.1}$$

ξ is the dimensionless longitudinal coordinate in the considered region between two consecutive rolls and the actual length of the strip in these regions is known. In order to get the absolute velocity of the strip $\dot{w}_{\mathrm{st}}(\xi,t)$ the transversal motions at the left-hand end $g_{\mathrm{l}}(t)$ and at the right-hand end $g_{\mathrm{r}}(t)$ of the region between two adjacent rolls, with the length of the strip l, are used:

$$\dot{w}_{\mathrm{st}} = \dot{w}^*(\xi,t) + \frac{1}{l}\frac{dw^*(\xi,t)}{d\xi}\dot{u}^*(\xi,t) + \xi g_{\mathrm{r}}(t) + (1-\xi) g_{\mathrm{l}}(t). \tag{17.2}$$

For the longitudinal motion $u^*(\xi,t)$ and the corresponding elastic deformation the Ritz approximation

$$u^+(\xi,t) = u_{\mathrm{l}}(t) + (1-\xi) u_{\mathrm{r}}(t) \tag{17.3}$$

is used, where $u_{\mathrm{l}}(r)$ is the longitudinal motion at the left-hand boundary and $u_{\mathrm{r}}(t)$ at the right-hand boundary of the region under consideration. Three regions are distinguished within the control volume. The first is between the coiling drum and the deflector roll, the second region is between the deflector roll and the pinch roll and the third region is between the pinch roll and the Steckel mill stand. In this third region only longitudinal oscillations are considered in the mechanical model. The potential and deformation energy of the strip is

$$U = \frac{1}{2}\int_0^l \int_A \sigma_{xx}\varepsilon_{xx} dA dx, \tag{17.4}$$

with

$$\varepsilon_{xx} = \frac{du}{dX} + \frac{1}{2}\left(\frac{dw}{dX}\right)^2 - z\frac{d^2w}{dX^2}. \tag{17.5}$$

The mechanical model of the coiling drum is a rotating roll with time-variable stiffness with respect to the body-fixed and rotating coordinate system. The actual stiffness is determined based on two effects. The first effect is that the coiling drum itself has different stiffness parameters in orthogonal body-fixed directions so that a coordinate transformation is necessary in order to get the equations of motion in the reference coordinate system, see Gasch et al. [6]. The second effect is that the orthotropic stiffness parameters additionally change with time due to the coiled strip length and have to be added to the stiffness of the bearings. It is assumed that the coiled strip is fixed at the coiling drum and does not have any relative movement when it is coiled so that we get a rigid coiled strip.

The coiling drum itself has three degrees of freedom and a variable radius, for which different evolutions can be assumed. In the present paper it is assumed that the radius varies linearly with the rotation angle of the coiling drum. Additionally, a thermal deflection of the coiling drum can be considered. A horizontal motion of the pinch roll is not allowed. As pure rolling conditions are assumed at all rolls, the coiled strip length can be computed using the radius of the roll and the rotation angle. The coiled strip length at the coiling drum has to be integrated numerically as the radius varies with time. The total length of the coiled strip is computed by the sum of the partial strip lengths in each time step.

The mechanical model of the deflector roll is shown in Fig. 17.2 in a plane perpendicular to the axis of the deflector roll. In this model actuators are considered so that control displacements

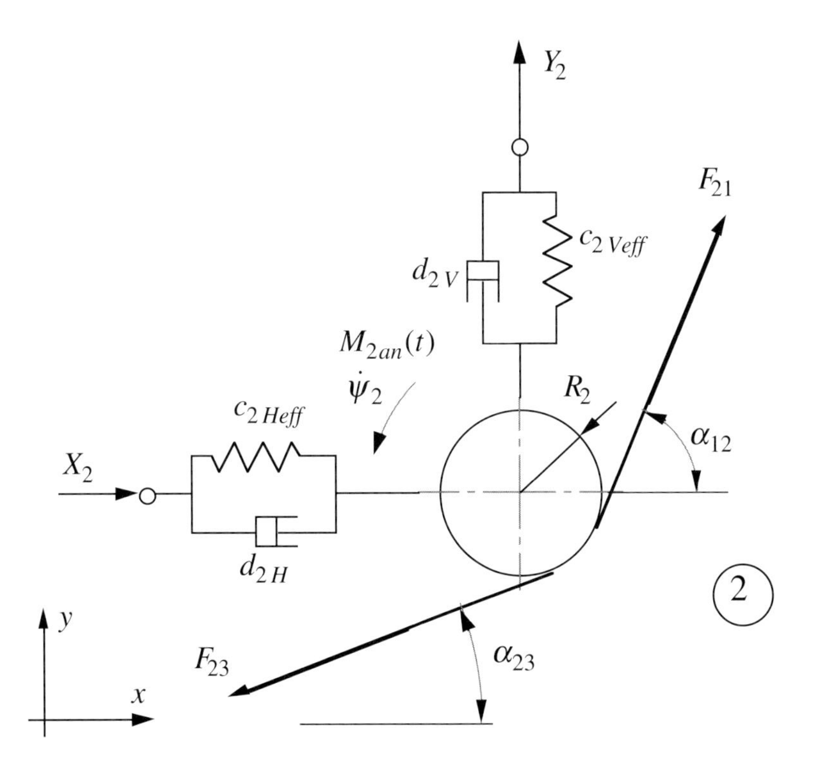

Figure 17.2: Mechanical model of the deflector roll.

X_2 and Y_2 can be defined as a function of time. The resulting effective stiffness in the horizontal and vertical direction $c_{2H\text{eff}}$ and $c_{2V\text{eff}}$ is computed based on the series connection of the stiffness of the roll and the stiffness of the bearings. In this model a thermal deflection of the rotating shaft of the deflector roll can be considered too. The tension forces of the strip on both sides of the deflector roll are defined in Fig. 17.2. In case of both the coiling drum and the deflector roll, the center of mass need not coincide with the shafts. A certain amount of eccentricity as well as a value for the thermal deflection can be defined for both.

The mechanical model of the pinch rolls is similar to that of the deflector roll and takes the stiffness of the rolls and the bearings into account. The horizontal deflection of the roll is locked and it is considered that one roll can be lifted vertically during operation. No control displacement is included in the model of the pinch roll.

The kinetic energy and the generalized forces are formulated for the mechanical system of Fig. 17.1. Having derived the kinetic energy and the generalized forces for the total system, they can be used in the Lagrange equations, which account for a variable mass. In the present case the Lagrange equations have to be formulated for a control volume for which the mass is not constant. A formulation of the Lagrange equations for the case of a non-material control volume has been developed by Irschik and Holl [1]. In [1] the equations of Lagrange have been derived for a non-material control volume by using a suitable extension of Reynold's transport theorem, where the surface is moving with a speed different from the material surface. The Lagrange equations then read:

$$\frac{d}{dt}\left(\frac{\partial_w T}{\partial \dot{q}_k}\right) - \frac{\partial_w T}{\partial q_k} + \int_S \frac{\partial_w T}{\partial \dot{q}_k}(\vec{v} - \vec{w})\vec{n}ds - \int_S T'\left(\frac{\delta \vec{v}}{\delta \dot{q}_k} - \frac{\delta \vec{w}}{\delta \dot{q}_k}\right)\vec{n}ds = Q_k, \quad (17.6)$$

with

$$T = \int_{v_w} T' dv_w = \int_{V_w} T' J dV_w = \int_v T' dv, \quad (17.7)$$

$$T' = T'' J^{-1} \tilde{\varrho} \quad (17.8)$$

and

$$T'' = \frac{1}{2}(\vec{v} \cdot \vec{v}). \quad (17.9)$$

The application of these modified Lagrange equations needs the kinetic energy T, the generalized forces Q_k, the k-th degree of freedom q_k, the normal vector $\vec{n}$ at the surface S of the non-material volume, the speed of the particles $\vec{v}$ at the surface of the boundary, the speed of the boundary of the control volume $\vec{w}$, the mass density in the reference configuration $\tilde{\varrho}$ and the Jacobian determinant J. The symbol δ means that the place is kept fixed during the differentiation. The operator $\partial_w T/\partial \dot{q}_k$, say, means that the partial derivative is taken with respect to the non-material control volume, i.e. studying the dependence of T on $\dot{q}_k$ for the non-material volume.

This form has to be applied in the present case. At the entrance the mass of the strip is moving into the control volume so that the flow of the partial derivatives through the control volume with respect to the degrees of freedom is considered in the Lagrange equations. Since we consider the entrance velocity to be prescribed, the surface integral in Eq. (17.6) does vanish, such that we are allowed to use the original form of the Lagrange equations formally for our non-material control volume. In [1] the non-material version of the Lagrange equations has been applied exemplarily to the equation of motion of a rocket and to the problem of a falling folded string. In the present contribution this equation is applied to the coiling of strips on a drum on elastic bearings, which results in

$$\mathbf{M}\ddot{\mathbf{q}} + (\mathbf{D} + \mathbf{G})\dot{\mathbf{q}} + (\mathbf{K} + \mathbf{N})\mathbf{q} + \mathbf{R}_{\mathrm{N}} = \mathbf{F}, \tag{17.10}$$

where $\mathbf{M}$ is the constant-mass matrix, $\mathbf{D}$ is the damping matrix, $\mathbf{G}$ is the gyroscopic matrix, $\mathbf{K}$ is the stiffness matrix, $\mathbf{N}$ is the matrix of the circulatory forces, $\mathbf{R}_{\mathrm{N}}$ is the vector of the nonlinear forces, which includes also the effects of the time-variable mass, is the vector of the excitation forces, including the excitation due to the eccentricity and the thermal deflection. As can be derived from the above equations, we also include quadratic and higher-order terms of the degrees of freedom of the mechanical model used. The terms which are linear with the degrees of freedom are inserted in the matrix $\mathbf{K}$. $\mathbf{q}$ is the vector of the unknown time-dependent degrees of freedom. The elements of $\mathbf{q}$ represent the generalized displacements of the mechanical parts and the transversal as well as the longitudinal motion of the moving strip as defined in Eq. (17.1). These resulting nonlinear time-varying equations of motion are solved numerically by a time-integration procedure, which has been discussed by Holl [4]. The general formulation of the resulting incremental scheme is an extension of the unconditionally stable version of the Newmark method, see [5]. In [7] and [8] these extensions are described in some detail, where numerical studies concerning the vibrations of a system with a defined (time-dependent) variable mass have been presented. The incremental formulation of the equations of motion

$$\mathbf{M}\Delta\ddot{\mathbf{q}} + (\mathbf{D} + \mathbf{G})\Delta\dot{\mathbf{q}} + (\mathbf{K} + \mathbf{N})\Delta\mathbf{q} + \Delta\mathbf{R}_{\mathrm{N}} = \Delta\mathbf{F} \tag{17.11}$$

is solved iteratively for each time step using the time evolution of the kinematic relations of the Newmark method, see [4, 5]. $\Delta\mathbf{R}_{\mathrm{N}}$ is the increment of the nonlinear force vector with respect to the previous iteration and $\Delta\mathbf{q}$, $\Delta\dot{\mathbf{q}}$ and $\Delta\ddot{\mathbf{q}}$ are the increments of the displacement, the speed and the acceleration vector with respect to the values of the previous iteration.

17.3 Results of the Simulation

For the simulation of the above-derived mechanical model of a Steckel mill reference parameters of an existing plant have been used. For the mechanical model of Fig. 17.1 a torque of $M_1 = 10\,\mathrm{kNm}$ is applied at the coiling drum after one second, corresponding to a ramp function, and this applied torque of the driving motor is kept constant afterwards. This assumption approximately reflects the result of the measurements of the current speed and the actual rotating speed at the driving motor. The simulated length of the strip is $370\,\mathrm{m}$, the width of the strip is $900\,\mathrm{mm}$ and the thickness of the steel strip is $4.22\,\mathrm{mm}$. (The case of a strip with 2.64-mm thickness was presented in a previous investigation [9].) For a simulation time of $65\,\mathrm{s}$ the results are shown in the following figures for some selected DOFs of the system and the tension force in the strip. The total coiled strip length, the speed profile and the corresponding acceleration profile are shown in Fig. 17.3. These speed profiles are the results

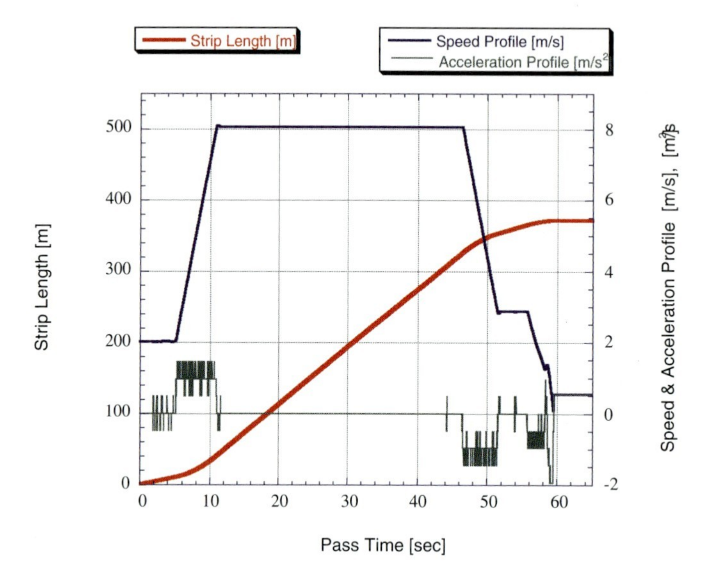

Figure 17.3: Speed profile of the strip in the Steckel mill.

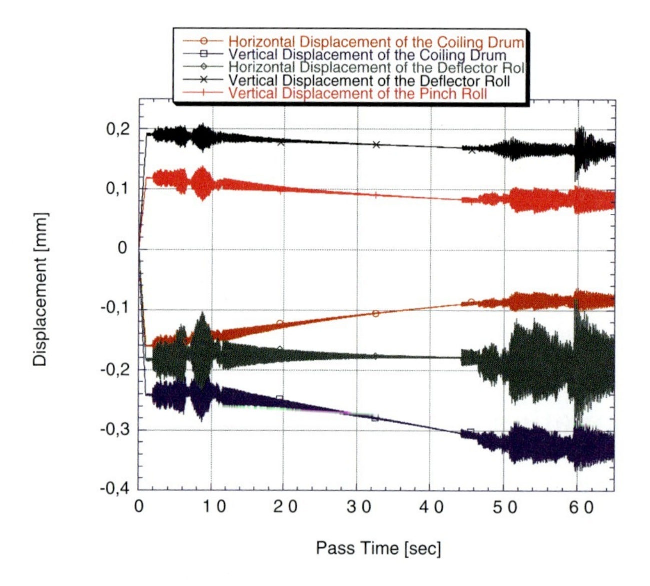

Figure 17.4: Positions of the centers of the rolls in the model of the Steckel mill.

of measurements during a production test and are used in the presented computation as input parameters.

In Fig. 17.4 the horizontal and vertical displacements of the centers of the shafts of the coiling drum, the deflector roll and the pinch roll are shown vs. pass time. It can be seen that

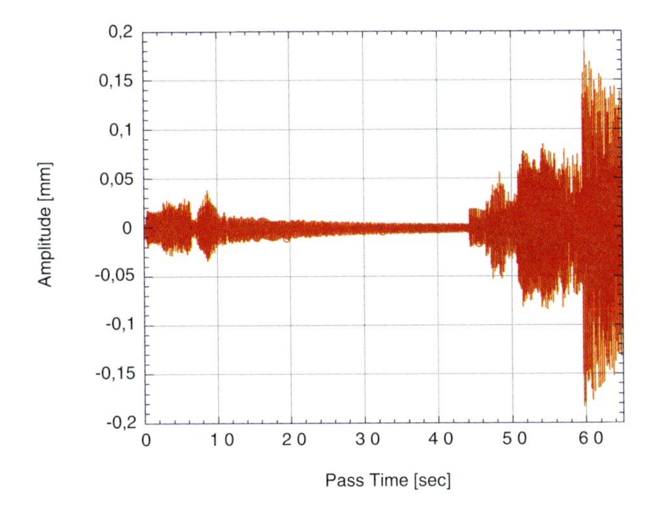

Figure 17.5: Transversal oscillations of the strip between the coiling drum and the deflector roll.

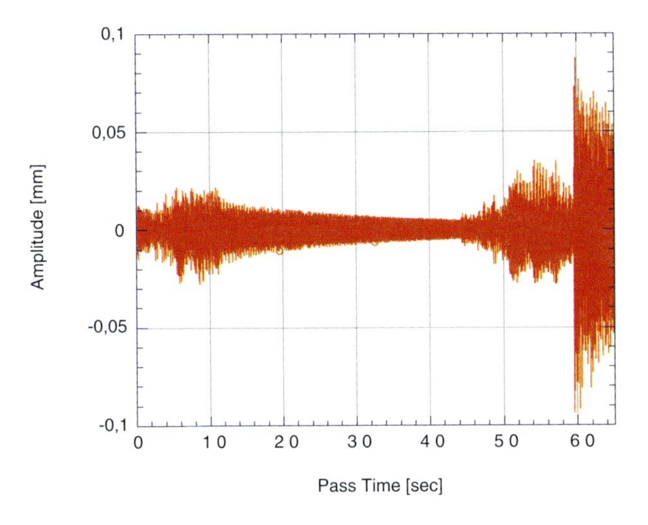

Figure 17.6: Transversal oscillations of the strip beween the deflector roll and the pinch roll.

the overall displacements mainly result from the torque, and the mean value of the vertical displacement results from the increasing mass of the drum. The amplitude function of the transversal vibrations of the strip is shown in Fig. 17.5 for the first bending mode in the area between the deflector roll and the coiling drum. It can be seen that a relatively small amount of oscillations results due to the high tension force in the strip. The vibration amplitude suddenly becomes high at the end of the process, when the speed of the coiling drum is slow and the end of the strip will be reached at the Steckel mill stand. From Fig. 17.3 it can be seen that after a period of 60 s the strip has only a very small speed, so that this time period is shown

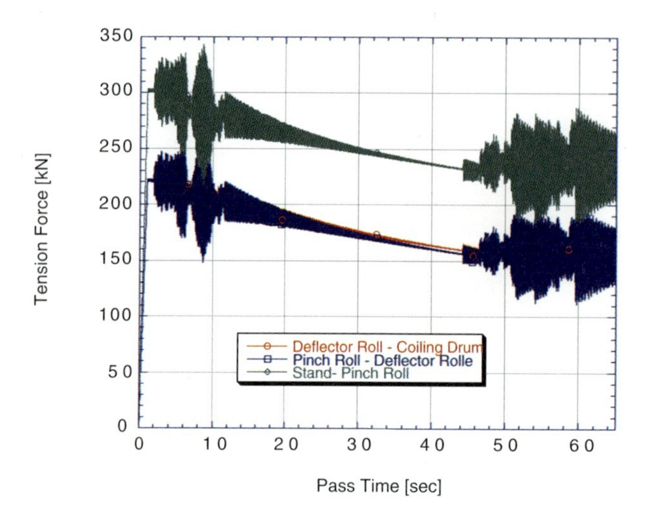

Figure 17.7: Tension forces in the strip.

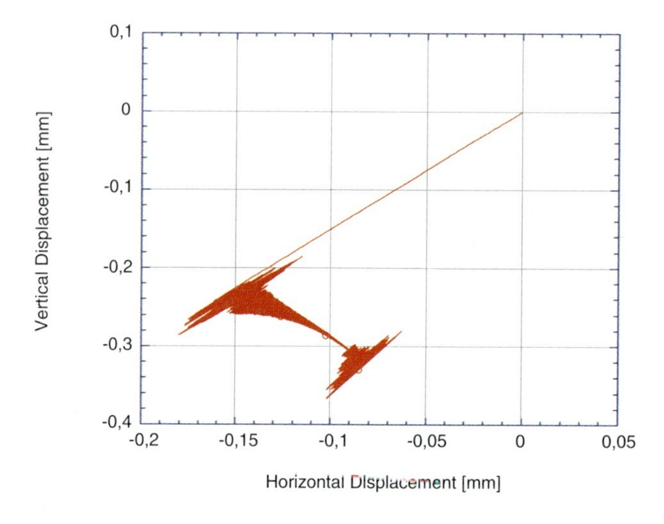

Figure 17.8: Position of the center of the coiling drum.

primarily for information but is not relevant for the judgment of the process and the correlated parameters.

In Fig. 17.6 the transversal oscillation amplitude in the region between the deflector roll and the pinch roll is shown. The amplitude also has a maximum when the strip moves with a minimal speed at the end of the process. Nevertheless, the total amplitude of the transversal oscillations in both regions is very small.

It can be seen in Fig. 17.7 that the tension force is decreasing during the process as the driving moment of the motor is kept constant and the radius of the drum is increasing linearly with the coiled strip length. The variations of the tension force in the strip result from the used measured data of the strip motion after the Steckel mill stand, which is the input for the simulation of the present pass. Additionally, the tension force results from the applied torque, which is pre-defined by the mentioned ramp function of one-second duration with a constant value of $M_1 = 10\,\mathrm{kNm}$ for the rest of the simulation.

For the assumed eccentricity of $1\,\mathrm{mm}$ of the coiling drum the result for the position of the center of the shaft of the coiling drum is shown in Fig. 17.8. A comparison with the case without eccentricity shows that there is only a negligible difference, as the speed of rotation is small. The vertical motion is determined by the mass of the coiled strip and the oscillating motion in the direction of the strip is due to the varying tension force in the strip, see Fig. 17.7. 10^7 time steps have been used for the computations for the presented cases, which led to converging results.

17.4 Conclusion

The computations for a time-variable system result in a high computational effort that could be successfully overcome by the present method. Based on the measurements the speed profile is pre-defined for the calculations. The results show a very small amount of transversal oscillations of the moving strip. The tension force in the strip strongly depends on the model of the growth of the radius of the coiling drum, where a linear dependence was applied.

Bibliography

[1] H. Irschik and H.J. Holl, *The equations of Lagrange written for a non-material volume*, Acta Mechanica **153** (3-4) (2002), 231–248.

[2] L. Cveticanin, *Dynamics of Machines with Variable Mass*, Gordon and Breach, The Netherlands, 1998.

[3] L. Cveticanin, *The influence of the reactive force on the motion of the rotor on which the band is winding up*, Journal of Sound and Vibration **167** (1993), 382–384.

[4] H.J. Holl, *An efficient semi-analytic time integration method with application to nonlinear rotordynamic systems*, Computational Mechanics **25** (2000), 362–375.

[5] H.J. Newmark, *A method of computation for structural dynamic*, Journal of Engineering Mechanics Division **85**, ASCE EM **3** (2094) (1959), 67–94.

[6] R. Gasch, R. Nordmann, and H. Pfützner, *Rotordynamik*, Springer, Berlin, 2002.

[7] H.J. Holl, *A time-integration algorithm for time-varying systems with non-classical damping based on modal methods*, Proceedings of the 15th International Modal Analysis Conference, 3–6 February 1997, edited by A.L. Wicks, Society for Experimental Mechanics, Orlando, 1997, 1558–1564.

[8] H.J. Holl, A.K. Belyaev, and H. Irschik, *A numerical algorithm for nonlinear dynamic problems based on BEM*, Journal of Engineering Analysis with Boundary Elements **23** (1999), 503–513.

[9] H.J. Holl, G. Finstermann, K. Mayrhofer, and H. Irschik, *Vibration simulation of the Steckel mill strip coiling process*, Proceedings of the Fifth World Congress on Computational Mechanics (WCCW V), edited by H.A. Mang, F.G. Rammerstorfer, and J. Eberhardsteiner, Vienna University of Technology, Austria, http://wccm.tuwien.ac.at, 2002.

Part III

Dynamics of Robots and Machines

Machine dynamics is a classical branch of applied classical mechanics and therefore nonlinear dynamic effects are ubiquitous and well known in this field. In this part we consider only aspects which are of high relevance for manufacturing processes. The sliding between surfaces and the possibility of loosing contact followed by impact processes are important scenarios. They play a large role e.g. in cutting processes as treated in the previous part II, where they can cause unwanted operating conditions such as machine chatter. But also under normal conditions friction and impact forces are essential for the functioning of machines and robots. These forces are special in so far as they provide two fundamental non-smooth nonlinearities in mechanical systems[1].

The first contribution of F. Peterka treats the simplest model system with impact, the one-degree-of-freedom-impact oscillator and its extension to the double-impact oscillator. Due to the impacts such driven harmonic oscillators turn into systems with a complex dynamic behavior as can be seen from the observed complicated bifurcation scenarios including transitions to chaos, and the occurence of hysteresis. Its practical relevance is demonstrated with the construction of a piercing machine based on the principles of the double-impact oscillator. The paper by G. Litak and M.I. Friswell provides another fundamental example where contact dynamics is an important phenomenon: the dynamic behavior of gear boxes is characterized by a periodically varying meshing stiffness, which can lead to repeated contact loss between the teeth of gear wheels. The resulting regular or chaotic vibrations provide a serious problem and means for their suppression are discussed.

Despite their ubiquitous presence friction processes are not understood in detail. This causes problems e.g. in robotics, where an accurate positioning of tools is required. Since there exists no universal friction law, which is applicable in every situation, the experimental identification of the friction law governing the application at hand becomes of central importance. The contribution of F. Al-Bender et al. reviews and discusses various approaches for measuring and identifying pre-sliding friction dynamics. The strongly nonlinear nature of friction becomes apparent in the observed slip-stick motion and in a complicated hysteretic behavior. Despite this an accurate modeling is possible as is shown in this paper.

The last contribution to this part by H. Nijmeijer and A. Rodriguez-Angeles deals with a dynamical mechanical problem of a quite different nature, which is of incrasing importance in modern industrial applications: the coordination of processes in general and especially the coordination of motions of mechnical robotic systems. The paper shows theoretically and experimentally how the control problem of position coordination for multi-robot systems can be solved without detailed knowledge of the physical system parameters using only information from position measurements.

[1] An elementary introduction and some complementary views on these topics may be found in the chapter "Nonlinear Dynamics with Impacts and Friction" of S.W. Shaw and B.F. Feeny, p.241-264, in the book "Dynamics and Chaos in Manufacturing Processes" (edited by F.C. Moon, Wiley, New York, 1998).

In memoriam František Peterka (1939–2003)

Dr. František Peterka died unexpectedly as result of an accident on August 30, 2003, only a few days after we received his final contribution to this book. We feel it is only appropriate to include in this book an obituary, so as to draw attention to some aspects of his scientific life.

František Peterka was born in Týn nad Vltavou (Southern Bohemia) on November 26, 1939. In 1957 he joined the Faculty of Mechanical Engineering at the Czech Technical University, Prague, and, in 1962, began his scientific activity at the Institute of Thermomechanics of the Czechoslovak Academy of Sciences. His research was focused on the dynamics of mechanical systems, particularly on problems of the dynamics of strongly non-linear systems with impacts and dry friction. In 1968 he submitted his PhD thesis "Theory of a dynamical impact damper with two degrees of freedom". He is the author of more then 30 reports, 180 papers and lectures and five books. He has solved analytically and by using analogue and hybrid computer simulations problems of periodic and chaotic impact motions. This allowed him to optimize the parameters of impact systems with viscous and dry friction in models of mechanical hammers, compacting and crushing equipments, as well as impact dampers. He is one of the scientists who discovered the existence and principle of chaotic oscillations in systems with inner impacts. In the last few years he had been working as the senior scientist dealing with problems of chaotic dynamics in mechanical vibro-impact systems.

Dr. Peterka was the Deputy Head of the Department of Systems Dynamics and the Head of the Laboratory Dynamics of Non-Linear Systems at the Institute of Thermomechanics, Academy of Sciences of the Czech Republic. He was a member of the Czech Society for Mechanics, the Euromech Society for Mechanics (EUROMECH), and of the International Federation for the Theory of Machines and Mechanisms (IFToMM), where he held the post of Chairman of the Czech National Committee and Secretary of the IFToMM Technical Committee "Non-linear Oscillations".

František Peterka was a calm man, very religious, and very honest. He was liked by everybody, within the Institute and also among the scientific community. He played the violin and was active in a choir for many years, but the majority of his free time he devoted to his family. He was happily married and has two sons, one daughter and four grandchildren. He will be sadly missed by those who knew him and his friendly nature.

Photography: František Peterka attending the 4th International Symposium on Investigations of Non-Linear Dynamic Effects in Production Systems in Chemnitz, April 2003.

18 New Type of Forming Machine

F. Peterka

The double-impact oscillator represents two symmetrically arranged single-impact oscillators, which correspond to commonly used spring hammers. It is the model of a forming machine, which does not spread the impact impulses into its neighborhood and has several other advantages. The anti-phase motion of this system has the identical dynamics as the single-impact oscillator and it is useful for practical application. The practically undesirable in-phase motion and the influence of asymmetries of the system parameters are studied using numerical simulations. Theoretical and simulation results are verified experimentally and they were applied for the construction of a piercing machine, which produces roller-chain details.

18.1 Introduction

The one-degree-of-freedom impact oscillator is one of the simplest strongly nonlinear mechanical systems. It consists of an elastically suspended and periodically excited mass (cf. one half of Fig. 18.1), which can impact against a rigid stop. Its dynamics has been thoroughly investigated theoretically, experimentally and using simulation methods (see references in [21] and [6]). The influence of the following parameters on the maximum velocity before impact during fundamental periodic impact motion was studied [6]: coefficient of restitution, viscous and dry friction damping, static clearance, amplitude and frequency of the excitation force. This fundamental motion is practically most important and is characterized by the repetition of one impact in every period of the excitation force. There exist also other periodic and chaotic impact motions. Each motion has a region of existence and stability in the space of system parameters. Regions can mutually penetrate and create hysteresis subregions. Several regimes of the system motion exist there. Motion initial conditions (basins of attraction) or other conditions decide which motion will appear (see the set of papers in [21, reference 16]).

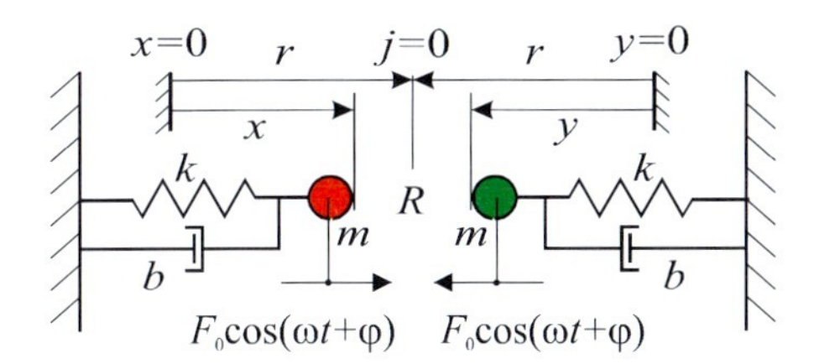

Figure 18.1: Scheme of the double-impact oscillator.

Nonlinear Dynamics of Production Systems. Edited by G. Radons and R. Neugebauer

ISBN 3-527-40430-9

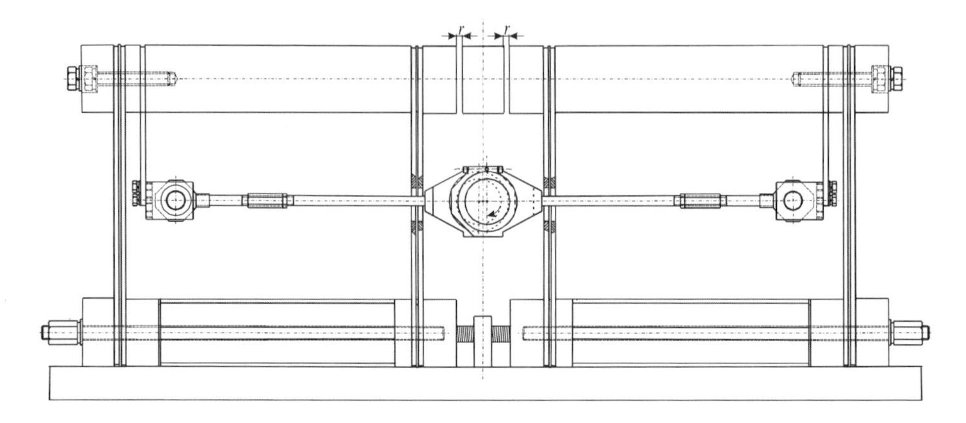

Figure 18.2: Structural drawing of the piercing machine.

Results obtained for the single oscillator are applicable for two symmetrically arranged impact oscillators – the double-impact oscillator (Fig. 18.1), which can be considered as the model of a new forming machine. Advantages of such a system, compared to the single-impact oscillator – the spring hammer, will be explained on the functional model of the piercing machine (Fig. 18.2) for the production of roller-chain details:

1. the construction of the machine is simple and allows us to increase the productivity, because the functional model has the piercing frequency 30 Hz (five times higher in comparison with current machines) and processing operation is completed during one impact of hammer heads. The quality of details is also higher, due to high speed of piercing,

2. the construction simplicity follows from the fact that hammers have the function of flywheels. The energy for the piercing at impact is accumulated during one period of excitation forces,

3. internal forces and moments are balanced due to the system symmetry and therefore the piercing machine does not influence the neighborhood by impact impulses,

4. impact impulses are transferred neither into the frame nor to the motor, and construction calculations correspond only to a non-impact load, which implies the decreasing of the machine weight,

5. the construction has also environmental aspects, because besides the elimination of impact pulses the machine can be covered by a shield as a noise eliminator,

6. the hammers' clamping using leaf springs does not require a special guide, the damping of impact-less motion is minimum and the efficiency of energy transfer from the motor into the forming is maximum.

High efficiency of the energy transfer is caused also by the identity of weight and form of hammers. It means that wave propagation and their reflection are also identical. The diffusion of energy in asymmetric impacting bodies is higher.

The second degree of freedom introduces asymmetry into the system motion, which is undesirable for the application. Differences in initial conditions, driving forces or natural frequencies of subsystems cause other asymmetries. Constructional or assembly imperfections can also introduce asymmetries. It has been shown using numerical simulations and experiments that small asymmetries do not affect the optimal operation of this system.

Rigid and soft impacts are assumed. The Newton model of impact with coefficient of restitution is used for the motion with rigid impacts; a special model of impact interactions is chosen for the motion with soft impacts, when the impact duration cannot be neglected.

The dynamics of a similar two-mass impact oscillator and its application in the forming is also deeply studied in [3].

18.2 Theoretical Analysis of Motion with Rigid Impacts

The theoretical analysis will arise here from the differential equations of the system impact-less motion and the relation between before- and after-impact velocities according to the coefficient of restitution. The solution of the simplest series of periodic impact motions and their stability are introduced in [21, 17, 18] for the symmetric system without viscous damping.

18.2.1 Symmetric Case

The impact-less motion is described by the system of transformed differential equations

$$X'' + 2\,\beta\,X' + X = \cos(\eta\tau + \varphi), \tag{18.1}$$

$$Y'' + 2\,\beta\,Y' + Y = \cos(\eta\tau + \varphi), \tag{18.2}$$

where $X = xk/F_0$, $Y = yk/F_0$ and $\tau = \Omega t \quad (\Omega = \sqrt{k/m})$ are transformations of displacements and the time, respectively, $\eta = \omega/\Omega$ and $\beta = b/\left(2\sqrt{km}\right)$ are the dimensionless frequency and viscous damping, respectively; $X' = dX/d\tau$, $X'' = d^2X/d\tau^2$.

Impacts occur when the following condition is met:

$$X + Y \geq 2\rho\,, \tag{18.3}$$

where $\rho = rk/F_0$ is the dimensionless static clearance.

It is assumed that impacts are described by Newton's theory of impacts. Let X'_-, Y'_- and X'_+, Y'_+ denote the before-impact and the after-impact velocities, respectively. Then:

$$\begin{aligned} X'_+ &= [(1-R)X'_- - (1+R)Y'_-]/2, \\ Y'_+ &= [(1-R)Y'_- - (1+R)X'_-]/2, \end{aligned} \tag{18.4}$$

where $R = -(X'_+ + Y'_+)/(X'_- + Y'_-)$ is the coefficient of restitution ($0 \leq R \leq 1$).

For plastic impacts ($R = 0$) the masses have equal after-impact velocities X'_p:

$$X'_+ = -Y'_+ = (X'_- - Y'_-)/2 = X'_\mathrm{p} \tag{18.5}$$

and there are two possibilities of the after-impact motion, according to the polarity of the after-impact relative acceleration $X'' + Y''$:

(a) When the condition

$$X'' + Y'' \leq 0 \tag{18.6}$$

is met, then no forces press the masses together and the after-impact motion is described by Eqs. (18.1), (18.2) as for the motion with elastic impacts, when $X'_+ + Y'_+ \leq 0$.

(b) When the condition (18.6) is not met, then the masses move together according to the equation

$$J'' + 2\beta J' + J = 0, \tag{18.7}$$

with initial conditions

$$J(0) = X(\tau_\mathrm{i}) - \rho \text{ and } J'(0) = X'(\tau_\mathrm{i}) = X'_\mathrm{p}\,, \tag{18.8}$$

where τ_i is the instant of the plastic impact and J is the displacement of the masses from the center ($j = 0$ in Fig. 18.1).

The time interval of such motion is named the after-impact dead zone of the masses relative motion. It ends at the instant τ_e, when the press force F_p disappears. This force is proportional to the fictive positive acceleration $X'' + Y''$, which is evaluated during numerical simulation of the motion (Fig. 18.3a).

The motion is described then by Eqs. (18.1), (18.2) with initial conditions

$$\begin{aligned} &X(0) = J(\tau_\mathrm{e}) + \rho\,,\; Y(0) = -J(\tau_\mathrm{e}) + \rho\,, \\ &X'(0) = -Y'(0) = J'(\tau_\mathrm{e}). \end{aligned} \tag{18.9}$$

The motion of the system with asymmetric initial conditions, plastic impacts and after-impact dead zones is introduced in Fig. 18.3. Figure 18.3a shows that the polarity of the force F_p (condition (18.6)) decides about the after-impact dead zone and its duration. The first plastic impact does not meet the condition for the joined motion of masses after impact. Dead zones appear after the remaining three impacts. The motion of the general impact–dry-friction pair of bodies is described in more detail in [8].

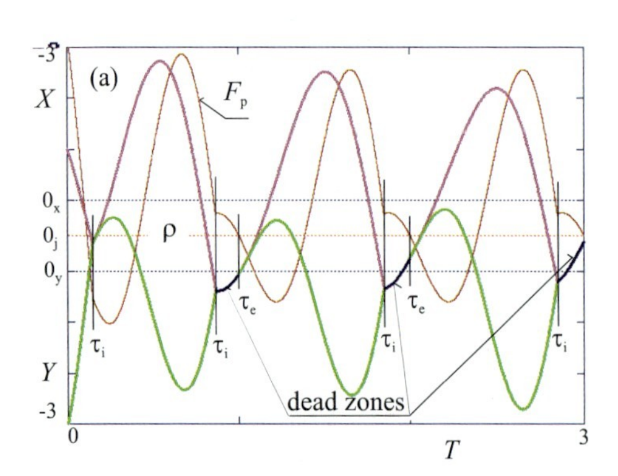

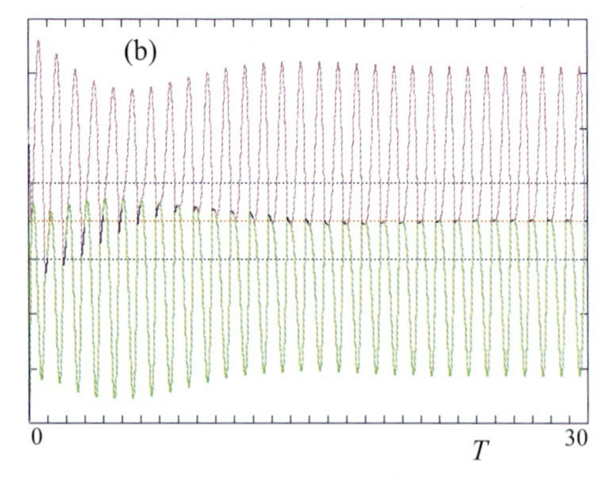

Figure 18.3: Time series of motion during three and 30 excitation periods T with asymmetric initial conditions $X(0) = -1$, $X'(0) = 1.5$, $Y(0) = -3$, $Y'(0) = 2.5$, $\varphi = \pi/4$ and dimensionless parameters $\eta = 0.95$, $\beta = 0.1$, $\rho = 0.7$ and $R = 0$.

Transformation of the System Coordinates

The transformation of coordinates X, Y into coordinates U, V according to the equations

$$U = X + Y\ ,\ V = X - Y \tag{18.10}$$

introduces a clear view of the double-impact oscillator behavior. U and V are coordinates of the anti-phase and in-phase motion of the system, respectively. This transformation is introduced in [4]. Equations (18.1), (18.2) are transformed into the equations

$$U'' + 2\beta\, U' + U = 2\cos(\eta\tau + \varphi)\,, \quad U'_+ = -RU'_-\,, \tag{18.11}$$

$$V'' + 2\beta V' + V = 0, \quad V'_+ = V'_-. \tag{18.12}$$

It follows from Eq. (18.11) that anti-phase motion corresponds to the single-impact oscillator motion with double excitation force amplitude. The in-phase motion (Eq. (18.12)) is the free damped vibration, which introduces the asymmetry into the system motion. The presence of viscous damping eliminates this asymmetry, as seen in Fig. 18.3b. Therefore the symmetric impact motion gradually stabilizes. It follows from this analysis that all known results obtained for the single-impact oscillator describe the behavior of the symmetric motion of the double-impact oscillator, but before-impact velocities should be doubled.

Regions of Existence of Different Impact Motions in the Plane of Parameters ρ, η

The dimensionless frequency η and static clearance ρ are important parameters of the system, which decide about the regime of motion. Therefore the regions of existence and stability of motions are usually evaluated in the plane η, ρ. The map of regions of motions with plastic impacts is shown in Fig. 18.4 for example.

Regions are labeled by the quantity $z = p/n$, which denotes the mean number of impacts in one excitation period T (p and n are the number of impacts and the number of periods T in the impact motion period, respectively). The value $z = 1_j$ or 2_j means that during the $z = 1$ or $z = 2$ impact motion there appear after-impact dead zones (see Fig. 18.3 for the motion $z = 1_j$). Values $z = 0 \div 1$ and $z = 0 \div 2$ denote regions of the periodic subharmonic and chaotic impact motions, which are named regions of beat impact motions.

The region of impact-less motion ($z = 0$) is bounded by grazing boundaries ρ', where a certain impact motion should arise. Boundaries ρ' are identical with the amplitude–frequency characteristics of the impact-less motion. The region of the periodic one-impact motion without after-impact dead zone ($z = 1$) is bounded by the period-doubling stability boundary s_1, where the system motion transits into the beat motion region between boundaries ρ' and s_1. Region $z = 1$ is also bounded by the saddle-node stability boundaries s_2, where one-impact motion suddenly transits into impact-less motion. Therefore two responses ($z = 0$ or $z = 1$) of the system exist in regions between boundaries ρ', s_2, named the hysteresis regions. The region of the one-impact motion with after-impact dead zones ($z = 1_j$) is bounded from below by the boundary $\rho_\mathrm{p} = -1$, where the dead zone increases to the whole excitation period T, because $F_\mathrm{p} = F_0$. The impact motion vanishes and masses remain joined all the time in the static position ($j = 0$ in Fig. 18.1). The system motion with plastic impacts is described in more detail in [18].

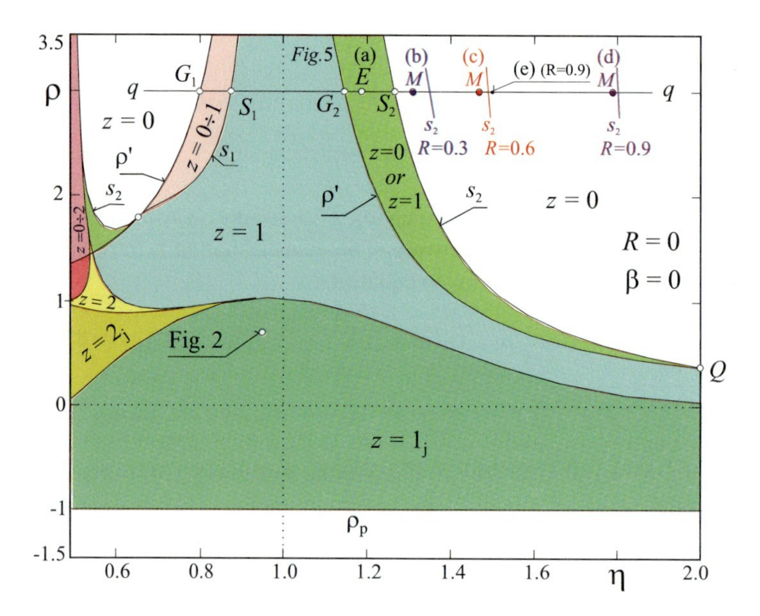

Figure 18.4: Regions of existence and stability of motions with plastic impacts.

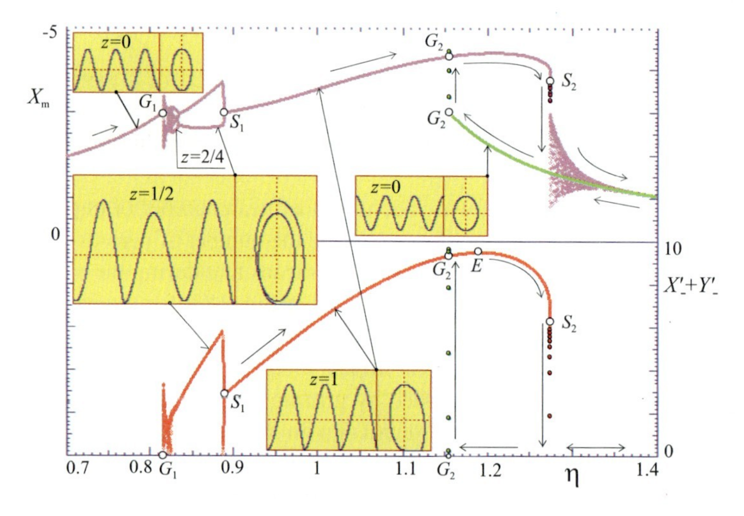

Figure 18.5: Bifurcation diagram and trajectories of motions along line q in Fig. 18.4.

Bifurcation Diagrams

The evaluation of the motion characteristics, e.g. along the section q of existence and stability regions in Fig. 18.4, offers a deeper view of the system motion behavior. The quasi-stationary amplitude characteristics $X_m(\eta)$ and the before-impact velocity characteristics $(X'_- + Y'_-)(\eta)$ are shown in Fig. 18.5. They were simulated numerically at increasing and decreasing frequency η (see arrows along characteristics). Figure 18.5 contains also time series and phase trajectories of typical motions, which exist along line q in Fig. 18.4 ($z = 0$ for $\eta = 0.8$ and $\eta = 1.2$, $z = 1/2$ for $\eta = 0.87$ and $z = 1$ for $\eta = 1$).

The impact-less motion transits in point G_1 into a narrow interval η of chaotic impact motion. Then the periodic impact motions $z = 2/4$, $z = 1/2$ and $z = 1$ gradually stabilize. The fundamental $z = 1$ motion is stable in the interval η between points S_1, S_2 – points of stability boundaries s_1, s_2 (Fig. 18.4). The impact-less motion, which arises in point S_2, slowly loses the component of a free vibration for the sake of a very small value $\beta = 0.002$ of the viscous damping. The frequency η decreases from the value $\eta = 1.4$ up to point G_2, where the $z = 1$ impact motion suddenly appears again. The extreme E of before-impact velocities exists in the hysteresis region between points G_2, S_2. This extreme regime can be attained by the quasi-stationary increase of frequency η from the region of the definite system response $z = 1$. If the extreme frequency $\eta = 1.19$ is constant, then it is necessary to choose the motion initial conditions from the basin of attraction of $z = 1$ motion, for its definite stabilization.

Basins of Attraction of Motions in Regions with Manifold Response of the System

The basin of attraction of the $z = 1$ impact motion in point E is shown in Fig. 18.6a. It was obtained under the assumption that $\varphi = 0$ and initial conditions of both subsystems are identical. The basin of attraction of impact-less ($z = 0$) motion is very small, so the periodic motion with plastic impacts can be attained simply, e.g. from the zero initial conditions. It has been shown in [18], that it does not depend on the initial phase φ of the excitation forces.

When the impacts become elastic and the coefficient of restitution R increases from zero to one, then regions of impact motions become more complex (see e.g. Figs. 18.8, 18.11). Hysteresis regions, as well as beat motion regions, enlarge. Figure 18.4 shows this schematically along line q for the hysteresis region. All stability boundaries s_2 shift right and touch the grazing bifurcation boundary ρ' in point Q. Points of the before-impact velocity extremes, denoted by $M$for $R = 0.3, 0.6, 0.9$ in Fig. 18.4, similarly shift and approach more and more the stability boundaries s_2. Therefore it is more and more difficult to attain these optimal regimes by the choice of motion initial conditions. This is graphically expressed in Fig. 18.6a–d. Figure 18.6 also shows basins of attraction (e) in the center of the hysteresis region of the almost elastic impact motion ($R = 0.9$).

18.2.2 Asymmetric Cases

Asymmetries in the double oscillator can considerably influence, in general, the system behavior. Asymmetries of

1. motion initial conditions,
2. amplitudes of driving forces,
3. natural frequencies

are considered as an example.

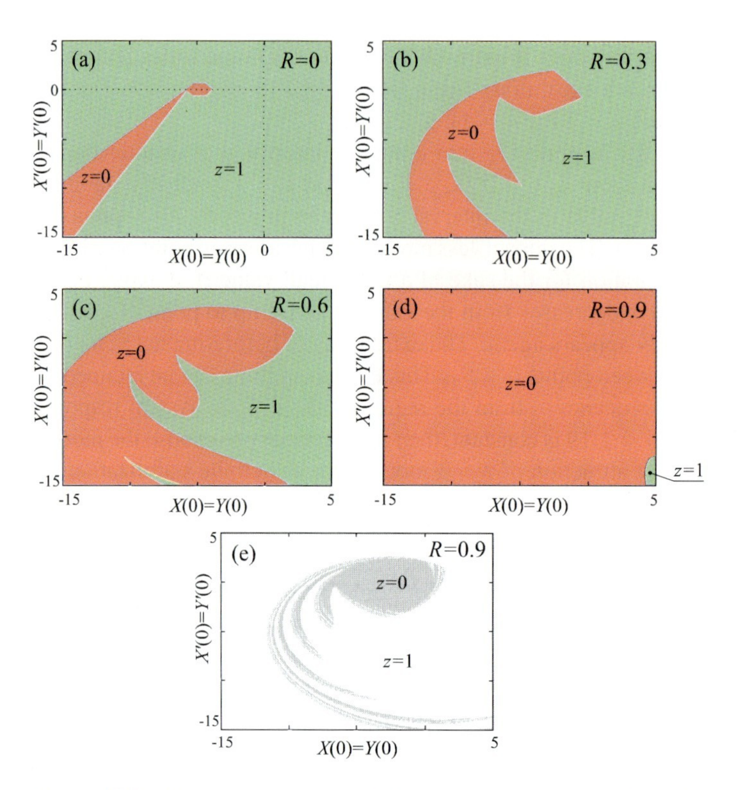

Figure 18.6: Basins of attraction of the impact motion ($z = 1$) and impact-less motion ($z = 0$) for parameters $\rho = 3$, $\beta = 0.02$ and $\varphi = 0$ for different coefficients of restitution R (a) $R = 0$, $\eta = 1.19$, $(X'_- + Y'_-)_{\max} = 9.26$; (b) $R = 0.3$, $\eta = 1.31$, $(X'_- + Y'_-)_{\max} = 9.53$; (c) $R = 0.6$, $\eta = 1.47$, $(X'_- + Y'_-)_{\max} = 11.9$; (d) $R = 0.9$, $\eta = 1.78$, $(X'_- + Y'_-)_{\max} = 28.16$; (e) $R = 0.9$, $\eta = 1.5$, $X'_- + Y'_- = 13.65$.

(ad 1) The oscillations become symmetric for asymmetric initial conditions, when viscous friction is present. When viscous friction is missing, then the initial asymmetry is preserved in spite of a considerable amount of energy losses during impacts, as has been discussed in Sect. 18.2.1.

(ad 2) Asymmetry of the driving force amplitudes introduces a systematic asymmetry of the motion, as follows from the transformation (18.10) of motion coordinates.

If a difference of 5% between excitation force amplitudes is assumed, then the differential equations of motion

$$X'' + 2\beta X' + X = \cos(\eta\tau + \varphi), \quad Y'' + 2\beta Y' + Y = 0.95\cos(\eta\tau + \varphi) \tag{18.13}$$

are transformed into

$$U'' + 2\beta U' + U = 1.95\cos(\eta\tau + \varphi), \quad V'' + 2\,\beta V' + V = 0.05\cos(\eta\tau + \varphi), \tag{18.14}$$

and the in-phase motion component V is present all the time.

(ad 3) Let us consider the equations of motion

$$X'' + 2\beta X' + X = \cos(\eta\tau + \varphi), \quad Y'' + 2\beta Y' + 0.95\,Y = \cos(\eta\tau + \varphi), \tag{18.15}$$

which express 2.5% asymmetry of natural frequencies. These equations are transformed into

$$U''+2\beta U'+U = 2\cos(\eta\tau+\varphi)+0.025(U-V), \quad V''+2\beta V'+V = -0.025(U-V). \tag{18.16}$$

It follows from Eqs. (18.16) that anti-phase and in-phase components of the system motion cannot be separated and the motion asymmetry is preserved again.

The influence of asymmetries on the optimum impact regime of the system (Fig. 18.5) has been investigated using bifurcation diagrams in the frequency interval $0.9 < \eta < 1.4$ (Fig. 18.7). The comparison of the behavior of asymmetric systems (Fig. 18.7c–f) with the symmetric case (Fig. 18.7a, b) shows that small asymmetries do not influence especially the before-impact velocity $X'_- + Y'_-$ in the neighborhood of the optimal frequency η, which corresponds to points E. The asymmetry manifests itself through resonance phenomena near the resonance ($\eta = 1$) of the impact-less motion of the single oscillator (see Fig. 18.7c, e and f) and it is more emphatic on the motion amplitude characteristics than on the before-impact velocity characteristics.

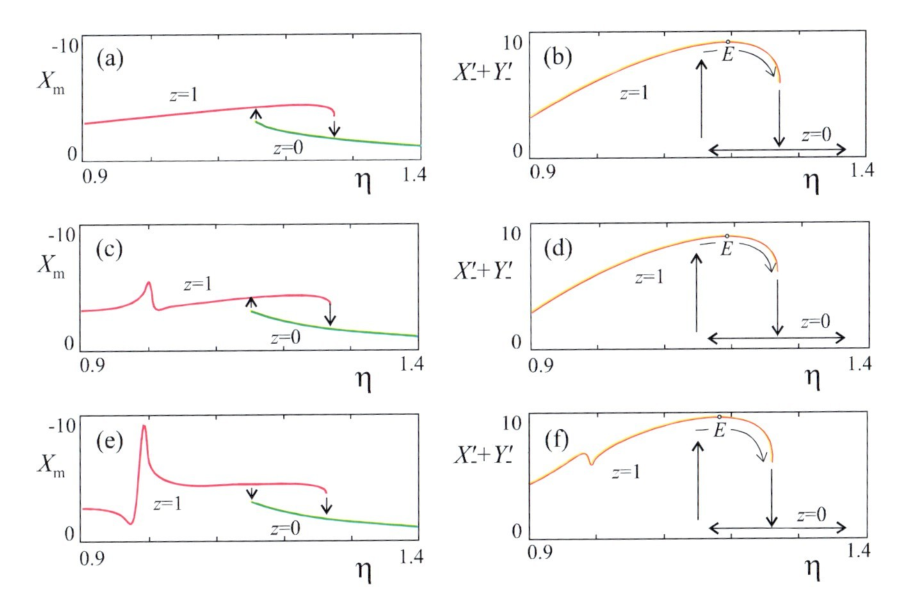

Figure 18.7: Frequency characteristics of the optimal impact motion ($z = 1$) and impact-less motion ($z = 0$) for symmetric subsystems (a), (b), asymmetric exciting force amplitudes (c), (d) and natural frequencies (e), (f).

18.3 Simulations

Thirty years ago, analogue simulations allowed us to obtain fundamental knowledge of the impact oscillator dynamics [5, 1]. The current numerical simulations offer more extensive both qualitative and quantitative investigations. The program NON-1-SIM [13] has been prepared especially for the investigation of the single-impact oscillator dynamics and it can be used also for educational purposes. This program was enlarged with the aim to simulate the motion with several kinds of soft impacts [9–11].

The most important problem of the dynamics of impact oscillators is the evaluation of regions of different kinds of impact motion regimes in the dependence on excitation frequency η and static clearance ρ. Boundaries of stability and existence regions (see e.g. Figs. 18.4, 18.8, 18.11, 18.14, 18.15) have diverse character and it is difficult to create a general program for their simulation. Therefore bifurcation boundaries should be evaluated interactively during the simulation.

Figure 18.8 shows, as an example, the diversity of boundaries of the chaotic impact motion regions. Four input boundaries (1)–(4) characterize different ways into the chaos. Three of them, (2)–(4), are specific to the impact motion. Ways (3) and (4) are caused by additional impacts, which appear during the development of the period-doubling and saddle-node instability of periodic subharmonic impact motions. The system behavior along all boundaries is described in [16, 7, 14].

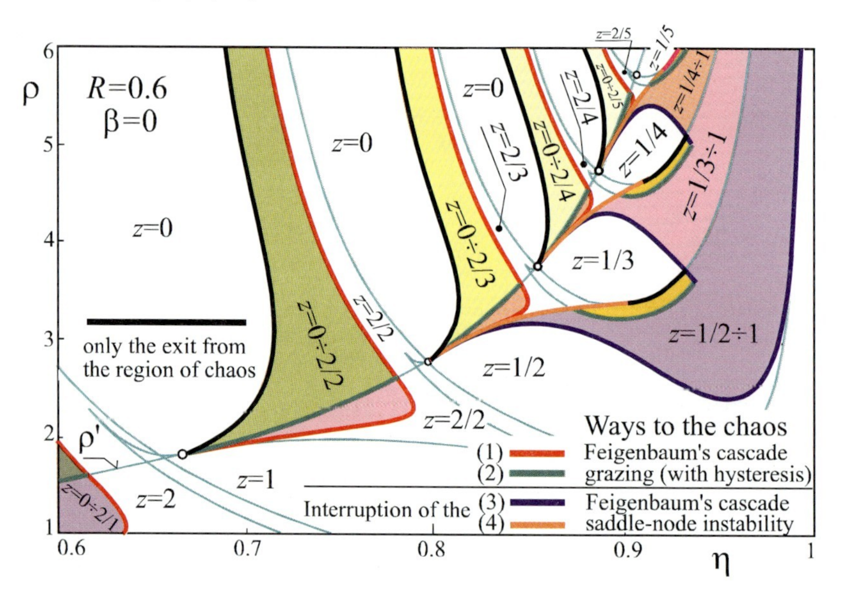

Figure 18.8: Input and output boundaries of chaotic motion regions.

18.4 Experiment

The experimental set of the double-impact oscillator is shown in Fig. 18.9. Its scheme and block scheme of electronic equipment controlling and evaluating the motion are given in Fig. 18.10.

Masses 1 are clamped using blade springs 2 to traversable frames 5, which also carry vibrators 4. These vibrators excite masses by rods 3. The static clearance $2r$ is controlled by the right–left screw 6 axially fixed to the foundation 7. Computer 10 determines both frequency and amplitude of the harmonic signal of the generator 12. 13 are amplifiers. Vibrations are measured by accelerometers 8 and integrating circuits 14. The velocity of one mass is measured by the laser vibrometer 15. One mass is electrically isolated and impacts switch on the circuit of the DC supply 16 for their indication. The computer stores measured signals during the time interval 1 s by the 16-channel AD converter 11, which samples signals with frequency 2 kHz. The natural frequency and viscous damping of separated systems are $\Omega = 12.6\,\mathrm{Hz}$, $\beta = 0.07$.

Figure 18.9: Model of double-impact oscillator.

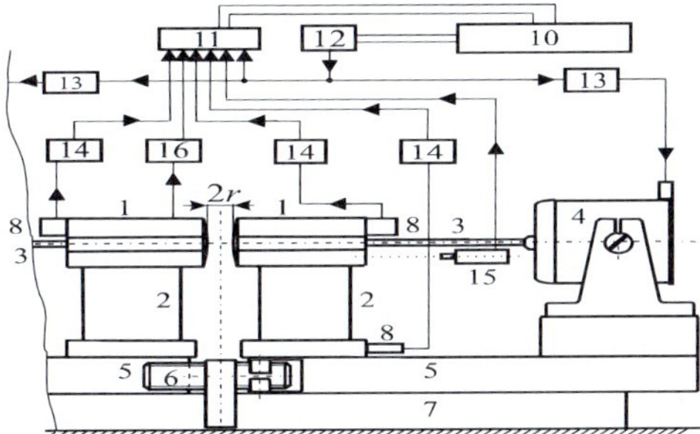

Figure 18.10: Scheme of experimental equipment.

18.5 Comparison of Simulation and Experimental Results

The results are compared by regions of existence of different system motions in the plane ρ, η (Fig. 18.11, where regions of all evaluated impact motions are denoted by a value of the impact number z) as well as by numerically simulated trajectories and experimentally measured quantities of typical periodic and chaotic impact motions (Figs. 18.12, 18.13). Figure 18.13 also shows the vibrations of the frame 5 (Fig. 18.10), which are of the order of μm without impact impulses.

The real value of the restitution coefficient $R = 0.9$ of impacting bodies was determined by the simulation of the stability boundary s_2 according to the experimentally obtained boundary s_2, because the course of the boundary s_2 depends expressively on R (see Sect. 18.2.1 and Fig. 18.4). There exist a series of $z = 1/n$-impact motions. Five of them ($n = 1$–5) are shown in Figs. 18.12A, B, D, F, H and 18.13A, B, D, F, H, corresponding to points (A), (B), (D), (F), (H) in Figs. 18.11–18.13. They also contain, for example, the periodic $z = 2/2$ impact motion (point (E)), which is one of other subharmonic motions. It arises by splitting of the $z = 1$ motion.

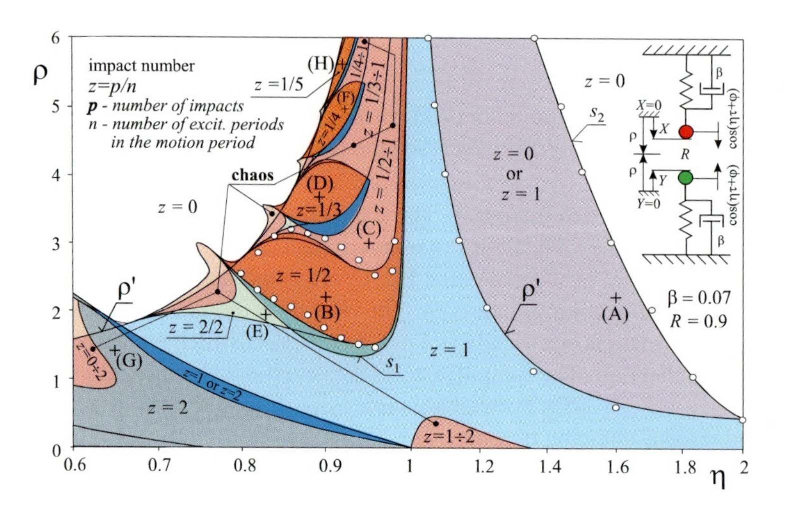

Figure 18.11: Impact motions $n \geq 2$ create groups of periodic subharmonic impact motions.

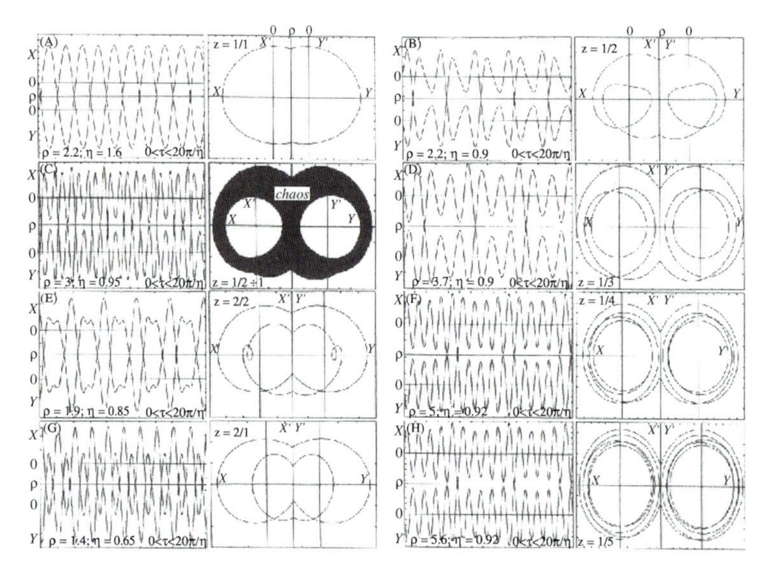

Figure 18.12: Time series ($0 < \tau < 20\pi/\eta$) and phase trajectories of numerically simulated impact motions in points (A)–(H) of Fig. 18.11.

The first motion from the group of more impact motions $z = p/n$ ($p \geq 2, n = 1$) is shown in Figs. 18.12G and 18.13G. The chaotic impact motion $z = 1/2 \div 1$ is in Figs. 18.12C, 18.13C.

Experimentally ascertained regions have the same structure as those attained numerically, but boundaries are shown only in a limited number of points denoted by circles along the hysteresis region of $z = 0$ and $z = 1$ motions and along the $z = 1/2$ region. The agreement of numerical and experimental results is good.

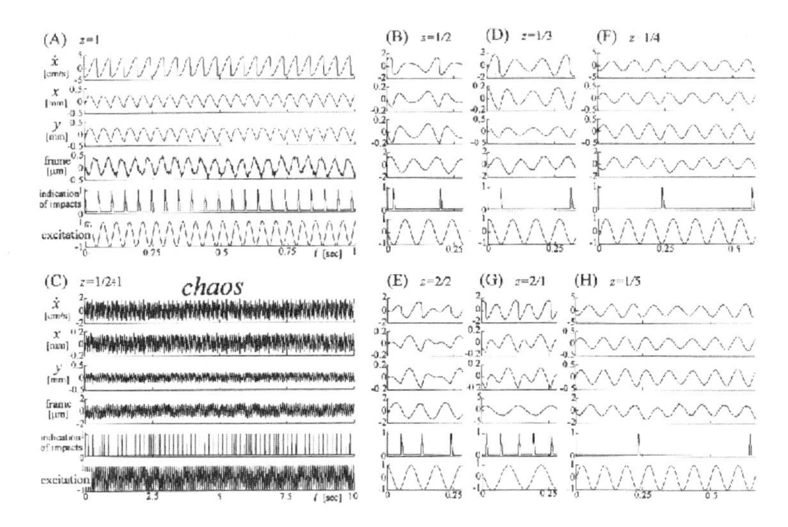

Figure 18.13: Time series of experimentally obtained motions in points (A)–(H) of Fig. 18.11.

18.6 Analysis of Motion with Soft Impacts

The theoretical analysis of an oscillator with soft impact is more difficult. Fundamental periodic impact motions and their stability were resolved in [2]. New phenomena in the dynamics of oscillators with soft impacts were obtained using numerical simulation and are explained in more detail in [19, 15, 10, 11, 22].

The theoretical analysis of the soft impact oscillator dynamics is more difficult. The exact solution of the fundamental group of periodic impact motions and their stability were resolved in [2] for the Kelvin–Voigt model of soft impact.

The global analysis by numerical simulations observed new phenomena, in comparison with dynamics of oscillators with rigid impacts, which are explained in more detail in [19, 9, 15, 22, 20].

Piercing forces at impacts can be modeled by a piecewise-linear (triangle) energetic loop, the scheme of which is shown in Figs. 18.14, 18.15. The triangle model of soft impact will be therefore used for the investigation of piercing machine motion.

Denoting the dimensionless deflection $X = x/x_{\rm st}$ ($x_{\rm st} = F_0/k_1$) and using the time transformation $\tau = \Omega t$, where $\Omega = (k_1/m)^{1/2}$, the oscillator motion is described by the equation

$$X'' + \beta X' + X + F = \cos \eta\tau, \tag{18.17}$$

where

$$X'' = \frac{d^2x}{d\tau^2}, \quad X' = \frac{dx}{d\tau}, \quad \beta = b_1/\sqrt{k_1 m} = 0.01,$$

$$F = \begin{cases} 0 & \text{for} \quad X \le \rho \; (\rho = r/x_{\rm st}), \\ (X-\rho)(k_2 + k_3 \operatorname{sgn} X')/k_1 & \text{for} \quad X > \rho \, . \end{cases}$$

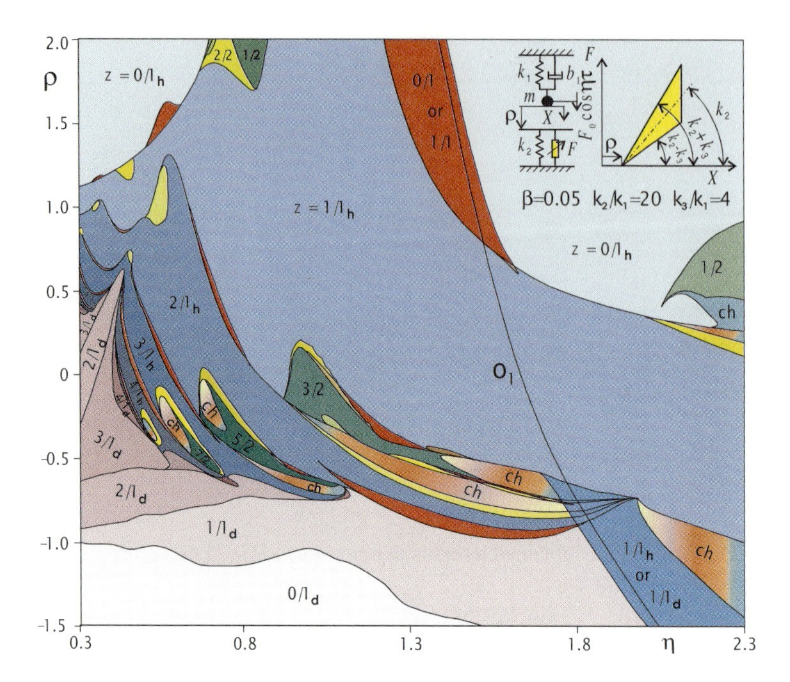

Figure 18.14: Regions of motion with triangle model of soft impact and small damping.

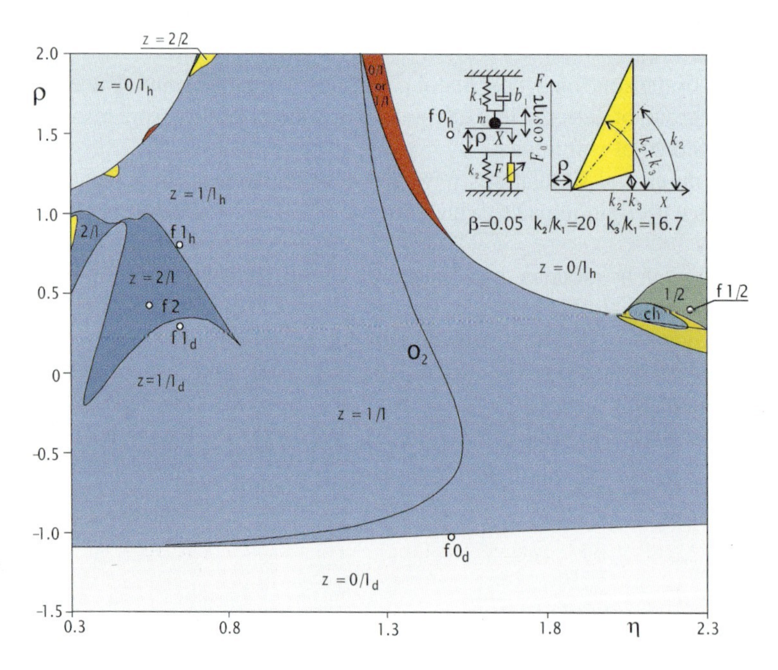

Figure 18.15: Regions of motion with triangle model of soft impact and large damping.

Results of the evaluation of regions of impact motions in plane η, ρ are shown in Figs. 18.14, 18.15. A more detailed explanation of Fig. 18.14 is in [12] and bifurcation diagrams of motion amplitudes and before-impact velocities along sections with constant clearance ρ are discussed in [23].

Figures 18.14 and 18.15 correspond to small and large energy losses at impacts (equivalent restitution coefficient $R_e = 0.816$ and $R_e = 0.3$), respectively. The less the energy dissipation at impacts the more complex is the diversity of periodic and chaotic impact motions. Nevertheless there exists a large region of practically most important $z = 1$ impact motion. Inside this region there exist local extremes of maximum before-impact velocities along curves O_1 and O_2, respectively. A similar evaluation was completed for the next series of equivalent restitution coefficients $R_e = 0.1, 0.5, 0.7, 0.8$ and results are shown in Fig. 18.16. The maximum impact velocities of the single oscillator for optimum combination of system parameters can be determined from this figure. Figure 18.17 shows, as an example, several phase trajectories and time series of impact-less and impact motions corresponding to points $f0_d$, $f1_d$, $f1_h$ and $f2$ in Fig. 18.15.

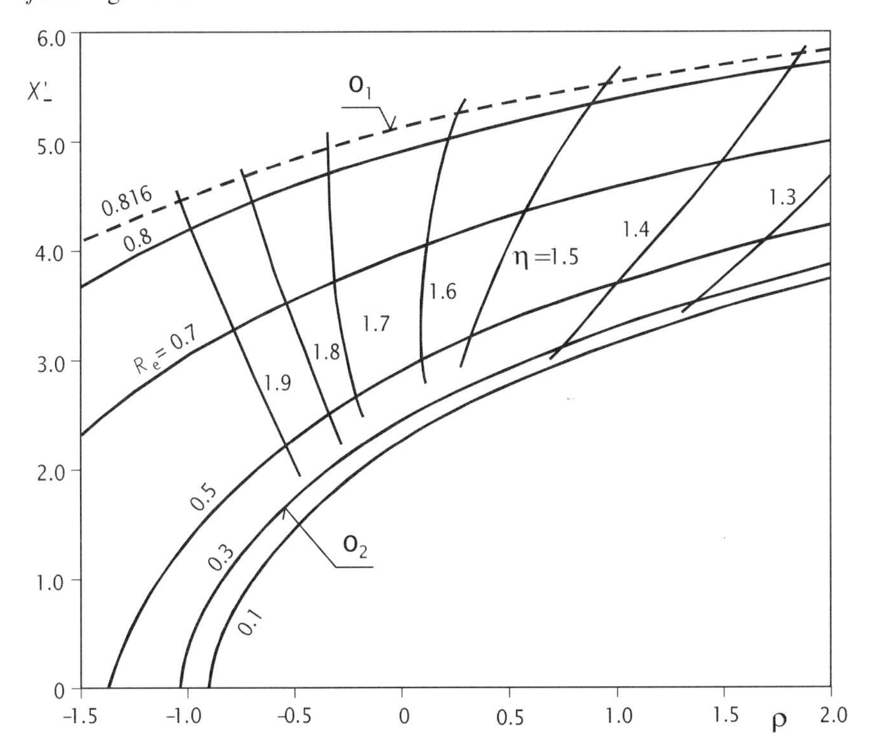

Figure 18.16: Optimum combination of system parameters for maximum before-impact velocity.

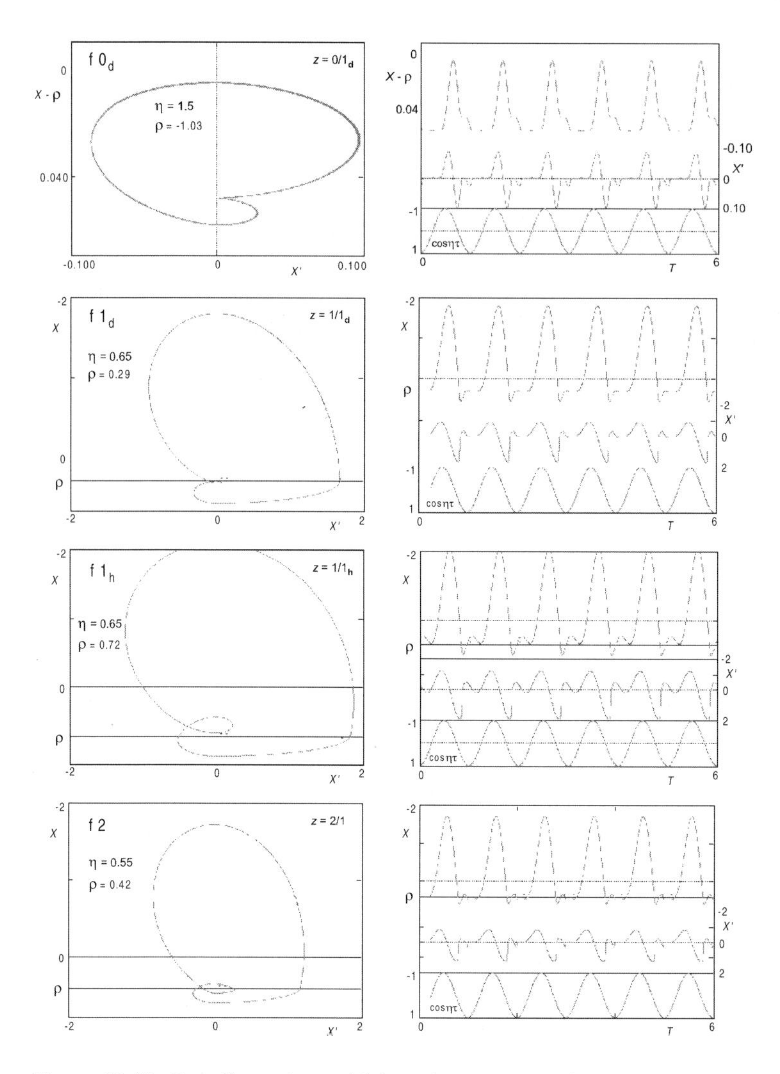

Figure 18.17: Periodic motions with large impact energy losses.

18.7 Conclusion

The correctness of the mathematical model of the double-impact oscillator and the numerical simulation of its motion have been successfully verified experimentally. Experiments also confirmed former results of the theoretical analysis and the simulation of the single-impact oscillator dynamics with respect to regions of periodic and chaotic motions and their structure in dependence on the static clearance ρ and excitation frequency η. The influence of energy losses at impact as well as both rigid and soft impacts was studied and optimum combinations of system parameters for maximum before-impact velocities were determined. Results obtained can make easier the construction of a new type of forming machine.

Bibliography

[1] T. Irie, G. Yamada, and H. Matsuzaki, *On the dynamic response of a vibro-impact system to random force*, Bulletin of the Faculty of Engineering, Hokkaido University **72** (1974), 13–24.

[2] L. Kocanda, Ph.D. thesis, Czech Technical University Prague, Faculty of Mechanical Engineering (2002) (in Czech).

[3] K. Koizumi, Doctoral dissertation, Tokyo University (1980).

[4] P. S. Landa, *Nonlinear Oscillations and Waves in Dynamical Systems*, Kluwer Academic Publishers, Dordrecht.

[5] F. Peterka, *Laws of impact motion of mechanical systems with one degree of freedom. Part II – results of analogue computer modelling of the motion*, Acta Technica ČSAV **5** (1974), 569–580.

[6] F. Peterka, *Introduction to Oscillations of Mechanical Systems with Internal Impacts*, ACADEMIA, Prague, (in Czech).

[7] F. Peterka, *Dynamics of the impact oscillator*, in: Proceedings of the IUTAM Symposium on New Applications of Nonlinear and Chaotic Dynamics in Mechanics, Ithaca, New York, USA, 27 July–1 August 1996, Kluwer Academic Publishers, Dordrecht.

[8] F. Peterka, *Analysis of motion of the impact-dry-friction pair of bodies and its application to the investigation of the impact dampers dynamics*, ASME 1999 Design Engineering Technical Conferences, MOVIC'99, Session Friction and Impact Induced Vibrations, paper DETC99/VIB-8350, Las Vegas, USA, 2–5 September, 1999.

[9] F. Peterka, *Simulation of the oscillator with soft impacts*, in: Proceedings of the 16th Computational Mechanics, Nečtiny, 30 October–1 November 2000, pp. 317–320.

[10] F. Peterka, *Program of numerical simulation of piercing process*, in: Proceedings of WG2 Workshop, Nonlinear Dynamics and Control Mechanical Processing, Budapest, 2001, pp. 29–30.

[11] F. Peterka, *Dynamics of oscillator with soft impacts*, Proceedings DETC'01, Pittsburgh, Pennsylvania, USA, 9–12 September 2001, paper DETC 2001/VIB-21609, CD ROM.

[12] F. Peterka, *Transition from impactless to impact motion in oscillators with soft impacts*, in: Proceedings Computational Mechanics 2002, Nečtiny, Czech Republic, 29–31 October 2002, pp. 349–356.

[13] F. Peterka and Formánek, *Simulation of motion with strong nonlinearities*, in: Proceedings of the CISS – First Joint Conference of International Simulation Societies, ETH Zurich, 22–25 August 1994, pp. 137–141.

[14] F. Peterka, S. Čipera, and T. Kotera, *Additional impact causes the intermittency chaos of unstable subharmonic motions of impact oscillator*, in: ICTAM 2000, Chicago, USA, 27 August– 2 September 2000, International Congress of IUTAM, Abstract Book, pp. 144–145.

[15] F. Peterka, L. Kocanda, and J. Veselý, *Simulation and experimental investigation of the oscillator with soft impacts*, in: Proceedings of VIII International Conference on the Theory of Machines and Mechanisms, Liberec, 5–7 September 2000, pp. 615–620.

[16] F. Peterka and T. Kotera *Four ways from periodic to chaotic motion in the impact oscillator*, Machine Vibration **5** (1996), 71–82.

[17] F. Peterka and O. Szöllös, *Dynamics of the opposed pile driver*, in: Proceedings of the IUTAM Symposium on Interaction between Dynamics and Control in Advanced Mechanical Systems, Eindhoven, The Netherlands, 21–26 April, 1996, Kluwer Academic Publishers, Dordrecht, The Netherlands, pp. 271–278.

[18] F. Peterka and O. Szöllös, *The stability analysis of a symmetric two-degree-of-freedom system with impacts*, in: Proceedings of EUROMECH – 2nd European Nonlinear Oscillations Conference, Prague, 9–13 September 1996.

[19] F. Peterka and O. Szöllös, *Influence of the stop stiffness on the impact oscillator dynamics*, in: Proceedings of the IUTAM Symposium Unilateral Multibody Dynamics, München, 3–7 August 1998, Kluwer Academic Publishers, Dordrecht, pp. 127–135.

[20] F. Peterka and A. Tondl, *Subharmonic motions of the oscillator with soft impacts*, in: Proceedings of Engineering Mechanics 2002, Svratka, Czech Republic, pp. 223–224, CD-ROM.

[21] F. Peterka and J. Vacík, *Transition to chaotic motion in mechanical systems with impacts*, Journal of Sound and Vibration **154**(1) (1992), 95–115.

[22] L. Püst and F. Peterka, *Impact Oscillator with Hertz's Model of Contact*, Meccanica, Aberdeen, EUROMECH 425 Colloquium (in print).

[23] L. Püst and F. Peterka, *Vibration of systems with clearances and prestress*, in: Proceedings of Computational Mechanics 2002, Nečtiny, Czech Republic, 29–31 October 2002, pp. 387–394.

19 Nonlinear Vibration in Gear Systems

G. Litak and M.I. Friswell

Gearbox dynamics is characterized by a periodically changing stiffness and a backlash which can lead to a loss of the contact between the teeth. Due to backlash, the gear system has piecewise-linear stiffness characteristics and, in consequence, can vibrate regularly or chaotically depending on the system parameters and the initial conditions. We examine the possibility of a non-feedback system control by introducing a weak resonant excitation term and through adding an additional degree of freedom to account for shaft flexibility on one side of the gearbox. We shall show that by correctly choosing the coupling values the system vibrations may be controlled.

19.1 Introduction

Gearbox dynamics is based on a periodically changing meshing stiffness complemented by a nonlinear effect of backlash between the teeth. Recently, their regular and chaotic vibrations have been predicted theoretically and examined experimentally [1–7]. The theoretical description of this phenomenon has been based mainly on single degree of freedom models [1–6, 8] or multi-degree models neglecting backlash [9, 10]. This paper examines the possibility of taming chaotic vibrations and reducing the amplitude by non-feedback control methods. Firstly, we will introduce a weak resonant excitation through an additional small external excitation term (torque) with a different phase [11, 12]. Secondly, we will examine the effect of an additional degree of freedom to account for shaft flexibility on one side of the gearbox [7]. This may also be regarded as a single-mode approximation to the torsional system dynamics, or alternatively as a simple model for a vibration neutralizer or absorber installed in one of the gears.

19.2 One-stage Gear Model

We start our analysis with modeling the relative vibrations of a single-stage transmission gear [6]. Figure 19.1 shows a schematic picture of the physical system. The gear wheels are shown with moments of inertia I_1 and I_2 and are coupled by the stiffness and damping of the teeth mesh, represented by k_Z and c_Z.

The equations of motion of the system may be written in terms of the two degrees of freedom ψ_1, ψ_2 that represent the rotational angles of the gear wheels. These angles are those that remain after the steady rotation of the system is removed [6]. Thus, if the backlash is initially neglected, we have

Nonlinear Dynamics of Production Systems. Edited by G. Radons and R. Neugebauer

ISBN 3-527-40430-9

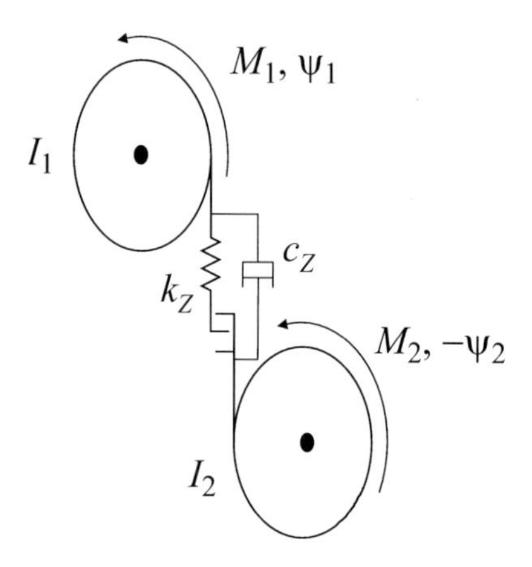

Figure 19.1: The physical model of a one-stage transmission gear system.

$$
\begin{aligned}
I_1\ddot{\psi}_1 + \Big[k_Z(r_1\psi_1 - r_2\psi_2) + c_Z(r_1\dot{\psi}_1 - r_2\dot{\psi}_2)\Big]\, r_1 &= M_1, \\
I_2\ddot{\psi}_2 - \Big[k_Z(r_1\psi_1 - r_2\psi_2) + c_Z(r_1\dot{\psi}_1 - r_2\dot{\psi}_2)\Big]\, r_2 &= -\,M_2,
\end{aligned}
\tag{19.1}
$$

where r_1 and r_2 are the radii of the gear wheels and the overdot represents differentiation with respect to time t.

In fact, it is possible to reduce the above two equations to one using the new relative displacement coordinate $x = r_1\psi_1 - r_2\psi_2$. That coordinate represents the relative displacement of the gear wheels at the teeth. The equation of motion for x is obtained by subtracting r_2/I_2 times the first equation from r_1/I_1 times the second (Eqs. (19.2)). Thus, in non-dimensional form, the equation of motion can be re-written as

$$
\frac{d^2}{d\tau^2}x + \frac{2\zeta}{\omega}\frac{d}{d\tau}x + \frac{k(\tau)g(x,\eta)}{\omega^2} = \frac{\bar{B}(\tau)}{\omega^2} = \frac{B_0 + B_1\cos(\tau + \Theta)}{\omega^2},
\tag{19.2}
$$

where the parameters are easily derived from Eqs. (19.2) and (19.2) as

$$
\begin{aligned}
&\tau = \omega t, \\
&2\zeta = c_Z\left[r_1^2/I_1 + r_2^2/I_2\right], \\
&\bar{B}(\tau) = r_1 M_1/I_1 + r_2 M_2/I_2.
\end{aligned}
\tag{19.3}
$$

ζ, $k(\tau)$, $g(x,\eta)$ and $\bar{B}(\tau)$ [6] and other symbols used are listed in Table 19.1 for convenience.

Note that the backlash and time-dependent meshing stiffness have been included by rewriting the meshing force as [6]

$$
k_Z\left[r_1^2/I_1 + r_2^2/I_2\right]x = k(\tau)g(x,\eta).
\tag{19.4}
$$

Equations (19.3) and (19.4) have assumed that the moments of the gear system are composed of a sinusoidal moment at frequency ω with a constant offset.

The meshing stiffness is periodic and the backlash is described by a piecewise-linear function. Figure 19.2 shows a typical time-dependent mesh-stiffness variation $k(\tau)$ [6] and the backlash is modeled for a clearance η as (Fig. 19.3)

$$g(x,\eta) = \begin{cases} x & x \geq 0, \\ 0 & -\eta < x < 0, \\ x+\eta & x \leq -\eta. \end{cases} \tag{19.5}$$

Table 19.1: Symbols and parameters used in the analysis.

Symbol	Description
I_1, I_2, I_3	Moments of inertia
ψ_1, ψ_2, ψ_3	Rotational angles
x, x_1, x_2	Relative displacements
v, v_1, v_2	Relative velocities
M_1, M_2, M_3	External torques
ω	Excitation frequency
τ	Dimensionless time
η	Backlash
ζ	Damping coefficient
$k(\tau)$	Meshing stiffness
k_S	Coupling stiffness
$g(x,\eta)$	Nonlinear stiffness function
$\bar{B}, B_0, B_1, D$	External excitations
A	Vibration amplitude
$\beta_1, \beta_2, \beta_3$	Constant parameters
$\lambda_i, i = 1, 2, ...$	Lyapunov exponents

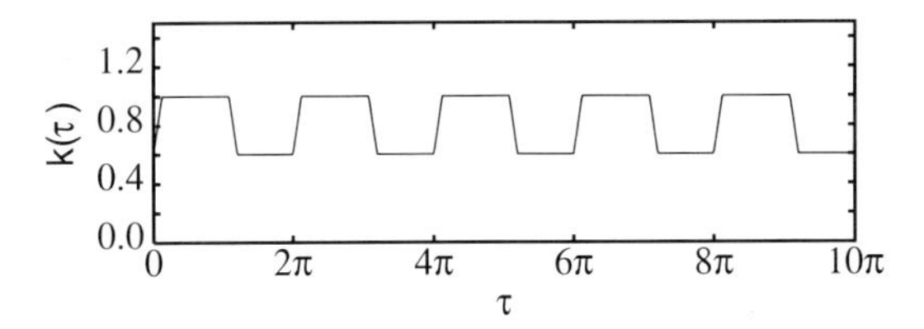

Figure 19.2: The meshing stiffness $k(\tau)$.

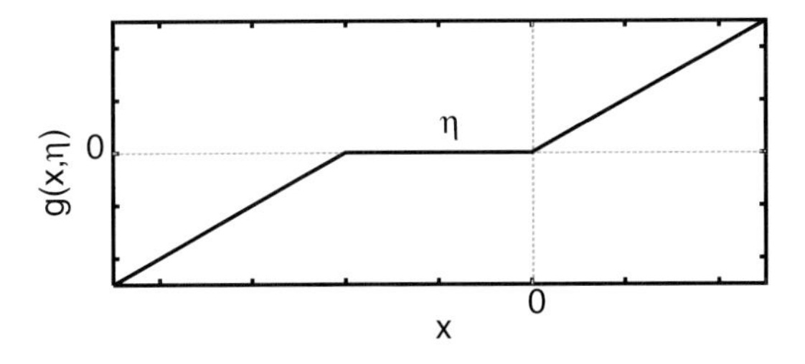

Figure 19.3: The nonlinear stiffness function $g(x,\eta)$.

19.3 Vibrations of a Gear System in Presence of a Weak Resonance Term

Now we are going to examine the gear system described by Eqs. (19.2)–(19.5) subjected to an additional external periodic excitation with a relatively small value D:

$$\frac{d^2}{d\tau^2}x + \frac{2\zeta}{\omega}\frac{d}{d\tau}x + \frac{k(\tau)g(x,\eta)}{\omega^2} = \frac{B_0 + B_1\cos(\tau+\Theta) + D\cos(\tau+\Theta')}{\omega^2}, \tag{19.6}$$

where the phase angle $\Theta' \neq \Theta$. Such inclusion has been discussed earlier in the literature as a suitable method of chaos control for Josephson-junction circuits [11] and Froude pendulum motion [12]. In those papers the authors claimed that the resonant term, if used adequately, can tame or include chaotic motion. We have performed numerical simulations of Eqs. (19.2) and (19.6) to highlight the effect of the addition of resonant term D on the basic equation of motion (Eq. (19.2)). In particular, we were interested in changes in vibration amplitude. The type of motion was analyzed using Lyapunov exponents obtained by the algorithm of Wolf et al. [13]. System parameters have been used as in [6]: $\omega = 1.5$, $B_0 = 1$, $B_1 = 4$, $\Theta = 0$, $\zeta = 0.08$, $\eta = 10$, while the additional excitation term in Eq. (19.6) was introduced as $D = 2$, $\Theta' = 0.1$.

Figure 19.4a shows the calculated vibration amplitude of the relative motions of the gear wheels, A, defined as

$$A = \left|\frac{x_{\max} - x_{\min}}{2}\right|. \tag{19.7}$$

The maximal Lyapunov exponent λ_1 vs. frequency ω, obtained by simulation, is plotted in Fig. 19.4b. Note that the initial conditions were assumed to be $x_0 = -2.0$ and $v_0 = \dot{x}_0 = -0.5$ for small starting ω ($\omega = 0.1$) and for each new ω_{i+1} calculations were performed for 400 excitation cycles, where the new initial conditions were the last pair of values of (x, v) for the previous frequency ω_i. Curves marked by '1' correspond to solutions of Eq. (19.2) while those marked by '2' correspond to Eq. (19.6). One can easily note that the amplitude of the modified system is slightly larger but simultaneously the system behaves more regularly. In most of the cases where the original system was in a chaotic state ($\lambda_1 > 1$ for the curve '1' in Fig. 19.4b) it vibrates regularly in the presence of the resonant term ($\lambda_1 < 1$ for the curve '2' in Fig. 19.4b). Unfortunately the taming of the chaotic motion was accompanied by additional features visible in Fig. 19.4a. Clearly, the curve '2' describing the amplitude of motion shows a number of jumps, i.e. for $\omega \approx 1.36$ and 1.70. These jumps may be associated with transitions to other solutions (attractors) with changing ω. This effect is typical for many nonlinear systems but it seems to be more transparent for $D \neq 0$. To explore further this effect we have also calculated Poincaré maps for many initial conditions chosen randomly. The results for frequencies $\omega = 1.5$ and 1.6 are shown in Fig. 19.5. In this figure, one can see that for $\omega = 1.5$ attractors (in gray colour) for $D = 0$ and (in black colour) for $D \neq 0$ are similar. Here, numbers '1' and '2' denote the regular and chaotic attractors, respectively. But if we move ω to a larger value ($\omega = 1.6$) the regular attractor for $D = 0$ disappears. The attractor remains for $D \neq 0$, but interestingly another chaotic attractor emerges. For some other frequencies additional attractors also exist, which supports the proposed explanation of the jumping phenomenon.

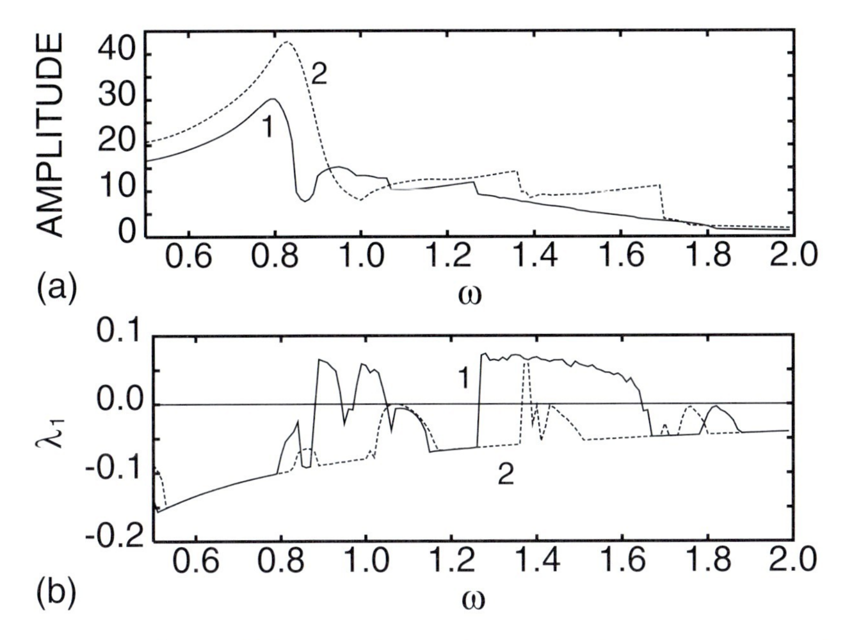

Figure 19.4: The amplitude of vibration (a) and the maximal Lyapunov exponent λ_1 (b) vs. frequency ω obtained by simulation. The initial conditions were assumed to be $x_0 = -2.0$ and $v_0 = \dot{x}_0 = -0.5$ for small ω ($\omega_0 = 0.1$) and for each new ω (ω_{i+1}) calculations were performed for 400 excitation cycles. Each time new initial conditions (x_0, v_0) were chosen as the last pair of values of (x, v) for previous ω_i ($(x_0, v_0)|_{\text{new}} = (x, v)|_{\text{old}}$). Curves '1' and '2' correspond to cases with $D = 0$ and $D = 2$, respectively.

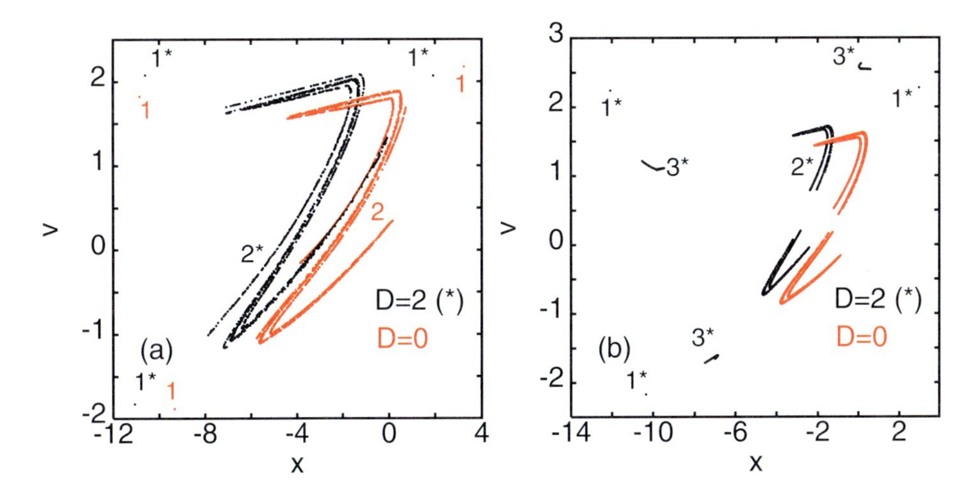

Figure 19.5: The Poincaré maps calculated for 10 randomly chosen initial conditions for $\omega = 1.5$ (a) and $\omega = 1.6$ (b). The gray color corresponds to simulations with respect to Eq. (19.2) while the black color corresponds to Eq. (19.6), respectively.

19.4 Vibrations of a Gear System with a Flexible Shaft

Here we examine the effect of adding an additional degree of freedom to the gear system (Fig. 19.6) by using a flexible shaft on one side of the gearbox [7].

Now the equations of motion of the system may be written in terms of the three degrees of freedom, ψ_1, ψ_2 and ψ_3, related to the rotational angles of the gear wheels and disk, and have the following form

$$
\begin{aligned}
&I_1\ddot{\psi}_1 + \left[k_Z(r_1\psi_1 - r_2\psi_2) + c_Z(r_1\dot{\psi}_1 - r_2\dot{\psi}_2)\right] r_1 = M_1,\\
&I_2\ddot{\psi}_2 - \left[k_Z(r_1\psi_1 - r_2\psi_2) + c_Z(r_1\dot{\psi}_1 - r_2\dot{\psi}_2)\right] r_2 - k_S(\psi_3 - \psi_2) = -M_2, \quad (19.8)\\
&I_3\ddot{\psi}_3 + k_S(\psi_3 - \psi_2) = -M_3.
\end{aligned}
$$

As in the previous case (Eqs. (19.2) and (19.2)) the above set of equations may be decoupled by using the new relative displacement coordinates $x_1 = r_1\psi_1 - r_2\psi_2$ and $x_2 = r_2(\psi_3 - \psi_2)$. Thus we reduce the number of equations from three to two:

$$
\begin{aligned}
&\frac{d^2}{d\tau^2}x_1 + \frac{2\zeta}{\omega}\frac{d}{d\tau}x_1 + \frac{k(\tau)g(x_1,\eta)}{\omega^2} + \frac{\beta_1 k_S x_2}{\omega^2}\\
&= \frac{B_0 + B_1\cos(\omega t + \Theta)}{\omega^2}, \qquad (19.9)\\
&\frac{d^2}{d\tau^2}x_2 + \frac{\beta_2 k_S x_2}{\omega^2} + \frac{2\beta_3\zeta}{\omega}\frac{d}{d\tau}x_1 + \frac{\beta_3 k(\tau)g(x_1,\eta)}{\omega^2} = 0,
\end{aligned}
$$

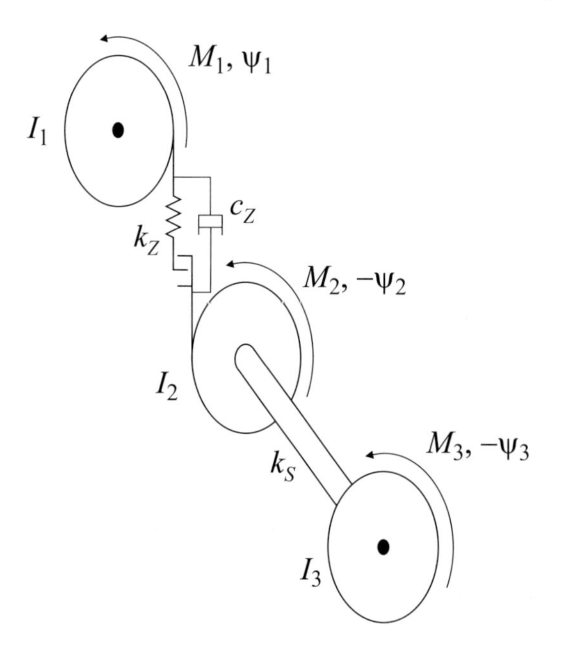

Figure 19.6: The physical model of the gear model with an additional degree of freedom introduced by a disk fixed by a flexible shaft to one of gear wheels.

where most of the parameters have already been defined in Eqs. (19.3) and (19.4). The rest of them are easily derived from Eqs. (19.8) and (19.9), as

$$\begin{aligned}
\beta_1 &= 1/I_2, \\
\beta_2 &= 1/I_2 + 1/I_3, \\
\beta_3 &= (r_2^2/I_2)/\left[r_1^2/I_1 + r_2^2/I_2\right], \\
0 &= r_2 M_2/I_2 - r_2 M_3/I_3.
\end{aligned} \tag{19.10}$$

The simulation results of the above system (Eq. (19.9), Fig. 19.6) are presented in Figs. 19.7 and 19.8. First of all we have repeated the calculations of the amplitude and Lyapunov exponents. The results for the amplitude A with various couplings (k_S) are summarized in Fig. 19.7a. Curves '1', '2' and '3' correspond to $k_S = 0$, 0.1 and 1.0, respectively. One can see that a suitable change to k_S may lead to the lowering of the amplitude. This effect is more efficient in the case of $k_S = 1.0$. Comparing the corresponding values of the maximal Lyapunov exponents one can investigate the chaotic nature of the vibrations (Fig. 19.7b). Namely, the case $k_S = 0.1$ (the curve '2') does not reduce the original chaoticity in the region around $\omega \approx 1.0$ and $\omega \in [1.24, 1.62]$ (the curve '1'), while $k_S = 1.0$ is large enough to tame the chaos in these regions as well as to make the vibration amplitude smaller.

It should be noted that for $k_S \neq 0$ there are four non-zero exponents to be examined. The positive value of the largest one, λ_1, is presented in Fig. 19.7b. It detects frequency regions of chaotic vibrations. But it is also interesting to see what has happened to the other exponents, especially for our multi-dimensional system. They were also calculated and are plotted for comparison in Figs. 19.7c and d. For a small coupling stiffness, $k_S = 0.1$, we have the surprising result that most of the chaotic regions are characterized by two positive exponents (Fig. 19.7c), which is a signal that the system is truly hyper-chaotic [14, 15, 10]. In this case two initially close trajectories escape exponentially in two different directions. The results of calculations with larger k_S ($k_S = 1.0$) are different (Fig. 19.7), where regions of chaotic motion with only one positive Lyapunov exponent were found, which signals typical chaotic behavior. To clarify this point we plot the phase portrait and Poincaré maps for the chosen frequency $\omega = 1.5$. Figure 19.8a shows the portrait plane and the Poincaré map of the chaotic attractor for $k_S = 0$ while, in Fig. 19.8b, for $k_S = 1.0$, the corresponding attractor is regular, synchronized with the excitation frequency ω. Figures 19.8c and d show the phase portrait and the Poincaré map of the hyper-chaotic attractor.

19.5 Conclusions

This paper has examined the effect of adding an additional small resonance excitation term or an additional degree of freedom to a simple model of gear vibration (Figs. 19.1 and 19.2). We have shown that the resonance term can successively reduce chaoticity in the dynamical system (Fig. 19.4b), although the vibration amplitude increases to a larger value than that for the basic system. We also noted several sudden jumps in the vibration amplitude value (Fig. 19.4a) which could destroy a real gearbox. The jumps are presumably caused by the creation of a larger number of attractors in the presence of the resonance term (Fig. 19.5). The extra degree of freedom (Fig. 19.6), which may represent a flexible shaft or a vibration neutralizer, also

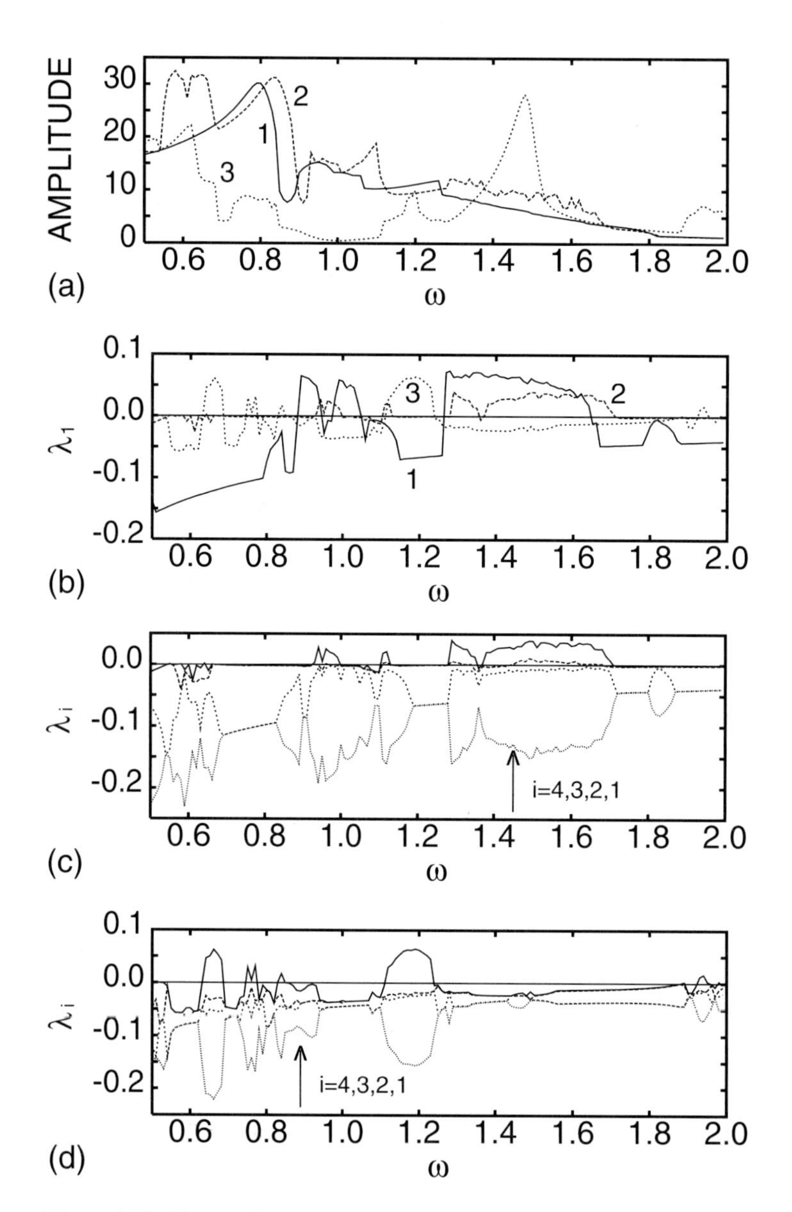

Figure 19.7: The amplitude of vibration (a) and the maximal Lyapunov exponent λ_1 (b) vs. frequency ω obtained by simulations. The initial conditions were assumed to be $x_{10} = -2.0$, $v_{10} = \dot{x}_{10} = -0.5$ and $x_{20} = 0$, $v_{20} = \dot{x}_{20} = 0$ for small ω ($\omega = 0.1$) and for each new ω_{i+1} calculations were performed for 400 excitation cycles, where the new initial conditions were the last pair of values of (x_1, v_1, x_2, v_2) for the previous ω_i. Curves '1', '2' and '3' were obtained for $k_S = 0, 0.1$ and 1.0, respectively. Four non-zero Lyapunov exponents λ_i ($i = 1, 2, 3, 4$) vs. frequency ω are shown for $k_S = 0.1$ (c) and $k_S = 1.0$ (d).

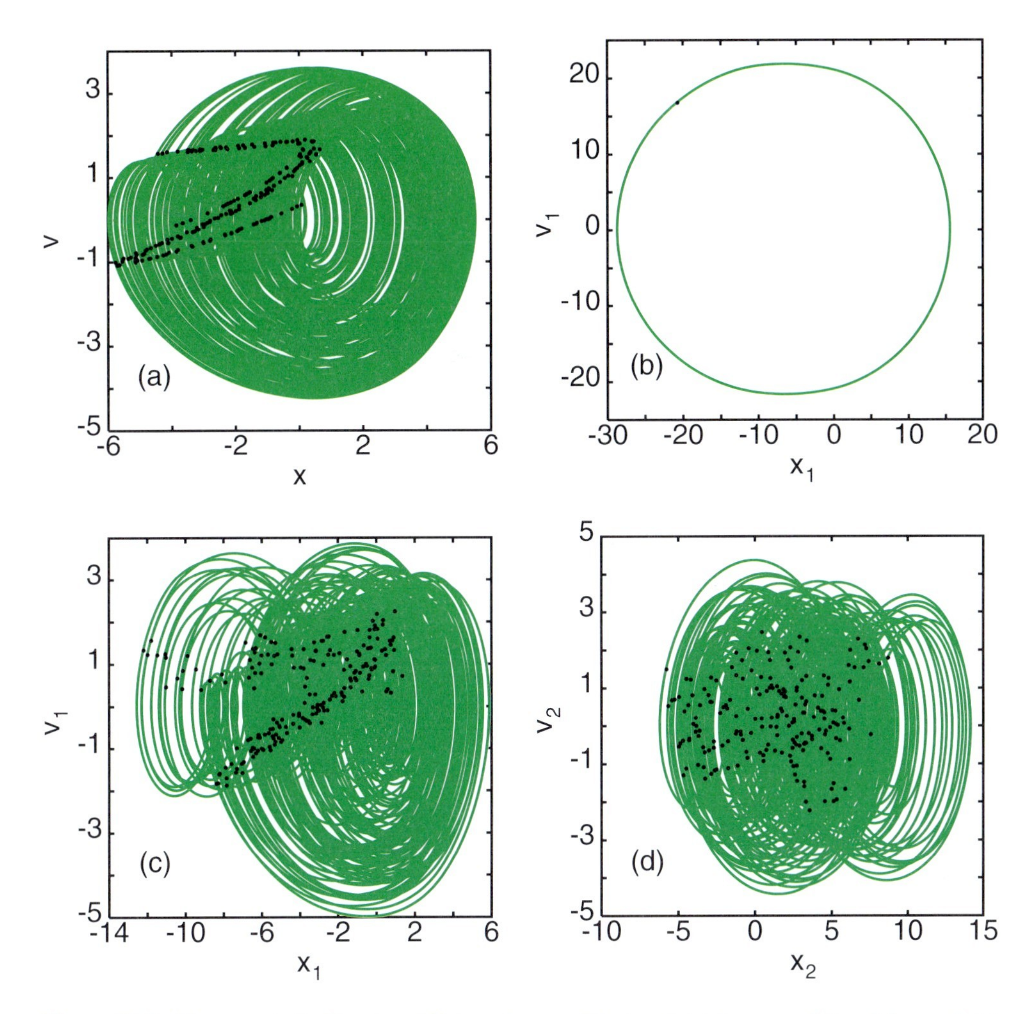

Figure 19.8: Phase portraits (lines) and Poincaré maps (black points) for $\omega = 1.5$ and $k_S = 0$ (a), $k_S = 1.0$ (b) and $k_S = 0.1$ (c and d). The initial conditions: $x_{10} = -2.0$, $v_{10} = \dot{x}_{10} = -0.5$ and $x_{20} = 0$, $v_{20} = \dot{x}_{20} = 0$.

has a considerable effect on the dynamics. We noted that the suitable use of the coupling value, the shaft stiffness k_S, can lead to regular attractors ($k_S = 1.0$) and, simultaneously, to a reduction in the vibration amplitude (Figs. 19.7a and b). Interestingly, we have found that for a small value of the coupling stiffness, $k_S = 0.1$, two of the four Lyapunov exponents were positive, leading to the hyper-chaos phenomenon. The study of the hyper-chaotic attractor will be reported in a separate article. Thus the coupling k_S is a proper bifurcation parameter giving a wide range of system behavior. The correct choice of this stiffness value can also be used to control the system vibrations.

Bibliography

[1] A. Kahraman and R. Singh, *Non–linear dynamics of a spur gear pair*, Journal of Sound and Vibration **142** (1990), 49–75.

[2] K. Sato, S. Yammamoto, and T. Kawakami, *Bifurcation sets and chaotic states of a gear system subjected to harmonic excitation*, Computational Mechanics **7** (1991), 171–182.

[3] G.W. Blankenship and A. Kahraman, *Steady state forced response of a mechanical oscillator with combined parametric excitation and clearance type non-linearity*, Journal of Sound and Vibration **185** (1995), 743–765.

[4] A. Kahraman and G.W. Blankenship, *Experiments on nonlinear dynamic behavior of an oscillator with clearance and periodically time–varying parameters*, ASME Journal of Applied Mechanics **64** (1997), 217–226.

[5] A. Raghothama and S. Narayanan, *Bifurcation and chaos in geared rotor bearing system by incremental harmonic balance method*, Journal of Sound and Vibration **226** (1999), 469–492.

[6] J. Warmiński, G. Litak, and K. Szabelski, *Dynamic phenomena in gear boxes*, in: M. Wiercigroch and B. De Kraker (eds.): *Applied Nonlinear Dynamics and Chaos of Mechanical Systems with Discontinuities*, Series on Nonlinear Science Series A, Vol. 28, World Scientific, Singapore, 2000, pp. 177–205.

[7] G. Litak and M.I. Friswell, *Vibrations in gear systems*, Chaos, Solitons and Fractals **16** (2003), 145–150.

[8] K. Szabelski, G. Litak, J. Warmiński, and G. Spuz-Szpos, *Chaotic vibrations of the parametric system with backlash and non-linear elasticity*, in: Proceedings of EUROMECH – 2nd European Nonlinear Oscillation Conference, Vol. 1, Prague, September 1996, pp. 431–435.

[9] G. Schmidt and A. Tondl, *Nonlinear vibrations*, Akademie-Verlag, Berlin, 1986.

[10] J. Warmiński, G. Litak, and K. Szabelski, *Synchronisation and chaos in a parametrically and self-excited system with two degrees of freedom*, Nonlinear Dynamics **22** (2000), 135–153.

[11] R. Chakon, F. Palmero, and F. Balibrea, *Taming chaos in a driven Josephson junction*, International Journal of Bifurcation and Chaos **11** (2001), 1897–1909.

[12] H. Cao, X. Chi, and G. Chen, *Suppressing or including chaos by weak resonant excitations in an externally-forced Froude pendulum*, International Journal of Bifurcation and Chaos (2003), in press.

[13] A. Wolf, J.B. Swift, H.L. Swinney, and J.A. Vastano, *Determining Lyapunov exponents from a time series*, Physica D **16** (1988) 285–317.

[14] T. Kapitaniak and L.O. Chua, *Hyperchaotic attractors of undirectionally coupled Chua's circuits*, International Journal of Bifurcation and Chaos **4** (1994) 477–482.

[15] J. Warmiński, G. Litak, and K. Szabelski, *Vibrations of a parametrically and self-excited system with two degrees of freedom*, in: Proceedings of the Second International Conference on Identification in Engineering Systems, Swansea, March 1999, edited by M.I. Friswell, J.M. Mottershead and A.W. Lees, University of Wales, Swansea 1999, pp. 285–294.

20 Measurement and Identification of Pre-sliding Friction Dynamics

F. Al-Bender, V. Lampaert, S.D. Fassois, D.D. Rizos, K. Worden, D. Engster, A. Hornstein, and U. Parlitz

One of the most important sources of nonlinear behavior in machines is the friction that is present in their moving components. Depending on its type, friction can be a real impediment to the effective servo control of machines: it can give rise to static errors, limit cycles or stick–slip motion. Although many models and model structures have been proposed in the literature, it is generally not possible to predict frictional behavior without resort to measurement and identification. This paper aims at the experimental identification and the modeling of pre-sliding friction, which is characterized by its hysteresis behavior in the displacement. After describing the basic characteristics of frictional behavior, and so situating the pre-sliding region, different test setups (and one data-simulation model) are described from which relevant friction data is acquired. This data can be obtained by applying any arbitrary input to the system. Thereafter, different identification techniques are employed to try to model the data. These range from the so-called "black-box" techniques; (N)ARMAX and support vector, through neural networks to physics-based ones; viz. Maxwell friction (MF) possibly in combination with a MISO FIR (finite-impulse response). The results obtained by the different techniques, although varying in their degree of accuracy, show good prediction capability of pre-sliding friction, thus indicating their effectiveness in modeling this type of friction behavior.

20.1 Introduction

Friction is defined as the resistance to motion when two objects are slid against one another. Depending on its type and on the application, friction can be a desirable thing or a drawback in a system. Examples of the first situation are brakes, clutches and friction drives; of the second, are bearings, slides and joints. One thing is common to both situations, however, and that is: one must be able to characterize the frictional behavior and, possibly, also be able to control it, in order to ensure proper functioning of a system. Except for viscous friction (which is proportional to the velocity), all other types of behavior exhibit strong nonlinearity in the displacement, velocity and time. Thus, rather than the presence of friction in itself, it is this nonlinear behavior that makes friction an impediment to system control [1]. A classical phenomenon that is frequently encountered in systems with friction is the so-called stick–slip motion, e.g. of a driven slider, which can appear in different guises. Although there are rule-of-thumb remedies for this phenomenon, e.g. increasing the drive stiffness and/or damping, the problem of accurate tracking/positioning in the presence of friction remains open. One

Nonlinear Dynamics of Production Systems. Edited by G. Radons and R. Neugebauer

ISBN 3-527-40430-9

thing however is sure: in order to deal effectively with friction, some sort of model or model structure is needed. But since friction is the result of extremely complex interactions between the contacting surfaces and lubricants, no accurate quantitative prediction of frictional behavior is yet possible based on the given material and surface properties: one must resort invariably to experimental determination. Understanding of the mechanisms involved in friction is, on the other hand, growing so that formulation of suitable models is becoming more advanced. The unknown parameters involved in such models should generally be estimated from identification tests of each given case. Notwithstanding this, here we try both "black-box" identification techniques that assume no *a priori* knowledge of the physics of the system, and physics-based techniques that are based on an assumed model structure with unknown parameters to be identified. The objective is to gauge the effectiveness of either of the approaches in identifying frictional behavior. This has important implications in regard to system identification and control as well as in validating the identification method itself in regard to universality and robustness.

In the next section, the general friction behavior will be described, together with an outline of the most important friction models and model structures. Thereafter, an overview is given of the various test setups used for obtaining friction data for the purpose of identification. Two following sections are dedicated to the two main classes of identification methods considered here, viz. regression and time-series modeling, including NARMAX, support vector, local and neural network models, and physics-based models that employ linear, nonlinear and dynamic linear regression on the Maxwell slip model structure. The results obtained by those methods are then compared and discussed. Finally, some conclusions are drawn.

20.2 Friction Characterization

Frictional behavior may be divided into two main regimes: pre-sliding and gross sliding. If we consider two objects in (frictional) contact, then there will always be a displacement resulting from an applied (tangential) force, unless the contact is infinitely stiff. (The same thing can be said of any, elasto-plastic, solid object that is subject to a force.) Now, below a certain force threshold, if the force is held constant, the displacement will likewise remain constant (except perhaps for creeping motion). When the force is decreased to zero, not all displacement will be recovered, i.e. there will, in general, be a residual displacement. This is the "pre-sliding" regime, in which, although there is relative motion, there are still points of unbroken contact and points of microslip on the two surfaces of the objects, resulting in hysteresis of the force in the displacement that marks the frictional behavior in that regime. Above that force threshold, the system will be critically stable, displacement will not remain constant for a constant applied force: the object will suddenly accelerate; all connections are broken, and we have true or gross "sliding". The term "friction force" is usually taken to mean "the resistance to the motion during true (or gross) sliding"; it usually has its maximum value at the commencement of motion (= static friction) and usually decreases with increasing relative velocity (= dynamic or kinetic friction). It has been observed and shown that the force is predominantly a function of the displacement in the pre-sliding regime showing quasi rate-independent hysteresis with non-local memory [5], and predominantly a function of the velocity (and its derivatives) in the true sliding regime, showing velocity weakening and lag behavior [1]. The borderline between the two regimes is obviously the "pre-sliding distance" and/or the "breakaway force" (= static friction force) threshold, which are not evident to determine (at least exactly) owing to many factors.

20.2.1 Friction Model Structures

Although this paper deals exclusively with pre-sliding friction identification, it may be instructive to give an overview of a representative state-of-the-art friction model structure.

Referring to [2, 3], friction force dynamics can be described, with the help of a state variable z representing average asperity deflection that satisfies a first-order nonlinear differential equation, as follows:

$$\frac{dz}{dt} = v\left(1 - \operatorname{sgn}\left(\frac{F_{\mathrm{h}}(z)}{s(v)}\right)\left|\frac{F_{\mathrm{h}}(z)}{s(v)}\right|^{n}\right), \tag{20.1}$$

$$F_{\mathrm{f}} = F_{\mathrm{h}}(z) + \sigma_1 \frac{dz}{dt} + \sigma_2 v. \tag{20.2}$$

Here, F_{f} is the friction force, which is seen to be composed of (i) a function $F_{\mathrm{h}}(z)$ that, for vanishingly small relative velocity v, corresponds to a hysteresis function with non-local memory (see further below); and for steady-state sliding, becomes equal to the Stribeck (or velocity weakening) behavior $s(v)$, and (ii) a viscous part in the state and in the velocity, characterized by the parameters σ_1 and σ_2, respectively.

During pre-sliding, the behavior will be dominated by that of F_{h} which will then tend to a hysteresis function in the relative displacement x. Since this has a direct bearing on the type of friction dealt with in the rest of this paper, we give here a more detailed sketch of pre-sliding hysteresis. Referring to Fig. 20.1(left), in a "virgin" contact, the motion starts at point $(0, 0)$ along $y(x)$, termed the "virgin" curve, which characterizes the hysteresis function. If the motion reverses at point 1, it will follow the path $(1-2-5)$, which is $y(x_m) - 2y((x - x_m)/2)$. If the motion should reverse again at point 2, it would follow $(2 - 3 - 3')$; if now there is a reversal at point 3, the trajectory will follow $(3 - 4)$. When point 4 (or 2) is reached, the shaded inner hysteresis loop is closed and the motion proceeds along $(1 - 2 - 5)$ as before. Points 2 and 3 of the closed loop are wiped out of the memory of the system, but all other reversal points must be kept in the memory (hence the term *non-local memory*). This type

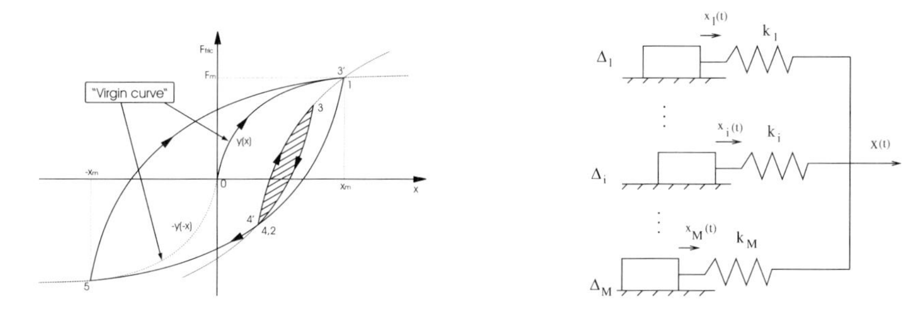

Figure 20.1: (Left) hysteresis function with non-local memory. (Right) Maxwell slip model representation.

of behavior may be modeled using either memory stacks [2], which is exact but difficult to implement, or by the easier Maxwell slip model [3, 4], which is however approximate. The latter is depicted in Fig. 20.1 (right): the hysteresis system is modeled as M massless elasto-slip elements in parallel (the system can thus have M reversal points at most). Each element i $(1 \leq i \leq M)$ is characterized by a stiffness k_i, the element's position $x_i(t)$, a spring deflection $\delta_i(t) = x(t) - x_i(t)$ and a maximum spring deflection Δ_i (before the element i starts to slip); the input displacement $x(t)$ is common to all elements. The total hysteretic force F_h is given by the summation of all operators' spring forces, $F_i(t)$:

$$F_\mathrm{h}(t) = \underbrace{k_1 \cdot \Delta_1 \cdot \bar{\delta}_1(t)}_{F_1(t)} + \ldots + \underbrace{k_M \cdot \Delta_M \cdot \bar{\delta}_M(t)}_{F_M(t)} \tag{20.3}$$

$$\underbrace{\text{if } |\delta_i(t)| < \Delta_i \text{ then } \begin{cases} \bar{\delta}_i(t) = \delta_i(t)/\Delta_i \\ x_i(t+1) = x_i(t) \end{cases}}_{\text{Stick}} \underbrace{\text{else } \begin{cases} \bar{\delta}_i(t) = \mathrm{sgn}(\delta_i) \\ x_i(t+1) = x(t) - \mathrm{sgn}(\bar{\delta}_i) \cdot \Delta_i \end{cases}}_{\text{Slip}} \tag{20.4}$$

Finally, the transition from pre-sliding to gross sliding, and back, can be quite a complex process and is thus outside the scope of this paper.

20.2.2 Acquisition of Friction Data

In order to obtain experimental friction data, be it for identification or model testing, several dedicated test setups have been built. The underlying principle of measurement, being the same for each setup, is depicted schematically in Fig. 20.2. The objective is to measure the friction force between the 'specimen' and the 'mating surface' and the relative displacement between the two as a function of time. The specimen and the mating surface may represent any friction system: plain bearing or rolling element bearing. The setup consists of a specimen holder that is attached, through a flexure, to a high-stiffness dynamometer for force measurement. The flexure allows only normal displacement and so ensures that the specimen conforms to

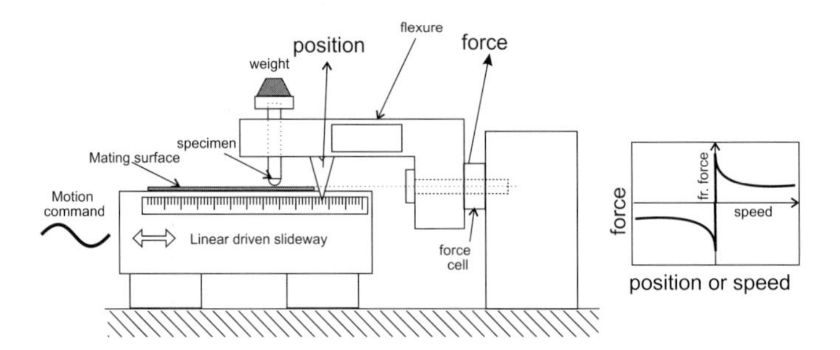

Figure 20.2: Schematic of friction test setup.

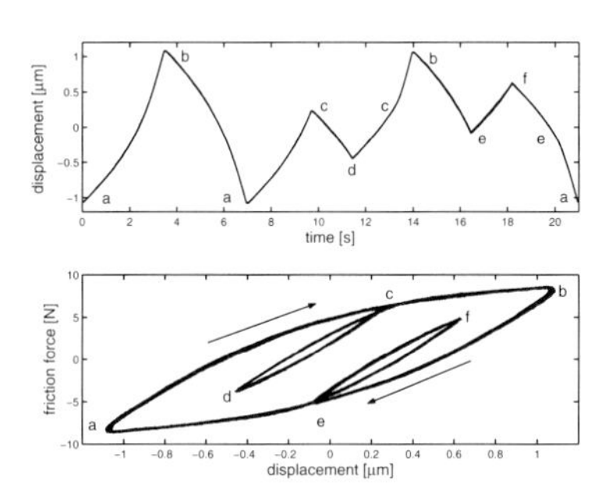

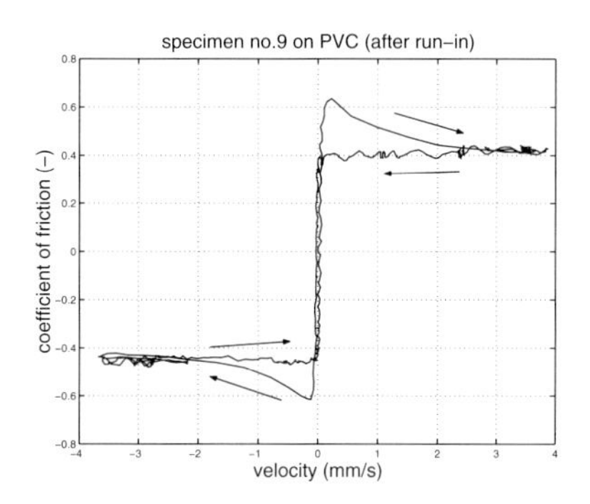

Figure 20.3: (Left) measured hysteresis friction (lower plot) in function of displacement input (upper plot). (Right) sliding friction measurement showing friction lag in the velocity.

the mating surface at the same normal load. The mating (counter) friction surface, on which the specimen rubs, is attached to the surface of a horizontal, linear slideway (table) that is driven by an electromagnetic actuator. (N.B. other means of driving the linear table are also possible depending on the required stroke, etc.). The relative position of the table with respect to the specimen may be measured by a variety of means ranging from linear encoder scales to interferometry. Not any arbitrary motion trajectory may be applied to the table, since that depends on the controllability of motion in the presence of friction. However, periodic motion can always be imposed.

Several variants of this setup have been built, notably (1) dry sliding friction test setup; (2) pre-rolling friction setup (short stroke and long stroke); (3) an ultra-accurate new test setup for investigating pre-sliding and the transition to gross sliding.

Typical measurement results are shown in Fig. 20.3.

20.2.3 Simulation of Friction Data

A novel generic friction model at asperity level has been developed at KUL/PMA [6]. This model simulates the interaction of a large number of idealized asperities subject to adhesion, creep and deformation. The input to the model is the displacement; the output is the total friction force that depends on the (idealized) contact process parameters: adhesion, creep and surface topography. So far, the model can generate friction data pertaining to low-velocity friction, which, to all intents and purposes, resembles real friction data. This is interesting in two ways. First, even if the model does not fully correspond with real friction behavior, "friction experiments" may be carried out on a desktop PC, i.e. much more easily than laboratory tests, with the added advantage that the process parameters can be changed at will, so that data for "black-box" identification techniques can easily be generated. If necessary, measurement uncertainties may be added to the data in the form of random noise. Second, if the model does prove to be able to fit real friction data, then it can be used to deduce suitable approximate macromodels for friction. This latter issue is still under investigation.

Typical pre-sliding friction data from this model that have been used for purposes of identification in this paper are shown in Fig. 20.4.

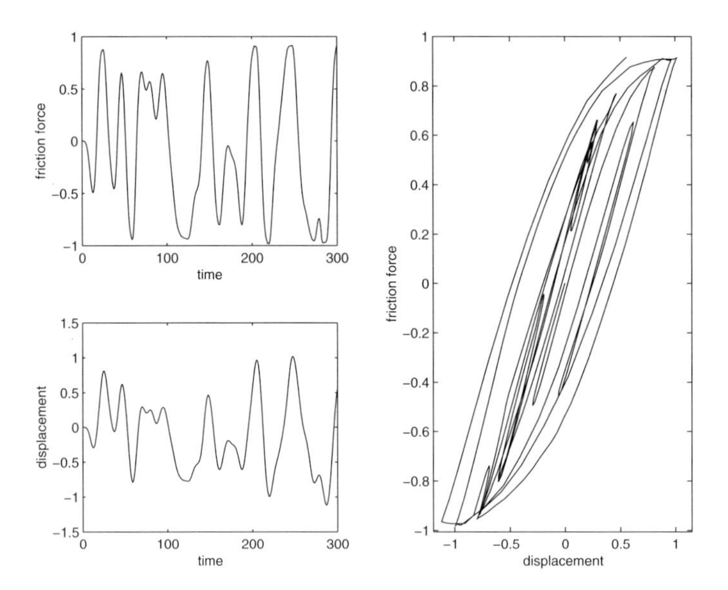

Figure 20.4: Sample of pre-sliding friction data generated by the generic asperity contact simulation model. All variables have arbitrary units.

20.3 Identificaton Methods and Results

The problem of system identification in its most general form is the construction of the functional $S[\cdot]$ which maps the inputs of the system to the outputs. The most basic approach would be to construct a complete 'black-box' representation on the basis of the data alone; this will be the first approach presented. The problem is simplified considerably by assuming that a model linear in the parameters with an appropriate structure can be used. An *a priori* structure is assumed (second approach), or clever structure detection is needed (third approach). In the remainder of this paper, the underlying theory of each identification approach will first be described, followed by its application to one and the same friction data set. This was generated by the simulation model outlined in the previous section, using a random broadband white-noise signal as input (displacement) that is sampled at 1000 Hz. In all applications, the first 8000 points were used as training set; the next 8000 as testing set. (For the neural network case, an additional 8000 points were used as validation set.) One standard, overall performance measure was used for all methods, namely the normalized mean-square error or MSE defined by

$$\mathrm{MSE}(\hat{y}) = \frac{100}{N\sigma_y^2}\sum_{i=1}^{N}(y_i - \hat{y}_i)^2, \tag{20.5}$$

where y is the output, σ_y^2 is its variance and the caret denotes an estimated quantity. This MSE has the following useful property: if the mean of the output signal $\overline{y}$ is used as the model, i.e. $\hat{y}_i = \overline{y}$ for all i, the MSE is 100. Experience shows that an MSE of less than 5.0 indicates good agreement while one of less than 1.0 reflects an excellent fit.

20.4 Regression and Time-Series Modeling

Given a data set $\{(\mathbf{x}_i, y_i)\}$, $i = 1, \ldots, N$, with vector inputs $\mathbf{x}_i \in \mathbb{R}^d$ and scalar outputs $y_i \in \mathbb{R}$, a regression problem consists in finding a mapping f from the inputs to the outputs. Usually this mapping depends on some finite number M of parameters θ_j:

$$\hat{y}_i = f(\mathbf{x}_i \,|\, \theta_1, \ldots, \theta_M) , \tag{20.6}$$

which are unknown and which have to be determined during the modeling procedure. There are two criteria for optimal parameters, accuracy and simplicity. The former guarantees that function outputs $\hat{y}_i$ deviate little from the true values y_i. The latter is a precaution to avoid overfitting phenomena for short and noisy data samples. In practice balance between the two criteria is found by using an optimization procedure which depends on the application at hand.

The regression problem can be easily extended to modeling dynamics from a time series $\{u_t\}$, $t = 1, \ldots, T$, by setting

$$y_i = u_t \quad \text{and} \quad \mathbf{x}_i = (u_{t-1}, \ldots, u_{t-d}) , \tag{20.7}$$

with $i = (d+1), \ldots, T$ and the embedding dimension d. For continuous systems an appropriate time delay τ may be taken into account: $\mathbf{x} = (u_{t-\tau}, u_{t-2\tau}, \ldots)$. If the actual time-series value u_t does not only depend on past values $u_{t-1}, \ldots, u_{t-d}$ but also on values of another time series $\{w_t\}$ these values can be included in the regression procedure by expanding the input vector $\mathbf{x}_i = (u_{t-1}, \ldots, u_{t-d}, w_t, w_{t-1}, \ldots)$.

With this formalism all other steps for finding a good model function are done analogously to the regression problem. So without any restrictions only the regression problem is considered in the next sections.

20.4.1 NARMAX Models

NARMAX models are nonlinear extensions of the well-known ARMAX[1] models ([7–10]). In the regression formalism the model form is described by some arbitrary nonlinear function $\hat{y}_i = f(\mathbf{x}_i \,|\, \boldsymbol{\theta})$. The input vector $\mathbf{x}_i$ includes past values of the time series $\{u_t\}$, values of the exogenous input $\{w_t\}$ and the noise values $\{e_t\}$ which correspond to measurement errors and model inaccuracies. Leaving out the noise terms results in a reduced form called the NARX model.

How the parameters $\boldsymbol{\theta} = (\theta_1, \ldots, \theta_M)$ enter into the model function is not defined. Practical aspects advise choosing f as a linear sum of so-called *basis functions* ϕ_j:

$$f(\mathbf{x}_i \,|\, \boldsymbol{\theta}) = \sum_{j=1}^{M} \theta_j \phi_j(\mathbf{x}_i) . \tag{20.8}$$

In this way the parameters θ_j contribute only quadratically to the sum of squared errors

$$V(\boldsymbol{\theta}) = \sum_{i=1}^{N} (y_i - f(\mathbf{x}_i \,|\, \boldsymbol{\theta}))^2 , \tag{20.9}$$

[1] AutoRegressive Moving Average with eXogeneous input.

reducing the least-squares fit to a simple convex minimization problem with a well-defined solution.

Theoretically any function type can serve as basis function. Nevertheless it should meet two requirements. First of all the basis functions have to be sufficiently flexible and complex. A combination of them should be able to approximate the potentially complex relationship between the inputs $\mathbf{x}_i$ and the outputs y_i. Obviously for nonlinear regression this implies that the basis functions have to be also nonlinear in some way.

The second requirement is contrary to the first and states that the basis functions should be as simple as possible. This is partly a matter of practicability as simple basis functions reduce computational efforts. However it also refines the control on the complexity of the model. Every basis function ϕ_j with a corresponding non-zero coefficient θ_j increases the complexity of the model by a small amount. Choosing an appropriate number M of basis functions effectively determines the trade-off between accuracy and complexity of the model (see forward orthogonal regression [7]).

In our models we have used *monomials* for basis functions:

$$\phi(\mathbf{x}) = \prod_{i=1}^{d} x_i^{p_i} \quad \text{for the input } \mathbf{x} \in \mathbb{R}^d, \tag{20.10}$$

with the maximum degree $p = \sum_i p_i$. Other popular choices are *radial basis functions*, *rational functions* and *wavelets*.

20.4.2 Support Vector Models

This section gives a short introduction to support vector regression (SVR), i.e. support vector machines used for regression. For a more in-depth discussion see [11], [12], [13] or [14]. Starting with the linear sum of basis functions

$$f(\mathbf{x}\,|\,\boldsymbol{\theta}, \theta_0) = \sum_{j=1}^{M} \theta_j \phi_j(\mathbf{x}) + \theta_0 \tag{20.11}$$

we want to find the optimal parameters $\boldsymbol{\theta}, \theta_0$. From another perspective this is equivalent to mapping the input vector $\mathbf{x} \in \mathbb{R}^d$ into a so-called *feature space* $\mathbb{R}^M$:

$$\mathbf{x} \mapsto \boldsymbol{\Phi}(\mathbf{x}) \equiv (\phi_1(\mathbf{x}), \ldots, \phi_M(\mathbf{x}))\,, \tag{20.12}$$

with $\boldsymbol{\Phi}$ as *feature vector*. In the feature space the output value y has an approximately linear relationship to the features:

$$y \sim \hat{y} = f(\mathbf{x}\,|\,\boldsymbol{\theta}, \theta_0) = \langle \boldsymbol{\theta} \,|\, \boldsymbol{\Phi}(\mathbf{x}) \rangle + \theta_0\,, \tag{20.13}$$

where $\langle \mathbf{v} \,|\, \mathbf{w} \rangle = \sum_i v_i w_i$ represents the inner product between two vectors. In most cases the dimension M of the feature space is set to much greater values than the dimension d of the original input space. Doing so results in a richer feature set and a potentially better quality of our approximation. However as a negative side-effect the computational costs are pushed toward, and sometimes beyond, bearable limits. The special ability of SVR is to lower these costs. Even more it enables the user to raise the dimension M to infinity.

There are many possible cost functions for the optimization of model parameters in SVR. Typically one minimizes

$$V(\boldsymbol{\theta}, \theta_0) = \frac{1}{2} \sum_{j=0}^{d} \theta_j^2 + C \sum_{i=1}^{N} |y_i - f(\mathbf{x}_i \,|\, \boldsymbol{\theta}, \theta_0)|_\varepsilon \,, \tag{20.14}$$

on the training set $\{(\mathbf{x}_i, y_i)\}$, $i = 1, \ldots, N$. The first sum is a regularization term. Small parameters θ_j produce simple flat functions, thus counteracting overfitting. The second sum penalizes model errors by using the robust *ε–insensitive function*

$$|x|_\varepsilon = \begin{cases} 0 & \text{if } |x| \leq \varepsilon, \\ |x| - \varepsilon & \text{otherwise,} \end{cases} \tag{20.15}$$

i.e. errors in a certain ε-bound are tolerated while all others raise the cost proportional to their distance from the bound. The constant factor C serves as a control parameter that regulates the balance between model complexity and model accuracy.

Via a Lagrange ansatz and a dual re-formulation the minimizing of the cost function (20.14) is expressed as a quadratic programming (QP) problem. Solving for the parameters yields the unique solution

$$\boldsymbol{\theta}^{\text{opt}} = \sum_{i=1}^{N} \lambda_i \boldsymbol{\Phi}(\mathbf{x}_i) \,. \tag{20.16}$$

The constant θ_0^{opt} is determined by constraints following from the QP formalism. The use of the ε-insensitive function (20.15) results in a sparse representation of the model parameters, i.e. most of the coefficients λ_i vanish. Our function now reads

$$f(\mathbf{x} \,|\, \lambda_1, \ldots, \lambda_N, \theta_0) = \sum_{i=1}^{N} \lambda_i \langle \boldsymbol{\Phi}(\mathbf{x}_i) \,|\, \boldsymbol{\Phi}(\mathbf{x}) \rangle + \theta_0 \,. \tag{20.17}$$

Note that the explicit knowledge of the former parameters $\boldsymbol{\theta}$ is not necessary. They are implicitly represented by the inner products in the feature space. However for high dimensions M the computation of these products is very time consuming. This is even more so for the QP problem, where they give rise to the main part of the computational burden.

SVR deals with this problem in an elegant way by using so-called *kernel functions*. These functions represent inner products in high- or even infinitely dimensional feature spaces but operate effectively on low- and finitely dimensional input spaces: $K(\mathbf{x}', \mathbf{x}) = \langle \boldsymbol{\Phi}(\mathbf{x}') \,|\, \boldsymbol{\Phi}(\mathbf{x}) \rangle$. The clue is that the complex mapping into the feature space $\boldsymbol{\Phi}$ and the tedious computation of inner products become superfluous. They are replaced by the kernel functions. Only in this way do the computations in infinitely high dimensional feature spaces become feasible.

Possible kernel functions are *polynomial kernels*, *Gaussian kernels* or *hyperbolic tangent kernels*. For our models we have used inhomogeneous polynomial kernels $K(\mathbf{x}', \mathbf{x}) =$

$(\langle \mathbf{x}' \mid \mathbf{x} \rangle + 1)^p$. They represent inner products in a polynomial feature space with p as the highest degree. The final form of a SVR model is

$$f(\mathbf{x} \mid \lambda_1, \ldots, \lambda_N, \theta_0) = \sum_{i=1}^{N} \lambda_i K(\mathbf{x}_i, \mathbf{x}) + \theta_0 \, . \tag{20.18}$$

20.4.3 Local Models

In contrast to the *global* models discussed so far, *local* models do not use any training data until queried with some point $\boldsymbol{x}$. A small neighborhood of $\boldsymbol{x}$ is located in the training set and a simple model using only the training points lying in this neighborhood is constructed. In statistical learning theory, local models are also referred to as *lazy learners* [17].

The most common choice for the neighborhood is to locate the k nearest neighbors $\boldsymbol{x}_{nn_1}, \ldots, \boldsymbol{x}_{nn_k}$ of $\boldsymbol{x}$ (*fixed mass*), i.e. the k points in the training set which have the smallest distance to the query point according to some arbitrary metric $\|\cdot\|$ (usually Euclidean). Alternatively, one can search for all points lying in some fixed neighborhood of the query point (*fixed size*). To find the nearest neighbors we use a fast algorithm called *ATRIA*, which relies on a binary search tree built in a pre-processing stage [16].

The model used in the neighborhood of the query point is usually fairly simple. A *locally constant* model computes a weighted average of the images of the nearest neighbors

$$\hat{f}(\boldsymbol{x}) = \frac{\sum_{i=1}^{k} w_i y_{nn_i}}{\sum_{i=1}^{k} w_i} \, . \tag{20.19}$$

Besides their speed of computation, locally constant models are very robust, as their predictions always remain in the data range given by the nearest neighbors. The weights w_i are usually drawn from a monotonically decreasing weight function, so that the influence of the furthest nearest neighbors is decreased. Otherwise, the model output becomes discontinuous, as shifting the query point $\boldsymbol{x}$ results in points suddenly entering or leaving the neighborhood.

A *locally linear* model fits a linear function

$$\hat{f}(\boldsymbol{x}) = \boldsymbol{a}^{\mathrm{T}} \cdot \boldsymbol{x} + a_0 = \tilde{\boldsymbol{a}}^{\mathrm{T}} \cdot \tilde{\boldsymbol{x}} \tag{20.20}$$

(with $\tilde{\boldsymbol{a}} = [\boldsymbol{a}; a_0]$ and $\tilde{\boldsymbol{x}} = [\boldsymbol{x}; 1]$) in the neighborhood of the query point by minimizing the weighted sum of squared errors

$$V(\boldsymbol{a}, a_0) = \sum_{i=1}^{k} w_i^2 (y_{nn_i} - \tilde{\boldsymbol{a}}^{\mathrm{T}} \cdot \tilde{\boldsymbol{x}})^2 \, . \tag{20.21}$$

The solution for $\tilde{\boldsymbol{a}}$ is given by

$$\tilde{\boldsymbol{a}} = (\boldsymbol{X}_W^{\mathrm{T}} \boldsymbol{X}_W)^{-1} \boldsymbol{X}_W^{\mathrm{T}} \cdot \boldsymbol{y}_W = \boldsymbol{X}_W^{\dagger} \cdot \boldsymbol{y}_W \, , \tag{20.22}$$

where $\boldsymbol{X}_W = \boldsymbol{W} \cdot \boldsymbol{X}$, $\boldsymbol{y}_W = \boldsymbol{W} \cdot \boldsymbol{y}$, $\boldsymbol{X} = [\tilde{\boldsymbol{x}}_{nn_1}^{\mathrm{T}}, \ldots, \tilde{\boldsymbol{x}}_{nn_k}^{\mathrm{T}}]^{\mathrm{T}}$, $\boldsymbol{y} = [y_{nn_1}, \ldots, y_{nn_k}]^{\mathrm{T}}$ and $\boldsymbol{W} = \mathrm{diag}([w_1, \ldots, w_k])$ [15]. The term $\boldsymbol{X}_W^{\dagger}$ denotes the pseudoinverse of $\boldsymbol{X}_W$, which can be calculated using the singular value decomposition $\boldsymbol{X}_W = \boldsymbol{U} \cdot \boldsymbol{S} \cdot \boldsymbol{V}^{\mathrm{T}}$, where $\boldsymbol{S} = \mathrm{diag}([\sigma_1, \ldots, \sigma_k])$ with the singular values σ_i. The pseudoinverse is then given by $\boldsymbol{X}_W^{\dagger} = \boldsymbol{V} \cdot \boldsymbol{S}^{-1} \cdot \boldsymbol{U}^{\mathrm{T}}$.

Locally linear models give usually more accurate estimations than locally constant ones, but they need an additional regularization method to secure stability. One popular approach for regularization is the *truncated principal component regression* (TPCR). During the calculation of the pseudoinverse $\boldsymbol{X}_W^{\dagger}$ small singular values in the diagonal matrix $\boldsymbol{S}$ are set to zero. This can be further improved by *soft thresholding*, where singular values lying in a specific interval $[s_1, s_2]$ are smoothly weighted down to zero using the weight function [15]

$$f(\sigma) = \begin{cases} 0 & s_1 > \sigma\,, \\ \left(1 - \left(\dfrac{s_2 - \sigma}{s_2 - s_1}\right)^2\right)^2 & s_1 \leq \sigma < s_2\,, \\ 1 & s_2 \leq \sigma\,. \end{cases} \tag{20.23}$$

For locally linear models, four types of parameters must be chosen: the number of nearest neighbors k and the metric used to locate these, the weighting function for the weights w_i and the regularization parameters s_1, s_2. For time-series prediction, one must also find good values for the embedding parameters dimension and delay.

By calculating the *leave-one-out cross-validation error* over the training set, one can use a simple cyclic optimization algorithm for finding good parameter values [15].

20.4.4 Neural Network Methods

Artificial neural networks have come into recent prominence because of their ability to learn input–output relationships by training on measured data and they appear to show some promise for the system identification problem. The most often used forms are the multi-layer perceptron (MLP) and radial basis function (RBF); the model used here will be the MLP. In order to form a model with a neural network it is necessary to specify the form of the inputs and outputs; in this case, the NARX functional form will be used,

$$y_i = F(y_{i-1}, \ldots, y_{i-n_y}; x_{i-1}, \ldots, x_{i-n_x}). \tag{20.24}$$

In the case of the MLP with a linear output neuron, the appropriate structure for a single-input–single-output (SISO) system is

$$y_i = s + \sum_{j=1}^{n_h} w_j \tanh\left(\sum_{k=1}^{n_y} v_{jk} y_{i-k} + \sum_{m=0}^{n_x-1} u_{jm} x_{i-m} + b_j\right), \tag{20.25}$$

where the w's, u's and v's are the weights, or if a nonlinear output neuron is used,

$$y_i = \tanh\left[s + \sum_{j=1}^{n_h} w_j \tanh\left(\sum_{k=1}^{n_y} v_{jk} y_{i-k} + \sum_{m=0}^{n_x-1} u_{jm} x_{i-m} + b_j\right)\right] \tag{20.26}$$

as a linear output neuron is often adopted for regression problems; the structure (20.25) is used here.

Some of the earliest examples of the use of neural networks for system identification and modeling are the work of Chu et al. [20] and Narendra and Parthasarathy [22]. Masri et al. are amongst the first structural dynamicists to exploit the techniques [21]. The last work is interesting because it demonstrates 'dynamic neurons' which are said to increase the utility of the MLP structure for modeling dynamical systems. The most comprehensive program of work to date is arguably that of Billings and co-workers, starting with [18] for the MLP structure and [19] for the RBF.

The neural networks described further were trained using a basic back-propagation algorithm.

The strategy for optimizing the neural network structure involves using a training set to establish weights and a validation set to fix the optimum hidden layer number, initial conditions, stopping time and number of lags. The programme is summarized by the following pseudo-code.

```
for different numbers of lags = 1 to 6
    {
    for number of hidden layer neurons = 1 to 30 step 2
        {
        for different random initial conditions = 1 to 10
            {
            train network on training data
            evaluate on validation data
            terminate training at minimum on validation set
            }
        }
    }
```

For simplicity the maximum number of x-lags was assumed equal to the maximum number of y-lags. During training, time-varying learning coefficients and momenta were used. These were initially set high to allow potential movement between local minima and then annealed to allow fine tuning. The networks updated the weights at each presentation of an input–output pair, i.e. an epoch of unity was used.

20.4.5 Numerical Results of Black-box Methods

In our models we estimated the current force value $F(t)$ by using the values of the displacement signal $x(t), x(t-1), x(t-2), \ldots$ but also by using past values of the original friction force signal $F(t-1), F(t-2), \ldots$. We refer to these estimations as 1-step predictions. There are situations where it is not possible to obtain the original force values. Then 1-step predictions are not applicable. However the models can still be applied by predicting the values $F(t)$ iteratively. Starting from some original value $F(t)$ we estimate $\hat{F}(t+1)$. With the estimation $\hat{F}(t+1)$ we estimate $\hat{F}(t+2)$ and so on. We refer to these estimations as free-run predictions.

As noted before we restrict ourselves to the regression case. This means our goal is to find a mapping from the input vectors $\boldsymbol{x}_i$ to the scalar outputs y_i. The optimal models were generated

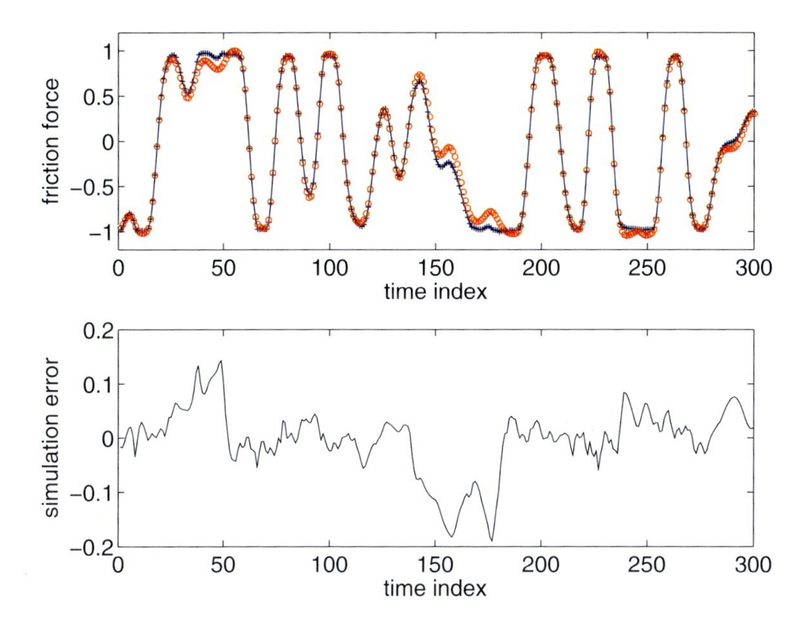

Figure 20.5: Top: interval of test set u_i (solid line, plus signs), free-run predictions of SV model $\hat{u}_i$ (circles). Bottom: Error $e_i = u_i - \hat{u}_i$, MSE $= 0.76$, $\max_i(\{|e_i|\}) = 0.230$.

as described in the previous sections. Here we specify the resulting parameters of the respective models. For all three models, $y_i = F(t_i)$, $\boldsymbol{x}_i = (F(t_i - 1), x(t_i), x(t_i - 1), \ldots, x(t_i - 6))$ with $t_i = i$. For the NARMAX model, the maximum polynomial degree is $p = 5$ and the number of monomials is 154. For the SVM, the maximum polynomial degree is $p = 5$ and the number of support vectors is 2643. For the local model, the number of next neighbors is 103 and the regularization is $s_1 = 9.86 \times 10^{-6}$, $s_2 = 1.42 \times 10^{-4}$.

Typical results obtained by the aforementioned methods are shown in Fig. 20.5. An overview of performance of all methods is given in Sect. 20.6.

For the neural network method, the data were divided into three disjoint sets of 8000 points each: a training set, a validation set and a testing set. The models were trained on the training and validation sets and finally judged on the basis of their prediction accuracy for the test set. The best results were obtained with a 6:2:1 network. This means that the maximum number of x- and y-lags was three and two hidden layer nodes were used. This structure gave a MSE for a free-run prediction of 5.79% on the training set and 6.56% on the validation set. The MSE on the final independent testing set was 6.49%. Figure 20.6 shows free-run predictions over the same interval.

20.5 Identification of Physics-based Models

Three different identification approaches, designated as linear regression (LR), dynamic linear regression (DLR) and nonlinear regression (NLR) are postulated and assessed. They are all physics-based, using the Maxwell Slip model, and are therefore capable of accounting for

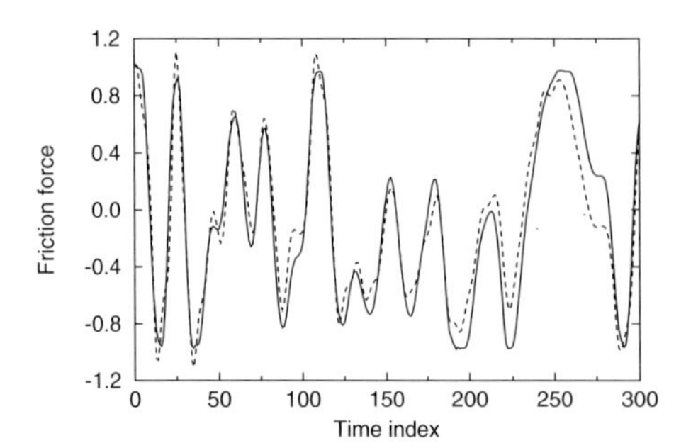

Figure 20.6: Interval of test set y_i (solid line), free-run predictions of MLP model $\hat{u}_i$ (dotted line). MSE = 6.49%.

the hysteretic, with non-local memory, relationship between the displacement (excitation) and pre-sliding friction force (response). Among their main advantages is simplicity and physical interpretation.

20.5.1 The Linear Regression (LR) Approach

The pre-sliding friction force is, in this approach, approximated via an LR(M) model from within the model class:

$$\text{LR}(M, \boldsymbol{k}): \qquad F(t) = k_1 \cdot \Delta_1 \cdot \bar{\delta}_1(t) + \ldots + k_M \cdot \Delta_M \cdot \bar{\delta}_M(t) + e(t). \tag{20.27}$$

This structure is obtained directly from the Maxwell slip model (see Sect. 20.2.1 above) with M superimposed elements, with $\bar{\delta}_i(t)$ $(i = 1, \ldots, M)$ designating the normalized (with respect to Δ_i) ith spring deflection and $e(t)$ the model error (assumed to be a zero-mean uncorrelated random sequence with variance σ_e^2). In this model structure the Δ_i's are pre-assigned as equally spaced within the range of the asperity deformation [3]; an apparently gross approximation. The model parameters to be estimated thus are the stiffnesses k_i $(i = 1, \ldots, M)$, collected into the vector $\boldsymbol{k}$. As the model structure is linear in the parameters, minimization of a quadratic function of the model error over a certain time range leads to a linear regression-type estimator for $\boldsymbol{k}$.

20.5.2 The Dynamic Linear Regression (DLR) Approach

In this approach the pre-sliding friction force is approximated via a DLR(M, n) model [23]. This is also based upon the Maxwell slip model with M superimposed elements, but the normalized spring deflections $\bar{\delta}_i(t)$, collected into the vector[1] $\bar{\boldsymbol{\delta}}(t) \triangleq \left[\bar{\delta}_1(t) \ldots \bar{\delta}_M(t)\right]^{\mathrm{T}}$, are now driven through a band of finite impulse response (FIR) filters (each one of order n) to produce the pre-sliding friction force. The DLR model class is thus of the form:

$$\begin{aligned}\text{DLR}(M, n, \bar{\boldsymbol{\theta}})\text{: } F(t) &= \boldsymbol{\theta}_0^{\mathrm{T}} \cdot \bar{\boldsymbol{\delta}}(t) + \ldots + \boldsymbol{\theta}_n^{\mathrm{T}} \cdot \bar{\boldsymbol{\delta}}(t-n) + e(t) \\ &= \sum_{j=0}^{n} \boldsymbol{\theta}_j^{\mathrm{T}} \cdot \bar{\boldsymbol{\delta}}(t-j) + e(t),\end{aligned} \tag{20.28}$$

[1] Bold-face symbols designate vector quantities.

with $\boldsymbol{\theta}_j$ $(j = 0, ..., n)$ designating the FIR filter band's jth coefficient vector, $\bar{\boldsymbol{\theta}}$ the composite FIR coefficient vector and $e(t)$ the model error. Note that, as in the previous case, the approximation of pre-assigned (equally spaced) Δ_i's is utilized.

The main advantage of this model structure is in the extra dynamics and complexity due to the FIR filter band, which may account for discrepancies between the previous LR model structure and the actual pre-sliding friction dynamics.

Finally, as the DLR(M, n) model structure is still linear in the parameters, minimization of a quadratic function of the model error over a certain time range leads to a linear regression type estimator for $\bar{\boldsymbol{\theta}}$.

20.5.3 The Nonlinear Regression (NLR) Approach

The nonlinear regression (NLR) approach [23] shares the basic Maxwell slip representation of the pre-sliding friction force, its important difference from the LR approach however being that the thresholds (Δ_i's) are no longer pre-assigned, but, instead, estimated along with the stiffnesses (k_i's). The NLR model class is thus of the form:

$$\text{NLR}(M, \boldsymbol{k}, \boldsymbol{\Delta}): \qquad F(t) = k_1 \cdot \Delta_1 \cdot \bar{\delta}_1(t) + \ldots + k_M \cdot \Delta_M \cdot \bar{\delta}_M(t) + e(t), \quad (20.29)$$

with M designating the number of superimposed elements, $\boldsymbol{k}$ the stiffness vector and $\boldsymbol{\Delta}$ the threshold vector.

The NLR approach thus corresponds to complete identification of the Maxwell slip model through elimination of the pre-assigned threshold approximation. This is naturally expected to lead to increased accuracy, yet, the price paid for it is that the linearity in the model parameters is now lost. As a consequence, minimization of a quadratic function of the model error now leads to a nonlinear regression-type estimator for the parameter vectors $\boldsymbol{k}$ and $\boldsymbol{\Delta}$.

NLR model parameter estimation is thus based upon a postulated two-phase, hybrid, optimization scheme. The first (*pre-optimization*) phase utilizes probabilistic genetic algorithm (GA)-based optimization [24] in order to explore large areas of the parameter space and locate regions where global or local minima may exist. The second (*fine-optimization*) phase utilizes the Nelder–Mead downhill simplex algorithm [25, p. 289] for locating the exact global or local minima within the previously obtained regions.

This two-phase scheme is capable of locating (with high probability) the true global minimum of the cost function and circumventing problems associated with local minima, which are quite common in this case [23]. Furthermore, the Nelder–Mead algorithm makes use of cost function evalutions but not of derivatives, which are not defined everywhere, as the cost function is discontinuous in the parameter space [23].

20.5.4 Model Order Selection and Assessment

LR(M) and NLR(M) model order selection (that is, selecting the number M of superimposed elements) is based upon the successive estimation (training) of models for increasing M and evaluation of the MSE criterion. On the other hand, DLR(M, n) model order selection is based upon the estimation (training) of models corresponding to various values of n for any given M. The final model is selected following consideration of various values of M.

Model testing and assessment is based upon the model's simulation performance within a subset of the data, referred to as the *testing set*, which was not used in estimation (*cross-validation principle*).

20.5.5 Identification Results

As in the previous method application, two sets of 8000 points each were used for training and testing respectively. Model order selection results are, for all three approaches, presented in Fig. 20.7, which depicts the mean square error (MSE) as a function of the model order. Figure 20.7a refers to the training set and Fig. 20.7b to the testing set. It is worth noting that, within the DLR approach, the FIR order $n = 2$ is selected (for every M) as the reduction of the MSE beyond this particular order was found to be practically insignificant.

As Fig. 20.7a indicates, the MSE criterion is, for all three methods, minimized for models including $M = 4$ operators; hence the LR(4), DLR(4, 2) and NLR(4) models are selected. The behavior of the models within the testing set is analogous (Fig. 20.7b), and they are thus selected as valid representations of the pre-sliding friction dynamics.

Although all three models exhibit low error, their examination indicates that the NLR(4) model achieves the minimum MSE, followed by the DLR(4, 2) and LR(4) models. In fact the NLR(4) model achieves a MSE that is less than half of that of the DLR(4, 2) model in both the training and testing sets (values in Table 20.1, Sect. 20.6). This achievement becomes even more significant when taking into account the fact that the NLR model is characterized by a parametric complexity (number of estimated parameters) of 8, versus 12 for the DLR(4, 2) model. An overview of performance of all methods is given in Sect. 20.6.

The superiority of the NLR(4) model over its LR(4) and DLR(4, 2) counterparts is also evident from Fig. 20.8, in which the model-based simulation is compared to the actual friction force for part of the testing set and each estimated model. The corresponding error signals are also presented.

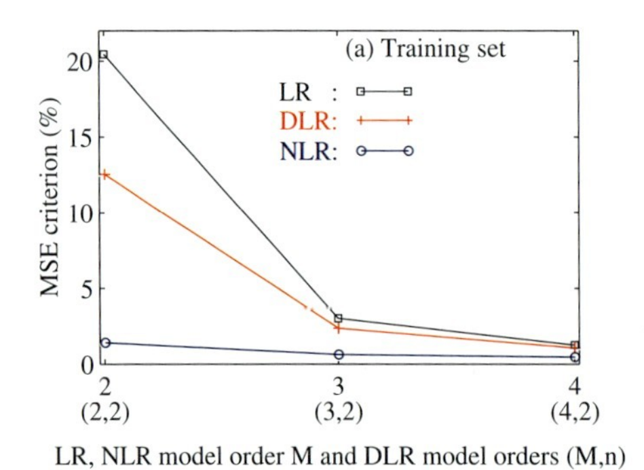

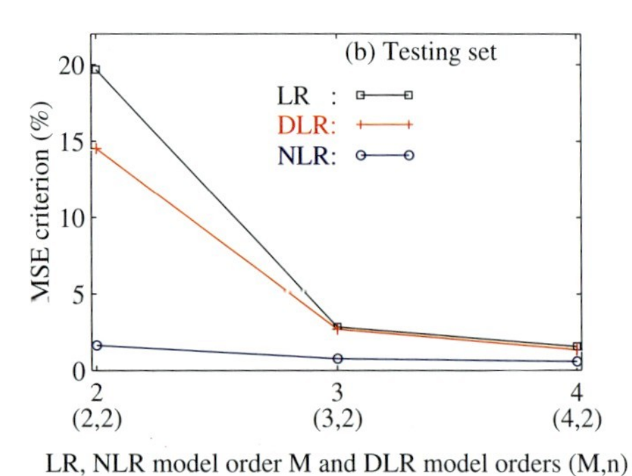

Figure 20.7: MSE criterion vs. model order: (a) training set, (b) testing set.

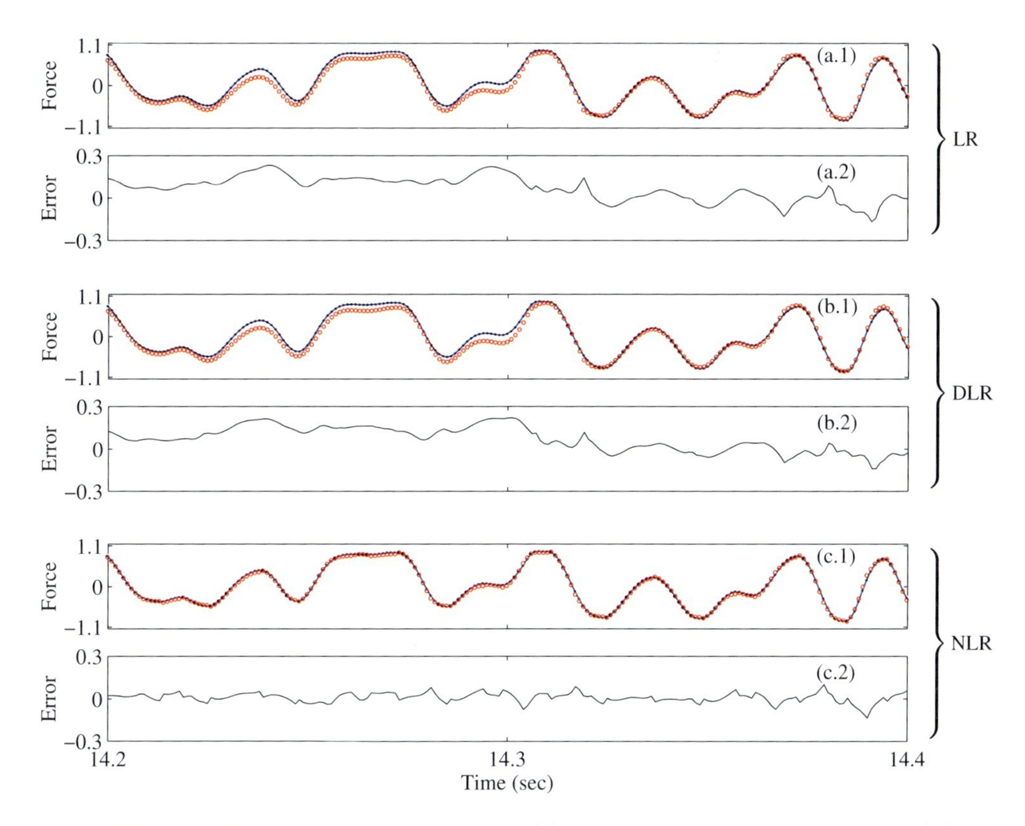

Figure 20.8: Model-based friction force simulation (o) contrasted to the actual friction force (−), and the corresponding error for each estimated model (part of the testing set).

20.6 Discussion and Conclusions

The identification results achieved by the various methods considered in this paper are in general very good and may suffice for control application purposes. There are some notable differences, however, in the performance of each method as reflected in Table 20.1. Firstly, using the performance criterion of Sect. 20.3, we conclude that the black-box (excepting neural networks) as well as the physics-based methods result in excellent fits; the neural network method is on the border of good agreement. Within the black-box methods, the NARMAX and SV score the best; and within the physics ones, the NLR. This is achieved, however, at the expense of both programming and calculation intensiveness. The number of unknown parameters to be estimated is not always a suitable measure for complexity but a good indicator thereof. Noteworthy is the fact that physics-based models, when we take into account their relative simplicity, would score the overall best. This is due, in our case, to the fact that the Maxwell slip model used as a basis for those methods accounts very closely for presliding friction force dynamics. Knowledge of a model (structure) representing a physical phenomenon, however approximately, is therefore a considerable asset in identification. In

Table 20.1: Characteristics of the estimated models: pre-sliding friction force simulations (testing set).

	Model class	Parametric complexity (number of estimated parameters)	MSE criterion (%)	Max error
Black box	NARMAX	154	0.66	0.23
	SV	2643	0.76	0.23
	Local model	N.A.	1.26	0.33
	Neutral network	22	6.49	0.46
Physics based	LR(4)	4	1.52	0.28
	DLR(4,2)	12	1.24	0.24
	NLR(4)	8	0.46	0.28

the absence thereof, black-box techniques prove very powerful tools for nonlinear system identification, not requiring any *a priori* knowledge of the physics, yet producing excellent fits.

In conclusion, this paper has presented an overview of friction force dynamics; its characterization, measurement and identification. Owing to its presence in most machine elements coupled with its highly nonlinear character, friction force dynamics has to be identified and modeled if effective control of machines is desired. This task is eased significantly by devising accurate measurement setups, constructing suitable models and developing efficient methods of identification, as this paper has clearly shown. An obvious next step is to extend this methodology from pre-sliding friction to gross sliding and the transition, and to apply it to experimental friction data of various sources, which will be the subject of future work.

Acknowledgment

The authors wish to acknowledge the financial support of the Volkswagenstiftung, under grant no. I/76938.

Bibliography

[1] B. Armstrong-Hélouvry, *Control of Machines with Friction*, Kluwer Academic Publishers, 1991.

[2] J. Swevers, F. Al-Bender, C. Ganseman, and T. Prajogo, *An integrated friction model structure with improved presliding behavior for accurate friction compensation*, IEEE Trans. on Automatic Contr. **45** (2000), 675.

[3] V. Lampaert, J. Swevers, and F. Al-Bender, *Modification of the Leuven integrated friction model structure*, IEEE Trans. on Automatic Contr. **47** (2002) 683.

[4] W. Symens, F. Al-Bender, J. Swevers, and H. Van Brussel, *Dynamic characterization of hysteresis elements in mechanical systems*, Proceedings of the American Control Conference, 2002, pp. 4129–4134.

[5] T. Prajogo, Ph.D. thesis, Werktuigkunde, Katholieke Universiteit Leuven, (1999).

[6] F. Al-Bender, V. Lampaert, and J. Swevers, *A novel generic model at asperity level for dry friction force dynamics*, Tribology Letters **16** (1) (2004), 81–93.

[7] S.A. Billings et al., *Identification of MIMO non-linear systems using a forward-regression orthogonal estimator*, International Journal of Control **6** (1989), 2157–2189.

[8] L.A. Aguirre and S.A. Billings, *Retrieving dynamical invariants from chaotic data using NARMAX models*, International Journal of Bifurcation and Chaos **2** (1995), 449–474.

[9] S.A. Billings and D. Coca, *Discrete wavelet models for identification and qualitative analysis of chaotic systems*, International Journal of Bifurcation and Chaos **7** (1999), 1263–1284.

[10] L.A. Aguirre et al., *Nonlinear multivariable modeling and analysis of sleep apnea time series*, Computers in Biology and Medicine **29** (1999), 207–228.

[11] V. Vapnik, *The Nature of Statistical Learning Theory*, Springer-Verlag, New York, 1995.

[12] A.J. Smola and B. Schölkopf, *A tutorial on support vector regression*, Technical Report NC-TR-98-030, Royal Holloway College, University of London, 1998.

[13] K.-R. Müller et al., *Predicting Time Series with Support Vector Machines* (Advances in Kernel Methods – Support Vector Learning) MIT Press, Cambridge, MA, 1999, pp. 211–242.

[14] A. Gretton et al., *Support vector regression for black-box system identification*, in: Proceedings of the 11th IEEE Workshop on Statistical Signal Processing, 2001.

[15] J. McNames, Ph.D. thesis, Stanford University (1999).

[16] C. Merkwirth et al., *Fast nearest-neighbor searching for nonlinear signal processing*, Physical Review E **62**(2) (2000), 2089–2097.

[17] C.G. Atkeson et al., *Locally weighted learning*, Artificial Intelligence Review **11** (1997), 11–73.

[18] S.A. Billings, H.B. Jamaluddin, and S. Chen, *Properties of neural networks with applications to modeling nonlinear dynamical systems*, International Journal of Control **55** (1991), 193–224.

[19] S. Chen, S.A. Billings, C.F.N. Cowan, and P.M. Grant, *Nonlinear systems identification using radial basis functions*, International Journal of Systems Science **21** (1990), 2513–2539.

[20] S.R. Chu, R. Shoureshi, and M. Tenorio, *Neural networks for system identification*, IEEE Control and Systems Magazine **10** (1990), 36–43.

[21] S.F. Masri, A.G. Chassiakos, and T.K. Caughey, *Structure-unknown nonlinear dynamic systems: identification through neural networks*, Smart Materials and Structures **1** (1992), 45–56.

[22] K.S. Narendra and K. Parthasarathy, *Identification and control of dynamical systems using neural networks*, IEEE Transactions on Neural Networks **1** (1990), 4–27.

[23] D.D. Rizos and S.D. Fassois, 4th International Symposium on Investigation of Nonlinear Dynamic Effects in Production Systems, Paper no. 21, Chemnitz, Germany, 2003.

[24] S.K. Pal and P.P. Wang, *Genetic Algorithms for Pattern Recognition*, 1st edn., CRC Press, Florida, 1996.

[25] W.H. Press, B.P. Flannery, S.A. Teukolsky, and W.T. Vetterling, *Numerical Recipes: The Art of Scientific Computing*, 2nd edn., Cambridge University Press, Cambridge, 1990.

21 Coordination of Mechanical Systems

H. Nijmeijer and A. Rodriguez-Angeles

A controller that solves the problem of position coordination of two (or more) mechanical systems is presented. The coordination controller works under a master–slave scheme, in the case when only position measurements are available. This work is focused in robot manipulators, which are mechanical systems widely used in industry applications nowadays. The controller consists of a feedback-control law and two nonlinear observers. It is shown that the controller yields semi-global ultimate uniform boundedness of the closed-loop errors, and a relation between the bound of the errors and the gains on the controller is established. Experimental results show, despite obvious model uncertainties, a good agreement with the predicted convergence.

21.1 Introduction

Nowadays the developments in technology and requirements on efficiency and quality in production processes have originated more complex and integrated systems. These integrated systems are a combination of many different disciplines such as mechanics, electronics, control, etc. The final goal of this synergy is to improve the performance, and in many cases to give rise to more flexible and robust systems.

In manufacturing processes, automotive applications and tele-operated systems there is a high requirement on flexibility and manoeuverability of the involved systems. In most of these processes the use of mechanical systems, particularly robot manipulators, is widely spread, and their variety in use is practically endless, e.g. assembling, transporting, painting, welding and grasping. All the mentioned tasks require large manoeuverability and manipulability from the robots, such that some of the tasks cannot be carried out by a single robot. In those cases the use of multi-robot systems, working under cooperative or coordinated schemes, has been considered as an option.

The cooperative schemes give flexibility and manoeuverability that cannot be achieved by an individual system, e.g. multi-finger robot hands, multi-robot systems, multi-actuated platforms [2, 12, 13, 15], vibro-machinery [1] and tele-operated master–slave systems [3, 5, 7, 11]. Typically robot coordination and cooperation of manipulators form important illustrations of the same goal, where it is desired that two or more mechanical systems, either identical or different, are asked to work in synchrony. In robot coordination the basic problem is to ascertain synchronous motion of two (or more) robotic systems. This is obviously a control problem that implies the design of suitable controllers to achieve the required coordinated motion.

Nonlinear Dynamics of Production Systems. Edited by G. Radons and R. Neugebauer

ISBN 3-527-40430-9

The problem of coordination of mechanical robotic systems can be seen as tracking between two (or more) robotic systems. Although it seems to be a straightforward extension of classical tracking controllers, this problem implies challenges that are not considered in the design of tracking controllers. These challenges arise by the interaction (inter-connection) between the robots. This inter-connection cannot be neglected, since it is precisely what generates the flow of information necessary to guarantee, and at the same time to determine, the coordinated behavior.

This work addresses the problem of coordination of mechanical systems, particularly robotic systems, under master–slave schemes. Since the pioneering work of Goertz [4], most of the master–slave robotic systems – if not all – are based on full knowledge of the dynamic model and joint variables (position, velocity and acceleration) of the master and slave robots [8, 10]. However, in practice, robot manipulators are equipped with high-precision position sensors, such as encoders, but very often the velocity measurements are obtained by means of tachometers, which are contaminated by noise. Velocity-measuring equipment is even omitted due to the savings in cost, volume and weight that can be obtained. New technologies have been designed for measuring velocities and accelerations, e.g. brushless AC motors with digital servo-drivers, accelerometers, etc. Nevertheless, such techniques are not very common in applications yet.

A coordination controller that solves the problem of position coordination of two (or more) robot systems is presented. The coordination controller works under a master–slave scheme, in the case when only position measurements of both master and slave robots are available. It is shown that the proposed coordination controller yields semi-global ultimate uniform boundedness of the closed-loop errors, and a relation between the ultimate bound of the errors and the gains on the controller is established. The coordination controller consists of a feedback-control law and two nonlinear observers. The observers re-construct the master and slave joint velocity and acceleration which are used in the feedback controller. Of course, presently other ways of estimating velocity and acceleration signals, like numerical differentiation or low-pass filters, are available, and in principle such alternatives could be used in the here-developed coordination scheme. These alternative techniques have the advantage of simplicity in implementation; however they present a reduced bandwidth and in general there is not (or it is too difficult to determine) an analytical method to guarantee stability of the closed-loop systems.

The setup here considered is as follows. Consider two rigid joint robots, such that the movement of one of the robots is independent of the other. This robot is the dominant one and will be referred to as the master robot. The master robot is driven by a control $\tau_{\rm m}(\cdot)$, which, in the ideal case, ensures convergence of the master robot joint positions and velocities $q_{\rm m}, \dot{q}_{\rm m}$ to a given desired trajectory $q_{\rm d}, \dot{q}_{\rm d}$. Then, the goal is to design inter-connections and a feedback controller for the non-dominant robot, hereafter referred to as the slave, such that its position and velocity $q_{\rm s}, \dot{q}_{\rm s}$ coordinate (synchronize) to those of the master robot $q_{\rm m}, \dot{q}_{\rm m}$. However, the input torque $\tau_{\rm m}$, the dynamic model and parameters of the master robot, as well as the velocity and acceleration variables $\dot{q}_{\rm m}, \ddot{q}_{\rm m}$, are not available for the design of the slave control law $\tau_{\rm s}(\cdot)$. Therefore for the design of the slave inter-connections and controller only master and slave angular positions $q_{\rm m}, q_{\rm s}$ are available by means of measurements.

Notice that the goal is to ensure coordination between the slave-robot trajectories $q_{\rm s}, \dot{q}_{\rm s}$ and the master-robot trajectories $q_{\rm m}, \dot{q}_{\rm m}$, and not to the master desired trajectories $q_{\rm d}, \dot{q}_{\rm d}$, which may not be realized due to model uncertainties or disturbances in the system, e.g. noise, unknown loads or friction.

Most of the master–slave robot systems are designed to interact with their environment, and thus force–position controllers are required [8, 10]. This work is focused only on the position-coordination problem. Nevertheless, in the case of a master–slave system interacting with its environment, passive compliance or end effector compliance models can be used in order to control the interaction forces between the slave robot and the environment.

The chapter is organized as follows. Section 21.2 presents the dynamic model of the master and slave robots. The proposed coordination controller is presented in Sect. 21.3. It includes a gain-tuning procedure that ensures the convergence properties of the coordination system. In Sect. 21.4 experimental results are presented and discussed. Section 21.5 presents general conclusions and some further extensions of the proposed controller.

21.2 Dynamic Model of the Robot Manipulators

Without loss of generality and considering that the friction phenomena can be compensated separately, it is assumed that the robots are frictionless. Consider a pair of fully actuated rigid robots, each one with the same number of joints, i.e. $q_i \in \mathbb{R}^n$, where $i = \mathrm{m}, \mathrm{s}$ identifies the master (m) and slave (s) robots; all the joints are rotational. This does not mean, however, that they are identical in their parameters (masses, inertias, etc.).

For each of the robots, the kinetic energy is given by $T_i(q_i, \dot{q}_i) = \frac{1}{2}\dot{q}_i^{\mathrm{T}} M_i(q_i)\dot{q}_i$, $i = \mathrm{m}, \mathrm{s}$, with $M_i(q_i) \in \mathbb{R}^{n \times n}$ the symmetric, positive-definite inertia matrix, and the potential energy is denoted by $U_i(q_i)$. Hence, applying the Euler–Lagrange formalism [17] the dynamic model of the robot is given by

$$M_i(q_i)\ddot{q}_i + C_i(q_i, \dot{q}_i)\dot{q}_i + g_i(q_i) = \tau_i, \qquad i = \mathrm{m}, \mathrm{s}, \tag{21.1}$$

where $g_i(q_i) = (\partial/\partial q_i)U_i(q_i) \in \mathbb{R}^n$ denotes the gravity forces, $C_i(q_i, \dot{q}_i)\dot{q}_i \in \mathbb{R}^n$ represents the Coriolis and centrifugal forces and τ_i is the $[n \times 1]$ vector of input torques.

21.3 Coordination Controller

As mentioned, it is assumed that only angular joint positions $q_{\mathrm{m}}, q_{\mathrm{s}}$ are measured. Therefore, the slave control τ_{s} can only depend on position measurements $q_{\mathrm{m}}, q_{\mathrm{s}}$. Thus estimated values for the velocities $\dot{q}_{\mathrm{m}}, \dot{q}_{\mathrm{s}}$ and accelerations $\ddot{q}_{\mathrm{m}}, \ddot{q}_{\mathrm{s}}$ are required to implement controllers based on velocity and acceleration feedback.

21.3.1 Feedback-Control Law

Under the assumption that the required estimated velocities and accelerations are available, and that the nonlinearities and parameters of the slave robot are known, the controller τ_{s} for the slave robot is proposed as

$$\tau_{\mathrm{s}} = M_{\mathrm{s}}(q_{\mathrm{s}})\widehat{\ddot{q}}_{\mathrm{m}} + C_{\mathrm{s}}(q_{\mathrm{s}}, \widehat{\dot{q}}_{\mathrm{s}})\widehat{\dot{q}}_{\mathrm{m}} + g_{\mathrm{s}}(q_{\mathrm{s}}) - K_d\widehat{\dot{e}} - K_p e, \tag{21.2}$$

where $\widehat{\dot{q}}_{\mathrm{s}}$, $\widehat{\dot{e}}$, $\widehat{\dot{q}}_{\mathrm{m}}$ and $\widehat{\ddot{q}}_{\mathrm{m}} \in \mathbb{R}^n$ represent the estimates of $\dot{q}_{\mathrm{s}}, \dot{e}, \dot{q}_{\mathrm{m}}$ and $\ddot{q}_{\mathrm{m}}$ respectively.

The coordination errors $e, \dot{e} \in \mathbb{R}^n$ are defined by

$$e := q_{\mathrm{s}} - q_{\mathrm{m}}, \qquad \dot{e} := \dot{q}_{\mathrm{s}} - \dot{q}_{\mathrm{m}}. \tag{21.3}$$

$M_{\mathrm{s}}(q_{\mathrm{s}})$, $C_{\mathrm{s}}(q_{\mathrm{s}}, \widehat{\dot{q}}_{\mathrm{s}})$ and $g_{\mathrm{s}}(q_{\mathrm{s}})$ are defined as in Eq. (21.1) and $K_p, K_d \in \mathbb{R}^{n\times n}$ are positive-definite gain matrices.

21.3.2 An Observer for the Coordination Errors $(e, \dot{e})$

Estimated values for the coordination errors $e, \dot{e}$ (Eq. (21.3)) are denoted by $\hat{e}, \widehat{\dot{e}}$; these estimated values are obtained by the full-state nonlinear Luenberger observer

$$\begin{aligned} \frac{d}{dt}\hat{e} &= \widehat{\dot{e}} + \Lambda_1 \widetilde{e}, \\ \frac{d}{dt}\widehat{\dot{e}} &= -M_{\mathrm{s}}(q_{\mathrm{s}})^{-1}\left[C_{\mathrm{s}}(q_{\mathrm{s}}, \widehat{\dot{q}}_{\mathrm{s}})\widehat{\dot{e}} + K_d \widehat{\dot{e}} + K_p \hat{e}\right] + \Lambda_2 \widetilde{e}, \end{aligned} \tag{21.4}$$

where the estimation position and velocity coordination errors $\widetilde{e}$ and $\widetilde{\dot{e}}$ are defined by

$$\widetilde{e} := e - \hat{e}, \quad \widetilde{\dot{e}} := \dot{e} - \widehat{\dot{e}}, \tag{21.5}$$

and $\Lambda_1, \Lambda_2 \in \mathbb{R}^{n\times n}$ are positive-definite gain matrices.

21.3.3 An Observer for the Slave Joint State $(q_{\mathrm{s}}, \dot{q}_{\mathrm{s}})$

Let $\hat{q}_{\mathrm{s}}, \widehat{\dot{q}}_{\mathrm{s}}$ denote estimated values for $q_{\mathrm{s}}, \dot{q}_{\mathrm{s}}$. To compute these estimated values, we propose the full-state nonlinear observer

$$\begin{aligned} \frac{d}{dt}\hat{q}_{\mathrm{s}} &= \widehat{\dot{q}}_{\mathrm{s}} + L_{p1}\widetilde{e}_q, \\ \frac{d}{dt}\widehat{\dot{q}}_{\mathrm{s}} &= -M_{\mathrm{s}}(q_{\mathrm{s}})^{-1}\left[C_{\mathrm{s}}(q_{\mathrm{s}}, \widehat{\dot{q}}_{\mathrm{s}})\widehat{\dot{e}} + K_d \widehat{\dot{e}} + K_p e\right] + L_{p2}\widetilde{e}_q, \end{aligned} \tag{21.6}$$

where the estimation position and velocity errors $\widetilde{e}_q$ and $\widetilde{\dot{e}}_q$ are defined by

$$\widetilde{e}_q := q_{\mathrm{s}} - \hat{q}_{\mathrm{s}} \qquad \widetilde{\dot{e}}_q := \dot{q}_{\mathrm{s}} - \widehat{\dot{q}}_{\mathrm{s}}, \tag{21.7}$$

and $L_{p1}, L_{p2} \in \mathbb{R}^{n\times n}$ are positive-definite gain matrices.

21.3.4 Estimated Values for $\dot{q}_{\mathrm{m}}, \ddot{q}_{\mathrm{m}}$

As stated, the master-robot variables $\dot{q}_{\mathrm{m}}, \ddot{q}_{\mathrm{m}}$ are not available; therefore estimated values for $\dot{q}_{\mathrm{m}}, \ddot{q}_{\mathrm{m}}$ are used in τ_{s} (Eq. (21.2)). From Eq. (21.3) and the definition of the estimated variables $\hat{e}, \widehat{\dot{e}}, \hat{q}_{\mathrm{s}}, \widehat{\dot{q}}_{\mathrm{s}}$, we can consider that estimated values for $q_{\mathrm{m}}, \dot{q}_{\mathrm{m}}, \ddot{q}_{\mathrm{m}}$ are given by

$$\begin{aligned} \hat{q}_{\mathrm{m}} &= \hat{q}_{\mathrm{s}} - \hat{e}, \\ \widehat{\dot{q}}_{\mathrm{m}} &= \widehat{\dot{q}}_{\mathrm{s}} - \widehat{\dot{e}}, \\ \widehat{\ddot{q}}_{\mathrm{m}} &= \frac{d}{dt}\left(\widehat{\dot{q}}_{\mathrm{s}} - \widehat{\dot{e}}\right). \end{aligned} \tag{21.8}$$

From the definition of the observers (Eqs. (21.4) and (21.6)) it follows that

$$\widehat{\ddot{q}}_{\mathrm{m}} = -(M_{\mathrm{s}}(q_{\mathrm{s}})^{-1}K_p + L_{p2})\widetilde{e} + L_{p2}\widetilde{e}_q,$$

which gives a clear insight of how $\widehat{\ddot{q}}_{\mathrm{m}}$ is re-constructed and why by increasing some appropriate gains, specifically K_p, L_{p2}, the closed-loop errors decrease in magnitude.

Remark 1. *Note that, in Eqs. (21.4) and (21.5), the estimate for $\dot{e}$ is given by $\widehat{\dot{e}}$, not by $\dot{\hat{e}}$. This definition introduces an extra correcting term in $\dot{\widetilde{e}}$, as it follows from Eqs. (21.4) and (21.5) that*

$$\dot{\widetilde{e}} = \dot{e} - \dot{\hat{e}} = \widetilde{\dot{e}} - \Lambda_1\widetilde{e}.$$

The term $\Lambda_1\widetilde{e}$ gives a faster estimation performance, especially during transients, but it has some negative effects on noise sensitivity, since it amplifies noise measurements on $\widetilde{e}$.

The same can be said for the observer (21.6) and the estimation position and velocity errors (21.7).

Figure 21.1 shows a schematic representation of the proposed coordination controller.

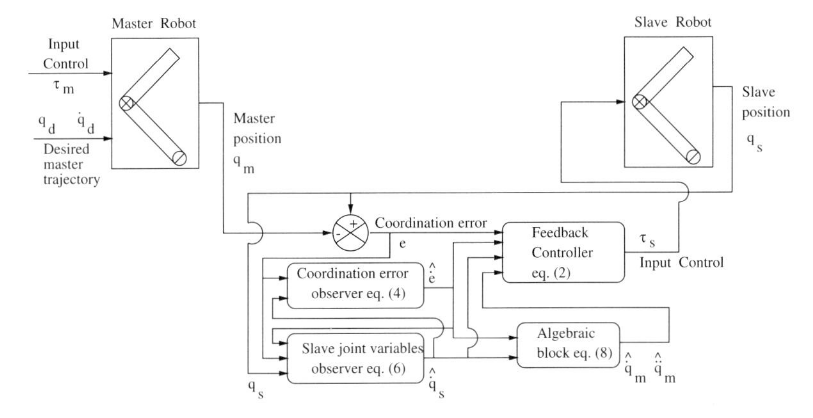

Figure 21.1: The coordination controller.

21.3.5 Ultimate Boundedness of the Closed-loop System

For the sake of simplicity in the stability analysis and without loss of generality the following assumptions are considered.

Assumption 1. *The gain matrices Λ_1, Λ_2 and L_{p1}, L_{p2} satisfy*

$$\Lambda_1 = L_{p1}, \qquad \Lambda_2 = L_{p2}. \tag{21.9}$$

Assumption 2. *The gains K_p, K_d, L_{p1}, L_{p2} are symmetric positive-definite matrices.*

Assumption 3. *The signals $\dot{q}_{\mathrm{m}}(t)$ and $\ddot{q}_{\mathrm{m}}(t)$ are bounded, i.e. there exist finite positive scalars V_M and A_M such that*

$$V_M = \sup_t \|\dot{q}_{\mathrm{m}}(t)\| , \qquad A_M = \sup_t \|\ddot{q}_{\mathrm{m}}(t)\| . \tag{21.10}$$

In practice, it is often not difficult to obtain on the basis of the desired motion $q_d(t), \dot{q}_d(t)$ and $\ddot{q}_d(t)$ of the master robot bounds on $\dot{q}_{\mathrm{m}}(t)$ and $\ddot{q}_{\mathrm{m}}(t)$, although, due to friction effects, tracking errors, etc., the actual motion of the master robot may differ from its desired motion.

Our main result can be formulated as follows.

Theorem 1. *Consider the master and slave robots that are described by Eq. (21.1), and the slave robot in closed loop with the coordination controller (21.2) and both observers (21.4), (21.6). Then there exist conditions on the gain matrices K_d, K_p, L_{p1}, L_{p2}, specifically on their minimum eigenvalues, such that the errors $\dot{e}$, e, $\dot{\widetilde{e}}$, $\widetilde{e}$, $\dot{\widetilde{e}}_q$, $\widetilde{e}_q$ in the closed-loop system are semi-globally uniformly ultimately bounded. Moreover, the bound of the errors can be made small by a proper choice of the minimum eigenvalues of the gains K_p, L_{p2}.*

Proof: The proof and the conditions on the gains K_d, K_p, L_{p1}, L_{p2} can be found in [13] and [14]. A sketch for the tuning gain procedure proposed in [13] and [14] is as follows.

1. Determine the bounds of the master trajectories $\dot{q}_{\mathrm{m}}, \ddot{q}_{\mathrm{m}}$, the physical parameters $M_{\mathrm{s}}(q_{\mathrm{s}})$, $C_{\mathrm{s}}(q_{\mathrm{s}}, \dot{q}_{\mathrm{s}})$ and their partial derivatives with respect to q_{s}.

2. Choose the weighting factors $\lambda_0 > 0$, $\varepsilon_0 > 0$, $\mu_0 > 0$ and $\gamma_0 > 0$ and a bound for the maximum eigenvalue of the gains K_p, K_d, L_{p1} and L_{p2}.

3. By considering the matrix Q_N (given by Eq. (49) in [13] and [14]) determine the minimum eigenvalue of the gain matrices K_p, K_d, L_{p1} and L_{p2}, such that Q_N is a positive-definite matrix. For this a set of nonlinear algebraic equations must be solved.

Remark 2. *Note that the conditions implied by Theorem 1 and the Assumptions 1–3 are only sufficient, but not necessary, to ensure stability and boundedness of the coordination system. Thus, for different values of λ_0, ε_0, μ_0 and γ_0, different minimum eigenvalues of the gains K_p, K_d, L_{p1} and L_{p2} would be obtained.*

21.4 Experimental Case Study

The proposed coordination controller has been implemented in a two manufacturing robot manipulator setup. The manufacturing robot is a Cartesian basic elbow configuration robot. It consists of a two-link arm which is placed on a rotating and translational base, and it has a passively actuated tool connected at the end of the outer link, see Fig. 21.2. The manufacturing robot is a pick and place industrial robot used for assembling. It has four degrees of freedom in the Cartesian space, denoted by x_{ci} $(i = 1, \ldots, 4)$, and seven degrees of freedom in the joint space, denoted by q_j $(j = 1, \ldots, 7)$, and is actuated by four DC brushless servo-motors. Although the robot has seven degrees of freedom in the joint space, three of them are kinematically constrained, with the set of constrained joints given by $\{q_3, q_6, q_7\}$. Therefore the robot

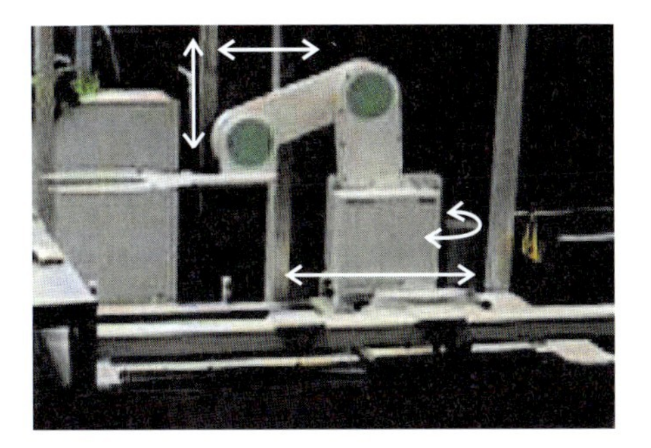

Figure 21.2: The CFT transposer robot.

can be represented in the joint space by four degrees of freedom $\{q_1, q_2, q_4, q_5\}$ actuated by four servo-motors. Although the shafts of the motors and the corresponding links are connected by means of belts, the servo-motor–link pair proved to be stiff enough to be considered as a rigid joint.

The four Cartesian degrees of freedom are rotation, up and down, forward and backward of the arm, forward and backward of the whole robot, see Fig. 21.2. The robot is equipped with encoders attached to the shafts of the motors with a resolution of 2000 ports per rotation, which results in an absolute accuracy of ± 0.5 [mm] in all motion directions. The tool connected at the end of the outer link is a kinematically constrained planar support. The tool is passively actuated and designed to remain horizontal at all times. A more detailed description of the structure of the robot can be found in [13] and [16].

For implementation of the controllers and communication to the robots, the experimental setup is equipped with a DS1005 dSPACE system, with a processor PPC750, a clock of 480 MHz and a bus clock of 80 MHz. Throughout the experiments the sampling frequency of the DS1005 dSPACE system was set to 2 kHz.

21.4.1 Joint Space Dynamics

The multi-robot system is formed by two structurally identical transposer robots, so that they have the same kinematic and dynamic model. However, the physical parameters of the robots, such as masses, inertias and friction coefficients, are different for both robots. The multi-robot system has been installed in the Dynamics and Control Technology Laboratory of the Department of Mechanical Engineering at the Eindhoven University of Technology.

Hereafter the notation q_i, for $i = \mathrm{m}, \mathrm{s}$, refers to the master or slave robot in the multi-composed system. According to [16] the dynamic model of the CFT robot is given by

$$M(q_i)\ddot{q}_i + C(q_i, \dot{q}_i)\dot{q}_i + g(q_i) + f(\dot{q}_i) = \tau_i, \qquad i = \mathrm{m}, \mathrm{s}, \tag{21.11}$$

where $f(\dot{q}_i)$ denotes the friction that is modeled as

$$f(\dot{q}_i) = B_{v,i}\dot{q}_i + B_{f1,i}\left(1 - \frac{2}{1 + e^{2w_{1,i}\dot{q}_i}}\right) + B_{f2,i}\left(1 - \frac{2}{1 + e^{2w_{2,i}\dot{q}_i}}\right), \tag{21.12}$$

with $q_i = [q_{i,1}\ q_{i,2}\ q_{i,4}\ q_{i,5}]^{\mathrm{T}}$ the vector of generalized coordinates of robot i, $M(q_i) \in \mathbb{R}^{4\times 4}$ the symmetric, positive-definite inertia matrix, $g(q_i) \in \mathbb{R}^4$ the gravity forces, $C(q_i, \dot{q}_i)\dot{q}_i \in \mathbb{R}^4$ the Coriolis and centrifugal forces and $\tau_i = [\tau_{i,1}\ \tau_{i,2}\ \tau_{i,4}\ \tau_{i,5}]^{\mathrm{T}}$ the vector of external torques.

The term $f(\dot{q}_i) \in \mathbb{R}^4$ corresponds to the forces due to friction effects, with $B_{v,i}$ the viscous friction coefficients and $B_{f1,i}$, $B_{f2,i}$, $w_{1,i}$ and $w_{2,i}$ parameters related to the Coulomb and Stribeck friction effects.

The entries and physical parameters in the matrices $M(q_i)$, $C(q_i, \dot{q}_i)$, the gravity vector $g(q_i)$ and the friction forces $f(\dot{q}_i)$ can be found in Appendix A. For further details of the modeling and identification of the dynamic model the interested reader is referred to [16].

21.4.2 Experimental Results

The desired trajectory for the master robot $q_{\mathrm{d}}(t)$ is obtained by transformation of a desired trajectory given in Cartesian coordinates $x_{cj,\mathrm{d}}(t)$, $j = 1, \ldots, 4$, which is given by

$$\begin{aligned} x_{cj,\mathrm{d}}(t) &= a_{0,j} + a_{1,j}\sin(2s_{f,j}\pi\omega t) + a_{2,j}\sin(4s_{f,j}\pi\omega t) \\ &+ a_{3,j}\sin(6s_{f,j}\pi\omega t) + a_{4,j}\sin(8s_{f,j}\pi\omega t), \end{aligned} \tag{21.13}$$

with the coefficients $a_{i,j}$, $i = 0, \ldots, 4$, $j = 1, \ldots, 4$ given in Table 21.1. The coefficients $a_{0,j}$ have been chosen as the middle value of the allowed displacements in the robots, while $a_{i,j}$, $i = 1, \ldots, 4$, were chosen to achieve the combination of maximum displacement and velocity allowed by the robots. This is done to generate a trajectory in amplitude that can be executed by the multi-robot system.

Table 21.1: Coefficients of the desired trajectory $x_{cj,\mathrm{d}}(t)$, $j = 1, \ldots, 4$.

$a_{i,j}$	$i = 0$	$i = 1$	$i = 2$	$i = 3$	$i = 4$	$s_{f,j}$
$j = 1$	-0.1343 [m]	-0.05 [m]	-0.015 [m]	-0.005 [m]	-0.01 [m]	1.0
$j = 2$	0.2766 [m]	0.05 [m]	0.03 [m]	-0.03 [m]	0.02 [m]	1.0
$j = 3$	2.4 [rad]	0.15 [rad]	0.05 [rad]	-0.03 [rad]	0.02 [rad]	1.0
$j = 4$	-0.265 [m]	0.2 [m]	0.1 [m]	-0.05 [m]	0.05 [m]	0.25

The fundamental frequency of the master robot's desired trajectory $x_{cj,\mathrm{d}}(t)$, given by Eq. (21.13), is set as $\omega = 0.4$ Hz.

The joint space desired trajectory $q_{\mathrm{d}}(t)$ is obtained by transformation of the desired Cartesian trajectories $x_{cj,\mathrm{d}}(t)$, $j = 1, \ldots, 4$, using the inverse kinematics [16].

The master robot is driven by PID controllers with control gains listed as in Table 21.2. After a series of experiments to decrease the coordination position error $e = q_{\mathrm{s}} - q_{\mathrm{m}}$, the gains on the slave robot controller (21.2) were set as listed in Table 21.2.

The initial position of the links and the initial conditions in the observers (21.4), (21.6) were chosen as in Table 21.3. The master and slave robots start from a steady state; therefore the joint velocity $\dot{q}(0)$, the estimated joint velocity $\widehat{\dot{q}}(0)$ and the estimated coordination error $\widehat{e}(0)$ are all equal to zero. The initial condition for the estimated coordination error $\widehat{e}(0)$ in observer (21.4) was set equal to zero.

Table 21.2: Control gains in the master (m) and slave (s) robots.

	K_p (m)	K_d (m)	K_I (m)	K_p (s)	K_d (s)	L_{p1} (s)	L_{p2} (s)
Joint q_1	11 000	50	2000	10 000	1200	500	100 000
Joint q_2	10 000	50	1000	8000	100	500	100 000
Joint q_4	40 000	600	1000	8000	100	500	100 000
Joint q_5	40 000	600	1000	8000	100	500	100 000

Table 21.3: Initial conditions for master and slave robots.

	$q_1(0)$ [m]	$q_2(0)$ [rad]	$q_4(0)$ [rad]	$q_5(0)$ [rad]
Master robot	-0.095	-0.4	-0.9615	2.1473
Slave robot	-0.079	0.0	-1.0355	2.1165
	$\widehat{q}_1(0)$ [m]	$\widehat{q}_2(0)$ [rad]	$\widehat{q}_4(0)$ [rad]	$\widehat{q}_5(0)$ [rad]
Slave robot	-0.07	0.1	-1.0	2.0

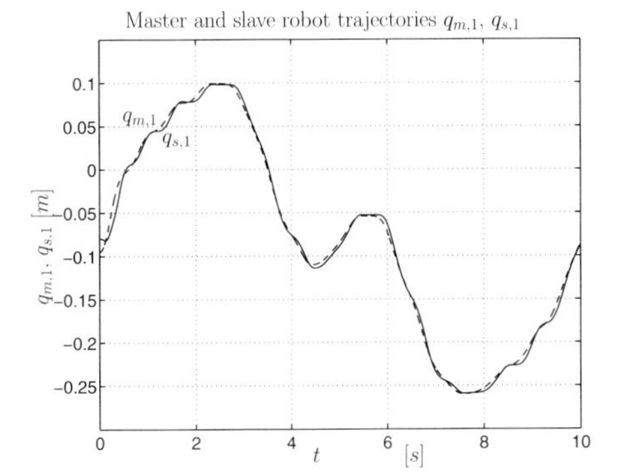

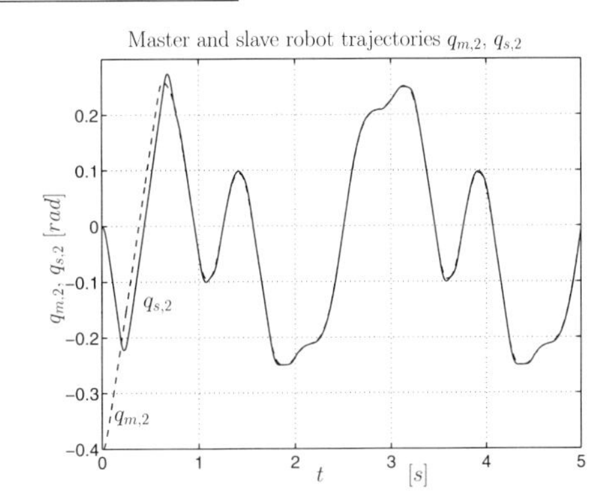

Figure 21.3: Master q_{m} (dashed) and slave q_{s} (solid) positions, joints q_1, q_2.

Figures 21.3 and 21.4 show the master $q_{\mathrm{m},j}$ (dashed) and slave $q_{i\mathrm{s},j}$ (solid) position trajectories for the joints $j = 1, 2, 4, 5$. The coordination errors $e = q_{\mathrm{s}} - q_{\mathrm{m}}$ after the transient period has finished are shown in Figs. 21.5 and 21.6.

From Figs. 21.3–21.6 it is evident that coordination between the master and slave robots is achieved, such that bounded coordination errors are obtained. Further experiments showed that the coordination errors can be decreased by increasing the gains K_p, which agrees with the result stated in Theorem 1.

The influence of the gains K_d and L_{p1} is related to the response to sudden changes on the system (transients, changes on trajectory, etc.). The increasing of the gains K_d and L_{p1} might result in large overshoots, amplification of noise in the measurements and saturation of the torques in the servo-motors.

Remark 3. *The conditions mentioned in Theorem 1 are very conservative. However, even without knowledge of the required physical bounds, the closed-loop system can be made uni-*

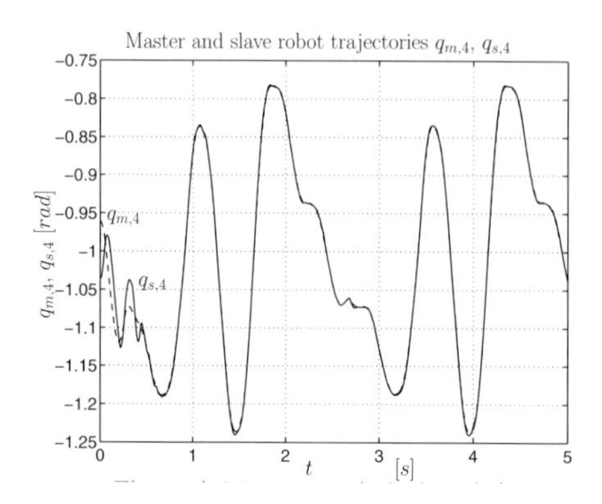

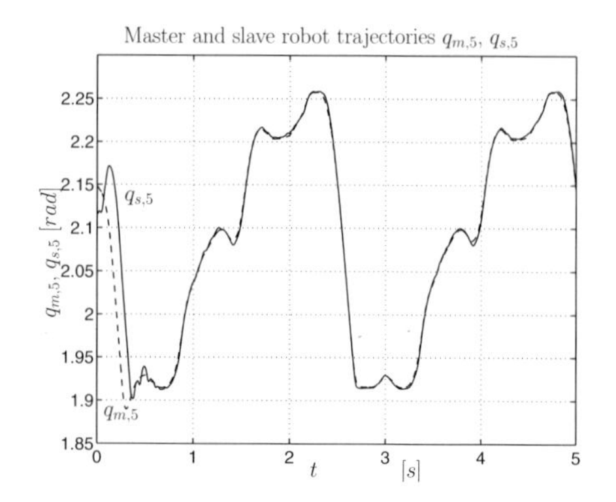

Figure 21.4: Master q_{m} (dashed) and slave q_{s} (solid) positions, joints q_4, q_5.

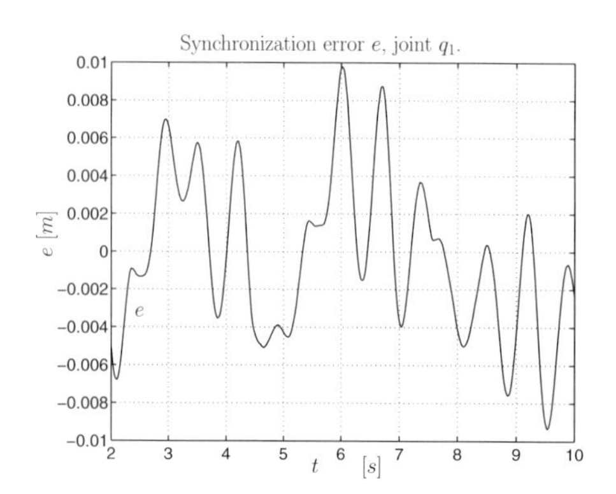

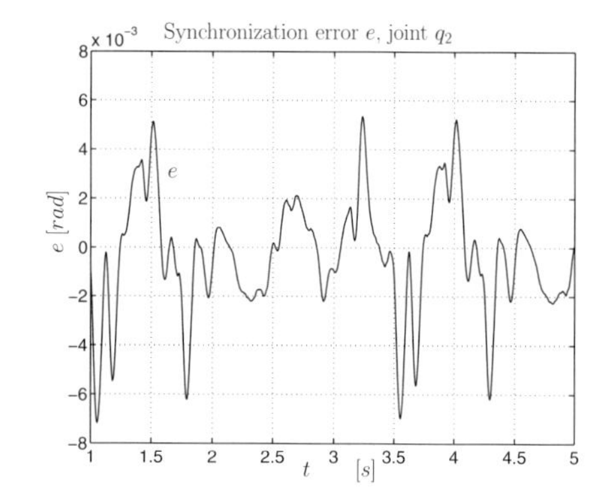

Figure 21.5: Coordination errors e, joints q_1, q_2.

formly ultimately bounded. This may be achieved by selecting the control gains large enough. However, such high-gain implementations are not always desirable in practical circumstances, since these may amplify unavoidable noise.

21.5 Conclusions and Further Extensions

A position coordination controller for multi-robot systems working in master–slave schemes has been presented. The proposed controller is independent of the master-robot dynamics and its physical parameters, and only requires position measurements. The coordination controller yields semi-global ultimate boundedness of the closed-loop errors. The bound of the errors can be decreased by a proper tuning on the controller gains. The proposed controller is robust against parametric uncertainties and noise in the position measurements.

Further extensions of the proposed coordination controller are the cases of flexible-joint and mixed rigid–flexible-joint robots. Also, some other coordination schemes different to master–slave are possible extensions, e.g. cooperative schemes.

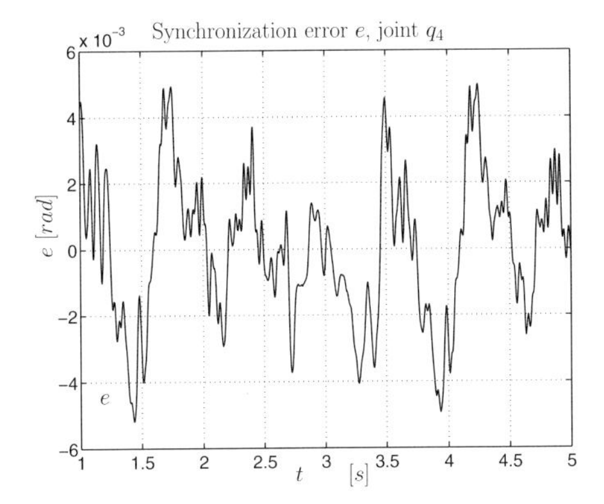

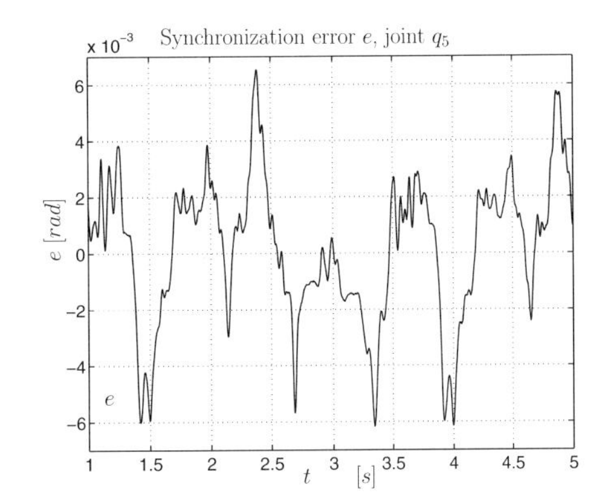

Figure 21.6: Coordination errors e, joints q_4, q_5.

Acknowledgment

The authors would like to thank D. Kostic and R. Hensen for very useful discussions about modeling and identification of robot manipulators, see also [6] and [9].

The second author acknowledges support from the CONACYT (National Council for Science and Technology), Mexico, Scholarship No. 72368. Support is also acknowledged to the Mexican Petroleum Institute, Mexico.

Bibliography

[1] I. I. Blekhman, *Synchronization in Science and Technology*, ASME Press Translations, New York, 1988.

[2] M. Brunt, Ph.D. thesis, Technical University Delft (1998).

[3] R. V. Dubey, T.F. Chan, and S.E. Everett, *Variable damping impedance control of a bilateral telerobotics system*, IEEE Control and Systems **17** (1997), 37–45.

[4] R.C. Goertz, *Manipulator systems development at ANL*, in: Proceedings of the 12th Conference on Remote Systems Technology, American Nuclear Society (1954), 117–136.

[5] G.S. Guthart and J.K. Salisbury, Jr., *The intuitive*™ *telesurgery system: overview and application*, in: Proceedings of the IEEE International Conference on Robotics and Automation (2000), 618–621.

[6] R.H.A. Hensen, G.Z. Angelis, M.J.G. v.d. Molengraft, A.G. de Jager, and J.J. Kok, *Grey-box modeling of friction: An experimental case-study*, European Journal of Control **6** (2000), 258–267.

[7] J.W. Hills and J.F. Jensen, *Telepresence technology in medicine: principles and applications*, Proceedings of the IEEE **86** (1998), 569–580.

[8] T. Komatsu and T. Akabane, *Control of a space flexible master–slave manipulator based on parallel compliance models*, in: Proceedings of the IEEE International Conference on Robotics and Automation (1998), 1932–1937.

[9] D. Kostic, R. Hensen, B. de Jager, and M. Steinbuch, *Modeling and identification of an RRR-robot*, in: Proceedings of the IEEE Conference on Decision and Control (2001), 1144–1149.

[10] K. Kosuge, J. Ishikawa, K. Furuta, and M. Sakai, *Control of single-master slave manipulator systems using VIM*, in: Proceedings of the IEEE International Conference on Robotics and Automation, Vol. 2 (1990), 1172–1177.

[11] H.K. Lee and M.J. Chung, *Adaptive controller of a master–slave system for transparent teleoperation*, Journal of Robotic Systems **15** (1998), 465–475.

[12] Y.H. Liu, Y. Xu, and M. Bergerman, *Cooperation control of multiple manipulators with passive joints*, IEEE Transactions on Robotics and Automation **15** (1999), 258–267.

[13] H. Nijmeijer and A. Rodriguez-Angeles, *Synchronization of Mechanical Systems*, World Scientific, Singapore, 2003.

[14] A. Rodriguez-Angeles and H. Nijmeijer, *Coordination of two robot manipulators based on position measurements only*, International Journal of Control **74** (2001), 1311–1323.

[15] A. Rodriguez-Angeles, Ph.D. thesis, Eindhoven University of Technology (2002).

[16] A. Rodriguez-Angeles, D. Lizarraga, H. Nijmeijer, and H.A. van Essen, *Modelling and identification of the CFT-transposer robot*, Technical Report 2002.52, Eindhoven University of Technology, Dynamics and Control Technology Group, 2002.

[17] M.W. Spong and M. Vidyasagar, *Robot dynamics and control*, Wiley, New York, 1989.

Appendix A: Dynamic Model of the Manufacturing Robot

In this appendix the dynamics of the manufacturing robot is given, and a set of physical dimensions and estimated physical parameters for the master and slave robots in the coordination setup are listed in Tables 21.4 and 21.5.

The dynamics of the master (m) and slave (s) robots in the multi-composed system are given by

$$M(q_i)\ddot{q}_i + C(q_i,\dot{q}_i)\dot{q}_i + g(q_i) + f(\dot{q}_i) = \tau_i, \qquad i = \mathrm{m},\mathrm{s}, \tag{21.14}$$

$$f(\dot{q}_i) = B_{v,i}\dot{q}_i + B_{f1,i}\left(1 - \frac{2}{1+e^{2w_{1,i}\dot{q}_i}}\right) + B_{f2,i}\left(1 - \frac{2}{1+e^{2w_{2,i}\dot{q}_i}}\right). \tag{21.15}$$

Since the master and slave robots in the multi-composed system are structurally identical, they are described by the same dynamic model, but they differ in the physical parameters. The dynamic model of the transposer robot (21.14), (21.15) includes 32 physical parameters, denoted by $\theta_{i,j}$, $i = \mathrm{m},\mathrm{s}$, $j = 1,\ldots,32$, where i identifies the robot and j the parameter. A detailed parametrization of the dynamics (21.14), (21.15) and physical interpretation of the parameters $\theta_{i,j}$ can be found in [16].

Entries of the Inertia Matrix $M(q_i)$.

The entries of the symmetric inertia matrix $M(q_i) \in \mathbb{R}^{4\times 4}$, as a function of the generalized coordinates $q_i = [q_{i,1}\ \ q_{i,2}\ \ q_{i,4}\ \ q_{i,5}]^{\mathrm{T}}$ of robot i, $i = \mathrm{m,s}$, and the parameters $\theta_{i,j}$, $j = 1,\ldots,32$, are given by

$$M_{1,1} = \theta_{i,1} + \theta_{i,11} + \theta_{i,12}$$

$$\begin{aligned}M_{1,2} &= (-\theta_{i,12}d_{2\_0'} - \theta_{i,11}d_{2\_0'} - \theta_{i,3})\sin(q_{i,2}) + (\theta_{i,2} + d_6\theta_{i,11})\cos(q_{i,2})\\ &+ \tfrac{1}{2}((L_4 - L_5)(\theta_{i,12} + \theta_{i,11}) - \theta_{i,9})(\cos(q_{i,5} + q_{i,2} + q_{i,4})\\ &\quad - \cos(q_{i,5} - q_{i,2} + q_{i,4}))\\ &+ \tfrac{1}{2}(\theta_{i,7} + \theta_{i,5} + \theta_{i,12}L_6)(\cos(-q_{i,2} + q_{i,4}) - \cos(q_{i,2} + q_{i,4}))\\ &+ \tfrac{1}{2}(\theta_{i,8} + \theta_{i,6})(\sin(q_{i,2} + q_{i,4}) - \sin(-q_{i,2} + q_{i,4}))\\ &+ \tfrac{1}{2}(-\sin(q_{i,5} - q_{i,2} + q_{i,4}) + \sin(q_{i,5} + q_{i,2} + q_{i,4}))\theta_{i,10}\end{aligned}$$

$$\begin{aligned}M_{1,3} &= \tfrac{1}{2}(-\theta_{i,5} - \theta_{i,7} - \theta_{i,12}L_6)(\cos(q_{i,2} + q_{i,4}) + \cos(-q_{i,2} + q_{i,4}))\\ &+ \tfrac{1}{2}((L_4 - L_5)(\theta_{i,12} + \theta_{i,11}) - \theta_{i,9})(\cos(q_{i,5} - q_{i,2} + q_{i,4})\\ &\quad + \cos(q_{i,5} + q_{i,2} + q_{i,4}))\\ &+ \tfrac{1}{2}(\theta_{i,8} + \theta_{i,6})(\sin(q_{i,2} + q_{i,4}) + \sin(-q_{i,2} + q_{i,4}))\\ &+ \tfrac{1}{2}(\sin(q_{i,5} + q_{i,2} + q_{i,4}) + \sin(q_{i,5} - q_{i,2} + q_{i,4}))\theta_{i,10}\end{aligned}$$

$$\begin{aligned}M_{1,4} &= \tfrac{1}{2}((L_4 - L_5)(\theta_{i,11} + \theta_{i,12}) - \theta_{i,9})(\cos(q_{i,5} + q_{i,2} + q_{i,4})\\ &\quad + \cos(q_{i,5} - q_{i,2} + q_{i,4}))\\ &+ \tfrac{1}{2}(\sin(q_{i,5} + q_{i,2} + q_{i,4}) + \sin(q_{i,5} - q_{i,2} + q_{i,4}))\theta_{i,10}\end{aligned}$$

$$\begin{aligned}M_{2,2} &= ((L_5 - L_4)(\sin(q_{i,5}) + \sin(q_{i,5} + 2q_{i,4})) - 2\cos(q_{i,4})d_{2\_0'})\theta_{i,8} + \theta_{i,4}\\ &+ (-2d_{2\_0'}\cos(q_{i,4} + q_{i,5}) - L_4\sin(2q_{i,5} + 2q_{i,4}))\theta_{i,10} + \theta_{i,12}d_{2\_0'}^2\\ &+ ((\tfrac{1}{2} - \tfrac{1}{2}\cos(2q_{i,5} + 2q_{i,4}))(L_5^2 + L_4^2) + 2L_4d_{2\_0'}\sin(q_{i,4} + q_{i,5}) + d_6^2\\ &+ d_{2\_0'}^2 + ((\cos(2q_{i,5} + 2q_{i,4}) - 1)L_4 - 2d_{2\_0'}\sin(q_{i,4} + q_{i,5}))L_5)\theta_{i,11}\\ &+ ((\cos(2q_{i,4}) - \cos(q_{i,5}) - 1 + \cos(q_{i,5} + 2q_{i,4}))L_4\\ &- 2d_{2\_0'}\sin(q_{i,4}))\theta_{i,5} - 2(\sin(q_{i,4} + q_{i,5})L_5 + \sin(q_{i,4})L_6)d_{2\_0'}\theta_{i,12}\\ &+ ((\cos(q_{i,5} + 2q_{i,4}) - \cos(q_{i,5}))(L_4 - L_5) - 2d_{2\_0'}\sin(q_{i,4}))\theta_{i,7}\\ &- \tfrac{1}{2}\theta_{i,12}(\cos(2q_{i,4}) - 1)L_6^2 - \tfrac{1}{2}(\cos(2q_{i,5} + 2q_{i,4}) - 1)(L_5^2 + L_4^2)\theta_{i,12}\\ &+ (-L_4(\sin(2q_{i,4}) + \sin(q_{i,5} + 2q_{i,4}) + \sin(q_{i,5})) - 2\cos(q_{i,4})d_{2\_0'})\theta_{i,6}\\ &+ ((\cos(2q_{i,5} + 2q_{i,4}) - 1)L_4 + (\cos(q_{i,5}) - \cos(q_{i,5} + 2q_{i,4}))L_6)L_5\theta_{i,12}\\ &+ ((\cos(2q_{i,5} + 2q_{i,4}) - 1)L_4 - 2d_{2\_0'}\sin(q_{i,4} + q_{i,5}))\theta_{i,9}\\ &+ (2\sin(q_{i,4} + q_{i,5})d_{2\_0'} - (\cos(q_{i,5}) + \cos(q_{i,5} + 2q_{i,4}))L_6)L_4\theta_{i,12}\end{aligned}$$

$$M_{2,3} = -\theta_{i,7}d_6\cos(q_{i,4}) + \theta_{i,8}d_6\sin(q_{i,4}) + \theta_{i,11}d_6(L_4 - L_5)\cos(q_{i,4} + q_{i,5})$$

$$M_{2,4} = \theta_{i,11}d_6(L_4 - L_5)\cos(q_{i,4} + q_{i,5})$$

$$
\begin{aligned}
M_{3,3} = {} & (((L_5 - L_6)L_4 + 2L_5L_6)\theta_{i,12} + L_5L_4\theta_{i,11} + (\theta_{i,9} - \theta_{i,5} - \theta_{i,7})L_4 \\
& + 2\theta_{i,7}L_5)\cos(q_{i,5}) + L_4(\theta_{i,6} + \theta_{i,8})\sin(2q_{i,4}) \\
& - L_4((\tfrac{1}{2}L_4 + L_6)\theta_{i,12} + \tfrac{1}{2}L_4\theta_{i,11} + \theta_{i,5} + \theta_{i,7})\cos(2q_{i,4}) \\
& - L_4((L_6 + L_5)\theta_{i,12} + L_5\theta_{i,11} + \theta_{i,9} + \theta_{i,7} + \theta_{i,5})\cos(q_{i,5} + 2q_{i,4}) \\
& + (\theta_{i,8} + \theta_{i,10} + \theta_{i,6})L_4\sin(q_{i,5} + 2q_{i,4}) + ((2L_5 - L_4)\theta_{i,8} \\
& - (\theta_{i,6} + \theta_{i,10})L_4)\sin(q_{i,5}) + (L_4^2 + (L_6 - L_5)L_4 + L_5^2 + L_6^2)\theta_{i,12} \\
& + ((\tfrac{1}{2}L_4 - L_5)(\theta_{i,12} + \theta_{i,11}) - \theta_{i,9})L_4\cos(2q_{i,5} + 2q_{i,4}) \\
& + (L_5^2 - L_5L_4 + L_4^2)\theta_{i,11} + (\theta_{i,7} - \theta_{i,9} + \sin(2q_{i,5} + 2q_{i,4})\theta_{i,10} - \theta_{i,5})L_4
\end{aligned}
$$

$$
\begin{aligned}
M_{3,4} = {} & \tfrac{1}{2}(\sin(q_{i,5} + 2q_{i,4}) - \sin(q_{i,5}))L_4\theta_{i,6} + \tfrac{1}{2}\theta_{i,12}(\cos(2q_{i,5} + 2q_{i,4}) + 1)L_4^2 \\
& + (\tfrac{1}{2}\cos(q_{i,5}) - \cos(2q_{i,5} + 2q_{i,4}) - \tfrac{1}{2}\cos(q_{i,5} + 2q_{i,4}) - 1)L_4\theta_{i,9} \\
& + (\tfrac{1}{2}\sin(q_{i,5} + 2q_{i,4}) + \sin(2q_{i,5} + 2q_{i,4}) - \tfrac{1}{2}\sin(q_{i,5}))L_4\theta_{i,10} \\
& + (L_5\cos(q_{i,5}) - \tfrac{1}{2}(\cos(q_{i,5}) + \cos(q_{i,5} + 2q_{i,4}))L_4)\theta_{i,7} \\
& + ((L_5 - \tfrac{1}{2}L_4)\sin(q_{i,5}) + \tfrac{1}{2}L_4\sin(q_{i,5} + 2q_{i,4}))\theta_{i,8} + \theta_{i,12}L_5^2 \\
& + (L_5^2 + \tfrac{1}{2}(\cos(q_{i,5}) - \cos(q_{i,5} + 2q_{i,4}))L_4L_5 + \tfrac{1}{2}(1 + \cos(2q_{i,5} + 2q_{i,4})) \\
& \times (L_4^2 - 2L_4L_5))\theta_{i,11} - \tfrac{1}{2}(\cos(q_{i,5}) + \cos(q_{i,5} + 2q_{i,4}))L_4\theta_{i,5} \\
& + (\cos(q_{i,5})L_6 - \tfrac{1}{2}(2 + 2\cos(2q_{i,5} + 2q_{i,4}) + \cos(q_{i,5} + 2q_{i,4}) \\
& \quad - \cos(q_{i,5}))L_4)L_5\theta_{i,12} - \tfrac{1}{2}\theta_{i,12}(\cos(q_{i,5}) + \cos(q_{i,5} + 2q_{i,4}))L_6L_4
\end{aligned}
$$

$$
\begin{aligned}
M_{4,4} = {} & ((\tfrac{1}{2}L_4^2 - L_5L_4)(\theta_{i,12} + \theta_{i,11}) - L_4\theta_{i,9})\cos(2q_{i,5} + 2q_{i,4}) \\
& + (\tfrac{1}{2}L_4^2 - L_5L_4 + L_5^2)(\theta_{i,12} + \theta_{i,11}) + (\sin(2q_{i,5} + 2q_{i,4})\theta_{i,10} - \theta_{i,9})L_4
\end{aligned}
$$

Entries of the Coriolis Matrix $C(q_i, \dot{q}_i)$.

The entries of the Coriolis matrix $C(q_i, \dot{q}_i) \in \mathbb{R}^{4\times 4}$, as a function of the generalized coordinates $q_i = [q_{i,1}\ \ q_{i,2}\ \ q_{i,4}\ \ q_{i,5}]^{\mathrm{T}}$ of robot i, $i = \mathrm{m}, \mathrm{s}$, and the parameters $\theta_{i,j}$, $j = 1, \ldots, 32$, are given by

$$
C_{1,1} = C_{2,1} = C_{3,1} = C_{4,1} = 0
$$

$$
\begin{aligned}
C_{1,2} = {} & \tfrac{1}{2}(\theta_{i,7} + \theta_{i,5} + L_6\theta_{i,12})((\dot{q}_{i,2} - \dot{q}_{i,4})\sin(q_{i,4} - q_{i,2}) + (\dot{q}_{i,2} + \dot{q}_{i,4}) \\
& \times \sin(q_{i,2} + q_{i,4})) - ((\theta_{i,12} + \theta_{i,11})d_{2\text{-}0'} + \theta_{i,3})\dot{q}_{i,2}\cos(q_{i,2}) \\
& + \tfrac{1}{2}(\theta_{i,9} + (L_5 - L_4)(\theta_{i,12} + \theta_{i,11}))((\dot{q}_{i,4} + \dot{q}_{i,2} + \dot{q}_{i,5}) \\
& \times \sin(q_{i,5} + q_{i,2} + q_{i,4}) + (\dot{q}_{i,2} - \dot{q}_{i,4} - \dot{q}_{i,5})\sin(q_{i,5} - q_{i,2} + q_{i,4})) \\
& - (\theta_{i,2} + d_6\theta_{i,11})\dot{q}_{i,2}\sin(q_{i,2}) + \tfrac{1}{2}(\theta_{i,8} + \theta_{i,6})((\dot{q}_{i,2} - \dot{q}_{i,4})\cos(q_{i,4} - q_{i,2}) \\
& + (\dot{q}_{i,2} + \dot{q}_{i,4})\cos(q_{i,2} + q_{i,4})) + \tfrac{1}{2}((\dot{q}_{i,2} - \dot{q}_{i,4} - \dot{q}_{i,5}) \\
& \times \cos(q_{i,5} - q_{i,2} + q_{i,4}) + (\dot{q}_{i,4} + \dot{q}_{i,2} + \dot{q}_{i,5})\cos(q_{i,5} + q_{i,2} + q_{i,4}))\theta_{i,10}
\end{aligned}
$$

$$C_{1,3} = \tfrac{1}{2}(\theta_{i,8} + \theta_{i,6})((\dot{q}_{i,4} - \dot{q}_{i,2})\cos(q_{i,4} - q_{i,2}) + (\dot{q}_{i,2} + \dot{q}_{i,4})\cos(q_{i,2} + q_{i,4}))$$
$$+ \tfrac{1}{2}(\theta_{i,9} + (L_5 - L_4)(\theta_{i,12} + \theta_{i,11}))((\dot{q}_{i,4} + \dot{q}_{i,2} + \dot{q}_{i,5})$$
$$\times \sin(q_{i,5} + q_{i,2} + q_{i,4}) + (\dot{q}_{i,4} - \dot{q}_{i,2} + \dot{q}_{i,5})\sin(q_{i,5} - q_{i,2} + q_{i,4}))$$
$$+ \tfrac{1}{2}(\theta_{i,5} + \theta_{i,7} + L_6\theta_{i,12})((\dot{q}_{i,4} - \dot{q}_{i,2})\sin(q_{i,4} - q_{i,2}) + (\dot{q}_{i,2} + \dot{q}_{i,4})$$
$$\times \sin(q_{i,2} + q_{i,4})) + \tfrac{1}{2}((\dot{q}_{i,4} + \dot{q}_{i,2} + \dot{q}_{i,5})\cos(q_{i,5} + q_{i,2} + q_{i,4})$$
$$+ (\dot{q}_{i,4} + \dot{q}_{i,5} - \dot{q}_{i,2})\cos(q_{i,5} - q_{i,2} + q_{i,4}))\theta_{i,10}$$

$$C_{1,4} = \tfrac{1}{2}((\dot{q}_{i,4} + \dot{q}_{i,2} + \dot{q}_{i,5})\cos(q_{i,5} + q_{i,2} + q_{i,4}) + (\dot{q}_{i,4} - \dot{q}_{i,2} + \dot{q}_{i,5})$$
$$\times \cos(q_{i,5} - q_{i,2} + q_{i,4}))\theta_{i,10} + \tfrac{1}{2}(\theta_{i,9} + (L_5 - L_4)(\theta_{i,12} + \theta_{i,11}))$$
$$\times ((\dot{q}_{i,4} + \dot{q}_{i,2} + \dot{q}_{i,5})\sin(q_{i,5} + q_{i,2} + q_{i,4})$$
$$+ (\dot{q}_{i,4} + \dot{q}_{i,5} - \dot{q}_{i,2})\sin(q_{i,5} - q_{i,2} + q_{i,4}))$$

$$C_{2,2} = -\tfrac{1}{2}(L_4\theta_{i,6} - L_5\theta_{i,8} + \theta_{i,8}L_4)((2\dot{q}_{i,4} + \dot{q}_{i,5})\cos(2q_{i,4} + q_{i,5})$$
$$+ \dot{q}_{i,5}\cos(q_{i,5})) - \dot{q}_{i,4}d_{2\_0'}(\theta_{i,5} + \theta_{i,7} + L_6\theta_{i,12})\cos(q_{i,4})$$
$$- (\dot{q}_{i,4} + \dot{q}_{i,5})(\theta_{i,9} + (L_5 - L_4)(\theta_{i,12} + \theta_{i,11}))d_{2\_0'}\cos(q_{i,4} + q_{i,5})$$
$$- \tfrac{1}{2}(\dot{q}_{i,4} + \dot{q}_{i,5})(2L_4\theta_{i,9} - (L_5 - L_4)^2(\theta_{i,12} + \theta_{i,11}))\sin(2q_{i,5} + 2q_{i,4})$$
$$- \tfrac{1}{2}\dot{q}_{i,4}(2L_4\theta_{i,5} - L_6^2\theta_{i,12})\sin(2q_{i,4}) - \dot{q}_{i,4}L_4\theta_{i,6}\cos(2q_{i,4})$$
$$+ \tfrac{1}{2}(L_4(\theta_{i,5} + \theta_{i,7}) + L_6\theta_{i,12}(L_4 - L_5) - L_5\theta_{i,7})(\dot{q}_{i,5}\sin(q_{i,5})$$
$$- (2\dot{q}_{i,4} + \dot{q}_{i,5})\sin(2q_{i,4} + q_{i,5})) + (\dot{q}_{i,4} + \dot{q}_{i,5})\sin(q_{i,4} + q_{i,5})d_{2\_0'}\theta_{i,10}$$
$$- (\dot{q}_{i,4} + \dot{q}_{i,5})\cos(2q_{i,5} + 2q_{i,4})L_4\theta_{i,10} + d_{2\_0'}\dot{q}_{i,4}(\theta_{i,8} + \theta_{i,6})\sin(q_{i,4})$$

$$C_{2,3} = \tfrac{1}{2}\dot{q}_{i,2}(L_6^2\theta_{i,12} - 2L_4\theta_{i,5})\sin(2q_{i,4}) - \dot{q}_{i,2}\cos(2q_{i,5} + 2q_{i,4})L_4\theta_{i,10}$$
$$+ \tfrac{1}{2}\dot{q}_{i,2}((L_5 - L_4)^2(\theta_{i,12} + \theta_{i,11}) - 2L_4\theta_{i,9})\sin(2q_{i,5} + 2q_{i,4})$$
$$+ (\dot{q}_{i,2}d_{2\_0'}\theta_{i,10} + d_6\theta_{i,11}(\dot{q}_{i,4} + \dot{q}_{i,5})(L_5 - L_4))\sin(q_{i,4} + q_{i,5})$$
$$- (((L_5 - L_4)(\theta_{i,12} + \theta_{i,11}) + \theta_{i,9})d_{2\_0'})\dot{q}_{i,2}\cos(q_{i,4} + q_{i,5})$$
$$+ \dot{q}_{i,2}((L_5 - L_4)(L_6\theta_{i,12} + \theta_{i,7}) - L_4\theta_{i,5})\sin(2q_{i,4} + q_{i,5})$$
$$+ (\dot{q}_{i,4}d_6\theta_{i,8} - \dot{q}_{i,2}(L_6\theta_{i,12} + \theta_{i,7} + \theta_{i,5})d_{2\_0'})\cos(q_{i,4})$$
$$+ (\dot{q}_{i,2}(\theta_{i,8} + \theta_{i,6})d_{2\_0'} + \dot{q}_{i,4}d_6\theta_{i,7})\sin(q_{i,4})$$
$$+ \dot{q}_{i,2}((L_5 - L_4)\theta_{i,8} - L_4\theta_{i,6})\cos(2q_{i,4} + q_{i,5}) - \dot{q}_{i,2}L_4\theta_{i,6}\cos(2q_{i,4})$$

$$C_{2,4} = ((\dot{q}_{i,4} + \dot{q}_{i,5})(L_5 - L_4)d_6\theta_{i,11} + \dot{q}_{i,2}d_{2\_0'}\theta_{i,10})\sin(q_{i,4} + q_{i,5})$$
$$+ \tfrac{1}{2}\dot{q}_{i,2}((L_5 - L_4)^2(\theta_{i,12} + \theta_{i,11}) - 2L_4\theta_{i,9})\sin(2q_{i,5} + 2q_{i,4})$$
$$- ((L_5 - L_4)(\theta_{i,12} + \theta_{i,11}) + \theta_{i,9})d_{2\_0'}\dot{q}_{i,2}\cos(q_{i,4} + q_{i,5})$$
$$- \dot{q}_{i,2}L_4\theta_{i,10}\cos(2q_{i,5} + 2q_{i,4})$$
$$- \tfrac{1}{2}\dot{q}_{i,2}(L_4\theta_{i,6} + \theta_{i,8}(L_4 - L_5))(\cos(q_{i,5}) + \cos(2q_{i,4} + q_{i,5}))$$
$$+ \tfrac{1}{2}\dot{q}_{i,2}((L_5 - L_4)(L_6\theta_{i,12} + \theta_{i,7}) - L_4\theta_{i,5})(\sin(2q_{i,4} + q_{i,5}) - \sin(q_{i,5}))$$

$$C_{3,2} = \dot{q}_{i,2}L_4\theta_{i,10}\cos(2q_{i,5}+2q_{i,4}) + \dot{q}_{i,2}L_4\theta_{i,6}\cos(2q_{i,4})$$
$$+ \dot{q}_{i,2}d_{2\_0'}(\theta_{i,9}+(L_5-L_4)(\theta_{i,12}+\theta_{i,11}))\cos(q_{i,4}+q_{i,5})$$
$$- \dot{q}_{i,2}d_{2\_0'}\sin(q_{i,4}+q_{i,5})\theta_{i,10} + \tfrac{1}{2}\dot{q}_{i,2}(2L_4\theta_{i,5}-L_6^2\theta_{i,12})\sin(2q_{i,4})$$
$$+ \dot{q}_{i,2}(L_4\theta_{i,5}-(L_5-L_4)(L_6\theta_{i,12}+\theta_{i,7}))\sin(2q_{i,4}+q_{i,5})$$
$$+ \tfrac{1}{2}\dot{q}_{i,2}(-(L_5-L_4)^2(\theta_{i,12}+\theta_{i,11})+2L_4\theta_{i,9})\sin(2q_{i,5}+2q_{i,4})$$
$$+ \dot{q}_{i,2}d_{2\_0'}(\theta_{i,5}+\theta_{i,7}+L_6\theta_{i,12})\cos(q_{i,4}) - \dot{q}_{i,2}d_{2\_0'}(\theta_{i,8}+\theta_{i,6})\sin(q_{i,4})$$
$$+ \dot{q}_{i,2}(L_4\theta_{i,6}+\theta_{i,8}(L_4-L_5))\cos(2q_{i,4}+q_{i,5})$$

$$C_{3,3} = \tfrac{1}{2}L_4(\dot{q}_{i,4}+\dot{q}_{i,5})((2L_5-L_4)(\theta_{i,12}+\theta_{i,11})+2\theta_{i,9})\sin(2q_{i,5}+2q_{i,4})$$
$$- \tfrac{1}{2}\dot{q}_{i,5}((2L_5L_6-L_6L_4+L_5L_4)\theta_{i,12}+(2L_5-L_4)\theta_{i,7}+\theta_{i,11}L_5L_4$$
$$+ L_4(\theta_{i,9}-\theta_{i,5}))\sin(q_{i,5}) + L_4(\dot{q}_{i,4}+\dot{q}_{i,5})\theta_{i,10}\cos(2q_{i,5}+2q_{i,4})$$
$$+ \tfrac{1}{2}L_4(2\dot{q}_{i,4}+\dot{q}_{i,5})((L_5+L_6)\theta_{i,12}+\theta_{i,7}+L_5\theta_{i,11}+\theta_{i,5}+\theta_{i,9})$$
$$\times \sin(2q_{i,4}+q_{i,5}) + L_4\dot{q}_{i,4}(\theta_{i,8}+\theta_{i,6})\cos(2q_{i,4})$$
$$+ \tfrac{1}{2}L_4\dot{q}_{i,4}((L_4+2L_6)\theta_{i,12}+2\theta_{i,7}+L_4\theta_{i,11}+2\theta_{i,5})\sin(2q_{i,4})$$
$$+ \tfrac{1}{2}L_4(2\dot{q}_{i,4}+\dot{q}_{i,5})(\theta_{i,8}+\theta_{i,6}+\theta_{i,10})\cos(2q_{i,4}+q_{i,5})$$
$$- \tfrac{1}{2}\dot{q}_{i,5}(L_4(\theta_{i,6}+\theta_{i,10})+\theta_{i,8}(L_4-2L_5))\cos(q_{i,5})$$

$$C_{3,4} = (\tfrac{1}{2}(\theta_{i,8}+\theta_{i,10}+\theta_{i,6})\cos(2q_{i,4}+q_{i,5})+\theta_{i,10}\cos(2q_{i,5}+2q_{i,4}))L_4$$
$$\times (\dot{q}_{i,4}+\dot{q}_{i,5}) - \tfrac{1}{2}(\dot{q}_{i,4}+\dot{q}_{i,5})(L_5L_4\theta_{i,12}+(2L_5-L_4)(\theta_{i,7}+L_6\theta_{i,12})$$
$$+ \theta_{i,11}L_5L_4+L_4(\theta_{i,9}-\theta_{i,5}))\sin(q_{i,5}) + \tfrac{1}{2}L_4(\dot{q}_{i,4}+\dot{q}_{i,5})((L_5+L_6)\theta_{i,12}$$
$$+ \theta_{i,7}+L_5\theta_{i,11}+\theta_{i,5}+\theta_{i,9})\sin(2q_{i,4}+q_{i,5})$$
$$+ \tfrac{1}{2}L_4(\dot{q}_{i,4}+\dot{q}_{i,5})((2L_5-L_4)(\theta_{i,12}+\theta_{i,11})+2\theta_{i,9})\sin(2q_{i,5}+2q_{i,4})$$
$$+ \tfrac{1}{2}(\dot{q}_{i,4}+\dot{q}_{i,5})((2L_5-L_4)\theta_{i,8}-L_4(\theta_{i,10}+\theta_{i,6}))\cos(q_{i,5})$$

$$C_{4,2} = \dot{q}_{i,2}d_{2\_0'}((L_5-L_4)(\theta_{i,12}+\theta_{i,11})+\theta_{i,9})\cos(q_{i,4}+q_{i,5})$$
$$+ \tfrac{1}{2}\dot{q}_{i,2}((L_5-L_4)(L_6\theta_{i,12}+\theta_{i,7})-L_4\theta_{i,5})(\sin(q_{i,5})-\sin(2q_{i,4}+q_{i,5}))$$
$$+ \tfrac{1}{2}\dot{q}_{i,2}(2L_4\theta_{i,9}-(\theta_{i,12}+\theta_{i,11})(L_5-L_4)^2)\sin(2q_{i,5}+2q_{i,4})$$
$$+ \dot{q}_{i,2}L_4\theta_{i,10}\cos(2q_{i,5}+2q_{i,4}) - \dot{q}_{i,2}d_{2\_0'}\theta_{i,10}\sin(q_{i,4}+q_{i,5})$$
$$+ \tfrac{1}{2}\dot{q}_{i,2}(L_4\theta_{i,6}-(L_5-L_4)\theta_{i,8})(\cos(q_{i,5})+\cos(2q_{i,4}+q_{i,5}))$$

$$C_{4,3} = \tfrac{1}{2}\dot{q}_{i,4}(L_4(\theta_{i,9}-\theta_{i,5}+\theta_{i,11}L_5)+(2L_5L_6-L_6L_4+L_5L_4)\theta_{i,12}$$
$$+ (2L_5-L_4)\theta_{i,7})\sin(q_{i,5}) + L_4(\dot{q}_{i,4}+\dot{q}_{i,5})\theta_{i,10}\cos(2q_{i,5}+2q_{i,4})$$
$$+ \tfrac{1}{2}\dot{q}_{i,4}L_4((L_5+L_6)\theta_{i,12}+\theta_{i,7}+L_5\theta_{i,11}+\theta_{i,5}+\theta_{i,9})\sin(2q_{i,4}+q_{i,5})$$
$$+ \tfrac{1}{2}\dot{q}_{i,4}L_4(\theta_{i,8}+\theta_{i,6}+\theta_{i,10})\cos(2q_{i,4}+q_{i,5})$$
$$+ \tfrac{1}{2}L_4(\dot{q}_{i,4}+\dot{q}_{i,5})((2L_5-L_4)(\theta_{i,12}+\theta_{i,11})+2\theta_{i,9})\sin(2q_{i,5}+2q_{i,4})$$
$$+ \tfrac{1}{2}\dot{q}_{i,4}(L_4(\theta_{i,6}+\theta_{i,10})+\theta_{i,8}(L_4-2L_5))\cos(q_{i,5})$$

$$C_{4,4} = \tfrac{1}{2}L_4(\dot{q}_{i,4}+\dot{q}_{i,5})((2L_5-L_4)(\theta_{i,12}+\theta_{i,11})+2\theta_{i,9})\sin(2q_{i,5}+2q_{i,4})$$
$$+ L_4(\dot{q}_{i,4}+\dot{q}_{i,5})\theta_{i,10}\cos(2q_{i,5}+2q_{i,4})$$

Table 21.4: Dimensions of the robot.

Dimension	Value [m]	Dimension	Value [m]
L_2, d_{1_2}	0.25	L_8	0.48
L_4	0.05	d_4, d_5	0.0
L_5	0.35	d_6	0.04
L_6	0.30	d_s	0.185
L_7	0.08	$d_{2\_0'}$	0.0916

Table 21.5: Estimated parameters for the master and slave robots.

Parameter	Master robot ($i = m$)	Slave robot ($i = s$)
$\theta_{i,1}$	147.0161	121.3049
$\theta_{i,2}$	2.1448	0.3107
$\theta_{i,3}$	−0.6363	4.1955
$\theta_{i,4}$	0.5931	1.7453
$\theta_{i,5}$	0.1701	0.8316
$\theta_{i,6}$	−0.0561	0.8687
$\theta_{i,7}$	0.8392	0.8105
$\theta_{i,8}$	1.9397	1.6721
$\theta_{i,9}$	−0.1428	−0.1879
$\theta_{i,10}$	1.7807	1.7850
$\theta_{i,11}$	0.1498	0.8759
$\theta_{i,12}$	4.4844	4.1328
$\theta_{i,13}$	83.2945	97.2600
$\theta_{i,14}$	11.2104	9.0999
$\theta_{i,15}$	16.6527	11.6257
$\theta_{i,16}$	13.6684	9.6229
$\theta_{i,17}$	−72.6918	−54.9912
$\theta_{i,18}$	−32.9333	18.4710
$\theta_{i,19}$	5.2337	−3.5232
$\theta_{i,20}$	−3.0435	−5.8564
$\theta_{i,21}$	−85.4138	−46.5915
$\theta_{i,22}$	−42.4819	11.1605
$\theta_{i,23}$	−4.3254	2.2684
$\theta_{i,24}$	5.5640	8.2304
$\theta_{i,25}$	149.9624	150.3190
$\theta_{i,26}$	142.7894	136.8945
$\theta_{i,27}$	8.6392	−35.3699
$\theta_{i,28}$	27.6979	36.0641
$\theta_{i,29}$	−100.2648	−98.9881
$\theta_{i,30}$	−142.2786	−170.4702
$\theta_{i,31}$	−1.8278	−89.3236
$\theta_{i,32}$	12.0224	16.2942

Entries of the Gravity Vector $g(q_i)$.

The entries of the gravity vector $g(q_i) \in \mathbb{R}^4$ as a function of the generalized coordinates $q_i = [q_{i,1} \ q_{i,2} \ q_{i,4} \ q_{i,5}]^{\mathrm{T}}$ of robot i, $i = \mathrm{m, s}$, the parameters $\theta_{i,j}$, $j = 1, \ldots, 32$, and the acceleration due to gravity $g = 9.81$ m/s^2, are given by

$$
\begin{aligned}
g_1 &= g_2 = 0 \\
g_3 &= -g(\theta_{i,9} + \theta_{i,12}L_5 + L_5\theta_{i,11})\sin(q_{i,4} + q_{i,5}) - g(\theta_{i,6} + \theta_{i,8})\cos(q_{i,4}) \\
&\quad - g(\theta_{i,5} + \theta_{i,12}(L_6 + L_4) + L_4\theta_{i,11} + \theta_{i,7})\sin(q_{i,4}) - g\theta_{i,10}\cos(q_{i,4} + q_{i,5}) \\
g_4 &= -g(\theta_{i,9} + \theta_{i,12}L_5 + L_5\theta_{i,11})\sin(q_{i,4} + q_{i,5}) - g\theta_{i,10}\cos(q_{i,4} + q_{i,5})
\end{aligned}
$$

Entries of the Vector of Friction Forces $f(\dot{q}_i)$.

The friction forces $f(\dot{q}_i) \in \mathbb{R}^4$ in the transposer robot are modeled by Eq. (21.12), such that the entries of $f(\dot{q}_i)$ can be written as a function of the generalized velocities $\dot{q}_i = [\dot{q}_{i,1} \, \dot{q}_{i,2} \, \dot{q}_{i,4} \, \dot{q}_{i,5}]^{\mathrm{T}}$ for robot i, and the parameters $\theta_{i,j}$, $i = \mathrm{m, s}$, $j = 1, \ldots, 32$:

$$
\begin{aligned}
f_1(\dot{q}_{i,1}) &= \theta_{i,13}\dot{q}_{i,1} + \theta_{i,17}\left(1 - \frac{2}{1 + e^{2\theta_{i,25}\dot{q}_{i,1}}}\right) + \theta_{i,21}\left(1 - \frac{2}{1 + e^{2\theta_{i,29}\dot{q}_{i,1}}}\right) \\
f_2(\dot{q}_{i,2}) &= \theta_{i,14}\dot{q}_{i,2} + \theta_{i,18}\left(1 - \frac{2}{1 + e^{2\theta_{i,26}\dot{q}_{i,2}}}\right) + \theta_{i,22}\left(1 - \frac{2}{1 + e^{2\theta_{i,30}\dot{q}_{i,2}}}\right) \\
f_3(\dot{q}_{i,4}) &= \theta_{i,15}\dot{q}_{i,4} + \theta_{i,19}\left(1 - \frac{2}{1 + e^{2\theta_{i,27}\dot{q}_{i,4}}}\right) + \theta_{i,23}\left(1 - \frac{2}{1 + e^{2\theta_{i,31}\dot{q}_{i,4}}}\right) \\
f_4(\dot{q}_{i,5}) &= \theta_{i,16}\dot{q}_{i,5} + \theta_{i,20}\left(1 - \frac{2}{1 + e^{2\theta_{i,28}\dot{q}_{i,5}}}\right) + \theta_{i,24}\left(1 - \frac{2}{1 + e^{2\theta_{i,32}\dot{q}_{i,5}}}\right)
\end{aligned}
$$

Physical Dimensions and Estimated Parameters.

The physical dimensions of the CFT robot are listed in Table 21.4; d_{i_i+1} and d_{s} denote the distance between the origin of the frames assigned to the links, and L_i denotes the length of the ith link.

The physical parameters $\theta_{i,j}$, $i = \mathrm{m, s}$, $j = 1, \ldots, 32$ of the master and slave robots have been estimated by using an extended Kalman filter and the least-square method. The estimated parameters are listed in Table 21.5; see [16] for a physical interpretation of the parameters $\theta_{i,j}$, $i = \mathrm{m, s}$, $j = 1, \ldots, 32$.

It is important to mention that there are no available physical parameters from the designer or constructor, such as inertia, masses, etc. Therefore, all the parameters have to be estimated. Note that both master and slave robots are structurally identical; however the parameter values listed in Table 21.5 present some deviations. This difference on the master and slave parameters might be due to several reasons, such as the excitation during the estimation and wear on the motors, servo-amplifiers, joints, etc.

Part IV

Non-conventional Manufacturing Processes

With classical manufacturing techniques, such as turning or milling, the material is usually cut with a solid tool typically made from specially treated steel or certain ceramics, etc. Problems that may arise and their possible solutions were treated in Part II. Nowadays there exist alternatives, which in several respects are superior to the classical methods. They use as tools, among others, beams of particles such as sand grains, ions, or electrons, or a coherent light beam, i.e. laser light. These techniques are often more flexible regarding the geometry of the cut and the choice of the material to be machined. Such beams are used for cutting, drilling, milling, deburring, and so on. Also with these non-conventional methods unwanted phenomena may occur, which find their explanation in the realm of nonlinear dynamics. For instance, a typical observation with beam-cutting techniques, at least for high cutting speeds, is the formation of ripples at the cutting edge, which can be explained with concepts from pattern formation in extended nonlinear dynamical systems.

The first contribution in this part considers ripple formation for abrasive water-jet cutting, where the eroding action of sand grains is used for cutting the workpiece. A phenomenological theory is provided, which in similar form is supposed to hold for other beam-cutting techniques, too. In addition strategies for the suppression of the striation formation are discussed on the basis of this theory and compared with experiments. The second paper by W. Schulz et al. treats laser cutting, where ripple formation is also one of the problems to be considered. In addition, the formation of adherent dross is a topic of concern. Since the physical mechanisms of laser–matter interaction are quite well understood, a more detailed mathematical description is possible. The authors' theoretical predictions and solution strategies are checked experimentally. Laser beams are not only used for separating material, but also for joining. The corresponding technique, laser welding, is investigated in the article by J. Michel et al., again by mathematical modeling and process monitoring. In laser cutting and welding one is mostly interested in the long-term behavior, where a reduction of the number of relevant variables occurs due to the dissipative nature of the process. The short-time dynamics of these processes is more complicated and is relevant e.g. for drilling with pulsed laser radiation. In the article by V. Kostrykin et al. this problem is treated mathematically. The following paper by R. Donner et al. considers the problem of laser beam melt ablation and the corresponding surface structure formation more from a practical point of view: based on measured time-series a minimal stochastic process model is suggested and analyzed. Also, the last paper in this Part IV deals with measuring and monitoring manufacturing processes. The authors suggest a dynamic characteristic quantity, a coarse-grained information rate, which allows to detect dynamic transitions between different process states. The method is applied to the transition from deep- to shallow-penetration welding with laser beams, and it is demonstrated that it works also for the automatic detection of the transition from chatter-free machining to machining with chatter in conventional turning and grinding operations.

22 Nonlinear Dynamics and Control of Ripple Formation in Abrasive Water-jet Cutting

G. Radons, T. Ditzinger, R. Friedrich, A. Henning, A. Kouzmichev, and E. Westkämper

We show experimentally that there are two sources of ripple formation in abrasive water-jet cutting. Ripples can be triggered externally by periodic pressure oscillations, e.g. due to the pump, or they can be generated intrinsically. The latter is due to a spontaneous front instability as described by the nonlinear evolution equations for the cutting front. By including oscillatory motions of the beam, or equivalently oscillations in the beam intensity, we are able to explain on the one hand resonance phenomena between intrinsic and external ripple-forming mechanisms. On the other hand the same approach leads to a strategy for ripple suppression if the corresponding frequencies are well separated.

22.1 Introduction

Beam-cutting techniques are for many applications an alternative to mechanical manufacturing techniques. Abrasive water-jet cutting is an advanced technology which has already today many industrial applications. Especially for hard-to-machine materials, possibly with complex geometry, conventional manufacturing processes can be replaced or complemented using the high flexibility and universality of this technique. For a further spreading of this innovative technology, also in new applications, precision and attainable cutting performance should be improved. The precision of the cutting process is limited by structure formation at the bottom cutting edge surface. With higher feed rates unwanted grooves and striations occur at the surface. To avoid these structures suboptimal cutting velocities are used at present. Reduction of structure formation at the cutting edge could to a large extent compensate the economic disadvantages of this technique and a further spreading could result. Profound investigations are necessary for a deeper understanding of the process and for its optimization. Despite much work in modeling the behavior of the water-jet cutting process over the last two decades (for a recent review see e.g. [1] and references therein), no real understanding of the origin of the striation-pattern formation was attained. The reasons seem to lie in the fundamentally nonlinear nature of the cutting process, which makes it difficult to gain insights by explorative parameter studies and by applying too-simplified models. This view is supported by the fact that ripple formation is a phenomenon observed with most beam-cutting techniques, such as laser-beam [2], flame, ion-beam or electron-beam cutting. The current understanding of ripple formation in the case of laser cutting is documented in [3] and [4]. For a review of modeling and applications of laser cutting see [5, 6].

Nonlinear Dynamics of Production Systems. Edited by G. Radons and R. Neugebauer

ISBN 3-527-40430-9

In this paper the formation of surface structures during the cutting process is analyzed using methods of nonlinear dynamics and synergetics, which were developed in studies of instabilities and pattern formation in many physical, chemical and biological systems [7]. The starting point is a coarse graining over microscopic structures such as atoms, molecules or grains, and temporal events (e.g. particle collisions or impacts). Mathematically the resulting continuum theories are specified by one or several nonlinear partial differential equations (PDE). Below we show that such an approach may shed new light also on the structure formation at the cutting edge during abrasive water-jet cutting.

22.2 Phenomenology of Ripple Formation

Figure 22.1a shows the typical configuration of an abrasive water-jet cutting tool, which can be used to cut all sorts of materials ranging from titanium and glass to ceramics and compound materials: in a focussed beam with diameter $\approx 1\,\mathrm{mm}$ abrasive particles (sand or garnet) are accelerated by water and air up to velocities of approximately $900\,\mathrm{m/s}$. These fast particles impinge onto the workpiece and remove material while the beam is moving with a constant feed velocity u (e.g. in the x direction) of several $\mathrm{cm/min}$, producing a cut with a depth of up to several centimeters. A problem with this technique is that at high feed rates ripples and striation patterns with wavelength of the order of the beam diameter are formed at the side walls of the cut, which degrade the quality of the cutting edge (see Fig. 22.1b). This fact is quantitatively captured by measuring the surface variations of the cutting edge, which is done experimentally with an autofocus sensor producing surface scans with a dense mesh of data points. The height variation is sampled every $0.02\,\mathrm{mm}$ in the feed direction (x direction in Fig. 22.1) over a length of $4\,\mathrm{cm}$. The result of such a scan can be regarded as a continuous signal $h_j(x)$. The index j counts the scans, which are repeated with increments $\Delta z = 2\,\mathrm{mm}$ in the z direction, thereby proceeding deeper into the bulk of the workpiece. These signals can be depicted as a waterfall plot as in Fig. 22.2.

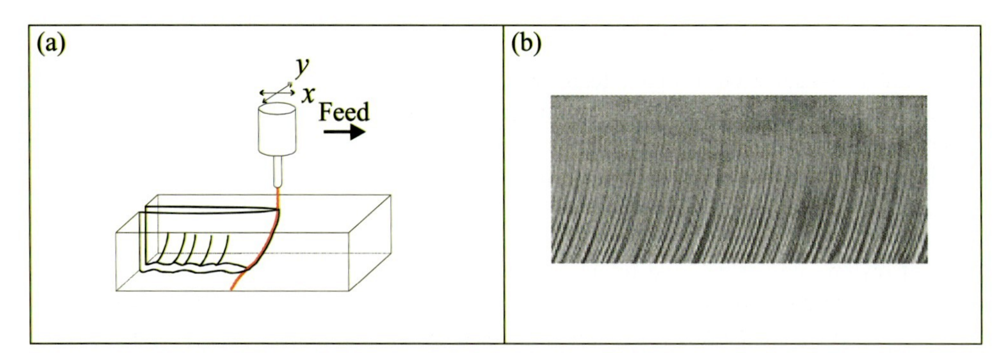

Figure 22.1: (a) Schematic view of the cutting process. The cutting head moves in the x direction, while the impinging grains in the jet cause a cut due to erosion processes at the workpiece. (b) At the side walls of the cut unwanted ripple patterns are generated.

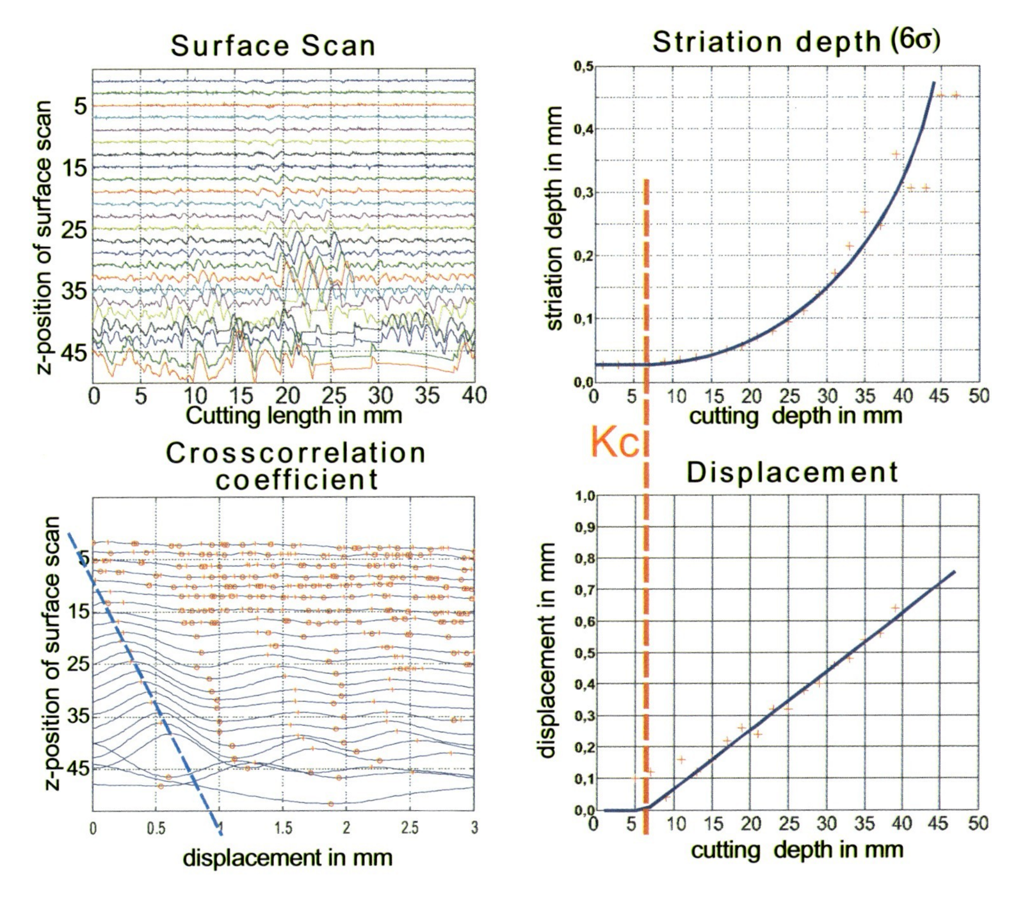

Figure 22.2: Upper left: waterfall plot of the surface variation signals as obtained by repeated scans in the cutting direction. Upper right: the increase of the surface roughness (mean square deviation σ of the signals) as function of the depth. Lower left: cross-correlation function of neighboring scans. The locations of first maxima measure the increments of the ripple lag as one proceeds into the workpiece. Lower right: the increments increase linearly over a wide range implying that the ripple form varies to a good approximation quadratically as a function of the depth. The observed behavior may be used to define the region $0 \leq z \leq K_\mathrm{c}$ of a quality cut.

22.2.1 Ripple Amplitude and Lag

A typical signature of the height variations is that they remain small and structureless near the top of the workpiece from $z = 0$ up to a depth of 5–8 mm (the region $z \lesssim K_\mathrm{c}$ in Fig. 22.2, which for the practitioner defines the quality cut region). With increasing depth the fluctuations measured by the standard deviation $\sigma = (\langle h_j^2(x)\rangle - \langle h_j(x)\rangle^2)^{1/2}$ ($\langle\ldots\rangle$ denotes an average over x over a length of 4 cm), increase approximately quadratically for $z \gtrsim K_\mathrm{c}$. With increasing depth the ripples in addition exhibit a typical lag which also increases with the depth. This may be quantified by detecting the first maximum ξ_j^* of the cross-correlation function $c_j(\xi) =$

$\langle h_{j+1}(x+\xi)\, h_j(x)\rangle$ depicted also in Fig. 22.2. As is seen the position of ξ_j^* increases over some range approximately linearly with j and thus with the depth $z = j\ \Delta z$ (implying a parabolic shape of the ripples in this range). For practical purposes this linear increase may also be used to define the value K_c of the onset of ripple formation by extrapolating to the j-value where $\xi_j^* = 0$. Note, however, that the determination of the ξ_j^* becomes ambiguous for small values, i.e. near K_c, which means that ripple formation not necessarily sets in at a sharp value of the depth, but it may also increase continuously starting from $z = 0$.

22.2.2 Ripple Wavelength

It is often stated in the literature [1] that the wavelength of the striation pattern is approximately independent of the cutting velocity. Such a behavior suggests that the origin of the ripples is of intrinsic nature, which means that it results from the interaction of the beam with the workpiece. This, however, is not a rigorously established fact. As in other beam-cutting techniques such as laser-jet cutting, one has to discuss the possibility that the striations are triggered by external perturbations, which in our case may be a result of oscillations in the cutting head or due to periodic pressure variations stemming from the pump. In the latter case one expects that the ripple wavelength increases linearly with the feed velocity. Our experimental investigations show that both scenarios are possible with the water-jet cutting technique. To see this we systematically determined the ripple wavelength for many experimental scenarios by local Fourier methods, i.e. by determining spectrograms $G_j(x,k) = \left|\int dx'\ \exp[-(x'-x)^2/(2\sigma_0^2)]\ h_j(x')\ \exp(i2\pi k x')\right|^2$ with suitably chosen fixed window width σ_0. Figure 22.3 shows typical results. There we depict measured signals $h_j(x)$ and the corresponding spectrograms $G_j(x,k)$. The spectrograms are plotted by gray coding the value G as a function of position x and wave number $k = 1/\lambda$ (λ is the wavelength). In this figure we also indicate by a dashed line the wave number which is associated with the ideal pump frequency $\nu_0 = 1/\tau_0$, where τ_0 is the time between two pressure maxima. The pressure which provides the acceleration of the abrasive particles is approximately constant, e.g. at an average pressure of $300\,\mathrm{MPa}$ the fluctuations are of the order $\pm 5\,\mathrm{MPa}$, i.e. less than 2%. Due to the construction, we use a two-piston plunger pump; the fluctuations are approximately periodic with period τ_0 in the range of 1.2–$3.1\,\mathrm{s}$. This period is determined under the idealistic assumption that both pistons act identically and exactly in anti-phase (see, however, the remarks below). If the pressure oscillations trigger the striation formation one expects a ripple wavelength $\lambda_0 = u\tau_0$, where u is the feed velocity, or in terms of the wave number $k\ k_0 = \nu_0/u$. This wave number k_0 is indicated in the spectrograms as a dashed line, and in addition the thin continuous line represents $k_0/2$. In the uppermost panel we see enhanced intensity exactly at the wave number k_0. This is observed for cutting stainless steel (1.4301) for a wide range of feed velocities in the range 18–$110\,\mathrm{mm/s}$ resulting of course in different values for k_0, and at a range of depths in the workpiece. Enhanced intensities at k_0 were observed also for constant feed under variation of the pumping frequency. The latter variation was a achieved by constructing a bypass with the only purpose of increasing the volume flow at the pump and therefore the pumping frequency without changing any other cutting parameter. In addition to experiments where the bulk of the workpiece was cut, we performed cuts along edges of the workpiece, where one half or some other portion of the jet was located outside the workpiece (polishing cuts). The ripple formation is much weaker

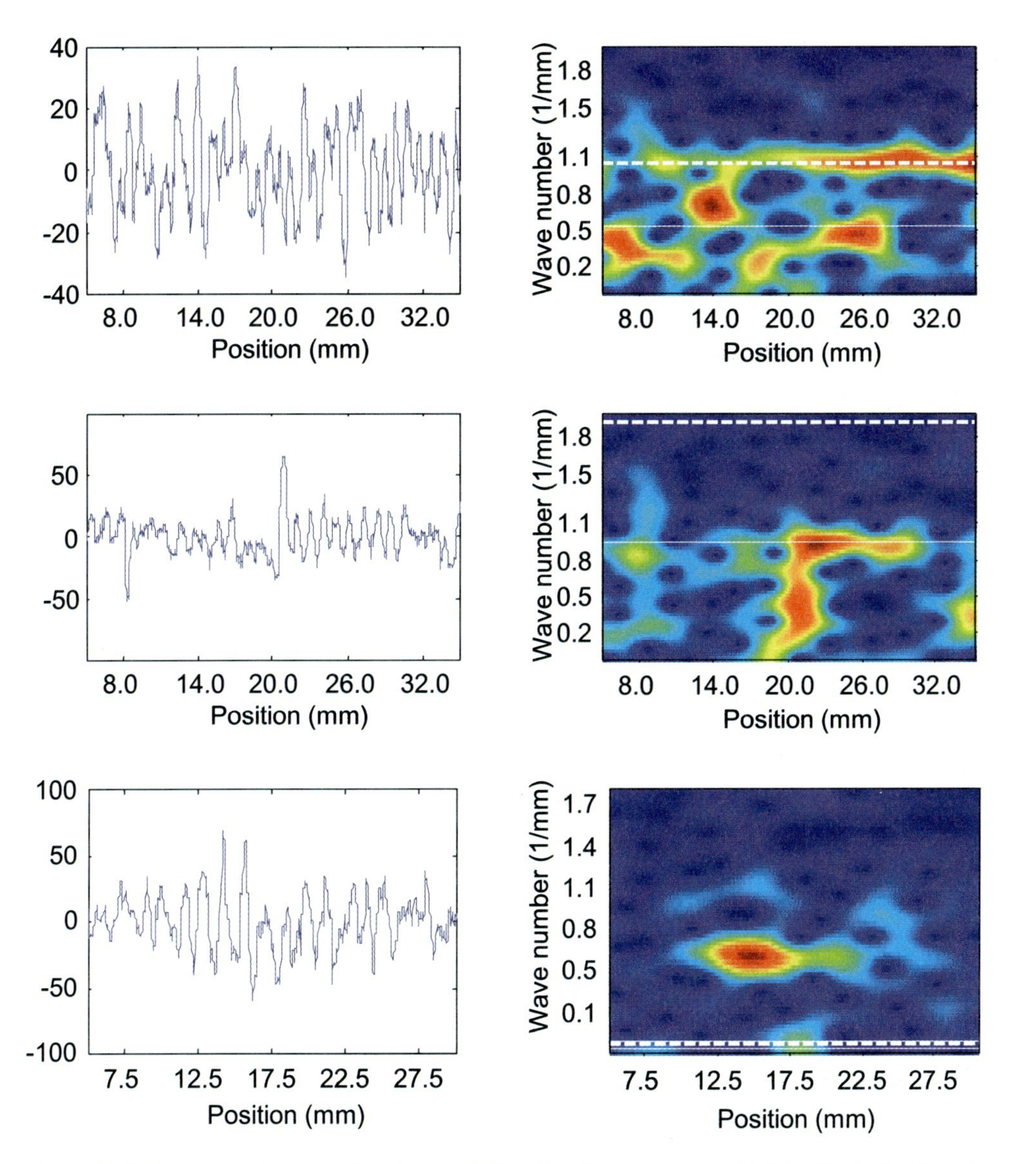

Figure 22.3: Typical surface variations in μm (left) and their spectrograms (right). The dashed lines in the spectrograms are at the spatial frequencies k_0 induced by the pressure oscillations (see Fig. 22.4), the dot-dashed lines correspond to $k_0/2$. Top: cut in stainless steel (feed velocity $u = 18\,\mathrm{mm/min}$, pumping period $\tau_0 = 3.15\,\mathrm{s}$, resulting wave number $k_0 = 1/(\tau_0 u) = 1.058/\mathrm{mm}$). Enhanced intensity at k_0 is seen in the spectrogram, but also some at $k_0/2$. Middle: same feed velocity as before but with enhanced pumping frequency ($\tau_0 = 1.74\,\mathrm{s}$ resulting in a wave number $k_0 = 1.916/\mathrm{mm}$). No intensity is found at k_0 but instead at $k_0/2$. Bottom: cut in PVC at a feed velocity $u = 500\,\mathrm{mm/min}$ ($\tau_0 = 2.42\,\mathrm{s}, k_0 = 0.050/\mathrm{mm}$). The attainable high feed velocity shifts the wavelength $\lambda_0 = 1/k_0$ to high values ($\lambda_0 = 20\,\mathrm{mm}$) so that intrinsic ripple-generating mechanisms decouple from the externally triggered ripples. At $k \approx 0.6/\mathrm{mm}$ one observes strong intensity, which is clearly separated from $k_0 = 0.050/\mathrm{mm}$, and therefore attributed to intrinsic mechanisms.

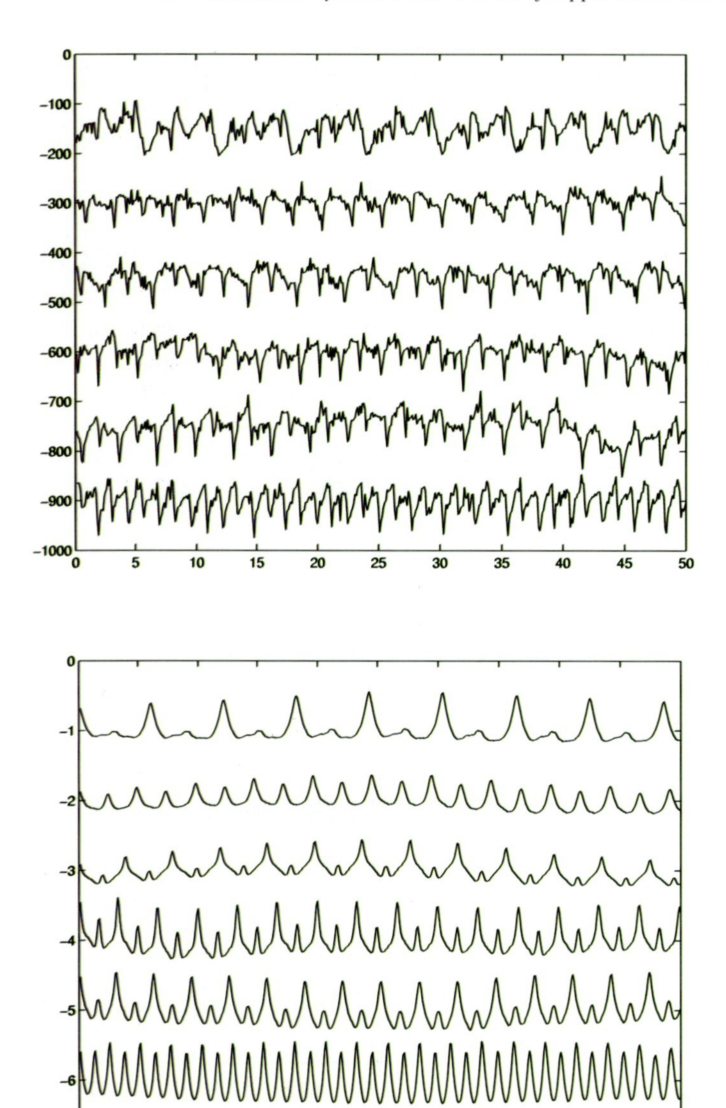

Figure 22.4: Top: water-pressure fluctuations (scale in bar) around the mean value of $3000\,\text{bar}$ ($300\,\text{MPa}$) as a function of time (in s) for various pumping frequencies. Note that the size of the fluctuations is only about 2% of the mean value. The uppermost curve corresponds to the cut shown in Fig. 22.3 (top), the fourth curve to Fig. 22.3 (middle). Bottom: autocorrelation vs. time (in s) of the signals above. In the ideal case all peaks of the autocorrelation functions (and of the pressure) should be equal (nearly fulfilled in the second and the last signal). Slight asymmetries in the two pistons of the plunger pump, which alternately produce the pressure maxima, lift the degeneracy which is ideally expected. The periods τ_0 indicated in Fig. 22.3 correspond to this ideal situation. The realistic period is in most cases equal to $2\tau_0$, which explains the occurrence of a "subharmonic" wavelength $k_0/2$, but not the absence of intensity at k_0 in Fig. 22.3 (middle).

in this case with only little lag, but also there we found intensities in the surface structures stemming from the influence of the pump. With all experimental setups we even found signs of higher harmonics occasionally, e.g. ripples at a wavelength $2k_0$, which can be assigned to higher harmonics of the pressure oscillations. All these experiments clearly showed signs of the influence of the pressure oscillations on the ripple-formation process, though with varying intensity. Quite astonishing at first sight is the finding in the middle panel of Fig. 22.3 obtained for the same material and the same feed velocity, but with the pumping frequency almost doubled. Here we find no sign of ripple intensity at k_0, but instead there is intensity at the subharmonic frequency $k_0/2$. Subharmonics, especially subharmonic bifurcations, are often considered as signs of nonlinear dynamics scenarios. In this case, however, the appearance of subharmonics is to a large extent due again to the pumping action. A closer look at the pressure variations, which we measured near the mixing chamber at the nozzle, showed that the pressure maxima produced by the two pistons are not quite equivalent. This is seen from the pressure as function of time (Fig. 22.4) for the different pumping frequencies (generated with the bypass technique) and even more clearly in the autocorrelation function of the pressure signal. These deviations from symmetric anti-phase motion imply that the true period is rather $2\tau_0$ instead of τ_0, and therefore we have a simple explanation of the corresponding subharmonic frequency $k_0/2$. A remarkable effect, however, is that such small variations in the pressure (compared to its average value) have so strong and drastic consequences with respect to ripple formation. These are clear indications that an instability is here at work which amplifies small perturbations. These amplification mechanisms appear to work especially well for perturbations with wavelengths close to the wavelength of the expected intrinsic ripple-forming mechanism. The latter is expected and repeatedly reported to lie near the jet diameter of about $1\,\mathrm{mm}$.

Our findings above raise the question whether or not such intrinsic mechanisms exist at all. To clarify this question we performed cuts in PVC which allows cutting velocities more than an order of magnitude higher than in steel. As a consequence ripple wavelengths induced by the pump get shifted into the cm regime and therefore are clearly separated from intrinsically generated ripples expected on the scale of the jet diameter, i.e. on a mm scale. At the bottom of Fig. 22.3 we see the result of cutting PVC with clear oscillations at $k \approx 0.6/\mathrm{mm}$ corresponding to a wavelength $\lambda \approx 1.6$–$1.7\,\mathrm{mm}$. We checked whether these ripples are observed also for other feed velocities. The results are depicted in Fig. 22.5 for velocities in the range 200–$1000\,\mathrm{mm/min}$. One observes that in all cases ripples with $\lambda \approx 1.6$–$1.7\,\mathrm{mm}$ are present. This independence of the feed rate assigns these ripples to intrinsic sources, i.e. to the interaction of jet and workpiece. The nature of the intrinsic mechanisms has not yet been fully clarified. We will shed new light on this phenomenon below by applying methods from pattern formation in nonlinear systems.

22.3 Cutting Processes and Pattern Formation

It is well known [7] that systems far from equilibrium can exhibit spontaneous formation of spatio-temporal behavior. Especially systems with interfaces show a rich variety of such phenomena. A prominent example is the behavior of a flame front: a planar front undergoes an instability resulting in complex behavior of the interface [8]. Other examples are moving fronts in chemical reactions [9].

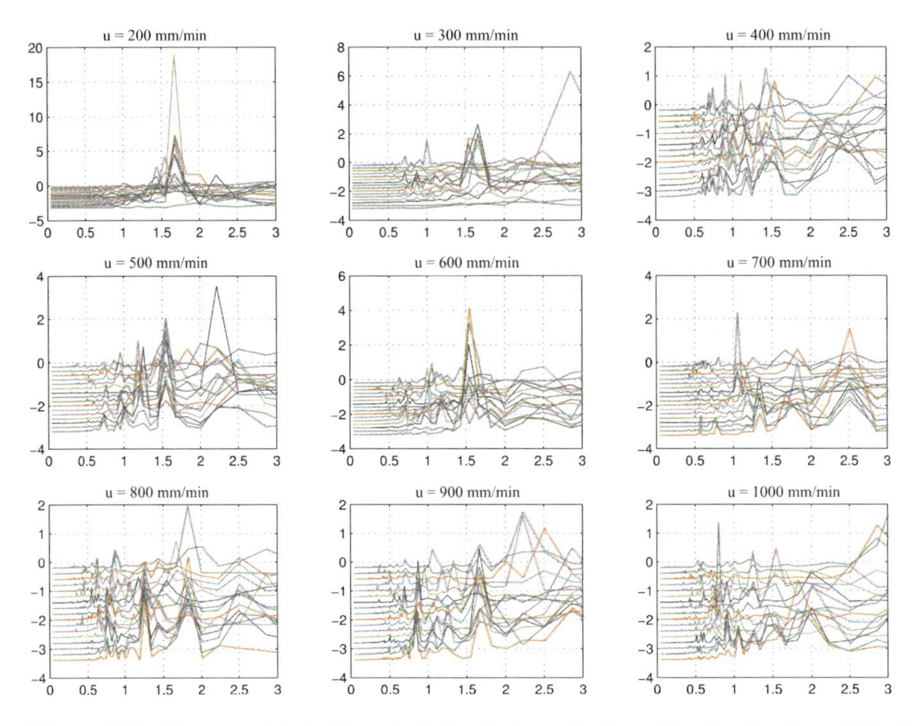

Figure 22.5: The spectral intensities of all signals (scans in the x direction for 16 or 17 depth values) in each cutting experiment for cutting velocities in the range 200–1000 $\mathrm{mm/min}$. They are plotted as a function of wavelength (in mm). One observes that in all cases enhanced intensities corresponding to ripples with $\lambda \approx 1.6$–$1.7\,\mathrm{mm}$ are present.

In the following we shall draw analogies between cutting processes like abrasive water-jet cutting or laser-beam cutting and such types of processes. The most important phenomenon here is the existence of a sharp cutting front. There are well-known situations where the cutting front spontaneously exhibits additional features in the form of ripples. There seem to be situations where these ripples are not enforced from the outside but are generated spontaneously due to interactions between beam and cutting surface. From the point of view of the theory of pattern formation in continuous media it is not surprising at all that the spontaneous emergence of patterns can be found in cutting processes since in such a process the system is in a state far from equilibrium due to the continuous impact of the beam. In the following we shall present a phenomenological theory for such processes which is based on an analogy with front instabilities in non-equilibrium systems: we shall put the evolution of a cutting front under a beam in close correspondence to the evolution of a flame front.

22.3.1 Pattern Formation by Front Instabilities

It is an appealing idea to study pattern formation by dynamic fronts from a unifying point of view. Such an approach is essentially based on the assumption that the dynamics of an interface or a sharp front at location $z = S(x, y, t)$ can be described by a closed evolution equation of the form

$$\frac{\partial}{\partial t} S(x, y, t) = M[S(x, y, t)], \tag{22.1}$$

where $M[S(x, y, t)]$ is a nonlinear functional of $S(x, y, t)$.

Let us consider a spatially extended system described by a state vector $\mathbf{q}(x, y, z, t)$ which fulfills an evolution equation of the form

$$\dot{\mathbf{q}}(x, y, z, t) = \mathbf{N}[\mathbf{q}(x, y, z, t)]. \tag{22.2}$$

Here, $\mathbf{N}$ denotes a nonlinear functional of $\mathbf{q}(x, y, z, t)$. The meaning of $\mathbf{q}$ depends on the system considered. For instance, for chemical reaction–diffusion systems the components of $\mathbf{q}$ describe the concentration of chemical species, whereas for cutting problems $\mathbf{q}$ includes the indicator function for the presence or absence of material of the workpiece at a given space–time point. It is interesting to consider the possibility to derive an evolution equation for a front (22.1) from this basic law. Without loss of generality one may split $\mathbf{q}$ into two parts:

$$\mathbf{q}(x, y, z, t) = \mathbf{q}_0(x, y, z - S(x, y, t)) + \mathbf{w}(x, y, z - S(x, y, t), t), \tag{22.3}$$

which decomposes the actual state into a part $\mathbf{q}_0$ which describes the change of the state due to a deformation of the *location* $S(x, y, t)$ of the front as it moves in one direction, e.g. the z direction, and a term which takes into account deviations of $\mathbf{q}$ from $\mathbf{q}_0$. Assuming now that these deviations can again be expressed as a functional of $S(x, y, t)$, i.e. $\mathbf{w} = \mathbf{w}[S]$, one arrives at the closed equation (22.1), as can be seen by differentiating Eq. (22.3) with respect to time.

The simplest case arises if the state vector $\mathbf{q}(x, y, z, t)$ describes a front solution moving in the z direction with constant velocity v:

$$\mathbf{q}(x, y, z, t) = \mathbf{q}_0(x, y, z - vt), \tag{22.4}$$

which arises for vanishing $\mathbf{w}$ and the corresponding functional $\mathbf{N}$.

The form of the evolution equation (22.1) can to some extent be fixed by symmetry considerations. For spatially homogeneous systems which allow for a shift of the interface $\mathbf{q}(x, y, z - S(x, y, t)) \to \mathbf{q}(x, y, z - z_0 - S(x, y, t))$, the evolution equation (22.1) has to be invariant with respect to the transformation

$$S(x, y, t) \to S(x, y, t) + z_0. \tag{22.5}$$

This shows that M cannot depend on $S(x, y, t)$ explicitly, but only on the spatial gradient ∇S and higher derivatives. Furthermore, one may perform a gradient expansion around the constantly moving front. In vectorial notation with $\mathbf{x} = (x, y)$ one obtains

$$\begin{aligned}\frac{\partial}{\partial t} S(\mathbf{x}, t) &= v + (\alpha\Delta + \beta\Delta^2 + ...)S(\mathbf{x}, t)\\ &+ [a\nabla S(\mathbf{x}, t) \cdot \nabla S(\mathbf{x}, t) + b(\nabla S(\mathbf{x}, t) \cdot \nabla\Delta S(\mathbf{x}, t)\\ &+ c\Delta(\nabla S(\mathbf{x}, t) \cdot \nabla S(\mathbf{x}, t)) + d\nabla\nabla S(\mathbf{x}, t)\nabla\nabla S(\mathbf{x}, t) + ...\end{aligned} \tag{22.6}$$

In principle, the coefficients v, α..., a, ... can be calculated from the basic equation by a standard perturbation approach [9].

In order to study the stability of the plane moving front $S(\mathbf{x}, t) = vt$ one performs a linear stability analysis with the ansatz:

$$S(\mathbf{x}, t) = vt + s(\mathbf{x}, t). \tag{22.7}$$

Neglecting higher-order derivatives leads to the following linear equation:

$$\dot{s}(\mathbf{x}, t) = [\alpha\Delta + \beta\Delta^2]s(\mathbf{x}, t). \tag{22.8}$$

A plane-wave ansatz $s = e^{\lambda(\mathbf{k})t}e^{i\mathbf{k}\cdot\mathbf{x}}$ yields a linear growth rate of the form

$$\lambda(k) = -\alpha k^2 + \beta k^4, \tag{22.9}$$

with $k = |\mathbf{k}|$. If $\alpha > 0$ and $\beta < 0$ the basic state is stable (the mode at $k = 0$ is neutrally stable since the plane front can be shifted in the z direction). However, if $\alpha < 0$, $\beta < 0$ the plane front is unstable ($\lambda > 0$) with respect to plane waves with $0 < |\mathbf{k}| < \alpha/\beta$. The mode corresponding to the maximal growth rate is obtained from $\partial\lambda(k)/\partial k = 0$ for

$$|k_{\mathrm{m}}| = \sqrt{\frac{\alpha}{2\beta}}. \tag{22.10}$$

Linear stability analysis indicates an instability of a planar front. The actual pattern which develops due to this instability is determined by the detailed structure of the nonlinear term. The evolution equation (22.6) can be viewed as an order-parameter equation for systems with front instabilities. It plays the same role as the generalized Ginzburg–Landau equations for instabilities in pattern-forming systems.

22.3.2 Phenomenological Theory of the Evolution of Cutting Fronts

For the case of abrasive water-jet cutting no mesoscopic theory formulated as a continuum theory exists. It is tempting, however, to develop a phenomenological theory for the location of the cutting front in the spirit of the Landau phenomenological theory of phase transitions. This theory should be in close analogy to the theory of front instabilities. Therefore, a starting point is the evolution equation (22.1), which, however, has to be modified in several respects [10].

1. Localized Beam

   The surface evolves under the influence of the beam. This fact is taken into account by introducing a beam function $v(\mathbf{x})$, which describes the intensity of the beam and which is, therefore, zero in regions where the beam is not active:

   $$\frac{\partial}{\partial t}S(\mathbf{x}, t) = v(\mathbf{x})M[S(\mathbf{x}, t)]. \tag{22.11}$$

   Usually, $v(\mathbf{x})$ is taken as a Gaussian centered at $\mathbf{x} = \mathbf{x}_0$:

   $$v(\mathbf{x}) = Ne^{-(\mathbf{x}-\mathbf{x}_0)^2/2\sigma^2}. \tag{22.12}$$

2. Feed Velocity

   The beam is moved with respect to the material. The process is best described in a co-moving coordinate frame. Therefore, we have to include a convective term with feed velocity $\mathbf{u}$:

$$\frac{\partial}{\partial t} S(\mathbf{x}, t) + \mathbf{u} \cdot \nabla_{\mathbf{x}} S(\mathbf{x}, t) = v(\mathbf{x}) M[S(\mathbf{x}, t)]. \tag{22.13}$$

   The feed velocity may depend on time t.

3. Removal Rate

   The nonlinear functional $M[S(\mathbf{x}, t)]$ can be considered as a removal rate, since $\partial S(x, y, t)/\partial t$ is proportional to the amount of material removed per unit time. The local removal rate predominantly depends (a) on the cut material, (b) on the composition of the water jet and (c) on the angle between beam direction and actual surface. As a result, the removal rate has to be a function of the gradient of $S(\mathbf{x}, t)$. Under isotropic conditions, the removal rate is a function of $(\nabla S(\mathbf{x}, t))^2$. A wide class of material (brittle material) is approximately described by the law [11]

$$M = \frac{1}{1 + (\nabla S(\mathbf{x}, t))^2}. \tag{22.14}$$

   Material removal is maximal when the surface is perpendicular to the beam. For ductile materials other relationships hold. The dependency of the removal rate on the material is well documented in the literature [11]. It has, however, been recognized already in [12] that such a law cannot describe ripple formation and that higher-order terms in the derivatives have to be included.

The next question thus concerns the possible dependency of the removal rate on the curvature of the surface as well as on higher derivatives. With regard to the Kuramoto–Sivashinsky equation we formulate the following phenomenological model:

$$\begin{aligned} &\frac{\partial}{\partial t} S(\mathbf{x}, t) + \mathbf{u} \cdot \nabla_{\mathbf{x}} S(\mathbf{x}, t) \\ &= v(\mathbf{x}) \left[\frac{1}{1 + (\nabla S(\mathbf{x}, t))^2} + \alpha \Delta S(\mathbf{x}, t) + \beta \Delta^2 S(\mathbf{x}, t) \right]. \end{aligned} \tag{22.15}$$

We mention that, in principle, the coefficients α and β may depend on $\nabla S(\mathbf{x}, t)$, i.e. on the angle between surface and beam. However, in the following we shall neglect such dependences. We assume $\beta < 0$, which assures stability, i.e. coarsening with respect to small-scale variations. However, we allow for negative α which indicates an instability, i.e. a roughening of the surface.

22.3.3 Solution of Model Equation

In Fig. 22.6 the results of a numerical simulation of the cutting process according to our model equation (22.15) are shown. The calculated surface $S(x, y, t)$ of the workpiece is

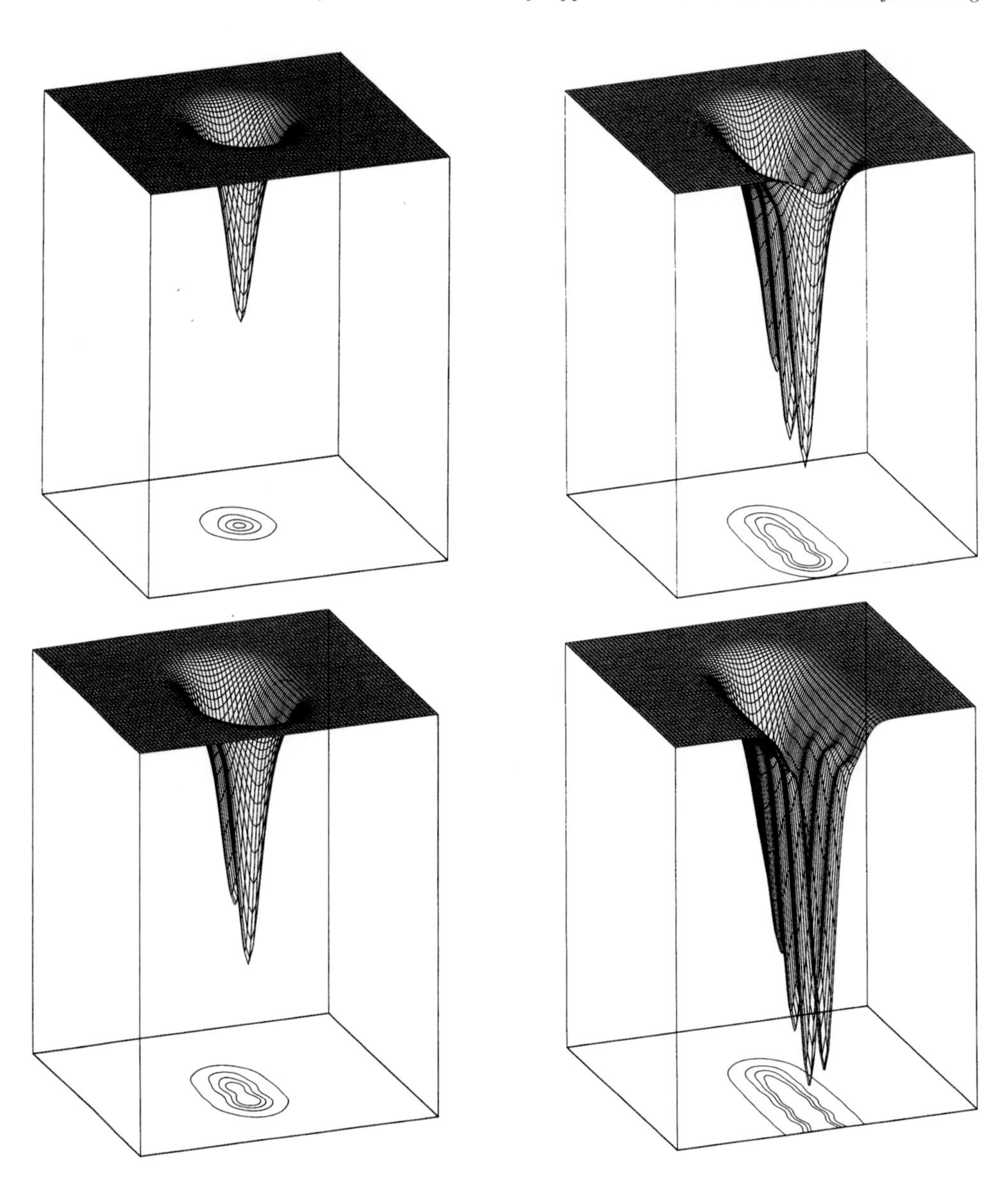

Figure 22.6: 3D simulation of the cutting process. For four subsequent times the surface $S(x, y, t)$ of the workpiece is plotted vs. the spatial coordinates x and y. In the co-moving frame used one has to think of material fed into the fixed beam from the back of the graphs. At the bottom and at the sides of the cutting surface the same spatial periodic patterns can be seen as in the experiment (cf. Fig. 22.1b).

plotted vs. spatial coordinates x and y at four different times. Note that kerfs similar to the ones in the experiment can be seen at the bottom of the cutting line. At the edge of the cut inhomogeneities in the surface evolve in close analogy to striation formation observed in the experiment (Fig. 22.1b). To better understand this result we consider in the following a one-dimensional geometry, i.e. we neglect the transverse degrees of freedom. For a smooth front $S(x, t)$ one can perform the approximation

$$\left|\frac{\partial S}{\partial x}\right| \gg \left|\frac{\partial^2 S}{\partial x^2}\right|, \left|\frac{\partial^4 S}{\partial x^4}\right|. \tag{22.16}$$

In that case we obtain a partial differential equation of first order which can be solved by the method of characteristics. Furthermore, an interpretation as a Hamilton–Jacobi equation allows one to make direct contact with classical mechanics [13, 14].

More detailed investigations of solutions of the model equation have to be done by numerical means. In the following we shall summarize the main results. However, an extensive investigation of the model equation for an extended parameter set has still to be performed. The results obtained so far, however, are of high relevance for the interpretation of the experimental results and for the design of optimal process strategies in order to suppress ripple formation and to improve the performance of abrasive water-jet cutting.

Analytical and Numerical Approach to the Solution of the Model Equation

Let us now consider stationary fronts, which have to fulfill the equation

$$uS_x = v(x)\frac{1}{1 + S_x^2}, \tag{22.17}$$

i.e. the spatial derivative S_x of the front is the solution of a third-order equation:

$$S_x^3 + S_x - \frac{v(x)}{u} = 0, \tag{22.18}$$

which allows for one real solution. Assuming $|v(x)/u| \ll 1$ one may neglect S'^3 to obtain the result

$$S(x) = S(-\infty) + \frac{1}{u}\int_{-\infty}^{x} dx' v(x'), \tag{22.19}$$

i.e. the cutting depth is given by $H = S(\infty) - S(-\infty)$:

$$H = \frac{1}{u}\int_{-\infty}^{\infty} dx' v(x'). \tag{22.20}$$

In the opposite case, $|S'| \gg 1$, one obtains

$$H = \frac{1}{u^{1/3}}\int_{-\infty}^{\infty} dx' v(x')^{1/3}. \tag{22.21}$$

Thus, for such material, one expects to observe a crossover in the cutting depth from $H \approx 1/u$ to $H \approx 1/u^{1/3}$.

Induced Ripple Formation

Let us first report results obtained for the case $\alpha > 0$. Here, a smooth stationary front develops which can be approximated by the expression (22.17) given above. Ripples, however, can also be generated numerically e.g. by a periodic modulation of the beam intensity $v(x)$. As we have seen in the first part this fact is highly relevant for the interpretation of experiments, since the beam intensity may vary periodically due to the action of the pump. Therefore, one has to distinguish between induced (extrinsic) and spontaneous (intrinsic) ripple formation.

22.3.4 Spontaneous Ripple Formation

In the case of an infinitely extended beam ($v(x) = \text{const.}$) an instability develops for the case $\alpha < 0$, $\beta < 0$. The wavelength L of the fastest growing disturbance is given by

$$L = \frac{2\pi}{k_\mathrm{m}} = \frac{2\pi}{\sqrt{\alpha/2\beta}}. \tag{22.22}$$

In order to obtain such an instability also for the case of a confined beam we take this wavelength (by a suitable choice of α, β) to be of the order of the beam thickness σ. For large values of the feed velocity a smooth front develops, whereas by lowering the feed velocity, i.e. by generating a deeper cut, an instability sets in (see Fig. 22.7). This instability is a convective instability [10].

22.3.5 Suppression of Spontaneous Ripple Formation by Periodic Modulation

It is tempting to consider periodic modulations of the parameters occurring in the model equations in order to predict phenomena which should be recovered in experiments. There are two ways to intervene in the cutting process from the outside:

(a) Modulation of the beam intensity:

Let us consider a modulation of the form

$$v(x,t) = v_0(x)(1 + A\sin\omega t), \qquad A < 1. \tag{22.23}$$

Introducing a new time variable τ:

$$\tau = t - \frac{A}{\omega}\cos\omega t, \tag{22.24}$$

the evolution equation can be re-cast into the form

$$\begin{aligned} &\frac{\partial}{\partial\tau}S(\mathbf{x},\tau) + \mathbf{u}(\tau)\cdot\nabla_\mathbf{x}S(\mathbf{x},\tau) \\ &= v_0(\mathbf{x})\left[\frac{1}{1+(\nabla S(\mathbf{x},\tau))^2} + \alpha\Delta S(\mathbf{x},\tau) + \beta\Delta^2 S(\mathbf{x},\tau)\right], \end{aligned} \tag{22.25}$$

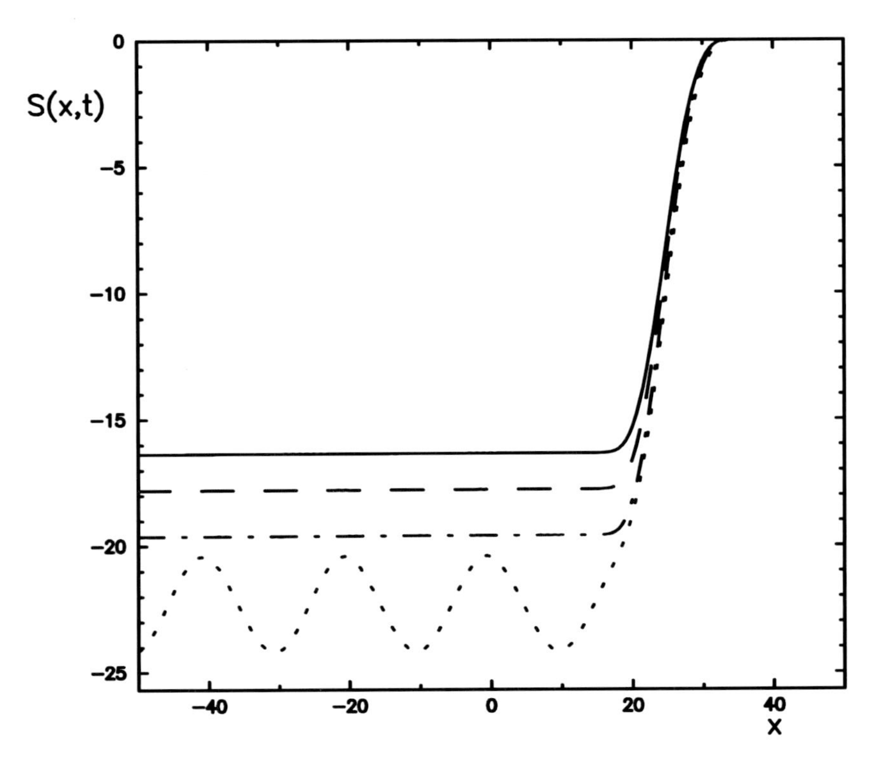

Figure 22.7: Top: the surface $S(x,t)$ at a late time $t = 20\,000$ for various feed rates $u = -0.061, -0.081, -0.101, -0.121$ (from bottom to top, parameters $\alpha = -1$, $\beta = -5.066$). For decreasing values of u the height of the front increases, leading eventually to a traveling wave behavior. Bottom: amplitude of the spatio-temporal modulation as a function of the feed velocity $|u|$ indicating the onset of an oscillatory instability of the stationary front through a Hopf-like bifurcation. The amplitude was determined from states at time $t = 10^6$.

where the feed velocity becomes time dependent:

$$\mathbf{u}(\tau) = \frac{\mathbf{u}}{1 + A \sin \omega t(\tau)}. \tag{22.26}$$

Therefore, a temporal modulation of the beam (described in the form $v(\mathbf{x})I(t)$) is dynamically equivalent to a modulation of the feed velocity.

(b) Modulation of the feed velocity:

$$u = u(t) = u_0(1 + A\ \sin\ \omega t). \tag{22.27}$$

This feed velocity is equivalent to the following time dependence of the position $x_0(t)$ of the beam:

$$x_0(t) = x_0 + u_0 t - u_0 \frac{A}{\omega} \cos \omega t. \tag{22.28}$$

For $|A| < 1$ the feed velocity is always positive. In the opposite case it can become negative, indicating that the beam propagates in the backward direction for a certain period of time.

It is quite interesting to consider periodic modulations of the feed velocity in the case where the front is unstable with respect to ripple formation. Since the evolution equation is nonlinear one expects resonance phenomena in the case where the modulation frequency is of the order of the frequency of the ripple formation defined by

$$\omega_0 = u_0 k, \tag{22.29}$$

where k is the wavenumber of the ripple. This appears to be exactly what we found in the experiments, which led to Fig. 22.3 in the case of cutting stainless steel. However, if the modulation frequency is higher than about four times the ripple frequency, a suppression of the instability can be observed for sufficiently high values of the modulation amplitude. In that case, however, the beam has to move back and forth [15]. The numerically observed suppression of the ripples is shown in Fig. 22.8.

Reduced Model for Ripple Formation

In order to study dynamical effects of ripple formation with respect to its suppression it is convenient to introduce the following phenomenological model [15]: close to onset of ripple formation we can expect the function $S(x, t)$ to be of the form

$$S(x,t) = S_0(x) + \psi(x,t)e^{ikx} + \text{c.c.} + \text{higher-order terms}, \tag{22.30}$$

where $S_0(x)$ denotes the basic, stationary front and the order parameter $\psi(x, t)$ describes the amplitude of the ripples, which, in the case of a convective instability, would take the form

$$\frac{\partial}{\partial t}\psi(x,t) + u\frac{\partial}{\partial x}\psi(x,t) = v(x)\left[\left[\epsilon + i\omega + D\frac{\partial^2}{\partial x^2}\right]\psi(x,t) - A|\psi(x,t)|^2\psi(x,t)\right]. \tag{22.31}$$

Here, ω denotes the ripple frequency, ϵ the linear growth rate and D the spatial diffusion. In order to take into account the finite extension of the beam, we again have introduced the beam function $v(x)$, which is localized in the beam region. Using this reduced model one can even analytically study the suppression of the instability by a modulation of the feed velocity $u(t) = u_0(1 + A\,\sin(\omega t))$. We mention that in order to study situations where the modulation frequency is close to the ripple frequency the inclusion of resonance terms is necessary.

22.4 Experimental Results for Ripple Suppression

The idea of an additional modulation of the feed velocity was tested by us in water-jet cutting experiments. In Fig. 22.9 the striation depth (6σ) is displayed for an oscillation parallel to the cutting direction. The cut without modulation is displayed with a dashed line for comparison. The effect does strongly depend on the choice of the modulation parameters. The best results

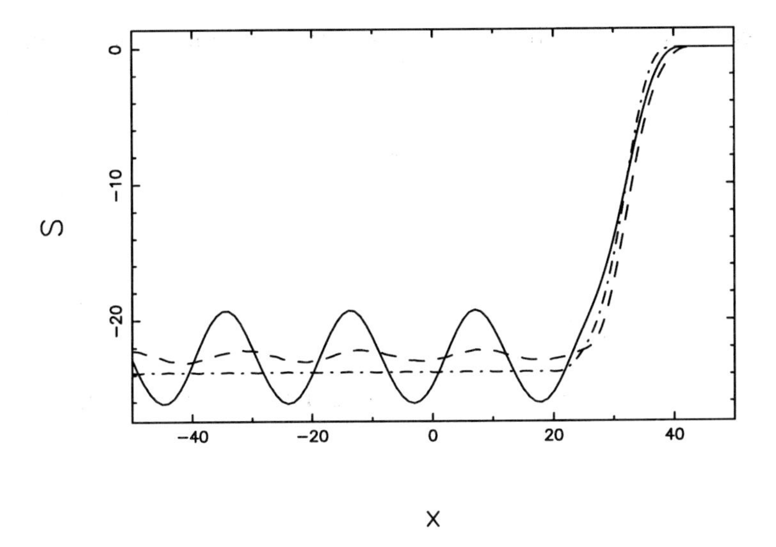

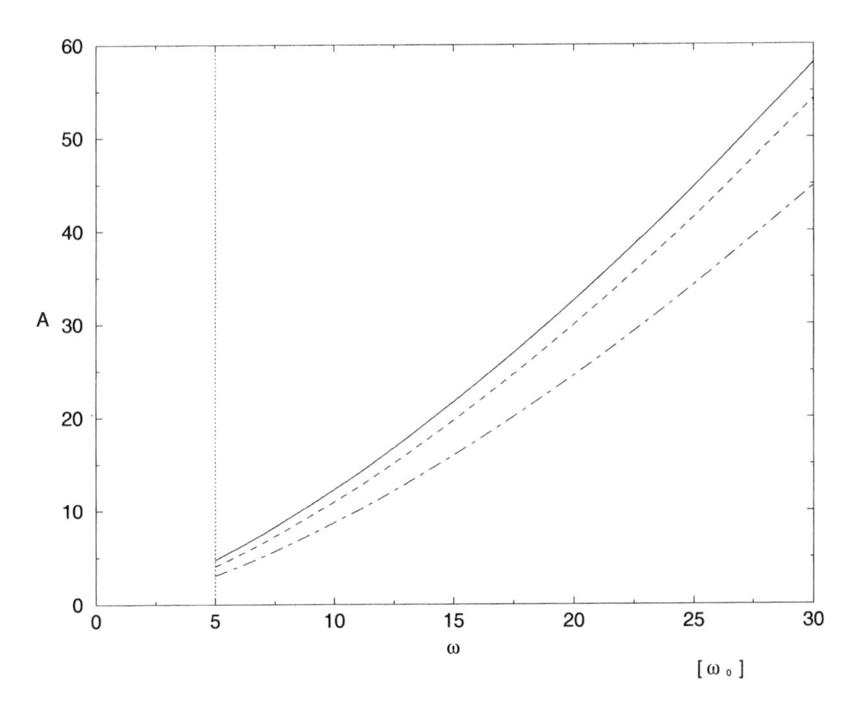

Figure 22.8: Top: the ripples (wavelength λ) in the surface $S(x, t)$ for modulation $A = 0$ (full), $A = 3.3$ (dashed) and $A = 8.3$ (dot-dashed) and frequency $\omega = 6\omega_0$ with $\omega_0 = 2\pi|u_0|/\lambda$. Bottom: phase diagram in the A–ω plane: above the lines corresponding to $|u_0| = 0.06, 0.065, 0.07$ respectively (from the top) the ripples are suppressed. For $\omega < 5\omega_0$ the occurrence of nonlinear resonances can lead to an enhancement of ripple formation, as we have observed in Fig. 22.3 for stainless steel. In the latter case the oscillations of frequency ω were due to the uncontrolled action of the pump.

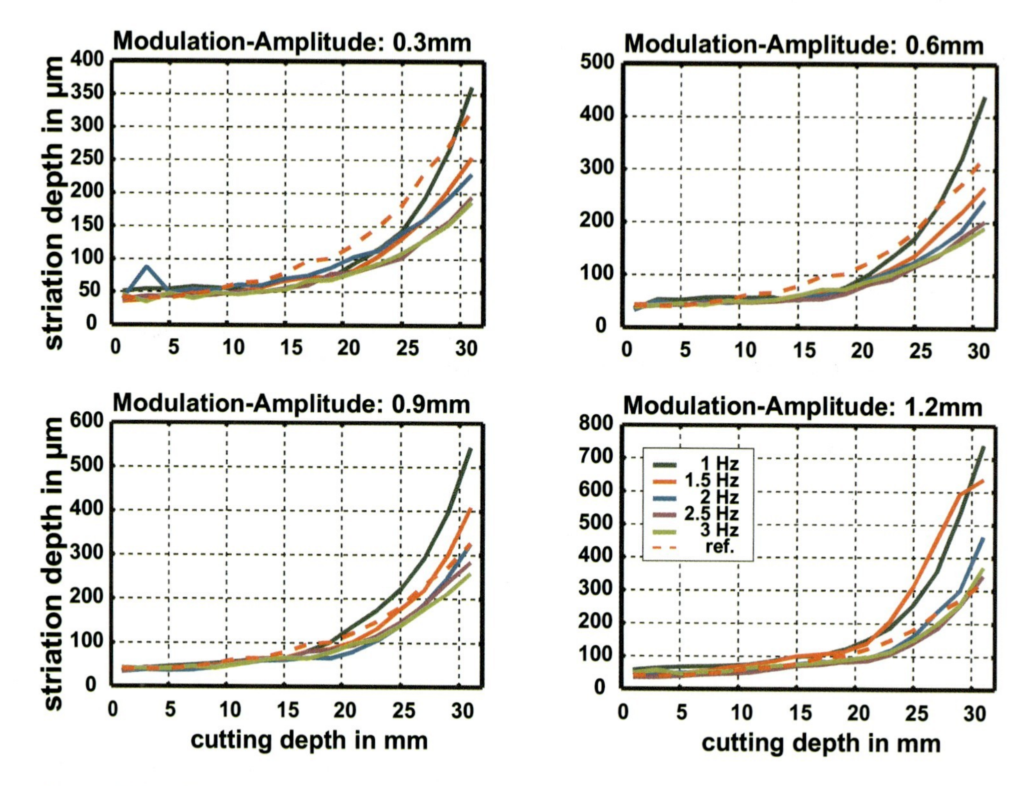

Figure 22.9: The striation strength as a function of depth is displayed for different amplitudes and frequencies of the feed-velocity modulation. In most cases a reduction of the ripple amplitude is observed.

are achieved by using small oscillation amplitudes $\overline{a} = u_0 A/\omega$ (see Eq. (22.28)) of $0.3\,\mathrm{mm}$ and $0.6\,\mathrm{mm}$. For these amplitudes the striation depth is reduced for nearly all used oscillation frequencies. The best results with respect to the modulation frequency $\omega = 2\pi\,\nu$ were achieved with $\nu = 2.5\,\mathrm{Hz}$. With $u_0 = 23\,\mathrm{mm/min}$ and $\lambda \approx 1\,\mathrm{mm}$ the oscillation frequency defined in Eq. (22.29) is obtained as $\omega_0 = 2\pi u_0/\lambda \approx 2\pi 0.4\,\mathrm{Hz}$. Thus for the experimentally observed optimal modulation one obtains a ratio $\omega/\omega_0 \approx 6$ and the dimensionless parameter $A = \overline{a}\omega/u_0$ of Eq. (22.27) is equal to 12.3 and 24.6, respectively. Thus according to the phase diagram of Fig. 22.8 a strong suppression of the ripples is expected. The experimental results, however, exhibit a reduction of the ripples, but not a full suppression. Similar results were obtained in [16] for mild steel under different cutting conditions and with a slightly different oscillation technique. As a perspective there is still room for a further improvement of the ripple-suppression strategy.

22.5 Conclusions

We have shown experimentally that ripples in abrasive water-jet cutting can be triggered externally by periodic pressure oscillations from the pump or they can be generated intrinsically.

The intrinsic mechanism is due to a spontaneous front instability as described by the nonlinear evolution equations for the cutting front, which we have developed. By including oscillatory motions of the beam, or equivalently oscillations in the beam intensity, we were able to explain on the one hand resonance phenomena between intrinsic and external ripple-forming mechanisms. On the other hand the same approach leads to a strategy for ripple suppression if the corresponding frequencies are well separated.

Acknowledgment

We gratefully acknowledge support from the Volkswagen Foundation within the priority area "Investigation of Nonlinear-Dynamic Effects in Production Systems".

Bibliography

[1] A. Momber and R. Kovacevic, *Principles of Abrasive Water Jet Machining*, Springer-Verlag, Berlin, 1997.

[2] Y. Arata, H. Maruo, I. Miyamoto, and S. Takeuchi, *Dynamic behavior in laser gas cutting of mild steel*, Transactions of the Japanese Welding Research Institute **8** (1979), 15–26.

[3] W. Schulz, V. Kostrykin, J. Michel, and M. Niessen, *Modelling and Simulation of Process Monitoring and Control in Laser Cutting*, Chapter 23, this book.

[4] R. Poprawe and W. König, *Modeling, monitoring and control in high quality laser cutting*, Annals of the CIRP **50**(1) (2001), 137–140.

[5] W. Schulz and C. Hertzler, *Cutting: modelling and data*, in: New Series Landolt-Börnstein, Group: Advanced Materials and Technologies, Volume VIII/1: Laser Physics and Applications, edited by M. Martienssen, R. Poprawe, H. Weber, and G. Herziger, Subvolume VIII/1C: Laser Applications, 2004, chapter 17.5.2. in press

[6] J. Franke, *Modellierung und Optimierung des Laserstrahlbrennschneidens niedriglegierter Stähle*, DVS-Berichte **161** (1994), 122–153.

[7] H. Haken, *Synergetics. An Introduction*, Springer-Verlag, New York, 1983.

[8] G. Sivashinsky, Acta Astronautica **4** (1977), 1177.

[9] Y. Kuramoto and T. Tsuzuki, Progress of Theoretical Physics **55** (1976), 356.

[10] R. Friedrich, G. Radons, T. Ditzinger, and A. Henning, *Ripple formation through an instability from moving growth and erosion sources*, Physical Review Letters **85** (2000), 4884.

[11] J.G.A. Bitter, *A study of erosion phenomena, Parts I and II*, Wear **6** (1963), 5–21, 169–190.

[12] I. Finnie and Y.H. Habil, *On the formation of surface ripples during erosion*, Wear **8** (1965), 60–69.

[13] T. Ditzinger, R. Friedrich, A. Henning, and G. Radons, *Nonlinear dynamics in modeling of cutting edge geometry*, in: Proceedings of the 10th American Waterjet Conference, edited by M. Hashish, Vol.1, Waterjet Technology Association, 1999, pp. 15–32.

[14] G. Radons, T. Ditzinger, R. Friedrich, A. Henning, and T. Kuebler, *Origin and control of ripple formation in beam cutting techniques*, in: Dynamics and Control of Mechanical Processing edited by G. Stepan and T. Insperger, The Publishing Company of Technical University of Budapest, 2001, pp. 23–28.

[15] T. Ditzinger, R. Friedrich, S. Merkel, A. Henning, and G. Radons, *Front instabilities and cutting processes*, in: Proceedings of the 3rd International Symposium "Investigation of Nonlinear Dynamic Effects in Production Systems", Cottbus, 2000.

[16] E. Lemma, L. Chen, E. Siores, and J. Wang, *Optimising the AWJ cutting process of ductile materials using nozzle oscillation technique*, International Journal of Machine Tools and Manufacture **42** (2002), 781–789.

23 Modeling and Simulation of Process Monitoring and Control in Laser Cutting

W. Schulz, V. Kostrykin, J. Michel, and M. Niessen

Laser cutting at moderate conditions is well established in industries. Both high cutting speed for efficient machining and low cutting speed during contour cutting change the dynamical properties of the process, which are related to quality features of the cut. To guarantee the product quality of laser processing the industry has tried to introduce monitoring and control systems. Understanding becomes crucial to develop robust and reliable laser processes and machines. The laser beam fusion cutting process is a Free Boundary Problem for the motion of two phase boundaries. In the long-time limit of such dissipative dynamical systems a reduction of the dimension in phase space occurs. The degrees of freedom are identified by methods of singular perturbation theory and integral methods are applied to derive the equations of motion. Applying spectral methods the solution can be calculated with arbitrary accuracy. As result, two mechanisms for the formation of adherent dross are revealed theoretically, identified by the monitoring system and can be avoided by modulation of the laser-beam power. The onset of evaporation and the increase of capillary forces are the two physical phenomena relevant for the build-up of adherent dross. The dynamic model predicts a modulation frequency for the laser power that leads to almost complete suppression of adherent dross in contour cutting.

23.1 Introduction

A variety of tools including flame [1], electric discharge [2], electron and laser beams are used for thermal processing. Each has a different mechanism for energy coupling, such as heat transition of the flame (Robin-type boundary condition), Joule heating of the current density (volume source and homogeneous Neumann-type boundary condition) as well as scattering and absorption of electrons and photons within a few tenths of an Ångström (Neumann-type boundary conditions). The quality of the processed material is controlled by similar dynamic features of the subsequent thermally induced processes. Process monitoring and identification of dynamic features are also investigated for other manufacturing processes like laser welding [3–7] and water-jet cutting [8].

Understanding and suppression of quality defects such as ripples and adherent dross become essential for advanced applications such as high-speed contour cutting because an extended range of processing parameters is involved. Analysis of specific experiments and monitoring of the technical process are combined with fundamental physical modeling, analytical calculus and numerical simulation to investigate the dynamic features of laser cutting including ripple formation and adherent dross.

Nonlinear Dynamics of Production Systems. Edited by G. Radons and R. Neugebauer

ISBN 3-527-40430-9

23.2 Diagnosis and Analysis of Dynamic Features

When a sheet is cut directly at the edge, thermal radiation emitted from the molten material and reflected light emitted from the cut edge can be observed simultaneously. Video recordings (Fig. 23.1a) of the cut edge show the ripples evolving at the borderline between the molten surface and the cut edge. The melt film separates from the sheet and is concentrated into a spatially localized melt thread. Depending on the processing parameters the melt thread can change between separating and wetting conditions while dross is formed. Monitoring the technical process using a CCD camera aligned coaxially to the laser-beam axis gives temporally and spatially resolved images of the whole interaction zone of laser beam and sheet metal (Fig. 23.1b). The thermal emission is sensitive to the movement of the cutting front and the dynamics of melt flow can be easily detected. Comparison of the video and CCD-camera images reveals that the position of the melt thread (brightest camera pixels) and its stability can also be observed with a CCD camera. A monitoring system can be used to identify the different relevant physical phenomena and their appearance at the CCD image.

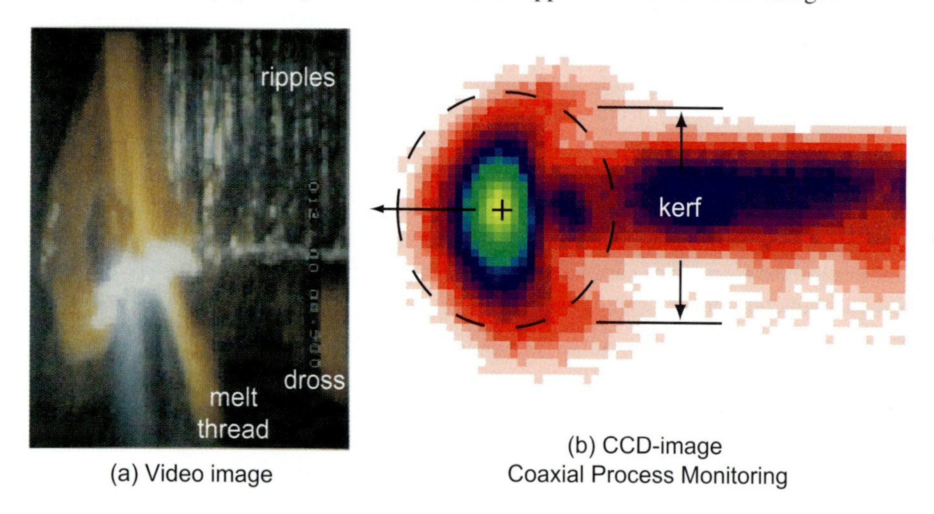

Figure 23.1: (a) Video recording during trimming of sheet metal allows us to observe the solid–liquid interface at the cutting front, the melt-flow separation into a single melt thread underneath the sheet metal as well as the formation of ripples and dross. (b) The thermal emission can be measured by a CCD camera, which is aligned coaxially with respect to the laser-beam axis. A melt thread forms where the kerf evolves.

Modeling the time-dependent interaction of the involved transport processes means formulating a free boundary problem [9–14] in terms of a system of partial differential equations. The moving boundaries at the solid/liquid and liquid/gaseous interfaces are the melting front and the surface of the melt film, respectively. In such dissipative dynamical systems a finite-dimensional inertial manifold exists which contains the attractor of the system [15–18]. The existence of a finite-dimensional inertial manifold means that the motion of a finite set of degrees of freedom can give a good approximation of the complete solution. Asymptotic methods [12] of singular perturbation theory are used to identify the degrees of freedom in terms of their corresponding characteristic dynamical variables. The characteristic variables of the reduced model and the corresponding spatial distributions reproduce the quasi-steady solution of the asymptotic equations. The general dynamics (τ) of spatially (ξ) distributed quantities like the temperature $\theta(\xi, \tau)$ in the solid is restricted to the movement of the characteristic dynamical variables. As a result, these variables appear to represent global quantities like the thermal energy content $E(\tau)$ as well as significant geometrical parameters of the free boundaries, like the position $A(\tau)$ and the radius $\alpha(\tau)$ of the melting front. Inserting the solution for the characteristic dynamical variables into the corresponding spatial distributions, e.g. Eq. (23.1), the spatial three-dimensional solution can be re-constructed (Fig. 23.3). As an illustrative example the spatial one-dimensional temperature $\theta(\xi, \tau)$ in the solid is considered and approximated by

$$\theta_{\text{app}}(\xi, \tau) = \exp\left[-\frac{\xi - A(\tau)}{E(\tau)}\right] , \quad \xi \geq A(\tau) , \tag{23.1}$$

where the position of the melting front $\xi = A(\tau)$ moves in the feed direction. The spatially integrated temperature equals the energy content

$$E(\tau) = \int_{A(\tau)}^{\infty} \theta_{\text{app}}(\xi, \tau) d\xi , \tag{23.2}$$

and can be interpreted as a penetration depth $\delta = E(\tau)$ (thickness of the thermal boundary layer) of the thermal energy.

Consider the spatial two-dimensional shape $\xi = A(\zeta, \tau)$ along the outflow direction described by ζ ($0 < \zeta < d$) where d is the sheet thickness. The melt film thickness d_m is determined by the movement of the melting front with the velocity $v_p = \dot{A}(\zeta, \tau)$ and the gas jet re-directs the melt to the outflow direction ζ. Also, the velocity component $v_m(\xi, \zeta, \tau)$ of the melt in the outflow direction is parameterized by its corresponding global quantity, namely the mass flow $m(\zeta, \tau)$:

$$m(\zeta, \tau) = [\bar{v}_m d_m]_{(\zeta,\tau)} = \int_{0}^{d_m(\zeta,\tau)} v_m(\xi, \zeta, \tau) d\xi. \tag{23.3}$$

Integral methods [14] are then applied to derive an approximate dynamical system [11, 12, 14] as a mathematical model of the Free Boundary Problem, which describes the motion of the characteristic dynamical variables. Typical features of dynamical systems, such as time-scale separation of the different dynamical variables depending on the processing parameters (cutting speed, beam radius, etc.) occur. As one consequence the model predicts increasing response amplitudes for the motion of the boundaries with respect to power modulation for smaller values of the cutting speed. The dynamical phenomena like ripple and dross formation follow directly and quantitatively.

23.3 Coupled Equations of Motion

We turn to the description of the equations of motion for the characteristic dynamical variables and in particular the coupling of different processes. For free moving boundaries the properties of coupling are closely related to the presence of boundary layer character for the solution. The thickness δ of a boundary layer can be compared with an external length scale like the spatial extent w_0 of the heat source or the radius R of curvature of the boundary, giving the parameter Péclet number $\mathrm{Pe} = w_0/\delta$ ($\mathrm{Pe}_R = R/\delta$). Boundary layers are a typical consequence from the counteraction of diffusive and convective transport when convection becomes dominant ($\epsilon = 1/\mathrm{Pe} \to 0$). As consequence, the diffusion of the transported quantity (thermal energy, momentum, etc.) becomes concentrated (singular) within a boundary layer of thickness $\delta = \kappa/v_0$ small compared to the radius $R \leq w_0$ of curvature of the boundary itself. The laser-beam axis moves with velocity v_0 (κ thermal diffusivity). In particular, the influence of curvature on the thickness of thermal boundary layers and the subsequent movement of the melting front as well as the movement of the melting front and the subsequent melt flow are discussed.

23.3.1 Axial Dynamics of the Melting Front

We turn to the description of the axial dynamics of the melting front. The influence of the thermal boundary layers at the top and the bottom of the sheet metal on the movement of the melting front will be discussed. To this end we restrict ourselves with the spatial two-dimensional model at the ceiling line of the front. The characteristic dynamical variables (position of the boundary Eq. (23.1), boundary layer thickness Eq. (23.2)) are now axially distributed and the inclination of the melting front becomes part of the solution. We start with the Free Boundary Problem for a single phase – which is the generic process [13] in cutting.

The absorbed intensity $q_a = \mu A(\mu)\gamma(\tau)f(\xi)$ depends on the angle ϑ of inclination ($\mu = \cos\vartheta$), the degree of absorption $A(\mu)$ and the incident laser intensity $\gamma(\tau)f(\xi)$ at the position $\xi = a(\zeta,\tau) - \mathrm{Pe}\tau$ of the melting front $a(\zeta,\tau)$ with respect to the moving ($\xi \propto \mathrm{Pe}\tau$) laser-beam axis. The laser intensity has the maximum value $\gamma = \gamma(\tau)$ and the distribution $f = f(\xi)$ ($0 < f < 1$). The absorbed intensity $\gamma A(1)f(\xi)$ at the top of the sheet heats the metal to the melting temperature and its value q_a at the melting front leads to cutting. The Free Boundary Problem for a single phase describing the axial motion of the melting front $a = a(\zeta,\tau)$ contains an additional smallness parameter w_0/d, which is the ratio of the laser-beam radius w_0 and the sheet thickness d.

In the non-dimensional variables $\xi = x/w_0$, $\zeta = z/d$ the problem reads

$$\frac{\partial\theta}{\partial\tau} = \frac{\partial^2\theta}{\partial\xi^2} + \left(\frac{w_0}{d}\right)^2 \frac{\partial^2\theta}{\partial\zeta^2},$$

$$q_a + \left[\frac{\partial\theta}{\partial\xi} - \left(\frac{w_0}{d}\right)^2 \frac{\partial\theta}{\partial\zeta}\frac{\partial a}{\partial\zeta}\right] = -h_m\frac{\partial a}{\partial\tau}, \qquad \theta|_{\xi=a} = 1,$$

$$\left.\frac{\partial\theta}{\partial\zeta}\right|_{\substack{\zeta=0\\ \xi\geq a(0,\tau)}} = -\left(\frac{d}{w_0}\right)^2 \gamma A(1)f(\xi), \qquad \left.\frac{\partial\theta}{\partial\zeta}\right|_{\substack{\zeta=1\\ \xi\geq a(1,\tau)}} = 0.$$

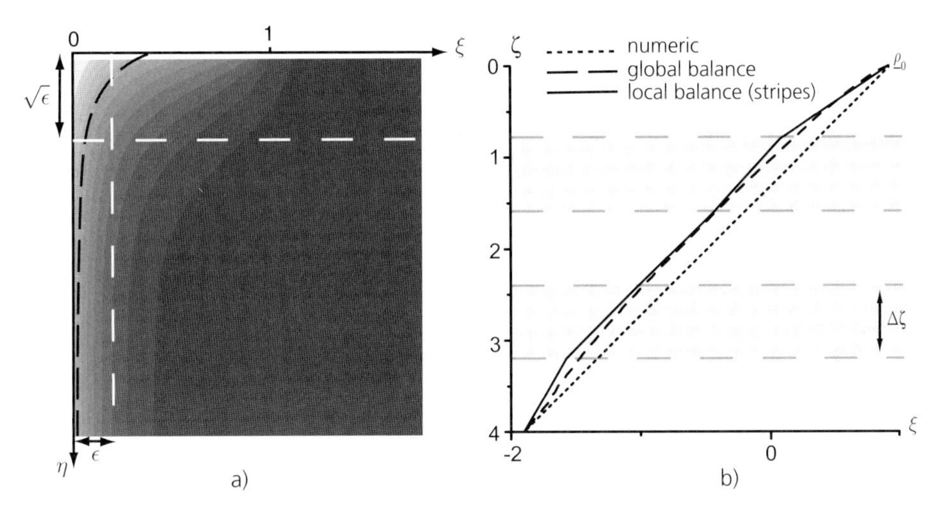

Figure 23.2: (a) The analytical temperature (23.4) for a rectangular shape of the boundary and constant intensity in the interval $\xi \in [0, 1]$. The analytical temperature is a singular perturbation showing the different length scales for the boundary layers at the top ($\delta_{\text{Heat}} \propto \sqrt{\epsilon}$) of the sheet by pre-heating and at the moving melting front ($\delta \propto \epsilon$) as well as the corresponding transition region. (b) Comparison of the geometrical shape of the cutting front from numerical calculation (dashed), global energy balance (dotted) across the entire sheet metal and the local energy balance (full) with respect to stripes of thickness $\Delta\zeta$. The gray regions indicate the thickness $\Delta\zeta$ of the balance volume (stripes) used for the local energy balance. The results are given for the absorbed intensity $\gamma A = 10.28$ and the inverse Péclet number $\epsilon = 0.33$.

The asymptotic analysis ($w_0 \ll d$) of this problem shows that everywhere in the material outside the boundary layers at the upper ($\zeta = 0$) and bottom ($\zeta = 1$) surfaces of the sheet metal the terms containing $(w_0/d)^2 \ll 1$ can be neglected. The dynamical behavior of the melting front is characterized by three boundary layers which become thinner with increasing velocity $O(\text{Pe})$ of the front. The smallness parameter $\epsilon = 1/\text{Pe}$ has to be considered. The boundary layer of the thickness δ_{Heat} appears through the pre-heating of the material on the not yet molten upper surface of the material. The influence of the boundary layer of thickness δ_{Insulate} on the bottom side of the workpiece is determined by thermal isolation from the environment.

Using the methods of singular perturbation theory Stier [19] derived the asymptotically ($\epsilon \to 0$) exact solution for a simplified problem (Fig. 23.2a) describing the thermal boundary layers near the moving boundary ($\xi = 0$) at the top edge ($\zeta = 0$) of the sheet metal. The material is moved in the negative ξ direction and at both boundaries – at the melting front $\Gamma_S = \{\xi = 0, \zeta\}$ and at the top $\Gamma_H = \{\xi, \zeta = 0\}$ – Neumann-type boundary conditions are posed. The solution θ_{app}

$$\theta_{\text{app}}(\xi, \zeta, \tau) = \theta_S\left(\frac{\xi - A(\zeta, \tau)}{\epsilon}\right) + \theta_H\left(\xi, \frac{\zeta}{\sqrt{\epsilon}}\right) + \theta_K\left(\frac{\xi}{\epsilon}, \frac{\zeta}{\sqrt{\epsilon}}\right) \tag{23.4}$$

is given analytically [19]. Three different regions with different thickness of the thermal boundary layer can be revealed by three different parts of the solution (23.4): two regions with the solutions θ_S, θ_H for the melting phase and the heating phase and a third region, where the transition takes place, are identified. Within the transition region both the influence of the curvature of the boundary and the change of the boundary conditions take place. In the transition region – where the different boundary conditions of heating and melting phases are co-existing – the angle φ between the direction of feeding (convective heat transport) and the surface normal of the boundary is changing together with the type (parabolic, elliptic) of the partial differential equations

$$\sin\varphi\frac{\partial}{\partial\nu}\theta - \epsilon\cos\varphi\frac{\partial}{\partial s}\theta + \frac{\partial^2}{\partial s^2}\theta = 0. \tag{23.5}$$

The distance ν from the boundary and the position s along the boundary as boundary fitted coordinates $\{\nu, s\}$ are introduced instead of $\{\xi, \zeta\}$. Hence, the solutions for the temperature and the thickness $(\sqrt{\epsilon}, \epsilon)$ of the thermal boundary layer undergo qualitative changes along the boundary.

As result, the axial dynamics of the melting front is determined dominantly by the spatial one-dimensional heat conduction in the direction of feeding [14, 20]. Effects of curvature, pre-heating by the laser beam and thermal isolation are additional effects which give corrections to the solution within the thermal boundary layers with thickness $\sqrt{\epsilon}$. By comparison of the analytical solution (Fig. 23.2a) with the numerical calculation it can be confirmed that the effects of curvature and changing boundary conditions remain localized in the vicinity of the top and bottom surfaces also for moderate values of the Péclet number. This result indicates that the mentioned additional effects for the movement of the melting front can be covered in the reduced cutting model by a refined description of the boundary layer properties.

An approximate solution for $(w_0/d)^2 \ll 1$ can be derived by solving the pre-heating problem separately, which yields the initial value for the position $\xi = a(\zeta, \tau)$ of the axially distributed melting front. Actually, the interaction of the boundary layers δ_{Heat} and δ can be calculated by boundary matching methods and introduces a correction for the position $a(\zeta, \tau)$ which vanishes for large values of the Péclet number.

23.3.2 Lateral Dynamics of the Melting Front

The reduced dynamical system for the spatial two-dimensional model including the lateral extent (width of the cut kerf) is explicitly given and discussed in [12, 13]. In particular, from comparison with numerical calculation the radius α of curvature of the melting front changes slightly with the azimuth angle and therefore the width of the cut kerf can be well approximated by the radius α. The lateral two-dimensional dynamical system for the case of large Péclet numbers takes the asymptotic form

$$\dot{\alpha} = \frac{1}{1+h_m}\left[\left(\gamma f_2 - \frac{b_2}{Q}\right) + \frac{1+h_m+b_2}{1+h_m-b_1}\left(\gamma f_0 - \frac{b_1}{Q}\right)\right],$$

$$\dot{A} = \frac{1}{1+h_m-b_1}\left(\gamma f_0 - \frac{b_1}{Q}\right), \quad \dot{Q} = b_1\left(\frac{1}{Q} - \dot{A}\right).$$

The dynamics of the spatial one-dimensional dynamical variables $\{A, Q\}$ is decoupled from the motion of the additional two-dimensional variable α. In the melting phase the thermal energy $E(\tau) = \theta_S(\tau)Q(\tau)$ equals the penetration depth $\delta/w_0 = Q$ of the heat. h_m denotes the inverse Stefan number ($b_1 = 3/5$, $b_2 = 1/10$). A Gaussian intensity distribution $f(\mathbf{x}) = \exp(-2\mathbf{x}^2)$ and the maximum intensity γ at the beam axis are considered. The intensity γf enters the equations of motion by its Taylor coefficients f_0, f_2:

$$f = f_0 + f_2\frac{\varphi^2}{2}, \quad f_0 = \exp(-2\bar{A}^2), \quad f_2 = -4\alpha(\alpha - \bar{A})\, f_0$$

at the vertex line, where the distance $\bar{A} = A - R_L$ between the position of the front $\xi = A(\tau)$ and the moving laser-beam axis $\xi = R_L(\tau)$ is introduced.

23.3.3 Melt Flow

Melt flow, heat transport in the molten material and in particular the temperature at the surface of the melt film are the physical mechanisms and quantities underlying the experimentally observable phenomena in cutting. The formation of ripples and adherent dross as well as the thermal emission from the cutting front are related to phenomena of melt flow.

The melt film of thickness d_m in front of the laser-beam axis occurs between the two free boundaries – the melting front and the absorption front. At the absorption front the melt film is heated by the absorbed laser radiation. Melting the solid material generates the melt that flows in the melt film across the melting front. The pressure gradient and the shear stress imposed by an external gas jet drive the melt downwards the melting front. The molten material separates from the sheet metal and the cut kerf evolves.

The typical length scales of the problem are the melt film thickness d_m, the radius w_0 of the laser beam and the sheet metal thickness d. The length scales are quite different ($d_m \ll w_0 \ll d$) resulting in boundary layer character for the heat and momentum transport. Moreover, according to the different length scales the corresponding times scales are separated. In the long-time limit of such dissipative dynamical systems a reduction of the dimension in phase space occurs. The integral formulation describes the motion of global characteristic variables. The proper scaled variables are the melt film thickness h, the mass flow m and the surface temperature θ_S. These variables are time-dependent parameters of the spatial distributions for the velocity of the molten material and its temperature (Eqs. (23.1), (23.2), (23.3)).

Within the scope of the integral formulation the comparatively fast changes of the spatial distributions towards their quasi-steady values will not be taken into account. The relaxation of the faster degrees of freedom will be treated by spectral methods.

The dynamics of the global dynamical variables is described by equations of motion derived from the balance equations for momentum, mass and thermal energy by spatial integration with respect to the small melt film thickness:

$$m = m(\zeta, \tau), \quad h = h(\zeta, \tau), \quad \zeta \in [0, 1], \quad \tau \in [0, \infty), \tag{23.6}$$

$$\mathrm{Re}\left(\frac{\partial m}{\partial \tau} + \frac{6}{5}\frac{\partial}{\partial \zeta}\left[\frac{m^2}{h}\right]\right) = -\frac{\partial \Pi_g}{\partial \zeta}h + 3\left(\frac{\Sigma_g}{2} - \frac{m}{h^2}\right), \quad \text{momentum} \tag{23.7}$$

$$\frac{\partial h}{\partial \tau} + \frac{\partial m}{\partial \zeta} = v_P, \qquad \text{mass} \qquad (23.8)$$

$$\frac{1}{2}\frac{\partial}{\partial \tau}\left[\theta_s h g_1\right] + \frac{5}{8}\frac{\partial}{\partial \zeta}\left[\theta_s m g_2\right] = \frac{1}{\mathrm{Pe_l}}\left[q_a - \frac{\theta_s}{h} g_3\right], \qquad \text{energy} \qquad (23.9)$$

$$m\Big|_{\zeta=0} = h\Big|_{\zeta=0} = \theta_s\Big|_{\zeta=0} = 0. \qquad \text{boundary values} \qquad (23.10)$$

At the surface of the melt film the energy density flow q_a is absorbed (23.9). The coefficients $g_j = g_j(\mathrm{Pe_l} h)$ in (23.9) are monotone decaying functions of the scaled melt film thickness $\mathrm{Pe_l} h$. The Reynolds number Re and the Péclet number $\mathrm{Pe_l}$ are the parameters of the solution. The dynamical system (23.7)–(23.10) for the melt flow is coupled to the gaseous phase by the pressure gradient $\partial \Pi_g / \partial \zeta$ and the shear stress Σ_g of the gas jet as well as to the solid phase by the velocity $v_p = v_p(\zeta, \tau)$ of the melting front. The coupled model is closed by the dynamical system for the movement of the melting front (Sect. 23.3.1).

23.4 Heat Convection Influences Ripple Formation

At least three different ripple patterns are observable, the so-called ripples of first, second and third types (Fig. 23.3b). They evolve subsequently with increasing sheet thickness and cutting speed. Especially at the top of the cut edge and up to feed rates of 4 m/min the ripples of the first type appear with an almost constant wavelength, i.e. their frequency increases nearly linearly with respect to the cutting speed up to a value of about 500 Hz at 4 m/min. Almost no re-solidified material can be observed by the metallurgical inspection of the cross section near the top of the cut edge (Fig. 23.3a). Ripples of second type are characterized by the doubling of the ripple frequency of the first type and with increasing cutting speed their onset is shifted towards the top surface of the sheet metal. Ripples of third type are characterized by the dominance of re-solidified molten material.

Ripples of first type are reproduced by the dynamical system as a result of direct heat absorption and diffusion (Fig. 23.3a). In the upper part of the sheet the influence of the melt flow on the movement of the melting front is negligible, since the heat contained in the melt film, its thickness and the mass flow of molten material remain fairly small. Heat convection in the melt film is involved when ripples of the second type are formed (Fig. 23.3b).

The mass flow of molten material increases linearly with the depth of the cut. The temperature in the molten material increases exponentially with the melt film thickness and the cutting speed. The two mechanisms of heat transport – diffusion and convection – contribute to the movement of the melting front and are different with respect to time scales and spatial distributions. Any ripple of first type causes an axial propagation of a melt wave. The subsequent heat transport – convection in the axial direction – leads to an additional and delayed motion of the melting front. The ripple frequency is doubled (Fig. 23.3b).

The correlations between the vertical focal position and the spatial distribution of the mass flow of molten metal at the cutting front form the basis to understand the mechanisms involved in subsequent formation of ripples. To suppress dross formation during CW cutting the focus is typically placed close to the bottom of the sheet. However, this working point has the disadvantage of increased ripple amplitudes of second type (Fig. 23.3b).

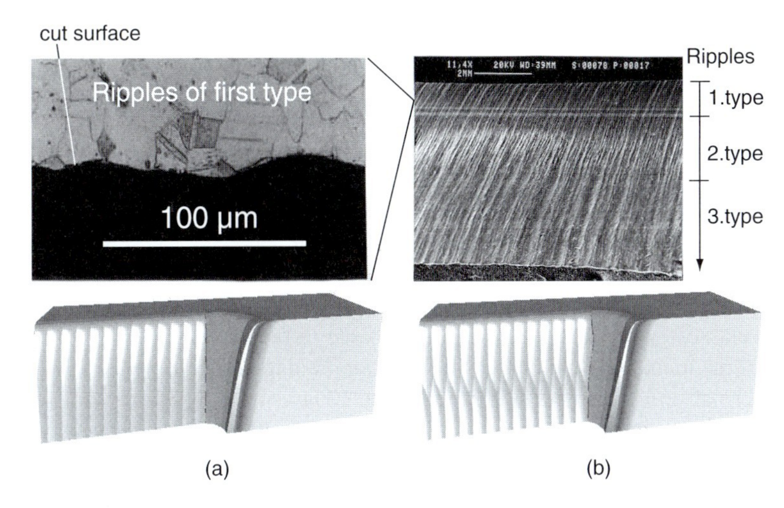

Figure 23.3: Ripples of first type (a) due to movement of the melting front. Ripples of second type (b) due to movement of the melting front and melt flow. (a) Ripples of first type occur at the top of the cut edge, where almost no re-solidified melt is observed. Direct absorption and heat diffusion mainly determine the movement of the melting front. (b) The downward-directed heat transport by the melt flow additionally moves the melting front and ripples of second type are formed (cutting speed 1 m/min, thickness 10 mm, N_2 14 bar, nozzle diameter 1.4 mm, CO_2 laser power 4.6 kW, stainless steel (1.4301)).

23.5 Observation of the Cutting Front

The imaging of the thermal radiation obtained with the CCD camera (Fig. 23.4a) can be interpreted by comparison with the thermal emission (Fig. 23.4b) as calculated from the dynamical system.

The CCD image as recorded during fusion cutting of stainless steel does not show the spatial shape of the cutting front. The different parts of the calculated interaction zone belonging to

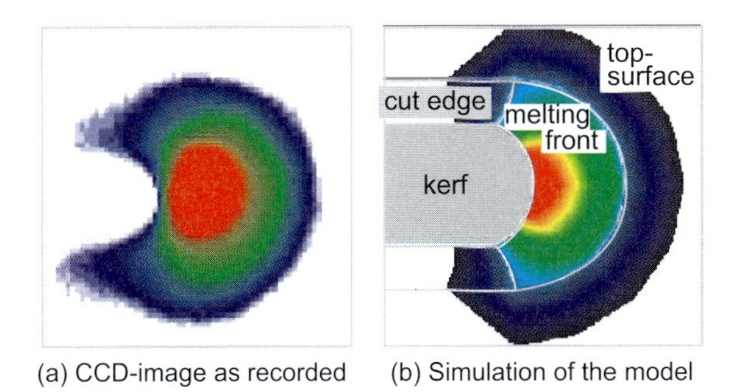

(a) CCD-image as recorded (b) Simulation of the model

Figure 23.4: (a) CCD image as recorded from experiments. Monitoring the cutting process with a CCD camera. (b) Simulation of the model. Comparison with the simulation of the dynamical model facilitates the interpretation of the measured data.

the pre-heated area at the top surface, the melting front and the kerf are re-constructed from the characteristic dynamical variables. The emission from the different geometrical parts (melt front, kerf, cut edge and top pre-heated area) as well as the position of the melt thread (Fig. 23.1) can be singled out and these findings for the quasi-steady state are in full accordance with the model predictions.

23.6 Quality Classes: Observation and Modeling

Introduction of quality classes by Arata et al. [21] is applicable to fusion cutting (Fig. 23.5, classes II, III, IV). The three main different processing domains become observable by online measurement of the thermal emission with a CCD camera. These quality classes can be correlated with the dynamical process domains and can be identified with the underlying physical mechanisms.

The ratio of the stagnation pressure $\rho \overline{v}_m^2$ and the counteracting capillary pressure σ / d_m is called the Weber number We (ρ and σ are the density and the surface tension of the melt). If the capillary forces become comparable with the inertia of the melt (We ≈ 1) then the separation of melt flow becomes unstable. The tendency to the formation of dross is estimated from the solution (Fig. 23.5, We) and occurs for slow cutting speeds. For increased cutting speeds the melt film starts to evaporate (Fig. 23.5, T_{V}). Finally the maximum cutting speed achievable (Fig. 23.5, E_{L}) is given by energy balance. The correlations between the processing domains (We, T_{V}), the characteristics of the CCD images and the formation of the different kinds of dross is striking. For low values of the Weber number, the CCD image shows an asymmetric position of the single melt thread and there is dross at one side of the kerf. With the onset of

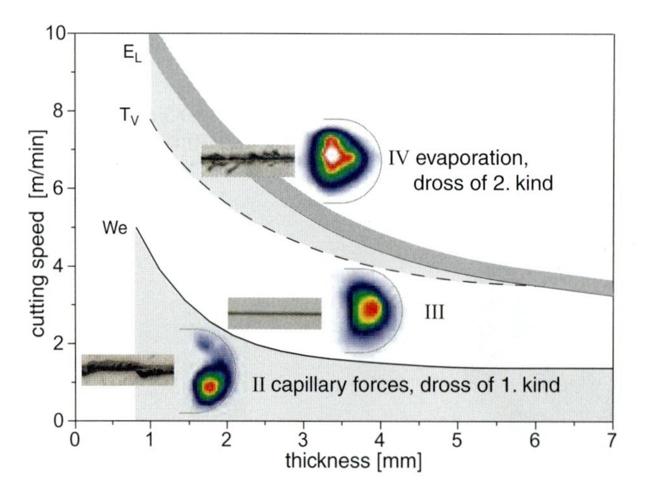

Figure 23.5: As a result of the modeling, quality degradation due to dross is related to the dynamics of the melt flow. Melt flow is interrupted by the dominance of capillary forces (We) or by the onset of evaporation (T_{V}). Cutting speeds achievable (E_{L}) are only limited by high material thickness or low laser power.

the evaporation pressure the CCD image shows a symmetric splitting of the melt thread and dross at both sides of the kerf appears.

23.7 Control

The capillary forces become dominant for slow cutting speeds and for CW operation. The melt flow does not separate completely from the sheet metal. Low cutting speeds are inherent in contour cutting and therefore a reliable control strategy is of interest. For this task it is crucial to differentiate between the cutting speed v_0 and the velocity $v_p(\tau)$ of the cutting front. From the results on the capillary forces (Fig. 23.5, We) the appearance of dross is related to small values of the Weber number We $\propto \langle v_{\text{melt}} \rangle^2 d_m$, which has to be kept at values larger than unity. Modulation of the laser power results in correspondingly modulated velocity of the melting front and hence the time-dependent mass flow $\overline{v}_m d_m = v_p(\tau) d \gg v_0 d$ can be significantly larger than its value $v_0 d$ averaged with respect to the cycle time (Eqs. (23.3), (23.7)). The effect of power modulation on the resulting cut quality then depends on the selection of the proper pulse duration and the duty cycle. These parameters can be chosen such that only one wave of ejected melt is produced during each laser pulse. The laser-pulse duration t_{on} has to be matched to the typical time $d/\overline{v}_m$ for the propagation of a melt perturbation along the whole melting front down the sheet thickness d. Also, the cycle time $t_{\text{on}} + t_{\text{off}}$ has to meet the time-averaged energy balance. Using these constraints the parameters for the pulse-mode operation, namely pulse duration and duty cycle, are determined. Figure 23.6 gives an example for a cut at a low cutting speed $v_0 = 1.2$ m/min performed with modulated laser power which remains dross-free. The CCD images show that the melt thread remains at the center of the cutting front.

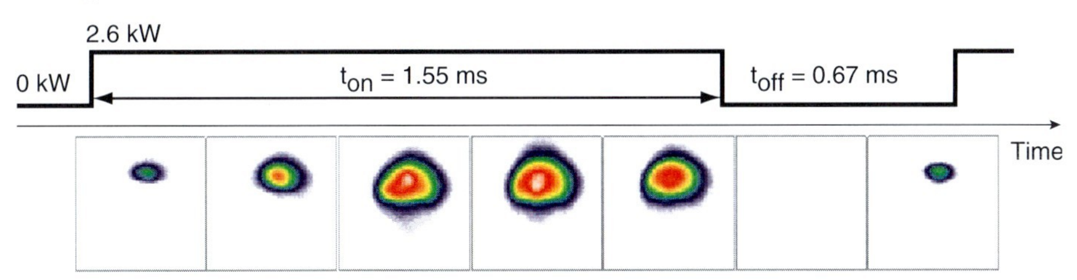

Figure 23.6: The cutting process can evolve periodically using modulated laser power. The recorded CCD images in this case show the typical properties of high-quality cuts of class III known from CW operation (cutting speed 1.2 m/min, thickness 3 mm, N_2 14 bar, nozzle diameter 1.4 mm, maximum CO_2 laser power 2.6 kW/pulsed, duty cycle 70%, pulse frequency 450 Hz, beam radius 150 μm, Rayleigh length 2.2 mm, focal position −2.4 mm, stainless steel (1.4301), CCD-camera frame rate 2.9 kHz).

23.8 Analysis Using Spectral Methods

The solution of the integral approximations discussed above describes the motion of global physical quantities (thermal energy content $E(\tau)$ (Eq. (23.2)), mass flow $m(\tau)$ (Eq. (23.3)), etc.). The exact quasi-steady solution often can be found in the asymptotic cases when for example the boundary layer character becomes dominant and the complexity of the problem is strongly reduced. As consequence, in the long-time limit after fast relaxation of the higher degrees of freedom also local properties (surface temperature θ_S (Eq. (23.9)), velocity of the melt, etc.) are well approximated. Using spectral methods the local corrections can be calculated with arbitrary accuracy [22]. Moreover, the spectral ansatz gives a constructive suggestion of how to choose the spatial distribution of additional degrees of freedom and what are the temporal and spatial scales which can be taken into account.

As an example, the heat transport in the melt film will be discussed. The method can be applied in a similar way to the melt flow itself and the motion of the melting front.

Mathematical Formulation. The heat transport in the melt film with thickness $h(z,t)$ is described by a partial differential equation for the temperature $\theta = \theta(x,z,t)$:

$$\frac{\partial \theta}{\partial t} + u\frac{\partial \theta}{\partial z} + v\frac{\partial \theta}{\partial x} = \mathrm{Pe}^{-1}\frac{\partial^2 \theta}{\partial x^2}, \qquad x \in [0, h(z,t)],\ z \in [0,1],\ t > 0 \tag{23.11}$$

subject to the boundary values

$$\theta|_{x=0} = 0, \qquad \left.\frac{\partial \theta}{\partial x}\right|_{x=h(z,t)} = \gamma(t) f(h(z,t)), \tag{23.12}$$

and the initial data $\theta(x,z,t=0) = g(x,z)$. At the melting front $x = 0$ – the phase front between solid ($x < 0$) and liquid ($x \geq 0$) – the temperature equals the melting point $\theta = 0$. At the surface of the melt film the intensity $\gamma(t) f(h(z,t))$ is absorbed. Consider the melt flow

$$v = 1, \qquad u = s_1(z,t)x + s_2(z,t)x^2,$$

where the coefficients $s_1(t,z)$, $s_2(t,z)$ as well as the melt film thickness $h(z,t)$ are determined in [13].

Degrees of Freedom Described by the Integral and the Spectral Methods. Consider the function θ_0 given by

$$\theta_0(x,z,t) = \frac{e^{\mathrm{Pe}\,x} - 1}{\mathrm{Pe}\, e^{\mathrm{Pe}\,h}}\gamma(t) f(h(t,z)), \tag{23.13}$$

which is the quasi-steady solution of Eqs. (23.11), (23.12) for $u = 0$.

To derive the integral formulation the global dynamical variable θ_s and the ansatz

$$\theta(x,z,t) = \frac{e^{\mathrm{Pe}\,x} - 1}{e^{\mathrm{Pe}\,h} - 1}\theta_s(z,t) \tag{23.14}$$

for the spatial distribution are introduced. The quasi-steady or instantaneous solution (23.13) of Eqs. (23.11), (23.12) is reproduced by the ansatz (23.14). The quasi-steady value of θ_s is

$$\theta_s(z,t) = \frac{e^{\mathrm{Pe}\,h} - 1}{\mathrm{Pe}\, e^{\mathrm{Pe}\,h}}\gamma(t) f(h(z,t)). \tag{23.15}$$

The equation of motion (23.9) for θ_s is derived by inserting the ansatz (23.14) into the boundary value problem (23.11), (23.12) and integration with respect to the melt film thickness. With the solution for θ_s a spatially distributed approximation of the temperature can be reconstructed using (23.14). The energy content $\int \theta(x,z,t)dx$ of the melt film with thickness $h(z,t)$ is described exactly and the dynamics of the spatial distribution is not taken into account.

For the spectral formulation the general ansatz

$$\theta(x,z,t) = \theta_0(x,z,t) + \theta_1(x,z,t)$$

is introduced. The correction θ_1 has the meaning of the difference between the exact solution and the instantaneous value of the approximate integral formulation. Compared to the integral formulation the spectral formulation additionally describes the changes of the spatial distribution of the temperature.

Solution Using the Spectral Method. A complete system of eigenfunctions $w_k(x; \mathrm{Pe}, h, u)$ and their eigenvalues λ_k can be given. The spectral ansatz (for the transformed temperature $S_1 = \theta_1 e^{-\frac{\mathrm{Pe}\,x}{2}}$)

$$\theta_1(x,z,t)e^{-\frac{\mathrm{Pe}\,x}{2}} = \sum_{k=0}^{\infty} a_k(z,t) w_k(x; \mathrm{Pe}, h, u). \tag{23.16}$$

leads to a system of coupled, inhomogeneous hyperbolic partial differential equations for the spectral coefficients $a_k(z,t)$:

$$c_k \frac{\partial a_k}{\partial t} + \sum_{j \neq k} c_{kj} \frac{\partial a_j}{\partial t} + \mathrm{Pe}\, \frac{\partial a_k}{\partial z} + \lambda_k a_k = H\left(\frac{\partial \theta_0}{\partial t}, \frac{\partial \theta_0}{\partial z}, a_i\right). \tag{23.17}$$

The velocity c_k of the kth mode, the coupling of the different modes c_{kj} and the inhomogeneity $H(\partial\theta_0/\partial t, \partial\theta_0/\partial z, a_i)$ can be given explicitly.

The initial-boundary value problem (23.11), (23.12) is reduced to an initial value problem (23.17). The spectral representation of the solution allows for a deeper understanding of the structure of the solution. Properties like the diffusive ($\lambda_k a_k$) and convective ($c_{kj}(\partial a_j/\partial t) + \mathrm{Pe}\,(\partial a_k/\partial z)$) contributions to the heat transport are visible. The eigenvalues $\lambda_k \propto k^2$ increase quadratically with the mode number k and hence a pronounced time-scale separation is present. Higher-order corrections undergo faster relaxation. After fast relaxation only a few spectral coefficients $a_k(z,t)$ contribute dominantly to the solution.

23.9 Conclusion and Outlook

A mathematical model describing the dynamical phenomena during cutting with laser radiation was formulated and analyzed. For the experimental observation of the cutting process a camera system was used to record the spatial and temporal properties of the thermal emission from the cutting front. Basic properties of the solution are assigned to the results from experimental observation of the cutting process and the quality features of the cut. The physical processes leading to ripple formation and adherent dross are identified by comparison of the camera images with the properties of the solution.

The processing parameters (laser power, laser intensity, etc.) cannot be kept ideally constant and depend on the condition of the machine, which could change during operation. With the observability of the dynamical state of the process a fundamental step towards quality assurance and system diagnostics is established.

Observability. The mathematical model is a Free Boundary Problem for the movement of two free phase boundaries (melting front and surface of the melt film). The geometrical shape at the top of the sheet metal, the transition of the melt film surface into the cut kerf and the position of the melt thread – where the melt separates from the sheet metal – are observable. With the camera images the appearance of different physical processes (pre-heating, melting, evaporation, outflow of the melt) can be distinguished.

Process Domains. The subsets of the processing parameters which lead to cuts with different types of adherent dross or cuts without dross are called ‘process domains’. The process domains can be distinguished by the quality features of the cut, by the properties of the camera images and by the type and number of physical processes involved. The solution of the model describes what are the physical processes which contribute to the different process domains.

Controllability. The transition between two different process domains depends on the dynamical control. As result, a dynamic control for the modulation of the laser-beam power is derived from the model and checked experimentally. By modulation of the laser-beam power the appearance of adherent dross can be avoided during contour cutting of sharp edges at low cutting speed.

Solvability. The mathematical model is a Free Boundary Problem for the motion of two free boundaries. The partial differential equations show boundary layer character, which was analyzed using methods from singular perturbation theory. Integral formulations of the equations of motion are derived which give approximate solutions. The dynamical system can be analyzed using spectral methods. The spectral correction to the approximate solution can be calculated with arbitrary accuracy.

Acknowledgment

The support by the Volkswagen Stiftung is gratefully acknowledged.

Bibliography

[1] J. Franke, *Modellierung und Optimierung des Laserstrahlbrennschneidens niedriglegierter Stähle*, DVS Berichte **161** (1994), 122–153.

[2] J.-P. Kruth, H.K. Tönshoff, and F. Klocke, *Surface and sub-surface quality in material removal processes for tool making*, in: International Symposium Electromachining ISEM XII 1998, Aachen, Germany, in: VDI Berichte, edited by F. Klocke and J.-P. Kruth, **1405** (1998), 33–64.

[3] E. Govekar, J. Gradisek, I. Grabec, A. Baus, F. Klocke, M. Geisel, and M. Geiger, *Dynamics based monitoring of manufacturing processes: detection of transitions between process states*, in: Investigation on Nonlinear Dynamic Effects in Production Systems, edited by G. Radons and R. Neugebauer, Vieweg Verlag, Chapter 27, this book, 2003.

[4] J. Michel, S. Pfeiffer, W. Schulz, M. Niessen, and V. Kostrykin, *Approximate model for laser welding*, in: Investigation on Nonlinear Dynamic Effects in Production Systems, edited by G. Radons and R. Neugebauer, Vieweg Verlag, Chapter 24, this book, 2003.

[5] M. Geisel, Dissertation, Lehrstuhl für Fertigungstechnologie, Universität Erlangen Nürnberg (2002).

[6] E. Govekar, J. Gradisek, I. Grabec, M. Geisel, A. Otto, and M. Geiger, Dissertation, Lehrstuhl für Fertigungstechnologie, Universität Erlangen Nürnberg (2002).

[7] A. Otto, Dissertation, Lehrstuhl für Fertigungstechnologie, Universität Erlangen Nürnberg (1997).

[8] G. Radons, T. Ditzinger, R. Friedrich, A. Henning, A. Kouzmichev, and E. Westkämper, *Nonlinear dynamics and control of ripple formation in abrasive waterjet cutting*, in: Investigation on Nonlinear Dynamic Effects in Production Systems, edited by G. Radons and R. Beugebauer, Vieweg Verlag, Chapter 22, this book, 2003.

[9] W. Schulz and C. Hertzler, *Cutting: Modelling and Data*, in: New Series Landolt-Börnstein, Group: Advanced Materials and Technologies, Volume VIII/1: Laser Physics and Applications, edited by M. Martienssen, R. Poprawe, H. Weber, and G. Herziger, Sub-volume VIII/1C: Laser Applications, Chapter 17.5.2, 2002, in press (preprint: Fraunhofer Institut Lasertechnik, Aachen).

[10] R. Poprawe, *Modeling, monitoring and control in high quality laser cutting*, Annals of the CIRP **50**(1) (2001), 137–140.

[11] W. Schulz and R. Poprawe, *Manufacturing with novel high power diode lasers*, IEEE Journal of Selected Topics in Quantum Electronics **6**(4) (2000), 696–705.

[12] W. Schulz and J. Michel, *Freie Randwertaufgaben der thermischen Materialbearbeitung: Inertiale Mannigfaltigkeiten und Dimension des Phasenraums*, in: Mitteilung des Curt-Risch-Instituts der Universität Hannover, '*Dynamische Probleme – Modellierung und Wirklichkeit – Systemüberwachung*', edited by W.-J. Gerasch, H. G. Natke, and U. Peil, **1** (2000), 271–288.

[13] W. Schulz, V. Kostrykin, M. Nießen, J. Michel, D. Petring, E.W. Kreutz, and R. Poprawe, *Dynamics of ripple formation and melt flow in laser beam cutting*, Journal of Physics D: Applied Physics **32** (1999), 1219–1228.

[14] W. Schulz, V. Kostrykin, H. Zefferer, D. Petring, and R. Poprawe, *A free boundary value problem related to laser beam fusion cutting: ODE approximation*, International Journal of Heat and Mass Transfer **40**(12) (1997), 2913–2928.

[15] R. Temam, *Infinite-dimensional Dynamical Systems in Mechanics and Physics*, Springer-Verlag, New York, 1988.

[16] R. Constantin, C. Foias, B. Nicolaenko, and R. Temam, *Integral Manifolds and Inertial Manifolds for Dissipative Partial Differential Equations*, Springer-Verlag, New York, 1989.

[17] L. Sirovich, B.W. Knight, and J.D. Rodriguez, *Optimal low-dimensional dynamical approximations*, Quarterly of Applied Mathematics **48** (1990), 535–548.

[18] J.C. Robinson, *Finite-dimensional behavior in dissipative partial differential equations*, Chaos **5** (1995), 330–345.

[19] B. Stier, Diplomarbeit in Physik, Lehrstuhl für Lasertechnik, RWTH Aachen (2002).

[20] V. Enß, V. Kostrykin, W. Schulz, C. Zimmermann, H. Zefferer, and D. Petring, *Thermische Materialbearbeitung mit Laserstrahlung: Schmelzschneiden*, in: *Mathematik – Schlüsseltechnologie für die Zukunft*, edited by K.-H. Hoffmann, W. Jäger, Th. Lohmann, and H. Schunk, Springer-Verlag, Berlin 1997, pp. 161–174 (in German).

[21] Y. Arata, H. Maruo, I. Miyamoto, and S. Takeuchi, *Dynamic behavior in laser gas cutting of mild steel*, Transactions of the JWRI **8**(2) (1979), 15–26.

[22] V. Kostrykin, M. Niessen, and W. Schulz, *Spectral method for free boundary problems*, submitted to Discrete and Continuous Dynamical Systems – Series B (preprint: Fraunhofer Institut Lasertechnik, Aachen).

24 Approximate Model for Laser Welding

J. Michel, S. Pfeiffer, W. Schulz, M. Niessen, and V. Kostrykin

The industrial application of laser welding implies a reliable and efficient production process. In order to facilitate this, modeling and simulation are used to reveal the crucial points of the welding process. The welding process is described by transport phenomena for mass, momentum and energy. The three involved phases (solid, liquid, gaseous) interact over the free-moving phase boundaries. The present boundary-layer characters allow us to reduce the dimension of these free boundary problems. However, the boundary-layer character of the melt flow is lost close to the stagnation point. This can be understood as a perturbation of the asymptotic solution in the boundary layer. This perturbation is identified to be singular. The description of the flow near the stagnation point leads to a modified Hiemenz problem which can be solved analytically. The resulting welding model reproduces the time-dependent spatially 3D-distributed welding process. Technically relevant prediction of seam width and depth are compared to experimental results. The reproduction of thermal monitoring signals from the interaction zone between laser and material will be approached.

24.1 Introduction

Laser welding is a well-established process in the industrial environment. In comparison to the so-called conventional welding processes, laser welding has the advantage of depositing the optical energy more precisely onto the workpiece. Thus, narrow weld joints with high aspect ratios (depth over width) at low thermal loads and distortions are achievable. The penetration is achieved by the formation of a plasma capillary within the sheet metal around the laser-beam axis: the absorbed heat flux heats the workpiece, which starts to melt and partially evaporates. The pressure induced by the evaporating particles drives a capillary of metal vapor into the molten material. The continuous evaporation keeps the capillary open. The melting material in front of the laser beam flows around the capillary in a thin melt film.

24.1.1 Technical Motivation and Physical Task

Today's development cycles of the metal processing industry are decreasing in duration. Today's trends towards globalization result in a need for flexible production in order to remain competitive. This flexibility allows companies to produce efficiently, even at small batch sizes. The adjustment of a laser welding process is complex: the number of process parameter is large; the dynamical process responds sensitively to a variation of parameters. Also, the parameter window for a robust process is small compared to the conventional welding processes. For

Nonlinear Dynamics of Production Systems. Edited by G. Radons and R. Neugebauer

ISBN 3-527-40430-9

example the gap width may not differ from the magnitude of the laser-beam radius (a couple of hundred micrometers); otherwise, the laser passes the gap between the joining partners without being sufficiently absorbed and the welding will be interrupted.

To guarantee the product quality of laser welding the industry tries to introduce monitoring and control systems [12]. Modeling as well as simulation of manufacturing processes become more important and could play a key role to qualify monitoring and control systems. While the simulation is often used to replace a time-consuming or expensive experiment, in laser welding modeling also supports the identification of the dominating phenomena of the process. This understanding will aid the development of robust reliable laser processes and machines.

The main effects of laser welding are based on phenomena that are dominated by the nonlinear dynamics of free-moving boundaries. The motion of the ablation front – the phase front between molten and vaporized material – feeds back to the motion of the melting front – the phase boundary between molten and solid material – and the dynamics of both free-moving boundaries are essential for the resulting quality of the weld seam. The dynamics in all three states of aggregation (solid, liquid, gaseous) are described by balance equations for energy, momentum and mass. Modeling the entire process as well as simulating the operation is not yet possible.

24.1.2 Asymptotic Methodology

The main aim of the model and the subsequent simulation is to support the interpretation of monitoring signals of the welding process. The laser welding process is too complex for a comprehensive modeling and simulation. Therefore, we use methods from singular perturbation theory. Thereby, the complex phenomena can be described by an approximate model which reproduces only the dominating phenomena of the process. All subordinated properties of the welding process that run on a faster time scale will not be taken into account. Since the omitted effects relax to their quasi-stationary values quickly, this approximate model is asymptotically accurate.

Many dissipative systems are characterized by the fact that their initially infinite-dimensional phase space shrinks to a low-dimensional subspace very fast, which could be reproduced by a few characteristic parameters [24]. These parameters – we call them generalized variables – often describe global properties of the dynamical system. In laser welding this typical property can be found in the form of a boundary-layer character for the transport phenomena of heat, mass and momentum [19, 20].

The approximation of the model is carried out by the use of integral and/or variational methods. Thereby, the degree of freedom is restricted by applying a parameterized distribution of the temperature or velocity field. The quality of this approximation is proven by application of pseudo-spectral methods [15]. They allow us to evaluate the solution for the full problem at an arbitrarily small error.

Current investigations on monitoring and control use interfaces between experimental and theoretical analysis to support the iterative refinement of reduced models. Iterative refinement aims to avoid unnecessary complexity and to reveal the essential features, i.e. the corresponding so-called process domains and characteristic appearance that support the analysis of the raw data from monitoring devices.

24.1.3 Former Works

Over the last decade we investigated approximate modeling of the thermal material processing by laser radiation. The first results have been achieved for the laser cutting process. The cutting model is able to identify the process domains that classify the product quality of the cut. The properties of the process domains could be related to the monitoring signals. Currently, the implementation to industrial cutting machines is in progress. Methodology and parts of the model can be transferred and adapted to laser welding.

Beside this, the presented results use the experiences and understanding that were developed by past research on parts of the welding process: initially, stationary heat-conduction models tried to reproduce the geometric shape of the weld seam [9]. The description of the melt flow around a given plasma capillary could refine their predications [10]. Additional physical phenomena of the welding process were taken into account, e.g. the Marangoni effect explained the extension of the melt pool [16]. The absorption of the laser radiation at metal surfaces was analyzed [11]. Stationary models are refined [30, 31] such that they allow process planning for special parameter ranges [8].

Dynamical models of these processes have been developed [14, 29, 2]. Recently developed models take the motion of the phase boundaries by melting or evaporating into account. The solution of most problems for these models have to be evaluated by time-consuming numerical methods.

24.2 Motion of the Melting Front

The phenomena in the solid material are effected by the heat conduction. The phase transition from solid to molten metal induces the motion of the melting front. This motion is identified to be the fundamental process of all thermal material processes. The description of the fundamental process is possible by a set of three ordinary differential equations applying asymptotical and integral methods.

24.2.1 Similarities of Thermal Material Processes

High-speed cutting can be seen as a transition from cutting to welding: with increasing feeding velocity the laser power has to be increased to guarantee the complete separation of the workpiece. Thus, the surface temperature of the ablation front reaches the evaporating temperature. A plasma capillary forms around the axis of the laser beam around which the melt flows. The jet of cutting gas drives the melt out of the cutting kerf behind the capillary (see Fig. 24.1). As a result of the formation of a plasma capillary, high-speed cutting represents a limit case that is very similar to deep-penetration welding.

This fact motivates the assumption that the results for the motion of the melting front in cutting are transferable to the welding process. Indeed, Schulz [28] has revealed that this motion represents a fundamental process for all thermal material processing (e.g. cutting, welding, drilling, metal removal).

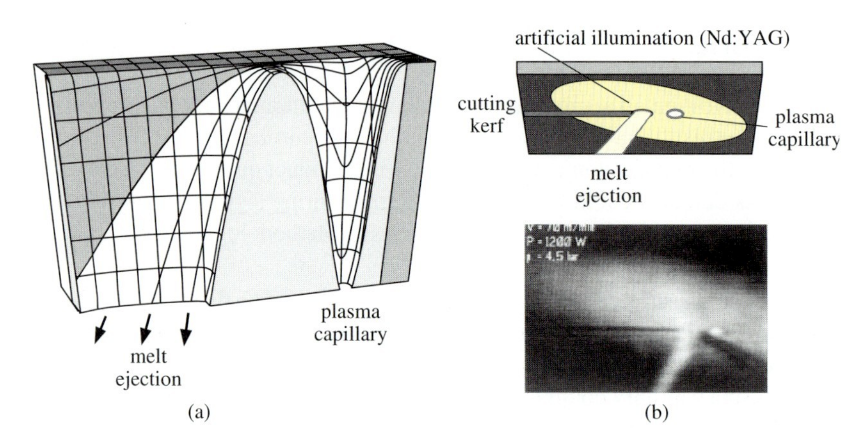

Figure 24.1: High-speed cutting: (a) at high feeding speeds ($> 30\,\mathrm{m/min}$) high laser power ($> 5 \times 10^7$ W) is used. The surface temperature reaches the evaporating temperature so that particles start to evaporate. The induced pressure towards the melt film drives the flow of melt in the azimuthal direction. A vapor capillary is formed. The melt that joins behind the capillary is driven out by the jet of cutting gas. (b) The offset of the position of the melt shoot in comparison to the position of the laser-beam axis can be experimentally observed [25].

24.2.2 The One-phase problem

The so-called one-phase problem [27, 28] describes the motion of the melting front. The laser radiation is absorbed at the melting front Γ_{m} and induces a heat flux q_{l} that heats the workpiece. When the surface temperature reaches the melting temperature T_{m} the melting front starts to move (see Fig. 24.2).

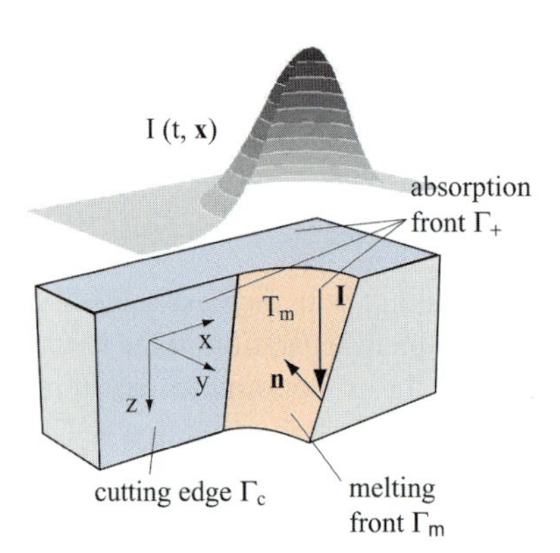

Figure 24.2: One-phase problem: the laser beam with intensity distribution I is absorbed at the absorption front Γ_+ and heats the workpiece. The part of the absorption front that has reached the melting temperature T_{m} is called the melting front Γ_{m}. The melting front is determined by the heat conduction and phase transition from solid to liquid material.

The motion of the melting front depends on the heat conduction within the solid material. Therefore, the temporal development of the temperature distribution $T(t, \mathbf{x})$ within the volume Ω_s of the solid material is determined. In the workpiece-fixed frame of reference the heat-conduction equation is

$$\frac{\partial T}{\partial t} = \kappa \, \Delta T, \qquad T = T(t, \mathbf{x}), \quad t \in [0, \infty), \quad \mathbf{x} \in \Omega_s, \tag{24.1}$$

where κ is the thermal diffusivity.

The temperature distribution is uniquely determined by specifying the boundary condition at all boundaries of the volume Ω_s. The part of the absorption front Γ_+ that is at the melting temperature T_m is called the melting front Γ_m. At this melting front Γ_m, the absorbed heat flux q_l heats the workpiece and moves the melting front by further melting. At the rest of the absorption front Γ_+/Γ_m the workpiece is heated by the absorbed heat flux. The bottom side Γ_- of the workpiece is isolated towards the ambient air. The border far away from the interaction zone between laser and material has ambient temperature T_a.

$$\mathbf{x} \in \Gamma_m: \qquad T = T_m, \qquad q_l - \lambda \nabla T \cdot \mathbf{n}_m = \rho_s \, H_m \, v_b, \tag{24.2}$$

$$\mathbf{x} \in \Gamma_+/\Gamma_m: \qquad q_l - \lambda \nabla T \cdot \mathbf{n}_m = 0, \tag{24.3}$$

$$\mathbf{x} \in \Gamma_-: \qquad - \lambda \nabla T \cdot \mathbf{e}_z = 0, \tag{24.4}$$

$$\forall \, |\mathbf{x}| \to \infty: \qquad T = T_a, \tag{24.5}$$

where λ is the heat conductivity, ρ_s the mass density of the solid material, H_m the melting enthalpy and v_b the velocity of the melting front in the normal direction. The normal vector of the melting front is given by $\mathbf{n}_m$; $\mathbf{e}_z$ represents the unit vector which points in the positive z direction.

The position of the melting front varies dynamically. Its velocity v_b is determined by the generalized Stefan boundary condition (24.2). Thus, the position of the melting front is part of the solution. This type of boundary is called a *free boundary*. Free boundary problems are nonlinear. The description of the heat-conduction equation (24.1) within the front-fixed frame of reference and insertion of the generalized Stefan condition (24.2) show formally the nonlinear character by the quadratic temperature gradient:

$$\frac{\partial T}{\partial t} + \mathbf{v}_b \cdot \nabla T = \kappa \, \Delta T \tag{24.6}$$

$$\overset{(24.2)}{\Rightarrow} \quad \frac{\partial T}{\partial t} + \frac{q_l}{\rho_s \, H_m} \nabla T \cdot \mathbf{n}_m - \frac{\lambda}{\rho_s \, H_m} (\nabla T \cdot \mathbf{n}_m)^2 = \kappa \, \Delta T \quad \forall \mathbf{x} \in \Gamma_m. \tag{24.7}$$

An explicit analytical solution of this free boundary problem does not exist. Even numerical evaluation of the solution requires the modification and extension of modern methods (level-set, hierarchical bases).

24.2.3 Approximate Equations of Motion

The motion of the melting front as a fundamental process for all thermal material processing has been thoroughly analyzed by Schulz [27, 28]. He has revealed that the difference of the typical length scales leads to a hierarchy of the spatial dimensions. By application of variational methods [5] the problem (24.1)–(24.5) can be reproduced approximately by an axially coupled system consisting of ordinary differential equations:

$$\dot{A} = \frac{1}{\ell_1}(\gamma_{01} - \ell_2\,\dot{\alpha}), \tag{24.8}$$

$$\dot{Q} = \frac{f_Q}{Q} - f_A\,\dot{A} - f_\alpha\,\dot{\alpha}, \tag{24.9}$$

$$\dot{\alpha} = \frac{1}{\ell_1\,\ell_4 + \ell_2\,\ell_3}(\ell_1\,\gamma_{21} + \ell_3\,\gamma_{01}). \tag{24.10}$$

The equations of motion describe the temporal development of global system variables like the position of the melting front A, the energy content Q and the radius α of the melting front. The variables γ_{01} and γ_{21} represent components of the absorbed heat flux. The coefficient functions f_X depend on the local Péclet number defined as α/Pe and l_x are numeric parameters [28].

24.3 Motion of the Capillary

The boundary of the capillary is the surface (liquid/gaseous) of the melt film. The dynamics within the melt film is described by transport phenomena of mass, momentum and energy. The boundary-layer character of the melt flow allows us to reduce the complexity of the description; the singularity of this asymptotic solution at the stagnation point is resolved by a separate description of the melt flow near the stagnation point.

24.3.1 Experimental Observation and Physical Analysis

Experimental observations are able to motivate suitable approaches for the approximate modeling of the complex physical relations. Especially, optical observation reveals crucial phenomena of the process and enables a process monitoring for industrial applications.

- **Thermography**
  The thermal radiation of the plasma and melt pool can be observed by a video camera coaxially aligned with the laser beam (see Fig. 24.3a). Since the plasma can be assumed to be optically thin, the coaxial alignment enables the view into the deep and narrow plasma capillary. The large dynamic range of modern CMOS detectors in today's digital cameras allows us to monitor the melt pool as well as the capillary simultaneously.

- **Artificial Illumination**
  The illumination of the interaction zone between laser and material by means of an artificial light source (diode laser) enables the observation of the geometrical shape of the scene (capillary, melt pool and solid material) [13]. Thereby, the thermal secondary radiation is filtered by a narrow band-pass filter for the wavelength of the almost monochromatic light source. The solid and molten material can be identified (see Fig. 24.3b) by the different reflection properties of both surfaces.

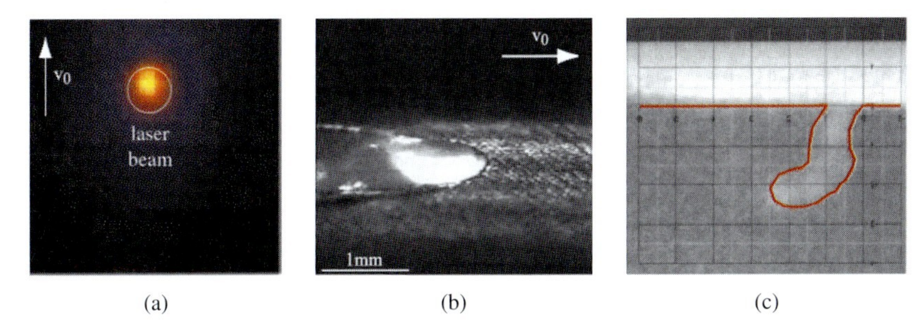

Figure 24.3: Temporally resolved optical monitoring of the laser welding process: (a) thermography: the thermal secondary radiation can be detected by a digital video camera aligned coaxially with the laser beam. Plasma capillary and melt surface are observable. (b) Artificial illumination: the geometrical shape of the interaction zone between laser and material can be revealed by artificial illumination. Solid, liquid and gaseous material can be differentiated [13]. (c) Radiography: by X-raying the sheet metal differences of mass density can be observed. The shape of the capillary and the melt pool as well as the mass flow in the molten material can be recorded [17, 22].

- **Radiography**
  The welding process can be transversely (perpendicular to laser-beam axis and feeding direction) observed by means of radiography [17, 22]. Thereby, the sheet metal is X-rayed during welding. The differences of the mass density in the workpieces are mapped to a fluorescent screen that is recorded by a high-speed video camera (see Fig. 24.3c). The radiographical records allow us to differentiate solid material, melt pool and capillary.

The experimental results reveal that the thickness of the melt film in front of the laser beam is small compared to the radius of the laser – and also of the melting front and the capillary. The melt flow is mainly laterally directed, i.e. no relevant axial transport of mass can be observed. The geometrical shape of the melting and absorption front in front of the laser beam is almost circular.

24.3.2 Mathematical Problem Formulation

The melt film in front of the laser beam axis is bordered by the melting front and the absorption front. Melting the solid material generates new melt which flows in the melt film across the melting front. At the absorption front the melt film is heated by the absorbed laser radiation. The surface temperature reaches the evaporating temperature, resulting in evaporation of material. The pressure gradient and the shear stress induced by the gaseous phase drive the melt film around the plasma capillary.

The molten material is assumed to be incompressible. Therefore, the transport of momentum and mass within the melt volume Ω_l can be described by the Navier–Stokes equation for incompressible fluids

$$\frac{\partial \mathbf{v}}{\partial t} + (\mathbf{v}\,\nabla)\,\mathbf{v} \quad = \quad -\frac{1}{\rho}\nabla p + \nu\,\Delta\mathbf{v} + g\,\mathbf{e}_z, \tag{24.11}$$

$$\nabla\cdot\mathbf{v} \quad = \quad 0, \qquad \mathbf{v} = \mathbf{v}(t,\mathbf{x}), \quad t \in [0,\infty), \quad \mathbf{x} \in \Omega_l, \tag{24.12}$$

where v is the velocity field, ρ the mass density and p the pressure within the fluid. ν represents the kinematic viscosity and g the acceleration due to the gravity.

The melt is driven by a pressure gradient and shear stress induced by the gaseous phase. The melt flow at the melting front is given by an in-stream – the melting front moves at the rate of v_p – and non-slip boundary condition

$$\mathbf{x} \in \Gamma_\mathrm{a}: \qquad \mathbf{n}_\mathrm{a} \cdot \left(\underline{\underline{\sigma}}\, \mathbf{n}_\mathrm{a}\right) = -p_\mathrm{g} + \sigma K, \tag{24.13}$$

$$\mathbf{t}_\mathrm{a} \cdot \left(\underline{\underline{\sigma}}\, \mathbf{n}_\mathrm{a}\right) = \tau_\mathrm{g} - \frac{d\sigma}{dT} \left(\nabla T \cdot \mathbf{t}_\mathrm{a}\right), \tag{24.14}$$

$$\mathbf{x} \in \Gamma_\mathrm{m}: \qquad \mathbf{n}_\mathrm{m} \cdot \mathbf{v} = \mathbf{n}_\mathrm{m} \cdot \mathbf{v}_P, \qquad \mathbf{t}_\mathrm{m} \cdot \mathbf{v} = \mathbf{t}_\mathrm{m} \cdot \mathbf{v}_P, \tag{24.15}$$

where p_g is the pressure and τ_g the shear stress within the gaseous phase. K represents the curvature of the absorption front Γ_a and σ its surface tension. The normal and tangential vectors of the melting front Γ_m and absorption front Γ_a are given by $\mathbf{n}_\mathrm{m}$, $\mathbf{t}_\mathrm{m}$ and $\mathbf{n}_\mathrm{a}$, $\mathbf{t}_\mathrm{a}$.

The thickness of the melt film and correspondingly the position of the absorption front is part of the solution for the dynamical system. The motion of the free surface is specified by the so-called kinematical boundary condition: a mass flux across the melt surface is barred – the loss of mass by evaporation is negligible – i.e. the streamlines run tangentially to the surface. The implicit representation $H(\mathbf{x}) = 0$ of the position of the absorption front Γ_a is described by a partial differential equation

$$\frac{\partial H}{\partial t} + \mathbf{v} \cdot \nabla H = 0. \tag{24.16}$$

The heat deposited in the volume Ω_l of the melt film is transported convectively by melt flow and diffusively by heat conduction. The temperature distribution is determined by the heat-transport equation

$$\frac{\partial T}{\partial t} + \mathbf{v} \cdot \nabla T = \kappa\, \Delta T, \qquad T = T(t, \mathbf{x}), \quad t \in [0, \infty), \quad \mathbf{x} \in \Omega_\mathrm{l}. \tag{24.17}$$

Its unique solution is determined by specifying the boundary condition at all boundaries of the volume Ω_l. At the absorption front Γ_a the melt is heated by the absorbed heat flux q_a. The melting front is defined to be at melting temperature T_m.

$$\mathbf{x} \in \Gamma_\mathrm{a}: \qquad -\lambda \nabla T = q_\mathrm{a}, \tag{24.18}$$

$$\mathbf{x} \in \Gamma_\mathrm{m}: \qquad T = T_\mathrm{m}. \tag{24.19}$$

An explicit analytical solution of this free boundary problem (24.11)–(24.19) does not exist. Even numerical evaluation of the solution requires the modification and extension of modern methods (level-set, hierarchical bases). The complexity of this model can be reduced by use of the present boundary-layer character that reveals the simple structure of the solution for this problem.

24.3.3 Boundary-layer Character of the Melt Flow

The experimental observations show that the values of typical length scales of the welding process change over several orders of magnitude: the thickness of melt film d_m (in the order of 10 μm) is significantly smaller than the radius of the laser beam w_0 (in the order of 100 μm) and the sheet thickness d (in the order of a millimeter).

$$d_\mathrm{m} \ll w_0, \qquad d_\mathrm{m} \ll d \tag{24.20}$$

$$\Rightarrow \qquad \epsilon := \frac{d_\mathrm{m}}{w_0} \ll 1, \quad \delta := \frac{d_\mathrm{m}}{d} \ll 1. \tag{24.21}$$

The differences in length scales indicate the boundary-layer character of the melt flow. The present transport phenomena link typical length scales with typical time scales via the velocities. The induced separation of time scales allows – as mentioned in Sect. 24.1.2 – a simplified description of the solution in the form of a reduced model for the long-term behavior [23, 21]. The problem (24.11)–(24.19) for the lateral plane – perpendicular to the laser-beam axis – given by polar coordinates (r, φ) is normalized to its typical scales

$$r = w_0\,\alpha - d_\mathrm{m}\,\eta, \quad \varphi\,\alpha = w_0\,\beta, \quad t = \frac{d_\mathrm{m}}{v_0}\,\tau \tag{24.22}$$

$$\Rightarrow \quad v_r = -v_0\,u_\eta, \quad v_\varphi = v_0\frac{w_0}{d_\mathrm{m}}\,u_\beta, \tag{24.23}$$

where the velocity field v is scaled towards u by the welding velocity v_0. The indices $(r, \varphi, \eta, \beta)$ denote the components of the velocity vector.

The smallness parameters (24.21) introduce a weighting of several phenomena among each other. The corresponding approximations for the melt flow in laser welding are:

- While the convective transport occurs mainly in the direction parallel to the melt film surface, the diffusive transport proceeds mainly in the perpendicular direction.
- The pressure stays constant in the direction perpendicular to the melt film surface.

The basis of the last approximation breaks down at the stagnation point. Here, the melt flow towards the stagnation point is decelerated by the pressure gradient. The reason for this contradiction stems from the rough assumption (24.23) that the typical length scale at the stagnation point in the tangential β direction will also be the radius of the laser beam w_0.

The approximation of the equations of motion has been analyzed by regular linear perturbation methods. The solution to the first order does not fulfill the boundary condition at the stagnation point. The perturbation of the boundary-layer character becomes singular at the stagnation point. Thus, the approximation induces not only a quantifiable error, but changes the qualitative behavior of the solution. In this case, singular perturbation theory instructs us to resolve both dynamics, in the inner and the outer areas. The inner solution (flow at the stagnation point) determines the initial conditions for the outer solution (flow around the capillary).

24.3.4 Flow at the Stagnation Point

The properties of the solution for the spatial two-dimensional flow (plane perpendicular to the axis of the laser beam) are discussed by means of a simplified model task. The value

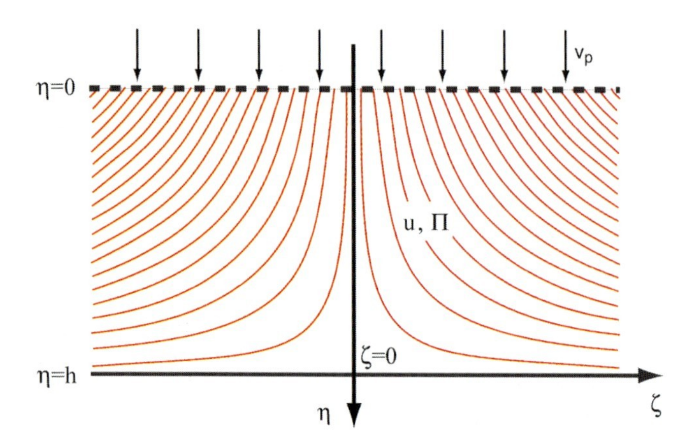

Figure 24.4: Model task for the velocity $\mathbf{u}$ and pressure distribution Π of the melt flow at the stagnation point: across the melting front ($\eta = 0$) new melt flows in a perpendicular direction. The thickness h of the melt film is constant and is given by the position of the frictionless absorption front ($\eta = h$) where a stationary pressure distribution $\Pi(\eta = h, \zeta)$ will be formed.

for the thickness of the melt film is small in comparison to the radius of the melting front. Curvature effects can be neglected and the thickness $h = h(t)$ of the melt film is independent of the location ζ (see Fig. 24.4). At first, the stationary problem is analyzed; subsequently, the dynamical equations of motion are presented.

The incompressible melt is driven by the pressure gradient $-\nabla\Pi > 0$ in the positive ζ direction. At the surface of the melt film (evaporation front) no shear stress exists.

$$\mathbf{u} = \mathbf{u}(\eta, \zeta), \quad \eta \in [0, h], \quad \zeta \in [0, \infty): \tag{24.24}$$

$$\mathrm{Re}\,(\nabla\,\mathbf{u})\,\mathbf{u} = -\nabla\Pi + \Delta\,\mathbf{u}, \qquad \nabla\cdot\mathbf{u} = 0, \tag{24.25}$$

$$u_\eta\Big|_{\eta=0} = v_p, \quad u_\zeta\Big|_{\eta=0} = 0, \qquad u_\eta\Big|_{\eta=h} = 0, \quad \frac{\partial\, u_\zeta}{\partial\,\eta}\Bigg|_{\eta=h} = 0. \tag{24.26}$$

The problem (24.24)–(24.26) is parameterized by the Reynolds number Re.

For the inertial limit case $1/\mathrm{Re} \ll 1$ the scalar velocity field $\nabla\Phi := \mathbf{u}$ is determined as a solution for the model task (24.24)–(24.26). In this limit case the solution can be analytically determined as

$$\Phi(\eta, \zeta) = \frac{1}{2}\frac{v_p}{h}\left(\zeta^2 - [\eta - h]^2\right). \tag{24.27}$$

To take the friction into account a generalized ansatz for the velocity $\mathbf{u}$ is introduced:

$$u_\eta = -f(\eta), \qquad u_\zeta = \zeta\, f'(\eta). \tag{24.28}$$

The solution for the problem without friction is covered by the ansatz (24.28) for the special case $f(\eta) = \eta$ (potential flow).

By the ansatz (24.28) the incompressibility condition $\nabla \cdot \mathbf{u} = 0$ is identically satisfied and the determination of the velocity $\mathbf{u}$ is decoupled from the determination of the pressure within the melt.

The task determining the function $f(\eta)$ represents a special case (Hiemenz case) of the Falkner–Skan problem [33] with modified boundary conditions:

$$\mathrm{Re}\left(-f\,f'' + f'^2\right) = \mathrm{Re} + f''', \qquad f(0) = -1, \quad f'(0) = 0, \quad f(h) = 0. \tag{24.29}$$

The modification is that the tangential boundary condition $f'(0) = 0$ is required at the melting front ($\eta = 0$) and not at the absorption front ($\eta = h$). Therefore, we call the problem above (24.29) a modified Hiemenz problem. The solution for the modified Hiemenz problem (24.29) is presented in the form of a convergent series [21, 23] and alternatively could be evaluated numerically by the shooting method (see Fig. 24.5).

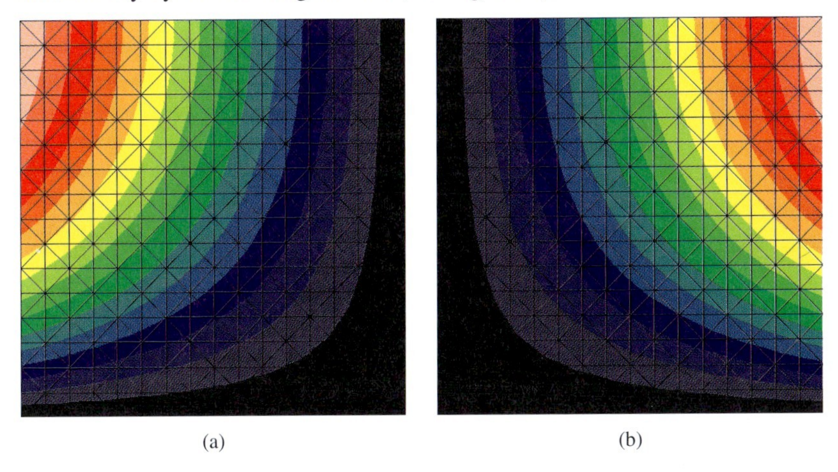

(a) (b)

Figure 24.5: Comparison of the solution from the modified Hiemenz problem (a) with the numerical solution (b, mirrored) of the full model task (24.24)–(24.26): visualization of the stream function. The deviations between both solutions are in the magnitude of the smallness parameter h/w_0 of the outer boundary layer flow.

The full model task (24.24)–(24.26) has been numerically solved by a finite-element method. The method used is based on the projection method of Chorin [6, 7]. Zienkiewicz and Taylor [34] combine this projection method with a characteristic Galerkin method that is characterized by stability over a large value range of the Reynolds number. The comparison with the potential and Hiemenz solutions shows that both limit cases contain crucial solution properties of the full model task. The deviations have the magnitude of the smallness parameter h/w_0 of the outer boundary layer flow.

24.3.5 Flow Around the Capillary

The singularity is only present in the neighborhood of the stagnation point at $\zeta = 0$. Its dimension has the magnitude of the melt film thickness. Outside this narrow neighborhood the melt flow shows the previously identified boundary-layer character (see Sect. 24.3.3).

We used integral methods exploiting the simplified structure of the solution to reduce the complexity of the model. The integration is done perpendicularly to the surface of the melt film where the diffusion processes will converge quickly to their quasi-stationary states. To enable the analytical evaluation of the integral terms in this formulation an ansatz for the flow profile is chosen. Therefore, the velocity distribution based on the analytical solution for the frictionless limit case is parameterized by a set of two generalized variables: the thickness h of the melt film and the mass flow m defined as

$$m(\beta) := \int_0^{h(\beta)} u_\beta(\eta, \beta)\, d\eta. \tag{24.30}$$

The equations of motion for the melt flow and its heat transport are a system of first-order partial differential equations. They fulfill the conservation equations of momentum, mass and energy

$$\mathrm{Re}\left(\frac{\partial m}{\partial \tau} + \frac{6}{5}\frac{\partial}{\partial \beta}\left[\frac{m^2}{h}\right]\right) = -\frac{\partial \Pi}{\partial \beta} h - \frac{3m}{h^2}, \tag{24.31}$$

$$\frac{\partial h}{\partial \tau} + \frac{\partial m}{\partial \beta} = v_p, \tag{24.32}$$

$$\frac{\partial}{\partial \tau}[\theta_\mathrm{s}\, h\, g_1] + \frac{5}{8}\frac{\partial}{\partial \beta}[\theta_\mathrm{s}\, m\, g_2] = \frac{1}{\mathrm{Pe}}\left[q_\mathrm{a} - \frac{\theta_\mathrm{s}}{h} g_3\right], \tag{24.33}$$

where v_p is the velocity of the melting front in the normal direction and the coefficients g_x are functions of the term $\mathrm{Pe}\, h$. θ_s represents the surface temperature of the absorption front and q_a is the absorbed heat flux. The dynamical system has a hyperbolic character: the transport of information occurs dominantly in the positive β direction. The velocities of propagation are identified by a linear stability analysis. The corresponding dispersion relation reveals that each perturbation induces one damped and one fanned response in the downstream direction [18, 19].

24.4 Evaporation

So far, the motion in the gaseous phase has not been described. To determine the driving forces of the gas towards the melt film the evaporating process is modeled. At a phase transition the hydrodynamical variables, e.g. the value of mass density or flow velocity, jump. The mass densities of liquid and gaseous phases differ typically by about three orders of magnitude, $\rho_\mathrm{g} \ll \rho_\mathrm{l}$. At the phase border a transition layer forms where deviations from the local thermodynamic equilibrium are observed. In evaporation this transition layer is called the Knudsen layer. Within the Knudsen layer the hydrodynamic approximation of the Boltzmann equation is not applicable.

The liquid and gaseous systems outside the Knudsen layer are described by hydrodynamic conservation equations of mass (density) ρ, momentum (density) $\rho\, u$ and energy E as well as the thermodynamic constitutive equations for pressure p and temperature T. Balancing these variables across the Knudsen layer we obtain

$$\rho_{\mathrm{l}}\,(v_p - u_{\mathrm{l}}) = \rho_{\mathrm{g}}\,(v_p - u_{\mathrm{g}})\,, \tag{24.34}$$

$$\rho_{\mathrm{l}}\,u_{\mathrm{l}}\,(v_p - u_{\mathrm{l}}) - P_{\mathrm{l}} = \rho_{\mathrm{g}}\,u_{\mathrm{g}}\,(v_p - u_{\mathrm{g}}) - P_{\mathrm{g}}, \tag{24.35}$$

$$E_{\mathrm{s}}\,(v_p - u_{\mathrm{s}}) - P_{\mathrm{s}}\,u_{\mathrm{s}} = E_{\mathrm{g}}\,(v_p - u_{\mathrm{g}})\,, \tag{24.36}$$

where the indices l and g denote the liquid and gaseous phases, leads to an under-determined system. The missing boundary values for the hydrodynamic variables are given by assumptions from the Knudsen layer that avoid a detailed analysis of the Boltzmann equation within the Knudsen layer. For minor evaporation Aden [1] assumes that the flux and the energy of evaporating particles remain constant across the Knudsen layer. He has found algebraic equations [1] that relate the surface temperature T_{l} of the fluid to the ablation pressure $P_{\mathrm{g}} + \rho_{\mathrm{g}}\,u_{\mathrm{g}}^2$ that acts upon the fluid surface:

$$\frac{P_{\mathrm{d}}}{P_{\mathrm{g}}} = \sqrt{\frac{T_{\mathrm{l}}}{T_{\mathrm{g}}}}\,\exp\left[-x^2\right] + \sqrt{\pi\,\frac{M\,k_{\mathrm{B}}\,T_{\mathrm{l}}}{2}}\,\frac{n_{\mathrm{g}}\,u_{\mathrm{g}}}{P_{\mathrm{g}}}\,\mathrm{erfc}\left[-x\right], \tag{24.37}$$

$$x^2 = \frac{M\,u_{\mathrm{g}}^2}{2\,k_{\mathrm{B}}\,T_{\mathrm{g}}}, \tag{24.38}$$

$$\frac{c_{\mathrm{v}}\,T_{\mathrm{l}}}{2} - \frac{3\,k_{\mathrm{B}}\,T_{\mathrm{g}}}{2\;\;M} = \frac{u_{\mathrm{g}}^2}{2}, \tag{24.39}$$

where P_{d} is the saturation vapor pressure, M the molar mass, k_{B} the Boltzmann constant, c_{v} the specific heat and n_{g} the particle density.

Subsequent works [26, 32] also take into account the re-condensation of gaseous material. For a more detailed description of the phenomena within the Knudsen layer the Boltzmann equation has to be analyzed [3].

24.5 Conclusion and Outlook

The approximate self-consistent modeling of the laser welding process has been presented. The crucial properties of the spatial three-dimensional welding process are reproduced by the motion of a few generalized variables. This welding model takes into account the phenomena in the solid and liquid materials as well as the driving evaporation of the melt film. The description of the re-condensation and the dynamics within the gaseous phase will be addressed in future work.

The simplified structure of the equations of motion enables an interactive process simulation by a standard personal computer. The model-based reproduction of the signals recorded by today's process monitoring systems has been addressed [4]. Further verification and refinement of the model will enable the use of model-based predictive control of the laser welding process.

Acknowledgment

The support by the Volkswagen Stiftung is gratefully acknowledged.

Bibliography

[1] M. Aden, Ph.D. thesis, RWTH Aachen, (1994).

[2] G. Albermann, Diploma thesis, RWTH Aachen (1998).

[3] K. Aoki and Y. Sone, *Gas flows around the condensed phase with strong evaporation or condensation – fluid dynamics equation and its boundary condition on the interface and their application*, in: R. Gatignol and Soubbaramayer (eds.): *Advances in Kinetic Theory and Continuum Mechanics*, Proceedings of a Symposium Held in Honor of Professor Henri Cabannes at the University Pierre et Marie Curie, Paris, Springer, Berlin, 1991, 43–54

[4] I. Bauer, Diploma thesis, RWTH Aachen (2002).

[5] M.A. Biot, *Variational Principles in Heat Transfer*, Oxford University Press, Oxford, 1970.

[6] A.J. Chorin, *Numerical solution of Navier-Stokes equations*, Math. Comp. **22** (1968), 745–762.

[7] A.J. Chorin, *On the convergence of discrete approximation to the Navier-Stokes equations*, Math. Comp. **23** (1969), 341–353.

[8] M. Dahmen, B. Fürst, S. Kaierle, E.W. Kreutz, R. Poprawe, and G. Turichin, *Model-based process planning and control-laser beam welding with CALAS*, in: XI International Symposium on Gas Flow and Chemical Lasers and High-Power Laser Conference, Proceedings of the SPIE **3092** (1997), 619–622.

[9] J. Dowden, M. Davis, and P. Kapadia, *The flow of heat and the motion of the weld pool in penetration welding with laser*, Journal of Applied Physics **57**(9) (1985), 4474–4479.

[10] J. Dowden, N. Postacioglu, M. Davis, and P. Kapadia, *A keyhole model in penetration welding with laser*, Journal of Physics D: Applied Physics **20** (1987), 36–44.

[11] M. Funk, PhD thesis, RWTH Aachen, (1994).

[12] S. Kaierle, P. Abels, C. Kratzsch, J. Michel, W. Schulz, and R. Poprawe, *New advances in process control for automotive laser applications*, in: Proceedings of the 31st course of Global Automotive Laser Applications Conference GALAC, Erice, Italy, 2001.

[13] G. Kapper, M. Dahmen, S. Kaierle, J. Michel, W. Schulz, K. Spielvogel, D. Petring, and R. Poprawe, *Coaxial process monitoring in heavy section laser beam welding*, in: Proceedings of LASER2001, Lasers in Manufacturing, Munich, Germany, 2001, 146–154.

[14] H. Ki, P.S. Mohanty, and J. Mazumder, *Modelling of high-density laser-material interaction using fast level set method*, Journal of Physics D: Applied Physics **34** (2001), 364–372.

[15] V. Kostrykin, M. Niessen, and W. Schulz, *Spectral method for free boundary problems* (2003), in preparation.

[16] C. Lampa, A.F.H. Kaplan, J. Powell, and C. Magnusson, *An analytical thermodynamic model of laser welding*, Journal of Physics D: Applied Physics **30** (1997), 1293–1299.

[17] A. Matsunawa, *Dynamics of keyhole and molten pool in high power laser welding*, in: Proceedings of the First International WLT-Conference on Lasers in Manufacturing, München, 2001.

[18] J. Michel, Diploma thesis, RWTH Aachen (1999).

[19] J. Michel, W. Schulz, M. Niessen, P. Abels, and S. Kaierle, *Advances in dynamical modeling, on-line monitoring and control in high quality cutting*, Proceedings of LASER2001, Lasers in Manufacturing, Munich, Germany, 2001, 112–122.

[20] J. Michel and W. Schulz, *Approximate Modelling of Dynamical Systems for Monitoring and Control*, Proceedings of Wesic2001, 3rd Workshop on European Scientific and Industrial Collaboration, Enschede, The Netherlands 2001, 395–405. (ISBN 90-365-12102).

[21] J. Michel, Ph.D. thesis, RWTH Aachen (2003), in preparation.

[22] M.G. Müller, B. Hohenberger, F. Dausinger, H. Hügel, T. Iwase, H. Sakamoto, H. Shibata, N. Seto, and A. Matsunawa, *An online monitoring system validated by visualization of laser welding phenomena*, in: Proceedings of LASER2001, Lasers in Manufacturing, Munich, Germany, 2001, 155–166.

[23] S. Pfeiffer, Diploma thesis, RWTH Aachen (2002).

[24] R. Poprawe and W. König, *Modeling, Monitoring and Control in High Quality Laser Cutting*, Annals of the CIRP **50**(1) (2001), 137–140.

[25] K.-U. Preissig, Ph.D. thesis, RWTH Aachen, (1995). (ISBN 3-8265-1041-0)

[26] J.W. Rose, *Accurate approximate equations for intensive sub-sonic evaporation*, International Journal of Heat Mass Transfer **43** (2000), 3869–3875.

[27] W. Schulz, V. Kostrykin, H. Zefferer, D. Petring, and R. Poprawe, *A Free Boundary Problem Related to Laser Beam Fusion Cutting: ODE Approximation*, International Journal of Heat Mass Transfer **40**(12) (1997), 2913–2928.

[28] W. Schulz, Habilitation, RWTH Aachen (1998).

[29] V.V. Semak, W.D. Bragg, B. Damkroger, and S. Kempka, *Transient model for the keyhole during laser welding*, Journal of Physics D: Applied Physics **32** (1999), L61–L64.

[30] P. Solana and J.L. Ocaña, *A mathematical model for penetration laser welding as a free-boundary problem*, Journal of Physics D: Applied Physics **30** (1997), 1300–1313.

[31] X. Ye and X. Chen, *Three-dimensional modeling of heat transfer and fluid flow in laser full-penetration welding*, Journal of Physics D: Applied Physics **35** (2002), 1049–1056.

[32] T. Ytrehus and S. Ostmo, *Kinetic theory approach to interphase processes*, International Journal of Multiphase Flow **22** (1996), 133–155.

[33] M.B. Zaturska, W.H.H. Banks, *A new solution branch of the Falkner-Skan equation*, Acta Mechanica **152** (2001), 197–201.

[34] O.C. Zienkiewicz and R.L. Taylor, *The Finite Element Method*, Vol. 3, Butterworth Heinemann, 2000. (ISBN 0-7506-5049-4)

25 Short-time Dynamics in Laser Material Processing

V. Kostrykin, W. Schulz, M. Nießen, and J. Michel

Material processings with laser radiation such as cutting, welding and drilling are established industrial processes. Their main advantages compared to conventional methods are the flexibility, precision and speed. The quality of processing results is determined mainly by the transient dynamics of the process. An application of asymptotic and spectral methods to the solution of the free boundary problem modeling the laser drilling of metals is presented. In the frame of this approach we obtain detailed information on the time and length scales involved. The approach can also be applied to study the transient dynamics of cutting and welding.

25.1 Introduction

Laser materials processing is established in industry. It provides the opportunity to achieve processing results that are difficult – if not impossible – to achieve using conventional methods. The dynamics of the process is mainly responsible for the resulting quality. As an example, in cutting of metals the dross formation at the bottom of the cutting kerf may require expensive mechanical post-treatment. The instabilities of the keyhole in welding may lead to high porosity of the weld seam, which drastically diminishes mechanical properties of the joint. In drilling the molten material solidified on the hole walls and at the crater entrance is a main factor decreasing the resulting precision. Therefore, the understanding of the dynamics of material processing with laser radiation is of considerable importance. Moreover, successful monitoring and control of the process require a detailed dynamical description of the process, thus making it possible to use full advantage of the flexibility, precision and speed of laser technology.

Any of the three mentioned processes (cutting, welding, drilling) includes a number of contributing physical subprocesses, in particular heat conduction, melt and gas flow, and plasma dynamics. A more or less complete mathematical model of the process is given by a multidimensional system of partial differential equations coupled at free boundaries. Even a qualitative analysis of such systems is a subtle problem. Their numerical simulation requires modern adaptive numerical methods and is unthinkable without high-performance computers.

From a practical point of view it is desirable to develop the simplest model of the process, e.g. by considering accurate approximate solutions. In contrast to the numerical solutions analytic approximations provide the detailed information on the dynamics of the process which in turn can be implemented in monitoring and controlling devices.

The long-time dynamics in laser material processing is rather well understood. It is well known that the long-time dynamics of dissipative partial differential equations is concentrated on finite-dimensional manifolds [1]. Based on this fact by means of integral, in particular,

Nonlinear Dynamics of Production Systems. Edited by G. Radons and R. Neugebauer

ISBN 3-527-40430-9

variational methods finite-dimensional dynamical systems modeling the cutting and welding processes have been developed [2–4]. These models are completely sufficient to answer a lot of practical questions. However, in a number of cases the understanding of short-time dynamics appears to be necessary.

The main aim of the present contribution is to settle an approach allowing us to describe the dynamics of the process on arbitrarily short time scales and arbitrarily small space scales. An approximate solution to the free boundary problem for the heat conduction equation modeling the drilling process is constructed by means of an eigenfunction expansion with respect to a suitably chosen time-dependent basis. This approach, originated from the classical Fourier method for solving parabolic partial differential equations (see e.g. [5]), is often called a *spectral* method. We refer to [1], [6], [7] and [8] for other applications of spectral methods to the analysis of nonlinear partial differential equations.

Cutting, welding and drilling have a lot of features in common. Actually, the underlying physical processes and their interaction are essentially the same. The knowledge gained from the modeling of the cutting process has been proved to be extremely useful for the understanding of dynamical features of welding and drilling. In the present work we will be concerned with the drilling of metals by pulsed laser radiation. We choose this process as representative mainly to illustrate the basic concepts of our approach. The implications to the other two processes (cutting and welding) will be presented elsewhere.

25.2 The Free Boundary Problem

We start with the formulation of the free boundary problem which models the drilling of a material with pulsed laser radiation. We consider an infinitely extended two-dimensional workpiece (see Fig. 25.1). The laser radiation is absorbed by the surface Γ of the workpiece. The induced heat flow heats the material. When the surface temperature reaches the melting point, the material melts. Evaporation from the surface of the melt film induces a large pressure gradient which pushes the melt to flow out of the drill hole. Therefore, the molten material will be assumed to be instantly removed from the drill hole. The additional effects introduced by the melt flow can be taken into account afterwards. Also, the extension of the model to the case of three dimensions is straightforward.

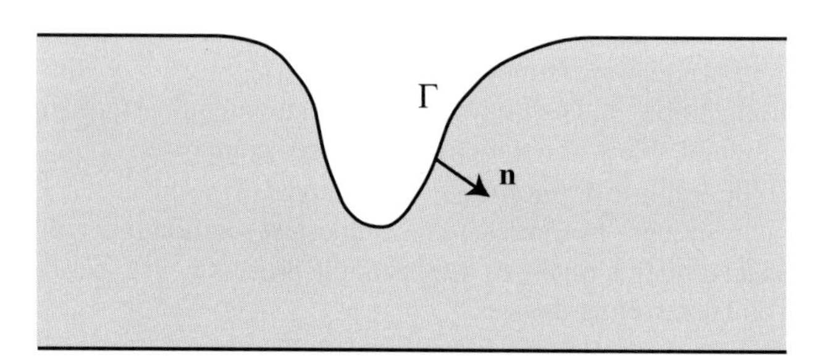

Figure 25.1: Free boundary problem.

Mathematically the drilling process can be described by a free boundary problem, where the geometry of the domain is unknown and has to be determined as a part of the solution. Denoting by $T(\mathbf{x}, t)$ the temperature we arrive at the following problem: Find the solution to the heat conduction equation

$$\frac{\partial T}{\partial t} = \varkappa \Delta T, \qquad t \in [0, t_{\text{pulse}}]$$

subject to the boundary conditions

$$I(t) f(\mathbf{x}) + \lambda \, \mathbf{n} \nabla T|_\Gamma = \begin{cases} 0 & \text{if} \quad T|_\Gamma < T_{\text{m}} \qquad \text{(heating phase)}, \\ H_{\text{m}} \overline{v}_p & \text{if} \quad T|_\Gamma = T_{\text{m}} \qquad \text{(melting phase)}. \end{cases}$$

Here $\varkappa$ is the thermal diffusivity of the material, λ its heat conductivity and H_{m} the specific latent heat of melting. The product $I(t)f(\mathbf{x})$ represents the laser intensity absorbed by the material at the point $\mathbf{x}$ of the surface Γ at time t. The function $I(t)$ describes the pulse form which has a duration t_{pulse}. The distribution $f(\mathbf{x})$, $0 \leq f(\mathbf{x}) \leq 1$, is an arbitrary rapidly decreasing function of the coordinate x parallel to the surface of the workpiece, e.g. $f(\mathbf{x}) = \exp\{-x^2/(2r^2)\}$. To simplify the discussion we ignore the dependence of the absorption degree on the angle of incidence of the laser radiation. The melting temperature of the material is denoted by T_{m}, $\mathbf{n}$ denotes the normal vector directed inwards and $v_p \geq 0$ is the velocity of the boundary Γ. On the lower surface of the workpiece we pose the homogeneous Neumann boundary conditions corresponding to perfect thermal isolation.

We will perform the asymptotic analysis of the problem using the fact that t_{pulse} is much smaller than the typical diffusion time $t_{\text{diff}} := r^2/\varkappa$. For this aim it is convenient to write the problem in the non-dimensional form. The available length and time scales are given by the beam radius r and associated diffusion time t_{diff}. In these non-dimensional variables the problem now takes the form (we use the same notation t and x for the non-dimensional variables)

$$\frac{\partial \theta}{\partial t} = \Delta \theta, \qquad t \in [0, \varepsilon] \tag{25.1}$$

with the boundary conditions

$$\varepsilon^{-1} \gamma(t) f(x) + \mathbf{n} \nabla \theta|_\Gamma = \begin{cases} 0 & \text{if} \quad \theta|_\Gamma < 1 \qquad \text{(heating phase)}, \\ h_{\text{m}} v_p & \text{if} \quad \theta|_\Gamma = 1 \qquad \text{(melting phase)}, \end{cases} \tag{25.2}$$

where

$$\theta = \frac{T - T_{\text{a}}}{T_{\text{m}} - T_{\text{a}}}, \qquad \gamma = \frac{r}{(T_{\text{m}} - T_{\text{a}})\lambda} \frac{t_{\text{pulse}}}{t_{\text{diff}}} I,$$

$$v_p = \frac{\overline{v}_p \, r}{\varkappa}, \qquad h_{\text{m}} = \frac{H_{\text{m}}}{c(T_{\text{m}} - T_{\text{a}})}$$

are the non-dimensional temperature, intensity, velocity of the free boundary and latent heat, respectively, T_{a} is the ambient temperature and c is the heat capacity. The parameter ε has a meaning of the non-dimensional pulse duration,

$$\varepsilon = \frac{t_{\text{pulse}}}{t_{\text{diff}}}.$$

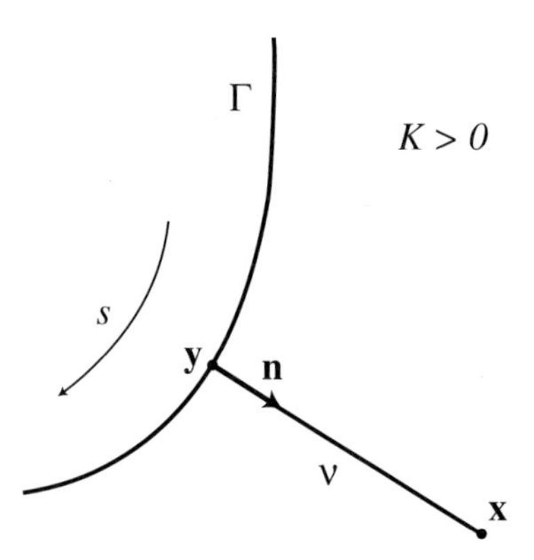

Figure 25.2: Boundary fitted coordinates.

The absorbed intensity in (25.2) scales inversely proportional to ε. This reflects the fact that with diminishing of the pulse duration the maximum laser intensity can be increased such that the pulse energy (which is proportional to the integral of $I(t)$ over the time) remains approximately constant.

It is easy to show that for $\varepsilon \to 0$ and fixed pulse energy the heating phase (the time when the material is only heated by the laser radiation and there is no melting) is of the order ε^2, i.e. much shorter as the pulse duration. Although the asymptotic analysis can also be performed for the heating phase, in what follows we will be concerned with the melting phase.

What we need for the asymptotic analysis of the problem are (local) boundary fitted coordinates. To introduce them, consider an arbitrary point $\mathbf{x}$ within the material such that there is a unique normal line to Γ passing through the point $\mathbf{x}$. Assume that this normal crosses Γ at the point $\mathbf{y}$ (see Fig. 25.2). Denote the distance between the points $\mathbf{x}$ and $\mathbf{y}$ by ν, i.e. $\nu = |\mathbf{x} - \mathbf{y}|$. The point y can be unambiguously characterized by the arc length s calculated from an arbitrarily fixed reference point on Γ.

In the coordinates ν, s the heat conduction equation has the form

$$\frac{\partial \theta}{\partial t} - v_p \frac{\partial \theta}{\partial \nu} - w \frac{\partial \theta}{\partial s} = \frac{1}{1+K\nu} \frac{\partial}{\partial s} \frac{1}{1+K\nu} \frac{\partial \theta}{\partial s} + \frac{\partial^2 \theta}{\partial \nu^2} + \frac{K}{1+K\nu} \frac{\partial \theta}{\partial \nu},$$

where $K = K(s)$ is the local curvature and w the tangential velocity given by

$$w = \frac{K}{1+K\nu} \frac{\partial v_p}{\partial s}.$$

Now we introduce the scaled time $\tau = t/\varepsilon$ and space $\widehat{\nu} = \nu/\varepsilon$ variables such that the pulse duration will correspond to $\tau = 1$. The velocity of the boundary then becomes $\widehat{v}_p = v_p/\varepsilon$. The problem takes the form

$$\begin{aligned} &\frac{\partial \theta}{\partial \tau} - \frac{1}{\varepsilon} \widehat{v}_p \frac{\partial \theta}{\partial \widehat{\nu}} - \widehat{w} \frac{\partial \theta}{\partial s} \\ &= \varepsilon \frac{1}{1+\varepsilon K \widehat{\nu}} \frac{\partial}{\partial s} \frac{1}{1+\varepsilon K \widehat{\nu}} \frac{\partial \theta}{\partial s} + \frac{1}{\varepsilon} \frac{\partial^2 \theta}{\partial \widehat{\nu}^2} + \frac{K}{1+\varepsilon K \widehat{\nu}} \frac{\partial \theta}{\partial \widehat{\nu}} \end{aligned} \tag{25.3}$$

with the boundary condition

$$\gamma(t)f(\mathbf{y}(s)) + \frac{\partial \theta}{\partial \widehat{\nu}}\bigg|_{\widehat{\nu}=0} = h_{\mathrm{m}}\widehat{v}_p. \tag{25.4}$$

The accurate analysis of the asymptotics of the solution to the problem (25.3) and (25.4) requires the application of singular perturbation theory. The detailed analysis will be presented elsewhere. Here we mention that outside those points of the free boundary where the curvature is large ($K \sim \varepsilon^{-1}$), all lower-order terms in Eq. (25.3) can be omitted. The resulting equation will then take the form of the heat conduction equation in one space dimension,

$$\frac{\partial \theta}{\partial \tau} - \frac{1}{\varepsilon}\widehat{v}_p\frac{\partial \theta}{\partial \widehat{\nu}} = \frac{1}{\varepsilon}\frac{\partial^2 \theta}{\partial \widehat{\nu}^2}. \tag{25.5}$$

In other words, in the limit $\varepsilon \to 0$ the heat conduction in the direction normal to the surface becomes dominant whereas that in the tangential direction becomes negligible. The tangential coordinate s enters the problem only parametrically via the boundary condition (25.4).

How can we analyze and solve the free boundary problem (25.5), (25.4)? We will be concerned with this question in the next section.

25.3 Finite-dimensional Approximations

In the previous section we have shown that the asymptotic analysis of the solution to the free boundary problem for $\varepsilon \to 0$ leads to the simplified free boundary problem for the heat conduction equation in one space dimension. Unfortunately, the general solution of this problem cannot be given in closed analytical form. Below we construct an approximate solution determined by a finite-dimensional dynamical system. A detailed account can be found in our paper [9].

Let $l(t)$ denote the distance between a point on the surface Γ and the lower surface of the workpiece measured along the normal to Γ (see Fig. 25.3). Performing the transformation of variables

$$x := l(t) - \widehat{\nu}, \qquad t = \tau/\varepsilon$$

we arrive at the following formulation of the free boundary problem (25.5), (25.4): Given a continuously differentiable function $\gamma(t) \geq 0$ find the functions $\theta(x,t)$ and $l(t)$ satisfying

$$\theta_t = \theta_{xx}, \qquad x \in [0, l(t)], \qquad t \in (0, T) \tag{25.6}$$

subject to the boundary conditions

$$\theta_x(0,t) = 0, \tag{25.7}$$

$$\theta(l(t),t) = 1, \tag{25.8}$$

$$\theta_x(l(t),t) - h_{\mathrm{m}}\dot{l}(t) = \gamma(t) \tag{25.9}$$

on the maximal time interval $[0, T)$ such that

$$\dot{l}(t) < 0 \quad \text{for almost all} \quad t \in (0,T) \quad \text{and} \quad l(t) > 0 \quad \text{for all} \quad t \in (0,T). \tag{25.10}$$

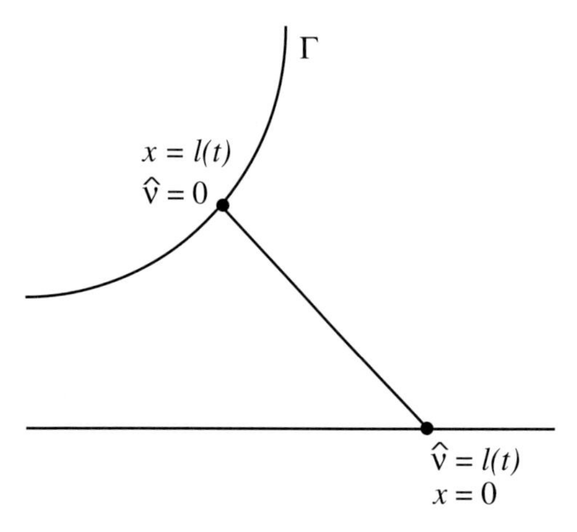

Figure 25.3: The coordinate transformation.

In other words $[0, T]$ is the maximal time interval on which melting is maintained. We assume the following initial conditions

$$l(0) = l_0, \qquad \theta(x, 0) = g(x),$$

where $g \in C^2(0, l_0)$ satisfies $0 \leq g(x) \leq 1$.

Looking for the solution of the problem (25.6)–(25.9) in the form

$$\theta(x, t) = 1 + u(x, t)$$

we obtain that the function u must satisfy the heat conduction equation

$$u_t = u_{xx},$$

and the conditions

$$u_x(0, t) = 0, \qquad u(l(t), t) = 0, \qquad u(x, 0) = g(x) - 1.$$

The free boundary $l(t)$ is determined by the equation

$$u_x(l(t), t) = h_{\mathrm{m}}\dot{l}(t) + \gamma(t). \tag{25.11}$$

Now consider the eigenvalue problem

$$w_k''(x; l) = -\lambda_k(l) w_k(x; l), \qquad w_k'(0; l) = 0, \qquad w_k(l; l) = 0, \qquad k \in \mathbb{N}_0.$$

Obviously, the eigenfunctions $w_k(x; l)$ satisfying the orthonormality condition

$$\int_0^l w_j(x; l) w_k(x; l) dx = \delta_{jk} \tag{25.12}$$

have the form

$$w_k(x;l) = (-1)^k \sqrt{\frac{2}{l}} \cos\left(\frac{\pi x}{2l}(2k+1)\right), \qquad \lambda_k(l) = \frac{\pi^2}{4l^2}(2k+1)^2, \qquad k \in \mathbb{N}_0.$$

Using the ansatz

$$u(x,t) = \sum_{k=0}^{\infty} a_k(t) w_k(x;l(t)), \tag{25.13}$$

we arrive at the following system of coupled ordinary differential equations:

$$\dot{a}_k(t) = -\lambda_k(l(t))a_k(t) - \frac{\dot{l}(t)}{l(t)} \sum_{j\in\mathbb{N}_0} Q_{kj} a_j(t), \tag{25.14}$$

where

$$Q_{kj} := l \int_0^l w_k(x;l) \frac{\partial w_j}{\partial l} dx = \begin{cases} 0 & \text{for } k = j, \\ \dfrac{(2k+1)(2j+1)}{2(k-j)(k+j+1)} & \text{for } k \neq j. \end{cases} \tag{25.15}$$

Using the fact that

$$\left.\frac{\partial \theta_1}{\partial x}\right|_{x=l(t)} = \sum_{k\in\mathbb{N}_0} a_k(t) \left.\frac{\partial w_k}{\partial x}\right|_{x=l(t)} = -\frac{\pi\sqrt{2}}{2l(t)^{3/2}} \sum_{k\in\mathbb{N}_0} (2k+1) a_k(t)$$

we obtain from (25.11) the additional differential equation for $l(t)$:

$$h_{\mathrm{m}} \dot{l}(t) + \gamma(t) = \frac{\pi\sqrt{2}}{2l(t)^{3/2}} \sum_{k\in\mathbb{N}_0} (2k+1) a_k(t). \tag{25.16}$$

The coupled (infinite-dimensional) system of differential equations (25.14), (25.16) yields the solution of the free boundary problem (25.6), (25.7)–(25.9), (25.10). Corresponding initial values are given by

$$a_k(t=0) = \int_0^{l(0)} g(x) w_k(x;l(0)) dx - q_k^{(0)}(l(0))$$

with

$$q_k^{(0)}(l) := \int_0^l w_k(x;l) dx = \frac{2\sqrt{2l}}{\pi(2k+1)}.$$

The truncated finite-dimensional equations (25.14) and (25.16),

$$\begin{aligned} \dot{a}_k(t) &= -\lambda_k(l(t))a_k(t) - \frac{\dot{l}(t)}{l(t)} \sum_{j=0}^{N} Q_{kj} a_j(t), \qquad k \in \{0,1,\ldots,N\}, \\ h_{\mathrm{m}} \dot{l}(t) + \gamma(t) &= \frac{\pi\sqrt{2}}{2l(t)^{3/2}} \sum_{k=0}^{N} (2k+1) a_k(t), \end{aligned} \tag{25.17}$$

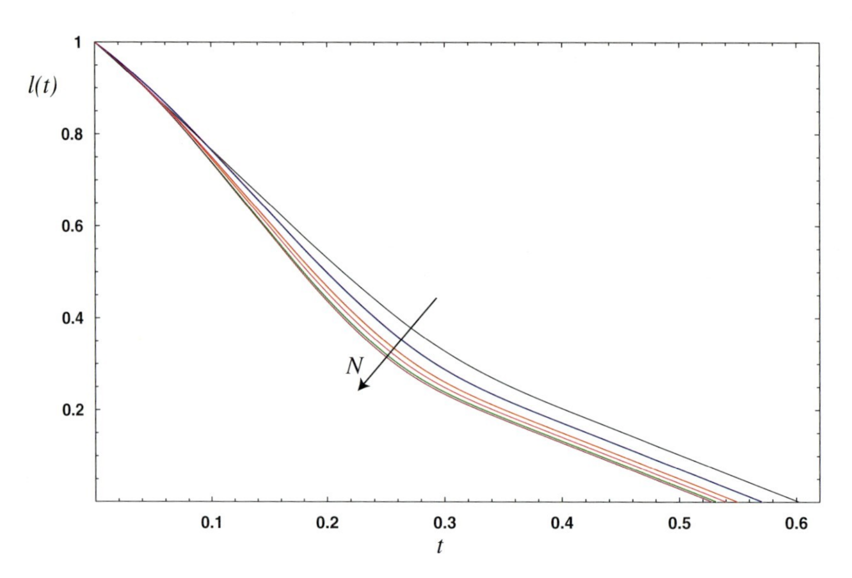

Figure 25.4: Convergence of approximations to $l(t)$ for increasing dimension of the truncated system (25.17). The plotted curves correspond to $N = 1, \ldots, 6$.

provide the approximate solutions

$$\theta_N(x,t) = 1 + \sum_{k=0}^{N} a_k(t) w_k(x; l(t))$$

to the problem (25.6), (25.7)–(25.9), (25.10). With increasing dimension N the solution converges towards the exact solution. The convergence is proven in [9]. Figure 25.4 illustrates the speed of convergence. The rapid convergence is mainly due to the fact that the eigenvalues $\lambda_k(l)$ are strongly separated.

The system (25.17) generates a finite-dimensional dynamical system which can be analyzed in detail. In particular, we obtain the following information on the behavior of the solution of the free boundary problem (25.6), (25.7)–(25.9), (25.10): Assume that $\gamma(t) = \gamma$ is constant and the solution of the free boundary problem (25.6), (25.7)–(25.9), (25.10) exists on the interval $(0, T)$ for some $T > 0$ such that

$$\lim_{t \uparrow T} l(t) = 0.$$

Then

$$\lim_{t \uparrow T} \dot{l}(t) = -\gamma / h_{\mathrm{m}}.$$

25.4 Conclusion

We performed the asymptotic analysis of the free boundary problem (25.1), (25.2) modeling the drilling of metals with pulsed laser radiation. We have shown that in the limit of small pulse duration the solution of this problem can be approximated by solutions of considerably more simple free boundary problems in one space dimension (25.5), (25.4). Furthermore, by means of the spectral method the latter problem has been approximated by a system of ordinary differential equations (25.17). Increasing the dimension N of the system (25.17) the solution of (25.5), (25.4) can be determined with an arbitrary accuracy. Thus, the infinite-dimensional dynamical system (25.1), (25.2) is reduced to a set of low-dimensional dynamical systems.

A modification of the spectral method to the case of a semi-infinite domain is developed in [10]. The spectral method described here is not restricted to free boundary problems for the heat conduction equation. Its application to free boundary problems for thin liquid films will be presented in [11].

Acknowledgment

The authors acknowledge the support by the Volkswagen Stiftung.

Bibliography

[1] P. Constantin, C. Foias, B. Nicolaenko, and R. Temam, *Integral Manifolds and Inertial Manifolds for Dissipative Partial Differential Equations*, Springer-Verlag, New York, 1989.

[2] W. Schulz, V. Kostrykin, H. Zefferer, D. Petring, and R. Poprawe, *A free boundary problem related to laser beam fusion cutting: ODE approximation*, International Journal of Heat and Mass Transfer **40** (1997), 2913–2928.

[3] V. Enß, V. Kostrykin, W. Schulz, H. Zefferer and D. Petring, *Thermische Materialbearbeitung mit Laserstrahlung: Schmelzschneiden* (Mathemathik – Schlüsseltechnologie für die Zukunft), edited by K.-H. Hoffmann, W. Jäger, T. Lohmann, and H. Schunck, Konferenzberichte zum "Statusseminar der anwendungsorientierten Verbundprojekte auf dem Gebiet der Mathematik des BMBF", München, 1995, Springer, Berlin, 1997, pp. 161–174.

[4] W. Schulz, V. Kostrykin, M. Nießen, J. Michel, D. Petring, E. W. Kreutz, and R. Poprawe, *Dynamics of ripple formation and melt flow in laser beam cutting*, Journal of Physics D: Applied Physics **32** (1999), 1219–1228.

[5] O. A. Ladyzhenskaya, *The Boundary Value Problems of Mathematical Physics*, Applied Mathematical Sciences **49**, Springer-Verlag, New York, 1985.

[6] C. Foias and G. Prodi, *Sur le comportement global des solutions non-stationnaires des equations de Navier-Stokes en dimension 2*, Rendiconti del Seminario Matematico dell' Università di Padova **39** (1967), 1–34.

[7] G. Ströhmer, About the approximation of steady viscous flows using the eigenfunctions of a certain differential operator, Mathematische Nachrichten **174** (1995), 283–304.

[8] G. Ströhmer, *About the approximation of unsteady flows by the eigenfunctions of a certain differential operator*, Mathematische Nachrichten **221** (2001), 151–168.

[9] V. Kostrykin, M. Niessen, and W. Schulz, *Spectral method for free boundary problems*, preprint (2003).

[10] H. te Heesen, Diploma Thesis, RWTH Aachen (2003).
[11] U. Eppelt, Diploma Thesis, RWTH Aachen (in preparation).

26 An Approach to a Process Model of Laser Beam Melt Ablation Using Methods of Linear and Nonlinear Data Analysis

R. Donner, A. Cser, U. Schwarz, A. Otto, and U. Feudel

As a non-contact process laser beam melt ablation offers several advantages compared to conventional processing mechanisms. During ablation the surface of the workpiece is melted by the energy of a CO_2-laser beam; this melt is then driven out by the momentum of an additional process gas. Although the idea behind laser beam melt ablation is rather simple, the process itself has a major limitation in practical applications: with increasing ablation rate the surface quality of the workpiece processed declines rapidly.

With different ablation rates different surface structures can be distinguished, which can be characterized by suitable surface parameters. The corresponding regimes of pattern formation are found in linear and nonlinear statistical properties of the recorded process emissions as well. While the ablation rate can be represented in terms of the line energy, this parameter does not provide sufficient information about the full behavior of the system.

The dynamics of the system is dominated by oscillations due to the laser cycle but includes some periodically driven nonlinear processes as well. Upon the basis of the measured time series, a corresponding model for the material ablation with a minimum complexity is suggested and analyzed. The deeper understanding of the process can be used to develop strategies for process control.

26.1 Introduction

In many industrial processes lasers have already taken the role of conventional machining [1–3], which is due to a major advantage of this new kind of machining: because of the lack of mechanical contact no tool wear exists and even extremely hard or brittle materials (e.g. tempered steel, glass or ceramics) can be processed [1, 4]. For example, during the last two decades the technologies of laser welding [5–7] and cutting [8–11] have been developed and found applications in a broad field.

In the case of a well-defined material removal from a metallic surface one distinguishes between evaporative microablation [12, 13], reactive ablation (chip removal) [14] and melt ablation. The last process is the only one to allow ablation rates high enough to be relevant for industrial manufacturing of three-dimensional surfaces. During processing, absorption of laser radiation leads to the melting of the material, and the momentum of an additional process gas is used to drive out the molten material from the laser–material interaction zone. Hence,

Nonlinear Dynamics of Production Systems. Edited by G. Radons and R. Neugebauer

ISBN 3-527-40430-9

melt ablation works with a higher energetic efficiency than evaporative ablation because of avoiding the phase transition to a vaporized state.

The idea behind the process of laser beam melt ablation is rather simple: intensive laser radiation enables the removal of a high amount of material. Unfortunately, a major limitation exists in practical applications: with increasing ablation rate, at about ablation rates relevant for industrial applications, the surface quality of the workpiece processed declines rapidly, which causes enormous difficulties in process control. Motivation of this work is to find a way to improve surface quality and parallely maintain high ablation rates to make this technology a competitive alternative in industrial applications.

The main aim of this contribution is to investigate some possibly suitable observables for radiation flux and surface quality that might be useful for the development of an online process control in the near future. Upon the basis of this analysis, a simple model for the ablation dynamics will be presented and discussed. After a short description of the experimental setup in Sect. 26.2, the detailed features of the process gained from experiments are summarized and statistically analyzed in Sect. 26.3. This reflects both the observables measured online and the final surface profiles. By using methods of linear and nonlinear data analysis, it is found that the dynamical behavior of the physical system is nonlinear, while this nonlinearity is hard to quantify with the quality and quantity of the previous experiments and the resolution of the corresponding data sets available at the moment. No significant local correspondence between the fluctuations of the observables and the surface structures is found, while global parameters calculated from quantities that can be measured online allow a quite good classification of the resulting process quality. The almost exponential distribution of ridges in the surface profiles indicates a fundamental importance of the gas–melt interaction that can be well approximated by a random process. According to this, a new stochastic approach to a time-continuous process model for the discrete ablation process is developed in Sect. 26.4 and described in more detail. Additional external forcing allows a regularization of the ablation process.

26.2 Experimental Setup

Two experiments from which all data were gained have been carried out using a five-axis, high-frequency CO_2 laser (Trumpf L5000) with a maximum output power of $2.2\,\mathrm{kW}$ [15]. The laser works in duty-cycle mode, which means that the pulse frequency as well as the ratio of pulse and pause can be varied. The pulse frequency was chosen as a constant value of $\omega_L = 10\,\mathrm{kHz}$. The material processed was commercial low-carbon steel (AISI 1008). For simplicity, a restriction to single-layer ablation was made that was carried out track by track with a horizontal offset of $0.1\,\mathrm{mm}$ while the focus diameter of the laser was kept constant at $0.3\,\mathrm{mm}$. The processing always took place in the same direction. Between two ablation tracks the laser was re-positioned while its power was reduced to $20\,\mathrm{W}$ so that no further ablation could take place. The process gas used in all experiments was compressed air with a pressure of $2\,\mathrm{bar}$. By varying the average laser power $\langle P \rangle$ and the feed rate v, the actual laser power $P(t_i)$ and the optical process emissions $F(t_i; N)$ (where N is the number of the current ablation track) were observed with the discrete times t_i sampled with a frequency of $25\,\mathrm{kHz}$. The final surface profiles were gained by a UBM laser triangulation device with depth values $d(x_i, y_j)$, where $x_i = x(t_i)$ is the discrete spatial coordinate along the ablation track belonging to $t = t_i$

sampled with a resolution of $10\,\mu\text{m}$, and y_j is a transverse coordinate sampled with $20\,\mu\text{m}$ in all measurements. The temporal and spatial coordinates are assumed to be directly related via the constant laser feed rate v via $x_i = vt_i$.

26.3 Linear and Nonlinear Data Analysis

In order to develop a process model of laser beam melt ablation a deeper understanding of the underlying physical mechanisms is necessary. Besides purely theoretical investigation, the analysis of data series recorded during processing allows us to prove the validity of empirical models and to develop new modeling strategies. Even though the measurement of only a few carefully chosen observables maps the dynamics from the possibly high-dimensional system of all dynamical variables to a much lower-dimensional phase space, the behavior of the corresponding low-dimensional models may give important information about the dominating physical processes.

Usually, data series gained from measurements are investigated by defining suitable (linear or nonlinear) statistical features. This strategy is valid for time series as well as for the resulting surfaces of the machining process. Moreover, the use of methods of statistical analysis allows us to define certain order parameters that can be used to classify the dynamics in the same way as the direct control parameters of the system can do. The physical meaning of such parameters and the possibility to quantify the relations between online observables and the final surfaces give information about the pattern-formation process and provide a key to gain knowledge about the full physical system necessary for designing appropriate controllers for industrial applications.

Surface Quality. It is found [15] that for high energy inputs (or low feed rates v for fixed laser power $\langle P \rangle$, because $E \sim v^{-1}$) into the workpiece the quality of the surfaces processed declines rapidly, as can be seen in Fig. 26.1. For a global classification, the quality of the processed surfaces can be evaluated using scalar parameters, such as the profile depth P_t and the arithmetic mean roughness S_a defined as follows [16]:

$$P_\text{t} = \max_{x_i,y_j} \{d(x_i,y_j)\} - \min_{x_i,y_j} \{d(x_i,y_j)\} \quad \text{and} \quad S_\text{a} = \frac{1}{n_i n_j} \sum_{x_i,y_j} d(x_i,y_j),$$

where n_i and n_j are the total numbers of sampling points in the x and y directions, respectively, and the sum, minimum and maximum have to be evaluated using all sampling points (x_i, y_j). These parameters which describe the quality of quite regular surfaces rather well lose their meaning for low feed rates (high energy inputs) since in such cases pattern formation becomes so dominant and the surface of the workpiece so poor that these cannot be applied any more. As is shown in Fig. 26.2, in dependence on v at least three different regimes can be found separated by some critical values of v (probably depending on the remaining set of parameter values, especially on $\langle P \rangle$): low profiles for higher feed rates $v > v_2$, deeper ones with not much increased roughness for intermediate values $v_1 < v < v_2$ and very rough ones for low values $v < v_1$. For $\langle P \rangle = 600\,\text{W}$, $v_1 \approx 2$–$3\,\text{m/min}$ and $v_2 \approx 4$–$5\,\text{m/min}$ seem to be reasonable estimates for these critical values.

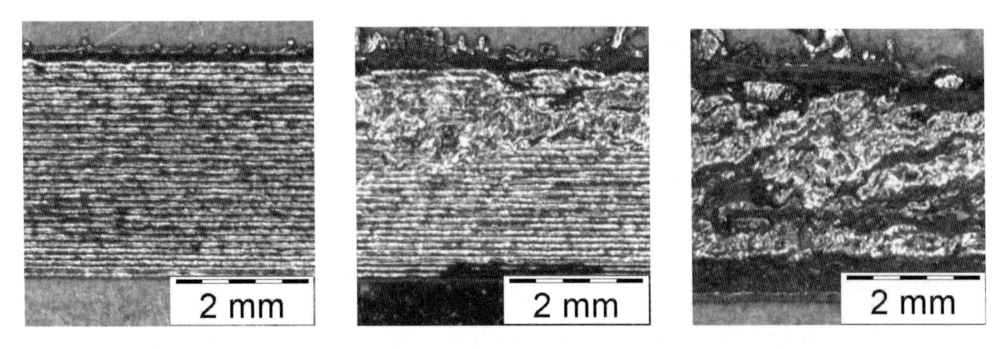

Figure 26.1: Surfaces processed with $\langle P \rangle = 600\,\mathrm{W}$ and $v = 5\,\mathrm{m/min}$ (left), $3\,\mathrm{m/min}$ (middle) and $2\,\mathrm{m/min}$. It can be clearly observed that the regular structures due to trackwise laser processing at high feed rates are successively replaced by growing irregular structures for increasing energy input.

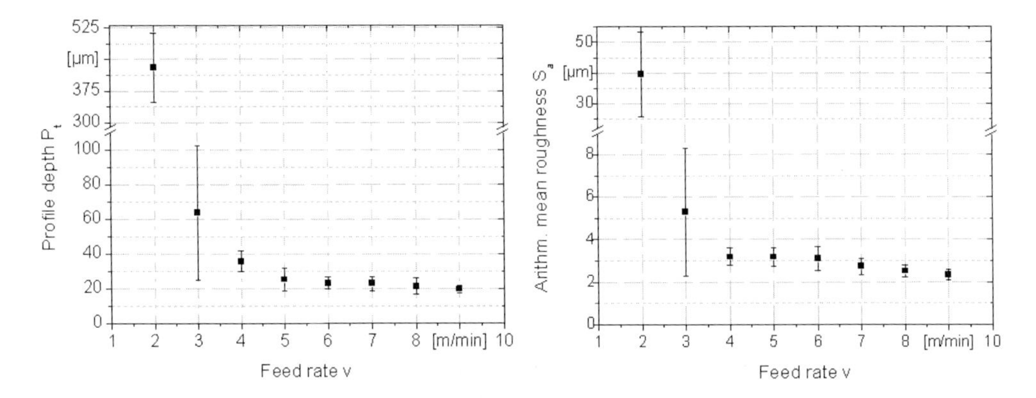

Figure 26.2: Statistical surface parameters and their variances found in experiments with $\langle P \rangle = 600\,\mathrm{W}$. The error bars are calculated from nine different parts of three different surfaces processed with the same set of parameters.

Time-series Analysis. The time series of process emissions and laser power are found not to give sufficient detailed information about the processes in the melt pool. The measured data can be approximated by a linear process with a simple oscillation driven by the laser cycle and a quite high noise level [15]. Both observables fluctuate almost simultaneously on small time scales and have strong linear correlations. Therefore, the assumption that the melt pool dynamics is directly triggered by the laser duty cycle seems reasonable. Nonlinearities of the physical system that can be found by performing a surrogate data analysis [17] or fitting autoregressive models onto the measured data [18] are included as dynamical noise. It is hard to say whether a further improvement of the different measurement techniques will allow us to get results that are significantly better and good enough to establish an online process control in this way.

To acquire additional information, it seems reasonable to try to subtract the laser dynamics from the optical emission data. A very simple approach to do this is to standardize the time series and consider the difference between $F(t_i)$ and $P(t_i)$ as follows:

$$r(t_i) = \frac{f(t_i) - f(t_{i-1})}{t_i - t_{i-1}} - \frac{p(t_i) - p(t_{i-1})}{t_i - t_{i-1}} \quad \text{with } x = \frac{X - \langle X \rangle}{\sigma_X},$$

where X stands for F and P, respectively, $\langle X \rangle$ is the average value and σ_X the corresponding standard deviation of the time series $X(t_i)$. The function $r(t_i)$ is called the (linear) residuum of the time series. Its linear properties are actually more similar to those of Gaussian white noise than those of the original data sets, but the oscillatory component is still occurring, indicating either that $\langle X \rangle$ and σ_X are not able to describe the data distributions completely due to differences from the Gaussian or that the transfer function between the laser input and the optical output of the melt pool has a nonlinear part. Probably, reality is best described by the combination of both cases.

Statistical Order Parameters. For a characterization of the different pattern-formation regimes a number of statistical parameters can be defined from the measured time series. Regarding such order parameters, it is possible to get a deeper insight into the physically relevant processes for different sets of control parameters. Besides the relative variances of the optical emission signals, the following linear parameters are evaluated as a function of the laser feed rate v and the track number N (with constant average input energy $\langle P \rangle = 600$ W):

$$\Gamma = \frac{\langle F(N, t_i) \rangle}{\langle F(N = 1, t_i) \rangle} - 1 \quad \text{and} \quad \gamma = \frac{\langle \dot{f}^2(t_i) \rangle}{\langle \dot{p}^2(t_i) \rangle}.$$

While the parameter Γ is designed to provide information about saturation behavior (and therefore about the stationarity of the physical process in some sense), γ measures the smoothing of the incident laser power in the optical process emissions. Additionally, the Ziv–Lempel complexity ZLC [19, 20] of the emission signals $F(t_i)$ was investigated. This measure quantifies the similarity of a given time series to a purely stochastic process. For its determination, the underlying time series is binary coded with the sign of the corresponding local time derivative. The number of different sequences in this symbolic representation is then divided by the maximally possible number to get a normalized and comparable feature.

The resulting features are shown in Fig. 26.3. As can be seen, there is a saturation behavior of the optical emissions for $N \geq 2$ found in the behavior of σ, Γ and ZLC. For high feed rates and therefore low energy inputs, the fluctuations of the optical emissions correspond directly to those of the current laser power as is shown by γ. The actual boundary conditions due to previous ablation tracks seem to act only via the profile depth $d(x_i, y_j)$ because the amplitude of $F(t_i)$ is probably correlated to d in some (maybe nonlinear) way. The optical emissions get saturated for $N > 2$ at high feed rates and for $N = 2$ at intermediate values of v. The high values of Γ for low-energy inputs indicate an increasing loss of laser radiation with increasing feed rate $v > v_2$ and therefore less and more regular material ablation, as found in Fig. 26.2. For small feed rates and therefore high energy inputs into the workpiece, no saturation of the process emissions occurs. This can be explained by the fact that the material has no possibility to cool down completely between two ablation tracks so that the surface temperature and therefore the emission amplitudes increase further and further. With this successive increasing, the emissions become more and more independent of the actual forcing

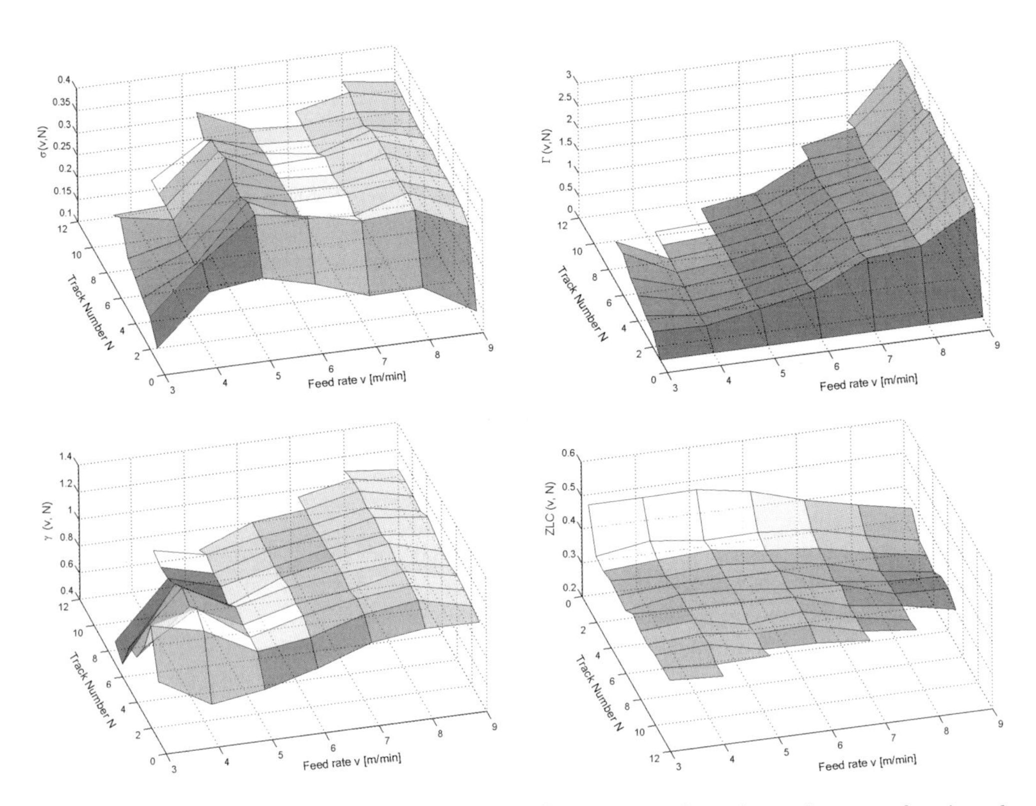

Figure 26.3: Statistical parameters describing the different pattern-formation regimes as a function of feed rate v and track number N. Upper figures: standard deviation of optical process emissions (left) and $\Gamma(v, N)$ (right). Lower figures: $\gamma(v, N)$ (left) and Ziv–Lempel complexity ZLC (right).

and seem therefore to be smoothed (low values of γ). The complexity of the optical process signals decreases almost monotonically with increasing N and v, indicating that the process is mostly stochastic (or mostly chaotic) for the first tracks and for the lowest energy inputs while becoming more regular with increasing energy density and track number.

Besides the different evaluation procedure for the ZLC with respect to [15], it has to be pointed out that an important premise occurring in the previous literature has to be corrected. For laser beam melt ablation, the combined parameter line energy, defined as $E_\mathrm{l} = \alpha\langle P\rangle/v$ with $\alpha \approx 0.1$ being the absorption coefficient of the system CO_2 laser–steel at the laser wavelength $\lambda = 10.6\,\mu\mathrm{m}$, does *not* provide sufficient information about the process as is the case for different laser machining mechanisms. This can be seen clearly by regarding the statistical parameters described above for constant line energy but different values of v and $\langle P\rangle$ as is shown in Fig. 26.4. Obviously, the values for different parameter sets yielding the same line energy are clearly outside the confidence intervals – estimated using the standard deviations – of each other. The reason for this is probably given by the highly competitive interaction of thermodynamical (described by $\langle P\rangle$) and (fluid-) mechanical (including v) processes.

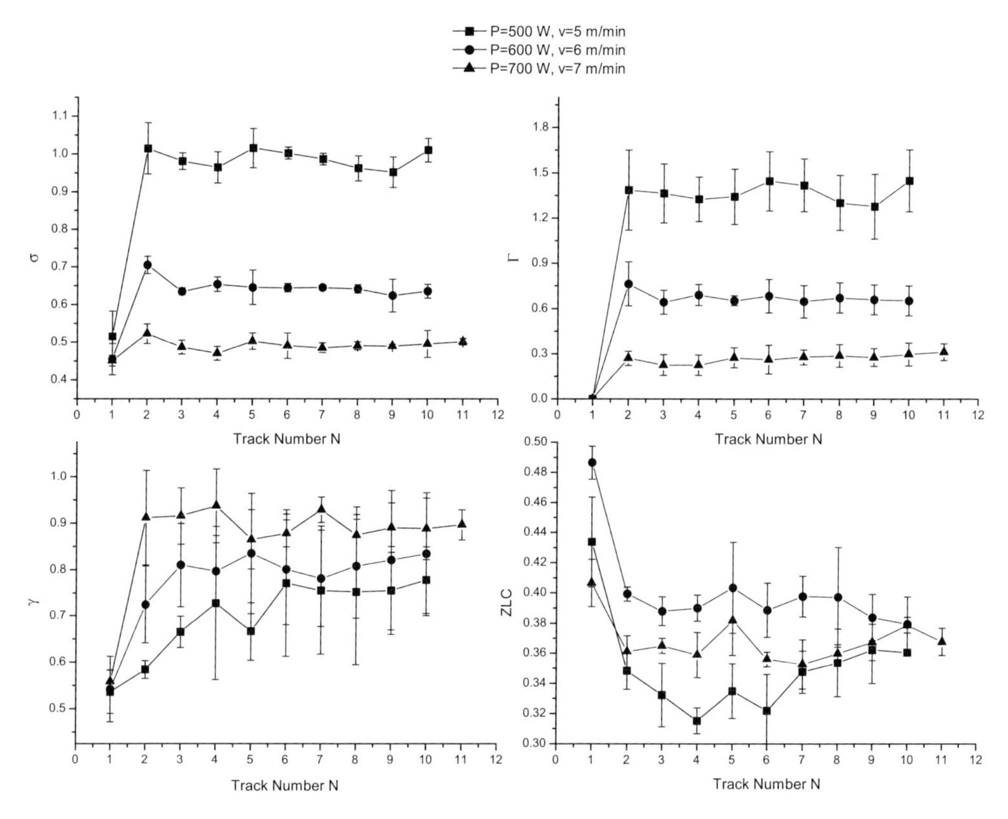

Figure 26.4: Statistical parameters for constant line energy $E_l = 600\,\mathrm{J/m}$ and varying parameter sets of feed rate v and average laser power $\langle P \rangle$. Error bars have been calculated from the variance of these parameters gained from three experiments with the same respective sets of process parameters.

Linear Correlation Analysis. As was shown, different regimes of process behavior can be found in the optical emissions $F(t_i; N) = F_{iN}$ during processing as well as in the final surfaces represented by the depth values $d(x(t_i), y_j) = d_{ij}$. Therefore it seems a promising idea to investigate the linear correlations within the ablation profiles $C(d_{ij}, d_{kl}; t_i - t_k)$, between the time series $C(F_{iN}, F_{jN'}; t_i - t_j)$ and between both kinds of data sets $C(d_{ij}, F_{kN}; t_i - t_k)$. It is found that there is indeed a weak dependence between different tracks in the profile data as well as between time series belonging to neighboring ablation tracks ($N' = N \pm 1$). Regarding the $F(t_i)$ signals, these correlations become more intensive with increasing track number N. In the profile data the correlations reach local maximum values at each integer multiple of the transverse horizontal offset of the laser, supporting the clear relationship between the dynamics of neighboring ablation tracks found in the optical emission signals. On the other hand, the corresponding absolute values have no sufficient statistical significance. The linear cross correlations between (normalized) emission amplitudes and profile depths $C(d_{ij}, F_{kN}; t_i - t_k)$ are even weaker. Therefore, it is an obvious conclusion that at least for the given sampling rates, the knowledge of the actual boundary conditions due to previous tracks and of the current emission amplitude probably does not give sufficient information for predicting the resulting ablation profiles. In this case, establishing an online process control by regarding only linear correlations (and so only linear processes) is not possible.

It is an important task for future work to determine whether a substantial increase of the spatio-temporal resolution of the online data and/or the investigation of nonlinear correlations between the possible observables allow for a significantly better understanding of the process dynamics.

Surface Profiles. By characterizing the surfaces only with scalar parameters, the fact that several overlapping ablation tracks have been performed to get a visible result is neglected. To get a deeper insight into the dynamics of the ablation process, it is necessary to statistically evaluate the full microscopic surface profiles by methods of data analysis. For this purpose, a laser triangulation device was used for measurements. By successive scanning parallel to the ablation track a three-dimensional map of the profile is gained. In the transverse direction with respect to the single ablation tracks, the offset between the neighboring tracks is clearly recovered. Along the different ablation tracks, the profile seems to be irregular. The end of any track is characterized by a positive height value indicating some remaining material that was pushed to each side by the gas pressure hitting the workpiece under an angle of $\phi = 90°$. The corresponding profile depth decreases almost linearly on a scale of $0.5\,\mathrm{ms}$. With a typical profile depth of $d \approx 50$–$100\,\mu\mathrm{m}$, one finds the inclination angle of the front to be clearly different from zero with $\vartheta \approx 10$–$20°$.

As is known from many previous investigations the ablation of fluid material from the melt pool is no continuous process but occurs in the form of discrete particles. On the other hand, the observed profiles in the direction of processing are very rough with relatively sharp ridges (local maxima of profile depth) as is shown in Fig. 26.5. As a hypothesis (that is hard to be proven) one can assume that any ridge along the ablation track corresponds to a single removal event of a single melt particle. In this case, one can investigate the statistical features of the distribution of the ablation points

$$\delta_k = \arg\left(\max_{\mathrm{loc}}(d(t_i; y_j))\right)$$

or, alternatively, of the times $\Delta t_k = \Delta x_k / v$ with $\Delta x_k = \delta_k - \delta_{k-1}$ between the removals of two melt particles forming one after the other. Together with the corresponding values of the profile depth in the interval $\{x \in [\delta_{k-1}, \delta_k]\} \times \{y = y_j\}$, Δx_k can be taken to estimate the dimensions of the ablated particles.

The distribution of the Δx_k for one particular experiment is shown in Fig. 26.5. One observes that there are only a few particles of smaller size. One particular reason for this is that the workpiece temperature increases more or less continuously due to the laser energy input. Hence, the melting of the material takes place continuously as well. On the other hand, material properties important for the ablation process (such as the surface tension of the melt pool) are probably strongly temperature dependent as well. Especially, smaller melt pools might be expected to have lower temperatures and higher surface tensions than larger ones. If this is correct, they are harder to remove from the processed surface. A second observation concerning the distribution function is that for higher values of the Δx_k, the corresponding frequency of occurrence decreases almost exponentially. This is a typical feature of Poissonian processes

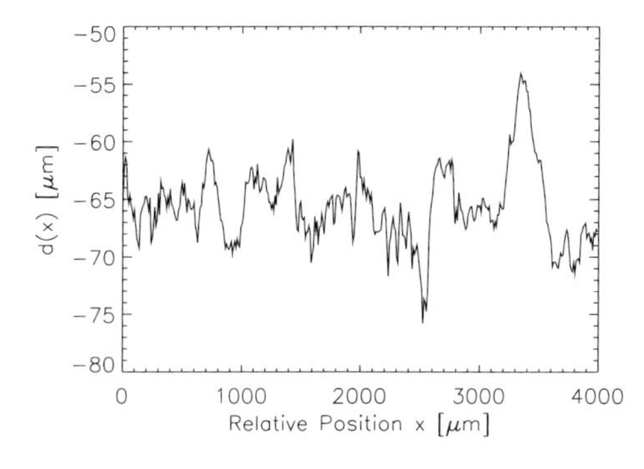

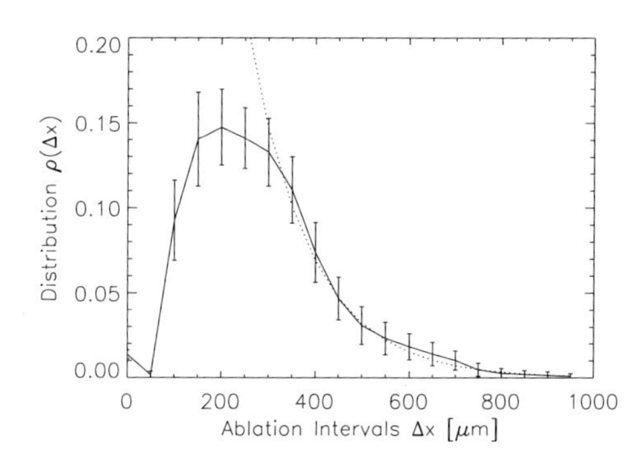

Figure 26.5: Left: part of a track profile in the processing direction. Right: distribution of the distances of neighboring ablation points found in experiments. Error bars are estimated by taking the variance over the histograms of 120 different tracks measured on one processed workpiece. The dotted line corresponds to an exponential fit.

where the single events are completely randomly distributed [21]. A possible explanation for this is that the ablation process itself is fundamentally controlled by the gas flow and the momentum transfer to the melt. As an outflow from a nozzle, this gas flow is spatio-temporally turbulent itself. The degree of turbulence is even enhanced due to its interaction with the rough surface. Therefore, finite-time samplings may arbitrarily well mimic a stochastic process. A strong dependence of the processing results of the nozzle design and gas pressure was previously found and described in several studies for laser cutting [22, 23] and melt ablation [14, 24].

26.4 A Stochastic Process Model

The results of the previous section give rise to the assumption that hydrodynamic processes, especially the interactions between melt pool and gas flow, have a dominating influence on the pattern formation via the ablation of discrete melt particles. To improve the knowledge about the ablation process (which is necessary in order to finally develop a strategy for an online process control) it is worth trying to describe the ablation dynamics by a simple conceptual model to extract the basic features. Assuming an approximately stochastically controlled material removal due to the action of the spatio-temporally turbulent gas flow, a model of minimum complexity would be given by a one-dimensional equation of motion including the different forces acting on the melt pool. Because thermodynamic processes seem to have less importance for the ablation dynamics itself, they are included only parametrically in the model. Especially, heat conduction will not be considered explicitly which is a major difference to melt-flow models of higher complexity as developed in previous studies for laser grooving and cutting [10, 11, 25, 26].

The heating due to the laser irradiation allows surface temperatures above the melting point of the workpiece. If the energy density coupled into the material is high enough, the phase transition from solid to liquid takes place in a certain surface layer. After a sufficiently long processing time allowing the physical system to get into a quasi-stationary equilibrium (the corresponding system is sketched in Fig. 26.6), the molten layer can be approximated as a

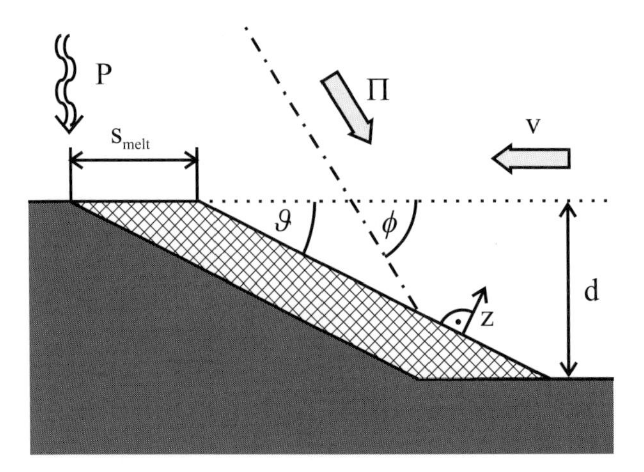

Figure 26.6: Sketch of the typical system parameters important for the ablation process in the membrane model described.

two-dimensional membrane, if its depth s_{melt} is sufficiently small compared to its transverse diameter $d/\sin\vartheta$, where d is the typical depth of the resulting ablation track. This membrane will be roughly modeled by a single massive point particle. To include some dimensional units for forces and particle mass, for simplicity a half-circular track profile is assumed (a different geometry will result in some constant factors in certain contributions to the model equations). In this case, $d \simeq d_{\text{f}}/2$, where d_{f} is the width of the track given approximately by the focus width of the laser for a Gaussian beam profile.

For small inclination angles ϑ of the melt front, it can be assumed that the laser forces the membrane to oscillate in first order normally with respect to the ablation front (z direction in Fig. 26.6). In this case, a one-dimensional formulation of this transverse motion is valid. An upper layer of the membrane is ablated if a certain threshold value for the oscillation amplitude is reached. If the typical time scale between two particle removals is small compared to the dynamical time scale s_{melt}/v, the size of this layer is $s \ll s_{\text{melt}}$. It seems reasonable that the average size $\langle s \rangle$ of the ablated particles is determined by the (temperature-dependent) surface tension of the material (for a given forcing). With respect to the constant laser feed rate v, s will be parametrized linearly in terms of the processing time t. Each ablation event is assumed to correspond to a ridge on the resulting surface. The exact form of the profile has to be described by thermodynamic relations as well. After any ablation event, the surface tension of the molten material causes the melt pool to form a new membrane with minimum surface on a very short time scale. Thus, melting and material removal can form a certain process cycle with irregularly distributed time intervals.

Due to the periodic exciation by the laser, the melt pool is forced to oscillate. In the limit of small amplitudes, these oscillations might be approximately harmonic. To describe nonlinearities and to allow for an ablation ($z \to \infty$) with a temperature-dependent threshold value, the oscillation potential $U(z)$ has to be metastable with a unique attractive region of finite size representing the melt pool. This can be achieved by adding a nonlinear term to a harmonic potential that models the influence of cohesion forces on the motion of the upper layer of the melt pool. The exact form of the nonlinear part is hard to determine; its physical significance is therefore neglected at this point. As long as the minimum and maximum of $U(z)$ are both unique, the qualitative features of the escape-time distribution (which is the statistical property easiest to be extracted from experimental surface data) should remain the same. As the possibly simplest and well-studied paradigmatic model [27–29], the potential $U(z)$ will be chosen to be of Helmholtz type, resulting in a quadratic nonlinearity of the potential force. Assuming linearly temperature-dependent activation energy $E_A(T)$ and membrane elasticity (surface tension) $k(T)$ (decreasing for increasing T for $T \gtrsim T_{melt}$), and setting $z = 0$ to be the equilibrium position of the membrane, the potential may have the form

$$U(z) = -\sqrt{\frac{k^3}{54E_A}} z^3 + \frac{k}{2} z^2 \text{ with } E_A(T) = E_0 \frac{T_{evap} - T}{T_{evap} - T_{melt}}$$

$$\text{and } k(T) = k_0 + \left.\frac{dk}{dT}\right|_{T=T_{melt}} T,$$

with $k_0 = k(T_{melt}) - dk/dT(T = T_{melt}) \cdot T_{melt}$ for $T_{melt} \lesssim T \ll T_{evap}$. This parametrization of $k(T)$ corresponds to a first-order Taylor expansion around the melting point T_{melt} of the material. $E_0 = E_A(T_{melt})$ is the activation energy necessary to overcome cohesion forces in the liquid layer at T_{melt}, and T_{evap} is the corresponding evaporation temperature at which the material is ablated instantaneously. For $T \gtrsim T_{melt}$, $U(z)$ is regular and provides a reasonable model if $s_{melt} \sim \mathcal{O}(z_0)$ with $z_0 = \sqrt{6E_A(T)/k(T)}$ representing the local maximum of the potential with $U(z_0) = E_A$. For weak amplitudes, oscillations occur with a temperature-dependent frequency $\omega(T) = \sqrt{k(T)/m(T)}$, where $m(T) = \rho_{fl}(T)\pi d^2 s_{melt}/2$ is the total mass of the molten layer and $\rho_{fl}(T)$ is the mass density of the molten layer that can be parameterized similar to $k(T)$.

The action of the periodic forcing enters the dynamics in two different ways. First, the laser photons cause a weak radiation pressure $\langle \Pi_{rad} \rangle$ resulting in a radiation force

$$F_{rad} \simeq \langle \Pi_{rad} \rangle \frac{\pi}{2} \frac{d^2}{\tan \vartheta} f(t)$$

on the melt pool, where $f(t)$ is the temporal profile of the laser power. Under general conditions, this term might be neglected. In a second way, the forcing also leads directly to a modulation of T (for example, $T(t) = T_0(1 + \epsilon g(f(t))$ where $g(\cdot)$ with $\max |g(\cdot)| \ll 1$ is a function transferring the modulations of the laser power into a time-periodic temperature signal), which results in a temporal modulation of almost all model parameters.

Beside the laser irradiation, the motion of the membrane is also driven by the turbulent contribution $\xi(t)$ of the gas flow at the surface of the workpiece and restored by the average gas pressure $\langle \Pi_{hyd} \rangle$ and by gravity. In physical terms, the gas flow acts against the cohesion force of the melt pool and causes at first a certain roughness (due to its shearing character) and second the removal of melt particles. This might be expressed by the following contributions:

$$F_{pres} = \xi(t) - \frac{\pi}{2} \frac{d^2}{\tan \vartheta} \langle \Pi_{hyd} \rangle \sin \phi \theta(\phi - \vartheta) \text{ and } F_{grav} = -mg \cos \vartheta,$$

where $\theta(\phi - \vartheta)$ is the Heaviside function indicating that the molten layer is influenced by the average gas pressure only if $\phi > \vartheta$ (see Fig. 26.6). The turbulent force contribution $\xi(t)$

reflects an ad hoc approximation of the fluctuating part of the gas flow at its interaction with the molten material in a rough and temporally variable boundary layer according to the results of the surface analysis that indicate an approximately stochastic ablation process. The transverse equation of motion for the membrane

$$m\frac{d^2z}{dt^2} + \frac{d}{dz}U(z) = F_{\text{pres}} + F_{\text{grav}} + F_{\text{rad}}$$

is thus given as a nonlinear stochastic differential equation of the following form:

$$\frac{\pi d^2 s_{\text{melt}}}{2}\left(\rho_0+\frac{d\rho}{dT}T\right)\frac{d^2z}{dt^2} + \left(k_0+\frac{dk}{dT}T\right)z - \sqrt{\frac{T_{\text{evap}}-T_{\text{melt}}}{6E_0(T_{\text{evap}}-T)}\left(k_0+\frac{dk}{dT}T\right)^3}\, z^2$$
$$= \xi(t) + \frac{\pi d^2}{2\tan\vartheta}\left(\langle\Pi_{\text{rad}}\rangle f(t) - \langle\Pi_{\text{hyd}}\rangle \sin\phi\theta(\phi-\vartheta)\right) - mg\cos\vartheta.$$

For a highly turbulent gas flow, the regular terms on the right-hand side of this equation may be neglected. Then, in dimensionless (dashed) quantities the equation of motion has the simple general form

$$\frac{d^2z'}{dt'^2} + \alpha(T)z' + \beta(T)z'^2 = \xi'(t'),$$

with certain temperature-dependent coefficients α, β. Therefore, T is a fundamental control parameter of this system if the corresponding dependences of k, ρ and E_A are prescribed. The fact that T is not regarded as a dynamic variable (as is actually the case) and is only parametrically modulated is a strong simplification of the model leading to certain restrictions of its validity. To be more exact, one also has to investigate the probably much more complex temperature dependences of the different model parameters experimentally. Moreover, an additional damping term should be included which is probably very small. If necessary, the dynamic behavior of the surface temperature can be prescribed by calculating the heat balance of the system. In this way the model can be extended easily. A possible further extension to three dimensions would allow us to investigate probably existing Marangoni-like surface instabilities of the molten layer that would contribute to the dynamics of the system in a certain way as well. Corresponding investigations have to be done in future work.

The statistical features of the dynamics of this kind of stochastically forced system have been investigated with an additional sinusoidal forcing [30] without parametric excitations. On the other hand, the dynamics of a purely deterministic Helmholtz-like oscillator system with parametrically modulated linear terms was the subject of different studies as well [31–33]. As a first step to further analysis, it is checked whether this kind of system actually shows statistical features of the escape-time distribution similar to the experimental observations. To achieve this task, ensembles of simulations with arbitrary initial conditions allow us to determine the typical escape times

$$\Delta t_k = \tau_k - \tau_{k-1} \text{ with } \tau_k = \min\arg(z(t_i)|z(t_i) \geq z_0)$$

of the model referring to the mean first passage times (FPT) at $z = z_0$ of single particles modeling the membrane oscillations with the initial condition $t_{\text{init}} = 0$. For simplicity, $\xi'(t)$ was

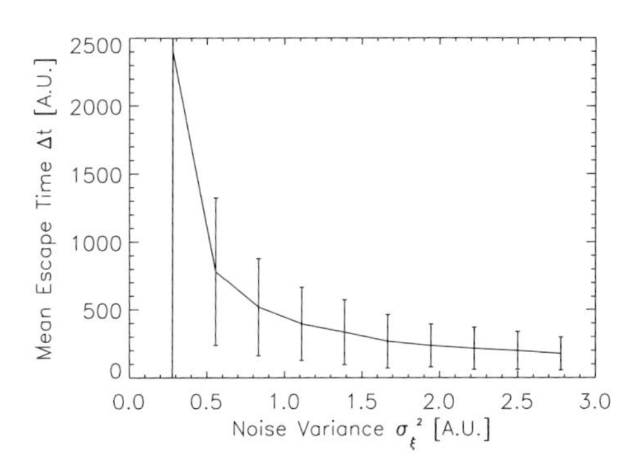

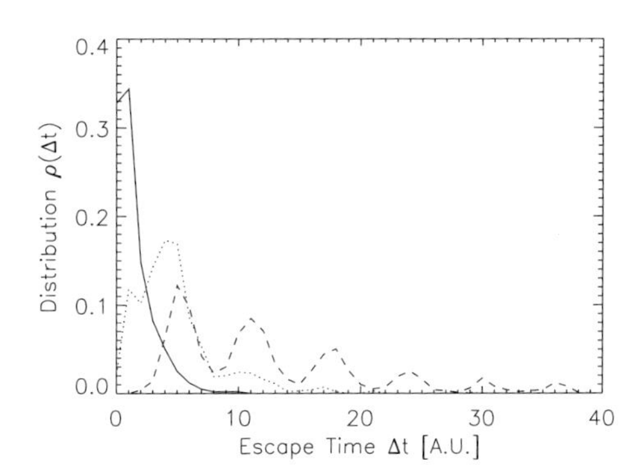

Figure 26.7: Left: mean first passage times and corresponding standard deviations (error bars) for the purely stochastically forced system. The values are estimated from 1000 simulation runs for each value of the noise amplitude (measured by its variance σ_ξ^2 – the corresponding units in the figure have been enlarged by a factor of 1000). Right: distribution of first passage times at $z = z_0$ in the periodically forced ($f(t) = \cos^2 \omega_L t$) stochastic model for $\sigma_\xi = 0.0045$ (solid), 0.0015 (dotted) and 0.0005 (dashed).

chosen to be normally distributed Gaussian noise with mean value $\langle \xi' \rangle = 0$. Moreover, because of $T \gtrsim T_{\text{melt}}$ the temperature and therefore all other model parameters were assumed to be constant in time. The simulations showed that the average escape times and the corresponding variances decrease monotonically with increasing amplitude of "noise", as is shown in Fig. 26.7. This result is well known from investigations of general escape processes [34, 35].

Additional periodic forcing terms are often able to stabilize stochastic processes in the sense of allowing more regular statistical features. For different kind of systems [36, 37] it is known that escape-like monotonically decaying temporal distribution functions become modulated with the frequency of the periodic forcing. A similar effect happens for the Helmholtz oscillator system without modulated parameters, as shown in Fig. 26.7 for a sinusoidal forcing. The strengths of periodic and stochastic forcings have to be balanced to allow a maximum regularity of escape. In particular, if the maximum of the non-modulated distribution function (depending on the variance of the stochastic forcing) coincides with the period of the additional forcing, a maximum number of particles escape after a time around this period, and the process is "more regular". For high "noise" amplitudes and high-frequency excitation, this may result in a substantial improvement of surface quality. As a suggestion for further experiments, this proposition should be tried to be proven in applications.

26.5 Discussion

As was shown in [15], the fluctuations of the optical emissions from the melt pool can be well described by an auto-regressive model of low order. Hence, the measured radiation flux may be interpreted as the original laser signal after a low-pass filtering that occurs due to a certain physical process in the melt pool that still has to be determined. Because of this, with the present temporal resolution the optical emissions may not be taken for the development of an online process control mechanism. The same statement can be assumed to hold for other possible thermodynamic observables as well.

The surface profile shows an exponential distribution for the distances between neighboring maxima. This observation suggests an approximately stochastically controlled material ablation when identifying ridges which are ablation points of discrete particles. Therefore, under certain assumptions a minimum model for the ablation dynamics might be given by a massive point particle evolving in a Helmholtz potential with stochastic external forcing. This particular model allows for an escape of (melt) particles from an attractive region (the melt pool). The attractivity is quantified in terms of the temperature-dependent surface tension of the molten layer. The highly turbulent gas flow at the rough boundary layer of the melt pool is the physical realization of the stochastic force.

As is observed, ablation can only take place in discrete portions due to the surface tension of the melt pool, so that an ideal flat surface profile cannot be reached with this method of processing. As the investigation of the stochastic model shows, there might be a possibility to get smaller and more regularly distributed ablation events (and therefore surfaces of a better quality) by taking a high-frequency external modulation of the process and a well-balanced magnitude of turbulent gas flow fluctuations to be controlled by arrangement and design of the gas nozzle. This suggestion has to be proven experimentally.

To reject this possibly too simple process model, more detailed measurements with higher spatial resolution are required. At the moment, neither the temporal resolution of the optical emission signals $F(t_i)$ nor the spatial resolution of the surface profiles are sufficient for developing a detailed description of the physical process. Such data with a higher spatial resolution are necessary to make all results even more reliable. For example, investigations of the exact forms of the ridges could help to understand more physical details contributing to the ablation mechanism. One would expect a certain smoothing of the profiles due to thermic and surface-tension effects. The minimum cutoff of the distribution function might be related closely to the actual surface tension of the molten material and could therefore provide a way to estimate the surface temperature. On the other hand, this property can be investigated as well pyrometrically or by measuring thermic spectra of the process emissions. Finally, it seems possible to detect ablation events in the $F(t_i)$ time series as well where ablated particles can be recovered in the form of sparks occurring as some very high and very short peaks in the emission amplitudes.

Besides further experimental investigations, it remains a task for the near future to analyze both models and data series in an advanced way. The use of new statistical modeling approaches such as unscented Kalman filtering (UKF) [38] may allow us to determine the amplitudes of the stochastic contributions and to estimate parameter values for the nonlinear differential equation of the process model. Hence, it can show reasonable approaches to make the stochastic model more reliable. This may be used for developing more detailed techniques for finally establishing an online process control.

Acknowledgment

The authors would like to gratefully acknowledge the financial support of the Volkswagen Foundation. Helpful discussions with Ekkehard Ullner provided many ideas realized in this work. Thanks are also due to an anonymous referee for his suggestions to improve this contribution substantially.

Bibliography

[1] G. Chryssolouris, *Laser Machining – Theory and Practice,* Springer, New York, 1991.

[2] W.M. Steen, *Laser Material Processing,* 2nd edn. Springer, London, 1998.

[3] N.B. Dahotre (ed.), *Lasers in Surface Engineering,* ASM International, Materials Park, 1998.

[4] S. Schuberth (ed.), *Präzisionsbearbeitung mit CO_2-Laserstrahlung (Abtragen),* VDI, Düsseldorf, 1996.

[5] M. Beck, *Modellierung des Lasertiefschweissens,* Teubner, Stuttgart, 1996.

[6] W.W. Duley, *Laser Welding,* Wiley-Interscience, New York, 1999.

[7] J. Michel, S. Pfeiffer, W. Schulz, M. Niessen, and V. Kostrykin, *Approximate model for laser welding,* Chapter 24, this book.

[8] M.F. Modest and H. Abakians, *Evaporative cutting of a semi-infinite body with a moving cw laser,* Transactions of the ASME – Journal of Heat Transfer **108** (1986), 602–607.

[9] J. Powell, *CO_2 Laser Cutting,* Springer, London, 1998.

[10] W. Schulz, V. Kostrykin, M. Niessen, J. Michel, D. Petring, E.-W. Kreutz, and R. Poprawe, *Dynamics of ripple formation and melt flow in laser beam cutting,* Journal of Physics D: Applied Physics **32** (1999), 1219–1228.

[11] W. Schulz, V. Kostrykin, J. Michel, and M. Niessen, *Modelling and simulation of process monitoring and control in laser cutting,* Chapter 23, this book.

[12] U. Sutor, Ph.D. Thesis, Aachen (1994).

[13] B. Lässiger, Ph.D. Thesis, Aachen (1994).

[14] D. Schubart, Ph.D. Thesis, Bamberg (1999).

[15] A. Cser, R. Donner, U. Schwarz, A. Otto, M. Geiger, and U. Feudel, *Towards a better understanding of laser beam melt ablation using methods of statistical analysis,* in: R. Teti (ed.): Proceedings of 3rd CIRP International Seminar on Intelligent Computation in Manufacturing Engineering (ICME 2002), Ischia, 2002, pp. 203–208.

[16] Surface parameters are defined according to the German norms *DIN V 32540: Thermisches Abtragen mit dem Laserstrahl* (Beuth, Berlin, 1997) and *DIN EN ISO 4287: Geometrische Produktspezifikationen (GPS) – Oberflächenbeschaffenheit: Tastschnittverfahren – Benennungen, Definitionen und Kenngrössen der Oberflächenbeschaffenheit* (Beuth, Berlin, 1998) applied to two-dimensional surface data.

[17] J. Theiler, S. Eubank, A. Longtin, B. Gladrikian, and J.D. Farmer, *Testing for linearity in time series: The Method of surrogate data,* Physica D **58** (1992), 77–94.

[18] W.W.S. Wei, *Time Series Analysis,* Addison-Wesley, Redwood City, 1990.

[19] A. Lempel and J. Ziv, *On the complexity of finite sequences,* IEEE Transactions in Information Theory **IT-22** (1973), 75–81.

[20] F. Kaspar and H.G. Schuster, *Easily calculable measure for the complexity of spatiotemporal patterns,* Physical Review A **36** (1987), 842–848.

[21] A. Papoulis, *Probability, Random Variables, and Stochastic Processes,* 3rd edn., MacGraw-Hill, Boston, 1991.

[22] H. Zefferer, D. Petring, and E. Beyer, *Investigations on the gas flow in laser beam cutting,* in: Vorträge und Posterbeiträge der 3. Internationalen Konferenz Strahltechnik. (DVS-Berichte **135**), Deutscher Verlag für Schweisstechnik, Düsseldorf, 1991, pp. 210–214.

[23] R. Edler, P. Berger, and H. Hügel, *Performance of various nozzle designs in laser cutting,* in: Proceedings of Laser Treatment of Materials 1992, DGM Informationsgesellschaft, Oberursel, 1992, pp. 111–116.

[24] D. Schubart, and A. Otto, *Process development of laser melt ablation,* in: M. Geiger and F. Vollertsen (eds.), Proceedings of Laser Assisted Net Shape Engineering 2 (LANE '97), Meisenbach, Bamberg, 1997, pp. 865–876.

[25] W. Schulz, W.: Ph.D. Thesis, Aachen (1992).

[26] W.C. Choi and G. Chryssolouris, *Analysis of the laser grooving and cutting processes,* Journal of Physics D: Applied Physics **28** (1995), 873–878.

[27] J.M.T. Thompson, *Chaotic phenomena triggering the escape from a potential well,* Proceedings of the Royal Society of London, Series A: Mathematical, Physical and Engineering Sciences **421** (1989), 195–225.

[28] J.A. Gottwald, L.N. Virgin and E.H. Dowell, *Routes to escape from an energy wall,* Journal of Sound and Vibration **187** (1995), 133–144.

[29] S. Lenci and G. Rega, *Optimal control of homoclinic bifurcation: theoretical treatment and practical rReduction of safe basin erosion in the Helmholtz oscillator,* Journal of Vibration and Control **9** (2003), 281–315.

[30] S. Linkwitz and H. Grabert, *Enhancement of the decay rate of a metastable state by an external driving force,* Physical Review B **44** (1991), 11901–11910.

[31] M.A.F. Sanjuán, *Homoclinic bifurcation sets of driven nonlinear oscillators,* International Journal of Theoretical Physics **35** (1996), 1745–1752.

[32] R. Chacón, F. Balibrea, and M.A. López, M.A.: *Role of parametric resonance in the inhibition of chaotic escape from a potential well,* Physics Letters A **235** (1997), 153–158.

[33] F. Balibrea, R. Chacón, and M.A. López, *Inhibition of chaotic escape by an additional driven term,* International Journal of Bifurcation and Chaos **8** (1998), 1719–1723.

[34] H.A. Kramers, *Brownian motion in a field of force and the diffusion model of chemical reactions,* Physica **7** (1940), 284–304.

[35] P. Hänggi, P. Talkner, and M. Borkovec, *Reaction-rate theory: fifty years after Kramers,* RevReviews of Modern Physics **62** (1990), 251–340.

[36] A.R. Bulsara, S.B. Lowen, and C.D. Rees, *Cooperative behavior in the periodically modulated Wiener process: noise-induced complexity in a model neuron,* Physical Review E **49** (1994), 4989–5000.

[37] J.J. Brey, J. Casado-Pascual, and B. Sánchez, *Resonant behavior of a Poisson process driven by a periodic signal,* Physical Review E **52** (1995), 6071–6081.

[38] A. Sitz, U. Schwarz, J. Kurths, and H.U. Voss, *Estimation of parameters and unobserved components for nonlinear systems from noisy time series,* Physical Review E **66** (2002), 016210.

27 Dynamics-based Monitoring of Manufacturing Processes: Detection of Transitions Between Process States

E. Govekar, J. Gradišek, I. Grabec, A. Baus, F. Klocke, M. Geisel, and M. Gieger

Monitoring of manufacturing processes often requires detection of a transition between different process states. If the transition is accompanied by a change of process dynamics it can be effectively detected using a dynamics-based characteristic. For this purpose the article proposes a nonlinear characteristic called the coarse-grained information rate (CIR) which measures predictability of process dynamics. Applicability of the CIR is demonstrated by detection of transitions in three manufacturing processes: turning, grinding and laser-beam welding.

27.1 Introduction

The fundamental task of process monitoring is recognition of a process state or detection of a transition from one state to another. In dynamic processes, transitions between different process states may or may not be associated with a detectable change of process dynamics. In manufacturing, both cases occur. Gradual wear of the cutting tool, for example, usually increases the mean cutting force but is only weakly reflected in the dynamics of force fluctuations until the tool is worn out. In contrast, the onset of chatter vibration in cutting not only increases the amplitude of machine tool–workpiece vibrations, but also significantly changes their dynamics. From the point of view of monitoring, a transition between process states accompanied by a change of process dynamics is a favorable situation, since such an event is most probably observable by a variety of monitoring methods and, consequently, easier to detect. When choosing the detection method for the transition at hand it is advantageous to consider methods based on process dynamics because they will be more robust and less influenced by the variation of process parameters which do not affect the transition. In fact, sometimes the detection method can be derived solely from the fundamental dynamics properties of the transition and independently of other process properties. Such a dynamics-based detection method can also be applied to other processes which exhibit dynamically similar transitions.

In this article, a dynamics-based detection method is applied to detect transitions between two operating regimes in three different manufacturing processes: turning, grinding and laser-beam welding. One of the two operating regimes is always favorable with respect to the workpiece quality and the other unfavorable, so that the transitions indicate improvement or deterioration of the workpiece quality. The transitions considered are: from chatter-free to chatter cutting regime in turning and grinding, and from deep- to shallow-penetration welding regime in laser-beam welding. Detailed analyses of process dynamics showed that the three

Nonlinear Dynamics of Production Systems. Edited by G. Radons and R. Neugebauer

ISBN 3-527-40430-9

transitions share a common property: they are all accompanied by an increase of predictability of the measured physical quantity [5, 2, 3]. The proposed detection method employs the information rate of the process as an indicator of the transition. Calculated from the recorded time-series data, a significant increase of the information-rate value indicates the transition in all three examples.

27.2 Information Rate

The information rate is a quantity defined in information theory. In the following, some basic definitions are briefly reviewed. More details can be found in [1].

Consider a discrete random variable X which can assume any value from a set $\Omega = \{x_1, x_2, \ldots, x_n\}$, with a probability distribution denoted as $p(x) = P(X = x)$. The entropy of the variable X is defined as:

$$H(X) = -\sum_{x \in \Omega} p(x) \log p(x) . \tag{27.1}$$

$H(x)$ quantifies the average amount of information gained from observing which x_i from Ω actually occurred in the experiment.

Consider now two random variables, X and Y, with corresponding sets of values Ω_X and Ω_Y, individual probability distributions $p(x)$ and $p(y)$, and a joint probability distribution $p(x, y) = P(X = x, Y = y)$. Analogously to Eq. (27.1), the joint entropy of X and Y is defined as:

$$H(X, Y) = -\sum_{x \in \Omega_X} \sum_{y \in \Omega_Y} p(x, y) \log p(x, y) . \tag{27.2}$$

The average amount of information about the variable Y contained in the variable X is measured by the mutual information $I(X; Y)$:

$$I(X; Y) = H(X) + H(Y) - H(X, Y) = \sum_{x \in \Omega_X} \sum_{y \in \Omega_Y} p(x, y) \log \frac{p(x, y)}{p(x)p(y)} . \tag{27.3}$$

In order to define the information rate, let us turn to dynamic systems for which the variables X and Y are time dependent, $X(t)$ and $Y(t)$. For many experimental dynamic systems, such as a machine tool–workpiece assembly, all variables describing their dynamics are neither known nor measurable. However, as shown by chaos theory [13, 10], it may often suffice to have access to a single variable which meaningfully reflects the system dynamics.

Now, let $X(t)$ denote the measured variable and $x(t)$ its value sampled at discrete times $t = i\Delta t$, $i = 1, 2, \ldots$. Replacing variables X and Y by $x(t)$ and its time-delayed value $x(t + \tau)$, respectively, the norm of the mutual information can be defined as [16]:

$$\| I\left(x(t); x(t + \tau)\right) \| = \frac{1}{\tau_{\max}} \sum_{\tau = \Delta\tau}^{\tau_{\max}} I\left(x(t); x(t + \tau)\right) \Delta\tau , \tag{27.4}$$

with $\Delta\tau = \Delta t$ as the usual choice for the delay increment $\Delta\tau$. The maximal time delay $\tau_{\max}$ should be chosen such that $I(x(t); x(t + \tau)) \approx 0$ for $\tau \geq \tau_{\max}$.

The norm $\|I(x(t); x(t+\tau))\|$ in Eq. (27.4) is called the "coarse-grained information rate" (CIR). The term "coarse-grained" stems from the fact that CIR is only a coarse and relative estimate of the exact information rate, which is difficult to estimate from experimental data [16, 14]. Nevertheless, both CIR and the exact information rate have the same physical interpretation as measures of predictability of process dynamics. Values of CIR are bounded to the interval $[0, M]$ where the upper limit M is finite and depends on the way of estimation of probability distributions (see Eq. (27.3)). CIR is close to zero for random processes which "forget" their past in a single time step Δt and are therefore not predictable at all. CIR is much higher for periodic processes which lose no information about their history and can be predicted for infinitely long time intervals.

For the results presented in this article, the probability distributions in Eq. (27.3) were estimated using histograms with $Q = 4$ equiprobable bins for $p(x)$ and $Q^2 = 16$ bins for $p(x, y)$. The upper limit of CIR was therefore $M = \log Q$. For monitoring purposes, it is convenient to normalize CIR by M so that its values are bounded to the interval $[0, 1]$. The normalized CIR is hereafter denoted by NCIR.

27.3 Examples of Transitions

In the following, the information rate CIR is applied to detect a transition between favorable and unfavorable operating regimes in three manufacturing processes: turning, grinding and laser beam welding.

27.3.1 Turning

Turning is an example of a cutting process in which a rotating workpiece is cut by a fixed tool with a single cutting edge (Fig. 27.1). In machining literature, two dynamically different cutting regimes are usually distinguished: chatter-free cutting and cutting accompanied by chatter. Chatter denotes self-excited large-amplitude vibrations of the machine tool–workpiece structure which detrimentally affect the workpiece quality and the machine tool itself. The principal cause of chatter is variation of the cutting force due to waviness of the cut surface, resulting from tool–workpiece vibration during previous cuts. As an unfavorable cutting regime, chatter has been studied intensively. Analyses of nonlinear models of cutting have revealed that the transition from chatter-free cutting to chatter in turning corresponds to a subcritical Hopf bifurcation from a stable fixed point to a stable limit cycle [20]. These analytical results have been confirmed experimentally [9]. Evidence in favor of such a description has been obtained also by time-series analysis of experimental data [7].

It is known that chatter can be achieved by a sufficient increase of the mean cutting force. One of the ways to do it is to increase the cutting depth. Figure 27.2 shows amplitude spectra of the cutting force from the experiment in which cutting depth was smoothly and slowly increased while all other cutting parameters were kept constant (Fig. 27.1). Two different cutting regimes can be clearly observed in the graph. Relatively flat spectra at low cutting depths correspond to chatter-free cutting, whereas spectra with a strong dominant peak and its higher harmonics at higher cutting depths correspond to cutting accompanied by chatter. Segments of the cutting

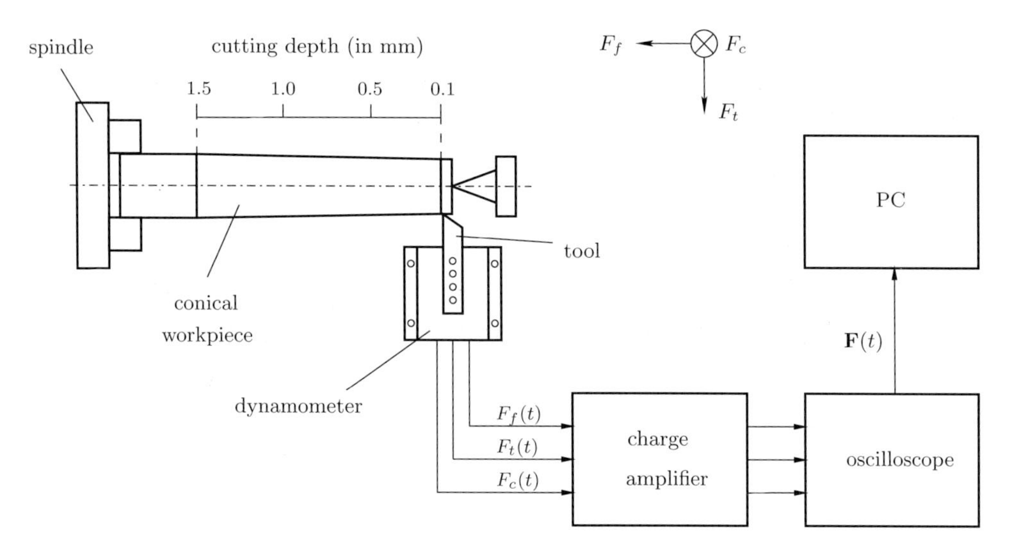

Figure 27.1: Schematic diagram of the experimental setup for turning. Details of the experiments can be found in [7].

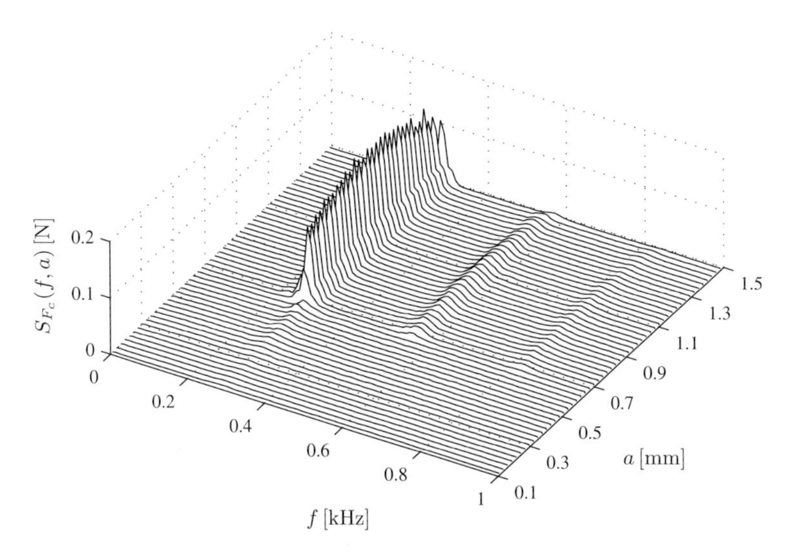

Figure 27.2: Dependence of amplitude spectra of the cutting component F_c on cutting depth a in turning. Two different cutting regimes can be observed.

force fluctuations recorded in the two cutting regimes are shown in Fig. 27.3. In the chatter-free regime, the cutting force fluctuates erratically with occasional bursts of high-frequency periodic oscillations, whereas in the chatter regime pronounced periodic fluctuations of the cutting force are observed with a dominant frequency markedly lower than in the chatter-free regime. The difference in dynamics of the two regimes can be described using a variety of

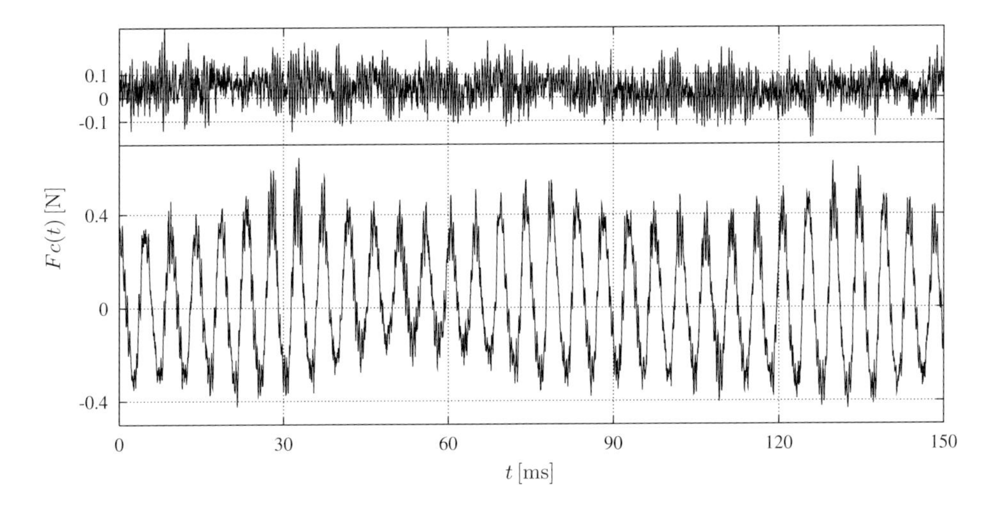

Figure 27.3: Cutting force fluctuations during chatter-free (top) and chatter cutting regimes.

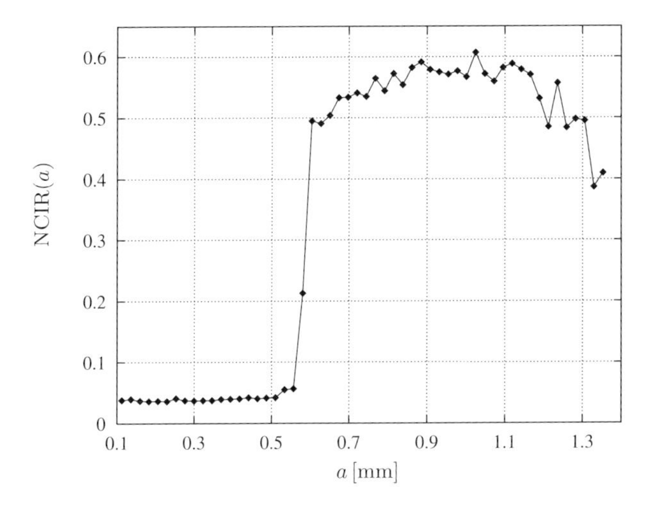

Figure 27.4: Dependence of NCIR on cutting depth a in turning.

characteristics [4] which could all be used as chatter indicators. However, coarse-grained characteristics were found to be the most appropriate for this task [5]. Figure 27.4 shows the dependence of the coarse-grained information rate CIR on cutting depth. As expected, a low CIR value is typical of chatter-free cutting while a much higher CIR value is typical of chatter. A sudden significant increase of the information rate value therefore clearly indicates the transition from chatter-free to chatter regime. A similar dependence of the information rate is generally found in dynamical systems at the points of bifurcation from irregular to periodic motion [15].

27.3.2 Grinding

Grinding is a cutting process with many tiny cutting edges, distributed randomly on the circumference of the grinding wheel. Grinding is often the finishing machining operation, providing high geometrical accuracy and a smooth surface of the workpiece. Similarly to turning, chatter in grinding denotes self-excited large-amplitude nearly harmonic vibration of the machine tool–workpiece structure. The principal cause of chatter is the same as in turning: variation of the cutting force. In diameter grinding, where both workpiece and grinding wheel rotate, the cutting force can vary due to a wavy circumference of either workpiece or grinding wheel [8]. In practice, chatter is most often caused by waviness of the wheel due to wear.

Figure 27.6 shows the amplitude spectra of the normal grinding force vs. specific material volume V'_{w} (cumulative volume of the ground material per unit width of the wheel) for the outer diameter plunge feed grinding experiment (Fig. 27.5). Flat spectra are typical of chatter-free grinding, while spectra with a strong dominant peak are typical of chatter. Slow growth of the amplitude of the dominant spectral peak indicates that wheel wear, which also progresses slowly, is the cause of chatter in this case. Segments of the normal force fluctuations from the two grinding regimes are shown in Fig. 27.7. As expected, low-amplitude aperiodic fluctuations in the chatter-free regime are in sharp contrast to large-amplitude periodic fluctuations in the chatter grinding regime. Dependence of the information rate CIR on specific material volume is shown in Fig. 27.8. Again, a low CIR value is typical of chatter-free grinding while a much higher CIR value is typical of chatter. A gradual increase of CIR correctly reflects the slow rise of chatter vibration. Comparing Figs. 27.6 and 27.8, one can observe that CIR attains rather high values already at $V'_{\mathrm{w}} = 200$–$400\,\mathrm{mm^3/mm}$, which corresponds to the weak, developing chatter. This shows that CIR can indicate chatter already in its early stages, when it is usually not yet recognized by the machine operator.

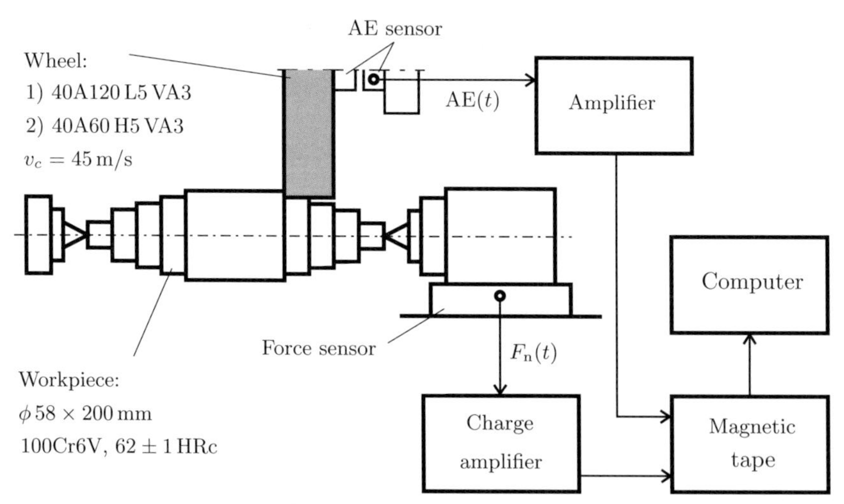

Figure 27.5: Scheme of experimental setup for grinding. Details of the experiments can be found in [2].

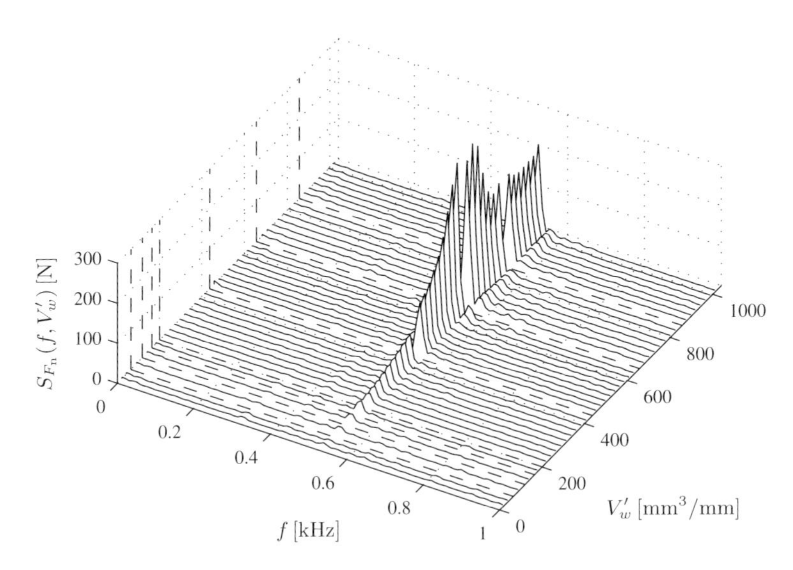

Figure 27.6: Dependence of amplitude spectra of the normal force F_n on the specific material volume V'_w in grinding. Slow growth of the vibration amplitude is typical of chatter caused by the wheel wear.

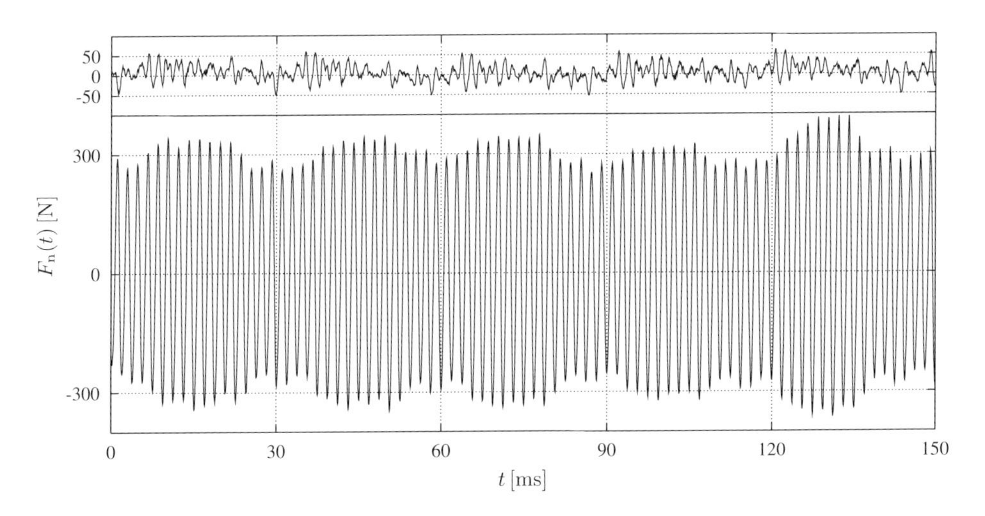

Figure 27.7: Normal force fluctuations during chatter-free (top) and chatter grinding regimes.

27.3.3 Laser-beam Welding

Laser-beam welding is a joining technique in which the components to be joined are locally melted by a laser beam. By varying the power density supplied to the workpiece by the laser beam, various welding regimes can be achieved [17]. In our study, two regimes of CO_2 laser-beam welding are considered: deep- and shallow-penetration welding. High power densities are required for deep-penetration welding. The energy supplied to the workpiece causes intense

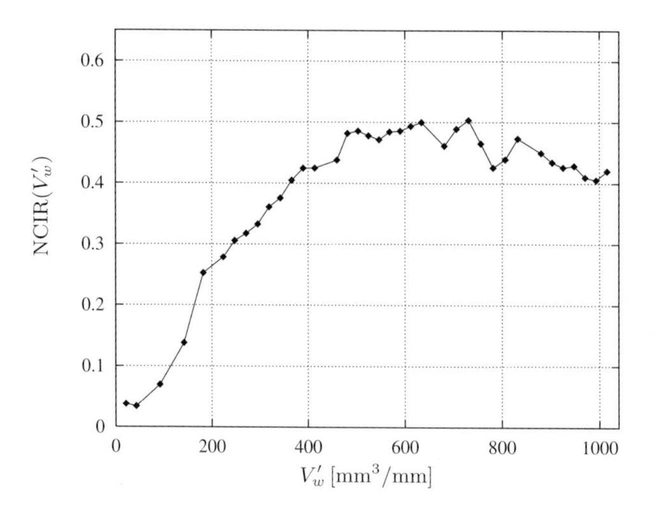

Figure 27.8: Dependence of NCIR on specific material volume V'_w in grinding.

vaporization of the material, such that a capillary known as a keyhole is created in the molten pool [21, 11, 12, 18]. The keyhole enables deeper penetration of the laser beam into the material, which deepens the molten pool. Shallow-penetration welding occurs at low power densities, where the material vaporization is less intense and no keyhole is formed. The penetration depth of the laser beam is therefore considerably smaller, and the molten pool shallower.

Transition from deep- to shallow-penetration welding can be achieved by a decrease of the power density supplied to the workpiece. This can be done by reducing the output power of the laser source, by increasing the feed velocity of the laser head or by increasing the focus position of the beam [12, 17, 18]. Figure 27.10 shows amplitude spectra of light-intensity fluctuations recorded during welding with smoothly increasing focus position (Fig. 27.9). A relatively flat spectrum is typical of deep-penetration welding, whereas in shallow-penetration

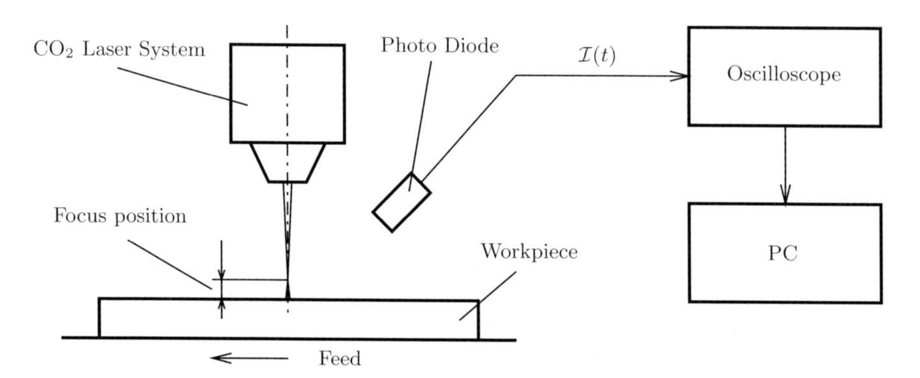

Figure 27.9: Experimental setup for CO_2 laser-beam welding. Details of the experiments can be found in [3].

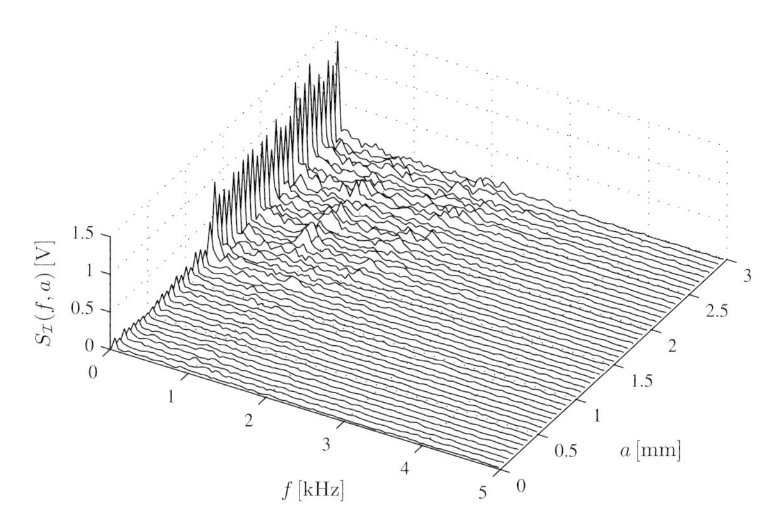

Figure 27.10: Dependence of amplitude spectra of the light-intensity fluctuations $\mathcal{I}$ on the focus position a. Two different welding regimes can be observed.

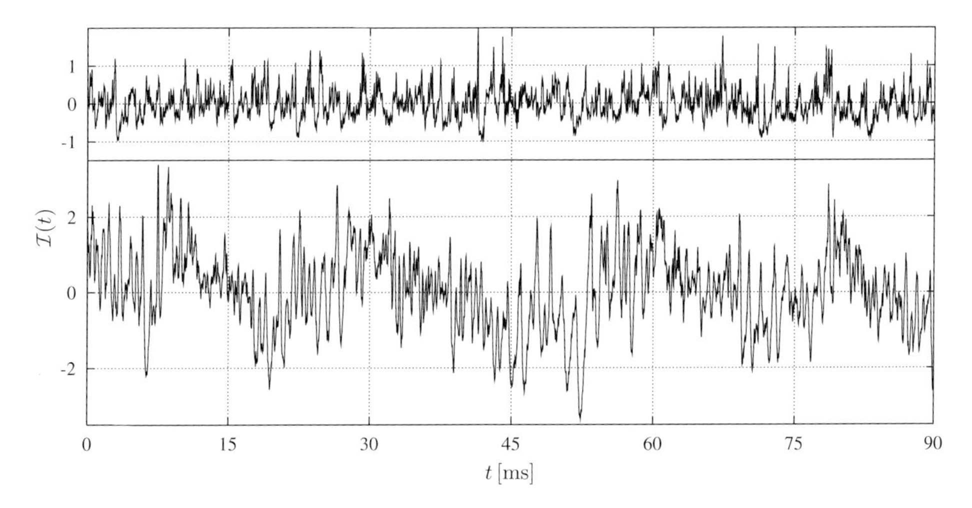

Figure 27.11: Fluctuations of emitted light intensity during deep- (top) and shallow-penetration welding regimes.

welding several peaks are found in the spectrum. The strongest peak lies at a frequency of $f \approx 66$ Hz. Segments of the light-intensity fluctuations recorded in the two welding regimes are compared in Fig. 27.11. In deep-penetration welding, irregular fluctuations of light intensity are observed. The occasional marked increases of intensity followed by significant drops presumably correspond to the keyhole activity and plasma effects in the molten pool. In shallow-penetration welding, irregular fluctuations of light intensity are interspersed with short segments of distinct periodic oscillations. Spiky patterns related to the keyhole activity are not observed. In the dependence of CIR on focus position, the transition between the two regimes is clearly indicated by a sudden increase of CIR value (Fig. 27.12). After the transition, CIR decreases, but it still retains a significantly larger value compared to the deep penetration welding regime. These results suggest that the difference in dynamics of the two regimes is not as distinct as in the chatter cases discussed earlier. However, the transition between deep- and shallow-penetration welding regimes can nevertheless be detected by CIR.

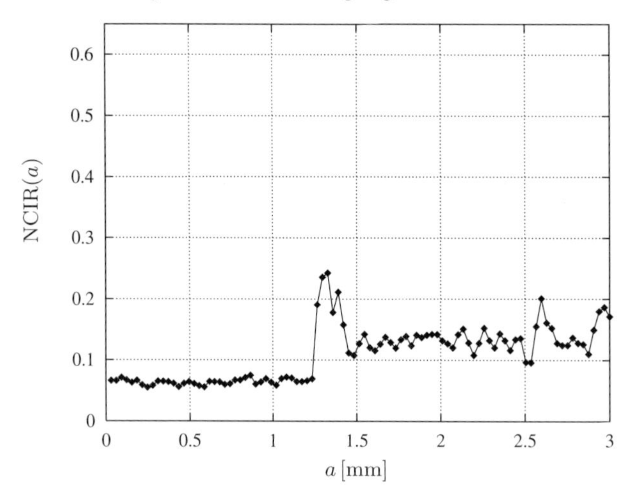

Figure 27.12: Dependence of NCIR on focus position a in laser-beam welding.

27.4 Discussion and Conclusions

The article discusses detection of transitions between process states associated with changes in process dynamics. The method of detection is based on a nonlinear characteristic called the coarse-grained information rate (CIR) which measures predictability of process dynamics. The method was applied to detect transitions from favorable to unfavorable operating regimes in turning, grinding and laser-beam welding. No adaptation of the method to a particular process was required.

The transitions from chatter-free to chatter regime in turning and grinding are dynamically similar. In both cases the transition represents a distinct change of process dynamics from random-like to nearly harmonic vibration of the machine tool–workpiece system, which is reflected in a marked increase of the information rate CIR. It is known from analytical modeling of turning that the transition from chatter-free to chatter regime is in fact a Hopf bifurcation from a stable fixed point to a stable limit cycle [20]. To our knowledge, no such analytical results exist for grinding. However, based on the observed similarity of the transitions it is reasonable to assume that chatter onset in grinding also represents a Hopf bifurcation. As shown in the article, this similarity can be of advantage in the case of monitoring, since the dynamics-based monitoring methods that were originally developed for turning can also be employed in grinding, and vice versa.

The transition from deep- to shallow-penetration welding is associated with a less distinct change of process dynamics than the two chatter transitions. The increase of information rate CIR at the transition nevertheless indicates that the transition might represent a phenomenon similar to a bifurcation [12, 19]. According to the stochastic time-series analysis [6], dynamics of light-intensity fluctuations in both welding regimes can be described as a randomly excited stable fixed-point motion, whereby the fixed point is a node in deep- and a focus in shallow-penetration regime. Although the change of the type of fixed point from node to focus is technically not a bifurcation, it does increase predictability of a process which can be detected by CIR.

In summary, the three detection examples show that there are at least two advantages of the dynamics-based detection methods over the heuristic ones. (1) The same method, i.e. the same indicator of a transition, can be applied to different processes which exhibit transitions associated with a change of dynamics. (2) Dynamics-based indicators can provide some insight into the physics of the process when the analytical models are not known.

Acknowledgment

The authors acknowledge support of the Volkswagen Foundation from Germany and EU COST Action P4. EG, JG and IG also acknowledge support of the Slovene Ministry of Education and Science.

Bibliography

[1] T.M. Cover and J.A. Thomas, *Elements of Information Theory* (Wiley Series in Telecommunications), John Wiley, New York, 1991.

[2] E. Govekar, A. Baus, J. Gradišek, F. Klocke, and I. Grabec, *A new method for chatter detection in grinding*, Annals of the CIRP **51**(1) (2002), 267–270.

[3] E. Govekar, J. Gradišek, I. Grabec, M. Geisel, A. Otto, and M. Geiger, *On characterisation of CO_2 laser welding process by means of light emitted by plasma and images of weld pool*, in: B. Scholz-Reiter (ed.), *Untersuchung Nichtlinear-Dynamischer Effekte in Produktionstechnischen Systemen*, 3. Symposium der Volkswagen Stiftung, Cottbus, Germany, 2000.

[4] J. Gradišek, E. Govekar, and I. Grabec, *Time series analysis in cutting: chatter versus chatter-free cutting*. Mechanical Systems and Signal Processing **12**(6) (1998), 839–854.

[5] J. Gradišek, E. Govekar, and I. Grabec, *Using coarse-grained entropy rate to detect chatter in cutting*, Journal of Sound and Vibration **214**(5) (1998), 941–952.

[6] J. Gradišek, E. Govekar, and I. Grabec, *Qualitative and quantitative analysis of stochastic processes based on measured data, II: applications to experimental data*, Journal of Sound and Vibration **252**(3) (2002), 563–572.

[7] J. Gradišek, I. Grabec, S. Siegert, and R. Friedrich, *Stochastic dynamics of metal cutting: bifurcation phenomena in turning*, Mechanical Systems and Signal Processing **16**(5) (2002), 831–840.

[8] I. Inasaki, B. Karpuschewski, and H.-S. Lee, *Grinding chatter – origin and suppression*, Annals of the CIRP **50**(2) (2001), 515–534.

[9] T. Kalmár-Nagy, J.R. Pratt, M.A. Davies, and M.D. Kennedy, *Experimental and analytical investigation of the subcritical instability in metal cutting*, in: Proceedings of the 17th ASME Biennial Conference on Mechanical Vibration and Noise, 1999 ASME Design and Technical Conferences, Las Vegas, Nevada (1999), pp. 1–9.

[10] H. Kantz and T. Schreiber, *Nonlinear Time Series Analysis* (Cambridge Nonlinear Science Series **7**,) Cambridge University Press, Cambridge, 1997.

[11] T. Klein, M. Vicanek, J. Kroos, I. Decker, and G. Simon, *Oscillations of the keyhole in penetration laser beam welding*, Journal of Physics D: Applied Physics **27**(10) (1994), 2023–2030.

[12] J. Michel, S. Pfeiffer, W. Schulz, M. Niessen, and V. Kostrykin, *Approximate Model for Laser Welding*, Nonlinear Dynamics of Production Systems, 2003, Chapter 24, this book.

[13] E. Ott, *Chaos in Dynamical Systems*, Cambridge University Press, 1993.

[14] M. Paluš, *Coarse-grained entropy rates for characterization of complex time series*, Physica D **93**(1–2) (1996), 64–77.

[15] M. Paluš, *On entropy rates of dynamical systems and Gaussian processes*, Physics Letters A **227**(5–6) (1997), 301–308.

[16] M. Paluš, V. Komárek, Z. Hrnčiř, and K. Štěrbová, *Synchronization as adjustment of information rates: detection from bivariate time series*, Physical Review E **63**(04-6211) (2001).

[17] R. Poprawe and W. Konig. *Modelling, monitoring and control in high quality laser cutting*, Annals of the CIRP, **50**(1):137–140, 2001.

[18] W. Schulz, J. Michel, M. Niessen, P. Abels, and S. Kaierle, *Modelling, Dynamical Simulation, and Online Monitoring in Laser Beam Welding*, Proceedings of Laser 2001, pp. 236–247.

[19] W. Schulz, V. Kostrykin, J. Michel, and M. Niessen, *Modelling and Simulation of Process Monitoring and Control in Laser Cutting*, Nonlinear Dynamics of Production Systems, 2003, Chapter 23, this book.

[20] G. Stépán and T. Kalmár-Nagy, *Nonlinear regenerative machine tool vibrations*, in: Proceedings of the 16th ASME Biennial Conference on Mechanical Vibration and Noise, 1997 ASME Design and Technical Conferences, Sacramento, California, 1997, pp. 1–11.

[21] C. Tix and G. Simon, *Model of a laser heated plasma interacting with walls arising in laser keyhole welding*, Physical Review E **50**(1) (1994), 453–462.

Part V

Chemical and Electro-chemical Processes

The dynamics of chemical and electro-chemical processes is often very complicated and not well understood due to the complex spatio-temporal behavior that may occur. Unfortunately this is true especially for processes of technical relevance. Correspondingly, in industrial applications a pragmatic approach is required which allows optimizing and controlling the processes of interest. Similar to the last paper in the previous part data driven process monitoring becomes important and sometimes is the only way to deal efficiently with the problem at hand. The first article by C.S. Daw provides a prototypical example for such an approach. In staged coal combustion one is interested in optimal burner respectively flame states. This problem is currently too complicated for using analytic and numerical approaches, because it involves complicated reaction–diffusion processes, turbulent multi-phase flow, unknown boundary conditions, and other unknown influences. The authors show that it is nevertheless possible to design a real-time monitoring tool for detecting dynamic state changes and for adjusting the system to optimal burning states. The underlying methods utilize ideas from nonlinear dynamics, especially symbolic dynamics, and statistics. The resulting monitoring system can serve as a paradigm for other industrial process applications.

The second paper by M. Mönnigmann et al. is very different in nature. It shows that in chemical engineering not only an understanding of given chemical processes with methods from nonlinear dynamics can be achieved, but that the optimal design of high-dimensional nonlinear systems and processes is possible. While the former methods rely on bifurcation theory and parameter continuation, the constructive design of high-dimensional systems employs normal vector-based constraints for achieving parametric robustness. The paper discusses and summarizes recent progress of such constructive nonlinear dynamics methods. The following contribution of P. Plath et al. presents and discusses some typical technical applications of concepts from nonlinear dynamics in the field of chemical engineering and for electro-chemical manufacturing processes. These examples include the separation of rare-earth elements by oscillatory extraction, the generation of mono-dispersive ceramic foams, surface structuring by electropolishing, and structure formation in etching processes. These examples show that the nonlinear dynamics of pattern formation is of fundamental importance in these applications. The remainder of this part is devoted to electro-chemical processes. The one by C. Gerlach et al. studies experimentally electropolishing of brass in different electrolytes, a widely used technical process. These nonlinear dynamical electro-chemical systems are identified from their bifurcation behavior as systems with negative differential resistance and further categorized into several classes and sub-classes for which mathematical descriptions are available. The subsequent paper by A. Mora et al. utilizes wavelets for the analysis of these electropolished surfaces. Consequently, the surfaces are characterized by their multi-fractal scaling behavior. In addition Markov properties in scale space are tested and evaluated. As a result surface features are related to specific dynamic processes and are therefore highly relevant for the optimization of these technical processes. The last article by U. Sydow et al. describes another very different and interesting application of nonlinear dynamics in electro-chemical systems. The state-of-charge and state-of-health evaluation of ubiquitously used lead–acid batteries is an outstanding problem of extremely important practical relevance e.g. for transportation systems and energy storage in general. The authors show how the monitoring of the spatio-temporal dynamics of local potentials via phase space and time series techniques can be used to evaluate the state of the battery. These new findings are relevant for both automotive and battery engineers and for those interested in such electro-chemical dynamical systems in general.

28 Real-time Monitoring of Dynamical State Changes in Staged Coal Combustion

C.S. Daw, C.E.A. Finney, T.A. Fuller, T.J. Flynn, and R.T. Bailey

We describe a system for monitoring the state of coal-fired utility burners using nonlinear dynamic characteristics of optical flame scanner measurements. Our signal analysis techniques are optimized for targeting characteristic dynamical patterns associated with global combustion instabilities that develop as operating parameters are changed. Various dynamical indicators are used, including the onset of temporal asymmetry and the emergence of characteristic symbol sequences associated with unstable periodicities. We show that careful application of such methods can accurately characterize the burner condition in terms of a range of known flame states. State-to-state transitions are interpreted in terms of global flame state bifurcations. We have implemented our monitoring methods in a hardware–software package designed for installation in utility boilers. This commercial package, referred to as the Flame Doctor® system, has been demonstrated in electric utility plant trials sponsored by the Electric Power Research Institute and participating utilities. Specific performance results from one plant trial are described. In addition, we discuss broader implications of our burner-monitoring results for other types of processes.

28.1 Introduction

Inefficient boiler control is responsible for wasting large amounts of fuel and releasing greenhouse gases (CO_2 and N_2O) and nitrogen oxide pollutants (NO_x). This is especially true in the United States where more than 84% of the energy consumed is produced by the combustion of fossil fuels [1]. More than one-fourth of this fossil fuel is coal, and coal is by far the largest source of energy for US electric power production. This situation has led to strong economic pressures and environmental regulations favoring the development of advanced management systems that efficiently control utility boilers.

It is now recognized that individual burner monitoring is essential for meeting the demands of advanced boiler management. Accurate monitoring is more important for so-called low-NO_x burners than conventional burners because these burners are more sensitive to changes in operating parameters and feed-system variations. Conventional combustion-monitoring systems provide information that has been averaged over many burners and long time scales (for example, measurements of excess air, coal feed or NO_x emissions at time scales of several minutes or hours). However, large NO_x and carbon burnout fluctuations can occur in individual burners over short time scales (i.e. time scales of tens of seconds down to small fractions of a second). These fluctuations produce widely different boiler performance for operating

Nonlinear Dynamics of Production Systems. Edited by G. Radons and R. Neugebauer

ISBN 3-527-40430-9

conditions that otherwise appear to be indistinguishable. Combustion diagnostics should thus reflect both long and short time-scale transients.

The burner-monitoring system described herein relies on the power of nonlinear time-series analysis for extracting information from complex signals. Specifically, nonlinear time-series analysis has been used to improve the extraction and utilization of information from existing burner sensors. Although the sensors in question were not originally intended to provide such information, it is now clear that they can be used in the new approach to burner diagnostics. Our research has demonstrated that this new type of monitoring system can be installed directly in existing power plants with minimal cost and hardware modification. This new system may serve as a paradigm for other industrial process applications.

28.2 Background

Our research has focused on the type of wall-mounted coal-fired utility burner illustrated in Fig. 28.1. This kind of burner is designed to achieve reduced nitrogen oxide emissions by staging the combustion so that the peak flame temperature is reduced. Pulverized coal is injected with a primary air stream (less than the stoichiometric amount required to combust the coal) into the central zone of the burner. Secondary air is injected in an outer annular zone with a tangential swirl imposed by vanes. The resultant outward flow from the burner consists of a central, fuel-rich core surrounded by a swirling annulus of oxygen-rich gas. The flame front develops at the interface between these zones, with a shape and volume depending on the relative flow velocities, temperature and concentrations of volatiles and devolatilized char particles. While the staging process typically extends the volume of the combustion region and thereby reduces peak temperature, it also makes the flame less stable. That is, the flame is closer to extinction and is sensitive to small changes in the local conditions due to turbulence or shifts in operating parameters. This instability makes better process control more important.

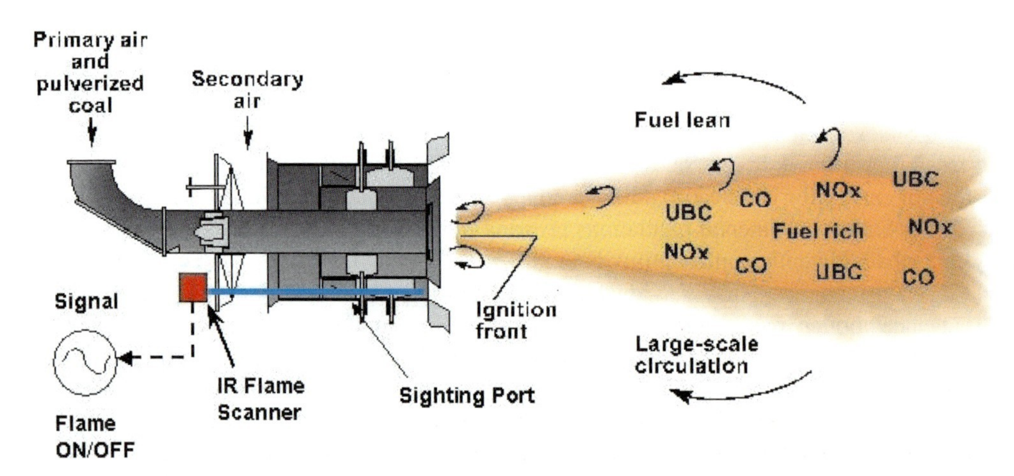

Figure 28.1: Schematic of a wall-fired, low-NO_x pulverized coal burner. Air staging reduces peak flame temperature and NO_x emissions but increases instability.

Utility boilers typically contain multiple wall-mounted burners. Depending on the boiler size, this could amount to a few burners or as many as a hundred. As shown in Fig. 28.2, these burners are usually arranged in rows on one or two walls. In the two-wall configuration, the burners are divided between opposite walls and directly face each other. For either single or opposing wall configurations, there is a very good opportunity for burner-to-burner coupling through direct flame interactions. There are also opportunities for coupling through shared air and fuel distribution systems.

The major visible characteristics of flames such as shape and ignition front location can shift radically as operating parameters are changed. In many cases, the flame states for a given burner design are at least empirically understood by combustion engineers. Some of the more common flame states are depicted in Fig. 28.3. A relatively steady flame beginning immediately at the central burner opening and extending outward with a cylindrical or slightly tapered shape is considered optimal (top example). This is frequently referred to as a 'stable attached flame' and usually has low emissions of both nitrogen oxides and unburned coal and carbon monoxide. Excessively high or low primary coal–air ratios can lead to a condition in which the flame front sporadically extinguishes at the burner face, rapidly moves further downstream and then rapidly returns to the burner face. This condition is referred to as 'intermittent detachment' (second example).

When the fuel and air imbalance becomes even more severe, a point is reached when the flame ignition front remains fully separated from the burner face. This condition is referred to as 'fully detached' and usually results in severe pollutant emissions (third example). Incorrect swirl or distribution of the secondary air can cause another type of flame instability characterized by side-to-side (or up-and-down) waving of the flame (fourth example). This 'flapping' condition can lead to large-amplitude oscillations in emissions. Excessively rapid primary–secondary mixing or high thermal feedback to the base of the flame can create a 'flared' condition such as that shown in the bottom example. This condition is of particular concern because it can create excessive heating of the burner face, causing severe damage to the hardware and ceramic insulation. The benefits of staging are also lost because of the higher peak temperature.

Each of the above burner states is characterized by a distinct variation in output light intensity, which is often referred to as 'flicker'. Figure 28.4 illustrates the scanner signals corresponding to three very distinct burner states. In the stable, well-attached case, most of the flicker is caused by combustion fluctuations associated with turbulent mixing of the coal and air. Because the shape and spatial position of the flame do not change much, this flicker is small in amplitude and does not reflect obvious characteristics of low-dimensional dynamics (that is, it appears to be largely random). The intermittent partial detachment condition involves a rapid back-and-forth movement of the ignition front due to reduced ignition stability, which creates large characteristic spikes in the scanner output. When the intermittent detachment becomes more severe, flicker becomes even more severe as the flame front jumps between two distinct spatial positions. This condition has similar attributes to a noisy two-well potential.

The essential concept behind our monitoring system is that the flicker fingerprint can be used to identify the flame state of each burner. The scanner signal differences are quite large in some cases and can be readily distinguished by eye. In other cases, however, the distinctions are much more subtle. Thus quantitative flicker recognition is challenging because of the complexity of the corresponding scanner signals. This becomes apparent if one attempts to

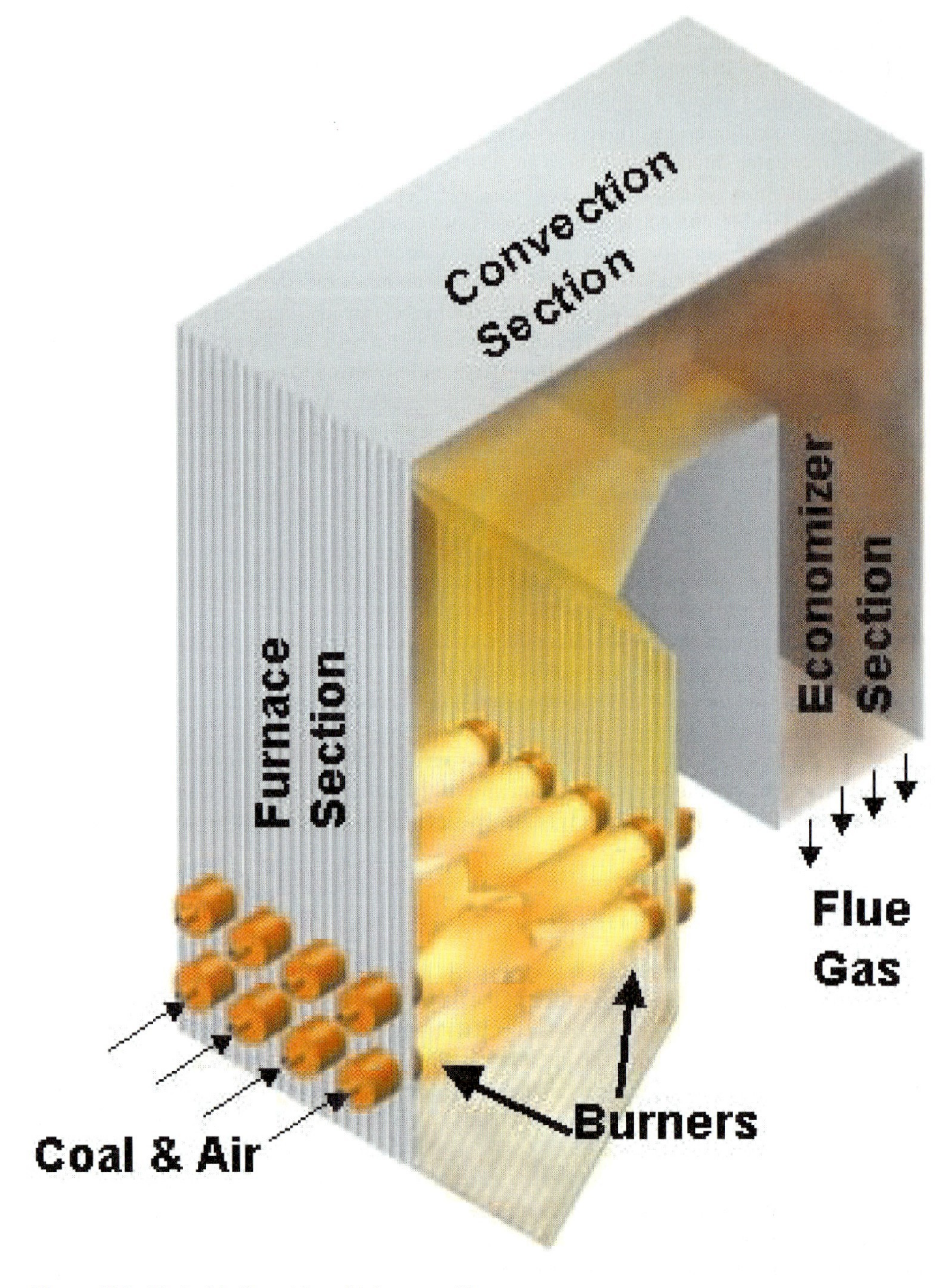

Figure 28.2: Typical boiler with multiple rows of burners.

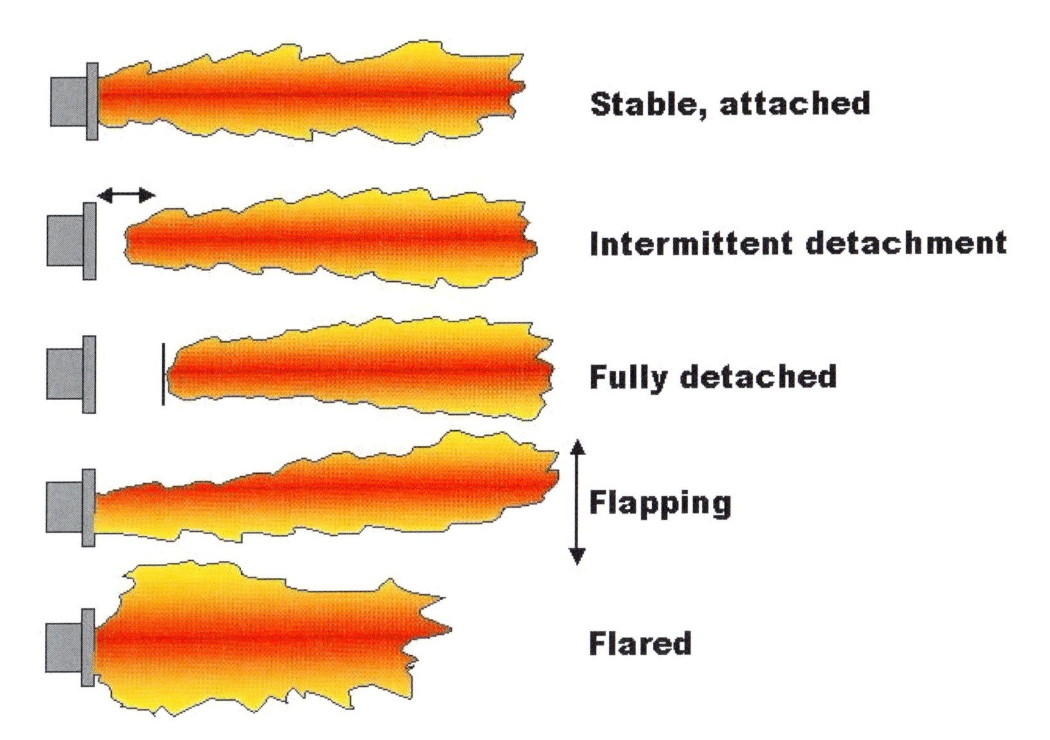

Figure 28.3: Examples of different characteristic flame states.

characterize scanner signals utilizing standard linear analysis methods such as Fourier decomposition. While some flame states can be distinguished in this way, many important differences become blurred due to the nonlinear instability. An example of this problem is illustrated by Fig. 28.5, in which the Fourier power spectra for three different flame signals are depicted. For all three cases, the burner parameters were carefully controlled in a pilot facility so that the burner state was well known. The two dashed lines represent very similar burner states where the flow ratio of primary air to coal (PA/C) was near 1.8. The solid line, on the other hand, represents a very different burner state where the same ratio was near 4. Both cases represent conditions where some degree of detachment occurs, but the physics of the detachment (and its causes) are considerably different and the flames are considerably different in appearance. Based on the Fourier power spectra, it would appear that the frequencies of the high-ratio case actually are a close match to one of the low-ratio cases but not the other. Since we know that the Fourier spectra comparison incorrectly matches these very different flame states, it is clear that such spectra are not very useful for achieving the desired discrimination.

The usefulness of alternate approaches for flicker characterization is illustrated in Fig. 28.6. This is a plot of the relative frequency of a range of possible flicker patterns extracted from the original signals used in Fig. 28.5. Initially, the original scanner signals were converted to corresponding symbol sequences by dividing the signal range into several equiprobable intervals. Symbol sequences (sometimes referred to as 'words') were then constructed by considering successive occurrences of these symbols. Figure 28.6 is then a plot of the relative word fre-

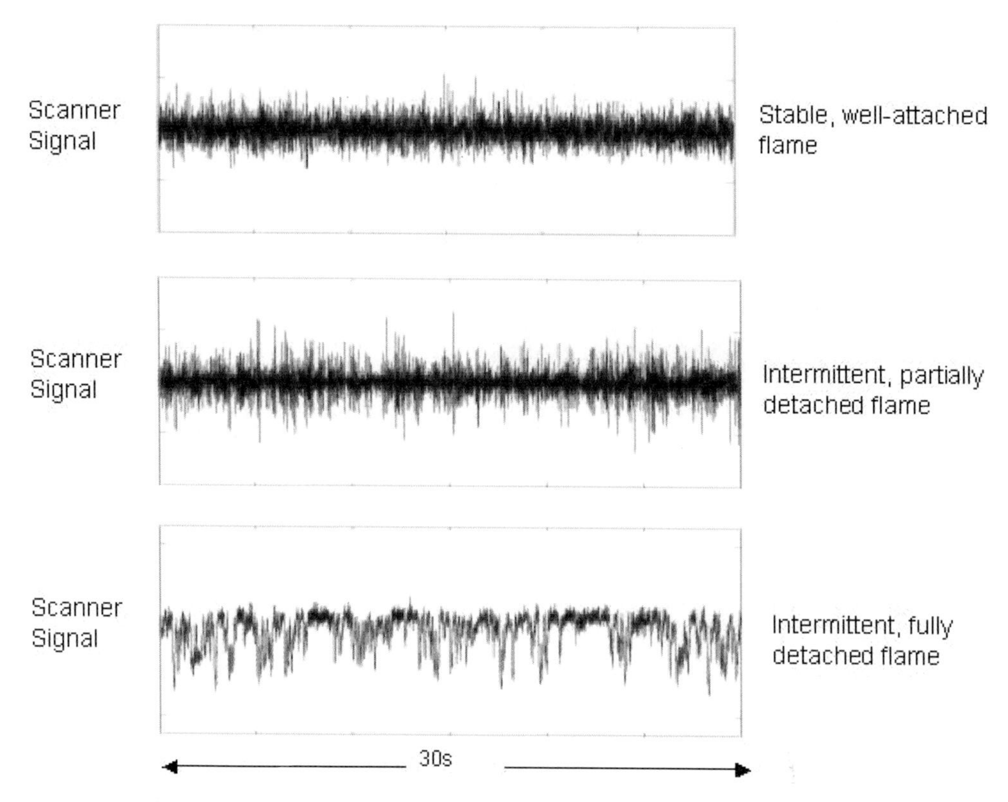

Figure 28.4: Example scanner signals for selected flame states. Some signal features can change with scanner type and location.

quencies (vertical axis) against each possible symbolic word (horizontal axis). The details of the symbolization process involve proprietary choices of parameters (e.g. time intervals between symbols and number of symbols per word) that are beyond the scope of this paper, but we describe the essential steps involved in this type of analysis in [3].

The point revealed by the symbol sequence histogram is that by extracting the coarse-grained features of the flicker signal, we are able to correctly match the two similar flame states and separate them from the third different state. In this case, the intermittent detachment occurring at the high PA/C condition produces high frequencies of particular symbolic events corresponding to unstable periodicities. Symbolization, like many other nonlinear time-series methods, is more appropriate for detecting such unstable periodicities. In particular, we are interested in the coarse-grained recognition of the global unstable periodicities in the noisy scanner signal.

Figure 28.7 is another example of how nonlinear statistics can be used to detect patterns that are invisible to linear diagnostics. In this case, we are using a measure of temporal asymmetry to distinguish two different flame states. Briefly, temporal asymmetry refers to time-series characteristics that appear different when viewed in forward and reverse time order. Temporal asymmetry cannot be measured using linear statistics such as Fourier power spectra and au-

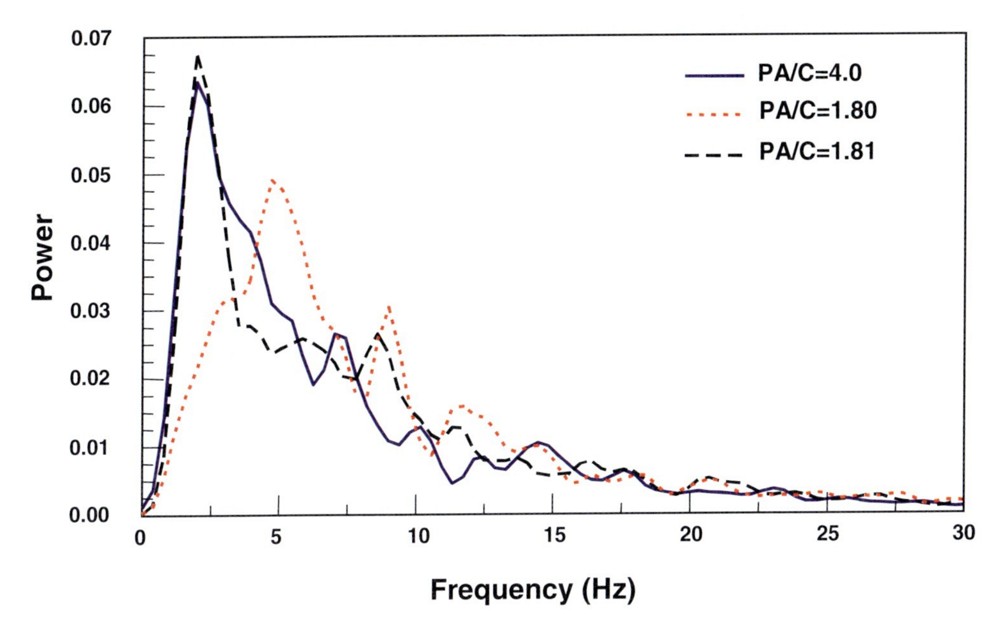

Figure 28.5: Fourier power spectra for three different flame conditions. Two of the flame conditions are known to be very similar (PA/C = 1.80 and 1.81), but their spectra have significant differences.

tocorrelation. However, temporal asymmetry is especially significant for nonlinear systems, and we provide additional background on this in [4]. The particular time-asymmetry statistic, $T(\tau)$, used here is evaluated as a function of differential time intervals and is determined by

$$T(\tau) = \sum_N (x(i+\tau) - x(i))^3 \Bigg/ \left[\sum_N (x(i+\tau) - x(i))^2 \right]^{3/2} .$$

The $x(i)$ are the observed scanner signal values τ is a time delay and N is the total number of time intervals evaluated. This type of function allows one to observe the variation in time asymmetry over different time scales. We observe in this case that the asymmetric time scales in the partially detached flame are below 10 milliseconds, corresponding to the rapid motion of the ignition front location during local extinction and re-ignition events. For other types of global bifurcations, the time scales associated with strong time asymmetries are much longer; e.g. when the flame length pulses in response to surging coal flow in the burner entrance.

Ideally, the best approach for identifying nonlinear dynamical signatures utilizes physical models to at least define the connections between operating parameter shifts and state bifurcations. Unfortunately, there are no existing burner models that accurately describe flame dynamics and the corresponding flicker patterns for coal-fired burners. We have had to rely, instead, on empirical mapping of the known burner conditions (based on pilot plant data and burner engineering expertise) using multi-variate statistical clustering. Figure 28.8 illustrates a qualitative conceptual map of the relationship among the various recognized burner conditions in parameter space. Transitions between states occur along characteristic boundaries as burner

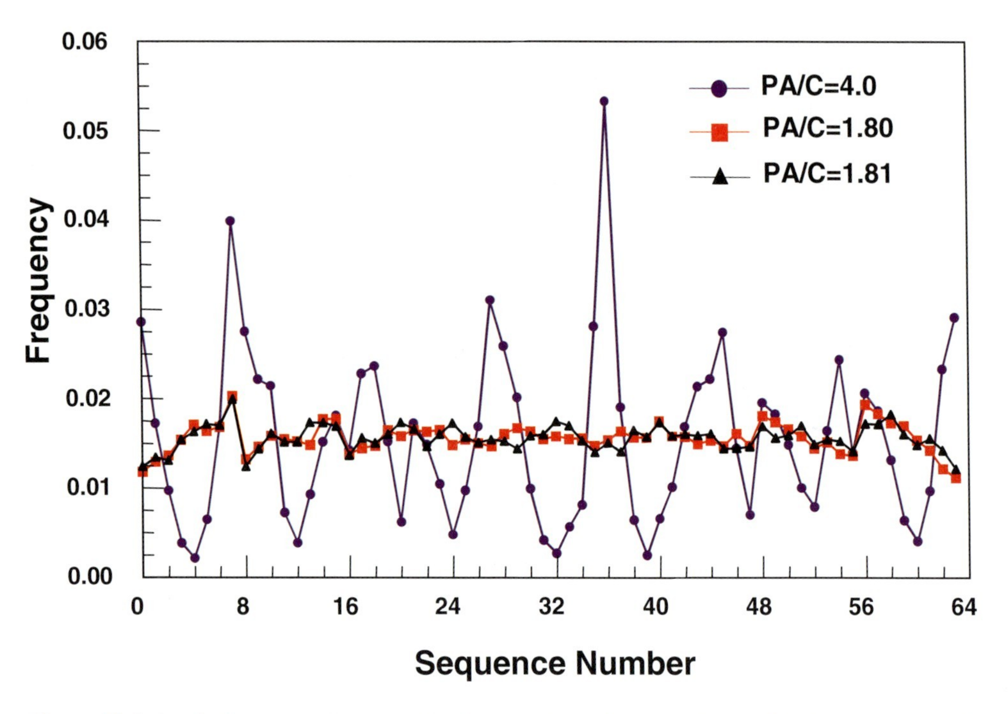

Figure 28.6: Symbol sequence histogram for the same three flame conditions. The histograms for the two similar conditions appear consistent.

parameters change, either due to deliberate control actions or to uncontrolled system drift. In Fig. 28.8, the various states are spaced outward from the center relative to their degree of deviation from optimal, with the farthest out being the worst. The worst states are also typically so unstable that they are on the verge of flameout. In general, our diagnostic approach begins by testing a set of several observed statistics for each burner against the null hypothesis that the burner is in an optimal state. Deviations from optimal in certain specific statistics then allow us to assign that burner to a particular reference state. We use an open-ended library of reference states, so the number of possible states can grow with experience. The system "learns" over time by adding new library members and/or discarding old members.

28.2.1 Practical Approach

We have constructed an embodiment of the above theoretical approach in a combined hardware and software system (referred to by the commercial name Flame Doctor®) for use in electrical generating plants. As illustrated in Fig. 28.9, the signals from existing optical scanners on each burner are collected either directly at the burners or (more typically) at a central relay cabinet where the scanner signals have already been routed for conditioning and further processing. Most scanner systems utilize such remote processing arrangements to generate the actual flame on/off indication so that the equipment actually located at the burner consists of only the sensor head and a local power supply. This allows the removal of the most sensitive electrical

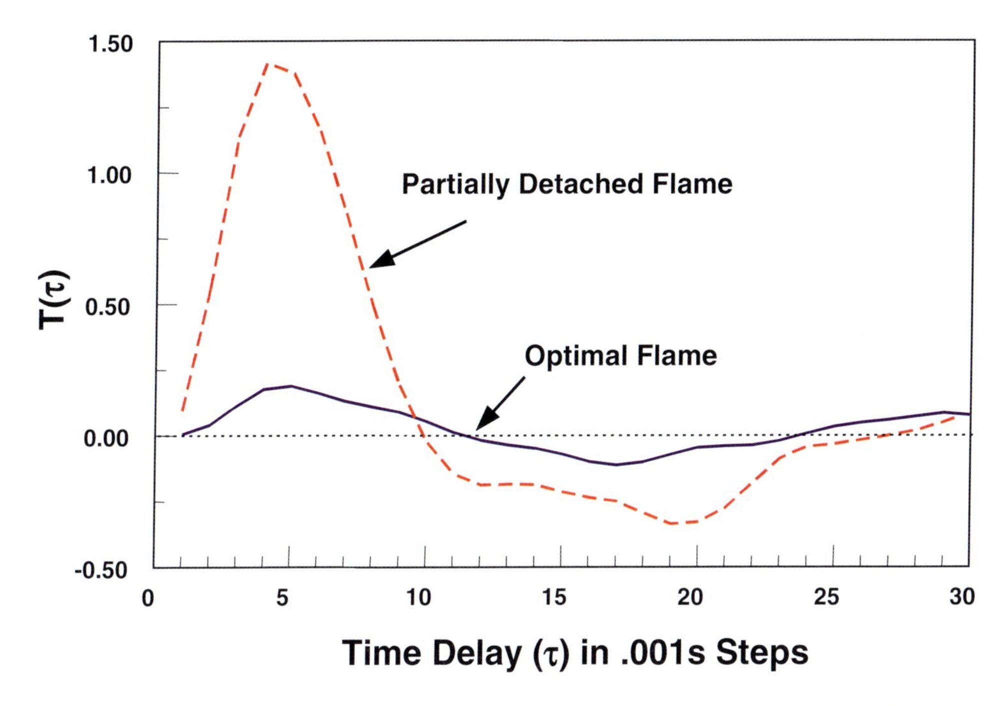

Figure 28.7: Time-asymmetry functions for two different flame states. The partially detached flame exhibits much more asymmetry at shorter time scales.

components to locations where the harsh environmental conditions around the boiler can be moderated. Because coal flames are typically full of particulates (and thus optically dense in the visible-light spectrum), most coal-boiler scanners also operate in the infrared and near-infrared frequency range, which is less attenuated by the particulates. For our system, we typically require that the scanner output signal be an analog voltage that has minimal filtering and sufficient amplitude so that it can be digitally re-coded by a multi-channel data-acquisition system. Digital data-acquisition units are now available in modular form so that it is possible to 'stack' as many as are needed to handle the required number of burners. Utility coal-fired boilers can contain anywhere from several to several dozen burners. The largest boiler we are aware of has ninety-six burners.

Through extensive pilot and full-scale testing, we have determined that it is necessary to record the output of each scanner for at least one to two minutes in order to collect a representative sample of flame dynamics. Shorter records are often insufficient to capture a statistically significant sample of the important long-time-scale periodicities in the burner. That is, we do not see good convergence of the statistics for samples of less than one minute. Sampling intervals longer than two minutes are usually impractical because of the need for frequent burner-status updates. Digital sampling rates also need to be sufficiently fast to observe the important small time-scale dynamics. As noted above regarding the temporal asymmetry example, these smaller time scales can be less than ten milliseconds. Another practical constraint is of particular importance for many existing commercial scanners; this is

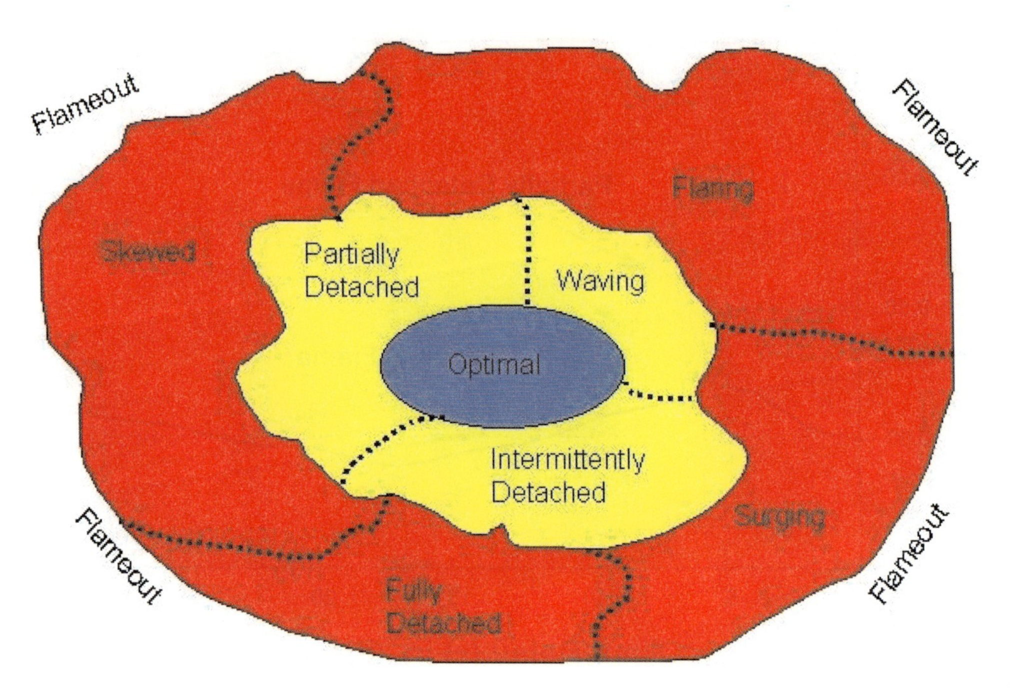

Figure 28.8: Conceptual map of flame states. Each state is associated with multiple dynamical features.

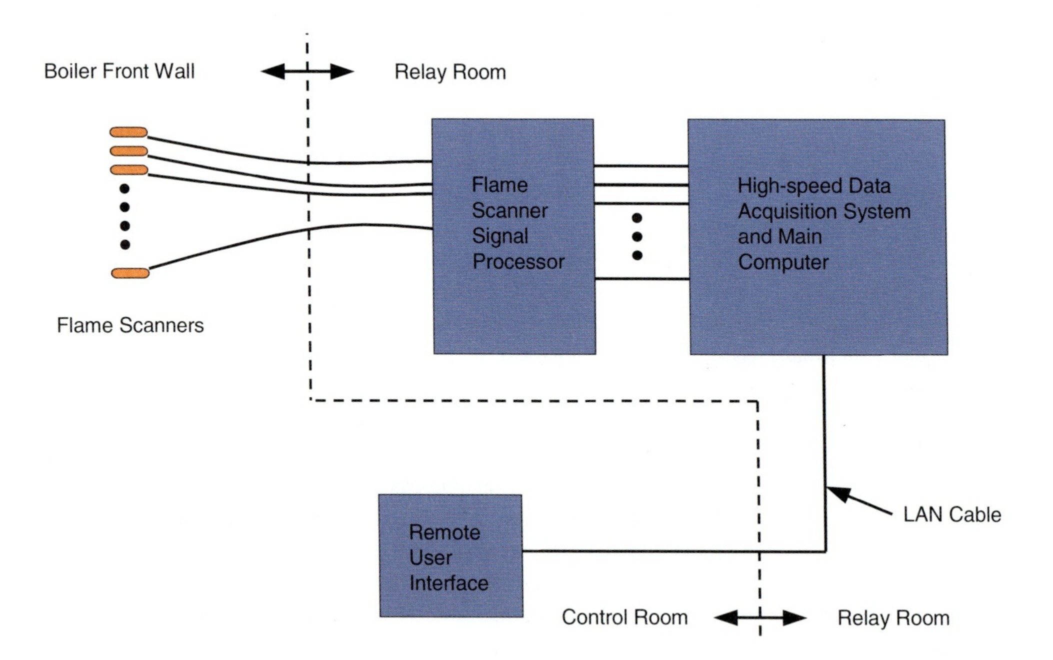

Figure 28.9: Schematic layout of the in-plant components for burner monitoring.

the so-called self-test interrupt. Such interrupts consist of an automated blocking of the light signal to the scanner or the signal coming from the scanner at regular intervals to produce an indication of signal loss at the processing unit. Under normal scanner operation, this simulated signal loss is used to determine whether or not the scanner is still functioning. For our needs, however, such interrupts create troublesome breaks in the signals being used to monitor flame dynamics. When scanners with short self-test interrupts are encountered, we make sure to extend the length of the self-test interval (e.g. by changing the scanner self-test timer setting) to be at least as long as the time required to collect representative output samples. We also have to be sure to exclude any data that are recorded during the interrupt periods from further analysis to prevent statistical bias.

Once the output signals have been digitally acquired, they are further processed on a standard Pentium-type desktop or laptop computer using proprietary software. The algorithms in this software produce several different statistical measures that are used to diagnose the condition of each burner relative to a set of reference states in the dynamic library. While the details of the diagnostic process are proprietary, the approach we use has many similarities to the approach used for diagnosing disease pathologies. Of course, instead of experience with diseases, we rely on the accumulated experience of burner engineers to identify non-optimal flame conditions and their underlying causes. The diagnostic results are displayed on a graphical user interface, which can be viewed at multiple remote locations including the boiler-control room. Multiple screens can be accessed via the interface so that it is possible to observe all the signal details coming from each burner that have been used to assess its condition. Historical trends for any of these features as well as the overall diagnosis for each burner and the boiler as a whole can also be viewed. In future versions, we expect to include recommended actions for burners in highly non-optimal conditions as well as condensed assessment outputs that can be fed as inputs to neural networks used for overall boiler management.

28.2.2 Example Application

An example full-scale application of our burner-diagnostics system was installed at the Ameren Union Electric Meramec generating plant near St. Louis, Missouri, in mid-2001. Specifically, the Flame Doctor system was installed on the Unit 4 boiler at Meramec, which produces 350 MW of electric power with three rows of six Babcock and Wilcox DRB-XCL front-wall-fired pulverized coal burners (eighteen burners total). The existing Coen flame scanners (Fig. 28.10) were used to generate the monitoring signal. All of the scanners were left in their original locations (adjacent to the primary air and coal line and angled toward the burner centerline), and no special connections or adjustments were made to the scanner head. The data-acquisition hardware and signal-processing computer were installed in a relay room located next to the existing flame scanner signal processing cabinet (Fig. 28.11). A display screen was also set up in the Unit 4 control room so that the graphical user interfaces could be observed from there.

According to normal procedure, time series were collected for each burner for a sufficient length of time to obtain two minutes of uninterrupted dynamics. The recorded signals were then subjected to a series of quality checks to ensure that they met acceptable criteria (e.g. checks were made to determine if there were unacceptably large components of electronic noise or if the signal digital resolution was too low). Once a signal passed quality checking, it was further

Figure 28.10: Typical Coen scanner arrangement at Meramec.

processed to identify the locations of self-test events, so that these could be discounted from the dynamical analyses. Any signals identified as having unacceptable quality were flagged on the graphical user interface and withheld from further processing. All signals passing the quality check were fully characterized with time-series algorithms and identified by their relative proximity to the signals in the reference library.

Example signals collected from the Meramec boiler and their corresponding symbol sequence histograms are illustrated in Fig. 28.12. For the optimal flame, there are few identifiable unstable periodicities, and the histogram is relatively flat. As the instabilities grow worse, how-

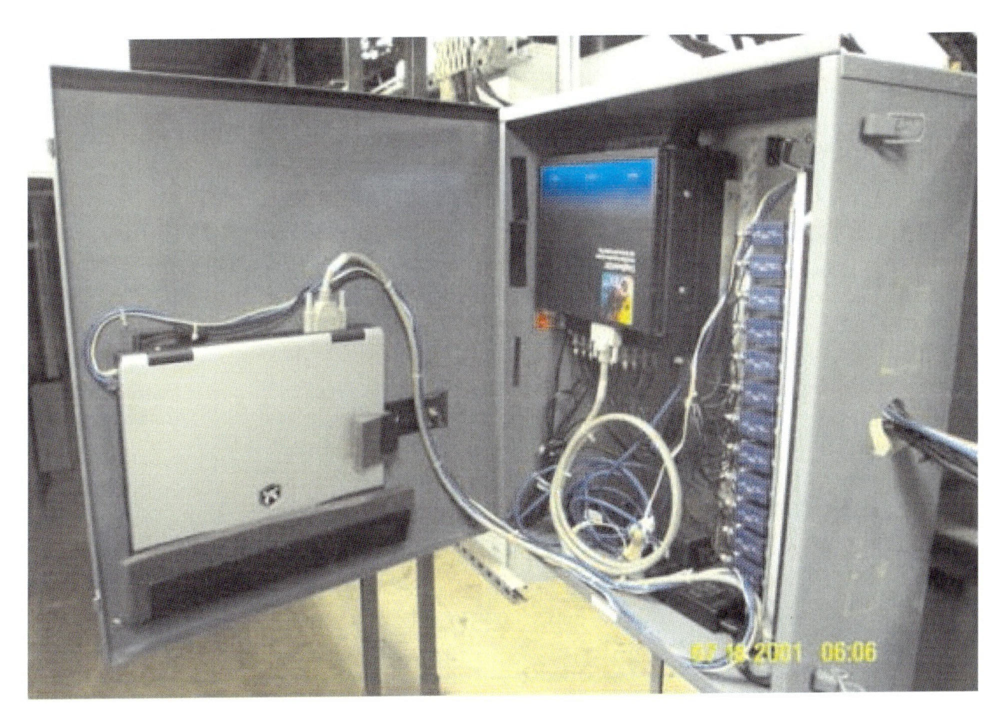

Figure 28.11: Data-acquisition and computer cabinet installed at Meramec.

ever, characteristic symbol sequence peaks appear and grow larger. The most severe reductions in flame quality correspond to conditions in which the flame becomes fully detached for significant periods.

The Flame Doctor graphical user interface includes a main status screen that summarizes the condition of all the burners together. Status is indicated by a colored icon and numerical quality rating for each burner. In the Meramec installation, the quality ratings ranged from eight to two, with eight being the optimum. A value of one meant that some type of scanner fault (i.e. bad signal quality) had been detected. Since the original Meramec installation, we have revised the quality scale to a numerical value ranging from 0 to 100 to increase discrimination. However, the basic approach for defining quality remains the same. In addition to providing a flame-quality assessment for each burner, the graphical user interface also provides an indication of the likely causes for poor flame quality. Many of the burners on Meramec Unit 4 exhibited varying degrees of flame detachment associated with an excessively high ratio of air to coal flow. Subsequent adjustments to reduce secondary air resulted in significant improvements in many of these burners.

A major recognized problem on Meramec Unit 4 has been the need to operate with excessively high air levels to avoid high carbon monoxide (CO) in the flue gas. This operation results in higher energy losses and difficulty in maintaining the lowest possible NO_x emissions. Using the indicated flame qualities as a guide, multiple burner adjustments were made with particular emphasis on several 'under-achieving' burners that were significantly lower in quality than the majority. These changes resulted in increasing the average burner assessment

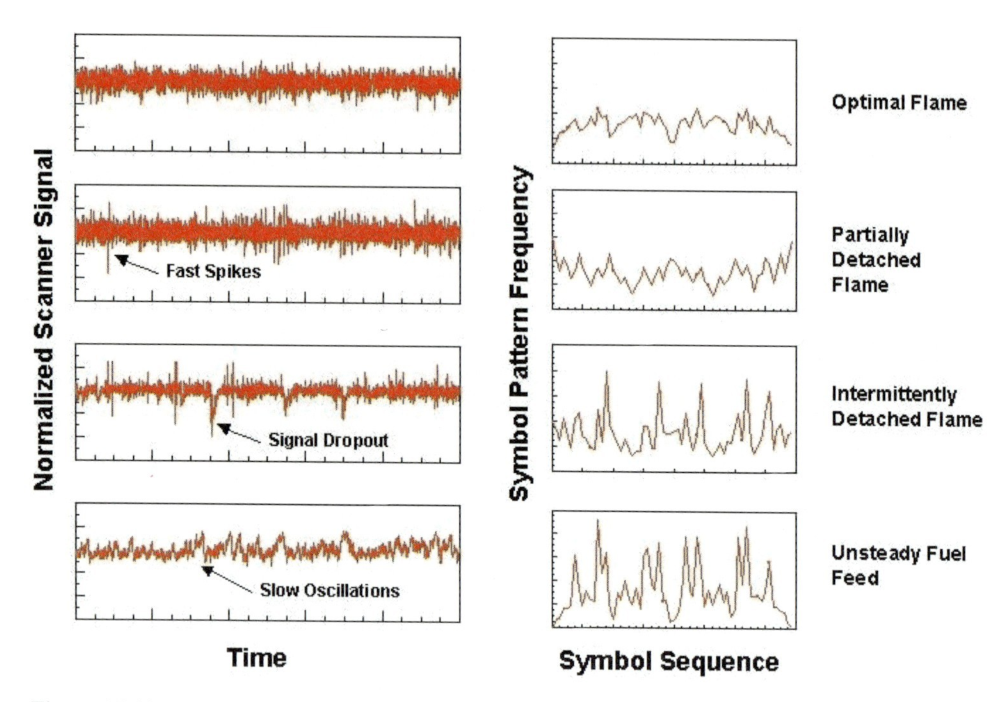

Figure 28.12: Example scanner output from Meramec showing signals for different flame states and corresponding symbol sequence histograms.

for the boiler from 6.5 to 6.7. The middle and bottom rows were most affected, increasing from 6.3 to 6.4 and 6.4 to 6.7, respectively. With this change in quality, the excess oxygen was reduced by 15% from what it had been previously to maintain acceptable CO and NO_x emissions. Further targeted adjustments on two individual burners increased the average boiler flame quality from 6.5 to 6.7 and decreased CO from > 1100 ppm (off scale on the instrument) to 600 ppm. The average NO_x level also dropped by 6%. Thus adjustments to just two burners had a disproportionate impact on overall boiler performance. Still further adjustments were able to reduce the CO levels to 300–400 ppm and lower NO_x by 10% from previous levels.

Key findings from the initial testing at Meramec Unit 4 can be summarized as follows:

- The Flame Doctor system correctly identified flame states associated with high primary and secondary air flows.
- There was typically a subset of burners (three or four) that deviated significantly from an optimal flame state. Adjustments to these burners resulted in disproportionate improvements in emissions.
- The subset of problem burners was not consistent over time (i.e. the burners with the most severe problems changed with time). Such shifts were especially likely when boiler load changed.
- Because of the complex connections between different parts of the air and coal supply systems, changes to individual burners frequently created unintended effects on other burners.

The last two observations are good illustrations of the need to have continuous burner monitoring. Even if problem burners are identified correctly at one point in time, this information can become quickly out of date as the burner state shifts due to load changes and control adjustments.

As a result of the positive results from the initial testing on Unit 4, Meramec plant engineers and test personnel continue to identify and correct poorly performing burners using the Flame Doctor system. Plans are now underway to connect the system to the plant's neural-net optimization system so that boiler control may be more fully automated.

28.2.3 Future Developments

Additional improvements to the Flame Doctor system are being pursued so that the power of dynamic burner monitoring can be further exploited. These improvements include:

- automated library updating;
- more detailed root-cause analysis;
- full integration with global plant control systems;
- extension to all types of coal burners;
- better utilization of numerical modeling; and
- direct incorporation into burner designs.

By automated library updating, we refer to the process by which the dynamic libraries can be automatically improved over time using artificial learning procedures such as adaptive pattern recognition. Such automated learning is needed to account for changes in burners due to ordinary equipment wear, fuel variations and changes in boiler operating or maintenance procedures. The need for flexibility and adaptability over time also applies to the root cause analysis component of diagnostics. We have already observed that information from the categorization of flame-scanner signals has led to improved understanding of burner engineering guidelines. In the future we expect that this improved understanding can be translated into more detailed feedback to boiler operators regarding appropriate actions to take in response to observed non-optimal behavior.

As described above in the Meramec example, it is expected that great benefits can be gained through connecting the burner-monitoring system to global boiler optimization systems such as neural networks. The single quality index produced for each burner is expected to be an ideal input for neural networks, so that training requirements are minimized. The neural-net inputs can be reduced by averaging groups of burners, but this also reduces sensitivity to individual problem burners. The existing monitoring system has been developed primarily for wall-fired pulverized coal burners, but there is no theoretical reason why flicker monitoring cannot also be used for cyclone and turbo-fired boilers. The main issue involves identifying appropriate flame-scanner locations and developing the required dynamic libraries. EPRI is currently funding projects in these areas.

To date, no adequate physical models have been developed to simulate the dynamics observed for coal-fired burners. Considerable effort has been expended in developing computer models, but these have been limited to time-average (e.g. Reynolds-averaged) descriptions of the combined effects of the fluid turbulence and combustion kinetics, without incorporating the actual dynamic behavior of the flame. We firmly believe that numerical computing power has now reached the stage where truly dynamic models of pulverized coal combustion are possible. One key feature of such models will be the strong nonlinear feedback between the main part of the flame and the ignition front. Once such models have been developed, it will be possible to understand the physics of flame bifurcations and the associated flicker with much greater certainty than before. This should greatly reduce the empirical work required to interpret flicker patterns and construct reference libraries. It should also lead to improved burner designs.

We believe that the burner industry is now on the verge of developing a new generation of burners with unprecedented performance. This will be achieved by combining feedback from measurements of flame flicker (or similar dynamic flame information) with new electromechanical, high-speed controls to create dynamic flame states previously unobtainable with conventional mechanical burners. Such a new generation of 'smart' burners will be critical to meeting the ever more stringent targets for emissions and energy efficiency.

28.2.4 Broader Implications

While the success of our burner-monitoring system has obvious value for utility boilers, we believe there are also broader implications for nonlinear process monitoring. One of the most important of these implications is that it appears that even highly complex, spatio-temporal systems can be effectively characterized in terms of global dynamics. Based on the observations reported here and previous experience with other combustion and multi-phase flow systems, we are inclined to describe such global dynamics as having the characteristics of noisy pseudo-low-dimensional systems. The low-dimensional features emerge from a high-dimensional background when the system approaches a global bifurcation point; i.e. a condition in which the global spatial and temporal features undergo large changes as some parameter is changed past a critical point. This implies that even though the inherent dynamics is high dimensional, only a few dimensions become dominant in the region of a global bifurcation. For any new system, one should seek out evidence that distinct dynamical operating regimes can occur with parametric changes. These known regimes should then form the basis for further experimentation and analysis.

Use of even crude dynamical signals should be considered for improving control of industrial processes. This is especially true where existing time-series measurements are already being made but are not currently utilized. It appears that techniques for characterizing noisy, nonlinear time series are quite useful for detecting and characterizing the pseudo-low-dimensional dynamics associated with global bifurcations. Specific analyses that seem particularly useful are symbol sequence analysis, measures of predictability (entropy) and time asymmetry. In our particular application, we have created specialized algorithms for singling out coarse-grain features of unstable periodicities that may be obscured by both measurement and dynamic noise. Such analyses can be used to generate multiple statistics that can identify the type of global bifurcation involved as well as its proximity to the current operating state. By using multiple statistics, it is possible to empirically group the observed dynamics into characteristic regions that coincide with qualitative maps of the possible global states. Thus, successful monitoring strategies can be developed even for spatio-temporal systems for which explicit models are not available.

A particular feature of boiler applications is that one is simultaneously monitoring many individual burners. This multiplicity of dynamical samples (both over space and time) has turned out to be of benefit, because it makes it possible to use burner-to-burner variability as a gauge for indicating level of significance. In other words, by considering measures of variability (e.g. standard deviation) between the dynamical features from burner-to-burner vs. variability for each burner over time, we are able to quickly determine appropriate significance values that distinguish important operating differences from noise. Such differences can also be used to make normalized comparisons between boilers to indicate the relative levels of control.

Even though there is considerable coupling among individual burners in large boilers, the use of single pilot burners to define the characteristics of burners in boilers has been surprisingly successful. This offers hope that such an approach can also work in other production or plant-scale systems where there are strong interactions between the basic dynamical units. Use of physically based models for the individual units will always be preferable when such models are available. However, in many industrial contexts, such models may be extremely crude or non-existent, and it will be necessary to rely on empirical or semi-empirical experience of engineers and/or plant operators. We recommend that such empirical knowledge be utilized as much as possible to supplement the dynamical information coming from measured time series.

Acknowledgment

Development of the Flame Doctor® burner-monitoring system has been sponsored by the Electric Power Research Institute (EPRI). The Meramec Unit 4 testing was sponsored by both EPRI and Ameren Union Electric Power.

Bibliography

[1] U.S. Environmental Protection Agency, *Inventory of U.S. Greenhouse Gas Emissions and Sinks: 1990–2000* (U.S. EPA Report No. EPA 430-R-02-003, 2002).

[2] *Steam: Its Generation and Use*, The Babcock and Wilcox Company, New York, 1963, Chapter 1.

[3] C.S. Daw, C.E.A. Finney, and E.R. Tracy, Review of Scientific Instruments **74** (2003), 915.

[4] C.S. Daw, C.E.A. Finney, and M.B. Kennel, Physical Review E **62** (2000), 1912.

29 Towards Constructive Nonlinear Dynamics – Case Studies in Chemical Process Design

M. Mönnigmann, J. Hahn, and W. Marquardt

Parameter continuation and numerical bifurcation analysis are two methods from Nonlinear Dynamics which have successfully been applied to chemical process models. While these methods have become established tools, a shift can be detected in the way these techniques are used in the chemical engineering community.

Early applications of applied Nonlinear Dynamics to chemical processes focused on gaining insight into the fundamental chemical and physical interactions and their impact on stability or on dynamics in general. In contrast, some more recent articles focus on obtaining information that is not merely meant to deepen the understanding of a given chemical process, but to aid process *design*. Information on the location of saddle-node, Hopf and other bifurcation points is valuable to a designing engineer, as it allows the avoidance of parameter regions that correspond to unstable or otherwise undesired process behavior. In this sense, a shift is taking place in the chemical engineering community from Nonlinear Dynamics analysis *for a deeper understanding* towards *analysis for design*. As an example of analysis for design, the present paper demonstrates the use of bifurcation analysis in controller design. By generalizing some of the results obtained for this case study, a procedure for bifurcation-based controller tuning can be derived.

Despite being very powerful, a numerical bifurcation analysis becomes tedious or even impossible for models with many parameters. The application of an analysis to guide the design of chemical processes is therefore limited to models with only a few parameters. To overcome this limitation, a new approach has been developed that incorporates information on the location of bifurcation points into optimization-based process design. This novel approach does not rely on analyzing, visualizing and interpreting the visualized information on the location of critical points in the process parameter space. The new approach is therefore in principle not restricted with respect to the number of process parameters. The present paper summarizes the progress made in the development of this constructive Nonlinear Dynamics method and gives two illustrative examples.

29.1 Nonlinear Dynamics Analysis in Chemical Engineering

Nonlinear dynamics analysis has a long tradition in chemical engineering research. While it is beyond the present article to attempt a historic review, it is noted that appreciable interest of chemical engineers in nonlinear dynamics can be traced back to the 1950s. Van Heerden [19],

Nonlinear Dynamics of Production Systems. Edited by G. Radons and R. Neugebauer

ISBN 3-527-40430-9

Bilous and Amundson [8] and Aris and Amundson [1–3] analyzed the most prominent example of a chemical engineering model to which nonlinear dynamics analysis has been applied, a continuously stirred tank reactor (CSTR) with a first-order exothermic reaction. Before numerical methods were available, these authors showed by analytical calculations that even the simple CSTR model can exhibit multiple steady states. In fact, earlier articles on the topic exist which had not, however, created persistent interest. Most notably, Liljenroth [24] reported the existence of multiple steady states in 1918 for a CSTR with a first-order exothermic reaction. In the Russian literature, Zeldovich and Zisin [36] reported the discovery of multiple solutions caused by isolated solution branches in the same classical CSTR example. For a review of the work until the 1980s the reader is referred to Razón and Schmitz [30].

Progress in the application of nonlinear dynamics analysis to chemical engineering problems can be accounted to both theoretical advances and the advent of powerful numerical methods.

Catastrophe theory [32] allows us to infer a lower bound on the number of multiple solutions in the vicinity of critical points. Similarly to catastrophe theory, singularity theory with a distinguished parameter [15, 14] predicts the topology of the steady state manifold in the vicinity of critical points. In both catastrophe and singularity theory, the term *critical point* refers to points on the steady state manifold of the dynamic system of interest, at which derivatives with respect to either state variables or the distinguished parameters vanish. By evaluating derivatives only at the critical point, these theories allow us to infer the existence of bifurcation diagrams. These bifurcation diagrams reveal information about the system behavior over an interval of a system parameter, as opposed to the point-like information contained in the derivatives. One strength of catastrophe and singularity theory lies in being able to predict the behavior of the system in a vicinity of the critical point based on only local information.

At first sight, catastrophe and singularity theory with a distinguished parameter are restricted with respect to their applicability, as they deal with scalar dynamic systems, i.e. models with a one-dimensional state space only. Historically, catastrophe and singularity theory were in fact first applied to chemical engineering models which can be reduced to a single nonlinear equation, such as the classic CSTR with a first-order exothermic reaction. By means of the Lyapunov–Schmidt (LS) reduction (see [14] for the LS reduction in the context of singularity theory), these results can be extended to models with more than one state variable. The use of the LS reduction became important in the context of numerical continuation, which is routinely conducted in the analysis of systems with several hundred state variables today.

Bifurcation theory [16, 23] addresses similar questions as catastrophe and singularity theory from a different perspective. In bifurcation theory, the focus is on changes of the phase portraits of a dynamic system under variations of process parameters. While in catastrophe and singularity theory the focus is on the existence and multiplicity of solutions, bifurcation theory also addresses the stability of these solutions.

The availability of numerical parameter continuation methods accounts to a great extent for the attractiveness of nonlinear dynamics analysis in chemical engineering. Parameter continuation has been used for the analysis of chemical engineering processes for more than two decades. Starting from the numerical bifurcation and singularity analysis of very simple process models [35], these methods have been successfully applied to the analysis of the nonlinear dynamics of industrially relevant processes. Examples can be found in the work of Ray and Villa on polymerization [29], or the more recent analysis of an industrial decanting reactor [18,

21]. Along with the application to larger, more realistic process models, implementations of continuation methods and methods for the detection of critical points from applied bifurcation theory have become more powerful. The software package auto2000 and its predecessors [11] has often been used by researchers in chemical engineering. Other software packages are available, for example CONTENT [22], or DIVA [25] which has successfully been used on large-scale chemical engineering systems.

In applying continuation-based bifurcation analysis, the focus has shifted from *analyzing* nonlinear dynamics of chemical engineering processes to using information on the nonlinear dynamics from such an analysis in *designing* a process [28].

Many early works which use continuation merely report the results of a numerical or analytical bifurcation or singularity analysis of a certain process model. These reports were meant to make the information on the location of critical points in the process parameter space available to the community for further research. Historically, research of this type was driven by the wish to identify the maximum number of multiple steady states, the maximum number of topologically distinct bifurcation diagrams or all types of topologically distinct phase diagrams for a given system. This type of investigation yields maps of the model parameter space which identify the regions of the parameter space in which the respective multiplicity or dynamic behavior exists. The early works by Van Heerden [19], Bilous and

Amundson [8] and Aris and Amundson [1–3], for example, identified three topologically distinct bifurcation diagrams and a maximum multiplicity of three. This work inspired other researchers to determine the maximum number of multiple steady states and topologically different bifurcation diagrams. Golubitsky and Keyfitz [13] in fact identified four more bifurcation diagrams. The work on the classic CSTR with a first-order reaction was followed by papers on consecutive reactions $A \to B \to C$ and parallel reactions $A \to B$, $A \to C$ for which 23 and 48 possible bifurcation diagrams were identified, respectively (see [30] and references therein).

In contrast to the early papers which concentrated on the analysis of the possible types of dynamic behavior, more recent applications of continuation and bifurcation analysis in chemical engineering focused on the use of information on the location of bifurcation points in the process parameter space for *process design* [7, 6, 5, 4, 31, 21, 18]. A typical approach is to (i) analyze the process using continuation, then (ii) visualize the results and finally (iii) design the process based on an interpretation of these visualized results. For ease of reference, an approach based on these three steps will be called analysis-based design in the following.

29.2 Analysis-based Process Design

A simple model of a continuous fermentation process is used to illustrate that a straightforward process optimization may fail if the dynamic behavior of the system is neglected. Subsequently, some results obtained by numerical bifurcation analysis and parameter continuation are presented. These results reveal the location of the critical points of the model which cause the loss of process stability.

29.2.1 Illustrative Example

A simple model of the fermentation process is given by

$$\begin{aligned} \dot{X} &= -\frac{F}{V} X + \mu(S)\, X, \\ \dot{S} &= \frac{F}{V}\, (S_F - S) - \sigma(S)\, X, \end{aligned} \tag{29.1}$$

where X [$\mathrm{kg\,m^{-3}}$] and S [$\mathrm{kmol\,m^{-3}}$] are the cell concentration and the substrate concentration, F [$\mathrm{kg\,s^{-1}}$], S_F [$\mathrm{kmol\,m^{-3}}$] and V [$\mathrm{m^3}$] denote the feed, the substrate concentration in the feed and the reactor volume, while

$$\begin{aligned} &\mu(S) = k\, S\, \exp(-S/K)\, [\mathrm{s^{-1}}] \quad \text{and} \\ &\sigma(S) = \mu(S)/(a + b\, S)\, [\mathrm{kg\,kmol^{-1}\,s^{-1}}] \end{aligned}$$

are the growth and substrate consumption rates, respectively. In the calculations that follow, the parameters $k = 1\,\mathrm{m^3 (kmol\,s)^{-1}}$, $K = 0.12\,\mathrm{kmol\,m^3}$, $a = 5.4\,\mathrm{kg\,kmol^{-1}}$, $b = 180\,\mathrm{kg\,m^3\,kmol^{-2}}$ and $V = 161\,\mathrm{m^3}$ are chosen. The substrate feed to the CSTR is a mixture of a concentrated feed stream with concentration $S_\mathrm{c} = 1.0\,\mathrm{kmol\,m^{-3}}$ and a diluted feed stream with concentration $S_\mathrm{d} = 0.3\,\mathrm{kmol\,m^{-3}}$. This imposes the inequality constraints

$$0.3\,\mathrm{kmol\,m^{-3}} \le S_F \le 1.0\,\mathrm{kmol\,m^{-3}} \tag{29.2}$$

on the process. Below it is assumed that the cost coefficients of the concentrated and diluted substrate feed streams are $c_\mathrm{c} = 200\,\$\,\mathrm{m^{-3}}$ and $c_\mathrm{d} = 10.8\,\$\,\mathrm{m^{-3}}$, respectively. For details of the process model we refer to an article by Brengel and Seider [9].

In a first attempt to find an optimal point for continuous operation of the fermenter model, an optimization with respect to the cost function

$$\phi(X, F, S_F) = c_\phi\, (c_1 X F - c_2(S_F) F) \tag{29.3}$$

is performed. The function ϕ models the profit resulting from produced cells diminished by the cost of the substrate in the feed. Note that due to the mixing of two substrate feed streams with different cost coefficients, the cost of the mixed stream is a function of S_F. Using the cost coefficients of the concentrated and diluted streams given above, $c_2(S_F)$ is given by

$$c_2(S_F) = -70.3\,\$\,\mathrm{m^{-3}} + 270.3\,\$\,\mathrm{kmol^{-1}}\, S_F. \tag{29.4}$$

In order to find a maximum of the profit function (29.3), the following nonlinear program is solved with NPSOL [12]:

$$\max_{X,\, S,\, S_F,\, F} \phi(X,\, F,\, S_F) \tag{29.5}$$

subject to

$$0 = -\frac{F}{V} X + \mu(S)\, X, \tag{29.6}$$

$$0 = \frac{F}{V}\, (S_F - S) - \sigma(S)\, X, \tag{29.7}$$

$$0 \le S_F - 0.3\,\mathrm{kmol\,m^{-3}}, \tag{29.8}$$

$$0 \le 1.0\,\mathrm{kmol\,m^{-3}} - S_F, \tag{29.9}$$

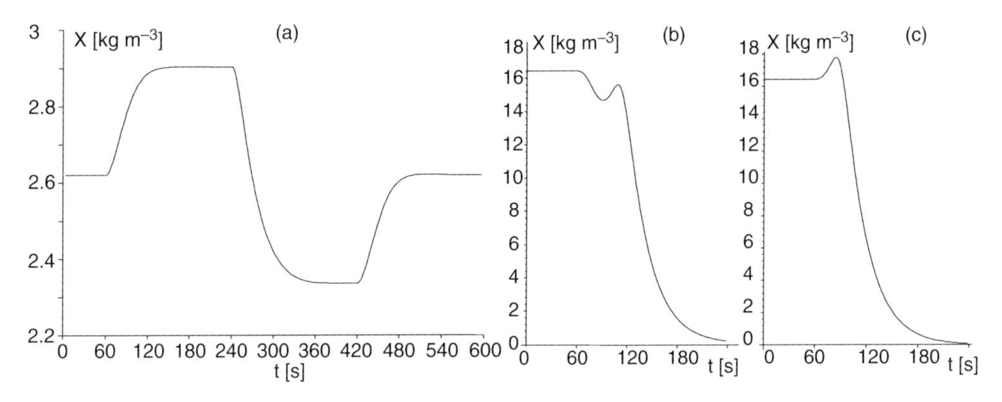

Figure 29.1: (a) Dynamics after disturbances to $S_F = (0.33, 0.27, 0.3)\,\mathrm{kmol\,m^{-3}}$ at $t = (60, 240, 420)\,\mathrm{s}$, respectively, for the stable solution (29.10). (b) Dynamics after a disturbance to $S_F = 0.97\,\mathrm{kmol\,m^{-3}}$ at $t = 60\,\mathrm{s}$ for the unstable optimum (29.11) and (c) after a disturbance to $S_F = 1.03\,\mathrm{kmol\,m^{-3}}$.

where the first two equality constraints result from requiring the model equations (29.1) to be at steady state, and the remaining inequality constraints result from Eq. (29.2). The optimization is started at the steady state solution

$$X = 2.6201\,\mathrm{kg\,m^{-3}}, \quad S = 0.022443\,\mathrm{kmol\,m^{-3}}, \tag{29.10}$$

for the parameter values $F = 3.0\,\mathrm{kg\,s^{-1}}$ and $S_F = 0.3\,\mathrm{kmol\,m^{-3}}$. A simple simulation suggests that the steady state given in Eq. (29.10) is stable, cf. Fig. 29.1a. This is confirmed by calculating the eigenvalues of the linearized process model at this steady state. The eigenvalues are $\lambda_{1,2} = -0.5363 \pm 0.24660\,i$.

Solving the nonlinear program (29.5) with NPSOL [12] results in an optimal point for continuous operation:

$$\begin{aligned} X &= 16.441\,\mathrm{kg\,m^{-3}}, S = 0.068003\,\mathrm{kmol\,m^{-3}}, \\ F &= 6.2184\,\mathrm{kg\,s^{-1}}, S_F = 1.0\,\mathrm{kmol\,m^{-3}}. \end{aligned} \tag{29.11}$$

Figure 29.1b and c clearly show that this point is not stable. The eigenvalues of the linearized process, $\lambda_{1,2} = 0.049605 \pm 0.079880\,i$, evaluated at this point confirm this result.

29.2.2 Continuation Analysis

Bifurcation analysis by numerical continuation reveals the cause for the loss of stability. Figure 29.2a shows the result for a variation of F while the parameter $S_F = 0.3\,\mathrm{kmol\,m^{-3}}$ remains fixed. Starting at a known steady state of the process model (29.1), a close-by solution can be found by approximating the curve of solutions by its tangent vector, and subsequently solving Eq. (29.1) at steady state for the new value of F. By making the step in F sufficiently small, the previous solution can be made an arbitrarily good starting point for e.g. a Newton solver. By repeated calculation of neighboring steady states, a curve of solutions like the curve in Fig. 29.2a can be built up.

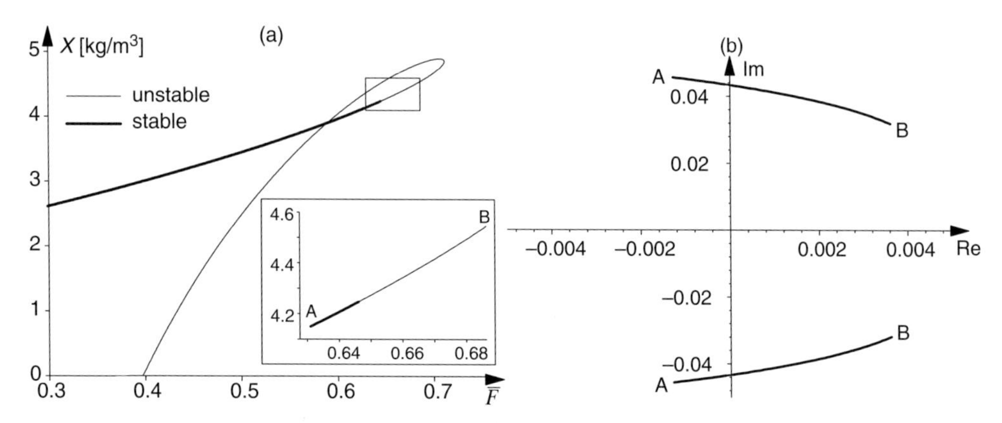

Figure 29.2: Continuation of equilibria (a) for $S_F = 0.3\,\mathrm{kmol\,m^{-3}}$ and eigenvalues along the enlarged part of the solution curve (b).

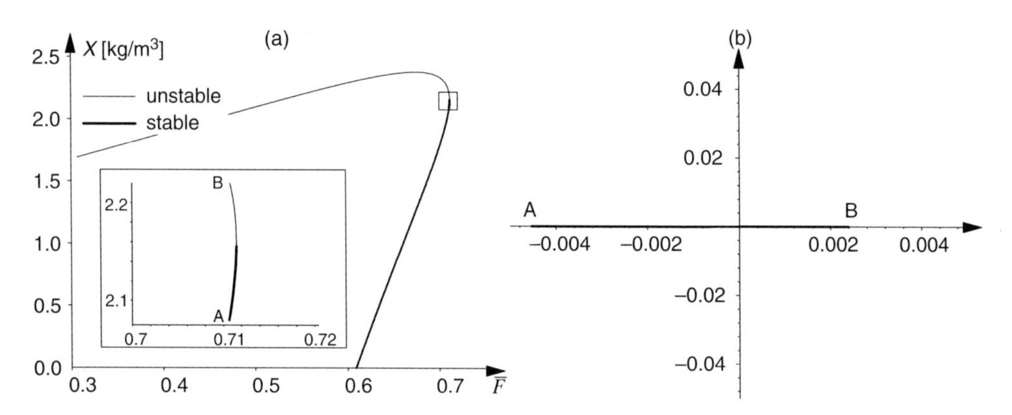

Figure 29.3: Continuation of equilibria (a) for $S_F = 0.2\,\mathrm{kmol\,m^{-3}}$ and eigenvalues along the enlarged part of the solution curve (b).

While continuation proceeds, the stability boundary of the model can be detected by monitoring the eigenvalues of the linearized model. Figure 29.2b shows eigenvalues for a part of the curve in Fig. 29.2a in the vicinity of the critical value of the parameter F at which stable steady states turn into unstable steady states.

A loss of stability can generally occur in two ways. Since the Jacobian of the process model is a real matrix which is in general not symmetric, both real eigenvalues and complex-conjugate pairs of eigenvalues may exist. Stability may therefore be lost as either a complex-conjugate pair of eigenvalues or a single real eigenvalue crosses the imaginary axis. The case shown in Fig. 29.2b corresponds to a Hopf bifurcation, while the case of a single real eigenvalue leads to a saddle-node bifurcation. An example of a saddle-node bifurcation for the fermenter is shown in Fig. 29.3.

Just as a curve of steady states was found by parameter continuation starting at a known steady state of the dynamic system (29.1), a curve of Hopf bifurcations or saddle-node bifurca-

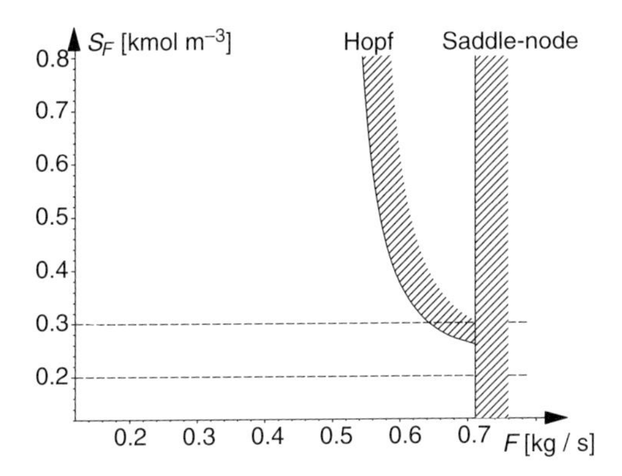

Figure 29.4: Curve of Hopf and saddle-node bifurcations in the (S_F, F)-parameter plane. Figures 29.2 and 29.3 were obtained by plotting X along the dashed lines at $S_F = 0.3\,\mathrm{kmol\,m^{-3}}$ and $S_F = 0.2\,\mathrm{kmol\,m^{-3}}$, respectively.

tions can be computed by parameter continuation starting from a known Hopf or saddle-node bifurcation, respectively. Note that this involves freeing a second model parameter. Figure 29.4 shows the result for the fermenter.

Figure 29.4 reveals the location of the stability boundary in the (F, S_F)-parameter plane. For parameter values to the left of the curve of saddle-node bifurcations and to the left and below the curve of Hopf bifurcations, only stable steady states exist. If the optimization of the fermenter sketched in Sect. 29.2.1 is restricted to this region, the optimum will be stable in contrast to the optimal point found in Sect. 29.2.1. In this sense, the continuation-based bifurcation analysis can provide information to the designer of which parts of the process parameter space should be avoided.

29.3 Analysis-based Control System Design

While bifurcation analysis has been extensively used to determine stability boundaries for uncontrolled processes, little work has been done on applying bifurcation analysis to systems under closed-loop control. Nevertheless, important information, e.g. what controller tuning is required for stability, can be gained from analyzing closed-loop systems. Towards this end, bifurcation analysis has been used as a tool to determine controller tuning parameters to achieve robust stability of a nonlinear process for certain forms of model mismatch. It is also possible to draw conclusions about the effect that certain types of model uncertainties have on the closed-loop system [17].

Application of bifurcation analysis to closed-loop systems is best illustrated by an example followed by a more general discussion.

29.3.1 Illustrative Example

Consider a continuously stirred tank reactor in which an exothermic, irreversible reaction, $A \rightarrow B$, is occurring [34]. Assuming constant liquid volume, the following dynamic model can be derived from a component and an energy balance:

$$\dot{C}_A = \frac{q}{V}\left(C_{A\mathrm{f}} - C_A\right) - k_0 \exp\left(-\frac{E}{RT}\right) C_A, \tag{29.12}$$

$$\dot{T} = \frac{q}{V}\left(T_\mathrm{f} - T\right) - \frac{\Delta H}{\rho C_\mathrm{p}} k_0 \exp\left(-\frac{E}{RT}\right) C_A + \frac{UA}{V\rho C_\mathrm{p}}\left(T_\mathrm{c} - T\right). \tag{29.13}$$

The values of the parameters and the nominal operating conditions for this process can be found in [20]. For this example the cooling fluid temperature, T_c, is chosen as the manipulated variable and the reactor temperature, T, is measured. This results in a single-input single-output system with two states. A bifurcation diagram of the open-loop system is shown in Fig. 29.5a, where the temperature of the cooling fluid is the bifurcation parameter. The equilibria consist of two stable branches and one unstable branch connecting the two stable ones. Under the assumption that all states are known, a feedback linearizing controller with integral action can be designed for this system:

$$u = \frac{-\frac{q}{V}\left(T_\mathrm{f} - T\right) + \frac{\Delta H}{\rho C_\mathrm{p}} k_0 \exp\left(-\frac{E}{RT}\right) C_A + \frac{UA}{V\rho C_\mathrm{p}} T}{\frac{UA}{V\rho C_\mathrm{p}}}$$

$$+ \frac{\frac{2}{\epsilon}\left(T_\mathrm{sp} - T\right) + \frac{1}{\epsilon^2}\int_0^t \left(T_\mathrm{sp} - T\right) d\tau}{\frac{UA}{V\rho C_\mathrm{p}}}, \tag{29.14}$$

$$T_\mathrm{c} = u. \tag{29.15}$$

This controller contains only one tuning parameter, ϵ, which corresponds to the closed-loop time constant. The integral action of the controller will eliminate offset between the set point and the measured output. Assuming that there is no mismatch between the plant and the model, the controller (29.14) results in a system that has a stable input–output behavior as well as a stable internal dynamics for any value of ϵ and any set point, T_sp, within the operating region. This can be observed in Fig. 29.5b where the unstable branch of equilibria has been stabilized. Unlike the open-loop case, the bifurcation parameter for the closed-loop bifurcation diagram is

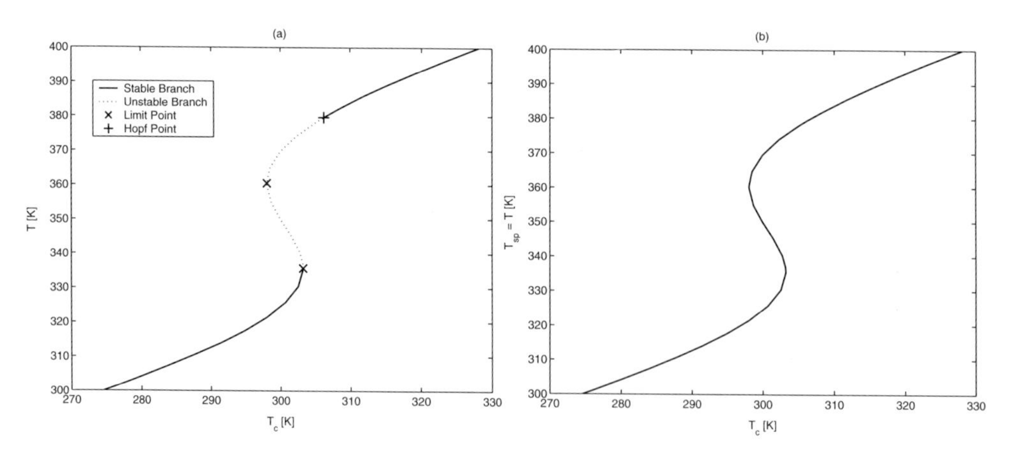

Figure 29.5: Bifurcation diagram of the open-loop system (a), closed-loop system (b).

chosen to be the set point for the reactor temperature. The reason for this is that the temperature of the cooling fluid cannot be used as a bifurcation parameter for the closed-loop case, since the controller will supply this variable as its output. At the same time the set point will be equal to the reactor temperature at steady state due to the integral action in the controller.

For the nominal case the value of ϵ can be chosen arbitrarily small, resulting in a very fast response. However, in reality there is always some mismatch between the plant and the model, which can lead to restrictions for controller tuning.

Parametric Uncertainty. One type of plant–model mismatch is parametric uncertainty. For example, uncertainty in the heat-transfer coefficient, the reaction-rate constants or the thermodynamic properties of the components can exist for this model. While uncertainties in all of these parameters have been investigated, only the results for the heat-transfer coefficient UA will be dicussed in more detail in this text. These investigations lead to the conclusion that there exists an upper bound for the value of ϵ when parametric uncertainty is considered. The larger the uncertainty in the parameter, the tighter the bound on the controller tuning parameter. Tuning the controller with a smaller value of ϵ than this bound will guarantee robustness against parametric uncertainty for the investigated case over the entire operating region [17].

Structured Uncertainty. Another form of plant–model mismatch to be considered is unmodeled dynamics. The effect of unmodeled dynamics on the achievable closed-loop performance can be investigated when bifurcation analysis is performed on a system that contains the most important part of the fast dynamics of a model, while a controller that has no knowledge about this dynamics is used to control it. For this case study, the plant model is augmented by

$$
\begin{aligned}
\epsilon_v \dot{T}_{\mathrm{c}} &= -T_{\mathrm{c}} + z, \\
\epsilon_v \dot{z} &= -z + u,
\end{aligned}
\tag{29.16}
$$

where ϵ_v corresponds to the time constant describing this fast dynamics. Equations (29.16) replace the original equation (29.15) for this investigation. The goal is to tune the controller such that robust stability is guaranteed for the closed loop even when the controller model does not include the dynamics described in Eqs. (29.16).

Investigation of the closed-loop system under the assumption of unmodeled dynamics given by Eq. (29.16), but no parametric uncertainty, reveals that the unmodeled dynamics results in a lower bound on the controller tuning parameter, ϵ, for any fixed value of ϵ_v. If ϵ is chosen to be greater than the critical value for a specific ϵ_v, then the system will be stable over the entire operating region. It is also possible to determine that the closed-loop system will remain stable for any value of ϵ_v that is smaller than the one that was used for the design [17].

Robust Controller Tuning. So far it can be concluded that there are upper and lower bounds for the controller tuning parameter ϵ for this case study. The upper bound results from uncertainty in the model parameters while the lower bound is caused by unmodeled dynamics. In a final step, the worst case which includes both parameter uncertainty and unmodeled dynamics is investigated. The results for this case are shown in the bifurcation diagram in Fig. 29.6, where Hopf curves separate the stable region for the controller tuning parameter from the unstable regions. Summarizing these results, it can be stated that the value of the controller tuning parameter ϵ for the worst-case scenario of

- uncertainty in the parameter UA of up to $\pm 10\%$
- unmodeled dynamics of the form of Eq. (29.16) with $\epsilon_v \leq 0.02\,\mathrm{min}$

can be determined from the diagram shown in Fig. 29.6. Any value of ϵ between the peak values of $0.0594\,\mathrm{min}$ and $7.20\,\mathrm{min}$ will result in robust stability of the closed-loop system over the entire operating region and for any plant–model mismatch as described. In order to achieve good performance in addition to robustness it is recommended to use a value of ϵ that does not lie directly on the stability boundary (e.g. $\epsilon = 0.5\,\mathrm{min}$ for this case).

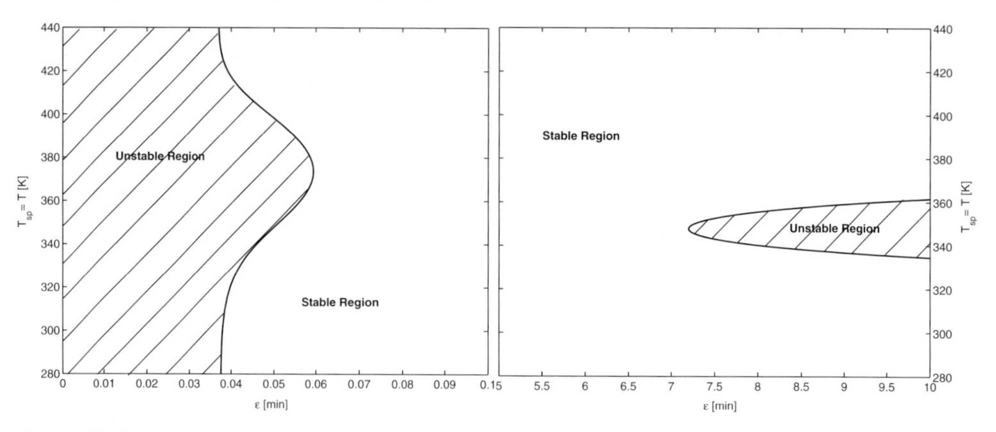

Figure 29.6: Regions of stability based upon the controller tuning parameter ϵ.

29.3.2 Controller Tuning Procedure

The previous subsection illustrated the controller tuning method by applying it to a specific example. However, the same tuning method has a more general applicability. No specific restrictions about the process to be controlled or the type of controller to be tuned exist.

For the special case of feedback linearizing controllers the following steps should be included in the controller tuning process:

1. Design the controller based upon the nominal model and implement it on the simulated process.

2. Investigate stability of the internal dynamics of the closed-loop system using bifurcation analysis. The internal dynamics should be stable at the operating point as well as over the entire operating region.

3. Determine a nominal performance requirement and tune the controller accordingly. This is trivial to do for the nominal case.

4. Identify the main sources of parametric uncertainty in the model. Subsequently, analyze the closed-loop system using bifurcation analysis under the assumption of parametric uncertainty. This can result in restrictions on the controller tuning parameters.

5. Include additional fast dynamics in the plant model but not in the model used for controller design. Perform bifurcation analysis on the closed-loop system. This investigation might place further bounds on the controller tuning parameters.

6. Investigate stability of the closed-loop process under the worst possible combination of parametric uncertainty and unmodeled dynamics over the entire operating region. Determine the restrictions that this places on the controller tuning parameters.

7. Check if the controller tuning parameters that satisfy the nominal performance requirement will also guarantee robust stability. If this is the case then the controller tuning process is complete. Otherwise, the controller has to be retuned in order to guarantee robust stability. It should be pointed out that it is desirable to use controller tuning parameters that do not lie close to a region of instability of the closed-loop process in order to also achieve good robust performance.

Following these steps will result in a controller that is tuned such that it meets nominal stability, nominal performance, as well as robust stability requirements [17].

29.4 Limitations of Analysis-based Design

While the examples discussed so far had a limited number of parameters, this is not the case for most models of chemical engineering processes. These systems often have a variety of parameters which can be varied in continuation, or become degrees of freedom in optimization. This section discusses how continuation-based analysis can be used to address high-dimensional parameter spaces. In the discussion, limitations will become apparent, which are addressed in Sect. 29.5.

Continuation can be used to analyze models with more than two parameters in two different ways: (i) analysis of high codimension critical points and (ii) analysis by collecting low codimension diagrams, as detailed in the sequel of this section. In practical applications both approaches are often combined.

Higher Codimension Critical Points. In Sect. 29.2, the analysis of the fermenter example started out with a steady state continuation, cf. Figs. 29.2 and 29.3. Along the curves of steady states obtained, saddle-node or Hopf bifurcations were detected. Subsequently, one-parametric continuations were started at these critical points to determine the saddle-node and Hopf locus as shown in the (F, S_F)-parameter plane in Fig. 29.4. Note that while the continuation of Hopf and saddle-node bifurcations in Fig. 29.4 yields a *one*-parametric curve, it involves the variation of the *two* model parameters F and S_F.

Along the curves of saddle-node and Hopf bifurcations, critical points of higher order can be detected. Figure 29.7b illustrates this for a cusp singularity detected along the curve of saddle-node bifurcations shown in Fig. 29.7a. The sketch of the one-parametric curve of cusp singularities in Fig. 29.7b now involves the variation of three model parameters, as opposed to two parameters in the saddle-node or Hopf bifurcation continuation, and as opposed to one parameter in the steady state continuation.

The emerging pattern is that higher-order critical points may exist on a curve of critical points. The subsequent continuation of critical points of higher order involves freeing an

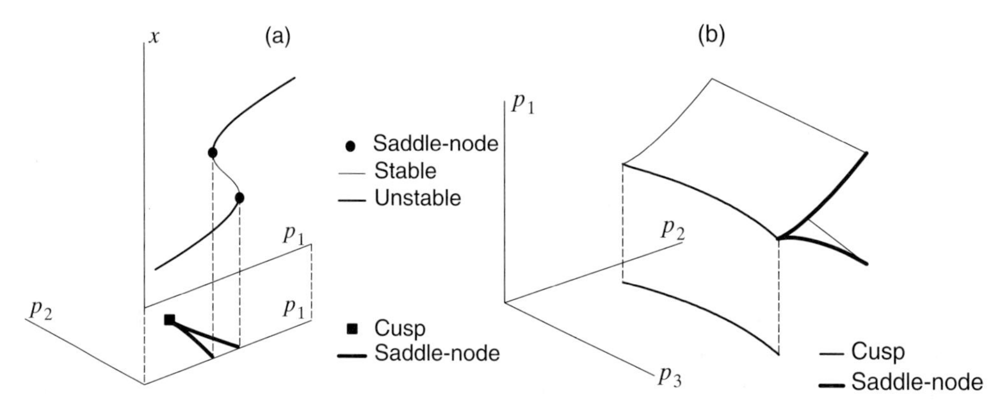

Figure 29.7: One-parameter continuation of steady states and two-parameter continuation of saddle-node bifurcation point (a), and three-parameter continuation of cusp singularities (b).

additional parameter. The number of free parameters needed for the continuation of a given critical point is denoted as its codimension. A precise definition of the term *codimension of a critical point* involves the notion of a universal unfolding (see, for example, Golubitsky and Schaeffer [14]). The term is used in a slightly different manner in catastrophe, singularity and bifurcation theory. In the present context, the definition based on the number of free parameters in continuation suffices.

The fact that continuation of higher codimension critical points involves more parameters can be exploited to some degree for an analysis of dynamic systems with a high-dimensional parameter space. The three-parameter continuation in Fig. 29.7b, for example, reveals how the saddle-node locus moves in the (p_1, p_2) plane as the third parameter p_3 is varied. Since the saddle-node locus is the stability boundary for the shown system, and since the cusp singularity bounds the saddle-node locus in Fig. 29.7, this information is valuable for a process designer who intends to avoid unstable process designs.

Categorizing critical points by their codimension reveals that a hierarchy of critical points exists. For saddle-node and Hopf bifurcation continuation, the model system is augmented by equations which ensure that a real eigenvalue or a complex conjugate pair of eigenvalues exists on the imaginary axis of the complex plane. For a cusp continuation, the saddle-node system is further augmented by additional equations which state that a particular function of second-order derivatives of the model system is zero. Points which fulfill the augmented system for cusp singularities therefore also fulfill the augmented system for saddle-node bifurcations. In turn, saddle-node bifurcations are steady state solutions of the model. As the analysis moves up in such a hierarchy, the number of model parameters to be varied increases. Therefore, hierarchies of this type allow the user to systematically analyze a dynamic system with a high-dimensional parameter space.

As the codimension increases in such an analysis, the obtained diagrams become more difficult to interpret. In fact, it is typically not possible to interpret a single, isolated diagram from a high codimension analysis, but diagrams across the levels of the hierarchy have to be interpreted jointly. The curve of cusp points in Fig. 29.7b is not useful by itself to a design engineer, as only the saddle-node points of codimension one reveal in which direction the unstable equilibria emerge. More importantly, only rather incomplete hierarchies of critical points exist. Golubitsky and Schaeffer [14] present hierarchies up to four-parameter continuation for those critical points related to saddle-node bifurcations. Clearly this still restricts the dimensionality of the parameter spaces which can be thoroughly analyzed. For those higher-order critical points emerging from Hopf bifurcations, the hierarchy is less complete [14].

While high codimension analysis of dynamic systems has been used in chemical engineering to analyze simple models such as the classic CSTR, it has apparently not been applied to larger, more realistic models.

Collection of Diagrams. Another approach to dealing with high-dimensional parameter spaces is to re-calculate curves after a variation of parameters which are fixed in the continuation. Figure 29.8 illustrates this idea. Figure 29.8a sketches how the repeated steady state continuation of a simple dynamic system results in a three-dimensional steady state manifold. Even though visualizations can comprise at most three dimensions, repeated steady state continuation for various values of the parameters p_i, $i \neq 1$, will in principle allow us to analyze a steady state manifold in high-dimensional parameter space. Similarly, repeated continuation of higher-order critical points is possible, cf. Fig. 29.8b.

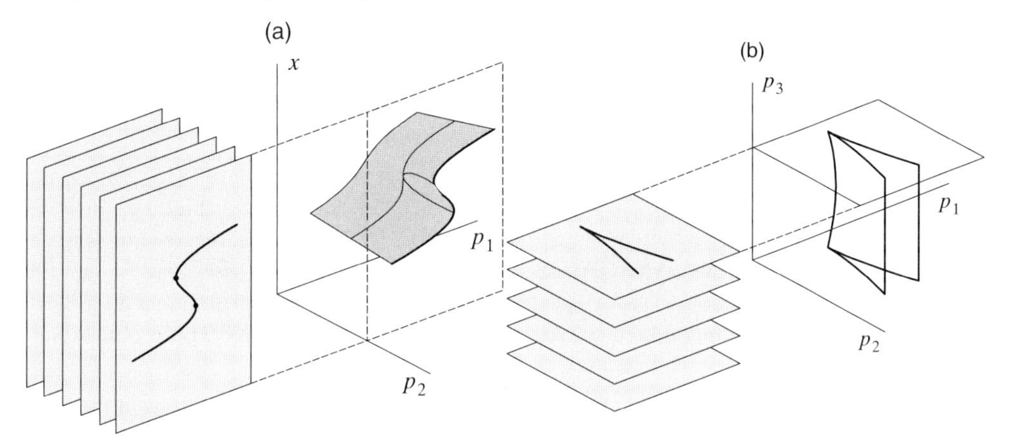

Figure 29.8: Repeated steady state continuation yields steady state manifolds (a), repeated continuation of the saddle-node locus yields a manifold of critical points (b).

While repeated continuation after variation of parameters allows us in principle to analyze a system with many parameters, the approach becomes tedious for more than three parameters. Clearly, the number of diagrams needed for a thorough analysis grows exponentially in the number of parameters to be varied and thus this approach is prohibitive for high-dimensional parameter spaces.

In practical applications, many curves are typically calculated and visualized in the course of analyzing a process model. Experienced users of continuation tools will usually not tackle the analysis by a brute-force continuation on a grid of parameter values as sketched in Fig. 29.8. Rather, they will apply continuation in an iterative manner. Curves that are calculated and visualized in one step can be used to guide a further analysis in the next step to those regions of the parameter space in which, loosely speaking, interesting dynamic behavior exists.

As the sketched approach of collecting diagrams involves the calculation of one-parametric curves only, there is the chance of overlooking regions of the parameter space with dynamic behavior of interest. Figure 29.9 shows examples of curves (a, b, c) which suggest that no multiplicities exist even though there are multiple steady states in the parametric vicinity of all bifurcation diagrams.

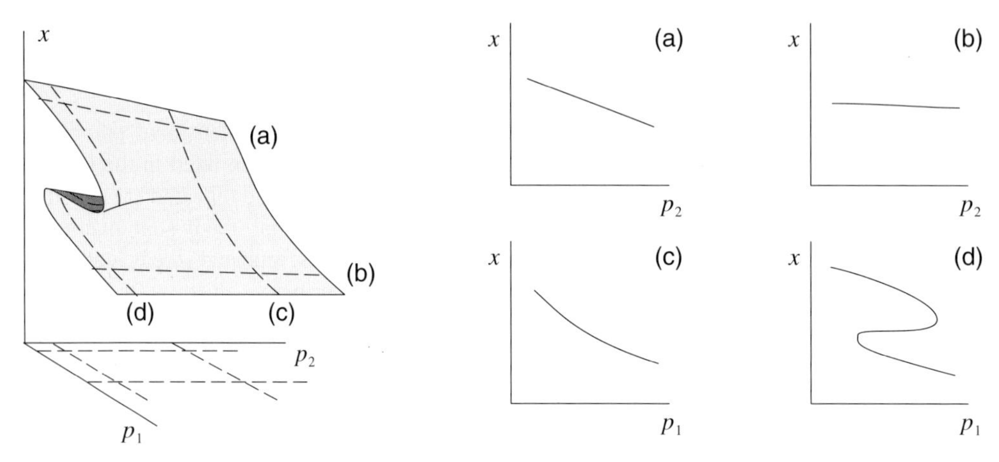

Figure 29.9: One-parametric curves in the vicinity of multiplicities and a cusp singularity. Only curve (d) reveals that multiplicities exist for the system.

29.5 Constructive Methods

While continuation and bifurcation analysis are powerful tools, limitations exist as pointed out in Sect. 29.4:

- A systematic continuation analysis of high codimension critical points is only possible for non-dynamic bifurcations up to four-parameter continuation. For dynamic bifurcations and combinations of dynamic and static bifurcations, hierarchies are less complete.
- A systematic analysis of high-dimensional parameter spaces is tedious as the number of necessary diagrams grows exponentially in the number of process parameters to be varied.
- As continuation provides one-parametric curves only, there is the danger of missing critical points in high-dimensional parameter spaces.
- In practical applications, a rigorous analysis of the entire parameter space is often avoided because it is too tedious or impossible for the above reasons. If one accepts the analysis to be premised on cycles of analyzing, visualizing, interpreting results and deciding on which parts of the parameter space to investigate further, continuation becomes an art. In such an approach, results will be the better the more experienced the user is in calculating, visualizing and interpreting the obtained data.

Ultimately, design methods which incorporate the information accessible via bifurcation analysis have to be developed. Ideally, these methods will avoid the complexity of a rigorous analysis and limit the necessity for experience-based decisions.

This section discusses research aiming at the development of *constructive* nonlinear dynamics methods. The central notions are (i) the concept of parametric distance in heterogeneous parameter spaces [27] and (ii) the normal space to manifolds of critical points which is used to locally determine closest critical points [10, 27]. After a brief introduction to these concepts, exemplary results are reported. References to publications which detail the concepts and results presented in this section are listed.

29.5.1 Normal Vector-based Constraints for Parametric Robustness

Instead of analyzing a candidate point of operation in the parameter space of a given process model to find the manifolds of critical points, it is often sufficient to know the *closest* critical points. Note that there will in general be more than one close critical point, since critical manifolds may arise due to physically different causes. There may, for example, be a close critical point for process stability due to a Hopf bifurcation and, independently, a close critical point due to a saddle-node bifurcation. Similarly, a critical manifold may not be convex, resulting in more than one locally closest critical point to a given point of operation.

The parametric distance between a candidate point of operation and the closest critical points is measured with the aid of parameter-space normal vectors. The concept of parameter space normal vectors is illustrated in Fig. 29.10. While this sketch uses a two-dimensional parameter space to illustrate the idea, the concept of a one-dimensional normal space generalizes to

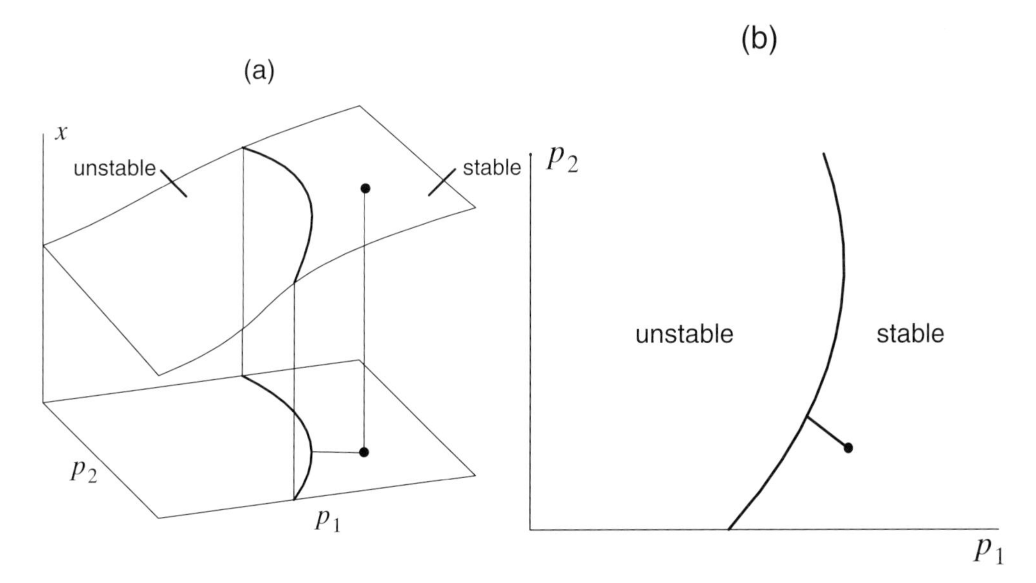

Figure 29.10: Visualization of minimal distance to critical manifolds. The closest connection from the marked candidate point of operation occurs along a direction which is normal to the critical manifold.

parameter spaces of arbitrary finite dimension. If another parameter was varied in Fig. 29.10b and a third dimension of the parameter space was added to the diagram, the critical curve would unfold to a critical surface in a three-dimensional parameter space. In general the critical points of codimension one will form an $(n_p - 1)$-dimensional critical manifold in an n_p-dimensional parameter space. Most importantly from a conceptual point of view, the normal space to the critical $(n_p - 1)$-dimensional manifold will be *one*-dimensional, regardless of the dimension of the parameter space. Note that the simple concept of identifying locally closest critical points by virtue of the *one*-dimensional normal space remedies the exponential growth of complexity of a continuation analysis in the number of process parameters. In some sense, the idea of exploiting the $(n_p - 1)$-dimensional *tangent* space to the critical manifold by parameter continuation has been replaced by the use of the *one*-dimensional *normal* space [27].

In fact this idea generalizes to higher codimension critical points. For critical points of codimension n_c, a continuation-based analysis of the $(n_p - n_c - 1)$-dimensional tangent space can be replaced by constraints in optimization based on *one*-dimensional normal vectors. We first focus on codimension one critical points such as the stability boundaries induced by the Hopf and saddle-node bifurcations in the fermenter example. A codimension two example will be discussed afterwards.

In models of engineering processes, some parameters p_i can usually not be fixed to precise nominal values, but they are only known up to a finite error Δp_i:

$$p_i \in \left[p_i^{(0)} - \Delta p_i, p_i^{(0)} + \Delta p_i\right], i = 1, \ldots, n_p. \tag{29.17}$$

This situation arises, for example, because some parameters, such as feed concentrations, slowly drift over time. Simlarly, bounds of the type (29.17) arise because physical constants of the model, such as kinetic or heat-transfer coefficients, have only been measured to within a finite precision. By re-scaling the model parameters according to

$$p_i \to \frac{p_i}{\Delta p_i}, \quad p_i^{(0)} \to \frac{p_i^{(0)}}{\Delta p_i} \tag{29.18}$$

the parameters become dimensionless and Eq. (29.17) translates to

$$p_i \in \left[p_i^{(0)} - 1, p_i^{(0)} + 1\right], \quad i = 1, \ldots, n_p, \tag{29.19}$$

cf. Fig. 29.11. The hyper-cube of side length 2 defined by Eq. (29.19) is over-estimated by an n_p-dimensional hyper-ball of radius $\sqrt{m}$. Note that in this re-scaled coordinate system, the usual inner product provides a meaningful measure of distance.

The requirement to keep a minimum distance to the manifold of critical points can now be visualized by a circle which is allowed to touch the critical manifold or stay off it. Note that while in the physical coordinates (29.17), the distance to the critical manifold cannot be visualized using the direction which is normal to the eye, the robustness ball in the rescaled parameters shown in Fig. 29.11b translates into an intuitive robustness ellipse in Fig. 29.11a.

The use of normal vectors to find the locally closest critical point to a fixed candidate point was first suggested by Dobson [10]. Mönnigmann and Marquardt [27] presented a general scheme for the derivation of systems of equations for the calculation of normal vectors on

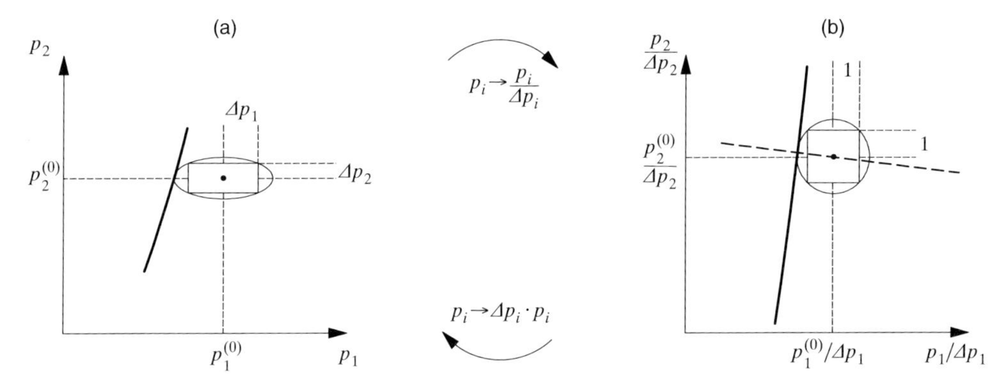

Figure 29.11: If the physical parameters vary within the ellipse in (a), the re-scaled parameters vary within the circle in (b). Since the parameters in (a) will generally have different physical dimensions, no meaningful measure of distance exists. After re-scaling, the parameters in (b) are dimensionless, and distances can be measured with the inner product.

manifolds of critical points of arbitrary finite codimension. Assuming that the model system has the form

$$\dot{x} = f(x, p), \tag{29.20}$$

where f maps into $\mathbb{R}^n$ and is smooth with respect to the state variables $x \in \mathbb{R}^n$ and the parameters $p \in \mathbb{R}^{n_p}$, the normal vector systems have the form

$$0 = F(x, \tilde{x}, p, r), \tag{29.21}$$

where $F : \mathbb{R}^n \times \mathbb{R}^{n_{\tilde{x}}} \times \mathbb{R}^{n_p} \times \mathbb{R}^{n_r} \to \mathbb{R}^{n+n_{\tilde{x}}+n_p+n_r}$ is smooth with respect to all variables. In Eq. (29.21), r denotes the $(n_r = n_p - \text{codim})$-dimensional normal vector to the manifolds of critical points of codimension codim, and $\tilde{x}$ denotes internal variables of the normal vector system of equations. For details of the derivation of normal vector systems of equations the reader is referred to [27].

Once the normal vector r to the critical manifold of interest is known, the requirement on the location of the critical hyper-ellipse with respect to the critical manifold in Fig. 29.11 can be stated as

$$\begin{aligned} &p^{(0)} = p^{(i)} + l^{(i)}\, r^{(i)}, \\ &l^{(i)}\, ||r^{(i)}||_2 \geq \sqrt{m}, \\ &l^{(i)} \geq 0, \qquad\qquad i = 1, ..., I, \end{aligned} \tag{29.22}$$

where the upper indices refer to the candidate point of operation 0 and to critical point number i, respectively. If conditions (29.22) hold for all locally closest critical points $i = 1, ..., I$, the process parameters will not cross any critical boundaries despite the parametric uncertainties (29.19). In this sense parametric robustness will be ensured.

The conditions (29.22) can now be used to state constraints for parametric robustness in optimization. Given a profit function $\phi(x, p)$, $\phi : \mathbb{R}^n \times \mathbb{R}^{n_p} \to \mathbb{R}$, the optimization problem has the form

$$\begin{aligned}
&\max_{x^{(0)},p^{(0)},l^{(1)},\ldots,l^{(I)},x^{(i)},\tilde{x}^{(i)},p^{(i)},r^{(i)}} \quad \phi(x^{(0)},p^{(0)}),\\
&\text{subject to} \quad 0 = f(x^{(0)},p^{(0)}),\\
&0 = F_i^{(j_i)}(x^{(i)},\tilde{x}^{(i)},p^{(i)},k^{(i)},\gamma^{(i)},r^{(i)}),\\
&p^{(i)} = p^{(0)} + l^{(i)}r^{(i)},\\
&0 \leq l^{(i)}||r^{(i)}||_2 - \sqrt{n_p},\\
&0 \leq l^{(i)},\\
&i = 1,\ldots,I.
\end{aligned} \tag{29.23}$$

The index j_i refers to the type of locally closest critical point number i. Optimization problems of the form (29.23) are solved in the next section for different types j_i of critical points. All optimizations are done with NPSOL [12].

29.5.2 Optimization with Robust Stability and Feasibility Constraints

The first example addresses robust stability and feasibility for a model of the polymerization of vinyl acetate [28, 26]. The process is carried out in a CSTR. The model consists of ordinary differential equations for the temperature and initiator, monomer and polymer concentrations. The gel effect and temperature-dependent densities and reaction rates are included. The reader is referred to [33] for details of the model and its experimental validation.

The polymerization process is optimized with respect to the cost function

$$\Phi = \left(-\phi_{\mathrm{m}}\rho_{\mathrm{mf}}C_{\mathrm{m}} - \phi_{\mathrm{s}}\rho_{\mathrm{sf}}C_{\mathrm{s}} - I_{\mathrm{f}}MW_{\mathrm{I}}C_i + v_{\mathrm{p}}\rho_{\mathrm{p}}C_{\mathrm{p}}\right)/\Theta \tag{29.24}$$

$$\approx \left(-\phi_{\mathrm{m}}\rho_{\mathrm{mf}}C_{\mathrm{m}} - \phi_{\mathrm{s}}\rho_{\mathrm{sf}}C_{\mathrm{s}} - I_{\mathrm{f}}MW_{\mathrm{I}}C_i + v_{\mathrm{p}}\rho_{\mathrm{p}}C_{\mathrm{p}}\right)\left(q_{\mathrm{out}}/q_{\mathrm{in}}\right)/\Theta, \tag{29.25}$$

where ϕ_{m} and ϕ_{s} refer to the monomer and solvent volume fractions in the feed, ρ_{mf}, ρ_{sf} and ρ_{p} are the densities of monomer in the feed, solvent in the feed and polymer in the reactor, MW_{I} and I_{f} are the molecular weight and the molar concentration of the initiator, Θ refers to the residence time, v_{p} denotes the volume fraction of polymer in the reactor and the C_i refer to the cost coefficients of the respective substances. The ratio of flow rates at the input and the output has to be taken into account as in Eq. (29.25), but the results are not affected if the simplified merit function (29.24) is instead used. For simplicity the process is optimized with respect to Eq. (29.24). The optimization results presented below were obtained with the relative cost coefficients $C_{\mathrm{m}} = 3C_{\mathrm{s}}$, $C_{\mathrm{p}} = 6C_{\mathrm{m}}$, $C_i = 20C_{\mathrm{p}}$.

Both Hopf and saddle-node bifurcations exist in the model. Apart from parametric robustness with respect to stability boundaries due to these bifurcations, the temperature constraint $T \leq 100°\mathrm{C}$ is imposed on the process to demonstrate how a feasibility constraint results in a critical manifold of codimension one.

For reference the model is first optimized *without* robustness constraints. The results are visualized in Fig. 29.12. Figure 29.12a shows that the optimal solution obtained without constraints for parametric robustness is in fact stable, but suffers from high parametric sensitivity. Clearly, an arbitrarily small deviation of the residence time θ towards lower values than the optimum will result in a hysteretic jump to the low branch in Fig. 29.12a. This is undesirable because the conversion from monomer to polymer is low on this branch.

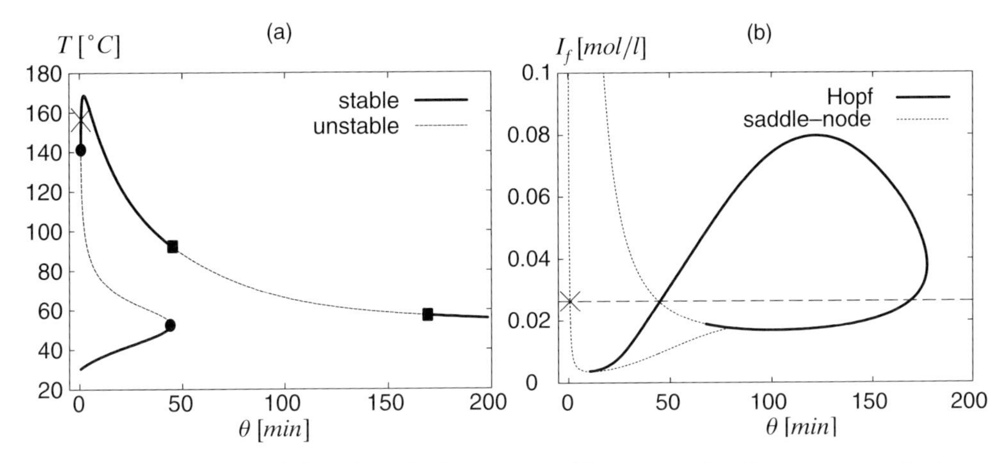

Figure 29.12: Optimum of the polymerization process without constraints for parametric robustness. The optimum is marked by the cross. Diagrams (a) and (b) show one- and two-parameter continuations, respectively. The dashed line in (b) marks the value of I_f for which diagram (a) was obtained.

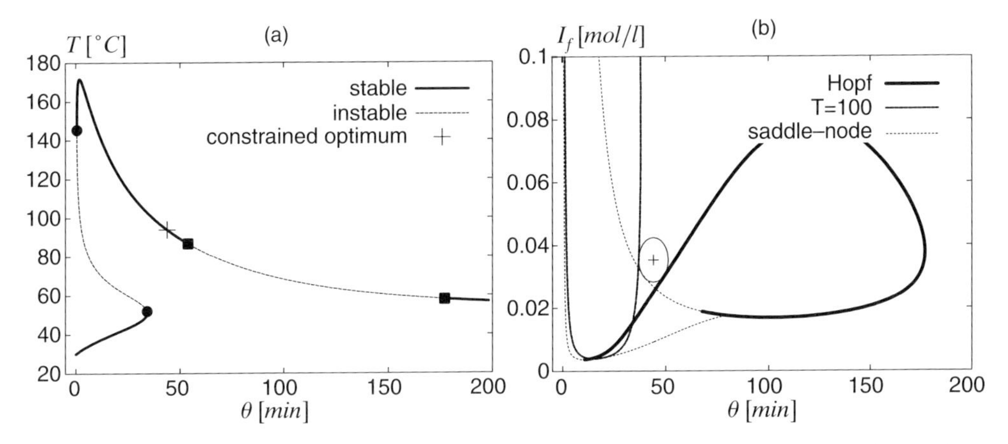

Figure 29.13: Result of the optimization with constraints on the location of Hopf bifurcations and the temperature constraint $T < 100°\mathrm{C}$.

Figure 29.13 visualizes the result of the optimization with constraints for parametric robustness with respect to saddle-node and Hopf bifurcations and the temperature constraint $T = 100°\mathrm{C}$. The parametric uncertainties were chosen to be $\Delta p_1 = \Delta\theta = 5\,\mathrm{min}$ and $\Delta p_2 = \Delta I_\mathrm{f} = 0.005\,\mathrm{mol/l}$. Figure 29.13b shows that the robustness ellipse guarantees parametric robustness with the three codimension one manifolds of critical points. Figure 29.13a shows the bifurcation diagram in θ passing through the optimal point. This diagram confirms that the optimum is at a parametrically safe distance to the saddle-node and Hopf bifurcations, as well as to the feasibility constraint $T < 100°\mathrm{C}$. For further details, the reader is referred to [26].

29.5.3 Optimization with Parametric Robustness with Respect to Hysteresis

The second example demonstrates the use of the normal vector-based constraints for a manifold of critical points of codimension two [27]. The process treated is the classical adiabatic CSTR with an exothermic first-order reaction $A \rightarrow B$. The CSTR is usually modeled in terms of dimensionless variables

$$\begin{aligned} 0 &= \frac{V}{Fc_0}\frac{dc_A}{dt} = (x_{10} - x_1) - \mathrm{Da}\exp\left(\frac{x_2}{1+gx_2}\right)x_1 =: f_1(x,\alpha), \\ 0 &= \frac{V}{gFT_0}\frac{dT}{dt} = (x_{20} - x_2) + B\mathrm{Da}\exp\left(\frac{x_2}{1+gx_2}\right)x_1 =: f_2(x,\alpha), \end{aligned} \tag{29.26}$$

where x_1, x_2 and Da are the dimensionless feed concentration, dimensionless temperature and Damköhler number, respectively [27].

The diagrams in Fig. 29.14 reveal that unstable steady states exist due to saddle-node bifurcations. Figure 29.14b shows that the two branches of saddle-node bifurcations meet in a cusp singularity. After freeing the dimensionless feed temperature x_{20}, the cusp point can be unfolded to a curve of cusp singularities in a three-parameter continuation. The projection of this curve into the (x_{10}, x_{20}) plane is shown in Fig. 29.15. The cusp locus splits this plane into a region in which hysteresis may occur due to the unstable solution branch framed by saddle-node bifurcations, and a region in which only stable solutions exist. The regions are labeled by sketches of the respective bifurcation diagrams inserted in Fig. 29.15. The objective in this example is to find an optimal point of operation which is parametrically robust with respect to the occurence of hysteresis. More precisely, while the dimensionless feed temperature x_{20} and the dimensionless feed concentration x_{10} are allowed to vary within the robustness ellipses shown in Fig. 29.15, the bifurcation diagram in the parameter Da is required not to contain any saddle-node bifurcation. Note that this task of guaranteeing parametric robustness with respect to codimension two shown in Fig. 29.15 amounts to guaranteeing traits of an *entire*

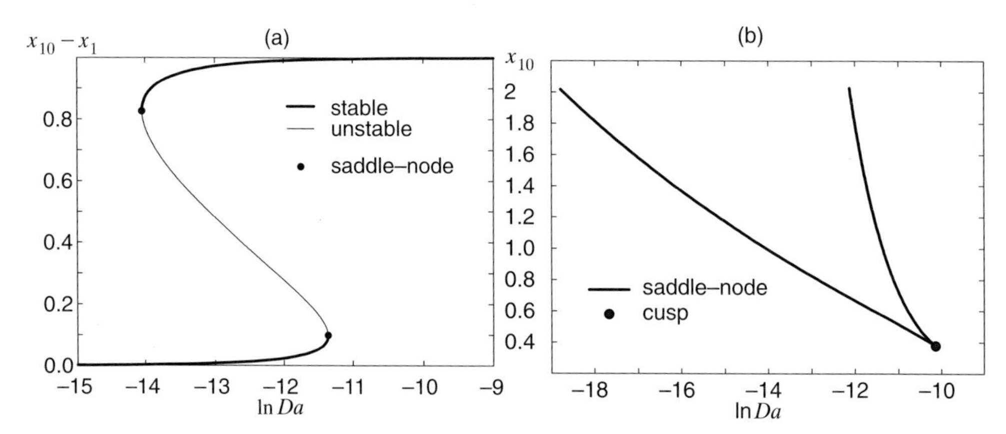

Figure 29.14: (a) Bifurcation diagram of the CSTR model (29.26) for $x_{10} = 1.0$ and $x_{20} = 10.0$. (b) Saddle-node locus obtained by two-parameter continuation.

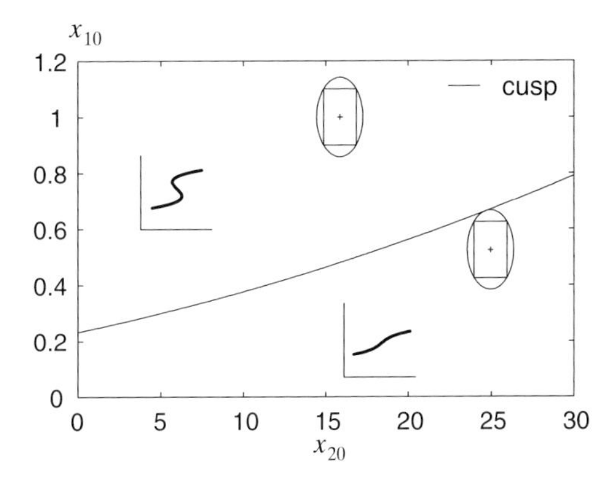

Figure 29.15: The curve of cusp points splits the (x_{20}, x_{10}) plane into a region in which bifurcation diagrams with hysteresis and unstable steady states exist and a region in which only stable steady states exist.

diagram. As opposed to the examples of parametric robustness of an optimal *point* of operation in Sect. 29.5.2, the application in this section guarantees traits for a one-dimensional manifold, i.e. the bifurcation diagram with respect to Da.

The CSTR is optimized with respect to the amount of B produced per unit time. The cost function is

$$\phi = (x_{10} - x_1)/\mathrm{Da}. \tag{29.27}$$

In the optimization, an upper bound on the residence time or, equivalently, a lower bound on the Damköhler number is imposed. Similarly, an upper bound on the feed concentration x_{10} is needed. In addition, the maximum temperature in the CSTR is restricted by bounding the adiabatic temperature rise $x_{2,\mathrm{ad}} = x_{20} + B\,x_{10}$. The bounds were chosen to be

$$\begin{aligned} &x_{2,\mathrm{ad}} \leq 35, \\ &-15 \leq \ln \mathrm{Da}, \\ &x_{10} \leq 1. \end{aligned} \tag{29.28}$$

For reference, the CSTR is optimized without a normal vector-based requirement on the minimal distance of the point of operation. The results of the optimizations with and without normal vector-based constraints for parametric robustness are shown in Fig. 29.15. While the optimization without these constraints results in a point of operation on a bifurcation diagram with two saddle-node bifurcations, the optimum obtained with constraints for robustness lies on a bifurcation diagram without an unstable branch. Therefore, for the optimum obtained with robustness constraints, no hysteresis can occur as the Damköhler number or, equivalently, the flow rate through the reactor is varied. Moreover, bifurcation diagrams which result for any point in the lower right robustness ellipse in Fig. 29.15 will not contain any saddle-node bifurcations.

29.6 Summary and Outlook

The chemical engineering community has been using methods from nonlinear dynamics for over fifty years. While first applications focused on the analysis of dynamic models aiming at a better understanding of the dynamic behavior of process models, a shift from, loosely speaking, an *analysis for a deeper understanding* towards *analysis for design* can be detected in the recent literature.

Since practically all chemical engineering processes involve a control system, any method used in the design stage must be applicable to closed-loop process models. Amazingly enough, little research seems to have addressed the application of methods from nonlinear dynamics, in particular of bifurcation analysis, to closed-loop systems. To this end, the present paper summarizes the results of a case study which shows that bifurcation analysis can be used to tune feedback linearizing controllers.

Despite being a very powerful tool, bifurcation analysis by continuation suffers from having to rely on interpreting visualized data. The present work summarized research which aims at exploiting the information, which can in principle be disclosed in a bifurcation analysis, in optimization-based process design. The use of constraints which guarantee parametric robustness with respect to critical manifolds has been demonstrated in examples. In the simplest application, these constraints can be used to guarantee parametrically robust stability in the design of continuous processes. Other applications allow us to exploit the information contained in the location of higher-codimension critical points. An example of constraints on the location of cusp singularities was used to illustrate this.

Bibliography

[1] R. Aris and N.R. Amundson, *An analysis of chemical reactor stability and control–i: the possibility of local control, with perfect or imperfect control mechanisms*, Chemical Engineering Science **7**(3) (1958), 121–131.

[2] R. Aris and N.R. Amundson, *An analysis of chemical reactor stability and control–ii: the evolution of proportional control*, Chemical Engineering Science **7**(3) (1958), 132–147.

[3] R. Aris and N.R. Amundson, *An analysis of chemical reactor stability and control–iii: the principles of programming reactor calculations. Some extensions*, Chemical Engineering Science **7**(3) (1958), 148–155.

[4] C. Bildea and A. Dimian, *Singularity theory approach to ideal binary distillation*, AIChE Journal **45**(12) (1999), S2662–S2666.

[5] C.S. Bildea and A.C. Dimian, *Stability and multiplicity approach to the design of heat-integrated PFR*, AIChE Journal **44** (1998), 2703–2712.

[6] C.S. Bildea, A.C. Dimian, and P. Iedema, *Nonlinear behavior of reactor-separator-recycle systems*, Computers and Chemical Engineering **24** (2000), 209–215.

[7] C.S. Bildea, A.C. Dimian, and P.D. Iedema, *Multiplicity and stability approach to the design of heat-integrated multibed plug flow reactor*, Computer and Chemical Engineering **25** (2001), 41–48.

[8] O. Bilous and N.R. Amundson, AIChE Journal **1** (1955), 513–521.

[9] D. Brengel and W. Seider, *Coordinated design and control optimization of nonlinear processes*, Chemical Engineering Communications **16** (1992), 861–886.

[10] I. Dobson, *Computing a closest bifurcation instability in multidimensional parameter space*. Journal of Nonlinear Science **3** (1993), 307–327.

[11] E. Doedel, A.R. Champneys, T.F. Fairgrieve, Y.A. Kuznetsov, B. Sandstede, and X.-J. Wang, *AUTO97: Continuation and bifurcation software for ordinary differential equations (with HomCont)*, Computer Science, Concordia University, Montreal, Canada, 1997.

[12] P. Gill, W. Murray, M. Saunders, and M. Wright, *User's Guide for NPSOL, version 4.0*, Systems Optimization Laboratory, Stanford University, Stanford, USA, 1986.

[13] M. Golubitsky and B.L. Keyfitz, *A qualitative study of the steady state solutions for a continuous flow stirred tank chemical reactor*, SIAM Journal on Mathematical Analysis **11** (1980), 316–339.

[14] M. Golubitsky and D. Schaeffer, *Singularities and Groups in Bifurcation Theory*, Vol. I, Springer, New York, 1985.

[15] M. Golubitsky and D.G. Schaeffer, *A theory for imperfect bifurcation theory via singularity theory*, Communications on Pure and Applied Mathematics **32** (1979), 21–98.

[16] J. Guckenheimer and P. Holmes, *Nonlinear Oscillations, Dynamical Systems, and Bifurcations of Vector Fields*, Springer, New York, 1993.

[17] J. Hahn, M. Mönnigmann, and W. Marquardt, *On the use of bifurcation analysis for tuning of feedback linearizing controllers*, Internal Report, Lehrstuhl für Prozesstechnik, RWTH Aachen University (2002).

[18] M.P. Harold, J.J. Ostermaier, D.W. Drew, J.J. Lerou, and L. Daniel Jr., *The continuously-stirred decanting reactor: steady state and dynamic features*, Chemical Engineering Science **51** (1996), 1777–1786.

[19] C. Van Heerden, *Authothermic processes: properties and reactor design*. Industrial & Engineering Chemical Research **45**(6) (1953), 1242–1247.

[20] M. Henson and D. Seborg, *Feedback linearizing control*, in: M. Henson and D. Seborg (eds.): *Nonlinear Process Control*, Prentice Hall, Upper Saddle River, New Jersey, 1997.

[21] J. Khinast, D. Luss, M. Harold, J. Ostermaier and R. McGill, *Continuously stirred decanting reactor: Operability and stability considerations*. AIChE Journal **44**(2) (1998), 372–387.

[22] Y.A. Kuznetsov, *CONTENT – Integrated environment for analysis of dynamical systems. Tutorial*. Dynamical Systems Laboratory, Centrum voor Wiskunde en Informatica, Amsterdam [ftp.cwi.nl/pub/CONTENT], 1998

[23] Y.A. Kuznetsov, *Elements of Applied Bifurcation Theory*, 2nd edn., Springer, New York, 1999.

[24] F.G. Liljenroth, Chem. Met. Engng. **19** 287–293, 1918. As cited in [30].

[25] M. Mangold, A. Kienle, E.D. Gilles, and K.D. Mohl, *Nonlinear computation in DIVA – methods and applications*, Chemical Engineering Science **55**(2) (2000), 441–454.

[26] M. Mönnigmann and W. Marquardt, *Chemical process optimization with guaranteed robustness of nonlinear dynamics characteristics*, Technical Report, Lehrstuhl für Prozesstechnik RWTH Aachen, Germany, 2000. Presented at AIChE Annual Meeting 2000, Los Angeles, 12–17 November 2000.

[27] M. Mönnigmann and W. Marquardt, *Normal vectors on manifolds of critical points for parametric robustness of equilibrium solutions of ode systems*, Journal of Nonlinear Science **12** (2002), 85–112.

[28] M. Mönnigmann and W. Marquardt, *Process optimization with guaranteed parametric robustness and flexibility*, accepted for publication in AIChE Journal, 2003.

[29] W.H. Ray and C.M. Villa, *Nonlinear dynamics found in polymerisation processes*, Chemical Engineering Science **55** (2000), 275–290.

[30] L. Razón and R. Schmitz, *Multiplicities and instabilities in chemically reacting systems – a review*, Chemical Engineering Science **42**(5) (1987), 1005–1047.

[31] L. Russo and B. Bequette. *Operability of chemical reactors: multiplicity behavior of a jacketed styrene polymerization reactor*, Chemical Engineering Science **53**(1) (1998), 27–45.

[32] P.T. Saunders, *An Introduction to Catastrophe Theory*, Cambridge University Press, 1980.

[33] F. Teymour and W.H. Ray, *The dynamic behaviour of continuous polymerization reactors – V. Experimental investigation of limit-cycle behavior for vinyl acetate polymerization*, Chemical Engineering Science **47** (1992), 4121–4132.

[34] A. Uppal, W. Ray, and A. Poore, *On the dynamic behavior of continuous stirred tank reactors*, Chemical Engineering Science **29** (1974), 967–985.

[35] A. Uppal, W.H. Ray, and A. Poore, *The classification of the dynamic behavior of continuous stirred tank reactors – Influence of reactor residence time*, Chemical Engineering Science **31** (1976), 205–214.

[36] Y.V. Zeldovich and U.A. Zisin, *On the theory of thermal stress. Flow in an exothermic stirred reactor, II. Study of heat loss in a flow reactor*. Journal of Technical Physics **11**(6) (1941), 501–508. As cited in [14].

30 Nonlinear Dynamics in Chemical Engineering and Electro-chemical Manufactory Technologies

P.J. Plath, M. Baune, M. Buhlert, C. Gerlach, A. Kouzmitchev, P. Thangavel, E. van Raaij, H. Mathes, S. Diaz Alfonso, and Th. Rabbow

We are convinced that nonlinear dynamics could be a useful tool to improve chemical process engineering. Usually pattern formation like oscillations or spatial-temporal patterns are not desired in chemical processes on an industrial scale. Therefore one likes to avoid self-organization of dissipative chemical systems by choosing special control parameters. Moreover one did not discuss such self-organisation phenomena wherefore one did not know enough to answer the question whether or not such phenomena could be helpful in production. On the other hand nonlinear science restricted themselves to the well-known examples like the Belousov-Zhabotinsky reaction (BZR) in order to improve their own tools. Here, we intend to give knowledge onto some technical important electrochemical processes like electropolishing, micro-structuring and etching which might be convincing examples for the advantage of the use of nonlinear dynamics in technical processes. Furthermore we present quite new processes like oscillating extraction/separation processes for rare-earth elements and the formation of mono-porous foams and ceramics which are based on structure formation far from equilibrium.

30.1 Introduction

A great variety of nonlinear phenomena are known in chemical dynamics. For example, one can find temporal as well as spatial pattern formation in the famous Belousov–Zhabotinsky reaction [1] for the catalytic bromation of malonic acid and related reactions [2]. Another very famous example is the heterogeneously catalyzed oxidation of carbon monoxide on support catalysts [3] as well as on single-crystal surfaces [4]. For a hundred years electro-chemical oscillations are well known during the anodic dissolution of metals [5]. But all these reactions are more of academic interest than of interest with respect to any kind of technical applications. Moreover, in chemical production techniques one likes to avoid oscillations or spatial pattern formation in general. But nonlinear dynamics may help to improve chemical technology just taking into account chemical oscillations and other pattern-formation phenomena. Here, we present various nonlinear dynamical systems which might be applicable for chemical engineering including electro-chemical manufacturing.

Nonlinear Dynamics of Production Systems. Edited by G. Radons and R. Neugebauer

ISBN 3-527-40430-9

30.2 Electropolishing

Electropolishing is a well-known electro-chemical method to achieve a brilliant and very smooth surface [6, 7]. Nonlinear methods – like for example driving the process with pulsating current – are used to improve the mass transport of the transpassive dissolution as well as the brilliance of the metal surface [8, 9]. Very often this process is executed in the vicinity of the transpassive region of the potential/current diagram (Fig. 30.1) in order to get a remarkable rate of removal of the metal surface.

In this region the anodic metal is dissolved while oxygen is formed at the same electrode. The oxygen forms small bubbles, which rise by buoyancy in the falling film of highly concentrated ionic solution of the dissolved metallic material. The whole process is highly nonlinear and very complicated and not at all understood up to now. Inspecting the potential/current curve close to the point where the oxygen formation sets in, one can find a small but negative differential resistance in case of brass. This is a strong indication for the occurrence of electro-chemical oscillations.

Indeed, Kouzmitchev observed chemical waves of very small oxygen bubbles spreading over the metal surface if he fixed the current galvanostatically to an appropriate value (Fig. 30.2). On the other hand Gerlach [9] proved that one can observe oscillations in the potential as well as in the production of oxygen bubbles under galvanostatic conditions. The upper level of the potential oscillations clearly correlates with the transpassive region, whereas the different minima correspond with the wide passive region (Figs. 30.1 and 30.3).

For technical purposes it is very important to use different electrolytes in order to achieve the desired surface structure on various metals. Usually extensive experimental tests are carried out to screen the high-dimensional parameter space in which electropolishing takes place.

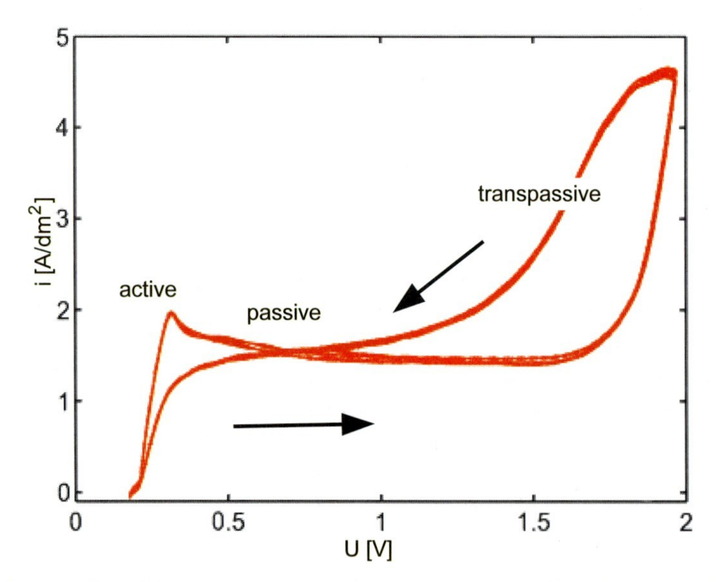

Figure 30.1: Cyclovoltammogram of brass in a standard electrolyte (60 vol% phosphoric acid (85%), 15 vol% 2-propanol, 15 vol% 2-butanol, 10% water); scanning velocity $v = 4$ mV/s.

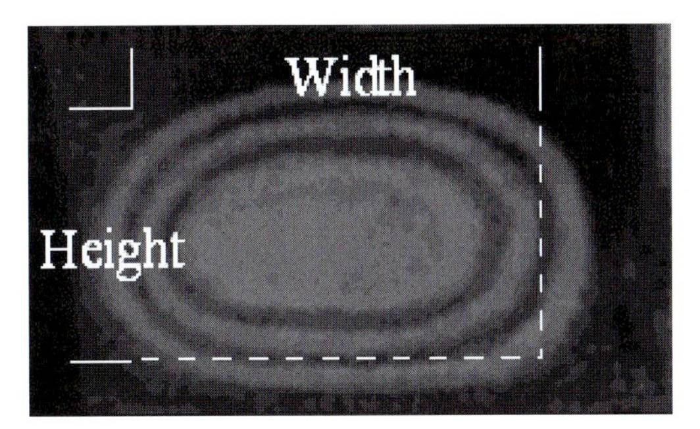

Figure 30.2: Spreading waves of oxygen bubbles in status nascendi on a sheet of brass at the insert of the transpassive region. The chemical constraints are comparable to those of Fig. 30.1.

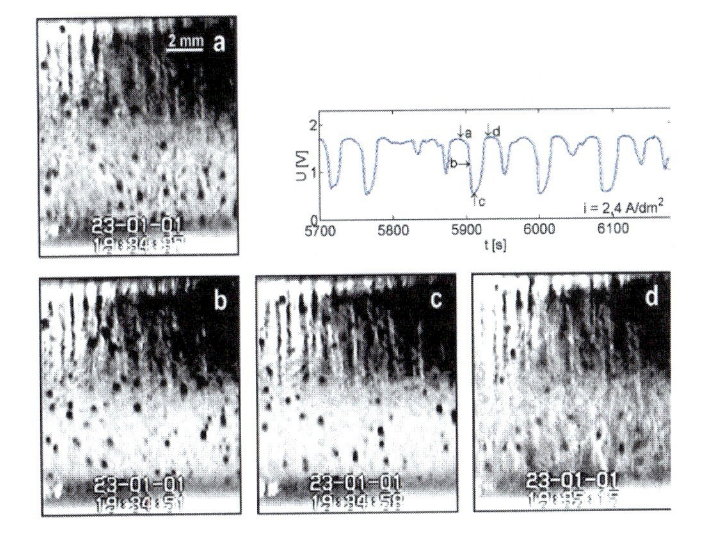

Figure 30.3: Correlation of the oxygen production with the potential oscillation during galvanostatic measurements in the transpassive region.

Surprisingly, even the use of different homologue alcohols creates different dynamical behavior (Fig. 30.4) and quite different surface structures (Fig. 30.5).

As mentioned above, the oxygen bubbles rise up close to the surface of the electrode creating deep valleys in the surface after a while. This kind of macroscopic structure formation has to be distinguished from the surface structure due to the chemical process and the composition of the electrolyte. Following the argument of Friedrich and Peinke [10–14], Kouzmitchev looked at the electropolished surface as to be structured by a Markov process (Fig. 30.6).

Using laser scanning equipment (UBM Co.) one can measure the surface structure on the micrometer scale, whereas atomic force microscopy allows us to characterize the surface structure on the nanometer scale. A very shiny surface can be very rough and mountainous on the micrometer scale if the same surface is very smooth on the nanometer scale. Just the opposite is true as well. A surface which looks very even and smooth on the micrometer scale

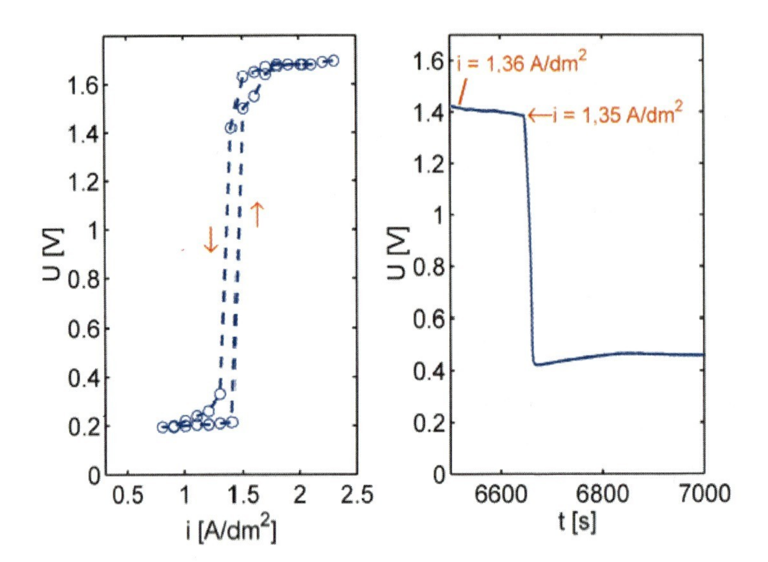

Figure 30.4: Electropolishing of brass in an electrolyte containing methanol 30 vol% (and in addition 60 vol% phosphoric acid and 10 vol% water): left-hand side: hysteresis behavior of the potential as a function of the current density; right-hand side: corresponding time series in the region of the hysteresis.

looks very dull if one can find a rough and mountainous structure on the nanometer scale. It is really not surprising that there does not exist a closed theory up to now for all the dynamic behavior interacting during electropolishing. But on the other hand there now exists a detailed overview of the various partial processes which contribute to the overall process [9]. It might be just the right time to put together all these experimental details and to start the development of a theory of electropolishing.

The height increments $\Delta h(x, r_0)$ and $\Delta h(x, r)$ in the direction x perpendicular to the valleys of the gas bubbles can be defined as: $\Delta h(x, r) = h(x, r) - h(x)$, where r is the fixed length scale (Fig. 30.6). Since we consider the system as a discrete stochastic process the system can be found in a countable number of states $s = 1, 2, \ldots, i, \ldots, j, \ldots, M$. Within this scale we define the state s of the system by the condition $\Delta h(x) \in (\Delta h_{s-1}, \Delta h_s)$, where $\Delta h_s = \Delta h_0 + (s/M)(\Delta h_M - \Delta h_0)$. Taking two different length scales r_0 and r with $r_0 < r$ and looking at the frequency of this situation $(\Delta h(x; r) = j) \cap (\Delta h(x; r_0) = i)$ the height increment at the position $(x; r)$ is of the size j whereas the height increment at (x, r_0) is i. The probability of this event is the conditional probability of j, if i has already entered. In the theory of Markov processes the conditional probability is defined as the joint probability for transitions. The most promising characterization of the stochastic process can be achieved if directly evaluated joint probabilities satisfy the Chapman–Kolmogorov (CK) equation:

$$p_{ij}(r, r_0) = \sum_k p_{kj}(r, r_1) p_{ik}(r_1, r_0) \text{ for any triplet } r_0 < r_1 < r.$$

Actually, the whole electro-chemical process is governed by two different stochastic processes: (a) a rough surface (before electropolishing) and (b) electropolished surface profiles. In order

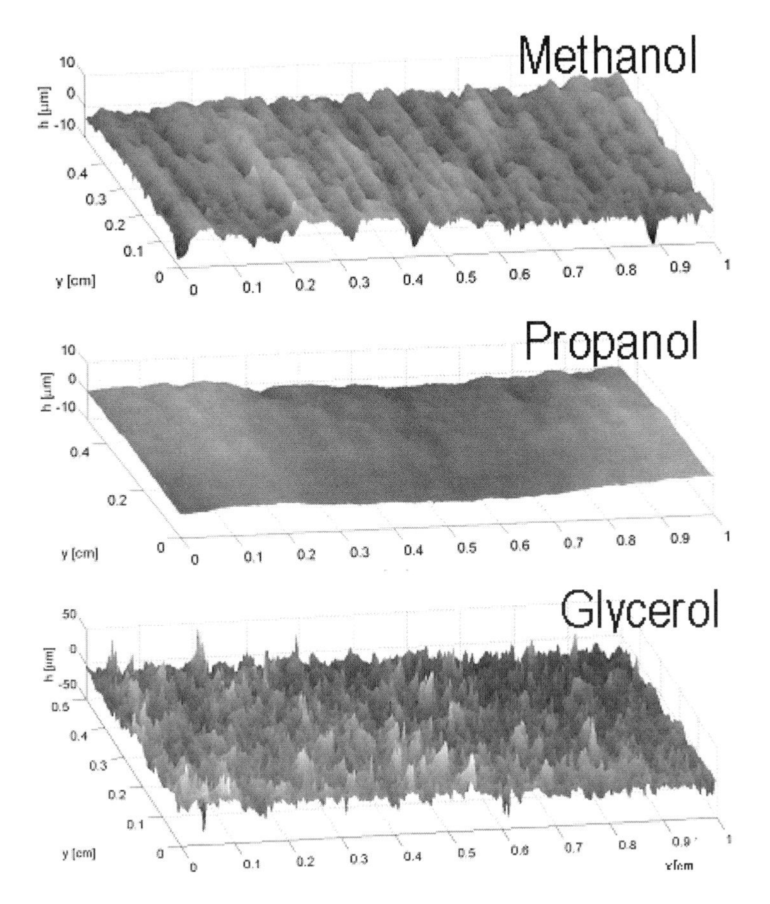

Figure 30.5: The use of different alcohols in an electrolyte containing phosphoric acid creates quite different surface structures.

to verify the Chapman–Kolmogorov equation both processes have to be considered separately. Figure 30.7 presents a comparison of directly evaluated probabilities with ones calculated by the CK equation. In both cases the CK equation is fulfilled in a restricted region ($\Delta h_1 \leq 3$ µm) quite satisfactorily (see Fig. 30.8). For other triplets $r_0 < r_1 < r$ the checking procedure gives similar results. The validity of the CK equation allows us to characterize correlations of the whole spectrum of length scales (r_0, r). However, it is only a necessary condition for Markovian processes. In common consideration a Markov process has no memory [14]. For the process running in space r, the Markovian property can be regarded as a local mechanism of the spatial pattern formation. But there are processes where the integration even over previous states of the system becomes necessary. The CK equation is indeed an integral equation, and that is why some processes with memory effects can satisfy this equation. The different local processes could be regarded approximately as independent from each other. But each of these localized processes has its own history. It is not a single event which may or which may not appear but it develops, wherefore this process possesses a memory.

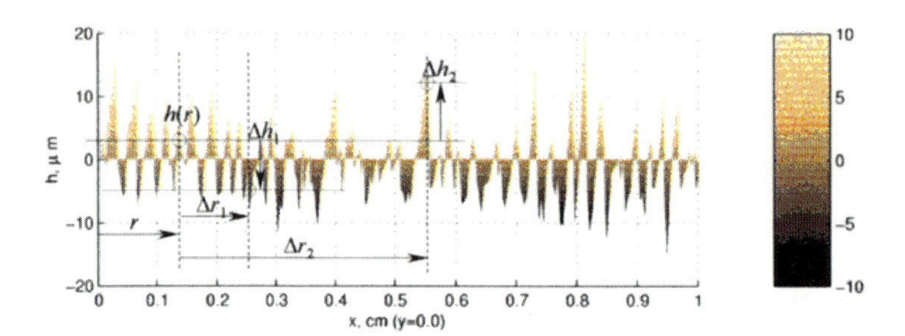

Figure 30.6: A cut of the electropolished surface perpendicular to the valleys which have been created by the gas bubbles. This cut can be used to estimate the Markov properties of the surface. One can take the surface structure as a stochastic process which can be measured for example by the height increments of the cut in the surface. The main idea of this ansatz is to consider the characteristic parameter of this system as a random variable abstracting from all the different physical and chemical forces which may be responsible for the resulting surface structure.

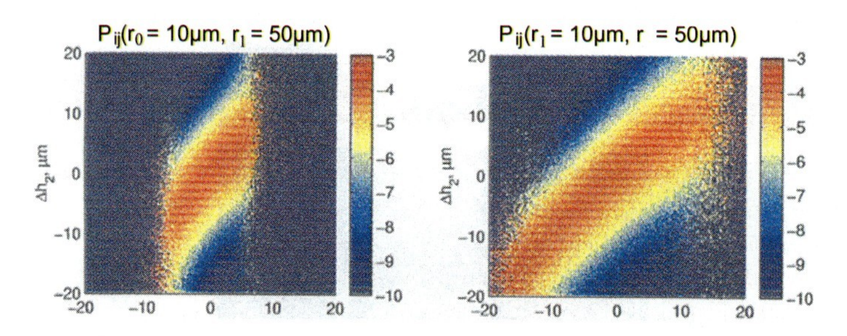

Figure 30.7: Verification of the Chapman–Kolmogorov equation for an electropolished surface. Left-hand side: joint probabilities directly evaluated from the experimental data. Right-hand side: joint probabilities calculated by the CK equation. For the triplet: $r_0 < r_1 < r$, where $r_0 = 0.01$ mm, $r_1 = 0.05$ mm and $r = 0.1$ mm.

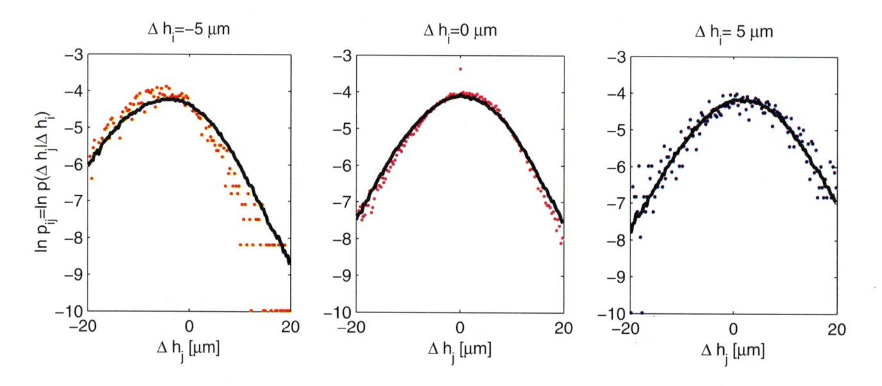

Figure 30.8: Logarithmic joint probabilities $p_{ij}(r, r_0)$ at fixed Δh_1 values. The set s of dots are taken from the experiment (Fig. 30.7). Solid lines are calculated from the CK equation.

30.3 Surface Structuring by Micro-electropolishing

Smooth, shining metal surfaces can be made by electropolishing. As a result, surfaces polished in this way are smooth, corrosion-resistant and buds adhere badly. Therefore, the manufacturing process is used in medical engineering. Metallic implants and medical instruments are electropolished for example. Venkatachalam et al. [15] and Labarga et al. [16] could show that surfaces get even smoother by use of pulsated direct current, while conventional electropolishing is achieved with direct current.

Many microdevices consist of microchannels. Microreactors are one example for such a microdevice [17]. These channels can be made by different microfabrication techniques like EMM (electro-micromachining) or PCM (photo-chemical milling). Most of the electrochemical removal processes need a resist to protect parts of the workpiece that should not be removed [18–23]. Only a few techniques are known to make channels without (photo-) resist. On the one hand there is the approach to create 3D structures by laser chemical etching [24–26]. On the other hand it can be done with the help of an atomic force microscope (AFM) used in STM mode [27–29]. In this case the removal results from charging the electro-chemical double layer.

Although theory of technical processes is not developed very often in a profound way, techniques work and new technologies are developing. Based on the process of electropolishing, Buhlert developed a technical application to get electropolished 3D micro-structures in a metallic surface. Taking the metal workpiece as an anode it can be electro-structured by means of a very tiny counter-electrode close to the anode.

From our knowledge on electropolishing with pulsed direct current [30–32], we developed a technique to electropolish spots (Fig. 30.9) and channels [33]. A little needle is placed a few micrometers in front of the workpiece. Pulsing the current between zero and a current density that allows a transpassive removal of the metallic workpiece on the opposite side, spots can be polished. Therefore a reasonable amount of current has to be used. Moving the needle electrode in front of the stainless sheet metal with a little xyz robot provides the possibility of making an electropolished channel structure.

First experiments have been done that show the possibility of producing channels in this way. The cathode should be positioned only a few micrometers away from the sheet. High currents allow us to work in the transpassive regime of the electro-chemical system where electropolishing takes place.

Figure 30.9: Micro-electropolished sheets of stainless steel. The microstructured hole has a width of 60 mm and a depth of 80 mm. The stainless steel has been electropolished in a mixture of sulphuric and phosphoric acids.

30.4 Etching Processes: Structure Formation on the Rotating Disk Electrode

In general, the rotating disk electrode is used to obtain exact experimental results pertaining to pure electro-chemical processes at the electrode by controlling the influence of the diffusion boundary layer. Classical electrochemistry postulates a diffusion current flowing in orthogonal direction through the laminarly flowing boundary layer at the rotating disk electrode. Due to the rotation of the disk, there is a constant vertical flow of solution from the bulk towards the electrode, as well as a horizontal flow from the center to the edge. The flow conditions in the solution beneath the electrode were well described by Kármán [34] and later by Levich [35] and Newman [36]. In these theoretical works, the flow towards the electrode is found to be independent of the radius (Fig. 30.10a), whereas the horizontal flow along the electrode is curved in the direction of rotation and follows spiral-like streamlines (Fig. 30.10b).

The main purpose of using the rotating disk electrode in electro-chemical investigations is to control a theoretically well-defined hydrodynamic flow along the working electrode. However, the flow and especially the idea of the fixed boundary layer at the electrode surface is not satisfactory if there is a coupling between the hydrodynamic flow and the dissolution of electrode material. Etching of stainless steel with concentrated $FeCl_3$ solutions is a widely used method for structuring and surface treatment. Nevertheless, only minor fundamental physico-chemical knowledge is available on the coupling of the electro-chemical etching process and the hydrodynamic flow close to the electrode. The rotating electrode offers the possibility to investigate just this coupling very precisely.

The strong light absorption of 3.5 M iron(III) chloride solution posed a major problem for realizing the videographic observation of the hydrodynamic flow beneath the electrode.

16 LEDs (light emitting diodes) with a wavelength of 605 nm and an overall power rating of 1.68 W were then arranged in a circle. The light intensity was 9500 mCd for each LED. The circular arrangement (Fig. 30.11) was used as a dark-field illuminator, which enhances the contrast of surface features such as laser embossed or engraved marks, or surface defects. This lighting source was embedded in an inert material (two-component epoxy resin, M9026, V. Höveling) and placed directly into the solution beneath the electrode at a distance of 2 cm (Fig. 30.12). Installing this special lighting source enabled sufficient light intensity for monitoring the hydrodynamics beneath the electrode.

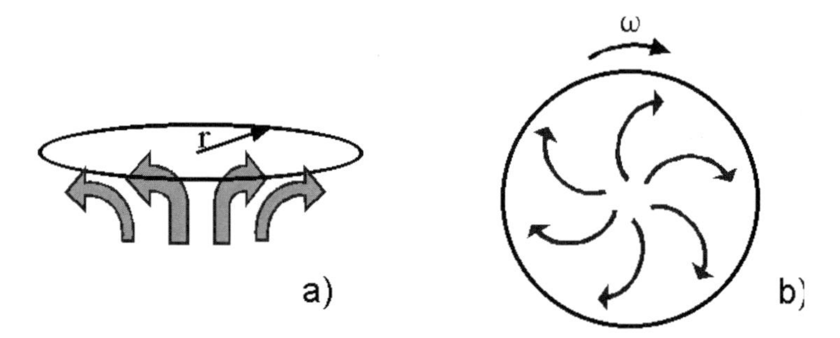

Figure 30.10: Flow of solution towards the electrode (a) and along the surface (b).

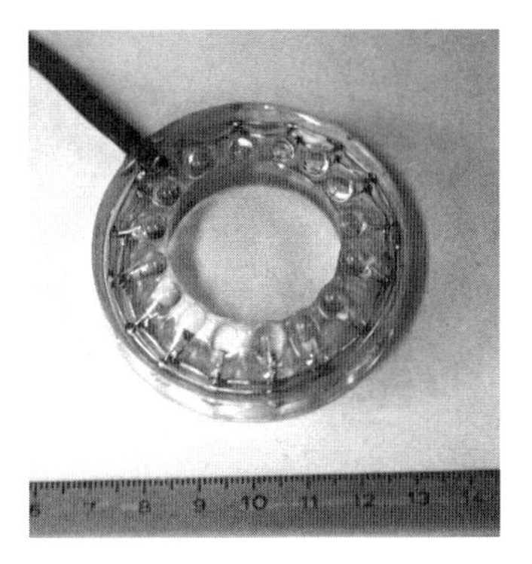

Figure 30.11: Special lighting source with 16 LEDs.

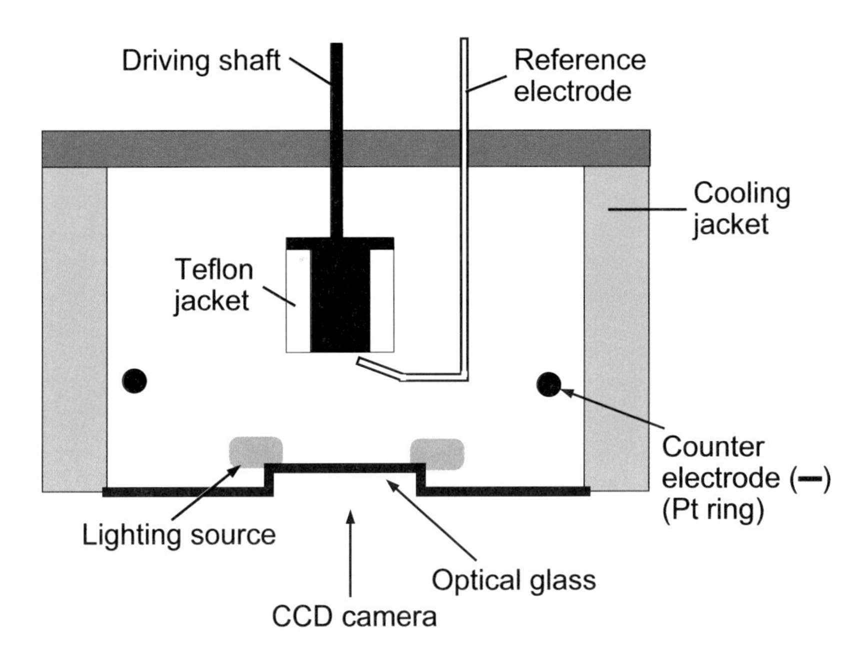

Figure 30.12: Schematic cross section of the reaction vessel with a three-electrode setup.

The low power of this construction also prevents the solution from unnecessary heating. An optical glass was integrated into the reaction vessel in order to avoid distortions while monitoring the processes at the working electrode and in the solution by means of the CCD camera (Fig. 30.12). The reaction vessel consisted of a glass cylinder, 14 cm in diameter and enclosed in a heating/cooling jacket (Fig. 30.12) that enabled the temperature to be controlled with an external thermostat (LAUDA UB40). By designing the experimental setup in this way, it was possible to maintain the temperature in the reaction vessel at levels between 15 and 50 °C.

The rotational speed of the working electrode could be set between 1 and 6000 rpm with an accuracy of 0.2% at 3000 rpm (according to the manufacturer's specification) by means of a rotation-control unit (JAISSLE R.S. PI-REGLER). A standard three-electrode setup connected to a JAISSLE potentiostat/galvanostat (IMP 83 PC-2) was used to control and measure both potential and current. The working electrodes (with a radius of 2.5 or 5.0 mm) consisted of chromium–nickel stainless steel (DIN 1.4301) with a Teflon jacket for protecting the side walls (Fig. 30.13).

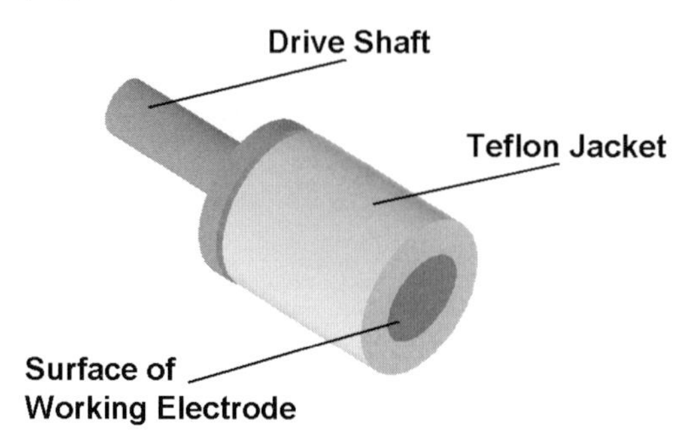

Figure 30.13: Working electrode.

A platinum wire with a diameter of 1 mm was used as the counter-electrode. It was placed circularly in the same plane as the working electrode, at a radial distance of 6 cm from the center. The circular design of the counter-electrode was chosen to guarantee radial distribution of current density. To avoid hydrodynamic pattern formation being disturbed, the reference electrode (standard Ag/AgCl METTLER TOLEDO 363-S7/120) was placed via a Luggin capillary next to the edge of the working electrode.

The onset of natural convection flow induced by a dissolution process for non-rotating as well as for slowly rotating disk electrodes causes the formation of etched patterns in the surface of the electrodes. For rotational speeds ranging between 0 and 50 rpm, the patterns in the surface were directly correlated with the convection flow beneath the electrode. Patterns generated at rotational speeds lower than 15 rpm are curved in the direction of rotation (Fig. 30.14, 10 rpm), because the processes which cause the formation of patterns are mainly influenced by the force of gravity. Processes at rotational speeds above 15 rpm, on the other hand, are mainly influenced by the centrifugal force, which results in topographical structures that are curved against the direction of rotation (Fig. 30.14, 20 rpm). A transition region at about 15 rpm was observed (Fig. 30.14, 15 rpm). This region indicates the point of balance between the two forces taking effect in opposite directions – the gravitational and the centrifugal forces.

The phenomenological model [37] of convection vortices in the boundary layer was developed to explain the formation of the different topographically structured surfaces as well as the occurrence of spatial bifurcation in the patterns on the electrode.

The behavior of the same system under conditions of fast rotating working electrodes (1000–6000 rpm) was investigated also [38]. Patterns with the shape of logarithmic spirals are generated at rotational speeds above 1000 rpm. When these spiral patterns were analyzed, an invariant curvature of the structures was found. This invariant behavior was identified for a broad range of external parameters, such as rotational speed, temperature, external current or the Teflon jacket enclosing the working electrode. Figure 30.15 shows the topography of a structured surface at 5000 rpm, with a superposed plot of two calculated spirals.

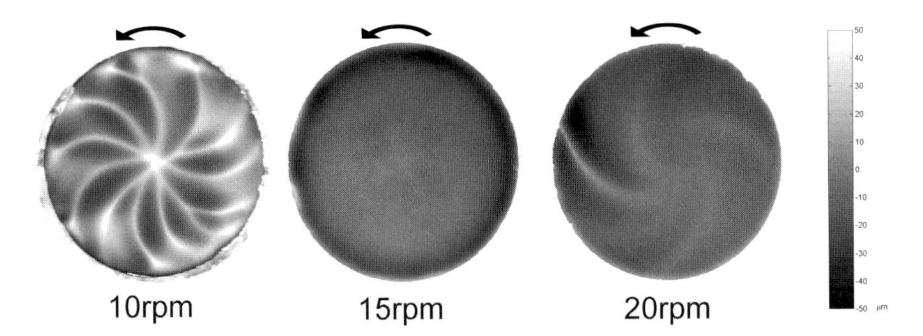

Figure 30.14: Generated patterns with a change in curvature direction at different rotational speeds (see the arrows) and a transitional region in between where no pattern is formed [37].

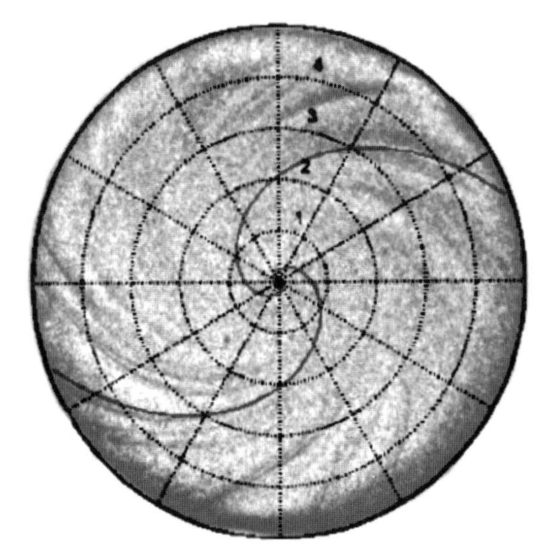

Figure 30.15: Topography of a structured surface at 5000 rpm and two superposed calculated spirals [38].

The shapes of the patterns were clearly determined as logarithmic spirals obeying the equation: $r = ae^{-b\Phi}$. For all the patterns formed at high rotational speeds the value for b in this equation is about -0.71; this value results from the constant ratio between tangential and radial flows of $1/\sqrt{2}$ that was discovered. In contrast to classical explanations of pattern formation at rotating disk electrodes, the Coriolis force was taken into account and the invariance of the logarithmic spiral patterns could be explained by a physical description that takes into consideration the balance between tangential velocity, centrifugal and Coriolis forces [37, 38].

This system with an electro-chemical induced hydrodynamic pattern formation exhibits also permanent galvanostatic potential oscillations, as known from many other examples for metals in aqueous solutions. Different regimes of current density were identified showing stationary, oscillatory and chaotic behavior.

The hydrodynamic vortex patterns in the boundary layer beneath the electrode and the electro-chemical process are coupled via the topographically structured surface and lead to a new type of superimposed potential oscillation [39]. The body of results obtained from these experiments provides a comprehensive analysis of moving workpieces in corrosive media, and of the coupling between electro-chemical and hydrodynamic instabilities as exemplified in a system comprising rotating steel disk electrodes in highly concentrated iron(III) chloride solution.

30.5 Oscillating BZ Reactors Coupled via Liquid Membranes

The Belousov–Zhabotinsky reaction (BZ reaction) is the most popular oscillating chemical reaction. Almost all details of this reaction are known because of the fact that this reaction has been investigated for decades by many famous groups [40–45]. Furthermore, the BZ reaction and the related Briggs–Rauscher reaction (BRR) serve as standard examples for studying coupling phenomena in oscillatory systems [46–49].

The different dynamic phenomena in this BZ reaction under CSTR (continuously stirred tank reactor) conditions have been studied both experimentally and theoretically. The three-variable Oregonator model (1974) [50] was the first famous model which has been studied extensively. In 1991 Györgyi and Field [51] proposed an 11-variable chaotic model which has almost matched the experimental behavior of the above reaction. Later [52–54] they have reduced the model to nine, seven, four and three variables. In this study we use the above four-variable model to study the control of chaos using flow rate as the control parameter. We used the following four-variable model [51] in order to describe well the possible chaotic behavior of the reaction with respect to the use of this reaction in the separation procedure:

This model leads to the following set of coupled kinetic equations:

$$\begin{aligned}
\frac{dx_1}{dt} &= -k_1x_1x_2 + k_2x_2 - 2k_3x_1^2 + 0.5k_4[\mathrm{H}^+]([\mathrm{Ce}]_{\mathrm{total}} - x_3)[\mathrm{BrO}_2^{\cdot}]_{\mathrm{eq}} \\
&\quad - 0.5k_5x_1x_3 - k_fx_1, \\
\frac{dx_2}{dt} &= -k_1x_1x_2 - k_2x_2 - k_7x_3x_4 + k_8[\mathrm{MA}^*]_{\mathrm{qss}}x_4 - k_fx_2, \\
\frac{dx_3}{dt} &= k_4[\mathrm{H}^+]([\mathrm{Ce}]_{\mathrm{total}} - x_3)[\mathrm{BrO}_2^{\cdot}]_{\mathrm{eq}} - k_5x_1x_3 - k_6x_3 - k_7x_3x_4 - k_{\mathrm{f}}x_3, \\
\frac{dx_4}{dt} &= 2k_1x_1x_2 + k_2x_2 + k_3x_1^2 - k_7x_3x_4 - k_8[\mathrm{MA}^{\cdot}]_{\mathrm{qss}}x_4 - k_{\mathrm{f}}x_4,
\end{aligned}$$

with the k_i values as defined above; k_{f} is the flow rate and

$$\begin{aligned}
[\mathrm{BrO}_2^{\cdot}] &= \sqrt{\frac{0.858x_1}{4.2 \times 10^7}}, \\
[\mathrm{MA}^*]_{\mathrm{qss}} &= \frac{-(2.4 \times 10^4x_4) + \sqrt{(2.4 \times 10^4x_4)^2 + 1.8 \times 10^9x_3}}{1.2 \times 10^{10}}, \\
\mathrm{H}^+ &= 0.26,
\end{aligned}$$

$$[\text{Ce}]_{\text{total}} = 8.33 \times 10^{-4},$$
$$x_1 = [\text{HBrO}_2],$$
$$x_2 = [\text{Br}^-],$$
$$x_3 = [\text{Ce(IV)}]$$

and

$$x_4 = [\text{BrMA}].$$

The model shows chaotic behavior at flow rates near 8.28×10^4; the corresponding phase portrait is shown in Figs. 30.16 and 30.17. Figure 30.16 shows the respective oscillations of the selected components. In Fig. 30.17 some controlled orbits of cerium are shown. It can be observed that the system admits a period-adding route to chaos. Further, it can also be seen that the system switches from chaotic to periodic and then again to chaotic oscillations.

Table 30.1: Model D from Györgyi and Field [51]

Reaction	Rate constant
$Br^- + HBrO_2 + H^+ \rightarrow 2\ BrMA$	$k_1 = 2.0 \times 10^6$
$Br^- + BrO3^- + 2H^+ \rightarrow 2\ BrMA + HBrO_2$	$k_2 = 2.0$
$2\ HBrO_2 \rightarrow BrMA$	$k_3 = 3.0 \times 10^3$
$0.5\ HBrO_2 + BrO_3^- + H^+ \rightarrow HBrO_2 + Ce(IV)$	$k_4 = 6.2 \times 10^4$
$HBrO_2 + Ce(IV) \rightarrow 0.5\ HBrO_2$	$k_5 = 7.0 \times 10^3$
$Ce(IV) + MA \rightarrow$	$k_6 = 0.3$
$BrMA + Ce(IV) \rightarrow Br^-$	$k_7 = 30.0$
$BrMA \rightarrow Br$	$k_8 = 2.4 \times 10^4$

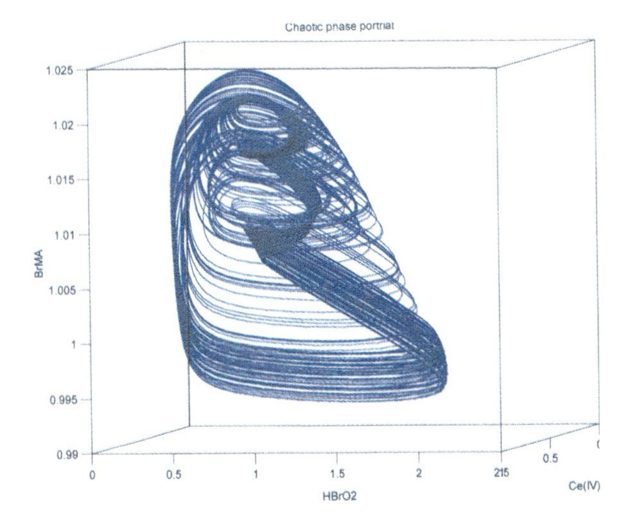

Figure 30.16: Three-dimensional subspace of the chaotic attractor of the four-variable model. The three variables shown here are: the concentration of the cerium(IV) ions, the hypobromic acid, and the bromomalonic acid.

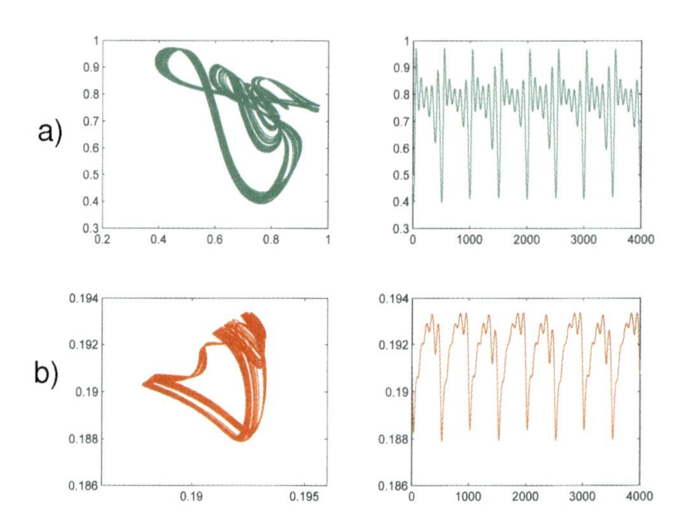

Figure 30.17: Attractors and time series of the oscillating cerium ions in the aqueous ((a), green) and the organic ((b), red) phases.

Afonin [55–57] proposed to use this reaction for extraction and separation of rare-earth elements under non-equilibrium constraints using fluid-membrane techniques. He used a batch reactor systems for his investigations. He claimed that one can enhance the extraction process using an oscillating reaction. For this purpose he carried out the BZR in a closed vessel in order to establish an oscillating chemical system. But such a system can never be stabilized which is why he started extraction experiments – sponsored by MIR-Chem GmbH and executed in their laboratories – in which the aqueous Ce^{3+}/Ce^{4+}–BZR systems formed an open low-flowing CSTR system, the oscillation in which could be stabilized in principle. As an extraction agent for the rare-earth metal ions he used n-decane loaded with the complexing agent TBP (tributyl phosphate). It was our assumption that the cerium ions – which are rare-earth metal ions as well – will also take part in the phase transfer. So, it would be unlikely to describe correctly the BZR with the classical three-variable model which mainly results in limit cycle behavior but not chaos as we expected for our system. Therefore we took the four-variable model of Györgyi and Field [51], in order to achieve chaos which could be controlled by choosing special flow rates. Moreover, it could be shown theoretically that we have to enhance the number of variables in order to include correctly the phase transfer of the cerium-metal ions between the aqueous phase and the organic TBP phase.

Moreover, he carried out preliminary experiments with two open Ce^{3+}/Ce^{4+}–BZR reactors, coupled via a fluid membrane (TBP), in order to look for an enhancement in the separation of neodymium and praseodymium. To achieve a reliable and efficient technical system for oscillating extraction of rare-earth elements, one has to form a cascade of oscillating reactors which are coupled via a fluid membrane in which only the catalyst and its chemically related rare-earth ions might be transported with variable time delays [58]. This is a fascinating idea but up to now there does not exist any reliable physical/chemical knowledge proved experimentally on such complex systems.

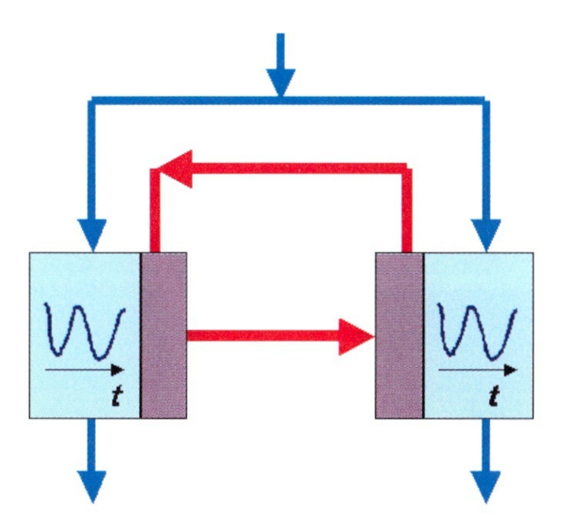

Figure 30.18: Schematic representation of two oscillating BZ reactors coupled via a cyclic pumped fluid membrane.

Therefore, Rabbow [59] studied in detail two oscillating BZ reactors which are coupled via a fluid membrane (see Fig. 30.18). He used the Ce^{3+}/Ce^{4+} redox system as catalyst and n-decane loaded with the complexing agent TBP (tributyl phosphate) as fluid membrane. Each of the two BZ reactors worked as an open low-flow CSTR system, whereas the fluid membrane was pumped in a cyclic way between the two reactors. The fluid membrane forms an interface with the non-miscible aqueous system of the BZ reactors. To achieve a high surface region for an optimal phase-transfer reaction of the catalyst between the aqueous and the organic phases the liquid membrane of each reactor was pumped through a frit into the aqueous oscillating phase of the other reactor.

Especially, the Ce^{4+} form complexes with the TBP, wherefore these ions become extracted from the aqueous phase into the organic one. When Ce^{4+} is formed during oscillations in the aqueous solution the concentration of these ions increases. This will force the phase-transfer reaction of the Ce^{4+} complexes from the aqueous into the organic phase. In this way an impulse of the Ce^{4+} concentration in the organic flow is produced. Depending upon the transportation time to the second reactor this impulse will interact with the oscillating system of the second reactor. Since both reactors are interacting in a cyclic way, both oscillating systems become coupled with a time delay only via the Ce^{4+} catalyst ions (see Figs. 30.18 and 30.19).

One can achieve a surprisingly good separation of the rare-earth elements praseodymium and neodymium if they are available in the aqueous phase. By oscillating extraction [58] the separation ratio of concentrations of both elements in the organic phase differs strongly from unity, if the absolute amount of the concentration is very small. This is a very interesting behavior but without a reasonable technical application because of the very difficult handling of the system and the great amount of chemicals needed for this process.

One can vary this method of oscillating extraction replacing the oscillating chemical system of the BZ reaction by an oscillating potential which serves as an external driving force [58]. This new procedure enables a promising technical application since one gets separation ratios of praseodymium and neodymium is of Pr/Nd = 1.4 to 1.5 even for concentration and time intervals of technical importance (see Fig. 30.20).

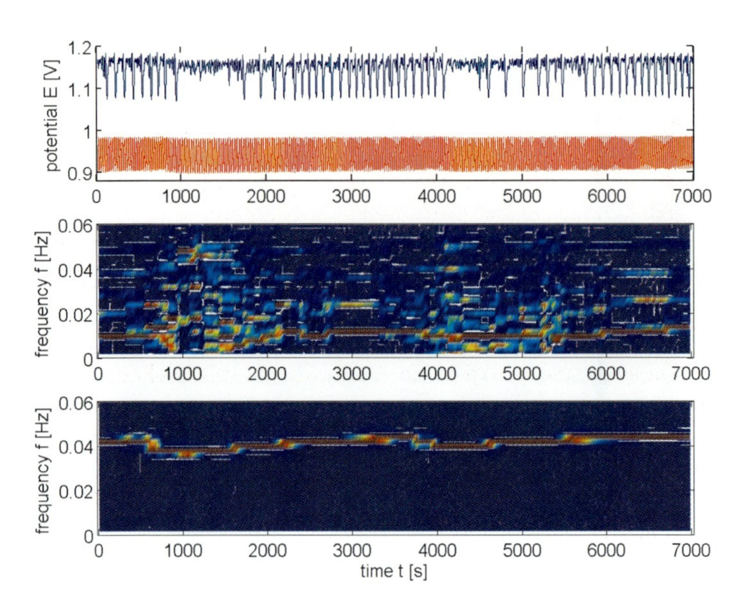

Figure 30.19: Oscillations of the coupled BZ reactions in the aqueous phases of the two phase reactors. The organic phase was pumped in a cyclic way with a pumping rate of 11.5 ml/min between the two reactors. Top: time series of the two reactors (blue and red curves); bottom: the temporal development of the correlated Fourier spectra (sonograms).

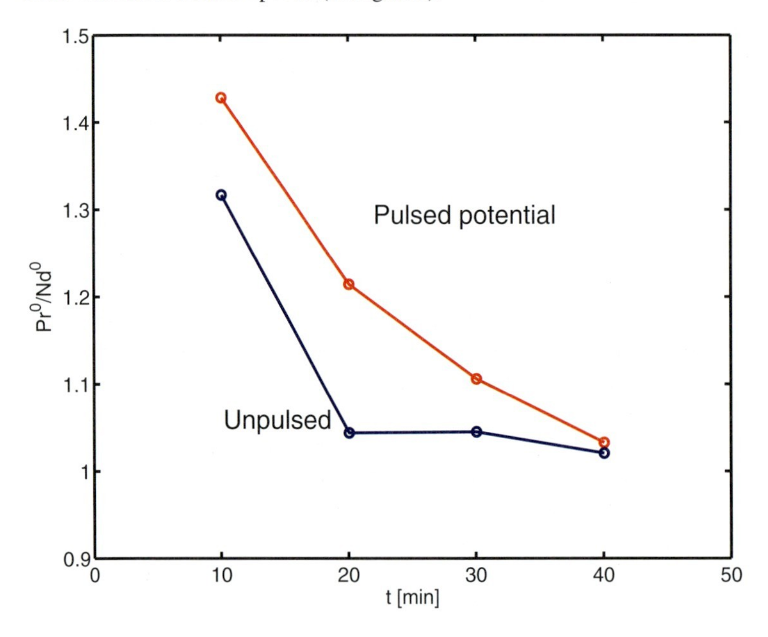

Figure 30.20: Temporal development of the Pr/Nd ratio in the stripped organic phase (TBP in kerosene).

30.6 Reaction in Mono-porous Foams

The rapid progress of developments in microstructuring techniques inaugurates a great variety of new possibilities for chemical process engineering. The major aim of this development should not be restricted to the miniaturization of the usual process units even though the handling of very dangerous and expensive chemicals as well would have enormous advantages. For example, it is urgently needed to understand the physico-chemical properties of mixing processes at the transition to smaller dimensions. On the other hand an optimal homogenized input stream of reactants makes any mixing devices inside of the reaction vessel unnecessary. A static micromixer has been designed at IMM in Mainz [60, 61] to separate filament-creating processes of mixing from the remaining diffusion process [62]. A microstructured inter-digital channel system combines two input streams into a laminated flow with wall thicknesses of a few micrometers [63]. Using a micromixer increases considerably the efficiency and quality of mixing [64]. The selectivity and the yield of chemical reactions and the efficiency of extraction processes have been improved by using these microstructured devices [65, 66]. Using a micromixer to deflocculate immiscible fluids was suitable for generating mono-disperse emulsions of silicon oil in water [67]. In practical application the energy demand of the dispersion process, which is needed to create the new interfacial area, was decreased considerably in comparison with conventional dispersing methods using inter-digital micromixers [68]. The flow rates of the microstructured mixers achieved a macroscopic dimension [60] of up to several liters per hour. Foaming experiments with glycerine/water solutions containing an anionic surfactant have advanced this work. It was possible to generate a so-called *hexagon flow* of highly ordered uniform bubbles by using several micromixers containing different mixing elements with various channel widths [69]. Continuously flowing foams with a regular cavity structure are obtained by using microstructured mixing devices. We examined the process by which such gas–liquid dispersions are generated and propose a simple method for optimizing the foaming conditions with regard to the foam quality. Foams generated with a narrow distribution of cavity sizes exhibit a particular stability owing to well-balanced Laplace pressures. Coalencence of the hollow structure generated is partially inhibited due to the uniformity of the entrapped gas volumina. Additionally, it is possible to create rather persistent foams of chemical solutions that are not usually intumescent. A continuous flow of a regular gas in liquid dispersion shows specific flow characteristics associated with a crystal-like arrangement of the uniform gas cavities. These mono-disperse foams are marked by a large and highly regular interface. Thus mono-disperse foams should be well suited for chemical gas–liquid reactions, particularly if the mass transport is a limiting process. The comparatively high gas fraction within the liquid phase is also advantageous for an effective reaction rate. Our interest is focused on the general cavity-formation mechanisms in microstructured devices, the generation process and the dynamics of flowing foams when containing a chemical reaction and the targeted exploitation of observable nonlinearities and pattern-formation processes.

Bubble formation under spatial restriction (see Fig. 30.21) and with simultaneous assistance from liquid flow is a practical and effective method for continuous production of mono-dispersive foams with controlled bubble size [70, 71]. Snapshots of the bubble-generation process show the uniformity of successively formed bubbles. Calculations derived from flow parameters and direct frequency measurements indicate that invariant bubble formation can be maintained to the order of magnitude of a few hundred bubbles per second, i.e. up to the kHz

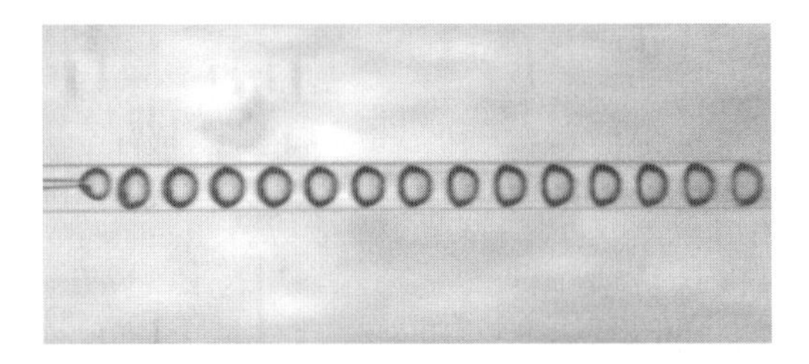

Figure 30.21: Central micromixing element for creating a mono-porous foam. One can easily detect the bubble formation (left-hand side) and the phase contacting in the capillary-in-capillary mixing channel system.

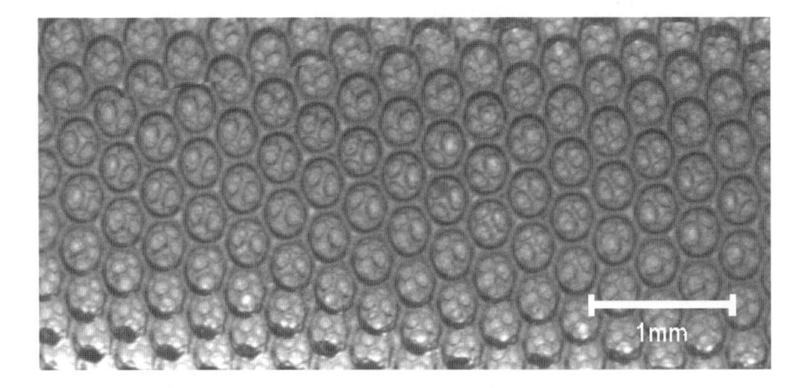

Figure 30.22: Mono-disperse foam in hexagonal closest sphere packing. Three-dimensional mono-dispersive foam: the lateral freeze image shows a continuously flowing foam inside a glass tube as it is used as a medium for a chemical reaction. The gaseous rich dispersion ensures educt supply, whereas the conversion takes place in the spatially structured liquid interstitial region.

region. Sizes of bubbles as well as the volume ratio of gas and liquid can be varied over a broadly extended range.

The static microchannel of the capillary-in-capillary mixing element setup used for the dispersion of gas and liquid results in a segmented, two-phase flow of successive, equally sized Taylor bubbles separated by uniform liquid plugs (Fig. 30.21). At the outlet of the mixer, the gas bubbles self-arrange into a loose crystal order with a close sphere package (Fig. 30.22), if the coalescence of bubbles can be suppressed by the presence of a surfactant. The close bubble package of a gaseous rich dispersion leads to a significant increase in the viscosity of the two-phase medium, which is unknown originally in either the gaseous or the liquid phase. Since the continuous liquid phase as a whole is turned into an immobilized matrix for embedding the gas-bubble assembly, the spatially fixed gas/liquid dispersion can be considered a structured two-phase medium.

Chemical reactions inside foams generally take place in the liquid-filled interstices between bubbles, which are spatially structured as a continuous lamellar system inter-connected by plateau borders. Obviously, the reaction space as a whole is a three-dimensional structure of a cohesive interface layer in a non-equilibrium situation.

Figure 30.23: Reaction zones during decoloring of *Remazol Black 5* in the vicinity of two gas bubbles (left- and right-hand sides) containing ozone.

Chemical conversion takes place on a molecular level. Consequently, two-phase reactions require a mass transfer of reactive compounds from one phase into the other, or straight contact at the phase boundary. Mass-transport processes inside thin layers are dominated by molecular diffusion. The reaction–diffusion dynamics of a fast chemical reaction cause a reaction front which moves from the phase boundary into the fluid (Fig. 30.23).

Figure 30.23 shows the reaction–diffusion areas of two bubbles being opposite to each other. These reaction–diffusion zones are produced within a few seconds but they remain stable even for minutes. The black border exactly at the gaseous side of the bubbles is due to the curved boundary layer of the two phases. The following lightened areas are correlated to the regions in which the ozone bleaches the dyestuff by oxidation.

A continuously flowing foam, homogeneously dispersed and characterized by an enlarged and initially well-structured interface region, a fixed contact area between phases and a high gas content is a very interesting medium for phase-transfer reactions under limited transportation. Mass transfer of reactive compounds out of the gaseous phase with small penetration depth into the fluid indicates some noteworthy effects in terms of yield, selectivity and rate of reaction, depending on reaction kinetics. In contrast to continuous gas-supply conditions as frequently used in phase-contacting equipment, a relative motion of phases in flowing foams is restricted to small values (e.g. small Reynolds numbers as well). Both lateral dispersion and back mixing inside foams are suppressed by the mechanical stiffness of the tightly packed bubble/lamellae structure (unlike the hydrodynamics of pipes or bubble columns).

Relative liquid motion occurs within the interstices of the bubble structure that are formed by bubble-separating films and the plateau border channel system [72]. Dynamic changes of

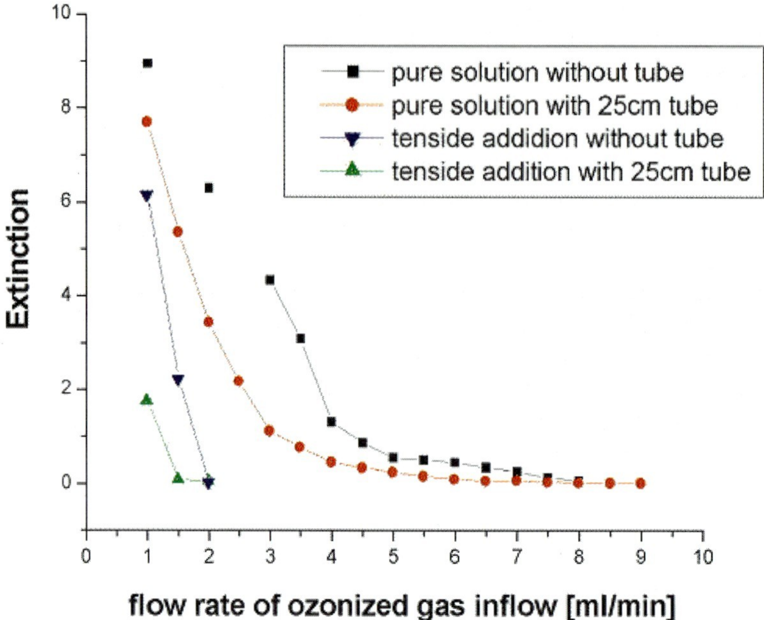

Figure 30.24: Kinetics of the reaction of indigo carmine with ozone in flowing systems. Comparison of different situations: black: without foam, without tube; red: without foam, in a tube; green: with foam, without a tube; blue: with foam in a tube.

bubble structure caused, for example, by bubble coalescence or due to lattice slippage in the bubble structure induce an additional factor to the local hydrodynamics.

Figure 30.24 shows a comparison of different situations of reactive flowing foam with a typical batch reactor or a tube reactor respectively. As can be seen easily, reactive flowing foams are much more effective than foamless tube reactors. Nevertheless, one has to take care of the positioning of the flowing foam with respect to gravitation. If the size of the tube is small enough gravitation can be ignored and a homogeneously flowing foam can be stabilized, which may serve as an ideal reaction medium.

In this regard, the use of foam as a chemical reactive medium opens up new pathways for the process engineering of two-phase reactions. Furthermore, there are potential applications for mono-dispersive solid foam in materials science.

Mono-dispersive Ceramics

Under appropriate circumstances dispersing of gas in a fluid with a glass-capillary micromixer leads to a continuous and homogenous flow of foam with a fixed aspect ratio of gas and fluid. The bubble-size distribution is a characteristic feature of a foam. If this distribution function possesses a very narrow half width, i.e. if all bubbles are almost of the same size, one may call the foam mono-dispersive.

Porous ceramics have been widely utilized in the industry for catalyst supports, filters and acoustic absorbents. In the environmental processes they have especially great potential as supports of microorganisms in biological treatment of waste water, and as biological filters for the treatment of toxic gases. On the other hand, the studies of viability have pointed out the employment of porous ceramics as filters for dust elimination from hot gases like NO_x and SO_x. All these applications are due to their characteristics such as low density, low thermal conductivity, thermal stability, specific high tension and high resistance to chemical attack [73].

Numerous techniques to prepare porous ceramics have been developed, for example mechanical and chemical foaming methods. The most important of them are a method of pre-sintering of granules or fibers, an aero gel or sol–gel method and the pyrolysis of several organic additives. The foaming method has some distinguishing characteristics since it allows an easier control of the pore structure. This control is important because, in most cases, it affects directly the thermal and mechanical properties and the operation of the porous ceramic. On the other hand, the foaming method involves serious problems because of the fact that the foam can be broken during the drying and because of the small mechanical resistance of the intermediate product. In the last decade, a significant part of this problem has been solved by means of the in situ polymerization, dehydration by means of freezing (freeze-drying) and thermal gelation.

At present, methods for obtaining micro- and mesoporous ceramic with cavities or uniform and predetermined pores exist; however, it is very difficult to obtain a macroporous ceramic with the same distribution of pore sizes, since the existing methods do not allow a high level of quality control of the ceramics, neither in the pore sizes nor in their distribution.

Lee and Park [74] outline that a great part of these problems can be solved by means of the in situ polymerization, freezing and thermal gelation; however, the authors use an agitation method for the foaming of a mixture with addition of foaming aid and a commercial gelling substance and they obtain a porous ceramic after drying and burning, but the porous system is very far from being uniform. Therefore, the methods of agitation are not the best way for obtaining uniform macroporous ceramics.

In the patent of Wood et al. [75], a ceramic foam is obtained by means of the reaction of poly-oxyethylene isocyanate with a great quantity of watery chemicals containing finely divided ceramic materials. The resulting foam is burned and a rigid ceramic structure is obtained. With this method it seems that the control of the pore size is very precise, but the authors do not give data of porosimetry. The limitations of this method are the same ones as those shown in the process of Lee and Park [74].

There are other methods that make use of the previously reported principles, with doubts regarding the uniformity of the pores of the ceramics. The porous ceramics can be obtained by means of foams of silicon carbide, polystyrene foams and many other types of foams. Additional information can be found in Brundage et al. [76], Tittizer [77], Schmidt et al. [78], Ryuichi [79], Bao et al. [73], Zhu et al. [80], Tulliani et al. [81], Murakata et al. [82] and Higuchi et al. [83].

The structure and geometry of the foams are extremely complex; they are composed by polyhedral bubbles that differ markedly in form and size. They are highly concentrated dispersions of gas (dispersed phase) in liquids (continuous phase). When they contain more than 90% of gas in volume, they are generally polyhedral, because the biggest fraction in volume for identical spheres is 0.74. This approach is applicable only to mono-dispersed systems; in poly-dispersed systems, the bubbles maintain the spherical form for higher fractions in volume

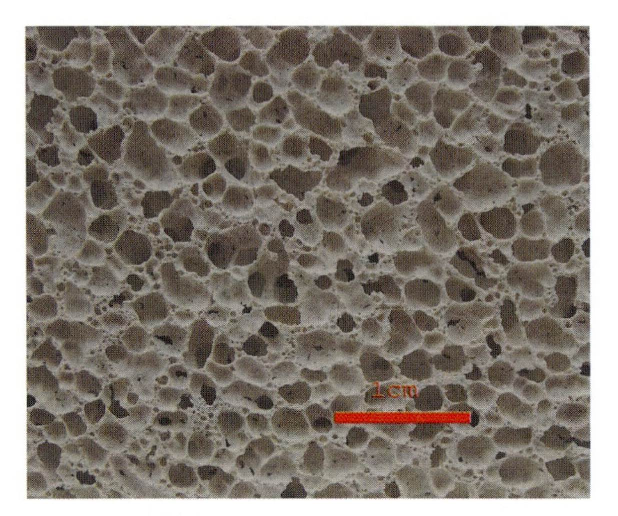

Figure 30.25: Porous structure of a ceramic foam.

because the smaller bubbles take places among the bigger ones. They are also metastable systems; they show a spontaneous tendency to separate into two different phases. Some foams can remain stable for just a few minutes and others for many days, depending on the generation conditions.

The starting pore-size distribution changes during lifetime of a foam. During this ageing process even a distribution of spherical bubbles transforms into an Apollonian structure (see Fig. 30.25) before ending up in polyhedral bubble shapes. The Apollonian structure is characterized by a few big spheres and a lot of essentially smaller spheres filling out the remaining space. Again there remains free space between the large and the smaller spheres, which can now be filled with much smaller spheres in a fractal way. Such a process can be observed very often in drying processes of foams as for example in beer or sol–gel procedures.

According to Bikerman [84], the foams, as colloidal systems, can be prepared by dispersion or condensation. In the dispersion methods, the dispersed phase is initially a great volume of gas, which is mixed with the dispersion phase (continuous phase, a liquid, in a foam). In the condensation methods, the dispersed material originally is presented as a solute, that is, as molecules dissolved in the liquid; when these molecules combine with big aggregates, the foam is obtained. The dispersion methods can be appreciated when a surfactant solution is agitated or when air is injected to this solution. A very common particular subdivision of these methods implies the injection of gas in a liquid through one or several holes.

Using a static micromixer Ehrfeld et al. [60, 61] showed that it is possible to create a hexagonal flow of bubbles in a tube if one presses air in a liquid flow. They argued that the special construction of the micromixer device causes this special formation of bubbles all of the same size. Mathes and Plath [70] confirmed these experiments. However, they proved that it is not the special construction of the inter-digital channel system of this IMM micromixer device which generates the bubbles but the construction of the slices in the outlet of this mixer device. Normally, this type of static micromixer generates bubbles with a wide bubble-size distribution.

Nevertheless, Mathes and Plath [70] were able to generate quite different bubble-flow systems while using this IMM micromixer system. These flows differ in the arrangements of the bubbles. They observed phase transitions among the different regimes of the bubble flows and they could estimate the region of flow rates and pressure in which these regimes give stable flows. Outside this range of parameters turbulent bubble flows are observable.

In a patent announcement Plath and Buhlert [85] stressed the idea to generate a mono-porous foam for the formation of a macroporous ceramic. For this purpose they proposed to vary the dispersion method taking into consideration the results of their micromixer experiments [70, 71]. Based on these ideas we developed a special glass-made microstructured capillary-in-capillary tool to generate a flow of silica sol in which air is pressed in a sufficient way so that one gets a silica-gel foam consisting of bubbles with a very narrow size distribution.

Mono-dispersive foams are distinguished by a higher persistence and coalescence stability in comparison to foams of equal chemical composition but wider bubble-size distribution. It was the aim of S. Diaz Alfonso to froth up a silicate sol by means of our specially developed glass-capillary-in-capillary-system creating a stable foam (Fig. 30.26). Afterwards the sol foam should undergo a sol–gel transformation resulting in a gel foam which should be dried and burned in order to create a mono-dispersive ceramic foam.

In order to get a stable bubble flow of the silical sol which remains stable even during gel formation we had to find a special surfactant which fits with the pH sensitivity of the silica sol. For obtaining the sol foam a solution with the silica sol and egg white was prepared; the proportion of the raw materials in the solution is the following: For 100 ml of solution: 70 ml of silica(IV)-oxide sol, purity 10%, pH = 6.9 (Merck) and 30 ml of egg white, purity 6% (SIGMA).

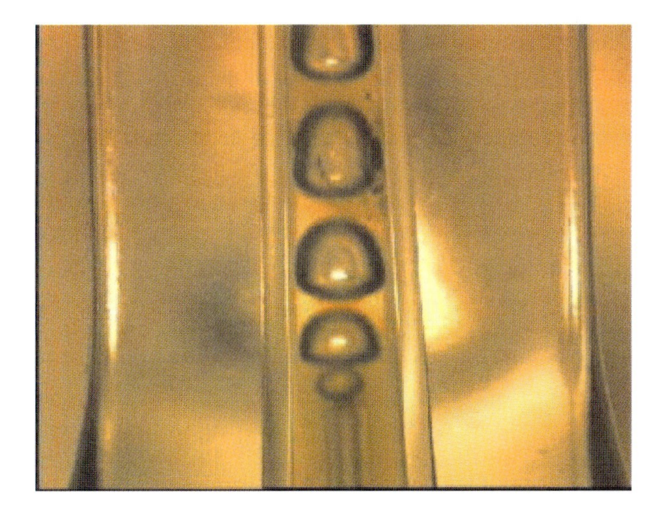

Figure 30.26: Generation of silica-sol bubbles of equal sizes by a system of capillaries. The bubbles develop at the outlet of the inner capillary. However, the size of the bubbles is determined by the radius of the medium capillary, which is also stabilizing the bubbles. In the outer capillary additional chemical components are transported for the reaction in the liquid surfaces of the bubbles at the outlet of the medium capillary.

With this quantity of surfactant the smallest proportion in the solution is guaranteed without affecting the stability of the foam. For bigger values of surfactant, the foam is very stable but after a short time the solution begins to present problems of solubility. On the other hand, for smaller values, the foam is not stable. The water added to the solution due to the dissolution of the egg white can be diminished, although in this case the experiments were carried out with the previously mentioned solution.

In order to get a ceramic from the sol–foam several problems have to be solved:

- To froth up to a foam a sol of high viscosity.
- To stabilize the sol foam.
- To carry out the sol–gel transformation of the stabilized sol foam.
- To dry the stable gel foam without destroying the intermediate product.
- To burn the dried gel foam without any additional friction getting a monolithic ceramic.

We have successfully figured out the special parameters to froth up the SiO_2 sols generating a mono-dispersive foam (Fig. 30.27).

It was very difficult to dry the gel foam. With the aid of a specially developed program S. Diaz Alfonzo could already manage to dry the gel foam in some cases. While slowing down the drying process, less destruction appears in the material (Fig. 30.28). However, up to now, the intermediate product was broken partially in all cases.

It is hardly possible to measure the bubble-size distribution of this broken piece of ceramic. But, without any doubt, no Apollonian structure is observable indicating a narrow size-distribution function. So we can conclude that it is possible to obtain mono-porous ceramics by a sintering method that does not seem to introduce new additional defects in the material. The obtained results demonstrate that one can obtain from a mono-porous sol foam macroscopic ceramics of almost the same pore-size distribution (Fig. 30.29). However, it is needed to achieve much better control of the drying and sintering procedures of this very sensitive system – or to choose another system.

Parameters		**Exp 7**
Pressure of air		0,20 bar
Flow of air		4,80 ml/min
Solution flow		4,02 ml/min
Outer capillary	$D_{ext.}$	0,80 mm
	$D_{int.}$	0,45 mm
Inner capillary	$D_{ext.}$	0,25 mm
	$D_{int.}$	0,18 mm
Results		
Bubble size distribution		0,42-0,49mm
Bubble diameter		0,45 mm

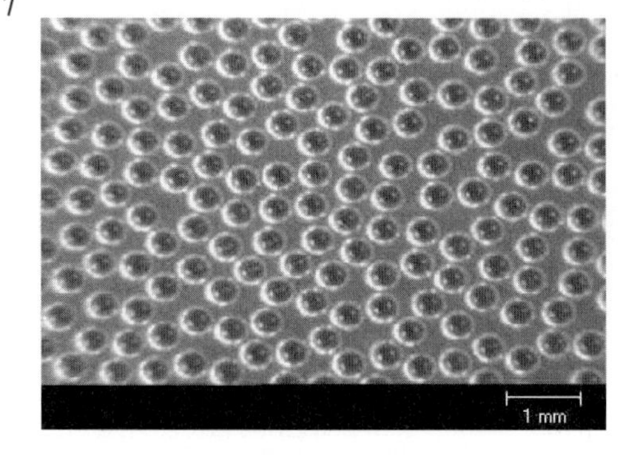

Figure 30.27: Two-dimensional view of the gel foam obtained with Exp. 7. The sol foam has been sampled on top of a sol film in a Petri dish before the sol–gel transition occurs.

Drying regimen 1	
Temperature (°C)	Time (min)
20	*
25	10
30	20
35	30
40	40
50	50
60	60
70	70
80	80
90	90
100	100
150	110

* drying until constant weight

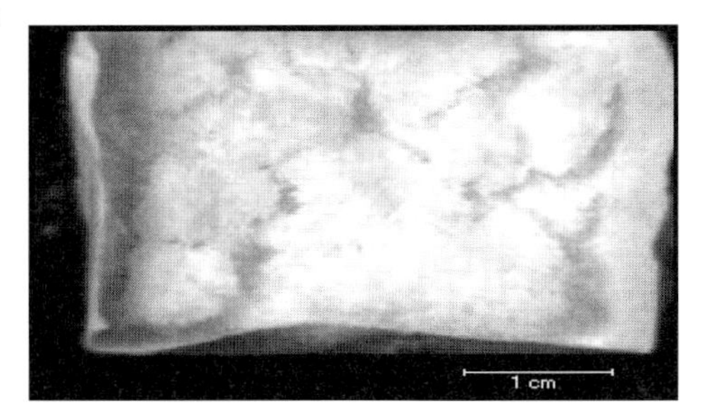

Figure 30.28: View of the foam after a slow drying process with the special drying regimen 1.

Figure 30.29: View of a ceramic obtained from a sol foam of $D = 0.45$ mm (Exp 7; see Fig. 30.27) after drying and sintering.

30.7 Conclusion

We have demonstrated the use of nonlinear dynamics in various applications in chemical engineering. We have shown that nonlinear dynamics plays a significant role for example in electro-chemical systems and that one has to take care of pattern-formation processes in electropolishing and etching of metals. But, especially in electro-chemical manufacturing, we used nonlinear behavior of the systems for example while driving the systems periodically in order to achieve very tiny electropolished microstructures. In case of the separation of rare-earth metals nonlinear processes using fluid membranes will enlarge the separation ratio for example of neodymium and praseodymium dramatically. The use of microstructured mono-dispersed foams for chemical reactions seems to be crazy on a first view. But we have proved that chemical reactions are much more efficient in that case compared with the traditional bubble column. The creation of microbubbles in spatially restricted flowing systems is a highly nonlinear process. The frequencies of the bubble formation are more than a hundred times higher than in classical bubble-formation processes. This offers the possibility of using effectively this special micro-bubble-formation process in various fields of chemical applications, like decontamination of waste water by means of ozone, the formation of polymeric hollow spheres as ultrasound absorbers in medicine and the formation of very new types of crystalline ceramics. We think that it is time to make use of all kinds of nonlinearities, structure formation and self-organization in order to develop new effective chemical processes and materials.

Acknowledgment

We gratefully thank the Volkswagen Stiftung for financial support of many parts of this work especially within the program *Untersuchung nichtlinear dynamischer Effekte in produktionstechnischen Systemen* (I/75 680 – micromixing and I/ 74515 – etching and electropolishing).

We are grateful to the Deutsche Forschungsgemeinschaft for financial support (PL 99/8-1 and PL 99/8-2). We also thank Prof. Visser (Fertigungstechnik, Universität Bremen) for his constant willingness for discussing our results and providing the microfocus scan (UBM). In addition we are indebted to Prof. R. Friedrich and his workgroup (Institute for Theoretical Physics, University of Münster) for some very inspiring discussions and Prof. W. Ehrfeld and Dr. St. Hardt (IMM-Mainz) for fruitful cooperation. Furthermore, we kindly thank the MIR-Chem GmbH for providing their laboratory and for additional funding.

We gratefully thank U. Sydow for converting our Word document to LaTeX.

Bibliography

[1] R.F. Field and F.W. Schneider, *Oscillating chemical reactions and nonlinear dynamics*, Journal of Chemical Education **66**(3) (1989), 195–204;
A.T. Winfree, *The prehistory of the Belousov–Zhabotinsky Oscillator*, Journal of Chemical Education **61** (1984), 661–663.

[2] I. Matsuzaki, T. Nakajima, and H.A. Liebhafsky, *Mechanism of the oscillatory decomposition of hydrogen peroxide in the presence of iodate, iodine etc.*, Faraday Symposium of the Chemical Society London **9** (1974), 55–65;
T.S. Briggs and W.C. Rauscher, *An oscillating iodine clock*, Journal of Chemical Education **50** (1973), 496;
S.D. Furrow and R.M. Noyes, *The oscillatory Briggs–Rauscher reaction I – examination of subsystems*, Journal of the American Chemical Society **105** (1982) 38–42.

[3] E. Wicke, *Dynamische Instabilitäten heterogen-katalysierter Reaktionen als Modelle für Informationsübertragung und biologische Strukturbildung*, Nova Acta Leopoldina **257**(55) (1983), 3–23;
N.I. Jaeger, K. Möller, and P.J. Plath, *Cooperative effects in heterogeneous catalysis Part I*, Journal of the Chemical Society, London, Faraday Transaction I **82** (1986), 3315–3330;
H. Herzel, P.J. Plath, and P. Svensson, *Experimental evidence of homoclinic chaos and type II intermittency during the oxidation of methanol*, Physica D **48** (1991), 340–352;
C. Ballandis and P.J. Plath, *A new discrete model for the non-isothermic dynamics of the exothermic CO-oxidation on palladium supported catalyst*, Journal of Non-Equilibrium Thermodynamics **25** (2000), 301–324;

M. Somani, M. Liauw, and D. Luss, *Hot spot formation on a catalyst*, Chemical Engineering Science **51** (1996), 4259–4269.

[4] G. Ertl, *Dynamik der Wechselwirkung von Molekülen mit Oberflächen*, Berichte der Bunsengesellschaft für Physikalische Chemie **99**(11) (1995), 1282–1294.

[5] W. Ostwald, *Periodische Erscheinungen bei der Auflösung des Chroms in Säuren I and II*, Zeitschrift für Physikalische Chemie **35** (1900), 33–76 and 204–256;
K.F. Bonhoeffer, *Über periodische chemische Reaktionen I*, Zeitschrift Elektrochemie **51**(1) (1947), 24–29;
U.F. Franck, *Chemical oscillations*, Angewandte Chemie **17**(1) (1978), 1–15;
J.L. Hudson and T.T. Tsotsis, *E*lectrochemical reaction dynamics: a review, Chemical Engineering Science **49**(10) (1994), 1493–1572;
M. Baune and P.J. Plath, *Chemically induced hydrodynamic pattern formation: slowly rotating disk electrodes under dissolving conditions and genesis of spatial bifurcation*, International Journal of Bifurcation and Chaos **12**(10) (2002), 2209–2217.

[6] H. Figour and P.A. Jacquet, French patent Nr. 707526 (1930);
P.A. Jacquet, *Electrolytic method for obtaining bright copper surfaces*, Nature **29** (1935), 1076;
P.A. Jacquet, *On the anodic behavior of copper in aqueous solutions of orthophosphoric acid*, Transaction of the Electrochemical Society **69** (1936), 629–655.

[7] M. Buhlert, *Elektropolieren und Elektrostrukturieren von Edelstahl, Messing und Aluminium. Untersuchung des transpassiven Abtragprozesses einschließlich unerwünschter Nebeneffekte*, Fortschritt-Berichte VDI, Reihe 2, No. 553, VDI-Verlag, Düsseldorf, 2000.

[8] M. Buhlert, Th. Hinte, P.J. Plath, and A. Visser, *Glanz und Rauhigkeit – Elektropolieren von Messing mit gepulstem Gleichstrom*, Galvanotechnik/mo Jahrgang **56**(2) (2002), 12–16.

[9] C. Gerlach, Doctoral Thesis, Universität Bremen (2002).

[10] R. Friedrich and J. Peinke, *Description of a turbulent cascade by a Fokker–Planck equation*, Physical Review Letters **78**(5) (1997), 863–866.

[11] R. Friedrich, J. Peinke, and Ch. Renner, *How to quantify deterministic and random influences on the statistics of the foreign exchange market*, Physical Review Letters **84**(22) (2000), 5225–5227.

[12] Ch. Renner, J. Peinke, and R. Friedrich, *Experimental indications for Markov properties of small scale turbulence*, Journal Fluid Mechanics **433** (2001), 383–409.

[13] M. Waechter, J. Peinke, F. Riess, and H. Kantz, *Stochastic analysis of road surface roughness*, Europhysics Letters (2003) (submitted).

[14] Ch. Renner, J. Peinke, and R. Friedrich, *Experimental indications for Markov properties of small scale turbulence*, Journal of Fluid Mechanics **433** (2001), 383–409.

[15] R. Venkatachalam, D. Kanagaraj, S. Vincent, and S. Mohan, *Electropolishing of stainless steel using pulse technique*, Transaction of the Metals Finishers Association of India **1**(4) (1992), 13–19.

[16] J.E. Labarga, J.M. Bastidas, and S. Fleiu, *A contribution to the study on electro-polishing of mild steel and alumnium using alternating current*, Electrochimica Acta **36**(1) (1991), 93–95.

[17] W. Ehrfeld, V. Hessel, and H. Löwe, *Microreactors*, 1st edn., Wiley-VCH, 2000; F. Völklein and S. Zetterer, *Einführung in die Mikrosystemtechnik*, Vieweg, Braunschweig 2000, pp. 229–244.

[18] A. Visser and M. Buhlert, *Theoretical and practical aspects of the miniaturization of lead frames by double-sided asymmetrical spray etching*, Journal of Materials Processing Technology **115** (2001), 108–113.

[19] M. Buhlert, *Elektropolieren und Eletrostrukturieren von Edelstahl, Messing und Aluminium*, Fortschritt-Berichte VDI, Reihe 2, No. 553, VDI-Verlag, Düsseldorf, 2000.

[20] D.M. Allen, *Electrolytic photoetching and its applications*, The Journal of the PCMI **27**(2) (1987), 10–12.

[21] D.M. Allen, *The Principles and Practice of Photoetching*, Adam Hilger, Bristol and Boston, 1986.

[22] P.-F. Chauvy, P. Hoffmann, and D. Landolt, *Electrochemical micromachining of titanium through a laser patterned oxide film*, Electrochemical and Solid-State Letters **4**(5) (2001), C31–C34.

[23] C. Madore, O. Piotrowski, and D. Landolt, *Through-mask electrochemical micromachining of titanium*, Journal Elec. Soc. **146**(7) (1999), 2526–2532.

[24] R. Nowak, S. Metev, and G. Sepold, *Laser-assisted chemical micromachining of metals and alloys*, Materials and Manufacturing Processes **9** (1994), 4;
R. Nowak, S. Metev, and G. Sepold, *Laser chemical etching of metals in liquids*, Mater. Manuf. Proc. **9** (1994), 429;
R. Nowak, S. Metev, and G. Sepold, *Laser-assisted wet chemical rtching of metals for microfabrication*, SPIE **2207** (1994), 633.

[25] A. Stephen, T. Lilienkamp, S. Metev, and G. Sepold, *Laser-assisted micromachining of large-area 3D metallic microparts*, in: Precision Engineering and Nanotechnology: Proceedings of the 1st Euspen Conference, 1999, Vol. 2, edited by P. McKeown et al., Shaker, Aachen, p. 20.

[26] A. Stephen and S. Metev, *Grundlegende Untersuchungen zur präzisen Bearbeitung von dreidimensionalen Strukturkörpern durch thermochemisches Laserabtragen, in Präzisionsabtragen mit Lasern*; Handbuchreihe Laser in der Materialbearbeitung, Band 12, Hrsg. VDI-TZ Physikalische Technologien, Bd. 2, S. 45, VDI – Herstellung und Druck, 2000.

[27] V. Kirchner, Dissertation, Freie Universität Berlin (2000).

[28] R. Schuster and V. Kirchner, Patent DE 19900173C.

[29] R. Schuster, Bunsen-Magazin **3**(5) (2001), 121–124;
R. Schuster, V. Kirchner, X.H. Xia, A.M. Bittner, and G. Ertl, *Nanoscale electrochemistry*, Physical Review Letters **80**(25) (1998), 5599–5602;
R. Schuster, V. Kirchner, P. Allongue, and G. Ertl, *Electrochemical micromachining*, Science **289** (2000), 98–101.

[30] A. Visser, M. Buhlert, and A. Rettinghaus, *Elektropolieren mit gepulstem Gleichstrom*, Galvanotechnik Saulgau **89**(3) (1998), 739–747.

[31] M. Buhlert, Proceedings of the Micro.tec 2000 Conference, Vol. 2., VDE World Microtechnologies Congress, VDE Verlag, Berlin, pp. 347–351.

[32] M. Buhlert, T. Hinte, P.J. Plath, A. and Visser, *Glanz und Rauheit. Elektropolieren von Messing mit gepulstem Gleichstrom*, Metalloberfläche Carl Hanser Verlag, München **52**(2) (2002), 12–16.

[33] M. Buhlert and P.J. Plath, Vorrichtung und Verfahren zum elektrochemischen Bearbeiten eines Werkstückes, Patent DE 10234122.2, June 2002.

[34] Th. v. Kármán, *Über laminare und turbulente Reibung*, Zeitschrift für Angewandte Mathematik und Mechanik **1** (1921), 233–252.

[35] V.G. Levich, *Physicochemical Hydrodynamics*, Prentice-Hall, Englewood Cliffs, NJ,1962, pp. 66–67.

[36] J. Newman, *Electrochemical Systems*, Prentice-Hall, Englewood Cliffs, NJ, 1973, pp. 280.

[37] M. Baune and P.J. Plath, *Chemically induced hydrodynamic pattern formation: slowly rotating disk electrode under dissolving conditions and genesis of spatial bifurcation*, International Journal Bifurcation and Chaos **12**(10) (2002), 2209–2217.

[38] M. Baune, V. Breunig-Lyriti, and P.J. Plath, *Invariant hydrodynamic Pattern formation: fast rotating disk electrode under dissolving Conditions*, International Journal Bifurcation and Chaos **12**(10) (2002), 2835–2845.

[39] M. Baune and P.J. Plath, *Coupling of chemical and hydrodynamic instabilities at the electrochemical dissolution of metals*, Z. Phys. Chemie (2003) (submitted for publication).

[40] A.M. Zhabotinsky, *Oscillating bromate oxidative reactions*, Berichte der Bunsengesellschaft für Physikalische Chemie **84** (1980), 303–308.

[41] J.J. Tyson, *Some further studies of nonlinear oscillations in chemical systems*, The Journal of Physical Chemistry A **58**(9) (1973), 3919–3930.

[42] K. Showalter, R.M. Noyes, and K. Bar-Eli, *A modified Oregonator model exhibiting complicated limit cycle behavior in a flow system*, The Journal of Physical Chemistry A **69**(6) (1978), 2514–2524.

[43] L. Kuhnert and K.-W. Pehl, *Oscillations in the Belousov–Zhabotinskii system (BZR) catalyzed by bis-bipyridine–silver complexes*, Chemical Physics Letters **84**(1) (1981), 155–158.

[44] A.K. Dutt and S.C. Müller, *Effect of stirring and temperature on the Belousov-Zhabotinskii reaction in a CSTR*, The Journal of Physical Chemistry A **97** (1993), 10059–10063.

[45] W. Hohmann, J. Müller, and F.W. Schneider, *Phase shifting in a chemical oscillator to encode analogue information*, Journal of the Chemical Society, London, Faraday Transaction **92**(16) (1996), 2873–2877.

[46] G. Dechert, K.-P. Zeyer, D. Lebender, and F.W. Schneider, *Recognition of phase patterns in a chemical reactor network*, The Journal of Physical Chemistry A **100**(49) (1996), 19043–19048.

[47] W. Hohmann, M. Kraus, and F.W. Schneider, *Recognition in excitable chemical reactor networks. Experiments and model-simulations*, The Journal of Physical Chemistry A **101**(40) (1997), 7364–7370.

[48] W. Hohmann, M. Kraus, and F.W. Schneider, *Learning and recognition in excitable chemical reactor networks*, The Journal of Physical Chemistry A **102**(18) (1998), 3103–31111.

[49] H. Mathes, Diplomarbeit, Universität Bremen (2000).
[50] R.J. Field and R.M. Noyes, The Journal of Physical Chemistry A **60** (1974), 1877.
[51] L. Györgyi and R.J. Field, *Simple models of deterministic chaos in the Belousov–Zhabotinsky reaction*, The Journal of Physical Chemistry A **95** (1991), 6594.
[52] L. Györgyi, S.L. Rempe, and R.J. Field, *A novel model for the simulation of chaos in low-flow-rate CSTR experiments with the Belousov–Zhabotinsky reaction. A chemical mechanism for two frequency oscillations*, The Journal of Physical Chemistry A **95**(8) (1991), 3159–3165.
[53] L. Györgyi and R.J. Field, *A three variable model of deterministic chaos in the Belousov–Zhabotinsky reaction*, Nature **355** (1992), 808.
[54] L. Györgyi, R.J. Field, Z. Noszticzius, et al., *Confirmation of high-flow-chaos in the Belousov–Zhabotinsky reaction*, The Journal of Physical Chemistry A **96** (1992), 1228–1233.
[55] A. Smirnov, M.A. Afonin, and V.M. Sedov, *Dynamics of cerium distribution during oscillatory extraction*, Radiochemistry **36** (1994), 282–288.
[56] M.A. Afonin and A.V. Smirnov, *Extraction of Ce and Eu under non-equilibrium conditions*, Radiochemistry **35** (1993), 676–681.
[57] M.A. Afonin, V.V. Romanovski, and V.A. Scherbakov, *Oscillatory extraction of uranium*, Solvent Extraction and Ion Exchange **16**(5) (1998), 1215–1231.
[58] MIR-Chem GmbH (P.J. Plath), *Verfahren und Vorrichtung zum Extrahieren und Separieren von Stoffen*, Patentanmeldung PCT/DE 02/01377 (2002).
[59] Th. Rabbow, Diplomarbeit, Universität Bremen (2002).
[60] W. Ehrfeld, V. Hessel, and H. Löwe, *Microreactors*, Wiley-VCH Verlag, Weinheim, 2000.
[61] W. Ehrfeld, (ed.), *Microreaction Technology – Industrial Prospects*, Proceedings of the Third International Conference on Microreaction Technology, Springer-Verlag, Berlin, 2000.
[62] H. Löwe, W. Ehrfeld, V. Hessel, Th. Richter, and J. Schiewe, Conference Preprints of IMRET 4, 4th International Conference on Microreaction Technology, personal communication, 2001.
[63] V. Hessel, W. Ehrfeld, V. Haverkamp, H. Löwe, and J. Schiewe, in: R.H. Müller and B. Böhm (eds.) *Dispersion Techniques for Laboratory and Industrial Production*, Wissenschaftliche Verlagsgesellschaft, Stuttgart, 1999.
[64] W. Ehrfeld, K. Golbig, V. Hessel, H. Löwe, and Th. Richter, *Characterization of mixing in micromixers by a test reaction: single mixing units and mixer arrays*, Industrial and Engineering Chemistry Research **38**(3) (2000), 1075.
[65] W. Ehrfeld, V. Hessel, S. Kiesewalter, H. Löwe, and Th. Richter, in: W. Ehrfeld (ed.) *Microreaction Technology – Industrial Prospects*, Proceedings of the Third International Conference on Microreaction Technology, Springer-Verlag, Berlin, 2000, pp. 165–170.
[66] K. Benz, K.-P. Jäckel, K.J. Regenauer, J. Schiewe, K. Drese, W. Ehrfeld, V. Hessel, and H. Löwe, *Utilization of micromixers for extraction process*, Chemical Engineering and Technology **24**(1) (2001), 11–17.
[67] V. Haverkamp, W. Ehrfeld, K. Gebauer, V. Hessel, H. Löwe, Th. Richter, and C. Wille, *The potential of micromixers for contacting of disperse liquid phases*, Fresenius' Journal für Analytische Chemie **364** (1999), 617–624.

[68] T. Bayer, H. Heinichen, and T. Natelberg, Proceedings of IMRET 4, 4th International Conference on Microreaction Technology, personal communication, 2000, pp. 167.

[69] V. Hessel, W. Ehrfeld, K. Golbig, V. Haverkamp, H. Löwe, and Th. Richter, Proceedings of the 2nd International Conference on Microreaction Technology, New Orleans, 1998.

[70] H. Mathes and P.J. Plath, *Generation of monodispered foams using a microstructured static mixer*, Proceedings of the Tunisian–German Conference – Smart Systems and Devices, 27–30 March 2001, Hammamat, Tunisia.

[71] H. Mathes P.J. and Plath, *Application of monodispersive foams as a spatially structured two-phase medium for chemical gas/liquid-reactions*, International Journal Bifurcation and Chaos (2003) (accepted for publication).

[72] S.A. Koehler, S. Hilgenfeldt, and H.A. Stone, *A generalized view of foam drainage: experiment and theory*, Langmuir **16** (2000), 6327–6334.

[73] X. Bao et al., Journal Material Science Letters **18**(3) (1999), 1003–1005.

[74] J.S. Lee and J.K. Park, *Preparation of porous ceramic pellet by pseudo double-emulsion method from 4-phase foamed slurry*, Journal Material Science Letters **20**(3) (2001), 205–207.

[75] L. Wood et al.; Patent US 3833386 (1972).

[76] K.R. Brundage, D.L. Hickman, and M. Lynn, *Production of porous mullite bodies*, Patent US 6238618 (2001).

[77] G. Tittizer, *Gas-static bearing of porous ceramic material*, Patent DE 3530448 (1987).

[78] H. Schmidt, D. Koch, G. Grathwohl, and P. Colombo, *Micro-/macroporous ceramics from preceramic precursors*, Journal American Ceramics Society **84**(10) (2001), 7899–7909.

[79] K. Ryuichi, *Method for producing porous ceramic product*, Patent JP 7002580 (1995).

[80] X. Zhu, D. Jiang, S. Tan, and Z. Zhang, *Improvement in the strut thickness of reticulated porous ceramics*, Journal American Ceramics Society **84**(7) (2001), 1654–1656.

[81] J.M. Tulliani, L. Montanaro, T.J. Bell, and M.V. Swain, *Semiclosed-cell mullite foams: preparation and macro- and micromechanical characterization*, Journal American Ceramics Society **82**(4) (1999), 961–968.

[82] T. Murakata et al., Journal Material Science Letters **27**(6) (1992), 1567–1574.

[83] T. Higuchi, K. Kurumada, S. Nagamine, A.W. Lothongkum, and M. Tanigaki, *Effect of addition of polymeric species with ether moieties on porous structure of silica prepared by sol-gel method*, Journal Material Science **35**(13) (2000), 3237–3243.

[84] J.J. Bikerman, *Foams*, Springer-Verlag, Berlin, Heidelberg, New York, 1973.

[85] MIR-Chem (P.J. Plath and M. Buhlert), *Vorrichtung und Verfahren zum Herstellen eines Schaums*, Patent PCT/DE 02/04260 (2002).

31 Galvanostatic Studies of an Oxygen-evolving Electrode

C. Gerlach, A. Visser, and P. J. Plath

Electropolishing is a technical process, which is carried out in the passive or transpassive region of a cyclic voltammogram. It is widely used to produce smooth metallic surfaces. The behavior of electropolishing of brass is representative for educt-limited systems. This article shows the temporal behavior of electropolishing of brass and it is shown that it can be described as a dynamical electro-chemical system with an NDR or an HNDR behavior.

31.1 Introduction

Many electro-chemical systems exhibit spatial, temporal or spatio-temporal behavior. Fechner provided one of the first descriptions in 1828 characterizing the periodic magnetic deflection caused by iron and silver in sulphuric acid and $AgNO_3$ [1]. More than one hundred years later, Franck et al. introduced a kinetic model for the oscillating dissolution of iron in phosphoric acid at the Flade potential. The main reason for these oscillations is the pH-dependence of the Flade potential, which leads to a periodic composition and decomposition of passivating oxide layers. They pointed out the correlation of oscillatory behavior with a negative differential resistance in the cyclic voltammogram [2–4]. Since then, many electro-chemical systems exhibiting temporal or spatio-temporal behavior were investigated, classified and modeled utilizing different approaches for modeling. Krischer explained that the appearance of dynamical behavior, such as fronts, waves and oscillations, is caused by diffusion and migration that lead to changes in the double-layer potential [5, 6]. Otterstedt et al. investigated the dissolution of cobalt in phosphoric acid. These authors pointed out that pH-differences are the main reason leading to passivating fronts [7]. Also, the dissolution of copper in acidic media is well examined and understood. Using this system, Glarum and Marshall showed that the dissolution is limited due to an educt limitation caused by a layer of used electrolyte in front of the electrode [8–10]. Electropolishing systems like the investigated one are educt-limited systems as well [11]. The work of Buhlert described the appearance of galvanostatic oscillations [12] in a technical electropolishing system, such as brass in a solution containing phosphoric acid and alcohol. Gerlach showed that oscillations occur due to insufficient water in front of the electrode [13]. Electro-chemical systems with dynamic behavior were recently reviewed by Hudson and Tsotsis [14] and Krischer [5]. All of these systems have the association with the appearance of negative differential resistance in common, which can be found in the I–ϕ plane. A classification of electro-chemical systems regarding different types of dynamical behavior and its correlation to negative differential resistance was provided by Koper [15, 16], Strasser et al. [17] and Krischer et al. [6].

Nonlinear Dynamics of Production Systems. Edited by G. Radons and R. Neugebauer

ISBN 3-527-40430-9

31.1.1 N-NDR and N-HNDR Behavior

Electro-chemical oscillators can be classified according to four categories [17], two of which, class III oscillators and class IV oscillators, exhibit an N-shaped I–U plot that is also referred to as a cyclic voltammogram.

Both class III oscillators and class IV oscillators may display current oscillations under potentiostatic conditions. However, galvanostatic conditions cause class III oscillators to show a strictly bistable behavior. An NDR (negative differential resistance) resides in the cyclic voltammogram of these oscillators. Class III oscillators can be described using a two-variable kinetic model with an activator and an inhibitor. The electrical field can be defined as activator, which is the fast variable. In contrast, the chemical step acts as inhibitor or slow variable.

In addition to potentiostatic current oscillations, class IV oscillators also exhibit galvanostatic potential oscillations. These oscillators show an HNDR (hidden negative differential resistance) that is located on a branch with a positive slope in the cyclic voltammogram. To model class IV oscillators, an additional variable is needed. This class of oscillators can be described using a three-variable kinetic model with one activator and two inhibitors. The electrical field again represents the fast variable and a second chemical step represents an additional slow variable. Class IV oscillators can be further divided into three subclasses depending on their respective bifurcation maps [15, 17, 6].

In this paper, we show that an electropolishing system containing brass in various electrolyte mixtures of phosphoric acid and alcohol can exhibit a class III or a class IV.3 oscillatory behavior. An experimental electro-chemical investigation of this system is presented. We exemplarily show that brass in a solution of phosphoric acid, water and methanol, referred to as METHANOL, displays N-NDR behavior. We also demonstrate that brass in BUTANOL, STANDARD or PHOSPHORIC ACID, the last being an electrolyte without additional alcohol, exhibits N-HNDR behavior. In the following, all electrolytes containing alcohol, phosphoric acid and water will be named according to their alcohol component which will be written in capital letters.

31.2 Experimental

The experimental setup is shown in Fig. 31.1. A basin measuring length $\times$width $\times$ height $=$ $180\,\mathrm{mm} \times 60\,\mathrm{mm} \times 30\,\mathrm{mm}$ was utilized as an electro-chemical cell. A three-electrode arrangement was employed, where the workpiece functioned as anode. Anode and cathode were made of brass (CuZn37) measuring $\mathrm{l} \times \mathrm{h} = 50\,\mathrm{mm} \times 50\,\mathrm{mm}$ each. The electropolished surface consisted of an area of $\mathrm{l} \times \mathrm{h} = 50\,\mathrm{mm} \times 20\,\mathrm{mm}$. Anode and cathode were positioned in the electrolyte vertically facing each other. The backsides of the sheets were shielded with Tesa adhesive tape. The distance between anode and cathode was approximately $14\,\mathrm{cm}$. All potentials measured were referenced to an Ag/AgCl electrode of $U = 0.222\,\mathrm{V}$. This reference electrode was connected to the electrolyte via a glass capillary filled with electrolyte solution and placed in the middle of the anode being electropolished. The distance between the anode and the tip of the glass capillary was approximately $4 \pm 1\,\mathrm{mm}$.

Many parameters can be optimized to improve electropolishing results. We compared different electrolytes, all of which contained the same amount of phosphoric acid and a varying

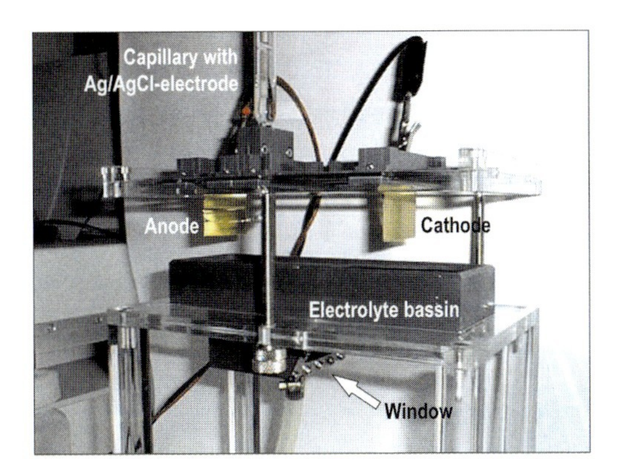

Figure 31.1: Experimental setup.

Table 31.1: Compounds of electrolyte mixtures.

Electrolyte	H_3PO_4	Alcohol	H_2O
METHANOL	60 vol.-%	30 vol.-% methanol	10%
ETHANOL	60 vol.-%	30 vol.-% ethanol	10%
BUTANOL	60 vol.-%	30 vol.-% 1-butanol	10%
STANDARD	60 vol.-%	15 vol.-% 2-propanol + 15 vol.-% 2-butanol	10%
PHOSPHORIC ACID	60 vol.-%		40%

component of alcohol. All electrolytes contained 60 vol.-% H_3PO_4 85%, 30 vol.-% alcohol and 10 vol.-% deionized water. The alcohols used in this study were methanol, ethanol, butanol and a mixture of 15 vol.-% isopropanol and 15.-% isobutanol referred to as STANDARD electrolyte. In addition, an electrolyte was utilized that consisted of a mixture of 60 vol.-% H_3PO_4 85% and 40 vol.-% deionized water, hereinafter referred to as PHOSPHORIC ACID. The electrolytes are referred to as the capitalized name of the respective parameter varied, which is the alcohol compound as shown in Table 31.1. All chemicals used in this study were of pure analytical (p.a.) grade. All electrolytes had pH = 0. The electrodes were rinsed with isopropanol to remove dust and grease. All experiments were performed at room temperature.

Table 31.2 displays the kinematic viscosity μ and electrolytic conductivity κ of all electrolytes. These physical parameters were determined using fresh solutions. It is evident that no differences in the order of magnitude regarding kinematic viscosity and electrolytic conductivity were detected between different electrolytes. Also, further experiments [13], which are not presented in this article, show that no correlation between and no tendency for these parameters and the dynamical behavior of the systems could be verified. Therefore, none of these physical parameters can account for the presented differences in dynamical behavior.

Table 31.2: Kinematic viscosity μ and electrolytic conductivity κ.

Electrolyte	μ [cSt]	κ [mS/cm]
METHANOL	13.6	86
ETHANOL	17.8	62
BUTANOL	23.3	57
STANDARD	21.7	58
PHOSPHORIC ACID	11.3	Out of range

31.3 Results

31.3.1 Cyclic Voltammogram

In electro-chemical systems, a cyclic voltammogram recorded under potentiodynamic conditions can be further analyzed. A representative cyclic voltammogram as shown in Fig. 31.2 can be divided into three regions where different electro-chemical processes occur: in the active region (I), the surface is etched. The second region (II) is the so-called passive region revealing a typical drop of potential. The transition from the active to the passive region is known as the Flade potential, which is also represented by the point of inflection of the graph. In this figure, the maximum of the graph was chosen because the point of inflection could not be clearly determined. For brass, this drop of potential is caused by transport limitation [8, 9] and not by formation of a hydroxide layer as described for other metals like steel and aluminum [18].

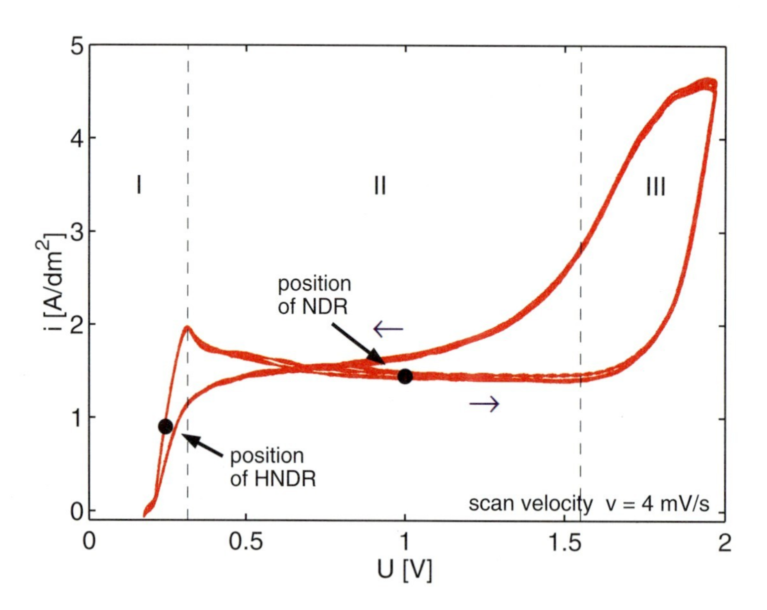

Figure 31.2: Representative cyclic voltammogram recorded using STANDARD electrolyte. Electrolytes containing alcohol reveal a cyclic voltammogram displaying a plateau in the forward scan and a loop in the backward scan caused by transport limitation.

In the transpassive region (III), a second electro-chemical reaction takes place, which is the hydrolysis of water leading to oxygen formation.

Dynamical behavior detected in an electro-chemical system is necessarily originated in negative resistance in the I–U plane. The cyclic voltammogram in Fig. 31.2 outlines typical positions of an NDR found in class III oscillators and an HNDR corresponding to class IV oscillators.

31.3.2 Methanol and Ethanol

The first system described here is a class III oscillator. It exhibits an N-NDR behavior under galvanostatic conditions. While electropolishing brass in METHANOL, this system reveals strict bistability displaying a hysteresis loop as shown in Fig. 31.3a. When current density is varied, a fixed point at a lower level of current densities $i \leq 1.4\,\mathrm{A/dm^2}$ and a second fixed point at an upper level of current densities $i \geq 1.5\,\mathrm{A/dm^2}$ can be detected. For $1.4\,\mathrm{A/dm^2} \leq i \leq 1.5\,\mathrm{A/dm^2}$, a hysteresis is observable. Figure 31.3b shows the jump from upper to lower level. The lower level was detected at a potential of approximately $U = 0.2\,\mathrm{V}$ and the upper level at a potential of approximately $U = 1.7\,\mathrm{V}$. Comparing this bifurcation map with the cyclic voltammogram in Fig. 31.2 reveals that the lower level corresponds to the active region of the cyclic voltammogram, where only metal dissolution is established. The upper level corresponds to the transpassive region, where, in addition to the dissolution of copper and zinc, oxygen production is established due to hydrolysis of water. This hydrolysis of water leads to gas bubbles rising in front of the electrode. Comparing the point of inflection of an assumed mathematical curve in the hysteresis loop with the cyclic voltammogram (Fig. 31.2) proves that this point of inflection lies on the negative I–U branch of the curve in the passive region

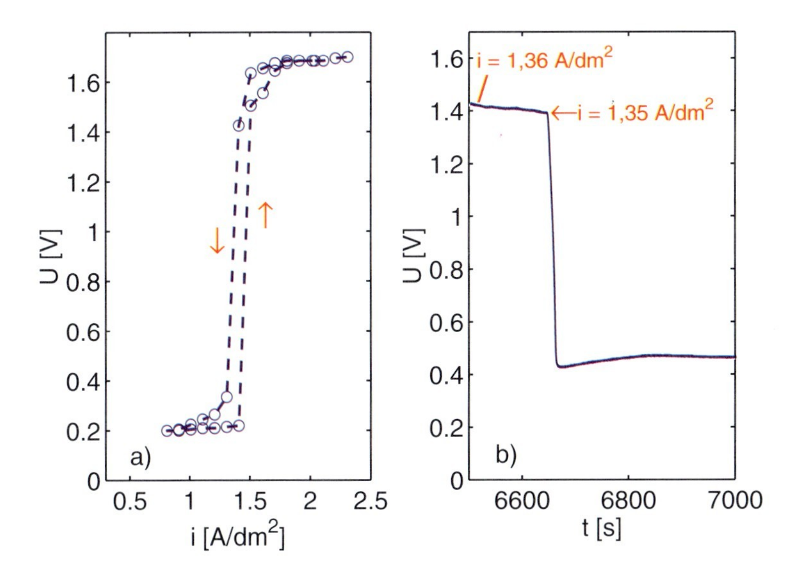

Figure 31.3: Brass in METHANOL. (a) Bistability with hysteresis loop; (b) changeover from the upper fixed point to the lower fixed point.

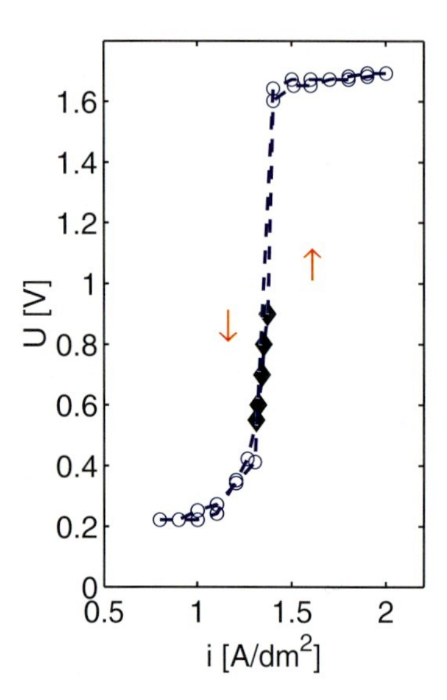

Figure 31.4: Bifurcation behavior of brass in ETHANOL. Circles: fixed points, diamonds: unstable oscillations.

(region II). This is in agreement with a class III oscillator exhibiting a negative differential resistance (NDR) that can be directly deduced from the cyclic voltammogram.

Class III oscillators can be described using a two-variable model including an activator step and an inhibitor step. In this case, the current density accounts for the activator. The presented electropolishing system is an educt-limited system. Therefore, the inhibitor is realized by insufficient water in front of the electrode. This water is needed to transport newly generated metal ions into the electrolyte.

Electropolishing brass in ETHANOL led to time series as shown in Fig. 31.4 revealing a qualitatively different behavior. Strict bistability was not observed under galvanostatic conditions. A fixed point was established for low current densities smaller than $i = 1.3\,\mathrm{A/dm^2}$ corresponding to potentials in the range of $0.2\,\mathrm{V} \leq U \leq 0.4\,\mathrm{V}$. These potentials were located on the active branch of the cyclic voltammogram. For values higher than $i = 1.4\,\mathrm{A/dm^2}$, a fixed point with $U = 1.65\,\mathrm{V}$ was found where oxygen was formed.

No hysteresis appeared when electropolishing was performed in ETHANOL. In contrast, when current density was varied upwards between $i = 1.3\,\mathrm{A/dm^2}$ and $i = 1.4\,\mathrm{A/dm^2}$ corresponding to potentials of $U = 0.5\,\mathrm{V}$ to $U = 1.5\,\mathrm{V}$, unstable oscillations with high amplitudes were found as displayed in Fig. 31.5. When lowering current density, no oscillations were found. The potential values correspond to the passive region of the cyclic voltammogram. The current densities leading to unstable oscillations are displayed as diamonds in Fig. 31.4. They are located at the slope of the curve.

The presented time series recorded while electropolishing brass in ETHANOL experimentally realizes the transition between an NDR and an HNDR behavior.

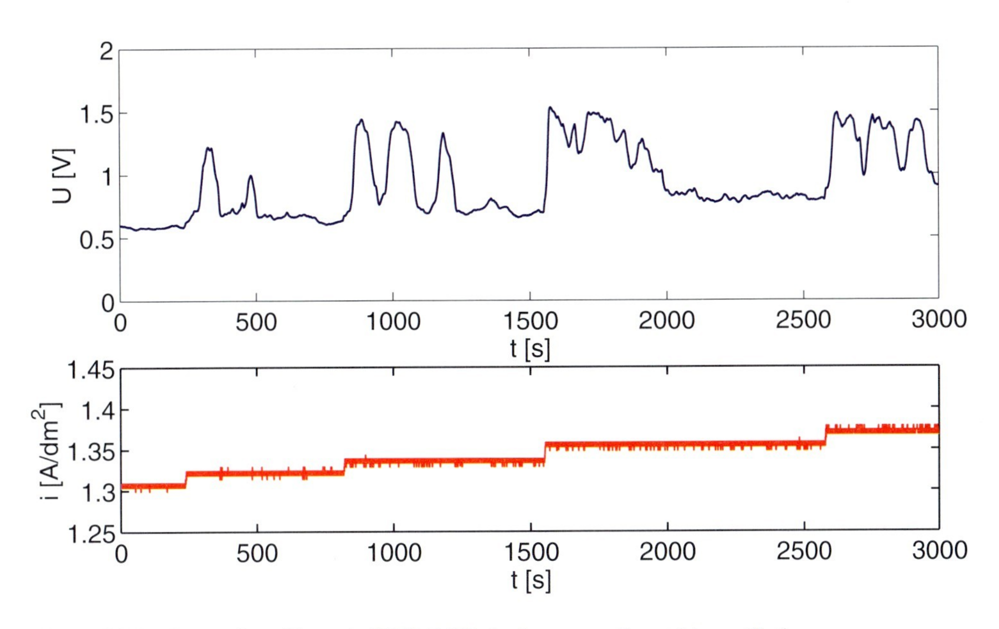

Figure 31.5: Time series of brass in ETHANOL in the range of unstable oscillations.

31.3.3 Butanol/Standard

The following subsection describes a change of electrolyte composition that modifies the system toward a class IV oscillator with an N-HNDR behavior. For several electrolyte solutions, an oscillatory regime can be observed under galvanostatic conditions as exemplarily shown for STANDARD, BUTANOL and PHOSPHORIC ACID (see next subsection). Figure 31.6 displays time series recorded under galvanostatic conditions. For BUTANOL and STANDARD, the latter being an electrolyte used for technical processes (for composition, see Table 31.1), potential oscillations with amplitudes of approximately 1 V were detected. The time series can be divided into three regions: a lower fixed point, an oscillatory regime and an upper fixed point. The oscillatory regime was detected between current densities of $i = 1\,\mathrm{A/dm^2}$ and $i \geq 3.5\,\mathrm{A/dm^2}$. For current densities lower than $i = 1\,\mathrm{A/dm^2}$, a fixed point with $U = 0.3\,\mathrm{V}$ was established. For high current densities, another fixed point with $U = 1.7\,\mathrm{V}$ was established, corresponding to the transpassive region with continuous oxygen production.

In the oscillatory regime, oscillations with high amplitudes of approximately 1 V were detected. Time series of different current densities are shown in Fig. 31.6. Nearly chaotic oscillations with one main frequency were observed for a current density of $i = 1.7\,\mathrm{A/dm^2}$. A period doubling was detected at $i = 1.9\,\mathrm{A/dm^2}$ representing a second main frequency. The frequencies were in the range of 10 mHz. At higher current densities, the amplitudes of the oscillations were more modulated as evident for $i = 2.1\,\mathrm{A/dm^2}$. Yet another behavior was observed at current densities of $i = 2.7\,\mathrm{A/dm^2}$. The time series exhibited oscillations with small amplitudes on a high potential level. Sometimes, these oscillations were interrupted by peaks of high amplitude. At this current density, the oxygen production was nearly continuous.

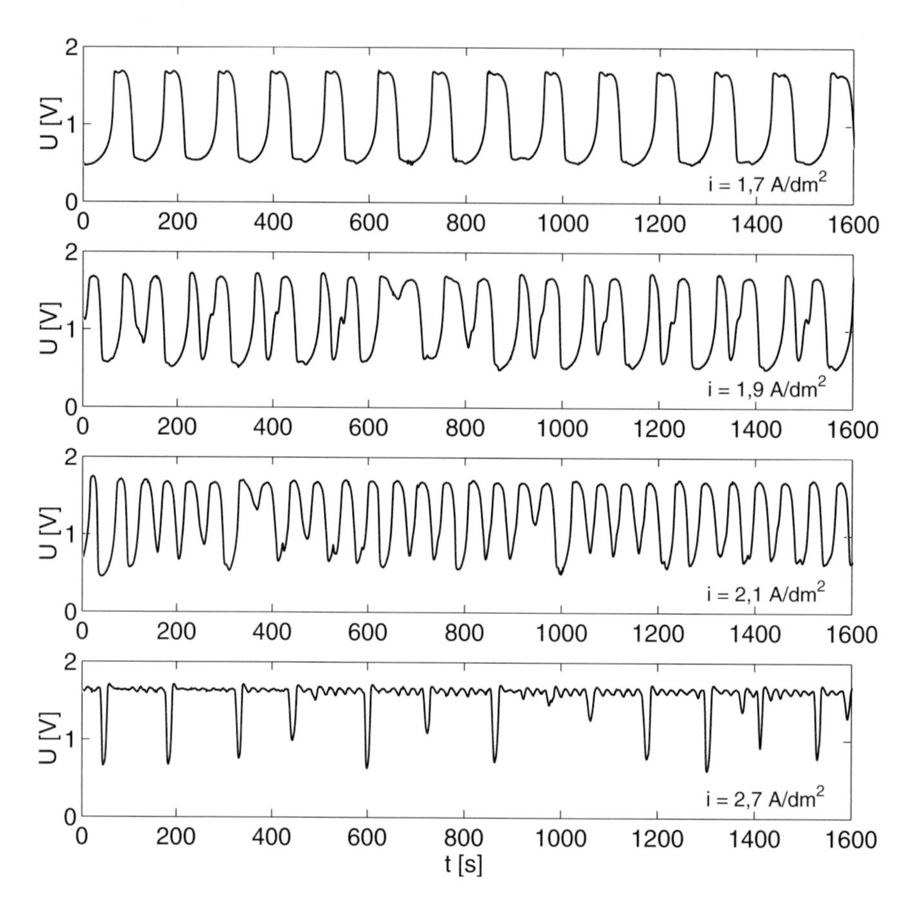

Figure 31.6: Time series of brass in BUTANOL for current density values $i = 1.7\,\mathrm{A/dm^2}$, $i = 1.9\,\mathrm{A/dm^2}$, $i = 2.1\,\mathrm{A/dm^2}$, $i = 2.7\,\mathrm{A/dm^2}$.

In the range of current density up to $i = 2.1\,\mathrm{A/dm^2}$, oxygen production was oscillating as well, corresponding to the potential oscillations. When the potential was at a low level, no hydrolysis of water occurred and no oxygen formed. Only dissolution of copper and zinc took place. As the potential increased, it eventually reached the upper level where hydrolysis started to occur and oxygen formed.

Figure 31.7 shows the period doubling observed between i = 1.7 A/dm^2 and i = 1.9 A/dm^2. The main attractor of the electropolishing system was discovered at i = 1.7 A/dm^2. Two domains were established: one domain corresponds to the dissolution of copper and zinc at the lower level and the second domain corresponds to potentials in the transpassive region, where two competing processes occur: metal dissolution and hydrolysis of water. A current density of i = 1.9 A/dm^2 revealed a second main frequency displayed as a loop in the delay map in Fig. 31.7. With increasing current densities, the lower domain of the attractor became more and more unstable and completely vanished at i = 2.7 A/dm^2, where the loop was dominating. This was accompanied by continuous oxygen formation.

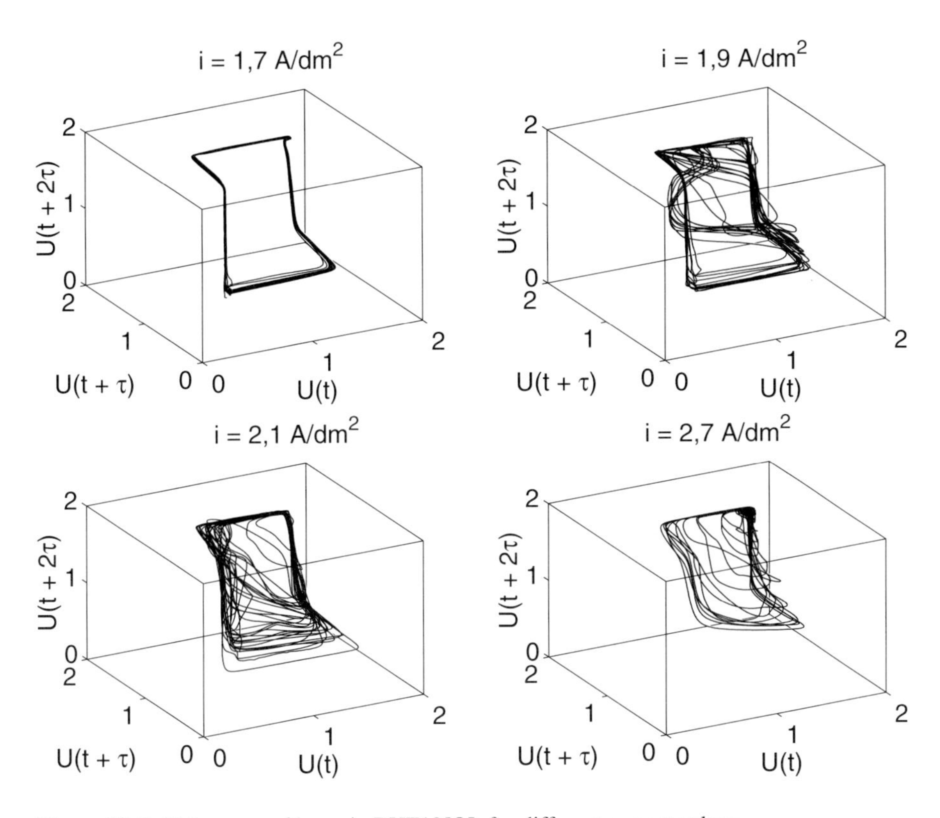

Figure 31.7: Delay map of brass in BUTANOL for different current values.

Potential oscillations appearing under galvanostatic conditions prove this system to be an oscillator of class IV. The existence of a hidden negative resistance (HNDR) in the I–U plot further supports classification as oscillator of class IV (see Fig. 31.2). The position of the HNDR corresponds to a current density lower than that for which the first oscillations were established. For BUTANOL and STANDARD, the transition to the oscillatory regime was detected at $i = 1\,\mathrm{A/dm^2}$. Therefore, the HNDR should be located at a current density lower than $1\,\mathrm{A/dm^2}$. Comparison with the cyclic voltammogram shown in Fig. 31.2 revealed that it corresponds to the active branch of the cyclic voltammogram with a positive slope.

The major classification of the present system indicated a class IV oscillator. Furthermore, it exhibited the behavior of a special subtype of this oscillator: class IV.3. Figure 31.8 shows the transition into the oscillatory regime for brass in STANDARD. The increase in current density from $i = 0.9\,\mathrm{A/dm^2}$ to $i = 1.0\,\mathrm{A/dm^2}$ led to a changeover into the oscillatory regime achieved by a hard transition. The oscillations start abruptly with an amplitude of $1\,\mathrm{V}$. Therefore, this transition at low current densities can be classified as a hard transition while, for high current densities, the oscillatory regime is left via a soft transition into a fixed point (compare Fig. 31.6 for $i = 2.7\,\mathrm{A/dm^2}$). This type of bifurcation indicates a class IV.3 oscillator. This particular system contains two independent current carriers, which are the dissolution of copper and zinc as well as the hydrolysis of water.

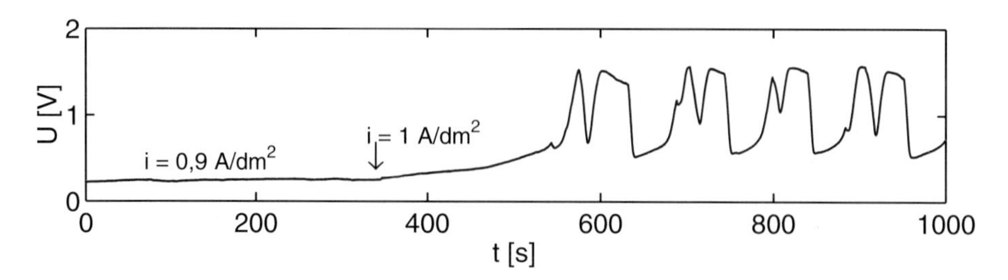

Figure 31.8: Transition into the oscillatory regime for low current densities shown for brass in STANDARD.

Modeling this electropolishing system can result in the description as a three-variable kinetic model with the current density representing one activator and the two independent current carriers as aforementioned, metal dissolution and hydrolysis of water, representing two inhibitors. Water is detected as a critical parameter for both current carriers.

The limiting factor of metal dissolution is the diffusion of water toward the electrode. This water is necessary to transport the products into the electrolyte. At high current densities, water is the limiting factor as well because, under these conditions, water itself acts as educt. If the electrolyte in front of the electrode is deprived of water, the potential breaks down and only dissolution of metal occurs.

31.3.4 Phosphoric Acid

Electropolishing brass in an electrolyte without any alcoholic compound produced a system that also exhibited an N-HNDR behavior with galvanostatic potential oscillations. Figure 31.9 displays the behavior of brass in PHOSPHORIC ACID. The parameter being varied was again the current density as shown in Fig. 31.9, rows 2 and 5. Rows 1 and 4 show the associated potential values and rows 3 and 6 show sonograms, which represent segmented Fourier analyses of the time series. Treating brass in PHOSPHORIC ACID led to current densities in the range of $i = 3\,\mathrm{A/dm^2}$ to $i = 12\,\mathrm{A/dm^2}$, which was three-fold higher as compared to using electrolytes with alcoholic components. Frequencies typically observed in this system were increased by a factor of approximately 10. However, its qualitative behavior was in general comparable to the behavior of the systems with alcoholic electrolytes.

Current densities lower than $i = 3\,\mathrm{A/dm^2}$ exhibited a fixed point with $U = 0.28\,\mathrm{V}$. An increase in current density to $i = 3.8\,\mathrm{A/dm^2}$ led to an oscillatory regime. This oscillatory regime revealed periodic oscillations and was stable up to current densities of $i = 11\,\mathrm{A/dm^2}$. The limit cycle displayed its low turning point at $U = 0.4\,\mathrm{V}$, while the high turning point was detected at $U = 1.6\,\mathrm{V}$. When providing an elevation of current density acting as a small perturbation, a reaction of the dynamical system was detectable. For current densities up to $i = 6\,\mathrm{A/dm^2}$, a small variation just led to an increase in the frequency of oscillation. However, for current densities higher than $i = 6\,\mathrm{A/dm^2}$, the system, upon perturbation, left the attractor for a short time, followed a trajectory towards a co-existing second solution in the phase plane, an unstable fixed point, and eventually remained in the limit cycle. Figure 31.10a shows the trajectory of the changeover drifting away from the unstable vortex into stable oscillations.

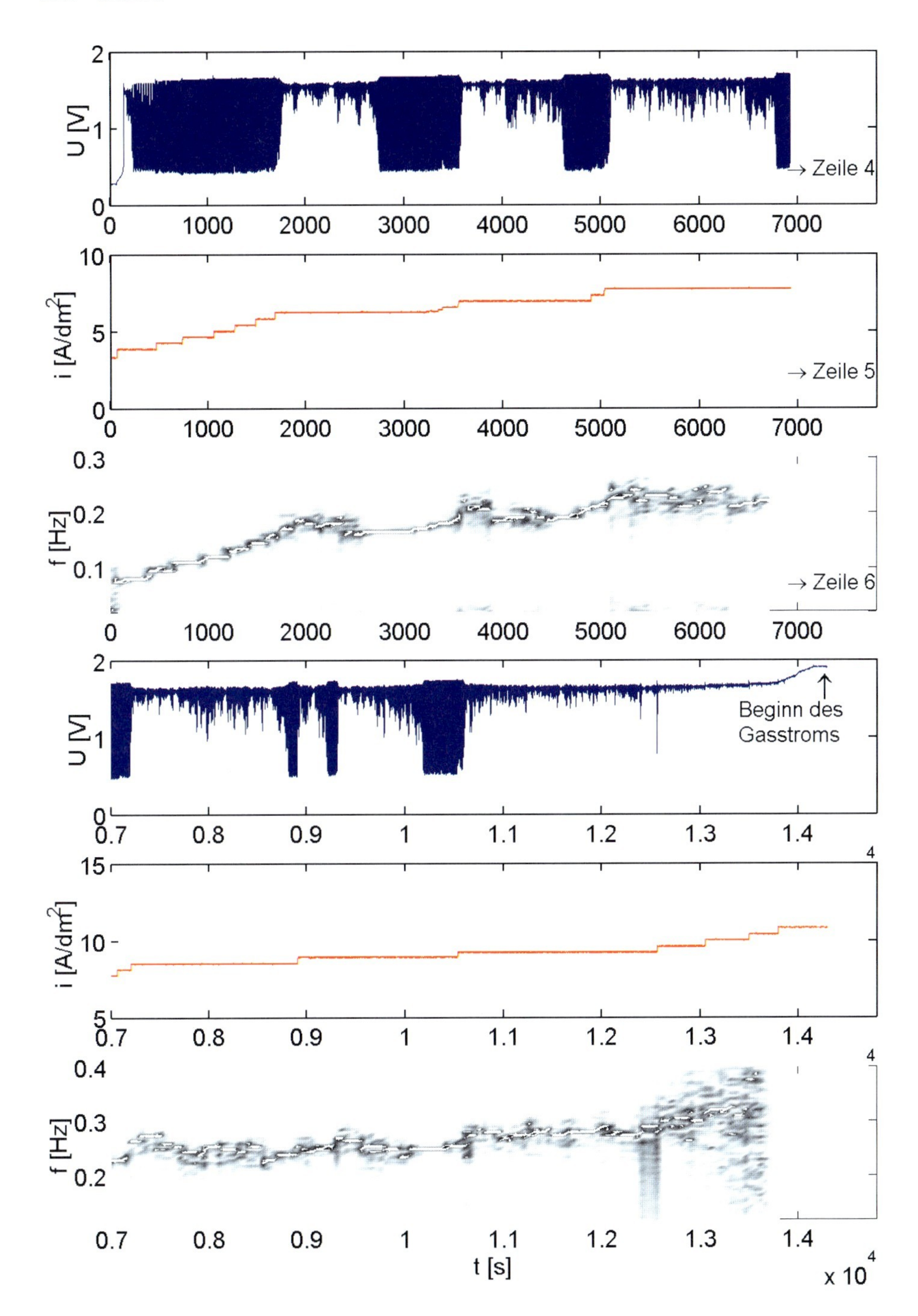

Figure 31.9: Brass in PHOSPHORIC ACID. Potential (rows 1 and 4), current (rows 2 and 5) and sonogram (rows 3 and 6).

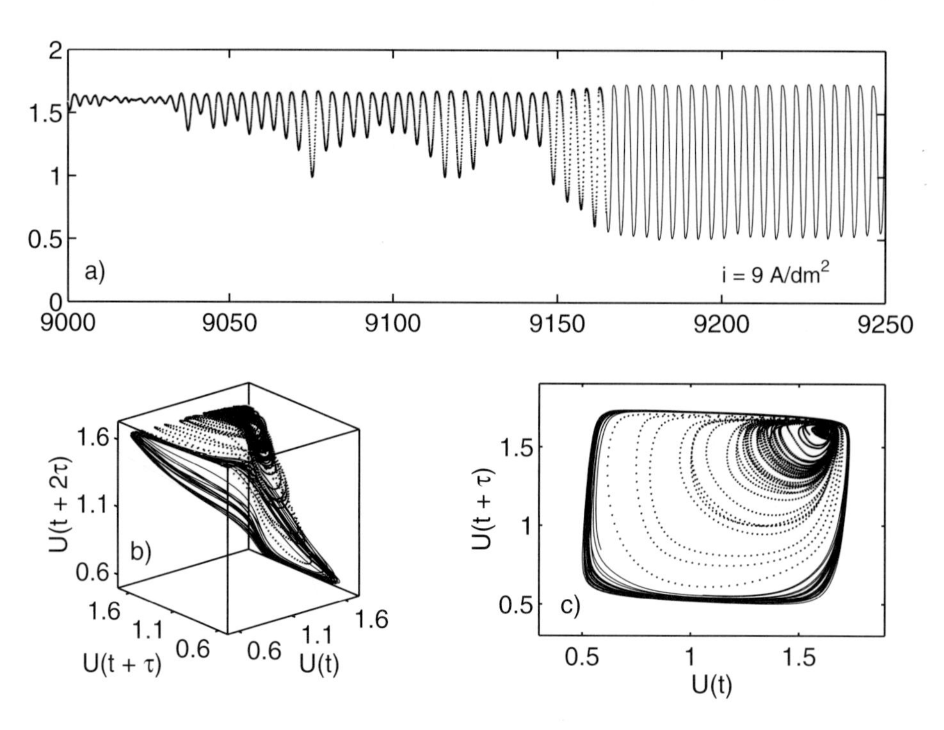

Figure 31.10: Brass in PHOSPHORIC ACID. Trajectory into the attractor.

Figure 31.10 also shows (b) a three-dimensional and (c) a two-dimensional phase portrait of the time series. The trajectory not corresponding to the limit cycle forms a cap over the limit cycle.

This transition was further supported by the sonograms in Fig. 31.9. When the oscillatory regime was stable, one stable frequency was observed. However, leaving the attractor caused amplitudes as well as frequencies to vary until the stable limit cycle was reached again. Within the limit cycle, frequencies were nearly linear with current density (Fig. 31.11).

At current densities higher than $9\,\mathrm{A/dm^2}$, amplitudes of the oscillations remained small. The system dropped off the oscillatory regime by displaying a soft transition.

At high current densities of $i \geq 11\,\mathrm{A/dm^2}$, a new fixed point was established with a potential of $U = 1.9\,\mathrm{V}$ corresponding to a strong ascent of gas bubbles in front of the electrode.

The transition into the oscillatory regime occurred via a hard transition as described above for BUTANOL and STANDARD. As evident in Fig. 31.12, the trajectory establishing the oscillatory regime also revealed the co-existence of a limit cycle and an unstable fixed point as reported above.

31.4 Conclusion

In the present paper we reported on a technical system: the electropolishing of brass in electrolytes containing phosphoric acid, water and alcohol. We have established that a variation of

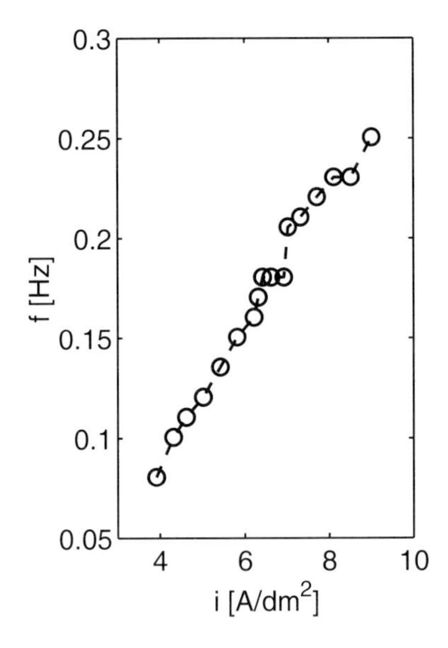

Figure 31.11: Frequencies of brass in PHOSPHORIC ACID.

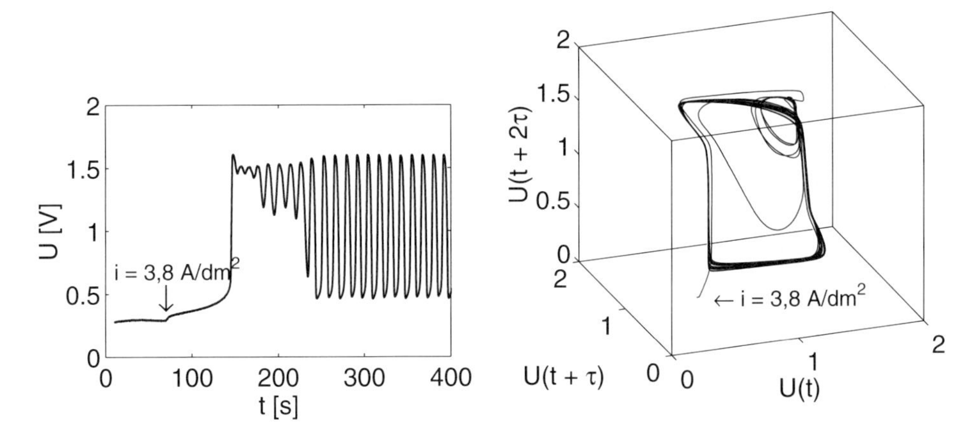

Figure 31.12: Brass in PHOSPHORIC ACID. Trajectory of the transition into the oscillatory regime.

electrolyte composition in this technical system can induce well-known dynamic behaviors of electro-chemical systems. These dynamic behaviors were classified accordingly. Electrolytes containing phosphoric acid and methanol revealed a behavior classified as class III oscillators with an N-NDR. The other systems described in the present study, such as BUTANOL, STANDARD and PHOSPHORIC ACID, were identified to belong to class IV.3 oscillators exhibiting an N-HNDR. The present investigation provides, for the first time, a possibility to describe this technical system theoretically. Our experiments clearly show that the established models, known for and previously applied to several electro-chemical systems, can also be employed for the technical system of electropolishing brass in electrolytes containing phosphoric acid, water and alcohol. In addition, we suggest that water may be the critical parameter of this system. Since the presented electro-chemical system has been classified according to class III and class IV.3 the mathematical equations of this system as well as its fast and slow variables are known.

Acknowledgment

We gratefully acknowledge the funding provided by the Volkswagen Foundation for project I/74151, in the priority area Investigation of Non-Linear Dynamic Effects in Production Systems.

Bibliography

[1] M.G.Th. Fechner, *Zur Elektrochemie. 1. Über die Umkehrung der Polarität in der einfachen Kette*, Jahrbuch der Chemie und Physik **9** (1928), 130–151.

[2] U.F. Franck, *Über die Aktivierungsausbreitung auf passiven Eisenelektroden*, Zeitschrift für Elektrochemie, **55**(2) (1951), 154–160.

[3] U.F. Franck, *Zur Stabilität von Elektrodenzuständen*, Zeitschrift für Physikalische Chemie – Neue Folge **3** (1955), 183–221.

[4] U.F. Franck and R. Fitzhugh, *Periodische Elektrodenprozesse und ihre Beschreibung durch ein mathematisches Modell*, Zeitschrift für Elektrochemie **65**(2) (1955), 156.

[5] K. Krischer, *Principles of Temporal and Spatial Pattern Formation in Electrochemical Systems*, Modern Aspects of Electrochemistry **51**, edited by B.E. Conway et al., Kluwer Academic/Plenum Publishers, New York, 1999.

[6] K. Krischer, N. Mazouz, and P. Grauel, *Fronten, Wellen, stationäre Muster in elektrochemischen Systemen*, Angew. Chem. **113** (2001), 842–863.

[7] R.D. Otterstedt, N.I. Jaeger, P.J. Plath, and J.L. Hudson, *Global coupling effects on spatiotemporal patterns on a ring electrode*, Chemical Engineering Science **54**, (1999), 1221–1231.

[8] S.H. Glarum and J.H. Marshall, *The anodic dissolution of copper into phosphoric acid, I. Voltammetric and oscillatory behavior*, Journal of Electrochemical Society **132** (1985),2872–2878.

[9] S.H. Glarum and J.H. Marshall, *The anodic dissolution of copper into phosphoric acid, II: Impedance Behavior*, Journal of Electrochemical Society **132** (1985), 2878–2885.

[10] M.R. Bassett and J.L. Hudson, *The dynamics of the electrodissolution of copper*, Chemical Engineering Communication, **60** (1987), 145–158.

[11] D. Landolt, *Fundamental aspects of electropolishing*, Electrochimica Acta **32** (1987), 1–11.

[12] M. Buhlert, *Elektropolieren und Elektrostrukturieren von Edelstahl, Messing und Aluminium. Untersuchung des transpassiven Abtragprozesses einschließlich unerwünschter Nebeneffekte*, Fortschritt-Berichte VDI, Reihe 2, Nr. 553, VDI-Verlag, Düsseldorf, 2000.

[13] C. Gerlach, Dissertation, Universität Bremen (2002). [http://elib.suub.uni-bremen.de/publications/dissertations/E-Diss427_Gerlach_2002.pdf].

[14] J.L. Hudson and T.T. Tsotsis, *Electrochemical reaction dynamics: a review*, Chemical Engineering Science **49** (1994), 1493–1572.

[15] M.T.M Koper, Advanced Chemical Physics **XCII** (1996), 161–298.

[16] M.T.M Koper, *Non-linear phenomena in electrochemical systems*, Journal of the Chemical Society, Faraday Transaction **94**(10) (1998), 1369–1378.

[17] P. Strasser, M. Eiswirth, and M.T.M Koper, *Mechanistic classification of electrochemical oscillators – an operational experimental strategy*, Journal of Electroanalytical Chemistry **478** (1999), 50–66.

[18] K.J. Vetter, *Elektrochemische Kinetik*, Springer-Verlag, Berlin, Göttingen, Heidelberg, 1961.

32 Wavelet Analysis of Electropolished Surfaces

A. Mora, C. Gerlach, T. Rabbow, P.J. Plath, and M. Haase

Electropolishing is a century-old technical treatment used to obtain bright and shiny surfaces by electro-chemical removal. The interplay between rising gas bubbles and a falling film of dissolved metal leads to complex surface structures. The choice of the electrolyte and the applied electrical potential have a significant influence on the surface structure on different length scales. The aim of the paper is to investigate the surface topography of electropolished surfaces and to classify the surface profiles resulting from different electrolytes. Wavelet techniques have now become well established for analyzing scaling properties of irregular functions and for investigating the roughness of surfaces. The wavelet transform modulus maxima method allows one to extract the scaling characteristics of irregular functions, which may even follow different power laws in different regions. While the continuous wavelet transform provides highly redundant information, the wavelet transform maxima lines contain the essential information about the evolution of scaling properties of irregularities across scales. Thus, they can be considered as a fingerprint of the signal. In addition, a preliminary stochastic study suggests that the statistics of the height profiles in some regimes can be described by a Markov process.

32.1 Introduction

Electropolishing is a widespread technology discovered in the 1920s and generally known since the 1930s [1]. It is a technical process for obtaining smooth, shiny surfaces by electro-chemical metal removal, i.e. the workpiece acts as an anode. For many metals and alloys, good electropolishing conditions can be found [2]. The operating point of electropolishing is either the so-called *passive region*, where metal is dissolved slightly, or the *transpassive region*, where two competing processes occur in the case of vertically arranged electrodes: the dissolution of metal leading to a falling film of spent electrolyte containing dissolved metal, and the hydrolysis of water, where oxygen is formed at the anode causing gas bubbles to rise. The interaction of the two processes leads to local hydrodynamic instabilities, which mainly influence the formation of the surface structures on different scales [3–5]. Common to all electropolished surfaces, which have been treated in the transpassive region of the cyclovoltammogram, is therefore an unwanted pattern on the micrometer scale, the so-called gas lines (Gasbahnen) [6].

The choices of the metal, the electrolyte and the applied electrical potential have a major influence on the surface structures. For aluminum a spatial pattern on a nanometer scale was found by Vignal et al. [7] and Yuzhakov et al. [8, 9]. While Vignal described hexagonal patterns reminiscent of convection cells occurring at free surfaces in the Rayleigh–Bénard

Nonlinear Dynamics of Production Systems. Edited by G. Radons and R. Neugebauer

ISBN 3-527-40430-9

convection, Yuzhakov postulated as the main reason for pattern formation a higher double-layer potential due to polarizable organic molecules. Pattern formation on a micrometer scale under electropolishing conditions for stainless steel, aluminum and brass was described by Visser and Buhlert [10, 6]. The scaling behavior of electropolished surfaces was investigated by Chauvy et al. [11]. They found that technical roughness measures depend on the interval length of the surface data. Sydow et al. [12] characterized the development of self-affine structures on a microscale of metal surfaces depending on the duration of electropolishing time. They pointed out that, at least after one hour of electropolishing, the fractal dimension of the spatial pattern is stationary.

In this study we investigate the surface topography of brass sheets, which have been electropolished in the perpendicular position in the transpassive region. They exhibit spatial pattern formation on the micrometer scale. The workpieces are scanned with a 3D-laser-focus scanner (UBM), where the height of a laser measuring head is controlled via a feedback loop to maximum reflectance [6]. The size of the focussed laser spot is about one micrometer, the lateral resolution of the scans is one micrometer and the vertical resolution is in the submicrometer range. Two different electrolyte solutions containing phosphoric acid and an amount of alcohol, denoted as methanol-elec. and glycerine-elec., are used. Although metal surfaces electropolished in methanol-elec. are very smooth, they show tiny ripples in the vertical direction. This anisotropy is due to the gas lines caused by rising gas bubbles (Fig. 32.1a). In contrast, glycerine-elec. leads to a rougher surface structure with higher and sharper peaks causing a more matt-finished appearance of the workpiece. Here, gas lines are not visible with the naked eye (Fig. 32.1b) [3, 4]. Our aim is a local characterization and classification of surface height profiles on different scales depending on the electrolyte used in order do get insight into the underlying interacting chemical and physical processes. This study is subdivided into three steps.

In the first step, global information is extracted, such as an estimate of the characteristic length scales introduced by the gas lines and an estimate of the scaling behavior in different regions. To this purpose, traditional Fourier spectral analysis is combined with more recently proposed wavelet power spectra and the continuous wavelet transform (CWT) based on complex progressive wavelets [13].

While power spectra are of great interest in finding regimes where scaling laws are valid, they can only provide estimates about a global roughness or a Hurst exponent H and thus about a self-affinity of the fractal surface height profile [14, 15]. Information on the spatial distribution of possibly varying local Hurst or Hölder exponents, i.e. fluctuations in the surface roughness resulting from multi-affine properties, is usually determined by means of the structure

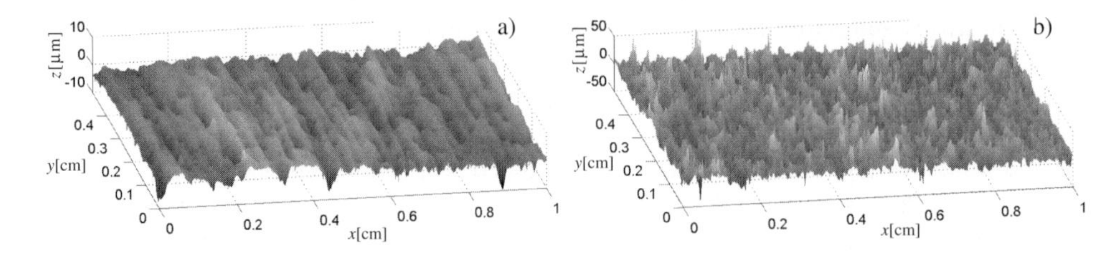

Figure 32.1: Laser scan of brass surfaces electropolished in (a) methanol-elec. and (b) glycerine-elec.

function method [16]. However, there are fundamental limitations in the structure function approach; in particular, irregularities in the derivatives of the function cannot be accessed. Wavelet techniques provide new tools for analyzing the regularity properties of functions. The recently developed wavelet transform modulus maxima (WTMM) method [17, 18] allows one to extract the local scaling behavior of singularities and provides a reliable method for the determination of singularity spectra [19]. The wavelet transform maxima lines contain the essential information about the evolution of scaling properties of irregularities across scales. Thus, they can be considered as a fingerprint of the scaling behavior of the function.

A characterization of the surface by means of singularity spectra is still insufficient, since possible correlations of the roughness measures on different scales are not taken into consideration [22, 23]. In a third step, we therefore study the evolution of conditional probability density functions (pdf) of surface height increments over scales. A preliminary analysis of multi-conditional pdf's suggests that the statistics of the height increments for scales larger than a certain threshold has Markov properties.

The paper is organized as follows. The experimental setup of the electropolishing is described in Sect. 32.2. In Sect. 32.3, a brief review of the continuous wavelet transform (CWT) and its scaling properties is given. We also report on another feature of the CWT when using complex progressive wavelets, namely to extract instantaneous frequencies or local wavelengths of a function, respectively. In Sect. 32.4, we study those scales where rising gas bubbles interact with the chemical removal of irregularities in the surface. Classical Fourier techniques and wavelet tools are combined to extract the length scales characterizing the gas lines. In Sect. 32.5, we focus on the wavelet transform modulus maxima (WTMM) method. After a short introduction the method is applied to the determinination of the multi-fractal spectra for surfaces electropolished in methanol-elec. and glycerine-elec., respectively. It can be seen that at least two processes are interacting, leading to different multi-fractal behavior on different scales. In Sect. 32.6, a preliminary stochastic analysis of the height profiles based on the theory of Markov processes [24] is given. Finally, Sect. 32.7 presents our conclusions and some perspectives of further investigations.

32.2 The Experimental Setup

As electro-chemical cells, basins of dimension length $\times$ width $\times$ height = 180 mm $\times$ 60 mm $\times$ 30 mm and $l \times w \times h$ = 180 mm $\times$ 60 mm $\times$ 60 mm were used, respectively. A common three-electrode arrangement was used, where anode and cathode were made of brass (CuZn37). The electrodes had the dimensions $l \times h$ = 50 mm $\times$ 50 mm and the electropolished surface was about $l \times h$ = 50 mm $\times$ 20 mm or $l \times h$ = 40 mm $\times$ 80 mm with an electropolished region of 40 mm $\times$ 50 mm, respectively. The electrodes were positioned face to face and vertically in the electrolyte. To avoid metal removal at the backside of the workpiece the anode was shielded with Tesa adhesive tape. The distance between anode and cathode was about 10 cm. An Ag/AgCl electrode was used as reference electrode; therefore all potential values are measured against U = 0.222 V. The reference electrode was fixed via a Haber–Luggin capillary in the middle of the electropolished area at the anode. The distance between the anode and the tip of the glass capillary was about 4$\pm$1 mm.

Two different electrolyte solutions were used. The composition of the electrolyte solution for one of the presented electrolytes contained 60 vol.-% H_3PO_4 85%, 30 vol.-% methanol and 10 vol.-% deionized water. This electrolyte solution is named methanol-elec. The other electrolyte contained 60 vol.-% H_3PO_4 85%, 30 vol.-% glycerine and 10 vol.-% deionised water. This electrolyte solution is denoted as glycerine-elec. All used chemicals were of pro analysi grade. For both electrolytes the pH-value was pH = 0. All experiments were carried out at room temperature.

The electrodes were rinsed with isopropanol to get rid of dust and grease. To ensure reproducible starting conditions and to be sure that the observed patterns are typical for electropolishing, each experiment started with a potentiodynamic scan, a so-called cyclic voltammogram. The important feature of this potential scan is to step into the transpassive region for a while. As soon as the potential in the cyclic voltammogram exceeds $U = 1.6$ V, oxygen production is initiated thereby cleaning the surface.

Afterwards, the workpieces were electropolished under galvanostatic conditions for longer than one hour to be sure that the pattern becomes stationary. For all experiments, the current densities were in a range between 1 and 5 A/dm^2.

The topography of the electropolished workpieces were measured with a 3D-laser-focus scanner (UBM). One- and two-dimensional measurements were made. The measurement resolution in x and y directions was 1 µm; the height resolution in the z direction was 0.01 µm.

32.3 Continuous Wavelet Transform

Recently, wavelet analysis has attracted much attention since it allows functions to be unfolded in space and scale. According to the definition

$$W_\psi f(a,b) = \frac{1}{a}\int_{-\infty}^{+\infty} f(x)\,\overline{\psi\left(\frac{x-b}{a}\right)}\,dx \qquad (a, b \in \mathbf{R},\ a > 0) \tag{32.1}$$

the continuous wavelet transform (CWT) decomposes the function $f(x) \in L^2(\mathbf{R})$ hierarchically in terms of elementary components $\psi\left(\frac{x-b}{a}\right)$ which are obtained from a single *mother wavelet* $\psi(x)$ by dilations and translations. Here, $\overline{\psi}(x)$ denotes the complex conjugate of $\psi(x)$, a the scale and b the shift parameter. The crucial point is to choose $\psi(x)$ so that it is well localized in both physical and Fourier space. A unique reconstruction of $f(x)$ is ensured if $\psi(x) \in L^1(\mathbf{R})$ has zero mean:

$$\int_{-\infty}^{\infty} \psi(x)\,dx = 0. \tag{32.2}$$

There is an infinite number of possible choices for the mother wavelets. The Gaussian family of real wavelets, which are obtained as derivatives of the Gaussian function, is especially suitable for detecting and characterizing irregularities in a function or even in its derivatives. For this purpose, we require $\psi(x)$ to be orthogonal to low-order polynomials:

$$\int_{-\infty}^{+\infty} x^k \psi(x)\, dx = 0, \qquad \forall k, \qquad 0 \le k < n_\psi. \tag{32.3}$$

The family of real Gaussian wavelets of nth order $\psi_n(x)$ is defined as [18, 25]

$$\psi_0(x) = e^{-x^2/2}, \qquad \psi_n(x) = \frac{d}{dx}\psi_{n-1}(x) \qquad (n \in \mathbf{N},\, n \ge 1). \tag{32.4}$$

As pointed out in the introduction, the regularity of a function at a point x_0 is characterized by the Hölder exponent $h(x_0)$, which is given by the largest exponent such that there exist a polynomial $P_n(x - x_0)$ of order $n < h(x_0)$ and a constant $C > 0$ such that for any point x in the neighborhood of x_0 the relation

$$|f(x) - P_n(x - x_0)| \le C|x - x_0|^{h(x_0)} \tag{32.5}$$

holds. Therefore, $h(x_0)$ characterizes, at least for non-oscillating (cusp) singularities [26], how irregular the function is at the point x_0: if $n < h(x_0) < n + 1$, then $f(x)$ is n times differentiable at x_0 but not $n + 1$ times. For instance, the Heaviside function ($H(x) = 1$ for $x \ge 1$ and $x = 0$ otherwise) corresponds to $h(0) = 0$, while the function $f(x) = |x|^{3/2}$ has a cusp singularity at $x_0 = 0$ with Hölder exponent $h(0) = 3/2$.

Assuming a cusp singularity with Hölder exponent $h(x_0) \in (n, n + 1)$ at x_0, the CWT scales like

$$|W_\psi f(a, x_0)| \sim a^{h(x_0)}, \qquad a \to 0^+, \tag{32.6}$$

provided the analyzing wavelet chosen has $n_\psi > h(x_0)$ vanishing moments [17, 18]. In contrast, if one chooses a wavelet with $n_\psi < h(x_0)$, the CWT scales with an exponent n_ψ:

$$|W_\psi f(a, x_0)| \sim a^{n_\psi}, \qquad a \to 0^+. \tag{32.7}$$

Thus, the faster the CWT of a function $f(x)$ decreases to zero around a point x_0, the more regular $f(x)$ is at x_0. It can be shown that this scaling behavior is also valid along the maxima lines of the modulus of the CWT, which point to the singularities [17]. Hence, for practical applications, these lines are used for extracting the Hölder exponents and can be regarded as *fingerprints* containing the complete information on the scaling behavior. A direct tracing of these maxima lines reduces the time-consuming calculation of the full redundant CWT, even so being sufficient for the characterization of the scaling properties of $f(x)$ [25].

In contrast with real wavelets, complex wavelets can separate amplitude and phase, enabling the measurement of instantaneous frequencies or local wavelengths and their temporal/spatial evolution [13, 27]. Let us introduce a family of complex progressive wavelets $\Psi_n(x)$ [28] for which the Fourier transform $\hat{\Psi}_n(\omega)$ vanishes for negative frequencies ω. They are defined as derivatives of the classical Morlet wavelet $\Psi_0(x)$ [30, 27]:

$$\Psi_0(x) = e^{-x^2/2}\, e^{i\omega_0 x}, \qquad \Psi_n(x) = \frac{d}{dx}\Psi_{n-1}(x) \qquad (n \in \mathbf{N},\, n \ge 1). \tag{32.8}$$

$\Psi_0(x)$ does not fulfill the admissibility condition, Eq. (32.2), in a strict sense. However, for practical purposes, because of the fast decay of its envelope towards zero, we can consider the Morlet wavelet $\Psi_0(x)$ to be admissible for $\omega_0 \geq 5$. On the other hand, all derivatives of $\Psi_0(x)$ are wavelets in a strict sense independent of ω_0.

The usefulness of this family of wavelets becomes evident if one considers, as an example, the wavelet transform of $f(x) = \cos \omega_1 x$. By inserting its Fourier transform $\hat{f}(\omega) = \pi[\delta(\omega + \omega_1) + \delta(\omega - \omega_1)]$ into Eq. (32.1), one obtains

$$W_\Psi f(a,b) = \frac{1}{2}\left[e^{ib\omega_1}\overline{\hat{\Psi}(a\omega_1)} + e^{-ib\omega_1}\overline{\hat{\Psi}(-a\omega_1)}\right], \tag{32.9}$$

which means that in general $|W_\Psi f(a,b)|$ is oscillating in the b direction. However, for progressive wavelets, these oscillations disappear, resulting in the relation

$$|W_\Psi f(a,b)| = \frac{1}{2}\left|\overline{\hat{\Psi}(a\omega_1)}\right|. \tag{32.10}$$

For the Morlet wavelet, the Fourier transform reads $\hat{\Psi}_0(\omega) = \sqrt{2\pi} e^{-(\omega-\omega_0)^2/2}$ and one observes a perfect localization of energy around the line $a = \omega_0/\omega_1$ [13, 29].

Here, we used the following convention for the Fourier transform $\hat{f}(\omega)$ of a function $f(x) \in L^2(\mathbf{R})$:

$$\hat{f}(\omega) = \int_{-\infty}^{+\infty} f(x)\, e^{-i\omega x}\, dx, \qquad x \in \mathbf{R}. \tag{32.11}$$

32.4 Characteristic Length Scales and Scaling Regions

The laser scan of a brass surface electropolished in methanol-elec. presented in Fig. 32.1a shows that the gas lines introduce a natural length scale into the surface structure. On the other hand, the brass sheet electropolished in glycerine-elec. (Fig. 32.1b) has a rough, fractal-like structure with high peaks, suggesting self-affinity. This, however, means the lack of a natural length scale in a certain range. Figure 32.2a, b show two typical surface height profiles transversal to the direction of the gas lines for methanol-elec. and glycerine-elec. In these one-dimensional profiles the fractal-like structure dominates in both cases, hiding a possibly existing waviness of the gas lines.

In order to find the characteristic length scales and scaling regions, the power spectral density $E(k)$ as a function of the wave number k for methanol-elec. and glycerine-elec. is plotted in Fig. 32.3, where we considered ensemble averages of profiles transversal to the direction of the gas lines and applied moving window averaging. For both power spectra, two regions with different decay can be seen, indicating the interplay of two different processes. For large scales the workpieces electropolished in methanol-elec. appear much smoother than those treated with glycerine-elec. Therefore, the methanol-elec. power spectrum displays a faster decay for large scales than the glycerine-elec. power spectrum. For small scales the behavior is just reversed. The latter observation is corroborated when we zoom into the profiles

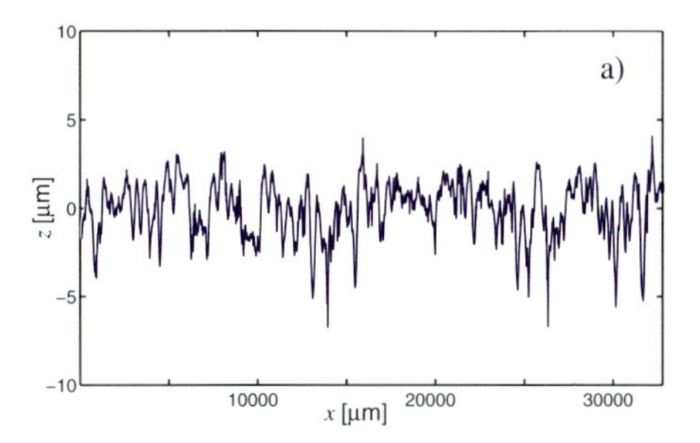

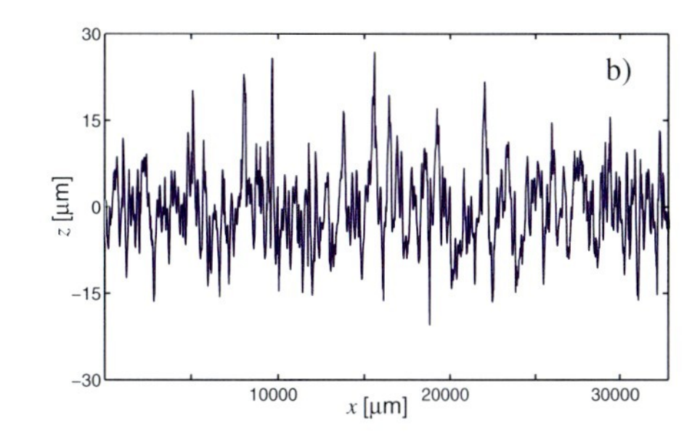

Figure 32.2: Height profiles $z(x)$ transversal to the direction of gas lines for surfaces electropolished in (a) methanol-elec. and (b) glycerine-elec. Note the different scales in the z direction.

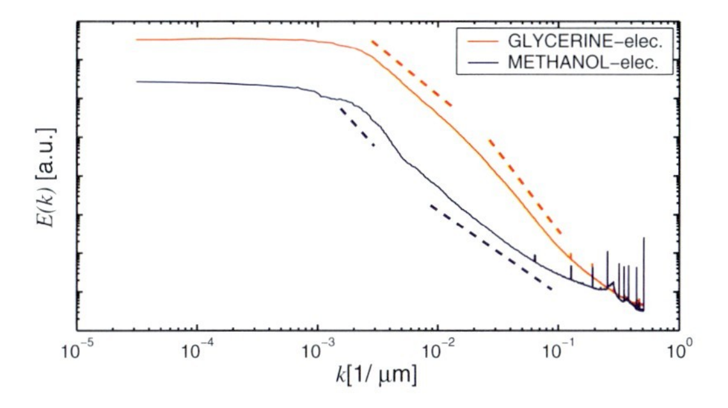

Figure 32.3: Power spectral densities $E(k)$ for ensemble averages of surface profiles transversal to the gas lines for (a) methanol-elec. and (b) glycerine-elec.

(see Fig. 32.4a, b). Some caution is advisable, because the scaling regions are rather small, covering at most one order of magnitude.

Since the characteristic length scales of the gas lines cannot easily be identified by means of the power spectral densities shown in Fig. 32.3, we measured the anisotropy introduced by the rising bubbles by calculating the power spectral density of ensemble averages transversal and parallel to the gas lines (see Fig. 32.5). As expected, the anisotropy is much more evident in the case of methanol-elec. From the ratio of the corresponding power spectra, the characteristic length scales can be readily extracted. For methanol-elec. we observe about 24 gas lines/cm. For glycerine-elec., the slight anisotropy is maximal for a frequency corresponding to about 38 lines/cm. However, it is not clear if one can speak about *gas lines* in this case.

Recently, complex progressive wavelets were introduced to decompose signals simultaneously in scale and space in order to extract local frequencies [13, 29]. We apply the CWT based on the Morlet wavelet Ψ_1 (Eq. (32.8)) to ensemble averages transversal to the gas lines for methanol-elec. (Fig. 32.6, top) and glycerine-elec. (Fig. 32.6, bottom). The graph of the respective height profile $z(x)$ (top), its Fourier power spectral density (PSD) (r.h.s.) and wavelet power spectral density (WPSD) [31] (l.h.s.) using discrete Meyer wavelets [30] is plotted together with the modulus of the CWT $|W_{\psi_1}z(a,b)|$ based on Morlet wavelets Ψ_1 defined in Eq. (32.6) for $\omega_0 = 5$. The wavelet power spectrum can be regarded as a smoothed Fourier power spectrum also allowing us to identify the scaling regions. The figures demonstrate the

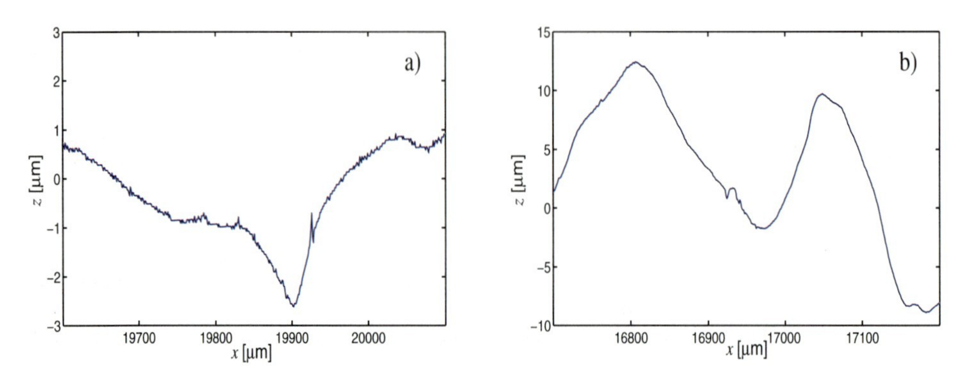

Figure 32.4: Zoom into height profiles transversal to the direction of gas lines for surfaces electropolished in (a) methanol-elec. and (b) glycerine-elec.

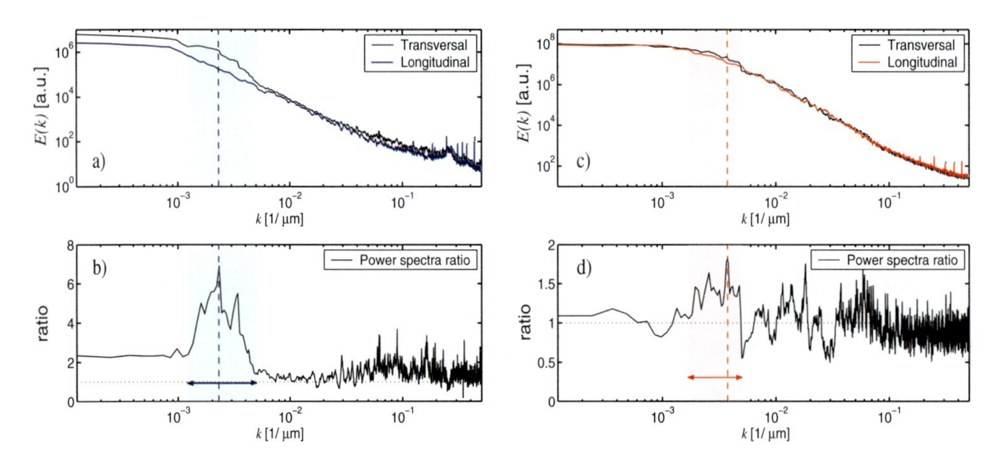

Figure 32.5: Comparison of power spectra of ensemble averages transversal and parallel to the gas lines for methanol-elec. (a) and glycerine-elec. (c). From the ratio of power spectral densities (b), (d) the range of length scales corresponding to the gas lines can be extracted.

ability of the CWT to unfold the information given in physical and Fourier space into a space–scale picture. In the case of methanol-elec. a band of peaks of varying frequency indicates the gas lines and their characteristic length scales (red arrow). For glycerine-elec. this band is slightly shifted to smaller scales and $|W_{\psi_1} z(a, b)|$ displays a topography, which is typically found for multi-fractals.

32.5 Multi-fractal Analysis

While the computation of the power spectral density $E(k)$ is of great interest in making conclusions about the existence of scaling laws, it gives only limited information about the mono- or multi-fractal properties of the surface roughness. It only allows for an estimation of a *global*

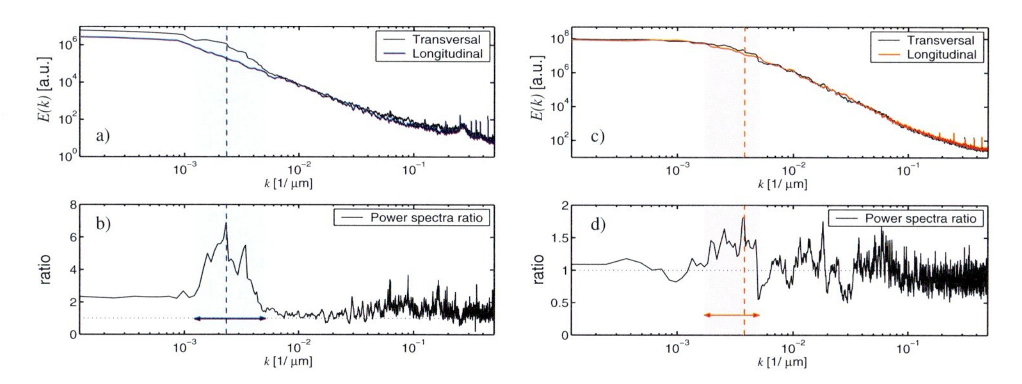

Figure 32.6: Space-scale decomposition of ensemble averages for methanol-elec. (top) and glycerine-elec. (bottom).

Hurst or Hölder exponent h via the relation $E(k) \sim k^{-1-2h}$. Local fluctuations in the degree of 'roughness' call for a location-dependent Hölder exponent $h(x)$.

The standard way [16] to extract multi-scaling properties, for example in a turbulent velocity field $v(x)$, is to study the scaling behavior of structure functions $S_q(r)$ of order q:

$$S_q(r) = < \delta v_r{}^q > \sim r^{\zeta_q} \tag{32.12}$$

of the longitudinal velocity increments $\delta v_r = v(x + r) - v(x)$. Multi-fractal behavior leads to a nonlinear scaling exponent ζ_q. By Legendre transforming the exponents ζ_q, one obtains an estimate of the spectrum $D(h)$ of Hölder exponents [16]

$$D(h) = \min_q (qh - \zeta_q + 1). \tag{32.13}$$

Here, $D(h)$ is the Hausdorff dimension of the subset of singularities, where the Hölder exponent is equal to h. A severe drawback of this method is that only exponents $0 < h(x) < 1$ are accessible, i.e. singularities in the derivatives of the function cannot be identified. In addition, negative moments ($q < 0$) may lead to divergencies in the structure functions $S_q(r)$, i.e. only a part of the spectrum of singularities can be described by the structure function method [18].

Wavelets provide alternative techniques which deserve special attention in the context of fractal analysis and synthesis because they can be used to extract local microscopic information about the scaling properties. Fractals are invariant under a group of self-affine transformations, i.e. they are similar to themselves when transformed by anisotropic dilations including translations and dilations, which are also basic operations for the generation of wavelets. Therefore, wavelets open a natural way to investigate irregular fractal functions. Based on the scaling properties of the wavelet transform described in Sect. 32.3, the wavelet transform modulus maxima (WTMM) method [18] allows in most cases an estimation of the full spectrum of singularities, i.e. a characterization of the multi-fractal behavior.

The WTMM method is a generalization of the classical multi-fractal formalism, which was originally developed for a statistical description of the scaling properties of singular measures [20, 21]. Instead of relying on box-counting techniques, wavelets are introduced as oscillating variants of box functions. Since the skeleton of maxima lines contains already the essential information on the singularities, one can define the following partition function [18]

$$Z(q,a) = \sum_{b_i \in \text{max.lines}} \left(\sup_{a' \leq a} |W_\psi f(a', b_i)| \right)^q . \tag{32.14}$$

For a given scale a, $Z(q,a)$ contains the qth moments of the contributions of $|W_\psi f|$ along the maximal lines, where the supremum in Eq. (32.14) is related to a Hausdorff-like covering with scale-adapted wavelets. Using this definition, divergencies due to negative-order moments are removed [18]. From the power-law behavior of the partition function

$$Z(q,a) \sim a^{\tau(q)}, \qquad a \to 0^+ \tag{32.15}$$

the whole spectrum of Hölder exponents $D(h)$ is obtained by Legendre transforming the scaling exponents $\tau(q)$:

$$D(h) = \min_q (qh - \tau(q)). \tag{32.16}$$

The spectrum of Hölder exponents $D(h)$ is used to characterize the fluctuations in the roughness of electropolished surfaces. We apply the WTMM method to profiles transversal to the gas lines for brass sheets electropolished in methanol-elec. and glycerine-elec. Figure 32.7a shows a typical height profile $z(x)$ together with the skeleton of maxima lines of the modulus of the CWT $|W_{\psi_2} z(a,b)|$ using the Mexican-hat wavelet ψ_2, see Eq. (32.4), shown in Fig. 32.7b. Blue lines indicate minima of the CWT $W_{\psi_2} z$ for $W_{\psi_2} z < 0$, red lines maxima of $W_{\psi_2} z$ for $W_{\psi_2} z > 0$. For each electrolyte we use an ensemble of five profiles transversal to the direction of the gas lines for the calculation of the partition function $Z(q,a)$ defined in Eq. (32.14). In accordance with the observation of two different scaling regions in the power spectra (Fig. 32.3) two regions with different power-law behavior occur, which lead to different distributions of the corresponding Hölder exponents. Up to 12 800 maxima lines are used for the evaluation of $D(h)$. The corresponding singularity spectra $D(h)$ for small and large scales are plotted in Fig. 32.8a, b. It should be noted that, due to the finite resolution of measurements and cross-over effects between small and large scales, the spectra depend to some extent on the scale interval selected for the evaluation of $D(h)$.

Before we discuss the results, some remarks may be helpful for the interpretation of the singularity spectra. Typical for multi-fractals is a single-humped shape of $D(h)$. The support of the spectrum then covers a finite interval $h_{\text{min}} \leq h \leq h_{\text{max}}$ indicating that the Hölder exponents fluctuate in this interval. For monofractal functions $D(h)$ degenerates to a single point. For large q-values, i.e. large positive moments of the contributions of $|W_\psi f|$ along the maxima lines (Eq. (32.14)), one obtains the strongest singularities, i.e. the sharpest peaks with the smallest exponent h_{min}. In contrast, the highest negative moments emphazise low values of $|W_\psi f|$ leading to the weakest irregularities with exponent h_{max}. The maximum of the spectrum $D(h_0) = D_0$ is obtained for the parameter $q = 0$. The corresponding value h_0 is the Hölder exponent which appears most frequently, because the fractal dimension of the subset of all singularities with exponent h_0 attains its maximal value D_0.

The spectra of Hölder exponents displayed in Fig. 32.8a, b show that the strengths of singularities for methanol- and glycerine-elec. are quite similar on small scales. To be more

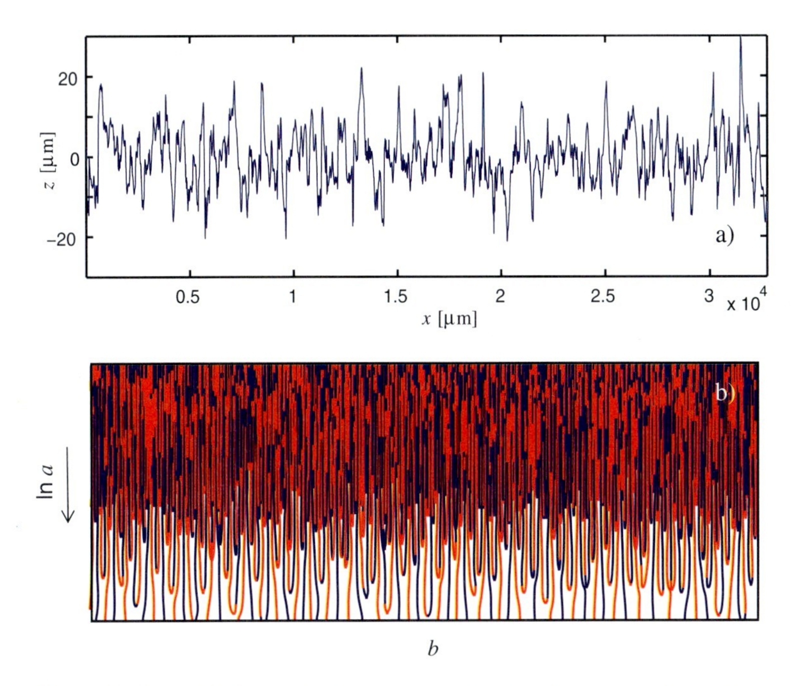

Figure 32.7: (a) Height profile $z(x)$ transversal to the direction of the gas lines for a surface electropolished in glycerine-elec. and (b) maxima lines of the modulus of the CWT $|W_{\psi_2} z(a, b)|$.

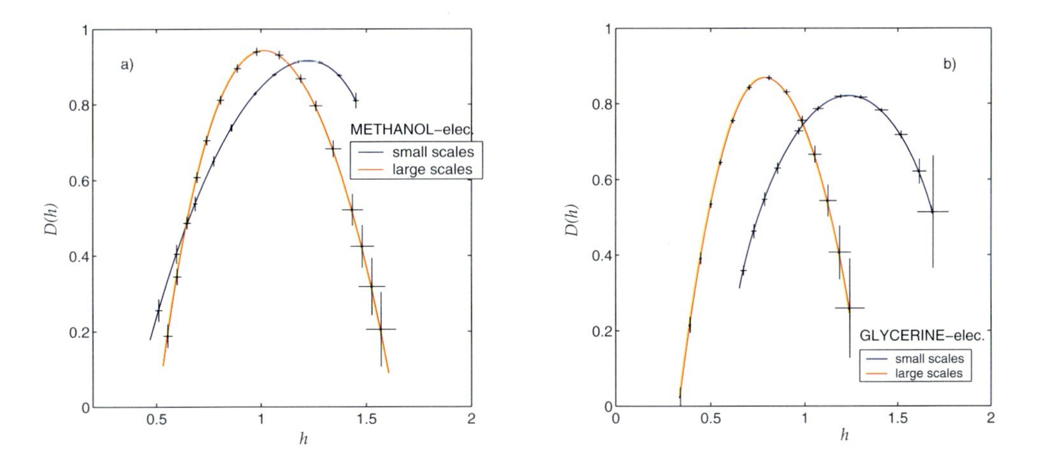

Figure 32.8: Spectra of Hölder exponents $D(h)$ for (a) methanol-elec. and (b) glycerine-elec. for different scaling regions.

specific: although the support of the $D(h)$ curve of methanol-elec. is slightly shifted to smaller values of h as compared to the $D(h)$ curve of glycerine-elec., the most frequent Hölder exponent h_0 is around 1.25 in both cases. A further difference is that $D(h_0)$ is smaller for glycerine-elec., indicating that the corresponding singularities are less frequent. Thus, in contrast to the visual impression received from Fig. 32.1, surfaces electropolished in glycerine-elec. are smoother on small scales than those treated with methanol-elec. (cf. Fig. 32.4). For large scales, however, the glycerine-elec. spectrum is clearly shifted to smaller h-values with $h_0 = 0.79$ as compared to $h_0 = 1.02$ for methanol-elec., which corroborates the observation that glycerine-elec. produces sharper higher peaks on large scales than the methanol-electrolyte. For instance, cusp-like irregularities of the type $|x - x_0|^{1/2}$ are almost absent for methanol-elec., while the subset of peaks with $h = 1/2$ has a fractal dimension $D(h = 1/2) \sim 1/2$ in the case of the glycerine electrolyte.

Let us mention that the power spectra displayed in Fig. 32.3 can only reflect the minimum global regularity [17] correlating to the sharpest peaks in the profiles. Thus, a smaller slope in the power spectrum corresponds to a smaller value of h_{min} in the singularity spectrum. The spectra $D(h)$ can therefore be considered as a quantification of the qualitative conclusions drawn from inspecting the profiles shown in Figs. 32.1, 32.2 and 32.4. The main difference in the surface topography of brass workpieces electropolished in different electrolytes seems to be caused by the rising gas bubbles, their magnitude, adhesion time and dynamics producing different patterns on large scales. This is due to the fact that the gas bubbles are generating a flow of fresh electrolyte etching the surface in the gas lines more effectively [3].

32.6 Stochastic Analysis

For rough surfaces displaying scaling behavior the singularity spectrum $D(h)$ can be regarded as a complete multi-fractal

characterization of the singularities [14, 15]. However, from a stochastic point of view, this characterization is still incomplete, since joint statistical properties of several height increments on different scales are not taken into account [32, 33]. In a series of papers a new approach for the stochastic analysis has been proposed, which allows one to extract the explicit form of the underlying stochastic process directly from experimentally measured data without making any assumptions, provided the process is Markovian [34, 24, 32]. Considering the height increment z_r of a surface profile $z(x)$

$$z_r(x) = z(x + r/2) - z(x - r/2) \tag{32.17}$$

as a stochastic variable in the length scale r, the aim is to describe the evolution of the conditional probability density function (pdf) as r is varied, where the conditional pdf $p(z_1, r_1|z_2, r_2)$ describes the probability for finding the increment z_1 on scale r_1 provided that the increment z_2 is given on scale r_2. In contrast to the left-sided increments used for the structure function (Eq. (32.12)), we refer to symmetrical increments in order to avoid the introduction of spurious correlations [22, 23]. A stochastic process is Markovian if the conditional probability densities fulfil the relations

$$p(z_1, r_1|z_2, r_2; ...; z_n, r_n) = p(z_1, r_1|z_2, r_2) \qquad \text{where} \qquad r_1 < r_2 < \cdots < r_n. \tag{32.18}$$

In this case, the conditional pdf fulfills a master equation. Expanding the distribution function into a Taylor series, the evolution equation can be written in the form [36, 24]

$$-r\frac{\partial}{\partial r}p(z_r, r|z_0, r_0) = \sum_{k=1}^{\infty}\left(-\frac{\partial}{\partial z_r}\right)^k D_k(z_r, r)\, p(z_r, r|z_0, r_0), \tag{32.19}$$

where the so-called Kramers–Moyal coefficients $D_k(z_r, r)$ can be directly estimated from experimental data. In the special case of $D_4(z_r, r) = 0$, Eq. (32.19) reduces to the Fokker–Planck equation

$$-r\frac{\partial}{\partial r}p(z_r, r|z_0, r_0) = \left\{-\frac{\partial}{\partial z_r}D_1(z_r, r) + \frac{\partial^2}{\partial {z_r}^2}D_2(z_r, r)\right\} p(z_r, r|z_0, r_0). \tag{32.20}$$

The partial differential equations (Eqs. (32.19), (32.20)) completely describe the underlying stochastic process. For details the reader is referred to [36, 24].

In what follows, we show some preliminary studies necessary for this stochastic approach. The height profile data were obtained from UBM measurements scanning an area of about $2\ \mathrm{cm} \times 3.3\ \mathrm{cm}$ of the electropolished brass workpieces. For both methanol-elec. and glycerine-elec., 201 profiles $z(x)$ (each containing 2^{15} points) in the x direction transversal to the gas lines were scanned at a distance of $\Delta y = 10\ \mu\mathrm{m}$ with a resolution of $1\ \mu\mathrm{m}$ in the x and $0.01\ \mu\mathrm{m}$ in the z direction, respectively. Polynomial trends of fourth order were removed from the profiles. In Fig. 32.9a, b the probability density functions of the z-increments are plotted for several scales r_i. The pdf's are normalized to their respective standard deviations σ_r and shifted in the vertical direction for clarity. For small scales the shapes of the curves deviate strongly from Gaussian distributions indicating pronounced intermittency effects. The asymmetry displayed in the case of glycerine-elec. is caused by sharp high peaks which are not balanced by grooves (cf. Fig. 32.2b).

In a next step we test the data for evidence of an underlying Markovian process. Since a general test of the condition, Eq. (32.18), for all sets of scales $r_1, r_2, ..., r_n$ and for all n is not possible, we test the validity of the following necessary condition

$$p(z_1, r_1|z_2, r_2; z_3, r_3) = p(z_1, r_1|z_2, r_2) \qquad \text{where} \qquad r_1 < r_2 < r_3. \tag{32.21}$$

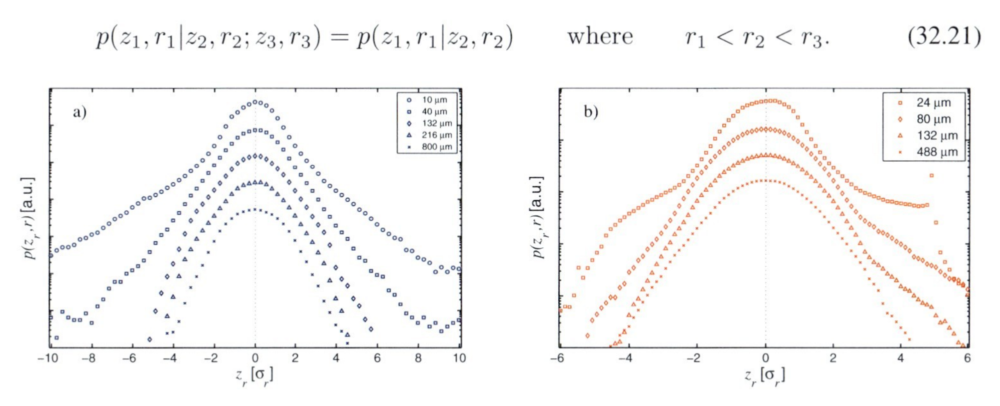

Figure 32.9: Probability density functions of height increments at different scales r_i for (a) methanol-elec. and (b) glycerine-elec.

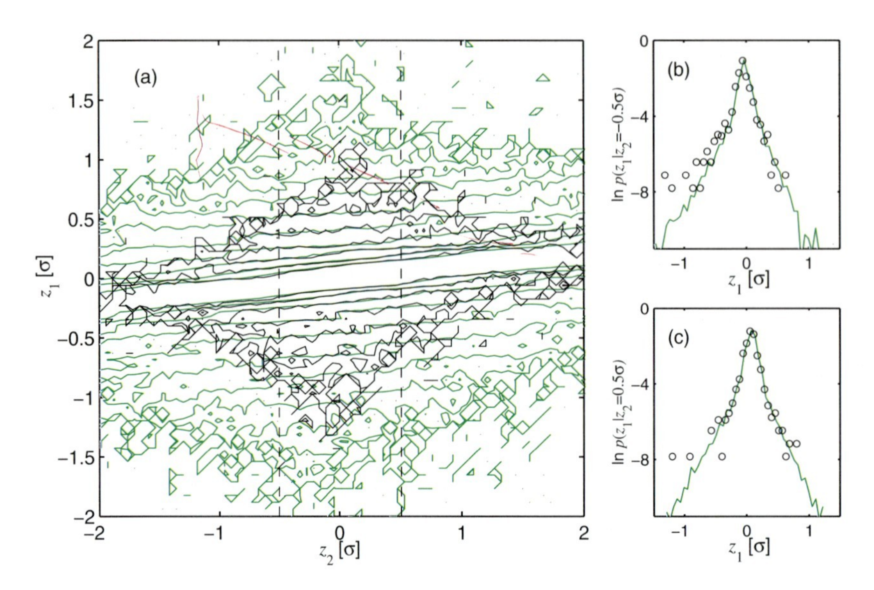

Figure 32.10: (a) Contour lines of the conditional pdf's $p(z_1, r_1|z_2, r_2)$ (green lines) and $p(z_1, r_1|z_2, r_2; z_3 = 0, r_3)$ (black lines) for $r_1 = 10\,\mu\text{m}, r_2 = 108\,\mu\text{m}, r_3 = 216\,\mu\text{m}$ for methanol-elec. (b), (c) Cuts through the conditional pdf's for $z_2 = \pm\sigma/2$.

The results for a brass sheet electropolished in methanol-elec. are presented in Figs. 32.10 and 32.11. In Fig. 32.10a, the contour plots of $p(z_1,r_1|z_2,r_2;z_3,r_3)$ (black lines) and $p(z_1,r_1|z_2,r_2)$ (green lines) for $r_1 = 10\ \mu\text{m}, r_2 = 108\ \mu\text{m}, r_3 = 216\ \mu\text{m}$ in units of the standard deviation of the z-data $\sigma \sim 1.52\ \mu\text{m}$ are shown. The rather good correspondence over several orders of magnitude is corroborated by two cuts for $z_2 = \pm\sigma/2$ displayed in Fig. 32.10b, c indicating the validity of the necessary condition (Eq. (32.21)). In Fig. 32.11a the same contour plots are presented for a different choice of scale increments $r_1 = 52\ \mu\text{m}, r_2 = 60\ \mu\text{m}, r_3 = 68\ \mu\text{m}$. From the two cuts for $z_2 = \pm 0.2\ \sigma$ shown in Figs. 32.11b, c it can be seen that the two sets of contour lines deviate strongly from each other, i.e. here a Markovian condition is not satisfied. Similar results are obtained for the glycerine-elec. case. A preliminary conclusion at this early state of our analysis is that the Markov properties are likely to hold for large enough differences in the scales r_i, but are violated for small differences. Further investigations including the Wilcoxon test [24] are necessary to estimate the range of validity of Markov properties.

Further work will comprise an estimation of the Kramers–Moyal coefficients D_1, D_2 and D_4, which allow us to set up Eq. (32.19) or Eq. (32.20), respectively, and thus to describe the stochastic process completely. Of special interest will be a comparison of the singularity spectra $D(h)$ calculated in the previous section with the scaling behavior of the height increments $\langle {z_r}^q \rangle \sim r^{\zeta_q}$, which can be derived from the master or Fokker–Planck equation, respectively [32]. According to Eqs. (32.12), (32.13) the Legendre transform of the scaling exponents ζ_q yields an estimation of $D(h)$.

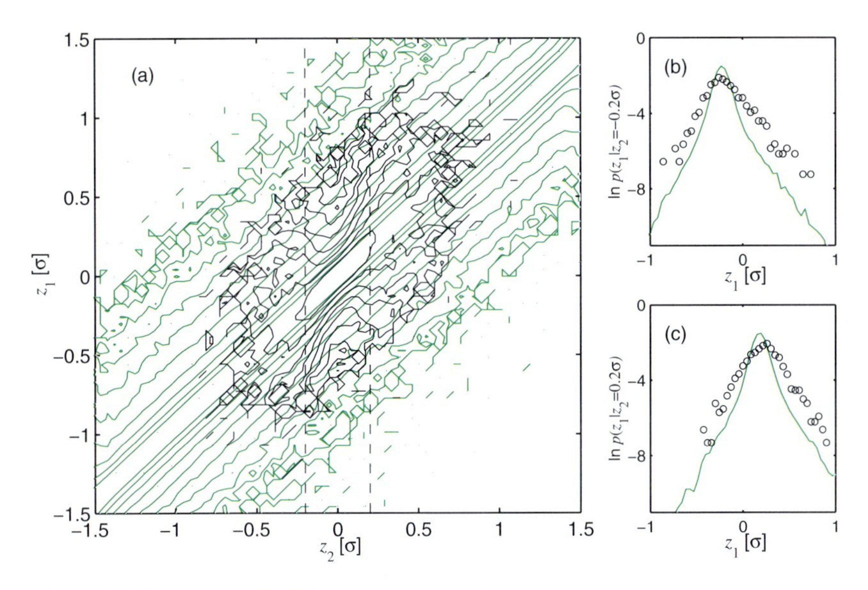

Figure 32.11: As Fig. 32.10, but for different scale increments: $r_1 = 52\,\mu\text{m}, r_2 = 60\,\mu\text{m}, r_3 = 68\,\mu\text{m}$.

32.7 Conclusions

In this study we presented various numerical techniques which are useful for the analysis of the surface topology of electropolished metals, for example brass sheets, with the objective to characterize and classify profiles resulting from different electrolytes. Classical Fourier techniques are combined with recently developed wavelet methods to search for scaling regions. The complete information about the scaling behavior is contained in the singularity spectra $D(h)$ which can be estimated by means of the wavelet transform modulus maxima method. For a complete stochastic description of the surfaces, some preliminary investigations in the framework of Markov analysis are presented. There are strong indications that the underlying processes fulfill the Markov conditions for certain scale differences.

As a result, it is possible to distinguish between the surface structures caused by the two interacting processes, namely the rising gas bubbles introducing an anisotropy into the topography and the dissolution of metal. This anisotropy is more pronounced in the case of methanol-elec. While workpieces electropolished in glycerine-elec. appear rougher on a large scale than those treated with methanol-elec., an opposite tendency is observed on the small scales. From the singularity spectra it can be concluded that the range of Hölder exponents of the singularities on the small scales, for which the dissolution of metal is mainly responsible, is almost the same for both cases, methanol-elec. and glycerine-elec. The main difference in the surface structure is caused by the specific peculiarity of the gas lines.

Still, there are many open questions which make further investigations necessary. Among them is the emergence of phase transitions in the singularity spectra resulting from a superposition of multi-fractal functions either with smooth C^∞ behavior [18] or with a second multi-fractal function [37] and the calculation of the corresponding $D(h)$ by means of the WTMM method. In addition, a possible occurrence of oscillating singularities should be taken into consideration [26]. Our main concern is to study whether or not one could benefit for the Markov analysis described in Sect. 32.6 from a replacement of the height increments by wavelet coefficients or the CWT values $W_\psi z(a, b_i)$ along the maxima lines, respectively. This work is in progress.

Acknowledgement

We gratefully acknowledge many stimulating discussions with R. Friedrich, J. Peinke, M. Wächter, A. Kouzmitchev, F. Kun and B. Lehle. This work was partially supported by the Volkswagen Stiftung grant I/77315: Nichtlineare Dynamik des lasergestützten elektrochemischen Jet-Verfahrens zur Ultrapräzisionsmikrostrukturierung von Metallen für optische Anwendungen and by grant I/74151: Untersuchung der Dynamik der Gasfahnen- und Kolkbildung bei der Elektrostrukturierung von Messing und Edelstahl.

Bibliography

[1] P.A. Jacquet, *Electrolytic and chemical polishing*, Metallurgical Reviews **1** (1956), 157–238.

[2] W.J.McG. Tegart, *The Electrolytic and Chemical Polishing of Metals in Research and Industry*, Pergamon Press, London, 1959.

[3] C. Gerlach, Dissertation, University of Bremen (2002) [http://elib.suub.uni-bremen.de/publications/dissertations/E-Diss427_Gerlach_2002.pdf].

[4] C. Gerlach, A. Visser, and P.J. Plath, *Galvanostatic studies of an oxygen evolving electrode*, in: *Nonlinear Dynamics of Production Systems, Chemical and Electro-Chemical Processes*, edited by G. Radons and R. Neugebauer, Wiley-VCH, Weinheim, 2003, Chapter 31.

[5] P.J. Plath, M. Baune, M. Buhlert, C. Gerlach, A. Kouzmitchev, P. Thangavel, E. van Raaij, H. Mathes, S. Diaz Alfonso, and T. Rabbow, *Nonlinear dynamics in chemical engineering and electrochemical manufactory technologies*, in: *Nonlinear Dynamics of Production Systems, Chemical and Electro-Chemical Processes*, edited by G. Radons and R. Neugebauer, Wiley-VCH, Weinheim, 2003, Chapter 30.

[6] M. Buhlert, *Elektropolieren und Elektrostrukturieren von Edelstahl, Messing und Aluminium*, Fortschritt-Berichte VDI, Reihe 2 **553** (2000), VDI-Verlag Düsseldorf.

[7] V. Vignal, J.C. Roux, S. Flandrois, and A. Fevrier, *Nanoscopic studies of stainless steel electropolishing*, Corrosion Science **42** (2000), 1041–1053.

[8] V.V. Yuzhakov, H.-C. Chang, and A.E. Miller, *Pattern formation during electropolishing*, Physical Review B **56**(19) (1997), 12608–12624.

[9] V.V. Yuzhakov, P.V. Takhistov, A.E. Miller, and H.-C. Chang, *Pattern selection during electropolishing due to double-layer effects*, Chaos **9**(1) (1999), 62–77.

[10] A. Visser and M. Buhlert, *Elektrostrukturieren von Messingoberflächen. Untersuchung erwünschter und unerwünschter Nebeneffekte*, Galvanotechnik **87**(5) (1996), 1454–1463.

[11] P.F. Chauvy, C. Madore, and D. Landolt, *Variable length scale analysis of surface topography: characterization of titanium surfaces for biomedical applications*, Surface and Coatings Technology **110** (1998), 48–56.

[12] U. Sydow, M. Buhlert, P.J. Plath, *Fractal characteristics of electropolished metal surfaces*, accepted for publication in Discrete Dynamics in Nature and Society.

[13] N. Delprat, B. Escudié, P. Guillemain, R. Kronland-Martinet, Ph. Tchamitchian, and B. Torrésani, *Asymptotic wavelet and Gabor analysis: extraction of instantaneous frequencies*, IEEE Transactions on Information Theory **38** (1992), 644–664.

[14] J. Feder, *Fractals*, Plenum Press, New York, 1988.

[15] A.-L. Barabási and H.E. Stanley, *Fractal Concepts in Surface Growth*, Cambridge University Press, 1995.

[16] U. Frisch and G. Parisi, *On the singularity structure of fully developed turbulence*, in: Turbulence and Predictability in Geophysical Fluid Dynamics and Climate Dynamics, Proceed. Intern. School of Physics 'E. Fermi', 1983, Varenna, Italy edited by M. Ghil et al., North Holland, Amsterdam, 1985, pp. 84–87.

[17] S. Mallat and W.L. Hwang, *Singularity detection and processing with wavelets*, IEEE Transactions on Information Theory **38** (1992), 617–643.

[18] J.F. Muzy, E. Bacry, and A. Arnéodo, *The multifractal formalism revisited with wavelets*, International Journal of Bifurcation and Chaos **4** (1994), 245–302.

[19] S. Jaffard, *Some open problems about multifractal functions*, in: Fractals in Engineering, edited by J. Lévy Véhel et al., Springer, London, 1997, pp. 2–18.

[20] T.C. Halsey, M.H. Jensen, L.P. Kadanoff, I. Procaccia, and B.I. Shraiman, *Fractal measures and their singularities: The characterization of strange sets*, Physical Review A **33** (1986), 1141–1151.

[21] J. Argyris, G. Faust, and M. Haase, *An Exploration of Chaos*, North-Holland, Amsterdam, 1994.

[22] A. Kouzmitchev, M. Wächter, and J. Peinke, *Violation of the Markov property for the Gauss-distributed height profile because of the height increment choice*, in preparation.

[23] A. Kouzmitchev, A. Visser, and P.J. Plath, *Stochastic characterization of rough and electropolished surfaces*, in preparation.

[24] C. Renner, J. Peinke, and R. Friedrich, *Experimental indications for Markov properties of small-scale turbulence*, Journal of Fluid Mechanics **433** (2001), 383–409.

[25] M. Haase and B. Lehle, *Tracing the skeleton of wavelet transform maxima lines for the characterization of fractal distributions*, in: Fractals and Beyond, edited by M.M. Novak, World Scientific, Singapore, 1998, pp. 241–250.

[26] A. Arnéodo, E. Bacry, S. Jaffard, and J.F. Muzy, *Oscillating singularities on Cantor sets: A grand-canonical multifractal formalism*, Journal of Statistical Physics **87** (1997), 179–209.

[27] R. Carmona, W.L. Hwang, and B. Torrésani, *Practical Time-Frequency Analysis*, Academic Press, San Diego, 1998.

[28] M. Haase, *A family of complex wavelets for the characterization of singularities*, in: Paradigms of Complexity, edited by M.M. Novak, World Scientific, Singapore, 2000, 287–288.

[29] M. Haase and J. Widjajakusuma, *Damage identification based on ridges and maxima lines of the wavelet transform*, International Journal of Engineering Science **41** (2003), 1423–1443.

[30] I. Daubechies, *Ten Lectures on Wavelets*, SIAM, Philadelphia, PA, 1992.

[31] J. Pando and Li-Zhi Fang, *Discrete wavelet transform power spectrum estimator*, Physical Review E **57** (1998), 3593–3601.

[32] M. Wächter, F. Riess, H. Kantz, and J. Peinke, *Stochastic analysis of surface roughness*, arXiv:physics/0310159 and forthcoming paper in Europhysics Letters.

[33] C. Renner, J. Peinke, R. Friedrich, O. Chanal, and B. Chabaud, *Universality of small scale turbulence*, Physical Review Letters **89** (2002), 124502.

[34] R. Friedrich and J. Peinke, *Statistical properties of a turbulent cascade*, Physica D **102**, 147–155.

[35] C. Renner, J. Peinke, and R. Friedrich, Journal of Fluid Mechanics **433** (2001), 383–409.

[36] H. Risken, *The Fokker-Planck Equation*, Springer Berlin, 1996.

[37] G. Radons and R. Stoop, *Superposition of multifractals: Generators of phase transitions in the generalized thermodynamic formalism*, Journal of Statistical Physics **82** (1996), 1063–1080.

33 Spatial Inhomogeneity in Lead–Acid Batteries

U. Sydow , M. Buhlert, E.C. Haß, and P.J. Plath

Local potential measurements in lead–acid batteries are realized in order to find reliable quantities for state-of-charge and state-of-health evaluation. It is known that in lead–acid batteries as an example for strongly coupled nonlinear electro-chemical systems, spatial inhomogeneities can be found like acid stratification due to an inhomogeneous current density distribution. To detect such inhomogeneities in situ we made spatially resolved potential measurements between the plates of a cell stack under charging/discharging conditions. The spatially resolved potential information from inside a cell stack gives additional information about the coupled electro-chemical system which cannot be obtained by potential and current monitoring at the outside poles of the battery, and is therefore suitable for developing a reliable prognosis of the battery's service time.

33.1 Introduction

In the light of a predicted overall growth of the world consumption of energy of around 40% over the next 20 years, energy storage will be a key technology for global energy sustainability [8]. There are four possibilities for energy storage: potential energy, kinetic energy, thermal energy and chemical energy. Chemical energy is generally stored as hydrogen or methanol for fuel cells, or as chemicals in batteries. Out of a variety of battery types, lead–acid batteries are still one of the most promising. Moreover, at present, lead–acid batteries are most likely to be used due to their acceptable cost. Lead–acid batteries are already used in conjunction with wind turbines and photovoltaic installations. Most widely, they are used in cars, etc., for engine starting and other duties. Future car electrical systems are demanding new power-optimized lead–acid batteries for starting and service [10, 14]. With increasing requirements from the car electrical system, a reliable prognosis of the state-of-charge (SOC) and state-of-health (SOH) of car batteries becomes more and more neccesary. A precise knowledge of the SOC and SOH is also of high significance for the maintainance of solar energy storage facilities. The increasing number of publications dealing with the capacity determination from the early 1990s on reflects the increasing requirements on lead–acid accumulators, especially from the automotive industry. Attempts to state an empirical equation for the capacity of lead–acid accumulators date back to the late 19th century, starting with the works of Peukert [15] and Liebenow [12]. The Peukert equation relates the discharge current I to the discharge time t for constant temperature (various discharge currents are applied until the same potential drop is established):

$$I^n t = K = \text{const.}, \tag{33.1}$$

Nonlinear Dynamics of Production Systems. Edited by G. Radons and R. Neugebauer

ISBN 3-527-40430-9

which can be linearized as:

$$\log t = \log K - n \log I, \tag{33.2}$$

providing an easy evaluation of the constants n and K. The Liebenow equation relates the amount of charge, usually referred to as the capacity of the battery for specific discharge current and specific temperature, $C = It$, to the discharge time t,

$$C = \frac{C_{\text{max}}}{1 + a/t^n}. \tag{33.3}$$

Liebenow found n to be always close to 0.5, so that the constants C_{max} and a remain to be determined. A detailed discussion of the Peukert and Liebenow equations, as well as a comparison of both equations with typical experimental data, is given by Compagnone [7]. The author also gives a new equation for the limiting capacity of the lead–acid accumulator. This new equation is an approximate solution for the diffusion problem for two planar, finite compartments of different lengths, simulating the porous electrode and the inter-plate space occupied by electrolytes of different apparent diffusion coefficients, D_1 and D_2.

Baikie et al. [1] gives an equation for the battery capacity including temperature effects. An overview of the different approaches to describe the battery capacity is given by Rydh [17]. Conventional attempts to determine the SOC are based on monitoring the battery current, the battery voltage and sometimes also the temperature of the battery. Constructing a model out of these data neglects the fact that the battery is not necessarily behaving as a homogeneous system. Even the measurement of the acid density [6, 13] will not give sufficient information about the SOC of a battery if e.g. it is measured only above the cell stacks. It is known that the bulk electrolyte density may vary by up to 20% during charge/discharge cycles, so that the local acid density cannot be expected to be homogeneous. Relaxation to homogeneous density distribution is affected by the diffusion coeffiecents. The diffusion coefficients of diluted sulfuric acid are also strongly dependent on the acid density. Armanta-Deu [9] gives values for the diffusion coefficient D in the range of 1×10^{-9} up to 20×10^{-9} m^2/s. So the process of acid density leveling by diffusion is a slow process. Since it is known that the current density between the plates is not necessarily homogeneous [2, 3, 4, 13] – and is influenced by the grid design [5] – it seems necessary to get spatially resolved information about the dynamics between battery plates under working conditions to get measures for a reliable prognosis of the battery's SOC and SOH.

Inhomogeneities due to acid stratification are particularly a problem of photovoltaic batteries which are more likely to undergo deep discharges during solar cycling. Firstly, we demonstrate the possibility of detecting spatial potential variation by statically measuring half-cell potentials with auxiliary Ag/AgCl electrodes. Secondly, we present here a new dynamic method for in situ measurement of spatial inhomogeneities between cell plates with auxiliary lead electrodes. In this dynamic method, the auxiliary electrode is connected periodically to either the positive or the negative battery plate.

33.2 Experimental

In a first set of experiments, static half-cell measurements with auxiliary electrodes (Ag/AgCl, Argenthal Typ 363-S7, Mettler-Toledo) between two battery plates have been done in order to check for the possibility of detecting inhomogeneities between the cell plates while charging/discharging. The electrodes were attached to capillaries small enough to be positioned between the plates. The capillaries were filled with the standard battery electrolyte.

A second set of experiments has been realized with nine Ag/AgCl electrodes between the two battery plates. The half-cell potentials have been recorded between the Ag/AgCl electrodes and either the positive or the negative plate. In this setup, vertical as well as horizontal potential inhomogeneities should be detectable. The positions of the auxiliary electrodes between the battery plates are shown in Fig. 33.1.

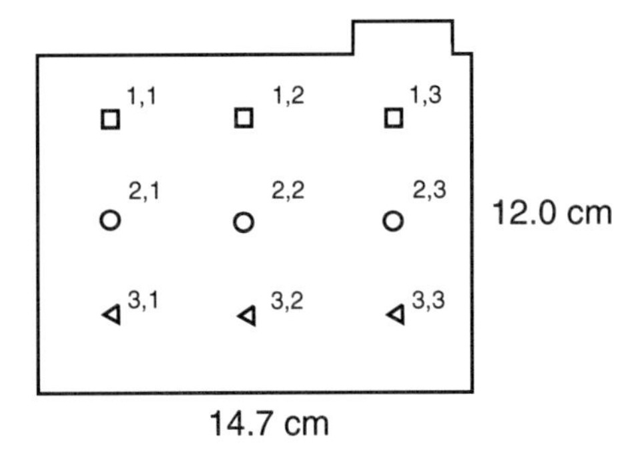

Figure 33.1: Positions of the capillaries of the auxiliary electrodes (Ag/AgCl or lead) in front of the PbO_2 electrode.

The plates were placed in a standard battery casing; the volume of the cell compartment was reduced with PVC plates to obtain an electrolyte volume between the plates comparable to the per plate volume of a whole cell stack. The standard electrolyte was sulfuric acid with a density of $d = 1.28$ g/cm. The used battery plates were pre-charged during tank formation. All measurements were carried out at room temperature (25°C).

Whereas in the above-mentioned experiments Ag/AgCl reference electrodes were used to statically record half-cell potentials, a third set of experiments was carried out with auxiliary lead electrodes as potential probes. The auxiliary lead electrodes have been dynamically switched between the positive and the negative battery plates in order to obtain alternating information about the positive and the negative battery plates. The battery has been discharged with a load resistance R, or charged/discharged with a laboratory power transformer.

33.2.1 Local Potential Measurements with Ag/AgCl Electrodes

The general setup is shown in Fig. 33.2. Two battery plates (negative and positive) out of a 110 Ah battery from VB Autobatterie GmbH were placed in a battery casing. PVC plates were placed in the casing to reduce the electrolyte volume. With that, the resulting electrolyte volume between the plates was comparable to the per plate electrolyte volume of a whole cell

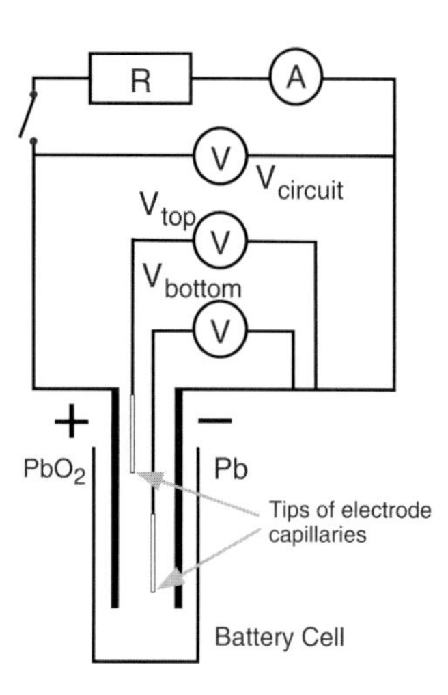

Figure 33.2: Sketch of circuit for local potential measurement between battery plates. The capillaries of the Ag/AgCl reference electrodes were placed in different positions between the battery plates. The potentials V_{top}, V_{bottom} and $V_{circuit}$ were measured with high-impedance voltmeters. The current was measured with an ammeter. The sampling rate was 0.0167 Hz.

stack. The distance between the plates (1 mm) was also comparable to the plate distances in the whole cell stack. The cell casing was filled with 500 ml sulfuric acid (d = 1.28 g/cm).

In an extended setup, nine Ag/AgCl reference electrodes were used. The potentials were recorded with an analog/digital converter card (ADLINK PCI-9112) at 0.1 Hz. The input impedance of the analog/digital channels was > 1 GΩ.

33.2.2 Local Potential Measurements with Auxiliary Lead Electrodes

Since Ag/AgCl electrodes are not easily implemented in lead–acid battery stacks, auxiliary lead electrodes have been developed. Lead electrodes are advantageous, since any chemicals, which are not normally present in lead–acid accumulators, and which may reduce the hydrogen over-potential, can be avoided in this way.

Lead electrodes were made by soldering a small piece of pure lead wire (Alfa Aesar puratronic 99.998%, 0.5 mm ⌀) to the end of an insulated copper wire (0.5 mm ⌀). The contact between the lead and the insulated copper wire was then coated as acid resistant. Only the tip of the lead wire remains uncoated to enable local potential mesurements.

In order to obtain knowledge of the behavior of both the negative and the positive battery plates, a modified experimental setup was chosen, in which the auxiliary electrodes could be switched periodically to either the positive or the negative cell plate with the help of a relay. This setup is shown in Fig. 33.3. Whereas in the experiments with Ag/AgCl electrodes only two battery plates were used, here the lead electrodes were placed in the middle of a whole cell stack consisting of eight Pb and seven PbO_2 plates. The cell compartment was filled with 1 L H_2SO_4, d = 1.28 g/cm.

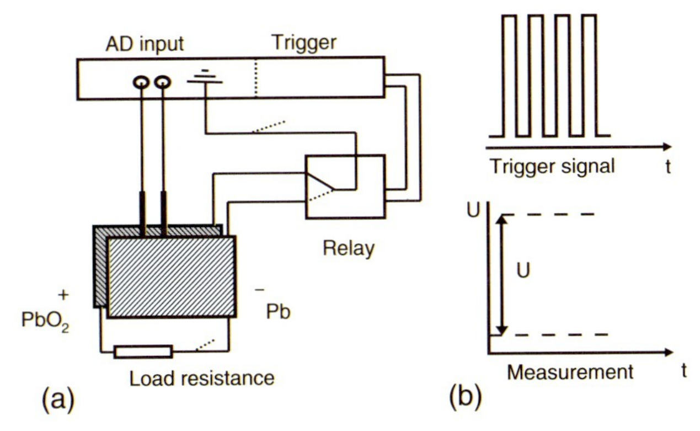

Figure 33.3: (a): Extended setup for local potential measurements between battery plates. The auxiliary lead electrodes could be periodically switched between the positive and the negative plate with a triggered relay. (b): The relay trigger frequency was set to 0.2 Hz. The sampling rate was 2 Hz. The input impedance of the analog/digital channels was > 1 GΩ.

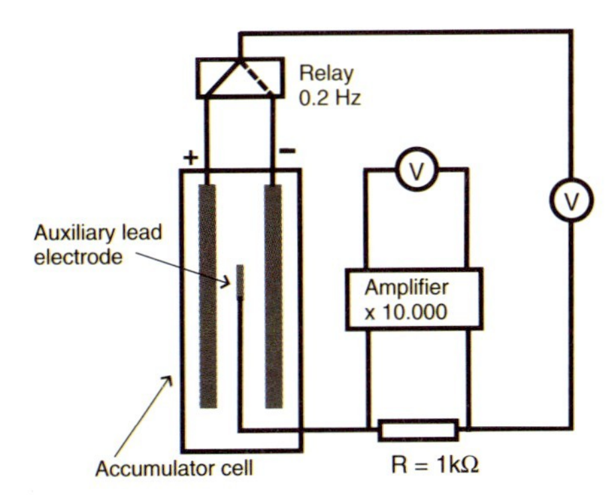

Figure 33.4: Setup for the measurement of the current flow through the auxiliary lead electrodes. The potential drop across the resistance R was recorded with an analog/digital card (ADLINK PCI-9114DG) at 2 Hz. The input impedance of the isolated analog/digital channels was $> 1\ G\Omega$.

The periodical polarization reversal induces a small current flow through the auxiliary electrodes. This current was measured with a modified setup, shown in Fig. 33.4.

33.3 Results and Discussion

33.3.1 Local Potential Measurements with Ag/AgCl Electrodes

Vertical Potential Gradient under Discharge Conditions with Ag/AgCl Electrodes

The first experiment was done in order to show the possibility of detecting potential inhomogeneities between the battery plates by means of auxiliary electrodes (here Ag/AgCl), see Fig. 33.5.

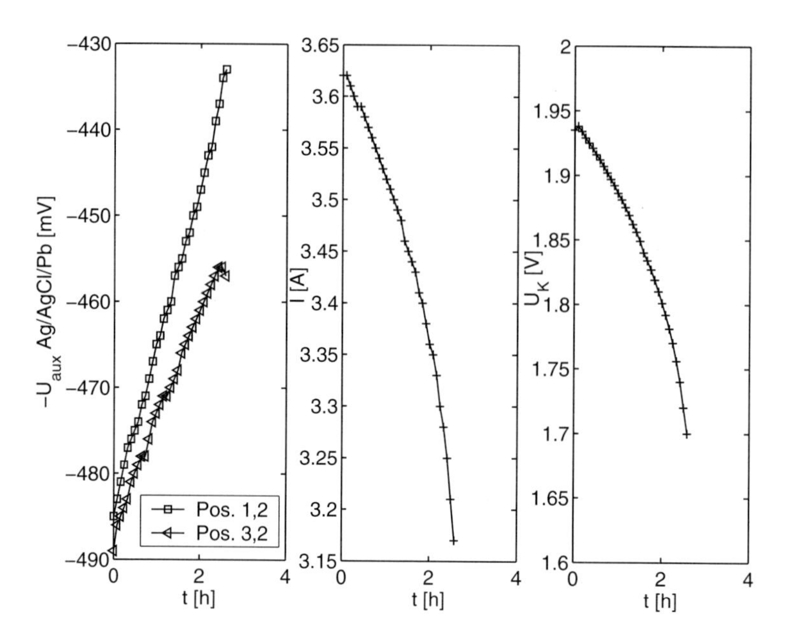

Figure 33.5: Half-cell potentials of Ag/AgCl electrodes against PbO_2 plate while discharging the cell over a $0.5\ \Omega$ resistance. The electrode positions are Pos. 1, 2 and Pos. 3, 2 from Fig. 33.1. Left: squares – electrode in upper position, triangles – lower position. Middle: current flowing through the cell. Right: cell potential.

The half-cell potentials between the auxiliary electrodes against the PbO_2 plate was recorded. Two electrodes have been applied in different vertical positions (Pos. 1, 2 and Pos. 3, 2, see Fig. 33.1). At the beginning of the measurement the cell had a rest potential of 2.114 V. During discharge, the potentials of the upper and the lower auxiliary Ag/AgCl electrodes drift apart (see Fig. 33.5, left), indicating differences in local current flow and acid density. The cell was discharged with a $0.5\ \Omega$ load resistance, so the ratio of the potential and the current (see Fig. 33.5, right and middle), U/I, is constant during the discharge.

The temporal behavior of the current (middle graph in Fig. 33.5) can be fitted by a polynomial ansatz:

$$I = \sum_{i=0}^{n} a_i t^i. \tag{33.4}$$

The best fit (see Fig. 33.6) was obtained for $n = 4$, with

$$a = (3.6194, -0.0358, -0.1244, 0.0798, -0.0202), \tag{33.5}$$

giving a regression coefficient R^2 of 0.952. A still reasonably good fit is obtained for $n = 2$, with $a = (3.6109, -0.0345, -0.0476)$.

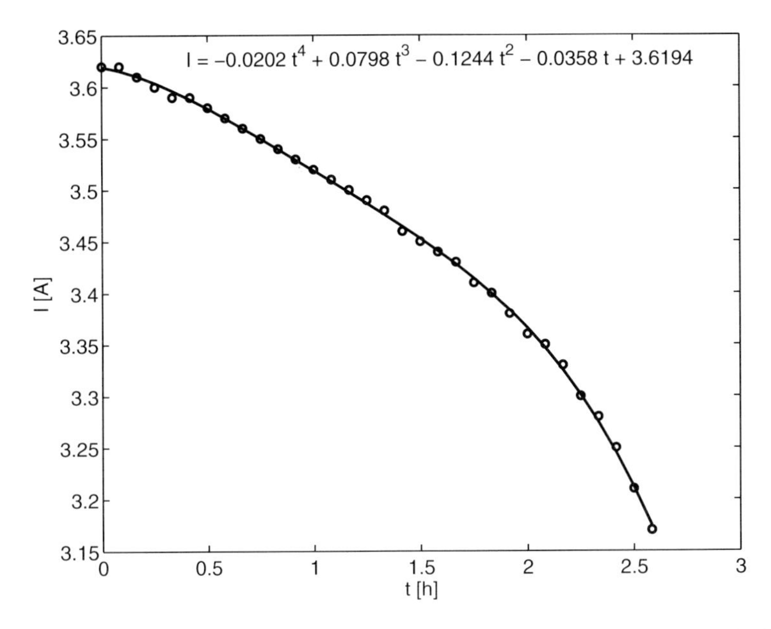

Figure 33.6: Fourth order polynomial fit of the temporal current behavior during discharge, see middle graph from Fig. 33.6.

Vertical and Horizontal Potential Gradients Under Charging/Discharging Conditions with Ag/AgCl Electrodes

Figure 33.7 shows the development of half-cell potentials of an array of Ag/AgCl electrodes against a PbO_2 plate (positive plate) near the end of the charging process, just before the onset of hydrogen and oxygen evolution as the remaining electrode reactions. The charging current was 2 A. In the shown time interval, the cell potential of the Pb/PbO_2 stack was slightly increasing from 2.194 V to 2.202 V. The general setup is shown in Fig. 33.2. The Ag/AgCl electrode positions are shown in Fig. 33.1. Clearly, the strongest variation of the local half-cell potentials is found at the upper electrode positions. The spatio-temporal variation of the local half-cell potentials, especially of the ones at the upper position of the plates, can be explained if one assumes an equalization of electrolyte compartments of different density through convection. That is, if a compartment of higher acid density moves downwards due to gravitation, the local compartment will be replaced with electrolyte of lower acid density and thus the local half-cell potentials will be influenced by the changing electrolyte density.

During charging, sulfuric acid of high density is produced at the cell plates, and moves slowly down by gravitation. We have here a strong indication that this effect is especially predominant in the upper part between the cell plates. During discharge, the acid density between the plates decreases. Due to its higher density and the low diffusion velocity for sulfuric acid, the electrolyte in the bottom part of the cell casing does not contribute to the reaction. Figure 33.8 shows the potential development during discharging. It can be seen that in this case no simple pattern results for the local potential development. Here, the half-cell potentials of the auxiliary Ag/AgCl electrodes against the Pb plate have been measured. The

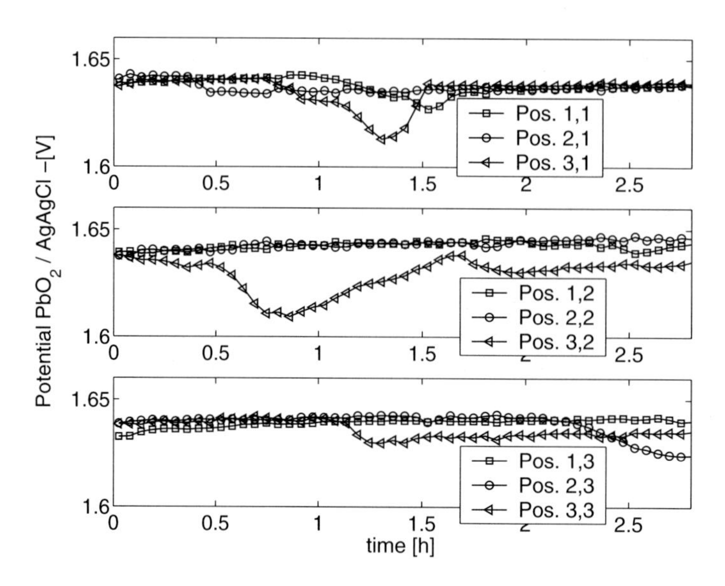

Figure 33.7: Half-cell potentials of Ag/AgCl electrodes against PbO_2 plate near the end of the charging process just before the onset of hydrogen and oxygen evolution as the remaining electrode reactions. The electrode positions are shown in Fig. 33.1.

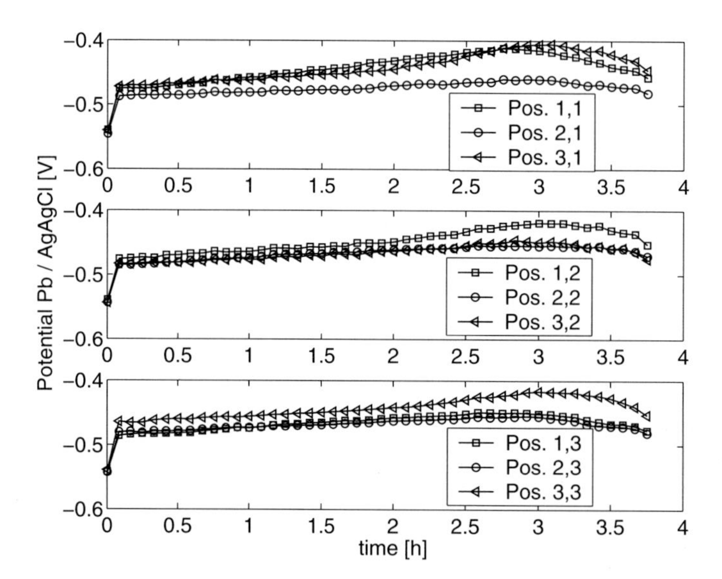

Figure 33.8: Half-cell potentials of Ag/AgCl electrodes against Pb plate while discharging over a 0.5 Ω, 25 W load resistance, corresponding to a current density of approx. 19 mA/cm^2 during the first 20 min of discharging. Discharging was stopped after 3 h. After this time the local potentials tend to approach a homogeneous state again.

potential drop due to discharge shows a small horizontal as well as a vertical variation. The discharging process was stopped after 3 h. After this time the local potentials tend to approach each other, thus reaching a homogeneous state again.

33.3.2 Local Potential Measurements with Lead Electrodes

The new dynamic measuring principle was introduced in the experiments with lead auxiliary electrodes. The lead auxiliary electrodes have been switched periodically to measure alternating the half-cell potential against the positive and the negative plates. The general setup is given in Fig. 33.3. The relay switching frequency was 0.2 Hz and the sampling rate was 2 Hz.

Local Potential Development During Cycling

A phase-space representation in two dimensions of a time series of the potential response of two auxiliary lead electrodes at positions 1, 1 and 1, 3 (see Fig. 33.1), placed in the middle of a cell stack, during cycling (40 min charging with max. 4.7 A/2.4 V and discharging 30 min with 4.7 A) is given in Fig. 33.9. Only the potential against the PbO_2 plate is shown. Clearly, both locally measured potential signals show a different dynamic development. The overall potential drop is much stronger for the upper auxiliary electrode (in Pos. 3, 1 from Fig. 33.1).

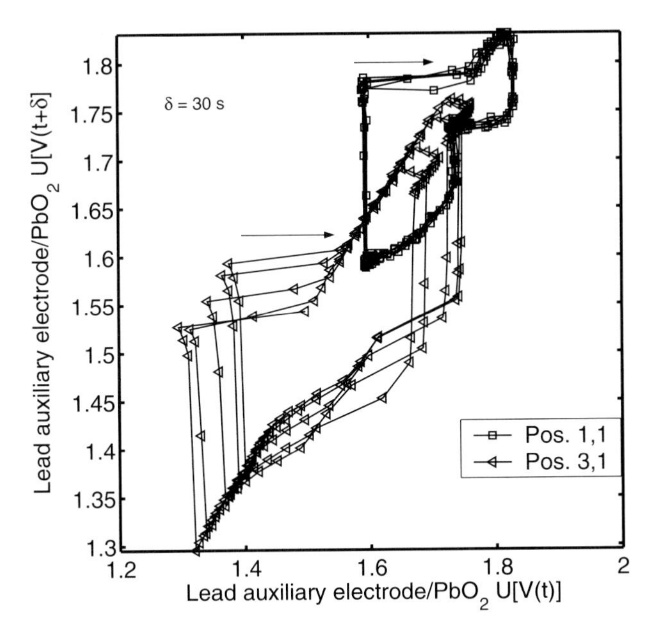

Figure 33.9: Two-dimensional phase-space representation of a time series of the potential response of two auxiliary lead electrodes at positions 1, 1 and 1, 3 (see Fig. 33.1). The auxiliary lead electrode in the upper position shows a markedly stronger potential drop, indicating a higher charge/discharge rate in the upper cell position.

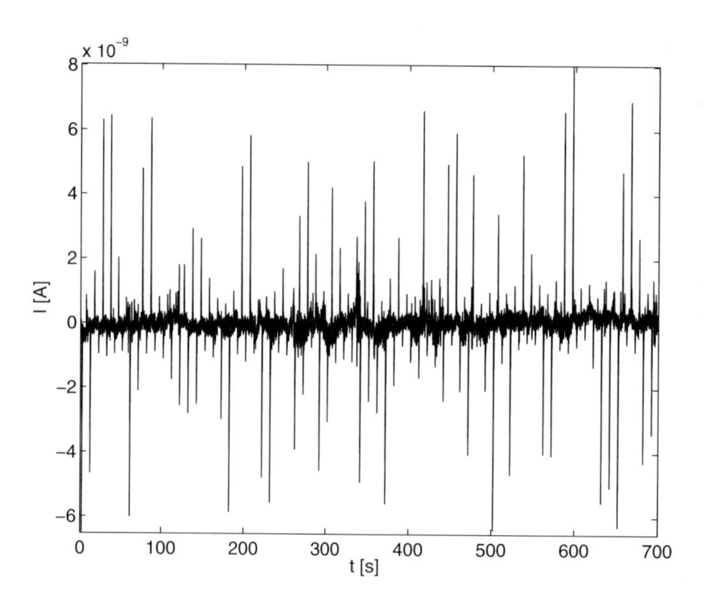

Figure 33.10: Current flowing through auxiliary lead electrode during periodic polarization reversal. The electrode was switched at 0.2 Hz between the positive and the negative plates. For the whole time series, the net current flow through the auxiliary lead electrodes is approx. zero.

Current Flow Through Lead Auxiliary Electrodes under Rest Conditions

The current flowing through the lead auxiliary electrode during periodic polarization reversal was monitored with the setup shown in Fig. 33.4. The current–response signal shown in Fig. 33.10 was recorded when the cell stack was at rest. The lead auxiliary electrode was at position (2, 2), see Fig. 33.1. The relay switching frequency was 0.2 Hz and the sampling rate was 100 Hz.

In this experiment it turned out that a periodical given signal (the switching of the auxiliary electrodes to either the positive or the negative plate) results in a chaotic response behavior of the system. The estimated largest Lyapunov exponent was found to be 0.990^{-s}, using the Kantz method [11] for embedding dimensions from 3 to 6. With the method of Rosenstein et al. [16] the largest Lyapunov exponent was found to be 0.680^{-s}. Both methods give positive largest Lyapunov exponents, indicating a chaotic system.

Two resolved peaks of intermediate size from Fig. 33.10 are shown in Fig. 33.11. The flowed charge has been determined according to Eq. (33.6):

$$Q = \int_{t_0}^{t_1} I dt = \frac{1}{fR} \int_{t_0}^{t_1} U dt, \tag{33.6}$$

with amplification factor $f = 1000$ and resistance $R = 1\ \mathrm{k}\Omega$ (see Fig. 33.4).

The net current flowing through the auxiliary lead electrode for the whole time series from Fig. 33.10 is almost zero. Nonetheless, the current peaks to the anodic and cathodic sides are of varying height and the absolute amount of charge varies between 0.5 nA s and 3 nA s; the average absolute amount of charge was found to be approximately 2 nA s.

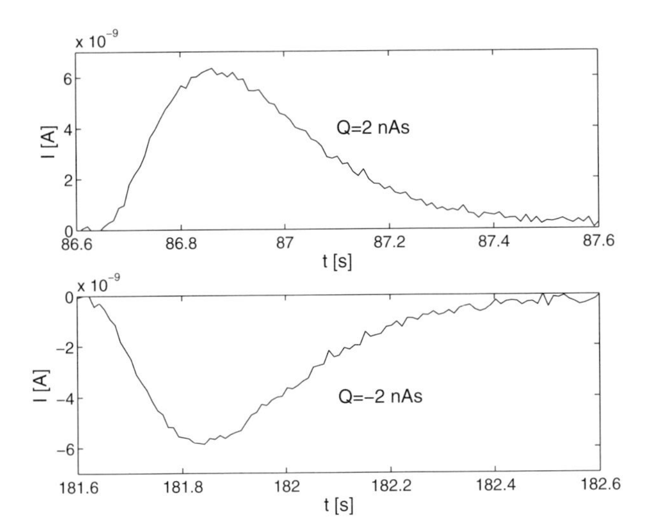

Figure 33.11: Close-up of two typical peaks at different times from the time series shown in Fig. 33.10. The height and the width of the peaks varied, so the absolute amount of charge for each peak varied between 0.5 nA s and 3 nA s. The average absolute amount of charge was found to be approximately 2 nA s.

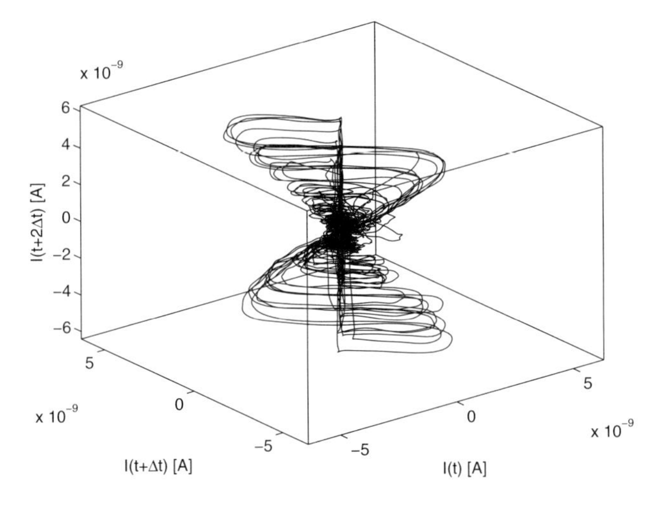

Figure 33.12: Slightly banded chaotic attractor of the time series shown in Fig. 33.10. The chosen time delay Δ was 0.25 s. Only the signal of one auxiliary lead electrode at position (2, 2) was recorded, so no spatially resolved system changes are observed here.

An attractor representation (Fig. 33.12) of the time series from Fig. 33.10 shows a slightly banded chaotic structure. The surprising (apparent) simplicity of the chaotic structure of the attractor implies that the dynamics at distinct positions is relative low-dimensional, although it is a spatio-temporal system. The current response of the periodically triggered auxiliary electrode appears to be very sensitive to its environment. Only the signal of one auxiliary lead electrode at position (2, 2) was recorded, so no spatially resolved system changes are observed here.

The electro-chemical situation of the auxiliary lead electrode in its specific surroundings determines its response behavior to the external trigger signal (periodic polarization reversal). Thus, we claim that one can detect the state-of-health (SOH) of a lead–acid battery while monitoring the changes of the chaotic response of the periodically forced auxiliary electrode.

33.4 Conclusions and Future Work

We showed that local potential measurements between cell plates under charge and discharge conditions revealed a spatially inhomogeneous temporal behavior between the cell plates ,indicating electro-chemical turbulence in lead–acid batteries. Moreover, it can be assumed that lots of different local states or potential distributions may be ascribed to the same set of certain values for the overall current I and the potential U, as measured at the poles of a battery. Therefore, it seems necessary to extend the phase space of variables describing the battery performance by locally measured ones. With such an extension of the phase space it is possible now to define regions in the phase space which can be correlated to the SOC and the SOH of the battery. This means that there will be regions in the phase space which, when visited by the vector of extended variables I, U, $U_{\text{local}_{xyz}}$ describing the battery performance, indicate a deterioration of the battery before this would be detectable from only monitoring I and U globally as measured at the battery poles.

Future work will be directed to answer the question of what is the minimal dimension of the phase space of the battery that allows for a reliable SOC and SOH prediction or how many local potential measuring points are necessary and what are the best positions for them.

Acknowledgment

We gratefully acknowledge VB Autobatterie GmbH for providing us with lead–acid batteries for our investigation.

Bibliography

[1] P.E. Baikie, M.I. Gillibrand, and K. Peters, *The effect of temperature and current density on the capacity of lead–acid battery plates*, Electrochimica Acta **17** (1972), 839–844.

[2] H. Bode and J. Euler, *Autoradiographische Untersuchung der Stromverteilung auf Platten von Bleiakkumulatoren – I*, Electrochimica Acta **11** (1966), 1211–1220.

[3] H. Bode and J. Euler, *Autoradiographische Untersuchung der Stromverteilung auf Platten von Bleiakkumulatoren – II*, Electrochimica Acta **11** (1966), 1221–1229.

[4] H. Bode, J. Euler, E. Rieder, and H. Schmitt, *Autoradiographische Untersuchung der Stromverteilung auf Platten von Bleiakkumulatoren – III*, Electrochimica Acta **11** (1966), 1231–1234.

[5] M. Calabek, K. Micka, P. Baca, and P. Krivak, *Influence of grid design on current distribution over the electrode surface in a lead–acid cell*, Journal of Power Sources **85** (2000), 145–148.

[6] J.M. Charlesworth, *Determination of the state-of-charge of a lead–acid battery using impedance of the quartz crystal oscillator*, Electrochimica Acta, **41**(10) (1996), 1721–1726.

[7] N.F. Compagnone, *A new equation for the limiting capacity of the lead/acid cell*, Journal of Power Sources **35** (1991), 97–111.

[8] R.M. Dell and D.A.J. Rand, *Energy storage – a key technology for global energy sustainebility*, Journal of Power Sources **100** (2001), 2–17.

[9] C. Armanta-Deu et al., *Determination of the diffusion coefficents for sulfuric acid in lea-acid batteries: Influence of the diffusion phenomenon on low-rate operation*, Journal of the Electrochemical Society **137**(4) (1990), 1030–1035.

[10] R. Friedrich and G. Richter, *Performance requirements of automotive batteries for future car electrical systems*, Journal of Power Sources **78** (1999), 4–11.

[11] H. Kantz and Th. Schreiber, *Nonlinear Time Series Analysis* (Cambridge Nonlinear Science Series), Cambridge University Press, Cambridge UK, 1997.

[12] C. Liebenow, *Über die Berechnung der Kapazität eines Bleiakkumulators bei variabler Stromstärke*, Zeitschrift für Elektrochemie, **4** (2) (1897), 58–63.

[13] F. Mattera, D. Desmettre, J.L. Martin, and Ph. Malbranche, *Characterization of photovoltaic batteries using radio element detection: the influence and consequences of acid stratification*, Journal of Power Sources **113** (2003), 400–407.

[14] E. Meissner and G. Richter, *Vehicle electric power systems are under change! Implications for design, monitoring and management of automotive batteries*, Journal of Power Sources **95** (2001), 12–23.

[15] W. Peukert, *Ueber die Abhängigkeit der Kapactität von der Entladestromstärcke bei Bleiakkumulatoren*, Elektrotechnische Zeitschrift **20** (1897), 287–288.

[16] M.T. Rosenstein, J.J. Collins, and C.J. De Luca, *A practical method for calculating largest Lyapunov exponents from small data sets*,. Physica D **65** (1993), 117.

[17] C.J. Rydh, *Bestämnig av kvarvarande kapacitet i ett batteri*, Oorganisk miljökemi, Chalmers tekniska högskola, Göteborg, 1997. http://homepage.te.hik.se/personal/tryca/battery/rydh_1997_restkap_littstudie.pdf,

Index